Parameter Extraction and Complex Nonlinear Transistor Models

For a listing of recent titles in the
Artech House Microwave Library,
turn to the back of this book.

Parameter Extraction and Complex Nonlinear Transistor Models

Günter Kompa

ARTECH
HOUSE

BOSTON | LONDON
artechhouse.com

Library of Congress Cataloging-in-Publication Data
A catalog record for this book is available from the U.S. Library of Congress.

British Library Cataloguing in Publication Data
A catalog record for this book is available from the British Library.

ISBN-13: 978-1-63081-744-2

Cover design by John Gomes

685 Canton Street
Norwood, MA 02062

10 9 8 7 6 5 4 3 2 1

To Anni, Lilli, and Janne

Contents

Preface

This book addresses the modeling of microwave field effect transistors (FETs) for circuit design with a focus on accurate measurement-based device characterization and physics-based consistent model parameter extraction covering a frequency range up to 120 GHz. Related topics have been treated and published in numerous papers and books. Therefore, the question may come up as to why to add a further reference book about device modeling.

Characterization and modeling of microwave devices has been one of the main research topics of the author and his research group over more than three decades, and many progressive contributions have been delivered during this time. Very early in the eighties, M. Schlechtweg introduced table-based FET models as an alternative to the analytic approach as to enhance the flexibility and accuracy of the modeling and circuit design process. In the second half of the eighties, F. van Raay performed pioneering work in the field of vectorial large-signal network analyzer (LSNA) development for error-corrected large-signal characterization of microwave field-effect transistors using initially a sampling scope in combination with a vector network analyzer. Later, at the beginning of the nineties, the proposed signal waveform measurement system could be drastically simplified by the availability of the microwave transition analyzer comprising both functions of the vector network analyzer and the sampling scope in one unit. Van Raay also developed an electronic load module (ELM) permitting cost-effective and very compact source-/load-pull measurement setups.

Using the established LSNA capabilities, original contributions were delivered by A. Werthof regarding direct model parameter extraction from signal waveform measurements. Alternatively to the classical bottom-up modeling approach with inherent model inconsistency, I. Schmale successfully studied the favorable features of a top-down modeling strategy concerning GaAs-based microwave devices. E. S. Mengistu applied this approach to GaN power HEMTs.

Continuously stringent requirements to today's digital communication systems concerning high linearity, high power efficiency, and low-intermodulation distortion of power amplifiers, initialized broad activities in our research group with respect to intermodulation distortion (IMD) analysis and appropriate measurement facilities for verification. E. R. Srinidhi delivered significant contributions concerning localization and identification of IMD sources in GaN power devices, and offered refined IMD measurement concepts to observe the sweet-spot occurrence under varying device operation.

Measurement setups have recently become more complex. In addition to output power and efficiency of the fundamental mode, higher-order harmonics terminations and two-tone and W-CDMA signals have become important to recognize the device potential. B. Bunz and W. Wittwer developed time-domain multiharmonic source- and load-pull measurement systems with envelope tuning capabilities. In addition A. Ahmed established frequency-domain source-and load pull measurement systems to analyze memory effects in high-power GaN amplifiers, investigating and verifying system behavioral models as a basis for predistortion strategies.

The FET device can be divided into an extrinsic linear part and an intrinsic active part. The nonlinear elements of the inner device are dependent on the inner control voltages, which are not directly accessible. Therefore, it is of fundamental importance to have accurate knowledge of the extrinsic parasitic network to determine the inner control voltages from the measured extrinsic ones by a deembedding procedure. Thus, it was of utmost concern to develop a robust and reliable straightforward extraction algorithm for the determination of the parasitics with physically meaningful element values. Several approaches have been investigated and analyzed over the years. Following optimization methods for model parameter extraction, starting values were initially roughly estimated from device technology and construction insofar as published by the manufacturer. This implied very large search spaces for the optimizer with high risk to be captured in a local minimum. M. Novotny modified the classical simplex algorithm with local search character in such a way that the algorithm was capable of escaping from a local minimum. Nevertheless, in spite of this effective leaping function, it could not be guaranteed that the global minimum was definitely found. Therefore we focused on optimization-based statistical extraction procedures. The idea was to sample the model parameter space around the (expected) global minimum of the error function. The distribution of the extracted model elements could give conclusive information about the location and size of model parameter values. In 1993 F. Lin proposed his bidirectional search method to increase computational efficiency. The number of optimization variables were drastically reduced focusing only on the extrinsic parasitics, whereby the intrinsic parameters were determined quasi-analytically.

At that time the starting parameter values for the optimization process had been defined independently from each other, although extraction results had already indicated some correlation between the model parameters. A completely new quality in the extraction procedure came up when these model parameter interdependencies were taken into account. W. N. Mwema provided a primary basis for reliable model parameter extraction using measurement-based correlated starting values for the model parameters. This idea meant an essential step forward with respect to physics-related element values. The extraction of the model parameter values follows in two steps. First the complex distributed equivalent circuit model is simplified by a pi topology, where each branch represents a composition of sub-branches of the original distributed model. The element values of the simplified pi model, also termed effective values, can be calculated directly from measured data. However, the extracted model elements are only valid for a lower frequency range where distribution of the topology (i.e., model elements) is not yet needed. It is the idea that the extracted effective model element values characterize a point in the model parameter space that is near the global minimum of the error function. However, for the higher-frequency range (or more complex topology structures as given with large-size multifinger GaN power devices), a further fine-tuning extraction step is needed satisfying the distribution of the effective model elements.

This successful modeling procedure was continuously refined by A. Jarndal focusing on GaN power HEMT devices with gate widths up to 4 mm. More recently, J. A. Zamudio

Flores has strengthened the model parameter extraction process by gaining additional well-approximated physics-related model element values for the optimization procedure directly from the surface layout structure of the considered FET. Moreover, he formulated a closed-form analytical approach to calculate the series inductance values. Thus, complex equivalent circuit structures comprising three admittance shells can be reliably determined without the need for optimization, thus avoiding any risk of getting stuck in a local minimum.

Circuit design issues on the basis of elaborated nonlinear device modeling approach were treated in subjects such as oscillator, frequency doubler and frequency tripler, both useful for the establishment of harmonic load-pull setups. A. Z. Markos tackled the more complex design task of a GaN power amplifier using the Doherty technique. R. Ma has demonstrated how to design an amplifier on first-pass success, employing partitioning design approach.

The modeling of devices may comprise the following different goals. The user of a device model may be less interested in the actual device measurements and modeling procedure than to rely on simulated data concerning device figure of merits such as output power, gain, power added efficiency (PAE), intermodulation distortion (IMD), adjacent channel power ratio (ACPR), and others. The model should be easily implementable in commercial software such as Advanced Design System (ADS) software and should operate and converge under complex working conditions as such appearing with currently used broadband digitally modulated stimulating signals. The accelerated evolution of modern communication systems demands for increased modulation bandwidth, multichannel mobile communication, and high maximum output power in base stations due to high peak-to-average power ratio (PAR). The complex digital modulation of carriers poses new challenges to the modeling process. This concerns the large modulation bandwidth, which coincides with the frequency dispersion region of microwave field-effect transistors, in particular, GaN technology based power devices. In addition, with high-power devices, temperature dispersion has to be taken into account as was well studied by S. Dahmani. Thus, device modeling has become very complex, and basic physical understanding of device behavior is needed to cope with the given subject.

In addition to the common circuit design goal, device modeling may also be utilized for device diagnostic purposes. Continuous enhancement of device performance is of utmost concern to device manufacturers. Currently, this is in particular related to simultaneous optimization of maximum achievable output power, high value of power-added efficiency within a large signal dynamic range, and high linearity at high power levels.

In a book titled *Basic Properties of III-V Devices—Understanding Mysterious Trapping Phenomena*, published with Kassel University Press in 2014, the author covered the technological issues in the fabrication of III-V devices for the third (3G) and fourth (4G) generation of mobile phone communication standards. Commonly observed anomalous physical effects in GaAs and GaN devices such as trapping, current collapse, gate-/drain lag, breakdown, and others, are illuminated and described. The reading of this book should contribute to a deeper understanding of the physical origin and properties of the parasitic effects, and may serve as an additional information source for respective device issues in this book.

The main focus of this book is devoted to the reliable and accurate small- and large-signal device modeling with particular focus on complex nonlinear transistor models and related model parameter extraction methods. It gives an overview and evaluation of diverse modeling approaches, and thoroughly analyzes model parameter extraction approaches known so far. It discusses extraction concepts that include the determination of physically meaningful values for a given microwave field-effect transistor. Extraction examples are discussed based on high-electron mobility transistors (HEMTs) on gallium arsenide and gallium nitride technology.

The author still remembers the 1996 IEEE MTT-S International Microwave Symposium in San Francisco when John Bandler organized a very interesting panel session called "Effective CAD: A Dilemma of Models!" Representatives of both passive device and active device simulators were joined in one session. Field theorists like Prof. Arndt from University of Bremen and Prof. Sorrentino from University of Perugia stated that one can always find a reliable CAD model for any passive structure. This depends only on numerical effort and time. After their talks, Mike Golio from Motorola stated at the beginning of his talk: "In my opinion all models (i.e., models of active devices) are wrong!" The participants of the panel session gave an enthusiastic applause.

I hope this book will highlight the significant progress in reliable, consistent, and accurate model parameter extraction of III-V devices with a focus on GaN power HEMTs that occurred during the last decades, and will give the reader a solid knowledge basis in the application of successful and reliable device modeling strategies.

The book is mainly addressed to microwave engineers dealing with device measurement, device modeling, and microwave circuit design. However, it should also be attractive for other researchers such as device physicists and device modelers designing III-V devices, or technologists responsible for material and device processing. The persistent evolution of mobile communication systems with ever-growing restrictive performance requirements makes strong cooperation necessary to benefit from the expertise of those in their disciplines.

Organization of This Book

The book is divided into 26 chapters, each of which explores a theme through several sections. Appendixes A to F complete the material under discussion and are related to respective chapters.

Chapter 1 provides an introduction. Chapter 2 is devoted to the most relevant microwave devices such as MESFET, HEMT, and HBT.

Chapter 3 gives an overview of the different classes of transistor models, focusing on numerical and analytical physical models, empirical and experimental models, and behavioral models, such as ANN-based and X-parameter-based models.

Chapter 4 is devoted to the classical Shockley model and its enhanced modifications including carrier velocity saturation. Two-region and short-channel saturation models are discussed as predecessors of today's modern modeling strategies; these basic analytical approaches may be even recognized in recent device modeling tasks.

In Chapter 5 the extrinsic part of the field-effect transistor is briefly addressed and the relationship of intrinsic and extrinsic control voltages discussed.

Chapter 6 focuses on the estimation of model parameter values needed as starting parameter values for the optimization procedure. Estimated values are derived from device construction and material, including its contact and interconnect structures.

Chapter 7 gives an overview on small-signal equivalent circuit topologies dependent on the range of operation frequency. Special attention is paid to GaN power devices with large gate-periphery and fabricated on silicon substrates.

Chapter 8 expands on the material in Chapter 6, describing how to obtain reliable model parameter values from low-frequency measurements. Various approximations are introduced to determine the intrinsic model parameters from the generic ones. The main target is to find parameter values that serve as a high-quality starting value vector in a final optimization process. The advantage of this approach is that the parameters are determined from a single set of S-parameters in a given bias point so that the intercorrelations between the variables are taken into account, which has strong influence on the extraction process.

Chapter 9 addresses static- and pulsed-DC measurement techniques, self-heating effects, thermal resistance characterization, and thermal and trapping time constant determination. The chapter closes with a thermal device model representation.

Chapter 10 summarizes traditional small-signal measurement technique using a vector network analyzer (VNA). After a short review of VNA evolution, specified uncertainties, error models, and calibration performance are described.

Chapter 11 discusses uncertainties in the device modeling process. After clarifying commonly used definitions such as accuracy, reliability, and validity, exemplarily observed measurement errors are illustrated and discussed. The chapter closes with some recommendations on how to proceed in the device modeling procedure.

Chapter 12 details a key part of the book thesis. Reliable model parameter extraction depends on some kind of optimization. Therefore it is reasonable to give a broad overview of optimization methods already in use for model parameter extraction purposes and those recently proposed and hereafter potentially applicable. The latter ones, in particular, concern combined local and global optimization strategies, also termed hybrid global optimizers. The author favors the leaping simplex algorithm developed and commonly used in his research group over years.

Chapter 13 gives a brief overview on the various model parameter extraction methods treated in the following chapters.

Chapter 14 reflects the beginning of the device model parameter extraction, taking into account all model parameter elements for optimization, mostly applying local search optimizers such as gradient-based methods. Good starting parameter values are derived from low-frequency S-parameter measurement and interrelations between intrinsic and generic model elements.

Chapter 15 presents decomposition-based extraction methods based on a experimental approach or on principal-component sensitivity analysis.

Chapter 16 details the bidirectional search method, where the error function is not defined as usual with respect to the external reference planes. The bidirectional optimization is done in the intrinsic reference planes, meaning that optimization is restricted to the extrinsic model parameters, whereas the intrinsic ones are quasi-analytically determined by least squares.

Definitions of starting values is not necessary. Multibias extraction is preferred to stabilize the extraction results.

Chapter 17 presents a parameter extraction method that relies on the pure analytical determination of the model parameter element values from four sets of measured S-parameters. Extraction stability is discussed in view of conditioning of the basic set of equations, measurement bandwidth, and measurement errors.

Chapter 18 addresses an analytical model parameter extraction using rational functions with the capability of direct model parameter extraction of all model elements.

The extraction methods in Chapters 19, 20, and 21 were thoroughly investigated by the author, having led to an extraction scheme, which from manifold experiences have be principally accepted and recommended, in particular, with respect to distributed small-signal equivalent circuits.

Chapter 19 proposes a repetitive random optimization approach with adaptive search space. At lower and medium frequencies effective branch capacitances and inductances are extracted from measurements. Then the distributed nature of the device model topology is considered. Repetitive starting value intervals of the individual model parameters are continuously decreased by discarding extraction results with larger error function values. The difference between mathematical and physics-based extraction results is highlighted.

Chapter 20 draws attention to the difficulty of extracting low-value series resistances as well as their bias dependence. It is shown that the proposed two-frequency scanning method can deliver reliable bias-dependent resistance values from S-parameter measurements.

Chapter 21 is a further key chapter, and describes the model parameter extraction approach with measurement-correlated parameter starting values. A detailed description of the various extraction steps is given. One main focus is put on the distribution condition of the capacitances in the gate-source, drain-source and gate-drain equivalent circuit branch. The total branch capacitances are obtained from low-frequency S-parameter measurements. Various approaches for inductive and resistive model parameter estimation based on estimated capacitances are discussed. In particular, a closed-form analytical calculation of the series inductances is presented. The high quality of extracted model element values of a distributed equivalent circuit is demonstrated. It is shown that the determined parameter values already fulfill accuracy requirement with respect toVNA uncertainty specifications. Nevertheless, further refinement of the extracted parameter values might be performed with the leaping simplex, resulting in excellent agreement of simulated and measured S-parameters in various bias points.

Chapters 22 through 26 treat large-signal modeling and large-signal measurement topics. Chapter 22 provides some basics in nonlinear device modeling. In Chapter 23 a nonquasi-static FET model is developed step by step. Extraction of the electrothermal drain current, including static and dynamic bias-dependent surface and barrier trap effects, are described.

Chapter 24 discusses high-power large-signal measurement techniques for device characterization and model verification. Setups in time domain and frequency domain with harmonic source- and load-pull facilities, including also the envelope frequency range, are presented, and special issues in IMD device characterization emphasized.

Chapter 25 discusses capabilities and limitations of standard device models on the basis of measured and simulated GaN power devices with respect to output power, power added efficiency, pulsed-DC, and intermodulation distortion.

Chapter 26 represents the finally proposed complete large-signal device model, its implementation in commercial software, and exemplary validation by experiments.

Appendixes A through F provide chapter supplements. Appendix A details the derivation of pi-type small-signal admittance parameters from the generic two-port admittance matrix. Appendix B gives a short overview about classical direct measurement techniques concerning series resistances. Appendix C is a supplement to the bidirectional search method in Chapter 16, deriving the relations for the quasi-analytical extraction of the intrinsic model elements. Appendix D is a contribution to the pure analytical model parameter extraction in Chapter 17, discussing the embedding of an impedance/admittance network into a two-port network. Appendix E details the derivation of the Ricatti equation in Chapter 21, which is solved for the calculation of the series inductances. Appendix F presents some details about the derivation of mathematical relations concerning the nonlinear n-port characterization in Chapter 23.

Who Should Read This Book

This book reflects the author's long experience over nearly four decades in model parameter extraction of microwave field-effect transistors and related device modeling and circuit design under large-signal operation. Therefore, it is expected that the contents of the book can serve the reader as a compass giving orientation in this difficult and complex field of activity.

The book considers the model extraction issues in a broad way, emphasizing the measurement and extraction conditions to gain good extraction and modeling results for successful active circuit design.

The contents of this book have been part of a microwave lecture offered at our university for international students in a postgraduate study program and supplemented by selected material from the aforementioned book *Basic Properties of III-V Devices—Understanding Mysterious Trapping Phenomena*. The lecture should prepare the students for research work involved in the preparation of their master and doctoral theses, respectively, completed by seminars and practical laboratory training.

Thorough knowledge and training in this field are regarded as very important factors for future vocational assignment of graduates as design engineers in research and development.

The book will be also useful for design specialists reading selective topics or using the clearly structured collected material as a reference book.

How to Use This Book

The book addresses several reader categories. The reader who wants to come into the field of transistor characterization and modeling for planar circuit design may read the book from the beginning to end.

The book is appropriate as a course book, primarily for graduate students or for undergraduate students in the last year of their studies. Basic knowledge of semiconductor physics and circuit theory of linear microwave networks (S-parameter concept) is necessary to understand the material.

Acknowledgments

Many people have contributed to the creation of this book. Without the motivation, creativity, and vital teamwork of my research coworkers, this strong field of model parameter extraction and modeling of microwave devices could not have been treated in the necessary breadth and depth over the years. Numerous attractive extraction algorithms have been published but frequently the ultimate question of physical relevance and physical meaningfulness remained. The author has great hope having closed this uncertainty gap substantially, in particular, with respect to complex, distributed equivalent circuits.

The generous provision of GaN power devices by the Fraunhofer Institute for Applied Solid State Physics (FhG-IAF), Freiburg, is greatly acknowledged. Dr. Michael Schlechtweg and Dr. Rüdiger Quay were always open to fruitful discussions. Gratitude is also given to Dr. J. Würfl, Ferdinand-Braun-Institut für Höchstfrequenztechnik (FBH), Berlin, providing us with GaN power devices. Fruitful discussions opened our eyes to III-V epitaxial and process issues.

I express my gratitude to Prof. Dr. G. Fischer (now with Friedrich-Alexander University Erlangen-Nürnberg (FAU)) and H. Schenkel, Lucent Technologies Network System GmbH, and Dr. W. Templ and D. Wiegner, Alcatel, for their great cooperation and integration of our research in their current industrial research projects, and thus participating in a national funding program announced by the BMBF (German Ministry of Technology).

I would like to express my deep gratitude to the many colleagues in the European TARGET(Top Amplifier Research Groups in a European Team) project. This international cooperation led to vital knowledge exchange among scientists in the area of gallium nitride technology.

I also thank my department successor, Prof. Dr. A. Bangert, letting me continue to work on running research projects so that they could be finished successfully.

I also thank Mrs. H. Nauditt from the department for her administrative support. Many thanks are also due to Mr. C. Sandhagen who was always accessible for solving software problems.

I acknowledge various support during the preparation of this manuscript; Dr. E. R. Srinidhi/NXP Semiconductors, Dr. F. van Raay/Fraunhofer Institute for Applied Solid State Physics (IAF), Prof. Dr. F. Lin/University of Science and Technology of China (USTC), Prof. Dr. E. Wasige/Glasgow University, Dr. M. Rui/Mitsubishi Electric Research Laboratories, Prof. Dr. M. Joodaki/Jacobs University (Bremen), Dr. J. Würfl/Ferdinand-Braun-Institut für Höchstfrequenztechnik (FBH), and Dr. W. N. Mwema, University of Nairobi. Then I also want to express my special thanks to Dr. J. A. Zamudi Flores, Scintec SG, for his proofreading of Chapter 5 concerning the regridding process.

I would also like to thank Artech House book company, especially Aileen Storry, for her support of the project. It was always a pleasure to work with the staff at Artech House. My thanks also go to the reviewers, who traditionally remain nameless, for their continuous useful commentary as the manuscript evolved.

I also want to thank all my friends, who kept me on their buddy list despite the long writing process.

Finally, I wish to express my immense gratitude and deep love to my wife, Pirkko Marjatta. Without her great empathy and generous support, this book could not have been written. Her patience has been endless.

Chapter 1

Introduction

Over the last decades, communication technologies have evolved dramatically. Microwave devices based on gallium arsenide GaAs were developed in the eighties and nineties. They were seen to be very competitive with silicon devices, and indeed, they very rapidly conquered higher frequency ranges up to about 100 GHz and beyond. This made it necessary to expand the topology structure of equivalent circuits. At low frequencies the topology of equivalent circuits might encompass a minimum number of only seven circuit elements [1]. With increasing frequency the complexity increased and has reached a large number of 22 circuit elements [2] for broadband operation. The operation mode of microwave devices was frequency modulation (constant amplitude modulation). Trapping effects were primarily related to the operating bias point.

The emergence of digital modulation (envelope modulation) at the turn of this century required additional demands for the modeling process. In this case the carrier of the device is operated in the RF region, but the information is an amplitude-varying signal in the baseband covering frequency bandwidths from 200 kHz to 5 MHz. Multicarrier UMTS operation extends the total modulation bandwidth up to 20 MHz. With the introduction of fourth generation (4G) mobile communication networks the modulation bandwidth largely extended to 100 MHz. Moreover, this will be further dramatically exceeded with the upcoming fifth generation (5G) communication standard specifying multiple carrier operation with a wideband single carrier bandwidth of 100 MHz.

As is known, dispersion effects (i.e., the dynamics of traps in the semiconductor device) have a significant impact on the modeling procedure [3]. The equivalent circuit must reflect both the low-frequency and high-frequency device properties. Because of the high output power of GaN-based microwave devices temperature dispersion is a further challenge in the modeling process.

It is common to divide the equivalent-circuit model into parasitic extrinsic and active intrinsic parts. It is well known that the extraction quality of the intrinsic model parameter values depends heavily on the accuracy of the extrinsic parameter values.

Essentially, there are two main extraction strategies: (1) direct model parameter determination, and (2) model parameter extraction based on optimization. A classical example for the direct approach refers to the standard 15-element small-signal equivalent circuit shown in Figure 1.1. It comprises eight extrinsic model parameters, namely, the pad capacitances C_{pg} and C_{pd}, the series inductances L_g, L_d, L_s, and the series resistances R_g, R_d, R_s.

As described in the classical paper of Dambrine et al. [4], concerning the modeling of a Ku–K-band low-noise GaAs MESFET with 0.3-μm long and 280-μm wide twofold driven

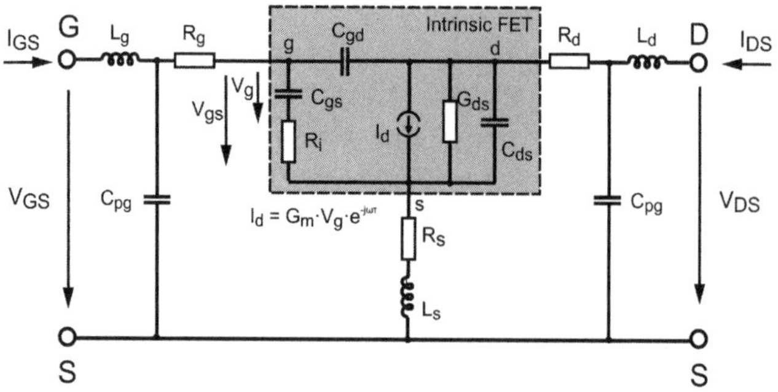

Figure 1.1 Standard 15-element small-signal equivalent circuit of a field-effect transistor. After Dambrine et al. [4].

gate (pi-gate), all parasitic model parameters can be determined sequentially in a direct way. With the extrinsic elements known, the intrinsic elements are determined with closed-form equations.

Various bias points were used to simplify the topology for model parameter extraction. Parasitic resistance and inductance parameters were extracted from cold measurements ($V_{ds} = 0$) under gate forward biasing. In this case, the current source and capacitances of the intrinsic transistor are negligible, so this part of the equivalent circuit is approximated as a resistive T-network that consists of fractions of the channel resistance. Then, the parasitic inductance and resistance parameters were determined from the imaginary and real parts of the Z-parameters, respectively, derived from measured S-parameters by matrix transformation. An additional relationship is required to find all the resistance parameters (four). It was proposed to obtain this relationship by the Fukui method [5], DC measurements, or analysis of the technological implementation.

Cold pinch-off biasing was used to determine the pad capacitance parameters. Under this biasing condition, the intrinsic transistor is reduced to a capacitive network. Then, at low frequencies, the inductance and resistance parameters are negligible and it is possible to find the values of the pad capacitance parameters by linear curve fitting.

With the values of the parasitic parameters known, the intrinsic-transistor parameters were extracted employing closed-form equations. This intrinsic-transistor parameter extraction was first proposed by Dambrine et al. [4] using frequency-constrained approximations to derive the required equations. Such approximations were later overcome by Berroth and Bosch [6], publishing more complete equations, leading to more accurate extraction results and better agreement between simulation and measurements at higher frequencies.

The standard 15-element model of Figure 1.1 is still widely used and accepted. It is robust, and the model elements can be reliably determined in a straightforward way. Frequently an additional resistance R_{gd} in series with C_{gd} is included to account for the basic device technological symmetry.

As previously mentioned the permanent increase of device operation frequency (focusing now on terahertz applications [7]) has made it necessary to expand the topology of equivalent circuits to account for distributed effects in the transistor layout [8]. Investigating

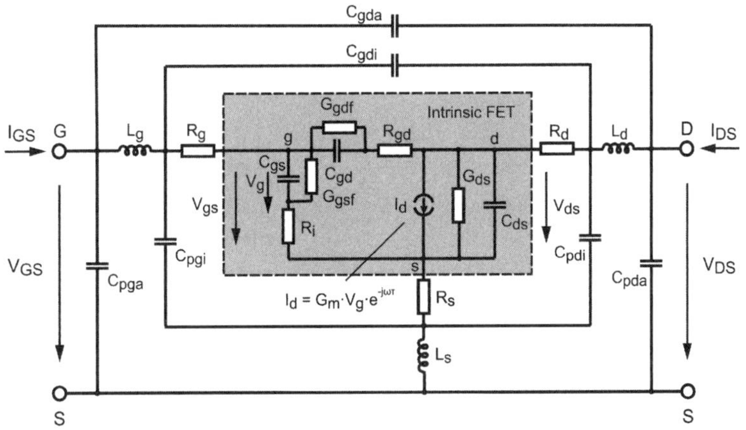

Figure 1.2 Distributed 22-element small-signal equivalent circuit of a field-effect transistor describing (1) large-size gate peripheries of multifinger power devices with interelectrode coupling, and (2) high-frequency operated small-size devices with strong stray capacitance effects.

sub-0.2 μm GaAs PHEMTs, the complex model topology shown in Figure 1.2 was proposed [2, 9]. The pad capacitance parameters in Figure 1.1 appear now as fractional parts, C_{pga} and C_{pgi} of C_{pg}, and C_{pda} and C_{pdi} of C_{pd}, respectively. The intrinsic part of transistor is modeled similar to Figure 1.1, but comprises an additional gate-drain resistance R_{gd}. Furthermore, conductance G_{gsf} models gate-leakage effects or describes forward-gate operation. Similarly, conductance G_{gdf} models drain-leakage current or breakdown effects.

Equivalent circuits with distributed capacitive effects have also become necessary to adequately model advanced multifinger power GaN HEMTs [10, 11]. Extended electrode bus bars contribute significantly to parasitic capacitance parameters, as well as interelectrode and crossover capacitive effects that appear due to source airbridges and field plates.

It is well known that the reliable and unique extraction of the model parameters of such a complex device model is very challenging. The working group of the authors has spent tremendous effort to tackle this problem.

Commonly, when the number of model elements becomes larger than eight, then it is not possible to extract the element values in a given bias point, as was already pointed out in the simple case of the standard 15-element model. In this case, using various bias points such as cold pinch-off and cold forward, the number of elements can be appropriately reduced so that direct determination from *S*-parameter measurements is still possible.

Regarding the complexity of the equivalent circuit with distributed capacitive effects shown in Figure 1.2, the use of the classical direct parameter extractions is inadequate. Rather, optimization approaches have been applied to find an acceptable (i.e., physically meaningful) solution set of the model parameter values. During the last five decades, a great amount of global and local optimization methods have been developed and applied in diverse fields. Typically, a desired network topology is chosen and tentative values are assigned to the network elements. Then, in each iteration, these values are altered in such a way that a preassigned error function is reduced. This iterative process is continued until the error function is minimized. A major disadvantage of optimization algorithms is that generally there is no guarantee that the iterations will converge to an optimal (i.e., physically meaningful) set of model parameters.

The optimization run is under the control of various conditions, such as choice of object function, starting values, termination criteria, definition of limited searching space of variables, suppression of unphysical model parameter values (notably concerning negative resistances), and provisions to escape from local minima.

Regarding model parameter extraction using optimization approaches, random search techniques [12, 13] with multistart options have been used frequently with predefined value ranges for every model element. As for the conventional gradient-based technique that is also frequently used in optimization approaches, which is principally a local optimizer, a satisfactory minimum can be found only with considerable expert guidance of the optimization runs. Even when these algorithms are restarted from several randomly chosen initial points, results are seldom satisfactory. Expert consideration of the equivalent-circuit element values related to a local minimum, with respect to the actual physics of the device, can be used to estimate a superior starting point for each subsequent gradient-based optimization run. This guiding process of the optimization algorithm is painstaking but can, at the end, result in realistic equivalent circuits for microwave devices [14].

The performance of gradient-based algorithms strongly depends on the initial values supplied. Several optimization runs with different initial guesses might be necessary if no a priori knowledge is available. There is a tremendous risk that the optimizer is stopped at a local minimum. Gradient-based optimizers do not yield a good result, although their initial progress is fast [15]. The major cost of gradient-based algorithms is the engineering time required to examine the extracted model parameter values, to decide on its merits, and to choose physically meaningful starting points [14].

Global optimization algorithms, such as genetic algorithm (GA) (e.g., [16]), simulated annealing (SA) (e.g., [17–20]) and tree annealing (TA) (e.g., [14, 21]), overcome the main disadvantage of the aforementioned local algorithms merely moving downhill. Probabilistic rules of decision permit to escape from local minima (hill-climbing). Nevertheless, although global optimization algorithms are conceptually simple, finding optimal algorithm settings for the parameters is by no means simple and straightforward [22]. Commonly the algorithm settings are problem-dependent and are best accomplished through trial and error. Global optimizers are rather time-consuming.

Decomposition methods for model parameter extraction that are based on parameter sensitivity analysis have been proposed. Model elements are sorted into different groups with stronger or weaker influence on the device response ([23–25]). As the model parameters interact with each other, the (sub-)problems cannot be solved independently but rather iteratively. Although classified as global-optimizer approaches, the relatively large spreads of extracted results indicate a high risk of the algorithms getting caught in local minima.

Global optimization algorithms are robust in finding solutions in larger search spaces without human intervention. However, as is known, the larger the search space (i.e., the larger the defined initial element values from the true value), the higher the risk to be trapped in a local minimum.

Over many years, the author's working group followed a different search strategy: How to find high-quality initial starting values for the model parameters, which already roughly reflect the solution of the optimization problem. In this case, a strict global search is no longer needed. Having generated high-quality initial model element values, it may be expected that local optimizers focusing on pure value refinement can deliver a reliable final result. Moreover, from a practical point of view, taking into account the measurement uncertainties of available vector network analyzers, optimization refinement may become redundant.

High accuracy in the measurement of microwave field-effect transistors (FETs) is a necessity for the development of physically based consistent device models. Device characterization and modeling are useful for better understanding the physical properties of the transistor and for studying the influence of small variations in geometry and material parameters on the electrical characteristics. However, with respect to the final device application, reliable device models are mostly developed for computer-based design purposes.

The measurement techniques for experimental characterization of microwave devices and model verification have also evolved dramatically in recent years [26, 27]. Beyond the classical S-parameter measurement with a vectorial network analyzer (VNA), nonlinear measurement concepts have been developed such as the large-signal network analyzer (LSNA) for large-signal device verification but also for direct large-signal model parameter extraction [28, 29]. Pulsed-DC measurements of microwave devices are necessary to characterize bias-dependent surface and barrier trapping effects as well as trapping dynamics [30–32].

Wide-bandgap semiconductor materials such as gallium nitride permit high breakdown electric fields, and are therefore virtually predestined for high power generation. Thus, large-size gate-periphery AlGaN/GaN power HEMT devices can deliver output powers of 100W and above [33]. This needs a corresponding performance adjustment of conventional soure/load-pull measurement systems. Starting from classical waveform measurement technique, the development of a 100W broadband source-/load-pull setup is described step-by-step. It is based on the IQ-modulator concept [34–37]. The key point of the presented setups is the split-signal type configuration. The forward and backward injected signals to the device-under-test are under control of a single microwave source, feeding all active load modules (IQ modulator plus frequency multiplier for higher harmonics). This is in contrast to the active source and load-pull system reported in [38], using a number of Agilent ESG series signal generators for signal magnitude and phase adjustment. Apart from high cost, the synchronization may be challenging (jitter) and may contribute to additional measurement errors.

In addition to high saturation output power, high linearity of the device is required. A frequency-domain measurement setup for intermodulation distortion characterization of GaN power devices including passive envelope load-pull testing is described. The issues in IMD measurements are illuminated. This covers in particular the bias tee configuration and the arrangement of measurement setup components that differs from conventional solutions.

The modeling of microwave active devices has become very complex. This relates in particular to AlGaN/GaN power transistors with large-size gate-periphery i.e. multifinger (N > 2) gate structures with strong interelectrode capacitive effects [39, 40]. Complex topologies may also be adopted for smaller-size transistors (N ≤ 2) at high frequencies (e.g., > 80 GHz) due to stray capacitance effects [8]. Having determined the extrinsic model parameters uniquely these can be eliminated from the complex model topology by the well-known de-embedding procedure. The remaining intrinsic active part is nonlinear and influenced by different factors such as bias-dependency, electrical and thermal dispersion, and gate- and drain-lag effects. Methods for characterization will be described, and implementation into the nonlinear device model discussed. Based on frequency- and time-domain large-signal measurement setups the developed nonlinear model is validated.

In addition to the investigated measurement-based modeling approach, analytical and behavioral modeling tasks will be illuminated, testing popular models on the basis of measured data of a 4-mm AlGaN/GaN power HEMT and giving a brief overview of the latest evolution.

REFERENCES

[1] Müller, J.-E.,"Investigation of GaAs MESFET Small Signal Equivalent Circuits for Use in a Cell Library," *Proc. European Microwave Conf.*, 1989, pp. 991–996.

[2] Kompa, G., "Reliable Extraction of Small-Signal Elements of a Generalized Distributed FET Model," *IEEE MTT-S Int. Microwave Symp. Dig.*, Baltimore, USA, 1998, pp. 291–294.

[3] Kompa, G., *Basic Properties of III-V Devices–Understanding Mysterious Trapping Phenomena*, Kassel, Kassel University Press GmbH, 2014.

[4] Dambrine, G. A., A. Cappy, F. Heliodore, and E. Playez, "A New Method for Determining the FET Small-Signal Equivalent Circuit," *IEEE Trans. Microw. Theory Techn.*, Vol. 36(7), 1988, pp. 1151–1159.

[5] Fukui, H., "Determination of the Basic Device Parameters of a GaAs MESFET," *The Bell System Technical Journal*, Vol. 58(3), 1979, pp. 771–797.

[6] Berroth, M., and R. Bosch, "Broad-Band Determination of the FET Small-Signal Equivalent Circuit," *IEEE Trans. Microw. Theory Techn.*, Vol. 3(7), 1990, pp. 891–895.

[7] O'Hara, J. F., S. Ekin, W. Choi, and I. Song, "A Perspective on Terahertz Next-Genearation Wireless Communications," *Technologies*, Vol. 7(2), Article 43 (18 pages), *https://www.mdpi.com.*

[8] Tasker, P. J., and J. Braunstein,"New MODFET Small Signal Circuit Model Rrequired for Millimeter-Wave MMIC Design: Extraction and Validation to 120 GHz," in *IEEE MTT-S Int. Microw. Symp. Dig.*, May 1995, pp. 611–614.

[9] Novotny, M., and G. Kompa, "Unique and Physically Meaningful Extraction of the Bias-Dependent Series Resistors of a 0.15 μm PHEMT Demands Extremely Broadband and Highly Accurate Measurements," in *IEEE MTT-S Int. Microw. Symp. Dig.*, June 1996, pp. 1715–1718.

[10] Schwantuschke, D., M. Seelmann-Eggebert, P. Brückner, R. Quay, M. Mikulla, and O. Ambacher, "A Fully Scalable Compact Small-Signal Modeling Approach for 100 nm AlGaN/GaN HEMTs," in *Proc. European Microwave Integrated Circuits Conf. (EuMIC)*, 2013, pp. 284–287.

[11] Nguyen, T. T.-L., and S.-D. Kim, "A Gate-Width Scalable Method of Parasitic Parameter Determination for Distributed HEMT Small-Signal Equivalent Circuit," *IEEE Trans. Microw. Theory Techn.*, Vol. 61(10), 2013, pp. 3632–3638.

[12] Kompa, G., "Small- and Large-Signal Characterization Methods," in *Proc. IEEE Workshop on Measurement Techniques for Microwave Device Characterization and Modelling*, Stuttgart, Germany, April 23, 1990, pp. 67–97.

[13] Kompa, G., and M. Novotny, "Highly Consistent FET Model Parameter Extraction Based on Broadband *S*-Parameter Measurements," in *IEEE MTT-S Int. Microw. Symp. Dig.*, June 1992, pp. 293–296.

[14] G. L. Bilbro, M. B. Steer, R. J. Trew, C. R. Chang, and S. G. Skaggs, "Extraction of the Parameters of Equivalent Circuits of Microwave Transistors Using Tree Annealing," *IEEE Trans. Microw. Theory Techn.*, Vol. 38(11). 1990, pp. 1711–1718.

[15] Heitzinger, C., T. Binder, and S. Selberherr, "Parallel TCAD optimization and parameter extraction for computationally expensive objective functions," *Proc. 15th European Simulation Multiconference: Modelling and Simulation (ESM 2001)*, Prague, Czech Republic, June 2001, pp. 534–538.

[16] Winter, G., *Genetic Algorithms in Engineering and Computer Science*, Wiley, 1995.

[17] Kirkpatrick, S., C. D. Gelatt, and M. P. Vecchi, "Optimization by simulated annealing", *Science*, Vol. 220, no. 4598, 1983, pp. 671–680.

[18] Corana, A., M. Marchesi, C. Martini, and S. Ridella, "Minimizing multimodal functions of continuous variables with the "simulated annealing" algorithm", *ACM Trans. on Mathematical Software*, Vol. 13(3), 1987, pp. 262– 280.

[19] Goffe, W. L., "SIMANN: A global optimization algorithm using simulated annealing," *Studies in Nonlinear Dynamics & Econometrics*, Vol. 1(3), 1996, 169–176.

[20] Ingber, L., "Adaptive simulated annealing (ASA): Lessons learned," *Control and Cybernetics*, Vol. 25(1), 1996, pp. 33–54.

[21] Bilbro and, G. L., W. E. Snyder, "Optimization of functions with many minima," *IEEE Trans. on Systems, Man, and Cybernetics*, Vol. 21(4), 1991, pp. 840–849.

[22] Shojaee, K., H. Shakouri, and M. Behnam Taghadosi, "Importance of the initial conditions and the time schedule in the simulated annealing," in: *Simulated Annealing, Theory with Applications* (Ed. Rui Chibante), 2010, pp. 217–233, (*http://www.intechopen.com/ books/simulated-annealing-theory-with-applications*).

[23] Kondoh, H., "An accurate FET modelling from measured *S*-parameters," in *IEEE MTT-S Int. Microw. Symp. Dig.*, June 1986, pp. 377–380.

[24] Patterson, A. D., V. F. Fusco, J. J. McKeown, and J. A. C. Stewart, "A systematic optimization strategy for microwave device modelling," *IEEE Trans. Microw. Theory Techn.*, Vol. 41(3), 1993, pp. 395–405.

[25] van Niekerk, C., P. Meyer, D. M. M.-P. Schreurs, and P. B. Winson, "A robust integrated multibias parameter-extraction method for MESFET and HEMT models," *IEEE Trans. Theory Techn.*, Vol. 48(5), 2000, pp. 777–786.

[26] Vye, D., "Fundamentally Changing Nonlinear Microwave Design," *Microwave Journal*, March 2010, pp. 22–39.

[27] Root, D. E., "Future Device Modeling Trends," *IEEE Microwave Magazine*, 2012, pp. 45–59.

[28] Werthof, A., F. van Raay, and G. Kompa, "Direct Nonlinear Power MESFET Parameter Extraction and Consistent Modeling," *IEEE MTT-S Int. Microwave Symp. Digest*, 1993, pp. 645–648.

[29] Schreurs, D., J. Wood, N. Tufillaro, D. Usikov, L. Barford, and D. E. Root, "The Construction and Evaluation of Behavioral Models for Microwave Devices Based on Time-Domain Large-Signal Measurements," *Int. Electron Devices Meeting* (IEDM), 2000, pp. 819–822.

[30] Filicori, F, G. Vannini, A. Mediavilla, and A. Tazon, "Modelling of Deviations Between Static and Dynamic Drain Characteristics in GaAs FETs," *23nd European Microwave Conference (EuMC)*, 1993, pp. 454–457.

[31] Jardel, O., F. D. Groote, C. Charbonniaud, T. Reveyrand, J. P. Teyssier, R. Quéré, and D. Floriot, "A Drain-Lag Model for AlGaN/GaN Power HEMTs," *IEEE MTT-S Int. Microwave Symp. Digest*, 2007, pp. 601–604.

[32] Jardel, O., F. D. Groote, T. Reveyrand, J. C. Jacquet, C. Charbonniaud, J. P. Teyssier, D. Floriot, and R. Quéré, "An Electrothermal Model for AlGaN/GaN Power HEMTs Including Trapping Effects to Improve Large-Signal Simulation Results on High VSWR," *IEEE Trans. Microw. Theory Techn.*, Vol. 55(12), 2007, pp. 2660–2669,

[33] Kikkawa, T., Mitani, E., Joshin, K., Yokokawa; S., and Tateno, Y., "An Over 100 W CW Output Power Amplifier Using AlGaN/GaN HEMTs," *https://csmantech.org/OldSite/Digests/2004/2004Papers/6.1.pdf*

[34] Poulin, D. D., J. R. Mahon, and J.-P. Lanteri, "A High Power On-Wafer Pulsed Active Load Pull System," *IEEE Trans.*, Vol. MTT-40, 1992, pp. 2412–2417.

[35] ATN Microwave Inc., *Product Feature: A Load-Pull System for Digitial Mobile Radio Power Amplifiers*, Microwave Journal, Vol. 38(3), March 1995, pp. 116–118.

[36] ATN Microwave Inc., *Product Feature: A Load-Pull System with Harmonic Tuning*, Microwave Journal, Vol. 39(3), March 1996, pp. 128–132.

[37] van Raay, F., and G. Kompa, "Waveform Measurements - The Load-Pull Aspect," *55th ARFTG Conference Digest*, Vol. 37, June 2000, pp. 14–21.

[38] Williams, D. J., J. Leckey, and P. Tasker, "A Study of the Effect of Envelope Impedance on Intermodulation Asymmetry Using a Two-Tone Time Domain Measurement System," *IEEE MTT-S Int. Microwave Symp. Digest*, 2002, pp. 1841–1844.

[39] Rudolph, M., C. Fager, and D. E. Root, *Nonlinear Transistor Model Parameter Extraction Techniques*, Cambridge: Cambridge University Press, 2011 (p. 69*ff.*)

[40] Luo, P., *GaN HEMT Modeling Including Trapping Effects Based on Chalmers Model and Pulsed S-Parameter Measurements*, Dissertation, Brandenburg University of Technology Cottbus-Senftenberg, 2018; published by Cuvillier Verlag, Göttingen, Germany, 2019.

Chapter 2

Transistor Concepts: MESFET, HEMT, and HBT

2.1 INTRODUCTION

This chapter gives a short overview on compound semiconductor-based transistor concepts, such as metal-semiconductor field-effect transistor (MESFET), high electron mobility transistor (HEMT), and heterobipolar transistor (HBT). The evolution of the III-V semiconductor devices is briefly reviewed. Construction and basic functioning of the various types are briefly described. Basic understanding of device physical structure and principal device operation is a crucial aspect in the device modeling process. For further information concerning III-V compound semiconductor materials, device physics, issues in device processing, and, in particular, trap-induced current collapse, the interested reader may be referred to [1].

2.2 EVOLUTION OF FET DEVICES

There are two types of microwave transistors: field-effect transistor and bipolar transistor. The main difference between these two types is that in field-effect transistors (FETs) the current flow is unipolar (i.e., the current is carried by one type of carriers only), whereas carriers with both signs are involved in bipolar devices. Regarding heterojunction bipolar transistors, hole current became negligible with the introduction of a wideband emitter. The resulting emitter-base heterojunction constitutes a large barrier for holes so that the holes from the base are back-injected.

2.2.1 Field-Effect Transistors

Table 2.1 illustrates the evolution of III-V transistors and their respective technological milestones. In 1952, Shockley [2] presented a first mathematical analysis of the operation of a junction field-effect transistor (JFET). Further details are given in Chapter 4. In 1966 Mead [3] proposed a Schottky barrier junction instead of a $p-n$ junction for the gate electrode, which provided higher isolation between drain and gate electrodes (less gate leakage current). The new device, termed MESFET (metal-semiconductor field-effect transistor), was first limited to small-signal low-noise applications. The first power MESFET was published in 1973 by Fukuta et al. [4,5].

A major breakthrough in FET device operation came in 1970 when Esaki and Tsu predicted the accumulation of carriers in a heterointerface [6]. However, it was not until eight

Table 2.1
Evolution of III-V Transistors and Respective Technological Milestones

Device	Type	Publisher	Year	Ref.
JFET	*n*-Ge, *p-n* junction gate electrode	W. Shockley	1952	[2]
MESFET (GaAs)	n-GaAs, Schottky barrier junction gate electrode	C. A. Mead	1966	[3]
2-DEG invention	Heterojunction	L. Esaki, R. Tsu	1970	[6]
Power-MESFET	GaAs, 1.6W output power @ 2 GHz, 5 dB gain	Fukuta et al.	1973	[5]
2-DEG confirmation	Heterojunction	R. Dingle et al.	1978	[7]
LM-HEMT	first report on AlGaAs/GaAs HEMT	T. Mimura et al.	1980	[8]
TEGFET	$Al_xGa_{1-x}As/GaAs$	D. Delagebeaudeuf et al.	1980	[9]
	Theoretical calculations on 2-DEG	D. Delagebeaudeuf et al.	1982	[10]
LM-HEMT on InP	$Al_{0.48}In_{0.52}As/Ga_{0.47}In_{0.53}As/InP$	C. Y. Chen et al.	1982	[13]
HBT concept	BJT with a wide-gap emitter	H. Kroemer	1957	[28]
HBT realization	AlGaAs/GaAs	P. M. Asbeck et al.	1981	[29]
HBT on InP	$Al_{0.48}In_{0.52}As/Ga_{0.53}As/InP$	Malik et al.	1983	[30]
PM-HEMT	$In_{0.2}Ga_{0.8}As/GaAs$ strained-layer superlattice	T. E. Zipperian et al.	1983	[14]
	$In_{0.15}Ga_{0.85}As/GaAs$ quantum well on GaAs	J. J. Rosenberg et al.	1985	[15]
	$In_{0.15}Ga_{0.85}As/Al_{0.15}Ga_{0.85}As/GaAs$, 0.25-μm gate	T. Henderson et al.	1986	[16]
MM-HEMT	$In_{0.30}Al_{0.70}As/In_{0.30}Ga_{0.70}As/GaAs$	P. Chu et al.	1988	[17]
	$Al_{0.5}In_{0.5}As/Ga_{0.5}In_{0.5}As/GaAs$	G.-W. Wang et al.	1988	[18]
AlGaN/GaN HEMT	first 2DEG confirmation	Khan et al.	1992	[19]
GaN HEMT	first AlGaN/GaN device (DC)	Khan et al.	1993	[20]
	first Microwave GaN MESFET (RF)	Binari et al.	1994	[21]
	first Microwave HFET (RF)	Khan et al.	1994	[22]
	doped channel AlGaN/GaN HEMT	Khan et al.	1996	[23]
GaN HBT	AlGaN/GaN	Ren et al.	2000	[31]

years later, in 1978, that this accumulation was demonstrated experimentally for the first time by Dingle et al. [7]. By using this concept, they achieved higher electron mobilities in GaAs than ever reported before. This initiated the development of a new type of field-effect transistor. In 1980 Mimura et al. at Fujitsu Laboratories reported on the first transistor ever based on a heterojunction structure [8]. Their new transistor had a gate length of 400 μm and showed high-speed microwave capabilities. Because of the high electron mobility in the device, the transistor was given the name high electron mobility transistor (HEMT). However, transistors of this type were under fabrication at other laboratories at the same time.

Emphasizing the epitaxial layer doping aspect with the new device, at AT&T Bell Labs they named their device selectively doped heterojunction transistor (SDHT) while the device fabricated by Thomson-CSF was given the name two-dimensional electron gas FET (TEGFET), whereby focusing on the electron distribution aspect [9]. Yet another name, modulation-doped FET (MODFET), was used for the transistor that was fabricated collaboratively by the University of Illinois and Rockwell International in the United States stressing the epitaxial layer doping aspect of the device. The latter device had a gate length of only 1 μm and was the first working device with reasonable microwave performance [9]. Thus, all these names describe various aspects of the device operation, which is commonly termed heterostructure field-effect transistor (HFET).

Initially the development of HFETs was mostly based on trial and error because a useful theory to describe the device characteristics was not available. However, in 1982 Delagebeaudeuf and Linh at Thomson-CSF developed such a theory [10]. Later the same year, Drummond and Morkoç refined this theory further [11]. Since then, a series of different material combinations have been tested to improve the transistor performance.

In principle lattice matching is required to grow a material on a substrate. But for thin layers, the growth of pseudomorphic material is possible, the thickness has to be lower than a critical thickness that depends on the lattice mismatch [1]. The main material system investigated was AlGaAs/GaInAs on GaAs showing a higher conduction band discontinuity than AlGaAs/GaAs.

The performance of the lattice-matched (LM) $Al_xGa_{1-x}As/GaAs$ MODFET was outperformed by the pseudomorphic (PM) heterostructure transistor such like the $Al_xGa_{1-x}As/In_yGa_{1-y}As$ MODFET on GaAs substrate. With growing fraction of indium in $In_yGa_{1-y}As$ the saturated velocity of free electrons in the 2DEG increases. In addition a larger conduction band discontinuity is found at the $Al_xGa_{1-x}As/In_yGa_{1-y}As$ interface in contrast to the $Al_xGa_{1-x}As/GaAs$ interface. This gives rise to higher density of free carriers in the channel. However, the $Al_xGa_{1-x}As/In_yGa_{1-y}As$ material system is not lattice-matched and, therefore, more difficult to control. The indium content and the $In_yGa_{1-y}As$ layer thickness can only be increased to some critical value. Beyond the critical thickness that depends on the indium content, disruptive dislocations occur in the crystal and worsen the electrical performance of the MODFET. Then the InGaAs is no longer pseudomorphic. A further increase of the saturation velocity cannot be expected in such metamorphic (MM) InGaAs layers [12].

Lattice-matched HEMTs on InP entail the advantages of high conduction band discontinuity, transport in unstrained $In_{0.5}Ga_{0.5}As$ providing high mobility and high intervalley separation and high electron velocity. The structure of the InP HEMT is similar to the AlGaAs one, except that an AlInAs/GaInAs heterojunction on an InP substrate is used. In this case the lattices are more closely matched, which allows a large amount of indium content. This results in increased mobility, which in turn results in increased electron velocity, increased conduction band discontinuity, increased 2DEG electron gas, and higher transconductance.

In 1982 C. Y. Chen et al. [13] reported about a first lattice-matched LM HEMT on InP. The conduction band discontinuity of InP-based MODFETs is much larger than those of GaAs-based MODFETs. Because of the high fraction of indium, a very high saturation velocity of the free electrons in the 2DEG can be obtained, increasing the cutoff frequency.

In1983, a pseudomorphic PM HEMT was reported by T. E. Zipperian [14]. For the first time, field-effect transistor action had been demonstrated in InGaAs/GaAs strained-layer superlattice SLS material. The samples, prepared by MBE, utilized 17 periods (repeat distance, 300Å) of alternating, modulation-doped, $In_{0.2}Ga_{0.8}As$ and GaAs layers to form the channel of the device.

In 1985 J. J. Rosenberg et al. [15] proposed a high electron mobility transistor HEMT utilizing a single $In_{0.15}Ga_{0.85}As/GaAs$ quantum well as conducting channel grown pseudomorphically on a GaAs substrate. The lattice-mismatched layer can be grown sufficiently thin so that the mismatch is accommodated entirely as elastic strain. In this situation the interface between the materials is essentially free from misfit dislocations and the thin layer is *pseudomorphic*. Thus, it had been shown that heterojunction systems other than $Al_xGa_{1-x}As/GaAs$ might be useful for devices provided that only a thin layer of one of the materials is needed.

In [16] Henderson et al. presented an $In_{0.15}Ga_{0.85}As/Al_{0.15}Ga_{0.85}As$ pseudomorphic modulation-doped field effect transistor (MODFET) (0.25-µm gate length) with excellent DC and millimeter-wave performance.

Further increase of indium fraction leads to metamorphic (MM) HEMT structures on GaAs. The high indium content InGaAs layers that provide these desirable characteristics are lattice-matched to InP but highly mismatched to GaAs substrates. In this case metamorphic buffer layer technology has been developed which provides a method through which high indium content layers may be grown on GaAs. The metamorphic buffer absorbs the strain of lattice mismatch and prevents the vertical propagation of dislocations, thereby maintaining the quality of the device active layers. In addition, metamorphic buffers allow the growth of practically any indium content layers, which is highly desirable for device optimization. Briefly, the metamorphic buffer layer is an AlGaInAs layer in which the indium concentration is linearly increased to expand the lattice constant from GaAs typically to that of InP.

In 1988, as reported by P. Chu et al. [17], $In_{0.30}Al_{0.70}As/In_{0.30}Ga_{0.70}As$ FETs were grown by molecular beam epitaxy MBE on GaAs substrates. At these compositions, the layers have an approximately 2.3% lattice mismatch relative to their GaAs substrates. In the same year, G.-W. Wang et al. [18] successfully grew $Al_{0.5}In_{0.5}As/Ga_{0.5}In_{0.5}As$ MODFET structures on lattice-mismatched GaAs substrates with a 3.8% difference of lattice constants.

Searching for new attractive III-V materials with superior properties such as higher breakdown and higher temperature for high-power applications of electronic devices, in 1992 Khan et al. [19] confirmed the presence of a two-dimensional electron gas (2DEG) in wide bandgap $Al_xGa_{1-x}N/GaN$ heterojunction samples, which then formed the basis of high electron mobility transistor devices in the $Al_xGa_{1-x}N$ material system. Already in 1993 the first DC functioning AlGaN/GaN device fabrication followed [20]. In 1994, Binari et al. reported on the first microwave GaN MESFET [21], and in same year, Khan et al. on a first GaN-based microwave heterojunction field-effect transistor [22]. In [23] Khan et al. reported about an AlGaN/GaN HEMT using a doped channel with increased cutoff frequency.

The extremely strong power amplification capability of GaN HEMTs has been demonstrated by the reports of a total output power of 230 W @ 2 GHz delivered by a single HEMT device (48-mm-wide recessed field-plated FET) [24] and a record high output power density of 40 W/mm at 4 GHz [25]. More recent efforts to expand the operation frequency of GaN HEMTs have realized W-band power amplification. In 2017 Fujisu announced the development of a gallium-nitride (GaN) high-electron mobility transistor HEMT power amplifier for use in W-band (75-110 GHz) transmissions [26]. 4.5W per millimeter of gate width have been achieved at 94 GHz, the world's highest output density in the W-band.

Table 2.2
MESFET Characteristics

Advantages	Drawbacks
Simple layer design	Electron transport in a doped layer
Implanted or epitaxial layers	→ low velocity
Simple optimization	→ low cutoff frequency
Low noise and power devices	
Mature technology	

Table 2.3
HEMT Characteristics

Type of HEMT	Advantages	Drawbacks
Conventional GaAs HEMT *n*-AlGaAs/GaAs	■ Electron separation ■ Better performance than MESFET ■ Same technology as MESFET ■ High electron mobility	■ Small conduction band discontinuity →low n_S (sheet carrier density) → poor charge control efficiency ■ DX center effects if Al content > 0.2 in AlGaAs
Pseudomorphic HEMT *n*-AlGaAs/InGaAs/GaAs	■ $\Delta E_C \sim 0.3$ eV ■ $n_S \sim 1.5 \times 10^{12}$ cm^{-2}	■ Transport properties in strained InGaAs similar to GaAs ones (strain effect)
Lattice matched HEMT on InP $Al_{0.5}In_{0.5}/In_{0.5}Ga_{0.5}As/InP$	■ $\Delta E_C \sim 0.5$ eV ■ $n_S \sim 3.0 \times 10^{12}$ cm^{-2}	■ InP substrate expensive, fragile ■ Low $In_{0.5}Ga_{0.5}As$ bandgap (0.8 eV) → impact ionization in the channel
Metamorphic HEMT on GaAs $In_{0.29}Al_{0.71}As/In_{0.3}Ga_{0.7}As/GaAs$	■ $\Delta E_C \sim 0.7 - 0.8$ eV ■ $n_S > 3.0 \times 10^{12}$ cm^{-2}	■ 2% mismatch between $Al_{0.5}In_{0.5}As/In_{0.5}Ga_{0.5}As$ and GaAs ■ Solution: Metamorphic buffer
Conventional GaN HEMT AlGaN/GaN	■ High breakdown voltage ■ High temperature operation	■ Current collapse

A comprehensive tutorial review on the HFET and its applications until 1992 was presented by Nguyen et al. [27]. Table 2.2 and Table 2.3 summarize advantages and drawbacks of discussed MESFET and HEMT devices (see e.g., [1]).

2.2.2 Heterojunction Bipolar Transistors

The idea of a heterojunction bipolar transistor concept was first proposed by H. Kroemer in 1957 [28]. The main goal was to improve the high-frequency performance of a conventional bipolar transistor by suppressing hole injection from the base into the emitter to increase the DC current gain and by speeding up transport across the base by reducing the base transit time. Kroemer had silicon and germanium in mind as the materials that would form the emitter/base heterojunction.

In fact, although the idea of HBT is as old as the homojunction transistor itself, HBTs had emerged as practical transistors only after molecular beam epitaxy MBE and metal-organic chemical vapor deposition (MOCVD) were developed to grow high-quality epitaxial layers in the 1980s. In 1981 the first heterojunction bipolar transistor (HBT) was fabricated in the AlGaAs/GaAs system by P. Asbeck et al. of Rockwell [29]. However, AlGaAs compound semiconductor was abandoned in the late 1990s in favor of InGaP. This possesses several unique material and fabrication advantages compared with AlGaAs. These advantages are its low DX center density [1], low surface recombination velocity, high ratio of valence-band-discontinuity (ΔE_V) to conduction-band-discontinuity (ΔE_C), and high etching selectivity between InGaP and GaAs [32]. The first InP-based HBTwas demonstrated at Bell Labs by Malik et al. in 1983 [30]. This device consisted of an InAlAs emitter and an InGaAs base and collector. Replacing *p*-GaAs base region by *p*-InGaAs results in larger conduction band discontinuity and improved charge transport. The (Al, In)As/(Ga,In)As layers were grown by MBE lattice-matched to a Fe-doped semi-insulating InP substrate. In 2000 Ren et al. reported about a first AlGaN/GaN HBT [31].

2.3 BASIC DEVICE STRUCTURES AND FUNCTIONING

This section briefly describes the physical structure and functioning of the most common microwave devices, such as MESFET, HEMT, and HBT.

2.3.1 MESFET

There are different classes of field-effect transistors (FETs): The metal-oxide-semiconductor FET and the junction FET (JFET) [1]. Regarding the junction FET, the junction may be a *pn* junction, which forms a *pn* JFET, or a Schottky barrier junction, which forms a metal-semiconductor FET, or MESFET. The MESFET was the working horse for microwave applications (wireless communications, radar) in the eighties, but today is of minor importance. Nevertheless, SiC MESFETs, with their wide bandgap, high thermal conductivity and radiation hardness compared to Si- and GaAs-based technologies, have undergone many structural modifications to enhance their performance [32]. Largely for educational reason the MESFET device, its physical structure, and its basic operation are briefly reviewed as follows.

2.3.1.1 Physical MESFET Structure

Figure 2.1 shows the cross section of the physical structure of a GaAs MESFET device. A highly pure semiconductor buffer layer is grown onto a semi-insulating (s.i.) GaAs substrate followed by an *n*-doped epitaxial layer (*n*-type active channel), which is then recessed to the final channel thickness by etching. This approach yields low bulk source and drain

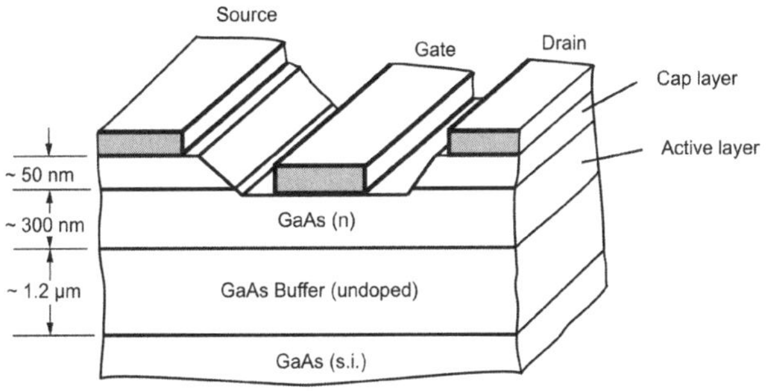

Figure 2.1 Simplified physical structure of a GaAs MESFET device (without passivation).

resistances. An epitaxially grown thin heavily doped cap layer (n^+-GaAs contact layer on the top of the active layer) facilitates the formation of low-resistance source and drain contacts. The buffer layer has the advantage that traps do not diffuse out of the substrate into the active epitaxial layer. The backside of the substrate is covered with a Au/Ge alloy, which is usually connected to the source terminal. Both drain and source are connected to the n-type layer through n^+ ohmic contacts. The carrier transport follows in the doped material, which reduces the drift velocity of the carriers and increases the noise. Field-effect transistors are typically covered by a passivation layer to reduce the influence of surface traps with respect to current dispersion [1]. Under typical operating conditions, the source contact of the MESFET is grounded. Assuming n-channel operation (the current is carried by electrons), the drain contact is positively biased with respect to the source. The Schottky gate contact can be either positively (forward) or negatively (reverse) biased with respect to the source. A MESFET has a negative temperature coefficient at high current levels, and hence the devices are thermally stable.

We distinguish between depletion and enhancement type of MESFET. Depletion devices are normally on devices requiring a negative gate-source potential to deplete the carriers through the channel region and turn the device off. In contrast, an enhancement MESFET, a positive gate potential must be applied in order to start drain-source conduction.

Figure 2.2 shows the band diagram of a GaAs MESFET in ON- and OFF-states. Assuming a homogeneously doped channel layer with doping concentration N_D and thickness a, the layer concentration n_s of free carriers in the channel (electron-density per unit area) can be estimated as (see Shockley model in Chapter 4),

$$n_s = N_D(a - h) \tag{2.1}$$

where

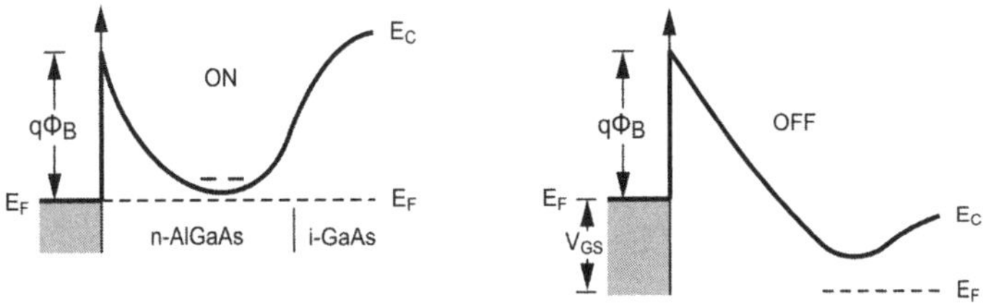

Figure 2.2 Band diagram in ON- and OFF-states of a GaAs MESFET device.

$$h = \sqrt{[2\varepsilon_s(V_{GS} + V_d)/qN_D]} \tag{2.2}$$

h is the thickness of the positively charged depletion region, ε_s is the dielectric constant of the semiconductor material, V_{GS} is the applied gate voltage, and q is the electron charge. Commonly the carrier concentration of MESFETS is related to unit volume.

Additional depletion regions are also formed under free surfaces due to surface states. These are caused by the abrupt interruption of the crystal structure and are sensitive to surface treatment. Surface states are partially filled with electrons that leave behind positive charges near the surface. The occupancy of surface states varies with the voltages applied to the MESFET. This mechanism is known to be frequency dependent and is one of the causes of the so-called frequency dispersion of the MESFET characteristics [1].

The wide bandgap property of SiC results in both high mobility and high breakdown voltage. A SiC MESFET can therefore have a frequency response comparable to that of GaAs MESFET, but breakdown voltages larger than silicon laterally diffused metal oxide semiconductor (Si LDMOS) transistors. This results in power densities of about 10 W/mm, which is ten times that of GaAs MESFET. The high thermal conductivity of the SiC substrate is particularly useful in high-power applications. The higher operating voltage and associated higher load impedance greatly simplify output networks and power combining.

2.3.1.2 Basic MESFET Operation

Regarding small-signal operation with fixed biasing, the DC quiescent point is under control of the DC drain and DC gate voltages. These voltages control the channel current by controlling the charge depletion region and the longitudinal electrical field in the channel. Three operation regions can be recognized for the $I_{DS}-V_{DS}$ characteristic curve of the MESFET: (1) the linear region, where I_{DS} is linearly proportional to V_{DS}, (2) the saturation region at higher V_{DS} values, where the current is almost constant, and (3) the pinch-off region, where the negative gate-source voltage is sufficiently high to suppress any channel current flow. Furthermore, breakdown and gate forward conduction must be considered under dynamic operation. Figure 2.3 illustrates the different modes of operation.

(1) *Linear operation region*: This implies very low drain-source voltages (e.g., $0 \leq V_{DS} \leq$ several 100 mV). The upper limit depends on the type of transistor. At very low drain voltage the MESFET behaves like a linear (voltage-controlled) resistor ($R = 1/[\sigma w(a-h(V_{GS}))]$ with the gate width w, electrical conductivity $\sigma = qN_D\mu_0$, and low-field electron mobility μ_0) as depicted in Figure 2.3(a). V_{DS} generates a longitudinal electrical field in the channel. As the voltage over the depletion region is larger at the drain side than at the source side ($V_{GS} < 0$), the thickness of the depletion layer increases toward the drain electrode. With increasing V_{DS} (and V_{GS} keeping constant) the cross section of the conductive channel reduces. With higher electrical field strength the velocity of the electrons increases, which causes a higher current. When V_{DS} is kept constant, and the negative gate control voltage is continuously increased, then the depletion layer grows and the conductive channel becomes smaller; the current decreases. When V_{GS} reaches the (extrinsic) pinch-off voltage, then the channel is totally pinched off; the drain current is zero, independent of the drain voltage. Most of the

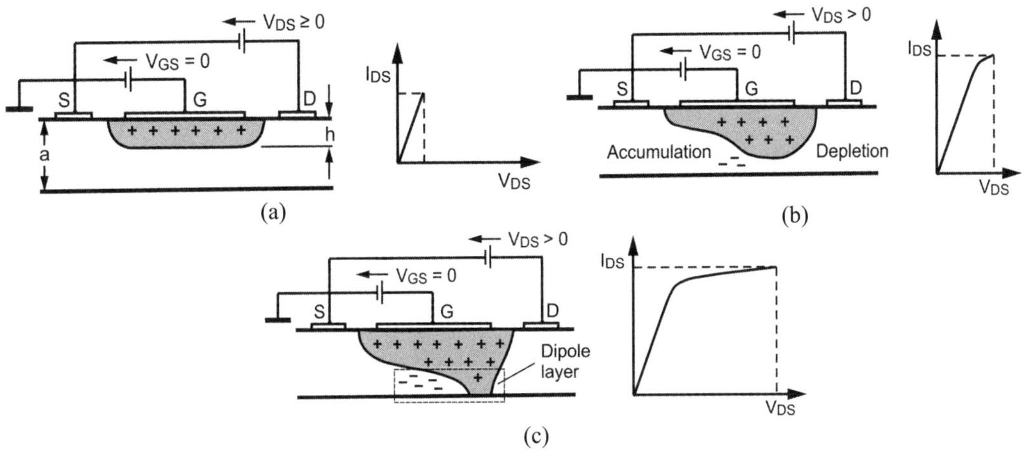

Figure 2.3 Space charge and $I_{DS}(V_{DS})$ characteristic of a MESFET at various bias conditions: (a) $V_{GS} = 0$, (b) pinch-off (onset of saturation), (c) saturation region.

depletion region is underneath the gate, which works as a reverse-biased Schottky diode.

(2) *Onset of saturation*: With growing V_{DS} the channel current increases; in particular, the voltage drop over the space-charge region at the drain side becomes higher (i.e., the channel cross section decreases strongly (Figure 2.3(b))). With respect to current continuity in the channel the electron drift velocity must grow toward the drain electrode. The electron velocity can reach a maximum value of the saturation drift velocity being approximately $v = 1.3 \times 10^7$ cm/s. When saturation velocity has been reached under the gate, then the electron density must increase to keep on fulfilling the condition for current continuity. As a consequence an accumulation of electrons is given at the drain side of the gate electrode. After leaving the gate electrode, the carrier density decreases due to the larger extension of the bulk semiconductor region so that an electron depletion region follows. The positive space-charge region originates from donors in the crystal. Due to V_{DS}, the electrical potential in the active channel increases from source to drain. Therefore, the channel-to-gate voltage is larger, and consequently, the depletion region is wider on the drain side of the gate.

(3) *Saturation region*: With further increase of V_{DS}, a continuously growing part of the voltage is dropped at the drain-sided end of the channel, which can be characterized as a stationary electrical double-layer, dipole layer, or sometimes it is called stationary Gunn domain (Figure 2.3(c)) [1]. Under this operating condition the electrons move with saturation drift velocity along most parts of the channel path.

The gate contact is constructed like a Schottky diode with mostly capacitive influence. At $V_{DS} = 0$ the gate channel capacitance is equally divided into the gate-source capacitance and gate-drain capacitance (device symmetry assumed). The capacitances can be derived from the incremental change of the gate space-charge region with respect to the change of the gate-source voltage and gate-drain voltage, respectively. In the saturation region, when the electrical dipole layer has been set up, then an increase of V_{DS} does not remarkably change the stored charge; that is, the gate-drain capacitance decreases to a value which relates to the remaining stray capacitance of the metal pattern of the electrode contacts. Conversely, the value of the gate-source capacitance increases in the saturation region and reaches approximately twice the value of that in the linear region, because it describes the complete gate capacitance with its space charge.

2.3.2 HEMT

The high electron mobility transistor HEMT is a field-effect device whose operation is very similar to that of the MESFET. The difference is that the HEMT utilizes a vertical structure, which comprises layers of different semiconductor materials. These form a device channel in which electrons are physically separated from their parent donors. This leads to large electron mobilities, as ionized impurity scattering is reduced. This results in improved high-frequency noise and gain characteristics compared to conventional GaAs MESFETs. The principle of operation in both MESFET and HEMT are similar. Differences are found in the charge control mechanism; that is, MESFETs are operated via channel thickness modulation and HEMTs via 2DEG carrier density modulation. Therefore, these devices can be described by similar equivalent circuit models.

2.3.2.1 Physical HEMT Structure

In Figure 2.4 a conventional AlGaAs/GaAs HEMT is shown. An undoped GaAs layer is epitaxially grown onto a semi-insulated GaAs substrate material. It follows a highly doped AlGaAs layer. A thin undoped AlGaAs is inserted to provide a clean interface between the AlGaAs and GaAs layer. Uppermost is a highly doped GaAs cap layer facilitating the formation of low resistance ohmic contacts to the device. After having grown the epitaxial HEMT layers, the wafers are passivated against environmental impact.

2.3.2.2 Basic HEMT Operation

In contrast to the MESFET, the carrier supply layer is now separated from the active channel providing superior electrical performance. Compared with the electron mobility in doped GaAs material, carriers that are separated from the positively charged donors exhibit higher mobility. Carrier separation is achieved by a heterostructure interface, where a wide bandgap

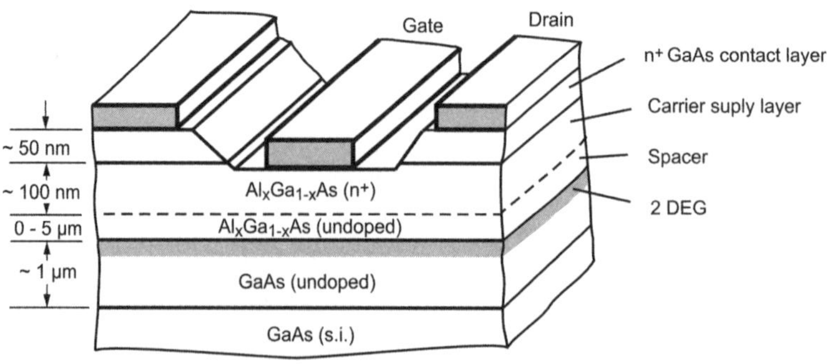

Figure 2.4 Physical construction of a conventional AlGaAs/GaAs HEMT.

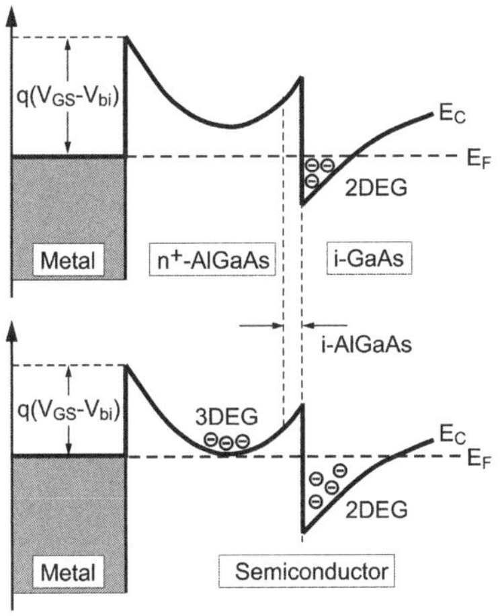

Figure 2.5 Energy band diagram of a HEMT: Normal operating mode above; additional parasitic conduction in the barrier (parasitic MESFET channel) at higher gate voltage below.

semiconductor is epitaxially grown onto a semiconductor material with smaller bandgap. At this interface the conduction bands of the paired semiconductors bend in such a way that a triangular (V-shaped) potential well occurs at the GaAs side, as illustrated in Figure 2.5. Due to the small extension of the V-shaped potential well at the AlGaAs/GaAs interface quantized subbands are formed (quantum well). As the layer with wider bandgap is doped (carrier supply layer), the free electrons diffuse into the potential well, forming the so-called two-dimensional electron gas 2DEG. The implementation of a thin undoped AlGaAs layer (spacer) between GaAs channel and doped AlGaAs enhances the mobility of the 2DEG. This results from reduced Coulomb scattering of the 2DEG at the positively charged donors near the AlGaAs/GaAs interface. The Coulomb potential of the donors diminishes with increased distance to the potential well. A further undoped layer under the gate would reduce the gate leakage current. The simplest HEMT structure comprises a Si doped AlGaAs layer grown onto an undoped GaAs substrate. Si atoms preferentially occupy donor sites in AlGaAs.

The current passing from drain to source through the 2DEG is determined by the sheet electron concentration beneath the gate contact, which can be modulated by varying the bias on the Schottky contact. In essence, the gate can be considered as a capacitor whose plates are formed by the gate metal and the 2DEG. The depleted AlGaAs donor and spacer layers form the dielectric of the capacitor. In the HEMT, the gate capacitance remains relatively constant with changing gate bias, as the plate separation is fixed by the thickness of the AlGaAs donor and spacer layers. Current modulation in the HEMT occurs as charge is added to or removed from the 2DEG in response to variations in voltage applied to the gate following the classical $Q = CV$ capacitor equation, where Q is the charge in the 2DEG, C is the gate capacitance and V is the applied voltage.

Under normal operation (Figure 2.5 (above)) the free electrons in the n^+-AlGaAs supply layer have moved to the lower-energetic potential well at the AlGaAs/GaAs interface forming there the 2DEG. The supply layer comprises a carrier-free positively charged space charge, which is separated from the 2DEG by a thin undoped spacer layer. Depending on the gate

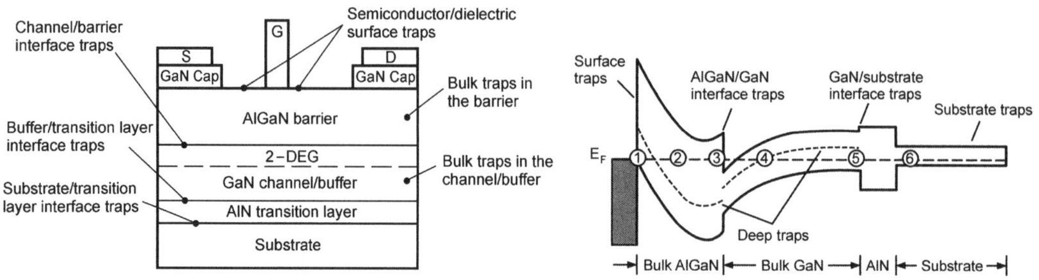

Figure 2.6 (a) Potential trap locations in a typical GaN HEMT and (b) band diagram of a doped AlGaN/GaN HEMT indicating possible trap sites [1].

control voltage, the position of the potential well is shifted up or down, influencing the electron density in the channel. At pinch-off (i.e., suffient high negative gate voltage) the V-shaped conduction band at the interface is entirely above the Fermi energy level. In this case, the channel is entirely empty. The current is zero.

With increasing gate voltage (less negative voltage value), the potential well is lowered more and more, until the conduction band edge of the AlGaAs carrier supply layer reaches the Fermi level. In this situation as shown in Figure 2.5 (below) carriers also occur in the wide bandgap layer leading to an additional channel path between source and drain. This conduction phenomenon is now similar to the MESFET operation principle, where part of the free carriers are propagating in a doped region with reduced electron mobility (parasitic MESFET channel).

2.3.2.3 Trapping Effects in GaN HEMTs

The presence of very large polarization fields in the GaN system is a fundamental difference between conventional III-V semiconductors and GaN. The large polarization fields lead to the formation of charged surface states, which are critical to the origin of the 2DEG in the HEMT structure in GaN [1].

Regarding GaAs-based field-effect transistors, the origin of traps and their densities are entirely related to the growth process of the device layers. With progress in material growth and devices, layer structure design results in ever-increasing suppression of trap densities and their effects on device performance. Meanwhile, GaAs technology is considered a mature technology. However, processing of GaN-based devices needs further improvement to exploit the outstanding properties of the wide bandgap material, such as high breakdown voltage, which is becoming increasingly attractive for high-power switching devices, and high output power, which, however, suffers from current collapse due to trapping phenomena. The possible mechanisms of the current collapse are (1) deep levels in the barrier layer under the gate metal, (2) deep levels in the interface and/or in the buffer layer, and (3) surface states.

Figure 2.6(a) illustrates the cross section of a AlGaN/GaN HEMT and indicates the various trap locations in the bulk materials and material interfaces. Possible trapping sites in a typical GaN HEMT are: (1) the semi-insulating substrate (influence of substrate traps may be negligible when the AlGaN/GaN interface is sufficiently far away from the substrate), (2) the interface between GaN and the substrate, (3) the GaN buffer layer, (4) the

AlGaN layer, and (5) the free surface of the AlGaN layer.

Figure 2.6(b) indicates possible trap sites from the perspective view of the band diagram of a doped AlGaN/GaN HEMT. Site 1 indicates the surface traps and site 2 bulk traps in the AlGaN barrier layer. Site 3 denotes AlGaN/GaN interface traps and site 4 bulk traps in the GaN buffer layer. Site 5 marks interface traps of transition from GaN buffer to the AlN thin AlN nucleation layer deposited on the substrate material [1].

2.3.3 HBT

The heterojunction-bipolar-transistor HBT is a further active device that has gained increasing attractiveness for application in the microwave and millimeterwave region. The concept of the HBT was suggested by H. Kroemer [28] who recognized that the usage of a wide bandgap emitter and a narrow bandgap base leads to band deformations that facilitates the electron injection into the base (*n -p-n* transistor), whereas the hole injection into the emitter is impaired as illustrated in Figure 2.7.

Table 2.4 indicates various HBT material systems and inherent band discontinuities [33]. Among all the HBT material systems, $Al_{0.3}Ga_{0.7}As$/GaAs HBTs have been widely studied and have also demonstrated good high-frequency performance. However, Table 2.4 shows that most of the bandgap discontinuity is due to the conduction band offset that reduces the emitter injection efficiency. In addition, Al in $Al_{0.3}Ga_{0.7}As$/GaAs HBTs generates DX centers [1], increasing the recombination currents further and therefore reducing the current gain.

$In_{0.5}Ga_{0.5}P$/GaAs has the most desirable band alignment out of the III-V compounds shown in Table 2.4 because most of the discontinuity in the band structure is in the valence band, which limits back injection of holes into the emitter. At the same time, the conduction band discontinuity is relatively small, which increases the emitter injection efficiency without the need for grading. This improves the current gain of the device. The problem related to DX centers in $Al_{0.3}Ga_{0.7}As$ are not present in $In_{0.5}Ga_{0.5}P$.

InP-based HBTs are attractive due to their compatibility with optical sources and detectors in optoelectronic integrated circuit applications. The smaller bandgap of $In_{0.53}Ga_{0.47}As$ (0.75 eV) compared to that of GaAs (1.42 eV) means that a lower base emitter voltage is required for operation, resulting in a low power consumption for these devices, which is advantageous for low power applications. Furthermore, HBTs with an $In_{0.53}Ga_{0.47}As$ base have higher peak electron velocity and lower surface recombination velocity compared to GaAs-based HBTs.

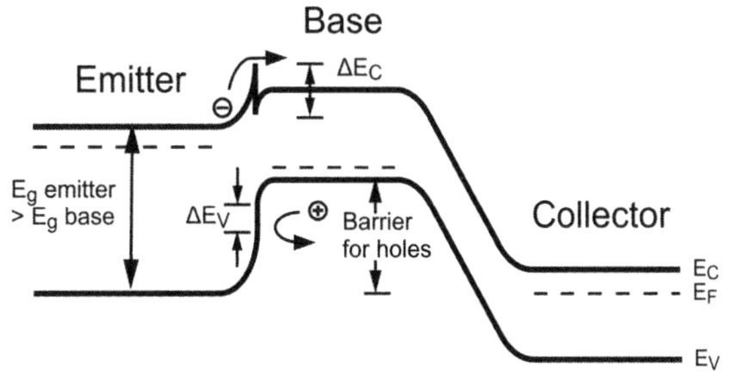

Figure 2.7 Schematic band diagram of a *n-p-n* abrupt heterojunction bipolar transistor.

Table 2.4
Band Discontinuities for Various HBT Material Heterostructures

Material Structure	ΔE_{g1} (eV)	ΔE_{g2} (eV)	ΔE_C (eV)	ΔE_V (eV)	ΔE_g (eV)
$Al_{0.3}Ga_{0.7}As$ / GaAs	1.86	1.42	0.28	0.15	0.43
$In_{0.5}Ga_{0.5}P$ / GaAs	1.92	1.42	0.12	0.38	0.5
InP / $In_{0.53}Ga_{0.47}As$	1.35	0.76	0.25	0.34	0.59
$In_{0.52}Al_{0.48}As$ / $In_{0.53}Ga_{0.47}As$	1.2	0.76	0.32	0.12	0.44
$In_{0.5}Al_{0.5}P$ / GaAs	2.35	1.42	0.31	0.62	0.93
Si / $Si_{0.75}Ge_{0.25}$	1.12	0.97	0.02	0.13	0.15

From [33].

These devices have a larger valence band offset when compared to the conduction band, which allows high injection efficiency with high doping. $In_{0.53}Ga_{0.47}As$ is also lattice matched to InP and therefore no grading of the base-emitter junction is required to remove the potential spike to improve the emitter injection efficiency [33].

One major problem associated with InP/$In_{0.53}Ga_{0.47}As$ HBTs, is the very low collector-emitter breakdown voltage, typically about 2.5V. This problem is less severe in DHBTs where a wide bandgap collector is used [34].

InP HBTs are ideal for submillimeter and THz power generation. In [34] a power amplifier with output power greater than 200 mW operating at a frequency greater than 200 GHz has been reported.

$In_{0.52}Al_{0.48}As/In_{0.53}Ga_{0.47}As$ has a higher value of conduction band offset compared to valence band offset, which implies low emitter efficiency. However, a large ΔE_C enables high-energy electron injection into the base, decreasing the base transit time and resulting in high-frequency operation and high bit rates.

An $In_{0.5}Al_{0.5}P$/GaAs material system is suitable for HBTs due to one of the largest values of ΔE_V, seen in any material: 0.62 eV. This provides excellent reduction of back injection efficiency and produces high current gains.

The Si/SiGe structure exhibits nearly ideal band discontinuities for a HBT. A high current gain can therefore be achieved from these devices. Due to the narrow bandgaps of both Si and Ge, they are not very suitable for high-temperature and high-power applications [33].

Figure 2.8(a) shows an initial HBT construction based on the AlGaAs/GaAs material system [29] and Figure 2.8(b) a more advanced InGaAs/InP HBT [35]. With an abrupt heterojunction, typically a spike-and-notch region in the conduction band at the surface is obtained (Figure 2.7). This energy spike increases the emitter-base turn-on voltage, limits the maximum current that can be driven through the device, and increases the emitter-base ideality factor. Increased emitter-base turn-on voltage undesirably increases power dissipation in the HBT. However, by grading the material composition of the heterointerface, the spike in the conduction band can be removed, resulting in a lowering of the barrier for electron injection into the base [1, 35].

In fact, the number of holes that transit from the base to the emitter is heavily reduced,

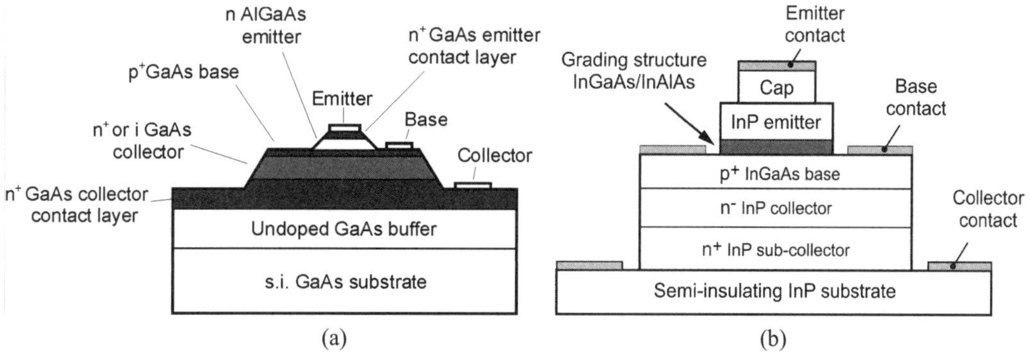

Figure 2.8 Schematic structure of (a) initial *n-p-n* AlGaAs/GaAs HBT construction [29] and (b) advanced InGaAs/InP HBTwith emitter-base grading structure [35] (passivation not shown).

since they have to pass the large potential energy step created at the base-emitter interface due to the heterojunction property. With this technique a greater gain, a higher breakdown voltage, and a higher operating frequency is obtained. The great advantage of HBTs over FETs is their greater linearity.

2.3.3.1 Physical HBT Structure

In the case of the AlGaAs/GaAs heterojunction bipolar transistor HBT (Figure 2.8(a)) the device is an *n-p-n* transistor with emitter up design. The collector comprises a heavily doped n^+-type contact layer and a lightly doped *n*- or *i*-layer to reduce the base-collector capacitance. The base is a heavily doped p^+-type layer in order to reduce the base resistance. It follows the low-doped *n*-type emitter (in order to reduce the base-emitter capacitance), which is capped with a heavily doped n^+-type GaAs layer. This layer shields the oxygen-sensitive emitter while serving as a contact layer. It would be rather difficult to form a low resistivity contact directly to AlGaAs. More tailored structures include compositional grading through the emitter layer to smooth out the potential spike at the base-emitter boundary that would otherwise hinder electron injection from the emitter. The epitaxial HBT structure is usually grown either using molecular beam epitaxy MBE or metal-organic vapor phase epitaxy (MOVPE).

The parasitic resistances and capacitances of the HBT can be far lower than known for the silicon bipolar transistor so that the RF performance is enhanced. The vertical design in combination with high power density significantly reduces the chip area required for a given output power.

2.3.3.2 Basic HBT Operation

The usage of a wide bandgap emitter and a narrow bandgap base leads to band deformations that facilitate the electron injection into the base (*n–p–n* transistor), whereas the hole injection into the emitter is impaired. These advantages also remain with a higher degree of doping of the base (at first Be and then C doping was applied) and low doping of the emitter.

The emitter injection efficiency of an HBT can be made very high. In this case, the transistor gain is limited by the recombination currents in the emitter-base space charge region, in the base region, and by surface recombination at the emitter edges. In devices where the overall recombination current (I_R) is small, the current gain β (β = $I_C/I_B \approx I_C/I_R$ could become very high (e.g., β ≥10.000). This increase in current gain can be used directly to enhance device performance or can be traded against increasing the base doping to reduce the base resistance, thus improving f_{max}. As a further example, the conduction band offset at the emitter-base junction means that the doping concentration of the emitter could be reduced, even to a level below that of the base, without compromising device operation. In this way, the emitter-base capacitance can be reduced.

The current flow in a HBT device is vertical so that surface imperfections have less effect on performance. Even though the base width in HBTs determines transit time across the base and, hence, device speed for the vertical device structure, in practical transistors the lateral dimensions of the emitter stripes are also very important. These dimensions determine parasitic capacitances, resistances, and power dissipation.

2.5 SUMMARY

Currently, a large variety of microwave devices exist that are based on different compound semiconductor material systems such as GaAs, InP, and GaN. Due to the continuous improvement of fabrication technologies such as molecular beam epitaxy, the device structures became more complex. Starting with homoepitaxial layers in MESFET devices, heterojunctions in HEMTs and HBTs opened the way for modern communication technology. Until now, regarding UMTS applications, the main focus was about 2 GHz, providing linear and high efficient power amplifiers. But, recently, within the scope of the 5G communication standard, higher frequencies are being attempted. This dynamic evolution of microwave devices, concerning growing design complexity, output power, linearity, efficiency, and operating frequency, is a permanent challenge not only for the device technologist but also for the device modeler who must now offer reliable device models for the circuit design just in time. The following chapters will discuss the various basic approaches to create a reliable field-effect transistor model for low- and high-power applications considering diverse parasitic effects such as temperature and frequency dispersion.

REFERENCES

[1] Kompa, G., *Basic Properties of III-V Devices–Understanding Mysterious Trapping Phenomena*, Kassel, Kassel University Press, 2014.

[2] Shockley, W., "A Unipolar Field-Effect Transistor," *Proc. I.R.E.*, 1952, pp. 1365–1376.

[3] Mead, C. A., "Schottky Barrier Gate Field Effect Transistor," *Proceedings of the IEEE*, 1966, pp. 307–308.

[4] Liechti, C. A., "Microwave Field-Effect Transistors—1976," *IEEE Trans.*, Vol. MTT-24, 1976, pp. 279–300.

[5] M. Fukuta, T. Mimura, I. Tujimura, and A. Furumoto, "Mesh Source Type Microwave Power FET," *1973 Int. Solid-State Circuits Conf.* (ISSCC 73), Dig. Tech. Papers, 1973, pp. 84–85.

[6] Esaki, L., and R. Tsu, "Superlattices and Negative Conductivity in Semiconductors," *IBM J. Res. Develop.*, Vol. 14, 1970, pp. 61–65.

[7] Dingle, R., H. L. Störmer, A. C. Gossard, and W. Wiegmann, "Electron Mobilities in Modulation-Doped Semiconductor Heterojunction Superlattices," *Appl. Phys. Lett.*, Vol. 33, 1978, pp. 665–667.

[8] Mimura, T., S. Hiyamizu, T. Fujii, and K. Nanbu, "A New Field-Effect Transistor with Selectively Doped GaAs/n-$Al_xGa_{1-x}As$ Heterojunctions," *Jpn. J. Appl. Phys.*, Vol. 19(5), 1980, pp. L225–L227.

[9] Delagebeaudeuf, D., P. Delescluse, P. Etienne, M. Laviron, J. Chaplart, and N. T. Linh, "Two-Dimensional Electron Gas M.E.S.F.E.T. Structure," *Electronic Lett.*, Vol. 16(17), 1980, pp. 667–668.

[10] Delagebeaudeuf, D., and N. T. Linh, "Metal-(n) AIGaAs-GaAs Two-Dimensional Electron Gas FET," *IEEE Electron Devices*, Vol. ED-29(6), 1982, pp. 955–960.

[11] Drummond, T. J., H. Morkoç, K.Lee, and M. Shur, "Model for Modulation Doped Field effect Transistor," *IEEE Electron Device Lett.*, Vol. EDL-3(11), 1982, pp. 338–341.

[12] Hülsmann, A., *Technologie zur Herstellung modulationsdotierter Feldeffektransistoren (MODFETs) unter Verwendung der Elektronenstrahl-Lithographie*, Doctoral Thesis, TU München, 1994.

[13] Chen, C. Y., A.Y. Cho, K.Y. Cheng, T. P. Pearsall, P. O'Connor, and P.A. Garbinski, "Depletion Mode Modulation Doped $Al_{0.48}In_{0.52}As$-$Ga_{0.47}In_{0.53}As$ Heterojunction Field Effect Transistors," *IEEE Electron Device Lett.*, Vol. EDL-3(6), 1982, 152–155.

[14] Zipperian, T. E., L. R. Dawson, G. C. Osbourn, and I. J. Fritz, "An $In_{0.2}Ga_{0.8}As$/GaAs, Modulation-doped, Strained-layer Superlattice Field-effect Transistor," *International Electron Devices Meeting (IEDM)*, 7-9 December, 1983, pp. 696– 699.

[15] Rosenberg, J. J., M. Benlamri, P. D. Kirchner, J. M. Woodall, and G. D. Pettit, "An $In_{0.15}Ga_{0.85}As$/GaAs Pseudomorphic Single Quantum Well HEMT," *IEEE Electron Device Lett.*, Vol. EDL-6(10), 1985, pp. 491–493.

[16] Henderson, T., M. I. Aksun, C. K. Peng, H. Morkoç, P. C. Chao, P. M. Smith, K.-H. Duh, and L. F. Lester, "Microwave Performance of a Quarter-Micrometer Gate Low-Noise Pseudomorphic InGaAs/AIGaAs Modulation-Doped Field Effect Transistor," *IEEE Electron Device Lett.*, Vol. EDL-7(12), 1986, pp. 649–651.

[17] Chu, P., C. L. Lin, and H. H. Wieder, "$In_{0.30}Al_{0.70}As$/$In_{0.30}Ga_{0.70}As$ Quasi-Insulating Gate Strained-Layer Field-Effect Transistors Grown by Molecular Beam Epitaxy," *Thin Solid Films*, Vol. 166, 1988, pp. 155–162.

[18] Wang, G.-W., Y.-K. Chen, W. J. Schaff, and L. F. Eastman, "A 0.1-µm $Al_{0.5}In_{0.5}As$/$Ga_{0.5}In_{0.5}As$ MODFET Fabricated on GaAs Substrates," *IEEE Trans. on Electron Devices*, Vol. ED-35(7), 1988, pp. 818–823.

[19] Khan, M. A., J. N. Kuznia, J. M. Van Hove, N. Pan, and J. Carter, "Observation of a Two-Dimensional Electron Gas in Low Pressure Metalorganic Chemical Vapor Deposited GaN-$Al_xGa_{1-x}N$ Heterojunctions," *Appl. Phys. Lett.*, Vol. 60, 1992, 3027–3029.

[20] Khan, M. A., J. N. Kuznia, A. R. Bhattarai, and D. T. Olson, "Metal Semiconductor Field Effect Transistor

Based on Single Crystal GaN," *Appl. Phys. Lett.*, Vol. 62, 1993, pp. 1786–1787.

[21] Binari, S. C., L. B. Rowland, W. Kruppa, G. Kelner, K. Doverspike, and D.K. Gaskill, "Microwave Performance of GaN MESFETs," *Electronic Lett.*, Vol. 30(15), 1994, pp. 1248–1249.

[22] Khan, M. A., J. N. Kuznia, D. T. Olson, W. J. Schaff, J. W. Burm, and M. S. Shur, "Microwave Performance of a 0.2 μm Gate AlGaN/GaN Heterostructure Field Effect Transistor," *Appl. Phys. Lett.*, Vol. 65, 1994, pp. 1121–1123.

[23] Khan, M. A., Q. Chen, J. W. Yang, M. S. Shur, B. T. Dermott, and J. A. Higgins, "Microwave Operation of GaN/AlGaN-Doped Channel Heterostructure Field Effect Transistors," *IEEE Electron Device Lett.*, Vol. EDL-17(1), 1996, pp. 325–327.

[24] Okamoto, Y, Y. Ando, K. Hataya, T. Nakayama, H. Miyamoto, T. Inoue, M. Senda, K. Hirata, M. Kosaki, N. Shibata, and M. Kuzuhara, "Improved Power Performance for a Recessed-gate AlGaN–GaN Heterojunction FET with a Field-modulating Plate," *IEEE Trans. Microwave Theory Tech.*, Vol. MTT-52(11), pp. 2536–2540.

[25] Y.-F. Wu, M. Moore, A. Saxler,T. Wisleder, and P. Parikh, "40-W/mm Double Field-plated GaN HEMTs," *64th Device Research Conf.*, 2006, pp. 151–152.

[26] Fujitsu Labs., Press Release 2017, *Fujitsu Achieves World's Highest Output Density with Power Amplifier for W-Band GaN Transmitters*, www.fujitsu.com.

[27] Nguyen, L. D., L. E. Larson, and U. K. Mishra, "Ultra-High-Speed Modulation-Doped Field-Effect Transistors: A Tutorial Review," *Proc. IEEE*, Vol. 80, 1992, pp. 494–518.

[28] Kroemer, H., "Theory of a Wide-Gap Emitter for Transistors," *Proc. IRE*, Vol. 45(11), 1957, pp. 1535–1537.

[29] Asbeck, P. M., D. L. Miller, R. A. Milano, J. S. Harris, Jr., G. R. Kaelin, and R. Zucca, "(Ga,Al)As/GaAs Bipolar Transistors for Digital Integrated Circuit," *International Electron Devices Meeting (IEDM)*, 7-9 December, 1981, pp. 629– 632.

[30] Malik, R. J., J. R. Hayes, F. Capasso, K. Alavi, and A. Y. Cho, "High-Gain $Al_{0.48}In_{0.52}As/Ga_{0.53}As$ Vertical *n-p-n* Heterojunction Bipolar Transistors Grown by Molecular-Beam Epitaxy," *IEEE Electron Device Lett.*, Vol. EDL-4(10), 1983, pp. 383–385.

[31] Ren, F., J. Han, R. Hickman, J. M. Van Hove, P. P. Chow, J. J. Klaassen, J. R. LaRoche, K. B. Jung, H. Cho, X. A. Cao, S. M. Donovan, R. F. Kopf, R. G. Wilson, A. G. Baca, R. J. Shul, L. Zhang, C. G. Willison, C. R. Abernathy, and S. J. Pearton, "GaN/AlGaN HBT Fabrication," *Solid-State Electronics*, Vol. 44, 2000, pp. 239–244.

[32] Hjelmgren, H., F. Allerstam, K. Andersson, P.-Å. Nilsson, and N. Rorsman, "Transient Simulation of Microwave SiC MESFETs With Improved Trap Models," *IEEE Trans. on Electron Devices*, Vol. 57(3), 2010, pp. 729–732.

[33] Khan, H. A., *Spectral Response Modelling and Analysis of Heterojunction Bipolar Phototransistors*, Doctoral Thesis, University of Manchester, 2010.

[34] Urtega, M., Z. Griffith, M. Seo, J. Hacker, and M.J.W. Rodwell, "InP HBT Technologies for THz Integrated Circuits," *Proceedings of the IEEE*, Vol. 105(6), 2017, pp. 1051–1067.

[35] Bahl, S. R., N. J. Moll, and M. Hueschen, *Heterojunction Bipolar Transistor (HBT) Having Improved Emitter-Base Grading Structure*, Patent Publication, US 2004/003/6082 A1, Pub. Date: Feb. 26, 2004.

Chapter 3

Classification of Transistor Models

3.1 INTRODUCTION

This chapter gives a short overview about the various types of transistor models. As shown in Figure 3.1, active device models may be classified as physical, empirical, experimental (measurement-based), and behavioral (black-box) models. Physical models may be further subdivided into numerical and analytical models. A model example for each type of transistor model is added.

3.2 PHYSICAL MODELS

Physical models have been developed with different levels of complexity, ranging from relatively simple models, based on Shockley´s gradual channel approximation (see chapter 4) and simplified carrier mobility description, to the highly sophisticated and numerically intensive two-dimensional models that can account for complex phenomena such as surface states, traps, and impact ionization. It is common to use the term physical model only for those complex models in which the partial differential equations (PDEs) are solved numerically. Sometimes this term is also used for simpler models that employ approximate analytical solutions to the PDEs. In order to distinguish them, the former will be called numerical physical models while the latter will be called analytical physical models [1].

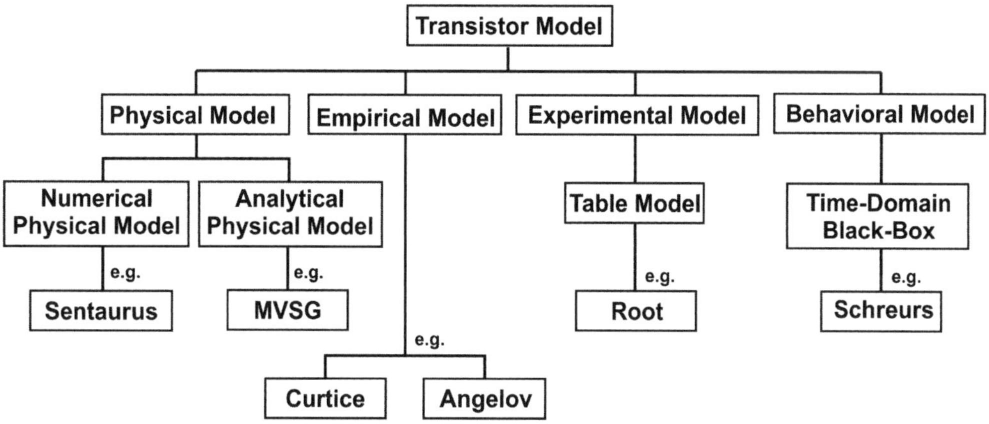

Figure 3.1 Classification of transistor models.

3.2.1 Numerical Physical Models

Initially, the research in physical modeling was almost entirely aimed at understanding the physical behavior of MESFETs to help fabricate better devices, and eventually, to obtain some qualitative information that could help the circuit designer. Physical modeling was therefore considered to concern primarily the device designer and to be somewhat remote from circuit design [1].

For device performance predictions, reliable data of physical parameters is necessary that describe the device geometry, material, and processing parameters (inner device structure). Such parameters include gate length, gate width, channel thickness, doping density, and temperature and field-dependent carrier mobility velocity. It may be impractical to obtain precise values of the physical parameter values such as the doping density distribution from the fabrication process.

In principle, physical models can predict circuit-level performance from process and material characteristics without first fabricating and measuring the devices. Process variations can be accounted for simply by changing the numerical values of the model parameters. Thus, the physical modeling approach permits a software-based development and optimization of new types of devices (virtual prototypes), and facilitates a direct feedback to the fabrication process for device characteristic optimization (tuning).

Physical models are mainly useful for deeper understanding of the functioning of a device and for the design of novel devices. They can predict behavior outside the range of known behavior or can determine the behavior of devices not yet fabricated. However, generally they are very complex, and commonly, the simulation time is very large.

The implementation of physical models in designer software requires a compromise between accuracy and acceptable computing time. In practice, numerical models have been used by device designers in order to optimize device properties. However, increasingly they also become important for the circuit designer. MMICs have already been designed with superfast computers that start on a purely physical design platform. Most of the physical models solve the semiconductor device equations using some form of numerical techniques such as finite differences or finite elements.

There are several software providers in the marketplace. The *Sentaurus™ Device* simulator supplied by Synopsis [2], for example, is capable of simulating electrical, thermal, and optical characteristics of silicon-based and compound semiconductor devices. These include (power) HEMTs and HBTs made from GaAs, InP, SiGe, SiC, AlGaAs, InGaAs, AlGaN, and InGaN. A real semiconductor device, such as a transistor, is represented in the simulator as a *virtual* device whose physical properties are discretized onto a nonuniform grid (or mesh) of nodes. The grid (or geometry) of the device contains a description of various regions; that is, boundaries, material types, and the locations of any electrical contacts. The data fields contain the properties of the device, such as the doping profiles, in the form of data associated with the discrete nodes. By applying a set of differential equations derived from Maxwell's laws onto this grid, the transport of carriers through the structure can be simulated. This means that the electrical performance of a device can now be modeled in DC, AC or transient modes of operation.

The *Victory Device™* Simulator from Silvaco [3] offers a similar set of features as discussed before. It allows the characterization and optimization of semiconductor devices for a wide range of technologies. This covers in particular III-nitride compound semiconductors.

3.2.2 Analytical Physical Models

Numerical physical models may deliver most accurate results for the electrical characteristics of semiconductor devices. However, calculations commonly take rather long time so that these models are not very useful for circuit design applications.

Analytical physical models offer an attractive compromise between simulation efficiency and quantitative accuracy. The first significant analytic model for field-effect transistors was presented by Shockley (see chapter 4). In this work Shockley presented the concept of the junction field-effect transistor JFET, along with a simple DC and AC incremental analysis.

Meanwhile physics-based models have reached a high level of development. This is true in particular for GaN HEMT devices. There are two models that have been recognized by the Compact Model Coalition (CMC) [4] as worldwide industry standard compact models. Members of the CSC organization are electronic design automation (EDA) vendors, foundries, identity management providers (IDMs), fabless companies, and research institutions [4, 5].

The Advanced Spice Model for GaN HEMT device (ASM-HEMT) has been developed at Kanpur Institute of Technology (IIT) [6]. The drain-current model is based on the drift-diffusion framework for carrier transport. The analytical and continuous expression for the drain current includes velocity-field relation and mobility degradation. The model ensures conservation of charge adopting Ward-Dutton partitioning for the source and drain charges [7]. It is worth adding that the reference planes of the intrinsic nonlinear part of the device refers to the gate- and drain-sided edges of the gate electrode. In this case the source and drain access resistances are dependent on the drain current $I_d(V_g, V_d)$. It is stated that, in GaN HEMTs, gate- and drain-access regions work as nonlinear resistance that limits maximum drain current. A simulation example in [8] exhibits an increase of R_d from 0.65Ω (I_d = 0.2A) to 0.95Ω (I_d = 1.4). For R_s the increase in the current interval is much smaller (i.e., from 0.05Ω to about 1.3Ω). In [9], for the same current interval, a larger increase from 1.5Ω to 5.5Ω was extracted for the differential source access resistor R_s. An accurate model of the series resistances, including temperature dependence, is of great importance to predict the drain current and transconductance for high-power as well as high-frequency GaN HEMTs [8] (cf. chapter 20).

The second relevant physics-based analytical model, which has been supported by the CMC and which has become an industry standard, is the MIT Virtual Source GaN HEMT (MVSG) compact model developed at the Massachusetts Institute of Technology (MIT). The model is well described in the thesis written by Radhakrishna under the supervision of D. A. Antoniadis [9]. This model comprises the intrinsic transistor and implicit-gate access regions at both sides of the gate electrode. The intrinsic transistor model is based on the virtual-source model in [10]. Carriers are injected into the source access region from source terminal which is a thermal-equilibrium reservoir. The carriers reaching the channel then face the potential energy barrier in the channel whose height is modulated by the gate voltage. With short gate lengths, this barrier peaks at the beginning of the channel at $x = x_0$ (gate-source edge at $x = 0$). The virtual-source charge density Q_{ix0} can be determined by electrostatic approximation. With the virtual-source velocity v_{x0} , which is related to the so-called unidirectional thermal velocity, the drain current can be written as $I_D = wQ_{ix0}\, v_{x0}$ with the gate width w. In this case, the transition of the gate-source access region and the channel region under the gate electrode is not really fixed. It is stated that Q_{ix0} and v_{x0} are almost independent of V_{DS}.

The access regions of the MVSG model are described by implicit-gate transistors. These are not actual gated transistors. The vertical field that terminates at the 2DEG in the access regions could originate from somewhere in the device; for example, donor states at the AlGaN surface or metal layers running above. In the limiting case one could assume that the field originates at infinity in which case the implicit-gate is at infinity. The location of the origin of this vertical field is the implicit-gate electrode. Since it is difficult to determine a location for the implicit-gate, the areal implicit-gate-capacitance is a fitting parameter in the model. The access region model delivers linear resistance behavior under very low drain bias conditions. However, the resistance value increases disproportionately with increasing drain current [9].

3.3 EMPIRICAL MODELS

In commercial microwave network simulation programs, mainly empirical FET device models have been implemented. In these models, the nonlinearities are described by closed-form analytical functions with formal parameters which are in general of no or less physical relevance. Classical analytical models are the Curtice [12], Curtice-Ettenberg [13], Statz [14], Tajima [15], and Materka-Kapczak models [16]. Originally, these had been developed as DC models, which were completed by capacitance models to improve AC properties. For the parameter extraction, generally simplified methods can be used, such as direct extraction methods.These simple models can still be found in foundry design kits [17] with possibly some modifications concerning, for example, frequency dispersion by adding an R–C series branch at the output.

Advanced compact models, such as the Angelov (Chalmers) model, have demontrated very successful results for a wide variety of device technologies using primarily DC and bias-dependent S-parameter data. A further widely accepted used model is the EEsof GaAs HEMT (EEHEMT) model, which is an extension of the Curtice model [18].

The main advantages of the empirical models are computational efficiency, automatic data smoothing, simplicity, short simulation time, and ability to simulate out of a measurement range. They are usually easy to implement in general CAD programs.Therefore, these types are widely used for circuit design.

In Chapter 25, for measured data of a 8×125 μm gate GaN HEMT, the performance of well-known standard models have been tested and evaluated concerning static-/pulsed-DC and large-signal (single-/two-tone) measurements.

The main disadvantages are limited accuracy often due to use of oversimplified expression, technology-dependent, difficulty in extraction of the model fitting parameters, and nonphysical meaning of the fitting parameters. The nonphysical behavior is often associated with the use of a function, such as a polynomial, to fit data over a specific range of voltages and subsequent application of the model to voltages outside this range. The function may not behave well outside the fitting range.

The recent evolution of new device types with permanently increasing stringent efficiency and linearity requirements necessitates more accurate modeling techniques. Traditional empirical modeling approaches, where a set of equations is used to present device characteristics, are still the preferred approach for many design environments. Equations are becoming more sophisticated in order to accurately handle different types of devices, such as PHEMTs, gallium nitride GaN devices, and enhancement-mode devices, over a wide range of voltages. Thus, recently developed models may comprise almost 100 parameters, which is large compared to the classical (standard) models. When 100 parameters are involved—most of them to be determined from measured data— model extraction requires a great deal of number crunching and extraction, which cannot be executed without the help of a smart extraction program [19].

3.4 EXPERIMENTAL MODELS

In comparison with analytical models the measurement-based (experimental) modeling approach is generally regarded as being very accurate, because the nonlinear model parameter description is not restricted to a predefined nonlinear analytical function, but rather directly derived from a database of measurement results. In this way the errors rising from the remaining mismatching of experimental data to empirical functions are avoided. Table-based data could be obtained from static-/pulsed-DC and small-signal S-parameter measurements at different bias conditions and/or signal waveform measurement analysis. The model topology is derived from physical and electrical device data.

All measured and calculated nonlinear model functions are stored in table form as functions of two independent controlling terminal voltages. Two-dimensional spline functions interpolate the tabulated functions during simulation. This model can therefore be classified as a lookup-table or table-based model. Lookup-table models, first introduced by M. Schlechtweg [20, 21], use data and interpolation (or approximation) techniques rather than analytical equations to define model port relations or equivalent-circuit element constitutive relations. They are also computing-time efficient.

Because of their formulation, measurement-based models should only be used for simulations of those measurement regions from which their parameters have been extracted. In certain situations this may be a significant limitation in comparison to physical or analytical models.

Table models can be both accurate and general — accurate, because the device's own data is used directly in the nonlinear model, and general, because the models are (largely) technology- and process-independent within device class such as FETs [22]. The HP Root model is one widely used table-based model.

However, the main drawback of table-based models has been the limitation of the interpolation algorithms used by the simulator to define the $I-V$ and $Q-V$ relations smoothly between discrete measured data points. Simulations of higher-order distortion products may become inaccurate when the input signal magnitudes are comparable to or smaller than the distance between neighboring data points. Recently artificial neural networks (ANNs) have become more attractive; they can replace the tables and interpolation schemes [23, 24].

3.5 BEHAVIORAL MODELS

Behavioral models do not require any knowledge of the actual internal device structure. Physical preknowledge is not required. Behavioral information is extracted from a set of selected tests based on the observation of the input–output signal relationships. Regarding small-signal analysis of microwave devices, the (*n*-port) S-parameter matrix could be presented as a (linear) behavioral model example. Because no circuit visualization of device elements is given, this kind of behavioral model is also termed black-box model. In the simple example of a two-port S-matrix, the model provides reflection and transmission coefficients for given bias condition and frequency.

Behavioral models do not deliver any information, which can be directly used for physics- and process-related device optimization. Instead this model approach aims at focusing on overall device characterization, which is mostly attractive for design tasks on the higher circuit and system level.

In recent years Schreurs et al. [25–29] have developed a technique to derive nonlinear behavioral models for microwave devices using time-domain large-signal measurements based on a limited harmonic bandwidth of 20 GHz [30]. The modeling procedure is based on nonlinear system identification using techniques developed in nonlinear time-series analysis (NLTSA) [31–33]. The two-port FET model is defined by representing the terminal currents as function of the terminal voltages and their derivatives. In a generalized form, the currents $I_1(t)$ and $I_2(t)$ of the two-port as function of the terminal voltages $V_1(t)$ and $V_2(t)$ can be written as

$$I_1(t) = f_1\left(V_1(t),\ V_2(t),\ \dot{V}_1(t),\ \dot{V}_2(t),\ \ddot{V}_1(t),\ \ldots,\ \dot{I}_1(t),\ \dot{I}_2(t)\right) \tag{3.1}$$

$$I_2(t) = f_2\left(V_1(t),\ V_2(t),\ \dot{V}_1(t),\ \dot{V}_2(t),\ \ddot{V}_1(t),\ \ldots,\ \dot{I}_1(t),\ \dot{I}_2(t)\right) \tag{3.2}$$

The objective of the modeling procedure is to find the functional relationships $f_1(.)$ and $f_2(.)$. The independent variables of the functions $f_1(.)$ and $f_2(.)$ are the state variables, which have to be determined. The approach comprises two steps. First a minimal set of independent variables needed to accurately predict the device dynamics is determined on the basis of a false nearest neighbors diagnostic test [26]. In a second step the functional relationships $f_1(.)$ and $f_2(.)$ are determined by fitting the measured time domain terminal currents using the independent variables determined in the preceding step. Multivariate polynomials have been used as fitting functions. Related coefficients are obtained by least-squares fitting. It has been pointed out that the proposed time-domain black-box modeling procedure is not restricted to the modeling of only weakly nonlinear phenomena unlike some frequency methods such as Volterra series analysis.

It is stated that the nonlinear behavioral model accurately predicts DC, small-signal, and large-signal behavior. However, it must be confirmed that the discussed modeling procedure restricts itself on the active inner part of the device. That is, it relies totally on a highly accurate extraction of the extrinsic parasitic parameters of the field-effect transistor performed previously. As is known, the values of the intrinsic nonlinear model parameters are sensitively dependent on the extraction reliability, which is extensively illuminated in the following chapters.

3.5.1 ANN-Based Models

Recently ANN-based FET models have become more prevalent [23]. ANN itself is a black box-model and hardly has any physical effect. However, knowledge-based ANNs are used quite often. This prior knowledge can be in the form of transfer function, empirical function, or equivalent circuit, which can make the modeling problem much easier because the prior knowledge has physical meaning in it.

3.5.2 X-Parameter-Based Models

X-parameter-based models also belong in the category of black-box models. No knowledge is needed of internal circuitry. X-parameter models capture the magnitude and phase information of the fundamental and harmonics [34].

The advantage of X-parameter-based models are their simplicity. For example, in the context of power amplifier (PA) design using a packaged transistor, nonlinear behavioral model extraction based on X-parameters will encompass not only the nonlinear behavior of the active die but also all the parasitic effects of wirebonds, package leads, and so forth. X-parameters in this sense help to protect the IP (intellectual property) of the device manufacturer and help the PA design house to develop a model faster compared to other modeling techniques.

Some drawbacks should also be mentioned: For X-parameter-based transistor modeling, there is a need for expensive, dedicated measurement equipment. X-parameters do not scale with device periphery. X-parameters, just like all black-box models, are only valid and useful in the environment (impedance, bias, frequency, etc.) under which the X-parameters were measured. Indeed, the model generality can be improved when X-parameters are extracted under different bias conditions, different frequencies, and/or power levels and different terminations.

3.6 SUMMARY

A variety of device models exist with strengths and weaknesses for a given application. Numerical physical device models are the most accurate ones because they are completely based on detailed device physics. However, geometry and technology data must be accurately known. Simulation execution time is very large and commonly not suited for circuit design. One-, two-, and three-dimensional device simulators are commercially available. Analytical device models, although based on semiconductor physics principles, comprise simplifications in the derivation of the physical device model. This leads to less model complexity and shorter simulation time. Various published models can be basically traced back to the classical but presently outdated Shockley model, which is nevertheless of general educational value. In the course of time the modeling activities have shifted from GaAs- to GaN-based modeling activities according to market challenges and progress in process and manufacturing technology. Recently, two nonlinear physics-based analytical GaN-based HEMTmodels, the ASM- and MIT-HEMT model, have become worldwide industry standard compact models.

Popular empirical models (e.g., Angelov and EEHEMT models) are widely used and found in current design software packages. They are the accepted standard for numerous applications in practice. They use analytical functions with formal parameters without any physical relevance. Implementation in circuit simulation programs is easy. They offer short computing time and comprise simplified model parameter extraction. Their accuracy is limited due to approximation errors.

Experimental (measurement-based) models are based on measured data. Electrical equivalent circuit topology is derived from semiconductor device physics. Table-based models reflect measured data. Two-dimension-spline interpolation is employed only between measured neighboring points. However, inaccuracies can occur with smaller input signal magnitudes. Recently it has become attractive to replace tables and interpolation schemes by artificial neural networks (ANNs). Dispersion effects are included, and models provide high simulation accuracy. Reliable model parameter extraction is not always easy. The development of a straightforward model parameter extraction strategy is the main topic of this book.

Behavioral (black-box) models focus on the device characterization with respect to their input-output terminals. No physical insight is available or used. They are empirical in nature, relying on measurements. It is therefore more a curve-fitting process than a modeling approach. With recent developments in large-signal vectorial time-domain/frequency-domain measurement techniques, attractive large-signal behavioral modeling techniques have been developed covering DC and small- and large-signal excitation.

REFERENCES

[1] Rodrigues, P. J. C., *Computer-Aided Analysis of Nonlinear Microwave Circuits*, Norwood: Artech House, Inc., 1998.

[2] Synopsys, *Sentaurus™ Device User Guide*, Version K-2015.06, June 2015, https://www.synopsys.com.

[3] Silvaco, Santa Clara, California: *Atlas User's Manual - Device Simulation Software*, August 26, 2016, https://www.silvaco.com.

[4] Green, K., *Compact Model Coalition*, http://www.mos-ak.org/washington_dc_2013/presentations/T4_Green_MOS-AK_WashingtonDC_2013.pdf.

[5] Dunn, M., and R. Sodhi, "RF GaN Modeling for 5G and Other Applications", *Electronic Design Innovation Concerence (EDICON)*, Beijing, China, 2019, http://www.mwjournalchina.com/edicon/Presentations/2019/EDICONChina2019_KeysightForum.

[6] Chauhan, Y. S., ASM-HEMT: *CompactModeling of GaN HEMTs for High Frequency and High Power Applications*, August 11, 2017, http://home.iitk.ac.in/~chauhan/ASM-HEMT-YSChauhan.pdf.

[7] Ward, D. E., and R. W. Dutton, "A Charge-Oriented Model for MOS Transistor Capacitances," *IEEE Journal of Solid-State Circuits*, Vol. SC-13(5), 1978, pp. 703–708.

[8] Palacios, T., S. Rajan, A. Chakraborty; S. Heikman, S. Keller, S. P. DenBaars, and U. K. Mishra, "Influence of the Dynamic Access Resistance in the g_m and f_T Linearity of AlGaN/GaN HEMTs," *IEEE Trans.*, Vol. ED-52(10), 2005, pp. 2117–2123.

[9] Radhakrishna, U., *A Compact Transport and Charge Model for GaN-Based High Electron Mobility Transistors for RF Applications*, Doctoral Thesis, Massachusetts Institute of Technology (MIT), 2013, http://hdl.handle.net/1721.1/82394.

[10] Radhakrishna, U., *MVSG-RF: Model Manual*, https://nanohub.org/resources/20180/download/MVSG_RF_1_0_0_manual.pdf.

[11] Khakifirooz, A., O. M. Nayfeh, and D. Antoniadis, "A Simple Semiempirical Short-Channel MOSFET Current-Voltage Model Continuous Across All Regions of Operation and Employing Only Physical Parameters," *IEEE Trans.*, Vol. ED-56(8), 2009, pp. 1674–1680.

[12] Curtice, W. R., "A MESFET Model for Use in the Design of GaAs Integrated Circuits," *IEEE Trans.*, Vol. MTT-28, 1980, pp. 448–456.

[13] Curtice, W. R., and M. Ettenberg, "A Nonlinear GaAs FET Model for the Use in the Design of Output Circuits for Power Amplifiers," *IEEE Trans.*, Vol. MTT-33, 1985, pp. 1383–1393.

[14] Statz, H., P. Newman, I. W. Smith, R. A. Pucal, and H. A. Haus, "GaAs FET Device and Circuit Simulation in SPICE," *IEEE Trans.*, Vol. ED-34(2), 1987, pp. 160–169.

[15] Tajima, Y., B. Wrona, and K. Mishima, "GaAs FET Large-Signal Model and Its Application to Circuit Designs," *IEEE Trans.*, Vol. ED-28(2), 1981, pp. 171–175.

[16] Materka, A., and T. Kacprzak, "Computer Calculation of Large-Signal GaAs FET Amplifier Characteristics," *IEEE Trans.*, Vol. MTT-33, 1985, pp. 129–135.

[17] Keysight, *ADS Design Kit*, http://edadownload.software.keysight.com/eedl/ads/2011/pdf/dkarch.pdf.

[18] Dunleavy, L., C. Baylis, W. Curtice, and R. Connick, "Modeling GaN: Powerful But Challenging," *IEEE Microwave Magazine*, October 2010, pp. 82–96.

[19] Tajima, Y., "Progress in RF Device Modeling: From MESFETs to GaN PHEMTs," *The Annual RF & Microwave Solutions Update*, pp. 40–44, 2006, www.rfglobalnet.com.

[20] Schreurs, D., *Measurement Based Modelling of Heterojunction Field-Effect Devices for Non-Linear Microwave Circuit Design*, Doctoral Thesis, Katholieke Universiteit Leuven (KU Leuven), 1997, p. 69.

[21] Schlechtweg, M., *Broadband Characterization and Modelling of GaAs-MESFETs and (AlGa)As/GaAs-MODFETs up to 40 GHz* (in German), Doctoral Thesis, Kassel University, 1989.

[22] Root, D., "Principles and Procedures for Successful Large-Signal Measurement-Based FET Modeling for Power Amplifier Design," *Agilent Technologies, Seminar: Gain Without Pain*, November 2000.

[23] Root, D. E., "Future Device Modeling Trends," *IEEE Microwave Magazine*, 2012, pp. 45–59.

[24] Wood, J., P. A Aaen, D. Bridges, D. Lamey, M. Guyonnet, D. S. Chan, and N. Monsauret, "*A Nonlinear Electro-Thermal Scalable Model for High-Power RF LDMOS Transistors*," *IEEE Trans.*, Vol. MTT-57(2), 2009, pp. 282–292.

[25] Schreurs, D., J. Wood, N. Tufillaro, D. Usikov, L. Barford, and D. E. Root, "The Construction and Evaluation of Behavioral Models for Microwave Devices Based on Time-Domain Large-Signal Measurements," *Int. Electron Devices Meeting* (IEDM), 2000, pp. 819–822.

[26] Schreurs, D., J. Wood, N. Tufillaro, L. Barford, and D. E. Root, "Construction of Behavioral Models for Microwave Devices from Time Domain Large-Signal Measurements to Speed Up High-Level Design Simulations," *2003 Wiley Periodicals, Inc.*, 2003, pp. 54–61.

[27] Schreurs, D., and J. Verspecht, "Large-Signal Modelling and Measuring Go Hand-in-Hand: Accurate Alternatives to Indirect S-Parameter Methods (Invited Paper)," *Int. J. RF and Microwave CAE*, Vol. 10, 2000, pp. 6–18.

[28] Schreurs, D., "Overview of Non-Linear Device Modelling Methods Based on Vectorial Large-Signal Measurements," *Proc. European Gallium Arsenide and Related III- V Compounds Application Symposium* (GaAs'99), 1999, pp. 381–386.

[29] Pedro, J. C., and T. R. Cunha, *Chapter 7: Behavioral Models for Microwave Circuit Design*, in *Microwave De-Embedding*. Eds.: G. Crupi and D. M. M.-P. Schreurs, Academic Press, 2013.

[30] Verspecht, J., P. Debie, A. Barel, and L. Martens, "Accurate On Wafer Measurement of Phase and Amplitude of the Spectral Components of Incident andf Scattered Voltage Waves at the Signal Ports of a Nonlinear Microwave Device," *1995 IEEE MTT-S Symp.*, 1995, pp. 1029–1032.

[31] Kantz, H., and T. Schreiber, *Nonlinear Time Series Analysis*, Cambridge University Press, New York, 1997.

[32] Hellerstein, J., "An Introduction to Modeling Dynamic Behavior with Time Series Analysis," *Performance Evaluation of Computers and Communication Systems*, Springer-Verlag, New York, 1993, pp. 203–223.

[33] Casdagli, M., "A Dynamical Systems Approach to Modeling Input-Output Systems," *Nonlinear Modeling and Forecasting, SFI Studies in the Science of Complexity*, Proceedings, M. Casdagli and S. Eubank (Eds.), Addison-Wesley, Reading, MA, Vol. XII, 1992.

[34] Dunleavy, L.,"Power Amplifier Design with X-Parameter Power Transistor Models," 2012, https://www.keysight.com/upload/cmc.../6Sept12_PAwebcast.pdf.

Chapter 4

Classical Shockley Model and Enhanced Modifications

4.1 INTRODUCTION

In this chapter we will discuss the classical Shockley model as an example for a physics-based DC FET model. The theoretical analysis of a junction field-effect transistor JFET was first proposed by Shockley in 1952 [1]. The theoretical analysis was performed under the assumption of long-channel field-effect transistors with gate lengths of several micrometers. Actually the ratio of the gate length L over the channel width a signifies the type of field-effect transistor. Following [2] a L/a ratio of 5.0 may be termed wide gate (long-channel), a value of 1.0 a narrow gate, and a value of 0.143 a very narrow gate device. Thus, regarding current short gate devices with gate lengths in the range of submicrometers (e.g., [3]), the question arises whether it is reasonable to treat this subject in a book that addresses the latest development in FET device characterization and modeling. In every way, the Shockley model marks the historical origin in transistor modeling, and it is useful to introduce a number of physical parameters. More interesting, it will be shown that the derived voltage-controlled relations for the drain current and related incremental device parameters, such as transconductance, output (channel) conductance, and gate capacitance, occur, although in modified form, frequently in empirical FET device models. The physics-based relations for the FET device parameters can also be used to derive estimated starting values for the model parameter optimizer-based extraction process (see Chapter 6).

4.2 LONG-CHANNEL (SHOCKLEY) MODEL

Figure 4.1(a) depicts the FET structure under discussion. The length and width of the gate are denoted as L and w, respectively. The channel thickness is marked as a. $h(x)$ is the height or opening of the channel at point x. Region I signifies the space-charge region, and region II the conductive channel. Furthermore, we have introduced a Cartesian coordinate system for the theoretical analysis. V_{DS} and V_{GS} are the applied drain and source voltages with respect to the source. V_{GS} is negative under common operation.

Figure 4.1(b) shows the cross section of the FET with the channel potential $V(x)$ at point x in the channel and the height of the space charge region $h(x)$. h_1 is the height of the space-charge region at the source-sided edge of the gate electrode, and h_2 the respective height at the drain-sided gate edge. $\varphi(h(x))$ is the space-charge potential (channel-to gate potential) at x.

At this moment of discussion we entirely neglect the external region of the FET, which

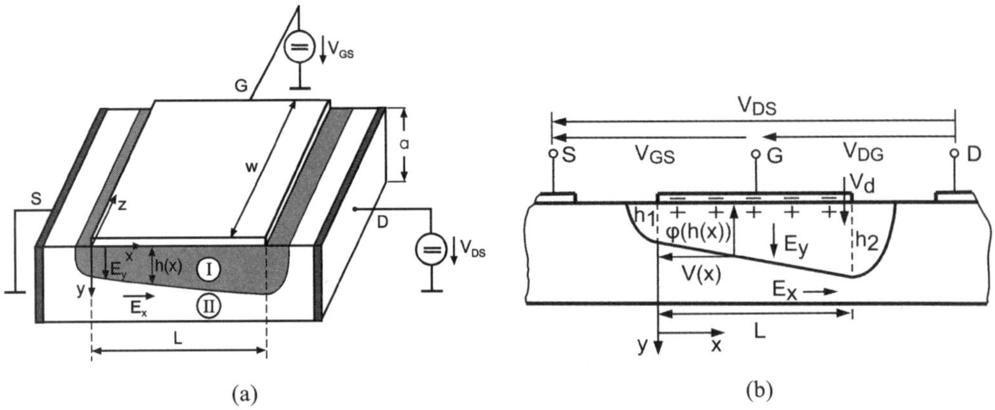

Figure 4.1 (a) Biased MESFET structure and coordinate system, and (b) definition of voltages and geometric quantities.

is outside the active region beneath the gate electrode and treated in Chapter 5. Thus, voltages and currents are referenced to the external electrodes that are accessible for device measurement. External voltages (and currents) are subscripted with capital letters. For example, V_{DS} denotes the external drain-source voltage, whereas V_{ds} signifies the internal voltage drop across the active channel below the gate electrode. Shockley's theory is based on the following simplifying assumptions: (1) *uniform charge distribution*: The epitaxial layer with thickness a is uniformly doped. The transition to the substrate is abrupt. (2) *gradual channel approximation*: The thickness of the space-charge region over the gate length is gradually widening. This is strictly true only for devices in which the channel thickness is much smaller than the gate length. (3) *Abrupt depletion layer*: The depletion layer under the gate ends abruptly. Under this assumption the field distribution under the gate electrode can be treated as a superposition of two one-dimensional fields, the longitudinal field E_x, corresponding to charge flow between the ohmic contacts, and the transverse field E_y, corresponding to charge depletion in the channel. (4) *Constant carrier mobility*: The mobility of carriers, μ, is field-independent; that is, the electron drift velocity, v_d, is proportional to the longitudinal electric field E_x ($v_d = \mu_0 E_x$).

For the uniformly doped n region, under the gradual channel approximation, the depletion-layer width h varies only gradually along the channel (x direction) and one can solve the one-dimensional Poisson equation in the y direction. Assuming a one-dimensional distribution of positive space charge $\rho(y) = qn(y) = qN_D(y)$ (all donors ionized), the Poisson equation reads

$$\frac{d^2\varphi(y)}{dy^2} = -\frac{dE_y(y)}{dy} = -\frac{qn(y)}{\varepsilon} = -\frac{qN_D}{\varepsilon} \tag{4.1}$$

where n, N_D, ε are the carrier density, donor concentration, and dielectric constant of the semiconductor material, respectively. Integrating (4.1), it follows for the electrical field $E_y(y)$

$$-E_y(y) = \frac{d\varphi(y)}{dy} = -\frac{q}{\varepsilon}\int_0^y N_D dy' + C_1 = -\frac{q}{\varepsilon}N_D y + C_1 y \tag{4.2}$$

By integration of (4.2) the potential $\varphi(y)$ is obtained,

$$\varphi(y) = -\frac{qN_D}{\varepsilon}\frac{y^2}{2} + C_1 y + C_2 \tag{4.3}$$

Defining the electrostatic potential at the surface ($y = 0$) as zero, the integration constant C_2 in (4.3) can be determined: $\varphi(0) = C_2 = 0$. At the space-charge edge the y-component of the electric field must vanish (i.e., from (4.2) we determine the integration constant C_1 to be $C_1 = -(q/\varepsilon)N_D h$). Then the electrostatic potential $\varphi(y)$ reads

$$\varphi(y) = \frac{qN_D}{\varepsilon}\left(-\frac{y^2}{2} + hy\right) \tag{4.4}$$

The current density $J_x(y)$ in the channel reads $J_x(y) = \sigma(y)E_x(x)$ with $\sigma(y) = qn(y)\mu_0$ where μ_0 is the low-field carrier mobility. The electric field in the conductive channel $E_x(x)$ is related to the incremental channel voltage drop as $E_x(x) = dV(x)/dx$. Then the channel current I_{DS} is calculated as

$$I_{DS} = \int_0^w \int_h^a J_x(y)\,dy\,dz = w\mu_0 q N_D(a - h)\frac{dV(x)}{dx} \tag{4.5}$$

Multiplying by dx and formal integration from $x = 0$ to $x = L$ delivers

$$\int_0^L I_{DS}\,dx = \mu_0 q w N_D \int_0^L (a - h)\frac{dV(x)}{dx}\,dx \tag{4.6}$$

As indicated in Figure 4.1(b) the applied external gate-source voltage V_{GS} is reduced by the internal diffusion voltage V_d ($V_d > 0$), also termed built-in voltage V_{bi}, so that the effective gate-source value $(V_{GS} - V_d)$ controls the space-charge region. The channel potential $V(x)$ and space-charge potential $\varphi(h(x))$ are related by

$$V(x) = \varphi(h(x)) + V_{GS} - V_d \tag{4.7}$$

The channel potential varies from zero at the source to the drain-source voltage at the drain. Differentiating (4.7) with respect to x [4],

$$\frac{dV(x)}{dx} = \frac{d\varphi(h(x))}{dh}\frac{dh(x)}{dx} = \frac{qN_D}{\varepsilon}h(x)\frac{dh(x)}{dx} \tag{4.8}$$

it follows

$$I_{DS} = \frac{w\mu_0 q^2 N_D^2}{\varepsilon L}\int_{h_1}^{h_2}(a - y)y\,dy \tag{4.9}$$

Evaluation of the integral yields

$$I_{DS} = \frac{w\mu_0 q^2 N_D^2 a^3}{6\varepsilon L}\left[\frac{3}{a^2}\left(h_2^2 - h_1^2\right) - \frac{2}{a^3}\left(h_2^3 - h_1^3\right)\right] \tag{4.10}$$

The drain current reaches a maximum with $h_1 = 0$ and $h_2 = a$. The latter condition ($h_2 = a$) implies that the conductive channel is entirely closed. Although in principle no carriers are present in the depleted channel (under this pinch-off condition a channel current flow should not be possible), carrier flow is assumed to exist in an infinitesimal sheet. Shockley has recognized this physical contradiction and further research had focused on its remedy.

Thus, it is important to note that there are two different definitions for the pinch-off voltage. We can define the (saturation) voltage V_{DSS} at which the $I_{DS}-V_{DS}$ characteristics saturate. This will be different from the gate-source (pinch-off) voltage ($V_{GS}=V_p$) required to reduce the drain-current to zero. In (4.10) a pinch-off channel current I_0 is defined as

$$I_0 = \frac{w\mu_0 q^2 N_D^2 a^3}{6\varepsilon L} \tag{4.11}$$

From (4.4) the electrostatic potential at the space-charge boundary $\varphi(y)|_{y=h(x)} = \varphi(h(x))$ reads

$$\varphi(h(x)) = \frac{qN_D}{\varepsilon}\frac{h(x)^2}{2} \tag{4.12}$$

Then with (4.7) the depletion region width $h(x)$ as a function of the control voltages can be calculated as

$$h(x) = \left[2\varepsilon/qN_D(V(x) - V_{GS} + V_d)\right]^{1/2} \tag{4.13}$$

For the source-sided width h_1 of the space-charge region it follows with $V(x=0)=0$

$$h_1 = \left[2\varepsilon/qN_D(-V_{GS} + V_d)\right]^{1/2} \tag{4.14}$$

For drain-sided width h_2 of the space-charge region one obtains with $V(x=L)=V_{DS}$

$$h_2 = \left[2\varepsilon/qN_D(V_{DS} - V_{GS} + V_d)\right]^{1/2} \tag{4.15}$$

We define V_{pi} as the intrinsic pinch-off voltage, which is the magnitude of reverse bias required to make the space charge penetrate the entire channel region at the drain side,

$$V_{pi} = qN_D a^2/2\varepsilon \tag{4.16}$$

As can be seen, the intrinsic pinch-off voltage V_{pi} includes material parameters such as doping concentration N_D and dielectric permittivity of the semiconductor ε, and the geometric parameter of channel thickness a. Therefore, its value is always positive. Frequently, in textbooks the relation (4.16) is simply denoted "pinch-off voltage." However, this may contribute to some confusion with the extrinsic pinch-off voltage denoted here V_p also known as threshold voltage V_{th} or V_T, which is practically measured (ideally) for zero channel current (see Section 21.2.3). In this case

$$V_{GS}(I_{DS} = 0) = V_p \qquad (V_{DS} \geq 0) \tag{4.17}$$

with

$$V_p = -V_{pi} + V_d \tag{4.18}$$

For example, with $V_d = 0.7$V and $V_{pi} = 5.1$V the extrinsic pinch-off voltage would turn out to be $V_p = -4.4$V (see also "Problems and Solutions" in Section 4.8).

We introduce the normalized voltage $u(x)$ defined as the normalized space-charge width $h(x)/a$,

$$u(x) = h(x)/a = (V(x) - V_{GS} + V_d/V_{pi})^{1/2} \tag{4.19}$$

Similarly we can write for the normalized quantities h_1 and h_2:

$$u_1 = h_1/a = (-V_{GS} + V_d/V_{pi})^{1/2} \tag{4.20}$$

$$u_2 = h_2/a = (V_{DS} - V_{GS} + V_d/V_{pi})^{1/2} \tag{4.21}$$

With the normalized voltages the channel current I_{DS} in (4.10) reads

$$I_{DS} = I_0[3(u_2^2 - u_1^2) - 2(u_2^3 - u_1^3)] \tag{4.22}$$

Figure 4.2 illustrates the flat depletion-layer boundaries beneath the gate electrode for $V_{DS} = 0$ and various gate-source voltages. For $V_{GS} = 0$ the channel is completely open (curve 1). With increasing reverse gate-source voltage the conductive channel narrows and is pinched off (curve 5) at large reverse gate potential ($V_{GS} \geq V_p$). The channel resistance can be approximately expressed by

$$R_{ch} = \frac{L}{\mu_0 q N_D w(a - h_2)} \tag{4.23}$$

The lowest value of channel resistance, $R_{ch} = R_{c0}$, occurs for $V_{GS} = 0$, when the space-charge thickness is $h_2 = 0$ (open channel resistance R_{c0}),

$$R_{c0} = \frac{L}{\mu_0 q N_D w a} \tag{4.24}$$

For $V_{GS} = V_p$, when the channel is totally pinched off (*i.e.*, $h_2 = a$) the resistance reaches an infinite value ($R_{ch} \rightarrow \infty$) (Figure 4.3).

With (4.14), (4.15) and (4.24) the drain current in (4.10) reads

$$I_{DS} = \frac{1}{R_{c0}}\left[1 - \frac{2}{3}\frac{V_{pi}}{V_{DS}}\left(\frac{V_{DS} - V_{GS} + V_d}{V_{pi}}\right)^{\frac{3}{2}} + \frac{2}{3}\frac{V_{pi}}{V_{DS}}\left(\frac{-V_{GS} + V_d}{V_{pi}}\right)^{\frac{3}{2}}\right]V_{DS} \tag{4.25}$$

Neglecting the V_d terms, a simplified equation is obtained for $V_{GS} = 0$,

$$I_{DS} = \frac{1}{R_{c0}}\left[1 - \frac{2}{3}\left(\frac{V_{DS}}{V_{pi}}\right)^{\frac{1}{2}}\right]V_{DS} \tag{4.26}$$

$$I_{DS}(V_{DS} \rightarrow 0) = V_{DS}/R_{c0} \tag{4.27}$$

For V_{GS} = constant, I_{DS} increases with growing V_{DS}. When the channel is pinched off at the drain side ($h_2 = a$) then the saturation current $I_{DSS} = I_{DS}(h_2 = a)$ flows. From (4.10), (4.14), and (4.15) we get

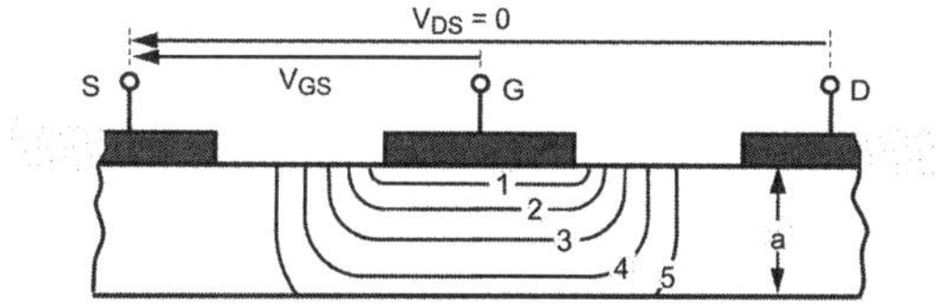

Figure 4.2 Expansion of the space-charge layer for $V_{DS} = 0$ and increasing reverse V_{GS} value.

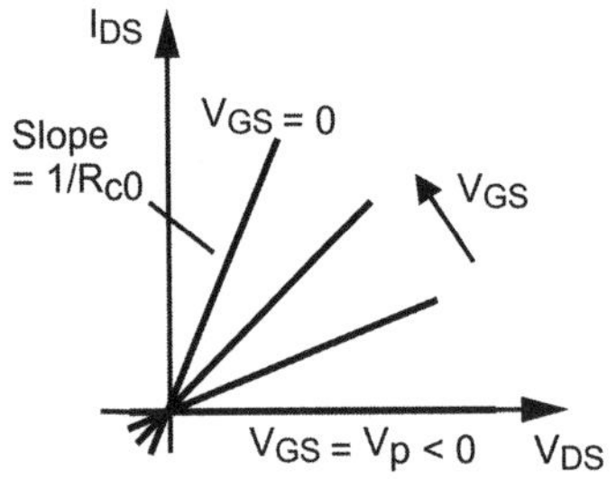

Figure 4.3 I_{DS}-V_{DS} characteristic for low V_{DS} values, indicating the decreasing channel resistance with increasing reverse gate-source voltage V_{GS}.

$$I_{DSS} = I_0\left[1 - 3\,\frac{-V_{GS} + V_d}{V_{pi}} + 2\left(\frac{-V_{GS} + V_d}{V_{pi}}\right)^{\frac{3}{2}}\right] \tag{4.28}$$

The drain-source saturation voltage V_{DSS} is the minimum drain-source voltage where pinch-off at the drain side occurs. From (4.13) with $V(x = L) = V_{DSS}$ and $h(x = L) = a$, we obtain

$$V_{DSS} = \frac{qN_D}{\varepsilon_s}\frac{a^2}{2} + V_{GS} - V_d \tag{4.29}$$

or

$$V_{DSS} = V_{pi} + V_{GS} - V_d \tag{4.30}$$

When the drain-source voltage exceeds the saturation voltage V_{DSS}, then it is assumed that the drain-source current remains constant ($V_{DS} > V_{DSS}$: $I_{DS} = I_{DSS}$ = constant).

Figure 4.4 illustrates the I–V characteristics based on Shockley's theory. The gradual channel approximation covers the lower drain-source voltage region (ohmic region) limited by the pinch-off locus, which describes the drain saturation current I_{DSS} versus the saturation drain-source saturation voltage V_{DSS}.

For gate voltages near the extrinsic pinch-off voltage V_p (threshold voltage V_{th}), (4.28) can be simplified. By replacing the diffusion voltage V_d by ($V_p + V_{pi}$) according to (4.18), the saturation current for a given V_{GS} value is given as

$$I_{DSS} = I_0\left[1 - 3\,\frac{-V_{GS} + V_p + V_{pi}}{V_{pi}} + 2\left(\frac{-V_{GS} + V_p + V_{pi}}{V_{pi}}\right)^{\frac{3}{2}}\right] \tag{4.31}$$

Applying the Taylor series expansion around the point $V_{GS} = V_p$ and taking the first three terms yields

$$I_{DSS} = I_0\frac{3}{2V_{pi}^2}\left(V_{GS} - V_p\right) = \frac{\mu_0 w\varepsilon}{aL}\left(V_{GS} - V_p\right)^2 \tag{4.32}$$

This quadratic dependence of the saturation current on the gate-source voltage has been used in most classical analytical MESFET circuit models. Equation (4.32) is usually written as

$$I_{DSS} = \beta\left(V_{GS} - V_p\right)^2 \tag{4.33}$$

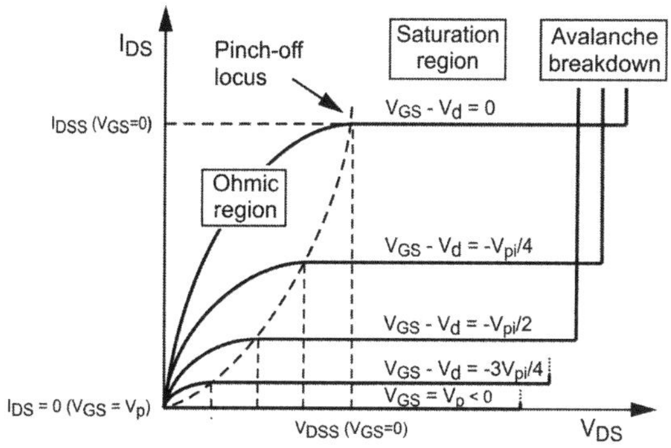

Figure 4.4 $I(V)$ characteristics based on gradual-channel approximation. The critical condition for pinch-off at the drain is indicated by the dashed line.

where β is called the transconductance coefficient.

Avalanche breakdown between drain and gate electrodes will occur when the drain-source voltage becomes sufficiently high ($V_{DS} \gg V_{DSS}$). For a given device construction, the breakdown voltage V_B, which signifies the utmost drain-gate voltage, is a constant value. It reads

$$V_B = V_{DG} = V_{DS} - V_{GS} = const. \tag{4.34}$$

Equation (4.34) indicates that the higher the reverse gate-source voltage, the lower the drain-source voltage for avalanche breakdown. This is also illustrated in Figure 4.4.

The most useful active parameter of a field-effect transistor is its (mutual) transconductance G_M. The transconductance G_M is defined as

$$G_M = \left. \frac{\partial I_{DS}}{\partial V_{GS}} \right|_{V_{DS} = \text{const.}} \tag{4.35}$$

Applying partial differentiation, one obtains G_M directly from (4.10)

$$G_M = \frac{w\mu_0 q N_D a}{L}(u_2 - u_1) \tag{4.36}$$

A maximum transconductance G_{M0} in the linear region can be deduced with $h_1 = 0$ and $h_2 = a$,

$$G_{M0} = \frac{w\mu_0 q N_D a}{L} \tag{4.37}$$

A high transconductance value is obtained with high carrier mobility and short channel length. Substituting G_{M0} into (4.36), the transconductance in the linear region reads in denormalized form:

$$G_M/G_{M0} = [(V_{DS} - V_{GS} + V_d)/V_{pi}]^{1/2} - [(-V_{GS} + V_d)/V_{pi}]^{1/2} \tag{4.38}$$

With $h_2 = a$ it follows for the transconductance in the saturation region ($I_{DS} = I_{DSS}$)

$$G_M/G_{M0} = 1 - [(-V_{GS} + V_d)/V_{pi}]^{1/2} \tag{4.39}$$

The dynamic output conductance of the drain terminal can be obtained from (4.10) in a similar manner. The channel conductance reads $G_{DS} = \partial I_{DS}/\partial V_{DS}$ with V_{GS} = constant. Applying partial differentiation one obtains G_{DS} from (4.10),

$$G_{DS} = w\mu_0 q N_D a/L(1 - h_2/a) \tag{4.40}$$

For the open channel conductance ($h_2 = 0$) we get

$$G_{D0} = qN_D\mu_0 aw/L \tag{4.41}$$

which corresponds to the reciprocal of the channel resistance (4.23). With G_{D0} it follows for the channel conductance in denormalized presentation

$$G_{DS}/G_{D0} = 1 - (V_{DS} - V_{GS} + V_d/V_{pi})^{1/2} \tag{4.42}$$

For very low drain-source voltages $V_{DS} \ll V_{DSS}$ it follows

$$G_{DS}/G_{D0} = 1 - (-V_{GS} + V_d/V_{pi})^{1/2} \tag{4.43}$$

With $h_2 = a$ the drain conductance in the saturation region is immediately obtained as $G_{DS} = 0$.

In short-channel FETs extremely high electrical field strength can exist, so that the carriers suffer from saturation effect. In this case, mostly the transconductance, the saturation current, and the saturation voltage are drastically reduced. To improve the theory after Shockley, more appropriate analytical relations for the steady-state field-dependent of the electron drift velocity were needed. In the following, different approaches are presented and their impact on the carrier transport along the channel discussed.

4.3 EXPERIMENTAL AND ANALYTICAL *v*(*E*)-CHARACTERISTICS

It has become apparent that a field-independent electron mobility cannot conveniently describe the carrier transport mechanism along the channel of a field-effect transistor at high electric field strength. At low electric field the carriers move with the low-field mobility μ_0 so that their drift velocity is proportional to the applied electric field, $v_d = \mu_0 E$. At higher electric field strength the drift velocity of the electrons saturates.

Figure 4.5 shows the steady-state field-dependent drift velocity $v_d = \mu(E)\cdot E$ for different semiconductor materials. Typically for many compound semiconductors like GaAs, InN, and GaN, the electron drift velocity increases with increasing applied electric field, reaches a peak value with the so-called critical field strength, and at higher applied electric fields it decreases until it saturates. With increasing electrical field strength, the electrons gain energy from the field and suffer from various kinds of scattering processes (inner-valley and inter-valley scattering, scattering at optical and acoustical phonons and at isolated traps). Beyond the critical field the electrons have acquired enough energy in order to transfer into an upper (satellite) valley. As the effective mass of the electrons in the upper valley is greater than that in the lower valley, the electrons in the upper valley will be slower. Finally, at high applied electric fields, the number of electrons in the upper valley saturates [5].

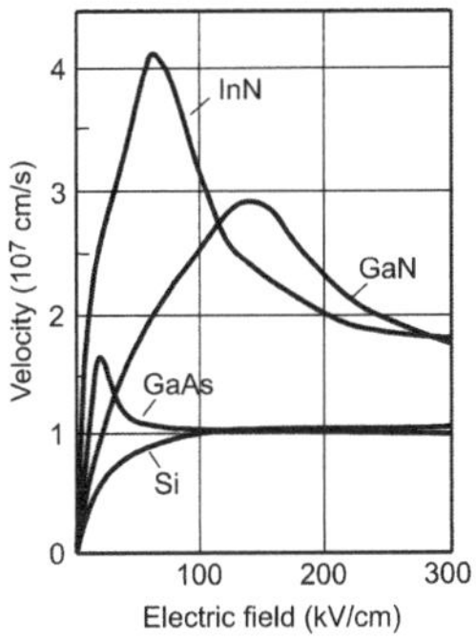

Figure 4.5 Electron drift velocity versus electrical field strength for different semiconductor materials ($N_D = 10^{17}$ cm^{-3}, $T = 300$K). See also [25].

At room temperature and doping concentration of 10^{17} cm^{-3} the maximum steady state drift velocities have been calculated using ensemble Monte Carlo simulations [6]. The peak drift velocity in GaAs is approximately 1.6×10^7 cm/s at a critical field strength $E_c \approx 4$ kV/cm. GaN reaches a maximum (peak) drift velocity value of approximately 2.9×10^7 cm/s at a high electric field of $E_c \approx 140$ kV/cm. InN achieves the highest steady-state peak electron drift velocity of 4.1×10^7 cm/s at an applied electric field of 65 kV/cm. For comparison, Si has been included. As can be seen, the saturation drift velocity v_s in Si is relatively low ($v_s \approx 1 \times 10^7$ cm/s) and is not reached until very high electric field strength values above 50 kV/cm.

The mobility of electrons can be estimated as follows ($N_D = 10^{17}$ cm^{-3}, $T = 300$K) ([6]: μ(GaN) ≈ 850 cm^2/(Vs), μ(AlN) ≈ 1250 cm^2/(Vs), μ(InN) ≈ 3450 cm^2/(Vs), μ(GaAs) ≈ 5300 cm^2/(Vs). In doped AlGaN-GaN heterostructures (with the electron sheet concentration $n_s \approx 10^{13}$ cm^{-2}) the measured room temperature mobility reaches 2019 cm^2/(Vs), which is much larger than for comparably doped silicon (approximately 500 cm^2/(Vs)) and half of that for a comparably doped GaAs (approximately 4,000 cm^2/(Vs)) [7].

The drift velocity depends both on doping concentration and temperature. Figure 4.6(a) shows the velocity-field characteristics for GaAs at 300K for various doping densities [8]. It is seen that doping densities greater than about 10^{15} cm^{-3} significantly reduce the peak drift velocity and the low-field mobility.

Figure 4.6(b) shows the calculated temperature-dependent electron drift velocity in intrinsic GaAs material as a function of electric field strength [8]. It is shown that the temperature of the material strongly reduces the electron drift velocity.

As the gate lengths of field-effect transistors have been reduced dramatically during the last decade, the electric field strength in the channel can reach very high values (i.e., that the electrons being injected at the source electrode are subjected to extreme acceleration and carriers travel with steady-state saturation drift velocity across most of the channel). Assuming a critical field strength of 3.3 kV/cm for GaAs and neglecting the voltage drop in the extrinsic bulk regions of the FET, the necessary applied drain-source voltage to reach the critical field strength is simply calculated as $V_{DS,c} = E_c \cdot L$ (Figure 4.7). For a 1-μm gate FET the necessary voltage is 330 mV, whereas in a short-gate FET with a gate length of 0.1 μm, the critical electrical field is already obtained with an applied drain-source voltage of only

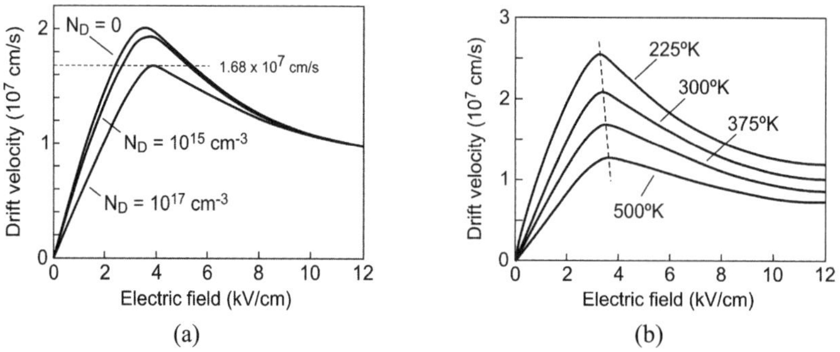

Figure 4.6 Electron drift velocity v_d in GaAs versus electrical field strength with (a) doping concentration as parameter (T = 300K), and (b) with temperature as parameter in intrinsic GaAs [8].

33 mV. The electrons propagate along the channel with saturation drift velocity.

There were various attempts to approximate the calculated or measured field-dependent $v_d(E)$ characteristic analytically to be suited for FET device modeling. Dacey and Ross [9] were the first to suggest that current saturation in FETs might be attributable to the decrease in mobility at high electric fields. They assumed that the mobility μ depends on the electrical field E according to $\mu = \mu_c(E_c/E)^{1/2}$, where E_c is the critical field at which μ begins to depart from a constant value and μ_c is the mobility at $E = E_c$. The drift velocity is then written as $v_d = \mu_c \sqrt{(E_c \cdot E)}$ for $E > E_c$. For electric fields greater than E_c, the mobility is proportional to $E^{-1/2}$ (i.e., that μ becomes zero only with $E \rightarrow \infty$). As the drift velocity increases monotonously with the field strength, a saturation velocity does not appear.

In [10] Trofimenkoff ascertained that an analytical expression of the form $\mu = \mu_0/(1 + E/E_0)$, where μ_0 is the low field mobility and E_0 is an empirically determined constant, could better approximate experimental measurements of the mobility as a function of the electric field, as shown in Figure 4.8.

Based on the results in [10] Lehovec and Zuleeg [11] have formulated the following analytical equation for the field-dependent drift velocity $v_d(E)$ (see Figure 4.9):

$$v_d = \frac{\mu_0 E}{1 + (\mu_0 E)/v_s} \tag{4.44}$$

This relation has been used frequently in FET device modeling, e.g. [12-14].

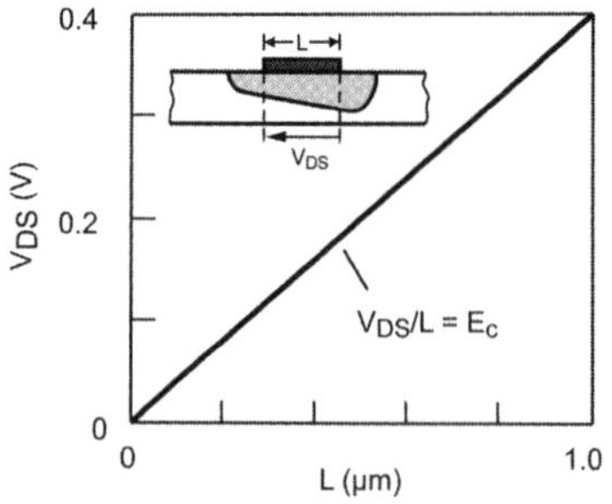

Figure 4.7 Drain-to-source voltage and respective gate length L to reach the critical field strength E_c in the channel ($E_c \approx 3.3$ kV/cm).

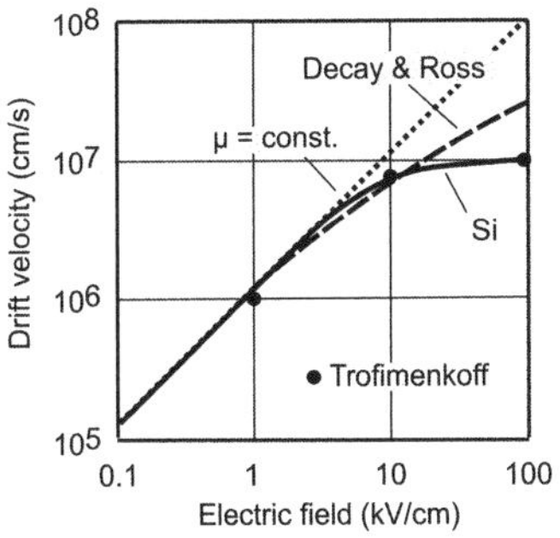

Figure 4.8 Approximation of the field-dependent drift velocity in Si (solid line) after Decay and Ross [9] and Trofimenkoff [10].

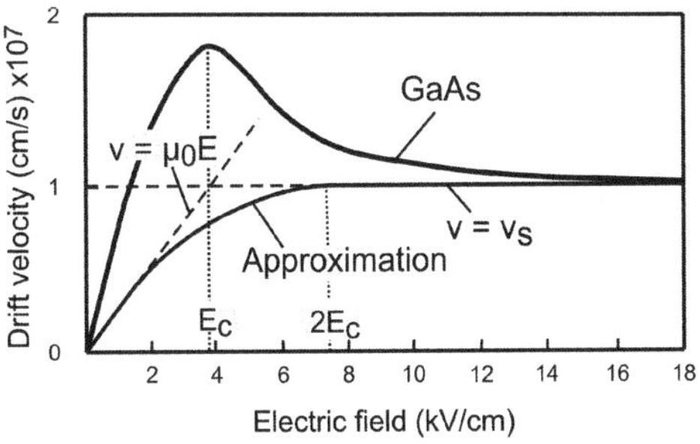

Figure 4.9 Measured drift velocity-electric field characteristic of epitaxially grown GaAs (layer thickness 30 μm, resistivity 1.5 Ω-cm, mobility 8,000 $cm^2/(Vs)$, f = 35 GHz) after Braslau and Hauge [15], parabolic approximation after Madjar and Rosenbaum [12].

4.4 IMPROVED SHOCKLEY MODEL INCLUDING CARRIER VELOCITY SATURATION

The Shockley model had to be improved by taking into account the carrier drift velocity saturation at high electrical fields. With the field-dependent relation (4.44) and the longitudinal electric field in the channel, $E_x(x) = \mathrm{d}V(x)/\mathrm{d}x$, (4.5) reads

$$I_{DS} = qN_D v_d(E_x)(a - h)w = qN_D \frac{\mu(E_x)E_x}{1 + \mu(E_x)E_x/v_s}(a - h)w \tag{4.45}$$

Integrating (4.45) between source $x = 0$ and a position x along the channel provides [16]

$$x/L = \frac{I_0}{I_{DS}}\left[3\left(1 - \frac{z}{3}\frac{I_{DS}}{I_0}\right)\left(u_2^2 - u_1^2\right) - 2\left(u_2^3 - u_1^3\right)\right] \tag{4.46}$$

At $x = L$ with $h = h_2$ (4.46) becomes [11, 16]

$$I_{DS} = \frac{I_0}{1 + z(V_{DS}/V_{pi})}\left[3\left(u_2^2 - u_1^2\right) - 2\left(u_2^3 - u_1^3\right)\right] \tag{4.47}$$

Comparing (4.47) and (4.10) shows that for given I_0, h_1/a, and h_2/a, the drain current is reduced by a factor of $(1 + \mu_0 V_{DS}/v_s L) = (1 + zV_{DS}/V_{pi})$ due to the field-dependent mobility.

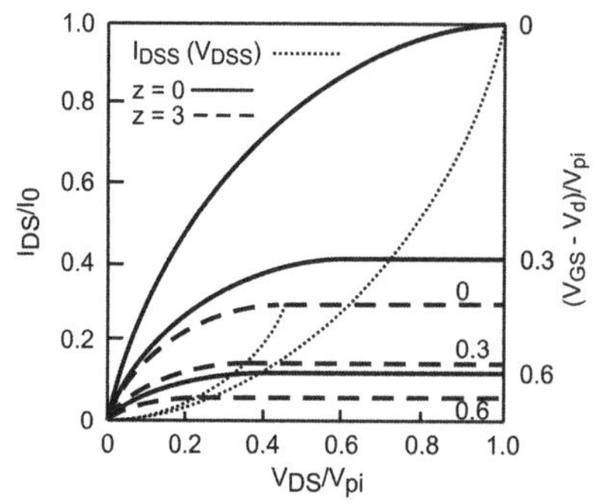

Figure 4.10 $I(V)$ characteristics with field-independent carrier mobility ($z = 0$) after Shockley and field-dependent carrier mobility ($z = 3$).

The saturation parameter z reads

$$z = \mu_0 V_{pi} / v_s L = V_{pi} / L E_c \tag{4.48}$$

which is the ratio between the small-field velocity extrapolated linearly to the field V_{pi}/L and the saturation velocity. In denormalized form (4.47) is written as

$$I_{DS} = \frac{I_0}{1 + z(V_{DS}/V_{pi})}\left[3\frac{V_{DS}}{V_{pi}} - 2\frac{\left(V_{DS} - V_{GS} + V_D\right)^{\frac{3}{2}} - \left(-V_{GS} + V_D\right)^{\frac{3}{2}}}{V_{pi}}\right] \tag{4.49}$$

Figure 4.10 illustrates the dramatic reduction of the drain current under assumption of a field-dependent carrier mobility (z = 3). The case (z = 0) indicates Shockley's theory. Then from the z-parameter definition (4.48) it follows $v_s \to \infty$. Furthermore it follows from (4.44) $v_d = \mu_0 E$ (i.e., $\mu = \mu_0 =$ constant). Similarly the channel conductance G_{DS} and transconductance G_M are reduced by the factor $(1 + zV_{DS}/V_{pi})$.

4.5 TWO-REGION MODEL

The theory described in the preceding section, including a field-dependent mobility, was successfully applied to field-effect transistors on the basis of silicon technology because the drift velocity is gradually increasing with growing electrical field strength, and does not saturate until a rather high value is reached. It can be seen from the illustrated function $v_d = v_d(E)$ for different semiconductor materials (Figure 4.5) that the saturation drift velocity of carriers in GaAs (and related GaAs-based materials) appears already at very low electrical field strength. Regarding GaAs MESFETs, the carriers are rapidly accelerated after entering the channel under the gate electrode, and then, after a very short transport length the saturation velocity is reached. Grebene and Gandhi [17] proposed a finite channel thickness to overcome the contradiction that a maximum saturation drain current flows even though the channel is pinched off. The channel region is divided into two parts (see Figure 4.11): (1) Region I with constant mobility of carriers, and (2) Region II with saturation drift velocity of carriers. Piecewise-linear approximation of the electron velocity as a function of the

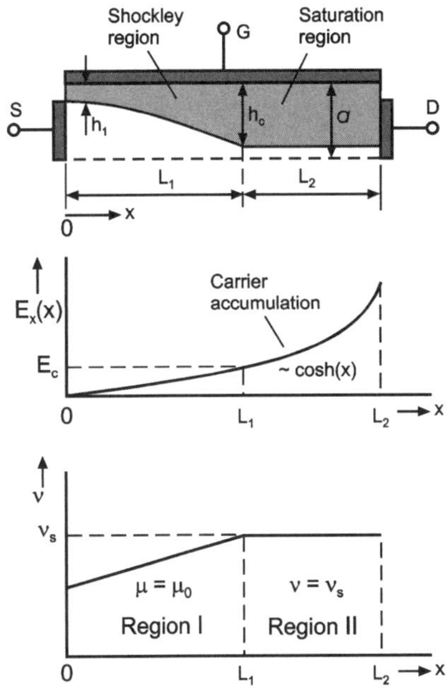

Region I: $0 \le x < L_1$, $E < E_c$
Shockley approximation with field-independent mobility:

$$\mu = \mu_0 = \text{const.}$$
$$v_d = \mu_0 E$$

Region II: $L_1 \le x \le L$, $E \ge E_c$
Saturation of carrier drift velocity in the high-field region:

$$v_d = v_s$$

Figure 4.11 Two-region model with space-charge contribution, electrical field, and carrier drift velocity along the channel.

electric field in the channel has been proposed [13]. The electron velocity is assumed to be proportional to the electric field until the saturation velocity v_s is reached at a critical field E_c, and then becomes constant (see Figure 4.9).

In Region I, in the vicinity of the source, the gradual-channel approximation is applied, and the depletion profile is estimated from a solution of the one-dimensional Poisson's equation. In Region II, near the drain, the two-dimensional Poisson's equation is solved to determine the potential distribution. The boundary values for the solution of Poisson's equation within Regions I and II are chosen to require the continuity of the potential distribution across the interface separating them. It is shown that the current flow is constrained in Region II to a narrow conductive filament (residual channel) along the channel. It is assumed that the interface between the two regions of the channel is determined by the onset of carrier drift velocity saturation.

After having passed the path L_1 the carriers have gained so much energy that they scatter into the satellite minimum of the material band structure, where their velocity decreases [5]. Thus, other hot electrons that follow catch up with these so that a high carrier density gradient comes up. That is why the electrical field strength strongly increases toward the drain-sided channel end. The transition from Region I to Region II is defined in such a way that the electrical field strength $E_x(x)$ reaches the critical value E_c:

$$E_x(x)\big|_{x=L_1} = E_c \tag{4.50}$$

We introduce a normalized layer thickness of the space-charge region at $x = L_1$:

$$u_c = h(L_1)/a = h_c/a \tag{4.51}$$

In the following the channel current is calculated in Region I and Region II, which is of the same value because of current continuity. Knowing the gate-source voltage and the drain-source current, the length L_1 of the acceleration region of the carriers (Shockley region) can be determined.

Region I: This (Shockley) region is described by drain-current relation (4.49). With $u_2 = u_c$, I_{DS} can be written in normalized form as follows:

$$I_{DS}^{I} = I_0(L_1)[3(u_c^2 - u_1^2) - 2(u_c^3 - u_1^3)] \tag{4.52}$$

with

$$I_0(L_1) = \frac{w\mu_0 q^2 N_D^2 a^3}{6\varepsilon L_1} \tag{4.53}$$

Region II: In this region the carriers drift with the saturation drift velocity v_s:

$$v_s = \mu_0 E_c \tag{4.54}$$

Then, with the current density $J_x = \sigma E_c$, the conductivity $\sigma = qN_D\mu_0$ and the current flow cross section $A = (a - h_c)w = a(1 - h_c/a)w$, the channel current reads

$$I_{DS}^{II} = J_x A = qN_D v_s w a(1 - h_c/a) \tag{4.55}$$

Defining the saturation current

$$I_s = qN_D v_s w a \tag{4.56}$$

equation (4.55) reads

$$I_{DS}^{II} = I_s(1 - h_c/a) = I_s(1 - u_c) \tag{4.57}$$

The saturation current I_s is flowing when the channel is entirely open and all carriers are moving with the same saturation drift velocity. Taking into account the derived formulas for the maximum transconductance G_{M0} (4.37) in the linear region, I_s can also be written as

$$I_s = G_{M0} L v_s/\mu_0 = G_{M0} L E_c \tag{4.58}$$

From the current continuity $I_{DS}^{I} = I_{DS}^{II}$ it follows

$$L_1 = zL\frac{(u_c^2 - u_1^2) - \frac{2}{3}(u_c^3 - u_1^3)}{1 - u_c} \tag{4.59}$$

with z in (4.48). Typical values for z lie in the order of 2 ... 20, whereas the higher values are given, for example, for microwave FETs operating at about 10 GHz. In this case $LE_c \ll V_{pi}$ ($z \gg 1$); then the Shockley model cannot be applied any longer. When the gate-to-source voltage V_{GS} and the channel current I_{DS} are known, then the length L_1 can be determined from (4.59).

The accumulation of electrons in Region II leads to a high field strength E_x in the channel, thus also an E_x component in the adjoining space-charge region. The electrical field strength

E_x is, as can be shown [16],

$$E_x^{II}(x,y) \approx -E_c \cosh\left[\frac{\pi}{2h_c}\pi(x-L_1)\right] \tag{4.60}$$

Through the practical exercises in Section 4.8 the reader should become familiar with typical physics-based device data and related terms, which will be strengthened in Chapter 6 focusing on the estimation of FET model element values from physical device parameters.

4.6 SHORT-CHANNEL SATURATION MODEL

When the gate length is very short, then it can be assumed that current saturation occurs over the whole channel region. Under consideration of the derivation of the drain current in the two-region model, the saturation current here reads [see (4.55)]

$$I_{DS} = qN_D v_s w(a - h) = qN_D v_s w a(1 - h/a) \tag{4.61}$$

With $h = h(V_{GS})$ (4.14),

$$h = [2\varepsilon/qN_D(-V_{GS} + V_d)]^{1/2} \tag{4.62}$$

the drain current reads

$$I_{DS} = qN_D v_s w\left[a - (2\varepsilon/qN_D)^{1/2}(-V_{GS} + V_d)^{1/2}\right] \tag{4.63}$$

that is, the current is directly modulated by the change of the thickness of the space-charge region controlled by the gate-source voltage. The validity of the relation could experimentally be verified for short-channel MESFETs with a gate length $L \leq 2$ µm. The transconductance G_M is calculated from (4.63) with $G_M = (\partial I_{DS}/\partial V_{GS})$, V_{DS} = constant, as

$$G_M = v_s w \sqrt{\frac{qN_D\varepsilon}{2}\frac{1}{\left(-V_{GS} + V_d\right)}} \tag{4.64}$$

For linear operation of a FET device, the transconductance G_M should be linearly dependent on the gate-source control voltage. This means that the transfer characteristic $I_{DS}(V_{GS})$ in Figure 4.12 should be linear above the pinch-off region. The transition from the linear range to pinch-off is smooth and frequently described by a quadratic function. The linear behavior can approximately be realized with special doping profiles [18]. Experimental investigations show that gradual or stepped doping profiles yield superior linearity.

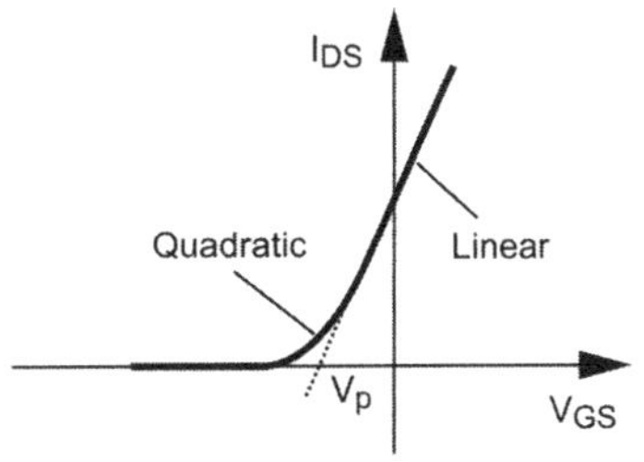

Figure 4.12 Linear transfer characteristic $I_{DS}(V_{GS})$.

The gate-source capacitance C_{GS} is calculated as follows. The charge of the space-charge region under the gate is

$$Q = qN_D wLh \tag{4.65}$$

Substituting h in (4.65) delivers

$$Q = wL[2\varepsilon qN_D(-V_{GS} + V_d)]^{1/2} \tag{4.66}$$

Then the gate-source capacitance reads

$$C_{GS} = \frac{\partial Q}{\partial V_{GS}} = wL\left(\frac{qN_D\varepsilon}{2}\frac{1}{-V_{GS} + V_d}\right)^{1/2} \tag{4.67}$$

From (4.64) and (4.67) we obtain the simple relation between transconductance and gate-source capacitance:

$$G_M = C_{GS}v_sL \tag{4.68}$$

The transconductance per unit length is calculated as

$$G'_M = C_{GS}v_s \tag{4.69}$$

4.7 RELATIONSHIPS BETWEEN MESFET AND HEMT DC CHARACTERISTICS

The output characteristic of a HEMT is very similar to that of a GaAs MESFET in that it has linear and saturated regions. In the following, based on rough approximations, simple relations of transconductance and gate capacitance of MESFET and HEMT are derived and compared, which may be useful for model parameter extraction purposes.

4.7.1 Transconductance

MESFET

According to (4.61) the DC drain current of a MESFET in a short-channel saturation model has been found to be

$$I_{DS} = qN_Dv_sw(a - h) = qN_D\mu_{3D}w(a - h)V_{DS}/L_{SD} \tag{4.70}$$

where w is the gate width, V_{DS} is the drain-source voltage, v_s is the carrier saturation velocity, $v_s = \mu_{3D}E = \mu_{3D}V_{DS}/L_{SD}$ with the carrier mobility μ_{3D} in the doped channel, and L_{SD} is the gate length. Note that in this chapter we do not consider the access regions on both sides of the gate electrode. Defining an effective sheet carrier concentration $n_{3D} = N_D(a-h)$ we get from (4.70)

$$I_{DS} = qn_{3D}(V_{GS})\cdot v_sw = qn_{3D}(V_{GS})\cdot\mu_{3D}wV_{DS}/L_{SD} \tag{4.71}$$

At pinch-off, with $(a-h = 0)$ the drain current becomes zero, $I_{DS} = 0$. Saturation current is obtained for $(h = 0)$ with $n_{3D,max} = N_da$. The rise of the drain current with V_{GS} follows a square root function.

HEMT

In a HEMT device, the electrons are transported in an undoped channel with higher drift velocity (see Figure 4.6(a)); that is, the carrier mobility in the 2DEG, μ_{2D} is much higher than in the doped MESFET channel, μ_{3D} . In the linear region, the drain-source current of a HEMT can be written as

$$I_{DS} = qn_{2D}(V_{GS}) \cdot \mu_{2D} w V_{DS}/L_{SD} \tag{4.72}$$

where n_{2D} (cm^{-2}) is the channel sheet carrier concentration under the gate. In the saturation region, the drain-source current can be written as

$$I_{DS} = qn_{2D}(V_{GS}) v_{eff} w \tag{4.73}$$

where v_{eff} is the effective channel velocity. v_{eff} denotes the electron velocity in HEMT devices, which can be much higher than the bulk saturation electron velocity v_s, for example, due to impact ionization, velocity overshoot and two-dimensional mobility [19]. Overshoot occurs in particular in nanometer-gate-width devices as described in the chapter "Short Channel Effect" in [5].

For an identical geometry device, the knee voltage for device saturation occurs at lower values of drain-source bias in a HEMT due to higher electron mobility.

The simple charge control model described in [20] can be used to derive an expression for the transconductance of a HEMT. The total charge under the gate can be written as

$$n_{2D}(V_{GS}) = \frac{\varepsilon_A}{q(d_A + \Delta d)}(V_{GS} - V_p) \tag{4.74}$$

with

$$\varepsilon_A = \varepsilon_0 \varepsilon_{rA} \tag{4.75}$$

Note that (4.74) describes the physical reality only roughly. Indeed, there is a linear increase for moderate gate-source voltages. However, with larger values sheet charge density saturates [5]. Furthermore, at larger gate voltage values a parasitic MESFET channel is formed in a HEMT structure, leading to a lower effective carrier mobility. Additionally, the gate-source capacitance increases. In (4.74) n_{2D} is the 2DEG sheet carrier concentration, ε_A and d_A are the AlGaAs permittivity and thickness, respectively, q is the electronic charge, and V_{GS} is the gate-source potential, which is normally negative for a depletion-mode device. Here Δd is a correction factor taking into account that the position of the 2DEG does not reside exactly at the GaAs-AlGaAs interface. Δd is the distance of the centroid of the 2DEG distribution from the AlGaAs/GaAs interface and is typically of the order of 80Å for $n_{2D} \sim 10^{12}/cm^2$. For room temperature, the following analytical function may be used [20]:

$$\Delta d/\text{Å} = 52.3(n_{2D}/10^{12}\,\text{cm}^{-2})^{-2/3} \tag{4.76}$$

From (4.73) and (4.75) we obtain

$$I_{DS} = \frac{\varepsilon_A v_{eff} w}{d_A + \Delta d}(V_{gs} - V_p) \tag{4.77}$$

The current control mechanism in the HEMT is control of the 2DEG density at the

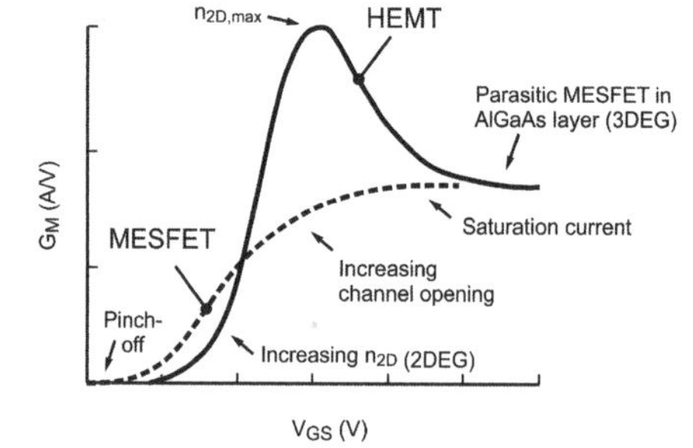

Figure 4.13 General dependence of MESFET and HEMT transconductances on gate-source voltage.

heterojunction by the gate voltage. Differentiating (4.77) yields

$$G_m = \frac{\partial I_{DS}}{\partial V_{GS}} = \frac{\varepsilon_A v_{eff} w}{d_A + \Delta d} \tag{4.78}$$

Therefore, the transconductance of a HEMT depends on the total gate-channel separation ($d_A + \Delta d$) and the effective carrier velocity v_{eff}. Relation (4.78) is used later to find approximated starting parameter values for a model parameter extraction process.

Figure 4.13 shows typical behavior of the transconductance of a HEMT in comparison with a MESFET device as a function of gate voltage. It can be seen that the transconductance of the MESFET is zero at pinch-off, then increases monotoneously and reaches its maximum at high gate-source voltages. However, the HEMT behavior is rather different. At pinch-off the 2DEG sheet carrier concentration n_{2D} is zero. With decreasing negative gate voltage, n_{2D} rises continuously until the carrier concentration saturates ($n_{2D} = n_{2D,max}$). Then, with further decrease of negative gate-source voltage additional parallel conduction (3DEG) in the doped AlGaAs layer occurs (parasitic MESFET effect), resulting in a lowering of the transconductance. The transconductance versus V_{GS} indicates a typical bell-shaped function.

4.7.2 Gate-Source Capacitance

MESFET

As derived in Section 4.6, the gate-source capacitance C_{GS} of a short channel device can be estimated by the following relation:

$$C_{gs} = wL(qN_D\varepsilon/2)^{1/2}(-V_{gs} + V_d)^{-1/2} \tag{4.79}$$

Frequently, the capacitance is normalized to the gate length L and gate width w. Then the capacitance per unit area reads

$$C'_{gs} = (qN_D\varepsilon/2)^{1/2}(-V_{gs} + V_d)^{-1/2} \tag{4.80}$$

As derived in section 4.6, normalized transconductance and capacitance are related to each other as

$$G'_m = C'_{gs} v_s \tag{4.81}$$

or in denormalized form,

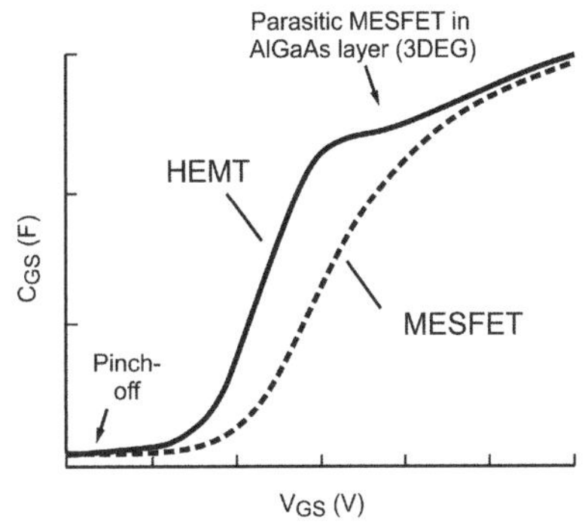

Figure 4.14 General dependence of MESFET and HEMT gate capacitance on gate-source voltage.

$$G_m = C_{gs} v_s / L \tag{4.82}$$

Regarding short-channel field effect transistors the fringing effect of the gate electrode must be taken into account that is, the effective gate length L_{eff} is simply approximated by

$$L_{eff} = L + 2d \tag{4.83}$$

with the extended gate length d, which is assumed to be equal to the height of the lateral depletion region of the zero-biased Schottky-gate. d is calculated as [1, 21]:

$$d = \left[\frac{2\varepsilon}{qN_D} (-V_{gs} + V_d) \right]^{1/2} \tag{4.84}$$

HEMT

The first HEMT charge control model was proposed by Delagebeaudeuf and Linh in 1982 [22]. The potential well at the AlGaAs/GaAs interface is approximated by a triangular well. The energy levels in this triangular well and the maximum 2DEG density, $n_{\text{2D,max}}$, can be calculated by solving the Schrödinger equation in the triangular well and Poisson equation in AlGaAs donor layers. For $0 < n_{\text{2D}} < n_{\text{2D,max}}$, the sheet charge density n_{2D} as a function of gate voltage V_{GS} can be expressed as

$$q n_{2D} A = C_{gs} (V_{gs} - V_p) \tag{4.85}$$

where C_{gs} is the 2DEG capacitance and is given with (4.74) by the following expression:

$$C_{gs} = \frac{A \varepsilon_A}{(d_A + \Delta d)} \tag{4.86}$$

The ratio $G_{\text{m}}/C_{\text{gs}}$ reads

$$G_m / C_{gs} = w v_{eff} / A \tag{4.87}$$

or with normalized parameters

$$G_m' = \frac{\varepsilon_A v_{eff}}{d_A} = C_{gs}' v_{eff} \tag{4.88}$$

G_{m}' is the transconductance per unit width and C_{gs}' the gate-source capacitance per unit area.

Figure 4.14 shows the differences in the gate-source capacitance behavior between MESFET and HEMT.

4.7.3 MESFET and HEMT Transconductance Comparison

With (4.64) and (4.62) the transconductance of a MESFET can be written as

$$G_m = \varepsilon_{GaAs} v_s w/h \tag{4.89}$$

where h is the height of the space-charge (depletion) region under the gate. Using (4.89) and (4.78) the ratio of the transconductance of HEMT and MESFET reads

$$\frac{G_{m,\,HEMT}}{G_{m,\,MESFET}} = \frac{\varepsilon_A v_{eff} w}{(d_A + \Delta d)} \bigg/ \frac{\varepsilon_{GaAs} v_s w}{h} \tag{4.90}$$

With $h \approx (d_A + \Delta d)$ (4.90) simplifies to

$$\frac{G_{m,HEMT}}{G_{m,MESFET}} = \frac{\varepsilon_{AlGaAs} v_{2DEG}}{d_{eff}} \bigg/ \frac{v_{n\text{-}GaAs} \varepsilon_{GaAs}}{h} = \frac{\varepsilon_{AlGaAs}}{\varepsilon_{GaAs}} \frac{v_{2DEG}}{v_{n\text{-}GaAs}} \frac{h}{d_{eff}} \tag{4.91}$$

With $\varepsilon_{AlGaAs} = 12.05 \times \varepsilon_0$ and $\varepsilon_{GaAs} = 12.9 \times \varepsilon_0$ the ratio $\varepsilon_{AlGaAs}/\varepsilon_{GaAs}$ is 0.934. The saturation velocity v_{2DEG} in the GaAs layer ($N_d = 0$) is about 2×10^7 cm/s, whereas in doped GaAs (9×10^{16} cm^{-3}...2×10^{17} cm^{-3}) the electron velocity is 1.8×10^7 cm/s [23]. Thus the ratio $v_{2DEG}/v_{n\text{-}GaAs}$ is approximately 1.11, and (4.91) reduces to

$$G_{m,\,HEMT}/G_{m,\,MESFET} \approx 1.04 \times a/d_{eff} \tag{4.92}$$

with $h \approx a$ [24]. Assuming a MESFET with typical channel layer thickness $a = 100$ nm and a HEMT with typical effective AlGaAs thickness $d_{eff} = 50$ nm, it is found that the transconductance of HEMT devices is approximately twice the estimated value for equivalent MESFET devices.

4.8 PROBLEMS AND SOLUTIONS

In this section some problems and their solutions are given concerning the discussed analytical models. This will allow the reader to become familiar with the physical parameters and their mathematical treatment.

Problem 4.1

Figure 4.15 shows the cross section of a GaAs MESFET. Assume that the doping concentration is $N_D = 7 \times 10^{16}$ cm^{-3} and that the dimensions are $a = 0.3$ µm, $L = 1.5$ µm, $w = 50$ µm (gate width), $\mu_0 = 4500$ cm^2/Vs (carrier mobility), and $V_d = 0.89$V (diffusion voltage). The saturated drift velocity for the given doping is assumed to be $v_s = 1.6 \times 10^7$ cm/s. In the operating point ($V_{GS} = 0$, $V_{DS} = 1$V) a drain current $I_{DS} = 14$ mA has been measured.

Calculate the channel parameter L_1 of the two-region model, and give a short comment on the validity of this model for the given device.

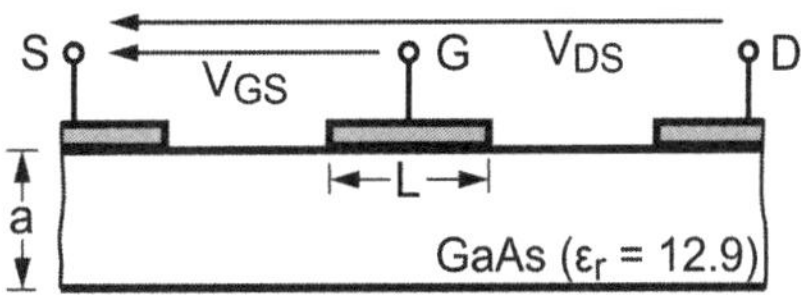

Figure 4.15 MESFET structure under investigation.

Solution:

The length L_1 can be calculated applying (4.16), (4.48) and (4.59). The factor qN_d/ε is calculated as

$$\frac{qN_d}{\varepsilon} = \frac{1.6\times10^{-19}\text{As}\times7\times10^{16}\text{cm}^{-3}}{12.9\times8.85\times10^{-14}\text{As/V/cm}} = 0.98\times10^{10}\text{V/cm}^2$$

Then the intrinsic pinch-off voltage is obtained from (4.16) as

$$V_{pi} = \frac{qN_d}{\varepsilon}\frac{a^2}{2} = 0.98\times10^{10}\text{V/cm}^2\times(0.3\mu\text{m})^2/2 = 4.41\text{V}$$

From (4.48) we then obtain

$$z = \frac{4500\text{cm}^2/\text{V/s}\times4.41\text{V}}{1.5\times10^{-4}\text{cm}\times1.6\times10^7\text{cm/s}} = 8.26$$

The normalized voltage u_1 is calculated as

$$u_1 = (V_d/V_{pi})^{1/2} = 0.89\text{V}/4.41\text{V} = 0.45$$

The saturation current I_s in (4.56) is obtained as

$$I_s = 1.6\times10^{-19}\text{As}\times7\times10^{16}\text{cm}^{-3}\times1.6\times10^7\ \text{cm/s}\times50\ \mu\text{m}\times0.3\ \mu\text{m} = 26.88\ \text{mA}$$

Solving in (4.57) for u_c yields

$$u_c = 1 - I_{DS}/I_s = 1 - 14\,\text{mA}/26.88\,\text{mA} = 0.52$$

Then the ratio L_1/L is calculated as

$$L_1/L = 8.26\times[(0.27 - 0.2025) - 2/3\times(0.14 - 0.091)]/0.48 = 0.59$$

or $L_1 = 0.59L = 0.88$ μm. Because $L_1 < L$, the two-region model can be applied.

Problem 4.2

A GaAs MESFET device as shown in Figure 4.15 has a gate length $L = 1.0$ μm, a channel width $a = 0.2$ μm, and a gate width w = 50 μm. The dopant density is $N_D = 1.5\times10^{17}$ cm^{-3}. The low-field carrier mobility is $\mu_0 = 5300$ cm^2/Vs. The diffusion voltage is assumed to be $V_d = 0.7$V.

Calculate the saturation current I_{DSS} for a gate-source voltage $V_{GS} = V_d$ and $V_{GS} = -1$V, respectively. Apply (1) the classical Shockley model, and (2) the improved one with field-dependent carrier mobility.

Solution:

We first calculate the intrinsic and extrinsic pinch-off voltages according to (4.16) and (4.18). With

$$\frac{qN_{\mathrm{d}}}{\varepsilon} = \frac{1.6\times10^{-19}\mathrm{As}\times1.5\times10^{17}\mathrm{cm}^{-3}}{12.9\times8.85\times10^{-14}\mathrm{As/V/cm}} = 2.1\times10^{10}\,\mathrm{V/cm}^2$$

we get for the intrinsic pinch-off voltage

$$V_{\mathrm{pi}} = 2.1\times10^{10}\,\mathrm{V/cm}^2\times(0.2\mu\mathrm{m})^2/2 = 4.2\mathrm{V}.$$

Then the external pinch-off voltage is obtained as

$$V_{\mathrm{pa}} = -4.2\mathrm{V} + 0.7\mathrm{V} = -3.5\mathrm{V}.$$

The pinch-off channel current I_0 in (4.11) is calculated as

$$I_0 = \frac{w\mu_0 q^2 N_{\mathrm{D}}^2 a^3}{6\varepsilon L} = \frac{V_{\mathrm{pi}}}{3}\,\frac{w\mu_0 q N_{\mathrm{D}} a}{L}$$

$$= \frac{4.2\mathrm{V}}{3}\,\frac{50\times10^{-4}\mathrm{cm}\times5300\,\mathrm{cm}^2/\mathrm{V/s}\times1.6\times10^{-19}\mathrm{As}\times1.5\times10^{17}\mathrm{cm}^{-3}\times0.2\times10^{-4}\mathrm{cm}}{10^{-4}\mathrm{cm}} = 178\,\mathrm{mA}.$$

I_{DSS} is calculated from (4.28) as $I_{\mathrm{DSS}} = I_0 = 178$ mA for $V_{\mathrm{GS}} = V_d$. Then, with (4.30) the respective saturation voltage reads $V_{\mathrm{DSS}} = V_{\mathrm{pi}} + V_{\mathrm{GS}} - V_d = V_{\mathrm{pi}} = 4.2\mathrm{V}$.
For the case $V_{\mathrm{GS}} = 1\mathrm{V}$, we first calculate the individual terms in (4.28):

$$3(-V_{GS} + V_d)/V_{pi} = 3\times1.7V/4.2V = 3\times0.4 = 1.2, \quad 2[(-V_{GS} + V_d)/V_{pi}]^{3/2} = 2\times0.4^{3/2} = 0.5$$

Then the saturation current I_{DSS} yields

$$I_{\mathrm{DSS}} = 178\,\mathrm{mA}\,(1 - 1.2 + 0.5) = 0.3\times178\,\mathrm{mA} = 53.4\,\mathrm{mA}.$$

The respective saturation voltage is obtained as $V_{\mathrm{DSS}} = 4.2\mathrm{V} - 1\mathrm{V} - 0.7\mathrm{V} = 2.5\mathrm{V}$.
Taking into account the field-dependent carrier mobility, the drain current is calculated from (4.49). Introducing the drain current after Shockley, it reads $I_{\mathrm{DS}} = I_{\mathrm{DS,Shockley}}/[1 + z(u_2^2 - u_1^2)]$. The z-parameter in (4.48) is calculated as

$$z = \frac{\mu_0 V_{\mathrm{pi}}}{L v_{\mathrm{s}}} = \frac{5300\,\mathrm{cm}^2/\mathrm{V/s}\times4.2\mathrm{V}}{10^{-4}\mathrm{cm}\times1.65\times10^7\,\mathrm{cm/s}} = 13.5.$$

Maximum drain-source current under pinch-off: $u_2^2 - u_1^2$ or $V_{\mathrm{DS}} = V_{\mathrm{GS}} - V_{\mathrm{d}} + V_{\mathrm{pi}}$. Then

$$I_{\mathrm{DSS}} = \frac{I_0}{1 + z(1 - u_1^2)}[3(1 - u_1^2) - 2(1 - u_1^3)] = \frac{I_0}{1 + z(1 - u_1^2)}(1 - 3u_1^2 + 2u_1^3) \tag{4.93}$$

For $V_{\mathrm{GS}} = V_d$, u_1 becomes zero so that

$$I_{\mathrm{DSS}} = \frac{I_0}{1 + z} = \frac{178\,\mathrm{mA}}{14.5} = 12.27\,\mathrm{mA}.$$

For $V_{\mathrm{GS}} = -1\mathrm{V}$, we have to evaluate (4.93): With

$$u_1^2 = (-V_{GS} + V_d)/V_{pi} = 1.7\text{V}/4.2\text{V} = 0.4$$

we obtain

$$I_{DSS} = \frac{178\ \text{mA}\times 0.3}{1 + 13.5\times 0.6} = 5.87\ \text{mA}.$$

4.9 SUMMARY

The Shockley model was the first analytical physics-based model that aimed to describe the electrical characteristics of (long-gate) field-effect transistors. It is based on simplified assumptions such as gradual-channel approximation and field-independent mobility and provides a closed-form expression for the channel current. The coefficients appearing in the equations of the model are material parameters, such as the dielectric constant ε and the doping density N_D, and geometrical parameters, such as the depth of the channel a and length of the gate L. Later on the model was improved by introducing electrical field-dependent carrier mobility. As the gatelength of transistors became continuously shorter for higher frequency applications, various modifications of the classical Shockley model were introduced creating, for example, the two-region and short channel saturation model. Popular analytical models frequently include mathematical expressions, which can be formally traced back to the Shockley model. Some relationships between HEMT and MESFET have been presented. Derived analytical expressions for some model parameters such as transconductance, output conductance, and gate-source capacitance may be useful for the estimation of model parameter values from given physical device data.

REFERENCES

[1] Shockley, W., "A Unipolar "Field-Effect" Transistor," *Proc. I.R.E.*, 1952, pp. 1365–1376.

[2] Kennedy, D. P., and R. R. O'Brien, "Computer Aided Two-Dimensional Analysis of the Junction Field-Effect Transistor," *IBM J. Res. Develop.*, 1970. 95–116.

[3] Leuther, A., A. Tessmann, P. Doria, M. Ohlrogge, M. Seelmann-Eggebert, H. Maßler, M. Schlechtweg, and O. Ambacher, "20 nm Metamorphic HEMT Technology for Terahertz Monolithic Integrated Circuits," *Proc. 9th European Microwave Integrated Circuits Conference (EuMA)*, Rome, Italy, 6 -7 Oct 2014, pp. 84 – 87.

[4] Bockemuehl, R. R., "Analysis of Field Effect Transistors with Arbitrary Charge Distribution," *IEEE Trans. on Electron Devices*, 1963, pp. 31–34.

[5] Kompa, G., *Basic Properties of III-V Devices–Understanding Mysterious Trapping Phenom ena*, Kassel, Kassel University Press, 2014.

[6] S. Kasap, and P. Capper (Eds.), *Springer Handbook of Electronic and Photonic Materials*, Springer Science+Business Media, Inc., 2006.

[7] Gaska, R., J. W. Yang, A. Osinsky, Q. Chen, M. Asif Khan, A. O. Orlov, G. L. Snider, and M. S. Shur, "Electron Transport in AlGaN-GaN Heterostructures grown on 6H-SiC Substrates," *Appl. Phys. Lett.*, Vol. 72(6), 1998, pp. 707–709.

[8] Ruch, J. G., and W. Fawcett, "Temperature Dependence of the Transport Properties of Gallium Arsenide Determined by a Monte Carlo Method," *Journal of Applied Physics*, Vol. 41, No. 9, 1970, pp. 3843–3849.

[9] Dacey, G. C., and I. M. Ross, "The Field Effect Transistor", *Bell Syst. Tech. J.*, Vol. 34, 1955, pp. 1149–1189.

[10] Trofimenkoff, F. N., "Field-Dependent Mobility Analysis of the Field-Effect Transistor," *Proc. IEEE*, Vol. 53, 1965, 1765–1766.

[11] Lehovec, K., and R. Zuleeg, "Voltage-Current Characteristics of GaAs J-FET's in the Hot Electron Range," *Solid-State Electronics*, Vol. 13, pp. 1415–1429.

[12] Madjar, A., and F. J. Rosenbaum, "A Large-Signal Model for the GaAs MESFET," *IEEE Trans.*, Vol. MTT-29(8), 1981, pp. 781–788.

[13] B andler, J. W., R. M. Biernacki, Q. Cai, S. H. Chen, S. Ye, and Q. J. Zhang, "Integrated Physics-Oriented Statistical Modeling, Simulation and Optimization," *IEEE Trans.*, Vol. MTT-40, 1992, pp. 1374–1400.

[14] Khatibzadeh, M. A., and R. J. Trew, "A Large-Signal, Analytic Model for the GaAs MESFET," *IEEE Trans.*, Vol. MTT-36, 1988, pp. 231–238.

[15] Braslau, N., and P. S. Hauge, "Microwave Measurement of the Velocity-Filed Characteristic of GaAs," *IEEE Trans.*, Vol. ED-17(8), 1970, pp. 616–622.

[16] Sze, S. M., *Physics of Semiconductor Devices*, New York: Wiley & Sons, Second Edition, 1981.

[17] Grebene, A. B., and S. K. Ghandhi, "General Theory for Pinched Operation of the Junction-Gate FET," *Solid-State Electronics*, Pergamon Press, Vol. 12, 1969, pp. 573–589.

[18] Williams, R. E., and D. W. Shaw, "Graded Channel FET's: Improved Linearity and Noise Figure," *IEEE Trans.*, Vol. ED-25, 1978, pp. 600–605.

[19] Quay, R., C. Moglestue, V. Palankovski, and S. Selberherr, "A Temperature Dependent Model for the Saturation Velocity in Semiconductor Materials," *Materials Science in Semiconductor Processing*, Vol. 3, 2000, pp. 149–155.

[20] Moloney, M. J., F. Ponse, and H. Morkoç, "Gate Capacitance-Voltage Characteristics of MODFET's: Its Effect on Transconductance," *IEEE Trans.*, Vol. ED-32(9), 1985, pp. 1675–1985.

[21] Han, J., "Electron Transport Characteristics of Sub-100 Nanometer GaAs MESFETs," *J. Korean Phys. Soc.*, Vol. 42, 2003, pp. 297–301.

[22] Delagebeaudeuf, D., and N. T. Linh, "Metal-(n) AlGaAs-GaAs Two-Dimensional Electron Gas FET," *IEEE Trans.*, Vol. ED-29(6), 1982, pp. 955–960.

[23] Masselink, W. T., "Electron Velocity in GaAs: Bulk and Selectively Doped Heterostructures," *Semicond. Sci. Technol.*, Vol. 4, 1989, pp. 503–512.

[24] Khemissi, S., and C. Azizi, "A Two-Dimensional Analytical Modeling of the Current-Voltage Characteristics for Submicron Gate-Length GaAs MESFET`s," *Int. J. of Engineering & Technology*, Vol. 12(4), pp. 27–33.

[25] O'Leary, S. K., B. F. Foutz, M. S. Shur, and L. Eastman, "Steady-State and Transient Electron Transport Within the III-V Nitride Semiconductors, GaN, AlN, and InN: A Review," *Journal of Materials Science: Materials in Electronics*, Vol. 17(2), 2006, pp. 587 –126.

Chapter 5

Extrinsic Transistor Network at DC

5.1 INTRODUCTION

As will be discussed later, the development of nonlinear transistor models requires bias-dependent small-signal model parameters dependent on the inner gate and drain control voltages V_{gs} and V_{ds}. However, extracted model parameters from measured scattering parameters are related to the outer bias voltages V_{GS} and V_{DS}. Thus, appropriate bias voltage correction must be performed. Figure 5.1 indicates the parasitic series resistances (i.e., source resistance R_s, drain resistance R_d, and gate resistance R_g), which cause a voltage drop and a shifting of the outer bias point.

5.2 INTRINSIC CONTROL VOLTAGES FROM RESISTIVE NETWORK DE-EMBEDDING

For the DC analysis we assume that the transistor is biased in a typical bias point so that the gate-source diode operates in the reverse direction. As well, avalanche breakdown is not taken into consideration. To determine the inner control voltages we regard the simplified DC equivalent circuit in Figure 5.2.

Applying Kirchhoff's loop rule, the following relations between the outer and inner bias voltages are obtained ($I_{GS} = I_{gs}$, $I_{DS} = I_{ds}$):

$$V_{gs} = V_{GS} - (R_g + R_s)I_{GS} - R_s I_{DS} \tag{5.1}$$

$$V_{ds} = V_{DS} - (R_d + R_s)I_{DS} - R_s I_{GS} \tag{5.2}$$

Or written in matrix form:

$$\begin{pmatrix} V_{gs} \\ V_{ds} \end{pmatrix} = \begin{pmatrix} V_{GS} \\ V_{DS} \end{pmatrix} - \begin{pmatrix} R_g + R_s & R_s \\ R_s & R_d + R_s \end{pmatrix} \begin{pmatrix} I_{GS} \\ I_{DS} \end{pmatrix} \tag{5.3}$$

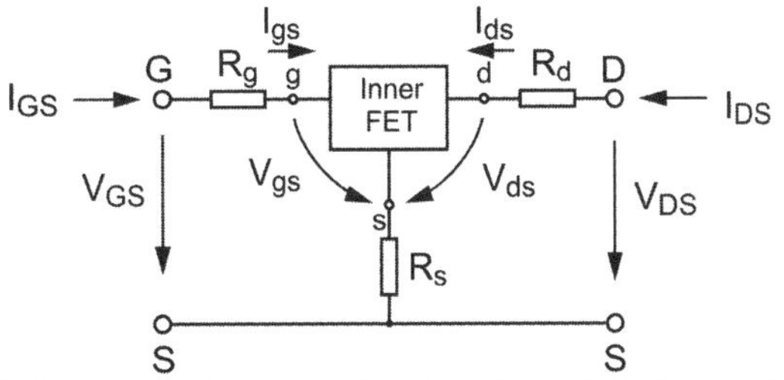

Figure 5.1 Extrinsic resistive network comprising R_g, R_d, and R_s has a strong impact on the inner transistor biasing.

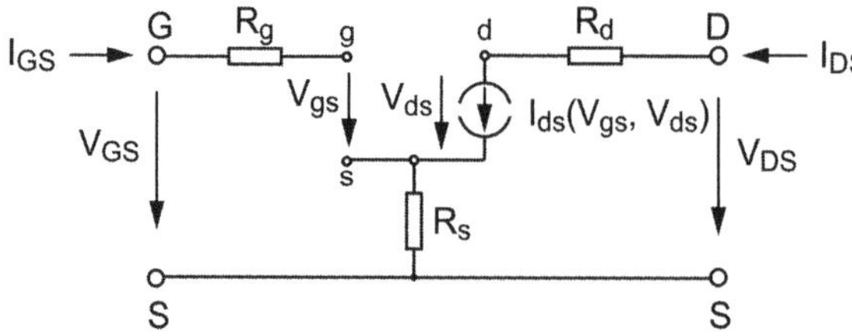

Figure 5.2 Outer bias voltages V_{GS} and V_{DS} transformed to inner bias voltages V_{gs} and V_{ds} via parasitic resistance network (R_s, R_d, R_g).

If $V_{GS} < 0$ the gate resistance R_g may be only relevant for alternating gate-source stimulus. Due to negative feedback the source resistance impairs the overall performance of the device. The drain resistance R_d is commonly higher than the source resistance R_s, which is particularly expected with high voltage devices with larger drain-gate distance to prevent breakdown appearance.

The inner gate-source control voltage is reduced by the voltage drop over the source resistor. Furthermore, in the ohmic region the $I_{ds}(V_{gs})$ characteristics is sheered with respect to the outer drain-source voltage by the voltage drop $(R_s + R_d)I_{DS}$. The modified characteristics affected by the bulk resistances are illustrated in Figure 5.3. It becomes obvious that for the same currents ($I_{DS} = I_{ds}$), the inner voltages (V_{gs}, V_{ds}) are lower than the outer voltages (V_{GS}, V_{DS}). The largest effect is on the knee voltage in the curve of the drain current versus drain voltage.

Table lookup models employ tables of parameter values, and each table has two control variables: the inner bias voltages V_{gs} and V_{ds}. The required conversion of outer to inner voltages complicates matters because the array of inner voltages is an irregularly spaced collection of values. If the desired grid of control voltages were the original grid of outer gate-source and drain-source voltages (i.e., the voltages applied in the scattering parameter measurements from which the model parameters were extracted) then any regridding process would be a straightforward interpolation. However, the values of the inner voltages V_{gs} and V_{ds} are necessarily not equidistant and require a nontrivial regridding process.

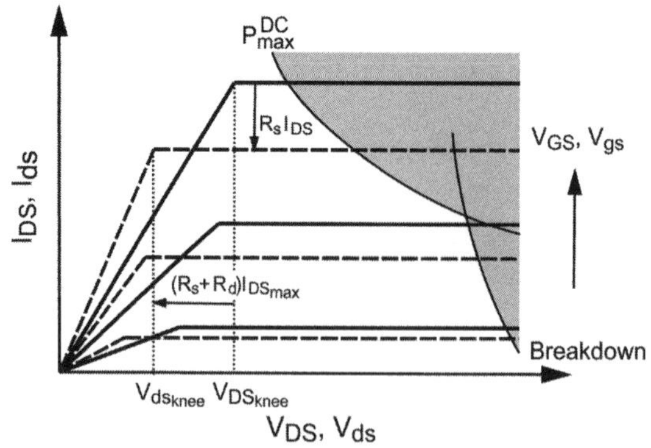

Figure 5.3 Schematic view of extrinsic and intrinsic field-effect transistor $I(V)$ characteristics with respect to inner and outer drain voltages. Solid lines refer to the outer FET terminals and dotted lines to the inner terminals (— $I_{DS} = I_{DS}(V_{GS}, V_{DS})$, - - - $I_{ds} = I_{ds}(V_{gs}, V_{ds})$). Limits of measurements: (1) maximum DC power limit, (2) breakdown.

A non-equidistant grid of control voltages complicates the intrinsic-element integration that is necessary for the extraction of the large-signal modeling of charge and current sources. In addition, table-based representations of model parameters with non-equidistant grids of control voltages are not convenient to handle in Advanced Design System's (ADS) simulator [1], in which many models are implemented.

A well-defined interpolation technique must be used to regrid each extracted bias-dependent parameter from the original outer voltage grid to an equidistant grid of inner voltages.

Note that for optimal modeling results, the density of DC operating points, on which scattering parameter measurements are performed, should be higher in regions of fast variation of the device current-voltage characteristics, and the points can be scarcer in regions with slower variation.

5.3 REGRIDDING OF NONORTHOGONAL INTRINSIC VOLTAGES

As reported in [2] numerical transformation of measured current-voltage data, or of an extracted model parameter, gridded on the outer voltages to an equidistant grid controlled by the inner voltages, may entail some practical problems that differ from a straightforward interpolation, and which could lead to flawed extraction results with dramatic effect on the derivation of final large-signal current and charge sources.

The outer-to-inner voltage conversion as described in (5.1) and (5.2) was evaluated with extracted values of R_g, R_d, and R_s of a 3.2-mm GaN power HEMT, using the measured voltages and currents from the static-DC I-V characteristics of the same device. The results of these calculations are illustrated in Figure 5.4, where the original V_{GS} - V_{DS} grid is marked by crosses, and the V_{gs} - V_{ds} grid is denoted by circles.

As observed in Figure 5.4, for $V_{ds} > 0$ and for V_{gs} values outside the pinch-off region (V_{gs}

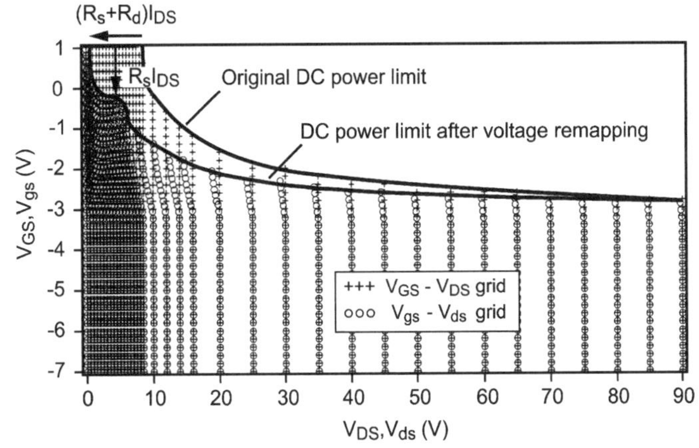

Figure 5.4 Outer control voltages (V_{GS}, V_{DS}) showing an orthogonal voltage grid (crosses) and inner bias voltages (V_{gs}, V_{ds}) with nonorthogonal voltage grid (circles). Example measurements referring to the drain current-voltage characteristics of a 3.2-mm GaN power device [2].

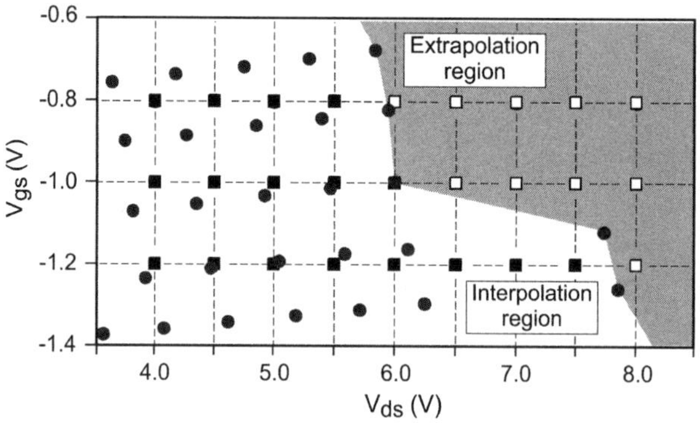

Figure 5.5 Partial illustration of the inner voltage grid that is shown in Figure 5.4. The dots are the nonorthogonal grid and the squares the orthogonal grid desired after the regridding process. Filled squares are the grid points that only require interpolation. Hollow squares denote grid points outside the actual measured data range, which imply extrapolation and should be considered out of the scope of the modeling strategy [2].

$> V_{th}, V_{th} = -3.3$V), the points on the inner-voltage grid appear shifted to lower voltage values with respect to the corresponding points in the outer-voltage grid. That observation agrees with the inspection of (5.1) and (5.2) that represent voltage drops on R_g, R_d, and R_s.

The absence of data with simultaneous high-V_{GS} and high-V_{DS} values in Figure 5.4 is due to the voltage limits used for S-parameters measurements, which are set by the maximum DC power limit of the device.

Figure 5.5 is an enlarged section of Figure 5.4, illustrating only the grid of inner voltages, using circle markers. The squares denote the orthogonal grid of the inner voltages at which that is desired (i.e., the goal of the regridding process). The filled squares are desired points that only require interpolation techniques, whereas the hollow squares extrapolation would be necessary. In reality, these grid points that required extrapolation should be considered out of scope of the modeling strategy, and the meaningfulness of the extracted model parameters on such grid points is doubtful or weak.

5.4 REGRIDDING ISSUES WITH MATLAB®

Zamudio [2] implemented the regridding process with a MATLAB routine developed in [3] that interpolates using a *b-spline* approximation technique. B-spline approximation techniques are preferred over other techniques, like polynomial or Bezier, for their high approximation accuracy to supplied data and the continuity of high-order derivatives of the received data [3]. On the other hand, as is pointed out in [2], using b-spline approximation techniques for extrapolation has been proven to produce unreliable values in many numerical implementations, mainly because the value threshold between the interpolation and extrapolation calculations has not been optimally defined. Figure 5.6 illustrates an example of the extracted bias-dependency of $C_{gs}(V_{gs}, V_{ds})$ showing with encircled zones unreliable

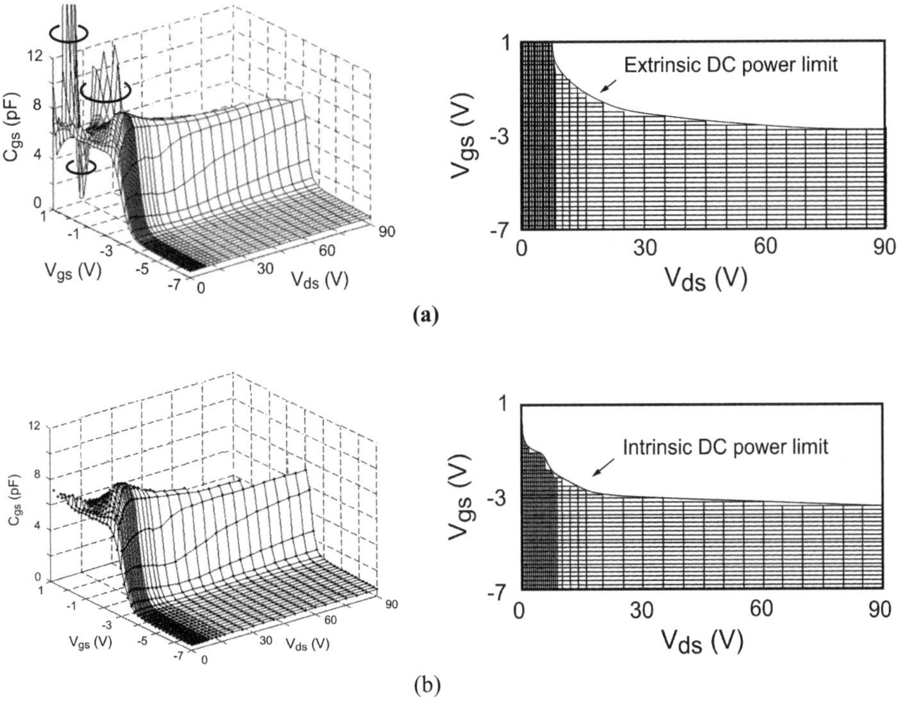

Figure 5.6 Extracted gate-source capacitance C_{gs} as function of V_{gs} and V_{ds} and associated voltage grid. (a) Results with the regridding process using the original DC power limit as threshold function, and (b) results with the regridding process using DC power limit after voltage remapping as the threshold function. (After [2].)

values. The ill-extrapolation must not always be as evident as in the encircled zones in Figure 5.6(a), in order to be unreliable. In this particular case the unreliability is probably so evident due to the nonlinearity of the C_{gs} bias dependency.

Thus, the key practical problem in the regridding process is defining a threshold to distinguish grid points that only must be interpolated and points that would need extrapolation (i.e., that should be discarded or at least marked as ill).

The original MATLAB routine developed in [3] applies indiscriminately the same b-spline approximation technique to interpolate and to extrapolate. In general, however, if the approximation technique applied for the regridding process can be reprogrammed, then it should be possible to inhibit unreliable data from extrapolation, for example, by assigning a specific marker to distinguish extrapolated grid points. In this way, only the grid points obtained by interpolation would be meaningful and all others would be marked as ill.

In [2], ill-marking was performed in the MATLAB environment, in which the model parameter extraction algorithms were programmed by assigning the marker NaN (not-a-number) to grid points that required extrapolation (NaN is a MATLAB preserved word that means that the value of a numeric variable is *not-a-number*, plus infinite, minus infinite, not defined, etc.). This ill-marking was also compatible with the ADS environment, in which the models were implemented.

In the MATLAB routine applied for the regridding process in [2], the ability to input a threshold function to perform the ill-marking, was added, and the appropriated function was created. The role of this function is to calculate a V_{gs} value for every V_{ds} value that it receives. For every point in the desired voltage grid, the V_{ds} value is input in the threshold function by the routine of the regridding process. If the value returned by the threshold function is lower than the V_{gs} value of the desired grid point, then the grid point is ill-marked.

Clearly, the original DC power limit of the device being modeled is an easy parabolic function to define, but it would not be a suitable threshold function for the ill-marking as implemented in the regridding process of [2], because this limit is deformed when the equation 5.3 is applied (see Figure 5.4).

In [2] the DC power limit after voltage remapping (for example as shown in Figure 5.4) is represented with a piecewise polynomial approximation $V_{gs} = \mathrm{P}(V_{ds})$ as part of the model extraction algorithm. Then, this approximation is provided as threshold function for the ill-marking in the routine of the regridding process.

Comparing the results shown in Figure 5.6(a) and Figure 5.6(b) demonstrates the improved numerical quality obtained of the C_{gs} voltage-dependencies when the routine of the regridding process uses the DC power limit after voltage remapping for the threshold function. The improvement has dramatic effect on the reliability of the integration results for the nonlinear gate charge source Q_{gs}. This shall apply equally to the bias-dependent drain-source and gate-drain capacitance and corresponding nonlinear charge sources.

5.5 SUMMARY

Extracted model parameters are a function of the extrinsic control voltages. However, for modeling purposes dependence on the intrinsic voltages is needed. Taking into account the voltage drop across the series resistances, the extrinsic orthogonal voltage grid is mapped onto an intrinsic nonorthogonal grid. Orthogonalization is based on interpolation. It has been shown that interpolation should be restricted to the known control voltage region limited by the intrinsic DC power limit to avoid the instabilities and inaccuracies frequently observed with extrapolation [4].

REFERENCES

[[1] Keysight Technologies, *Advanced Design System (ADS)*, www.keysight.com.

[2] Zamudio Flores, J. A., *Device Characterization and Modelling of Large-Size GaN HEMTs*, Doctoral Thesis, Kassel: Kassel University Press GmbH, 2012.

[3] Jarndal, A., and G. Kompa, "Large-Signal Model for AlGaN/GaN HEMT Accurately Predicts Trapping- and Self-Heating Induced Dispersion and Intermodulation Distortion," *IEEE Trans.*, Vol. ED-54, 2007, pp. 2830–2836.

[4] Rudolph, M., C. Fager, and D. E. Root, *Nonlinear Transistor Model Parameter Extraction Techniques*, Cambridge: Cambridge University Press, 2011.

Chapter 6

Estimation of Model Element Values Based on Device Physical Data

6.1 INTRODUCTION

The application of optimizer-based model parameter extraction procedures needs the definition of initial model parameter values. It is well known that the initial estimation of the model elements is very important for optimization efficiency and reliability. It is clear that the probability of finding a global solution (if that at all exists) depends, in addition to the definition of a suitable equivalent electrical network for the device under consideration, strongly on the choice of good initial values of the model elements. Besides that, well-estimated parameter values may help inspire confidence in extracted model elements.

Parasitic model elements of the transistor device, such as contact pad capacitances and inductances, may also be directly determined by measurement using adequate test structures. They may also serve as good estimates in a final optimization process. However, since such a technology resource is not always accessible to the device modeler, this issue will not get further consideration.

In this chapter model parameter values are estimated from typical device physical data currently available from device manufacturers and relevant publications. This encompasses the device material parameters, such as the dielectric constant of semiconductor layers, doping concentration, and carrier density, and geometric parameters, such as source, gate, and drain electrode layout and their spacing. Model parameter estimation is discussed both for GaAs- and GaN-based field-effect transistors.

6.2 RESISTANCES

The electrical contacting of microwave devices comprises ohmic contacts and bulk semiconductor resistances. Although a necessary part of the device, these are regarded as parasitics that should be kept as small as possible to ensure low voltage drop across the device contacts. Minimization of the ohmic contacts of a field-effect transistor provides high source-drain saturation current, low saturation and knee voltages, and maximum transconductance. In particular, low resistive ohmic contacts are essential for the efficient operation of high-power electronic devices to keep heat dissipation to a minimum to allow high-temperature operation. High drain currents may also cause serious voltage drop across the source resistance causing undesired feedback effects on the device performance. Notice the resistance values of high-power devices are commonly very small. However, because of high currents, the voltage drop may become significant with respect to device performance.

6.2.1 Ohmic Contact Resistance

Ohmic contacts for GaAs- and GaN-based devices have been treated in detail in Section 2.3 in [1], and thereto only some results are recalled here that are useful for a more quantitative description of contact resistances in field-effect transistors leading to estimated model element values for model parameter extraction purposes.

In an ohmic contact of infinite length, most of the current will flow from the semiconductor to the contact over a finite length near the edge of the contact. A finite transfer length l_t is defined in which the current in the semiconductor has fallen 1/e of its maximum value. The transfer length l_t reads [1]

$$l_t = (\rho_{c,norm}/R_{sh,con})^{1/2} \tag{6.1}$$

where $\rho_{c,norm}$ (Ω-mm^2) is the (specific) contact resistivity; that is, the resistance of the metal-to-semiconductor interface layer (barrier) per unit area. This is to distinguish from the bulk resistivity ρ (Ω-cm). $R_{sh,con}$ (Ω/sq) indicates the sheet resistance of the material under the ohmic contact, which is frequently approximated by the sheet resistance of the unplated (bulk) semiconductor layer $R_{sh,bulk}$ (Ω/sq). The sheet resistance under the contact is commonly lower due to the fact that most alloyed or sintered contacts introduce a heavily doped region directly under the contact. In addition, spiking of the contact can further degrade the assumption of a uniform conducting region under the contact. This is depicted in Figure 6.1 for planar and diffused contacts on highly doped GaAs material [2]. The Au/Ge/Ni alloyed system is the most widely used process for ohmic contacts to GaAs because it is capable of providing very low contact resistance in the order of 0.1 Ω-mm.

Contact resistances and contact resistivities are frequently measured using the transmission line method (TLM) [1]. In this case the total resistance R_{tot} of adjacent contact pads are measured, varying the distance between them. Figure 6.2(b) shows schematically the measured curve. It is a straight line with a slope $R_{sh,bulk}/w$ where w is the width of the ohmic contact. $R_{sh,bulk}$ denotes the bulk sheet resistance of the semiconductor layer. The contact resistance can be taken directly from the intersection with the y-axis.

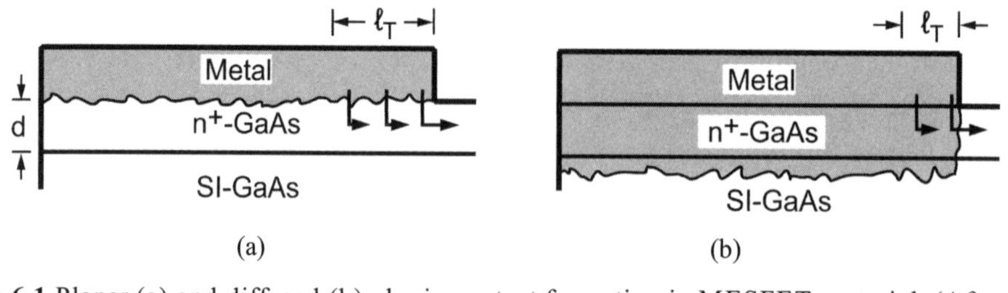

Figure 6.1 Planar (a) and diffused (b) ohmic contact formation in MESFET material. (After [2].)

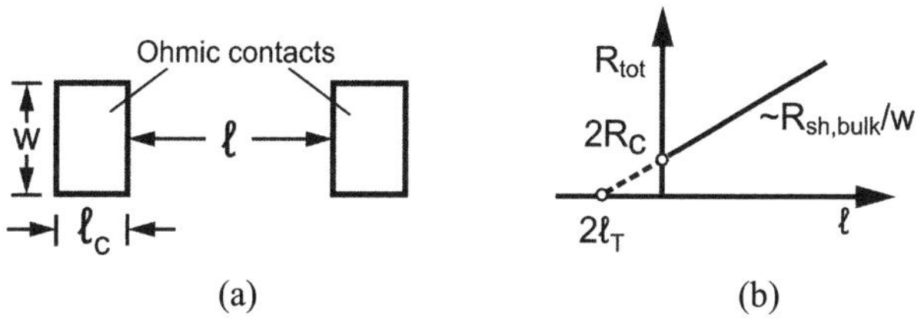

Figure 6.2 Schematic view of TLM method to measure the ohmic contact resistance R_c and the contact resistivity ρ_c. (a) TLM pattern with varying distance l between contact pads, and (b) measured total resistance versus spacing l.

The method most frequently used to achieve good ohmic contacts on the n^+-GaAs layer is to deposit a gold/germanium/nickel layer by evaporation and then alloying this layer with the GaAs [3]. The ohmic contact resistance is denoted by R_c (Ω). Frequently, regarding scaling rules, the contact resistance R_c is normalized to the gate width; thus, $R_{c,norm}$ (Ω-cm) is introduced as ohmic contact resistance per unit width. For planar ohmic contacts in MESFETs, $R_{c,norm}$ commonly has a value of less than 0.1 Ω-mm [4].

The following analysis example is derived from [2]. The intersection of the measured TLM curve with the y-axis delivers $2\,R_c = 9\,\Omega$ (i.e., the contact resistance of each contact pad is $R_c = 4.5\Omega$). Assuming a contact width $w = 65$ µm, then the normalized (specific) ohmic contact resistance $R_{c,norm}$ is calculated as $R_{c,norm} = R_c w = 4.5\Omega \times 65\ \mu\text{m} = 0.292\ \Omega\text{-mm}$.

We assume that the slope $R_{sh,bulk}/w$ has been measured as 2.1×10^5 Ω/cm. Then, with $w =$ 65 µm the bulk sheet resistance $R_{sh,bulk}$ yields 1.365×10^3 Ω/sq. The normalized contact resistance $R_{c,norm}$ is related to the specific contact resistivity $\rho_{c,norm}$ (eq. 2.138 in [1]),

$$R_{c,norm} = (\rho_{c,norm} R_{sh,con})^{1/2} \Rightarrow \rho_{c,norm} = R^2_{c,norm}/R_{sh,con} \tag{6.2}$$

With $R_{sh,con} \approx R_{sh,bulk}$ the specific contact resistivity $\rho_{c,norm}$ is obtained from (6.2) as 6.3×10^{-7} Ω-cm². With the normalized contact resistivity $\rho_{c,norm} = 6.3 \times 10^{-7}$ Ω-cm² and approximated contact sheet resistance $R_{sh,con} \approx R_{sh,bulk} = 1.365 \times 10^3$ Ω/sq, the transfer length l_t in (6.1) is calculated as 0.2 µm. In general, l_t typically varies between 0.2 and 0.6 µm [1].

The contact resistivity ρ_c depends primarily on the metal-semiconductor work function, doping density, and effective mass of carrier. As seen in Figure 2.43 in [1], the specific contact resistance for AuGeNi contacts on n-GaAs decreases to a minimum value of 10^{-6} Ω-cm² for dopant concentration above 4.7×10^{17} cm^{-3}.

Ohmic contacts to two-dimensional carrier gases as given with HEMT devices impose additional technological problems that are not encountered in ohmic contacts to *bulk* material. These problems result from the fact that in heterostructures more than one barrier (the metal semiconductor barrier) is present: two additional heterobarriers exist between the metal and the 2DEG, as shown in Figure 6.3. It is assumed that both the cap layer and the doping layer contain neutral (conducting, nondepleted) regions that allow current to flow parallel to the surface. The barriers at the heterojunction increase the resistance between metal and 2DEG, which cannot be tolerated for device applications. If the barriers are not too high and too thick, tunneling currents (I_{tun}) may flow. In the case of AlGaAs/GaAs

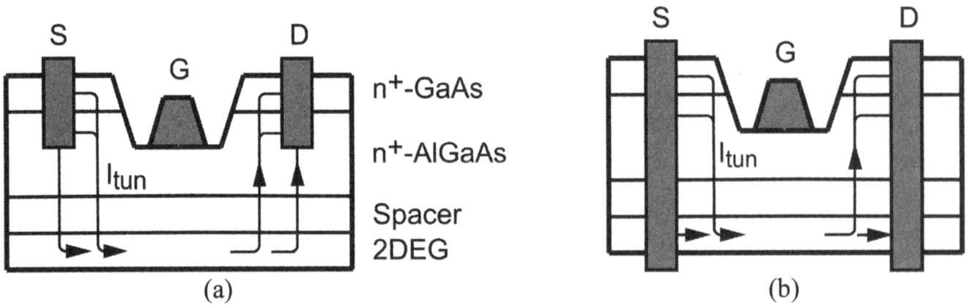

Figure 6.3 Ohmic contacts to recessed gate structures using highly doped cap layers, schematically. (a) Ohmic contact penetrates into the dopant layer only. (b) Ohmic contact penetrates into the buffer layer and/or substrate.

heterostructures it was shown that a thick highly conductive GaAs cap layer considerably reduces the source-drain resistance (which is composed of both the contact resistance and resistances of the semiconducting layers outside the gate region).

The contact resistance can be reduced, among other ways, as follows. The metal is alloyed such that the contact material penetrates into the doping layer as indicated in Figure 6.3(a), thus eliminating the barrier between cap and doping layer, or, as shown in Figure 6.3(b), penetrates into the channel layer or even into the substrate so that the doping/channel barrier is eliminated. It is evident that the two-dimensional gas will be destroyed inside the contact region,and probably outside this region, too, over a distance comparable to the Debye length.

The typical HEMT structure has a highly doped n^+-GaAs contact layer. The AlGaAs (or AlGaN) layer in HEMTs presents a high-energy barrier to the flow of current from the contact layer to the channel. The alloy material is required to penetrate this barrier to reach the 2DEG layer and the channel for useful conduction to take place. Even though the thickness of this AlGaAs (or AlGaN) layer is in the range of a few hundred angstroms (a few ten of nanometers), the difficulty in penetration of the alloy material is quite high owing to higher bond energies, especially of Al with As (or Al with N), which are stronger than those between Ga and As (or Ga and N). Difficulty of penetration of alloy material increases with the thickness of device layers, especially the AlGaAs (or AlGaN) spacer layer. In order to surmount this, sufficient thermal energy must be provided during the alloying proces. R_c values routinely obtained for different GaAs-based HEMT structures are mostly in the range of 0.09−0.2 Ω-mm. In [5] very low and consistent R_c values in the range 0.05−0.06 Ω-mm have been obtained with AuGe/Ni/Au ohmic contacts. For a contact layer width $w = 100$ μm the contact resistance would be 0.5Ω − 0.6Ω.

Table 6.1 summarizes published contact resistances on AlGaN/GaN material system for various ohmic contact configurations. In many cases the electrical properties of ohmic contacts on semiconductors are strongly determined by the formation of characteristic intermetallic phases near the metal-semiconductor interface [6]. Low contact resistance per unit width of 0.38 Ω-mm was measured by Selvanathan et al. [7] on Ti/Al/Mo/Au based contact. Nebauer et al. [6] developed a Ti/Al/Ti/Au/$WSiN_x$ scheme delivering low normalized contact resistance of 0.55 Ω-mm. In [8, 9] almost similar values of contact resistance ranging from 0.3-0.5 Ω-mm from various metallization schemes were obtained. Jacobs et al. [10] reported for n-type AlGaN/GaN FET structures very low contact resistance

Table 6.1
Ohmic Contacts to AlGaN/GaN Heterostructures

Metallization System	Layer Thicknesses (nm)	R_c (Ω-mm)	ρ_c (Ω-cm^2)	Reference
Ti/Al/Ni/Au	300/1800/400/1500	0.2	7.3×10^{-7}	[10]
Ti/Al/Ti/Au/WSiN$_x$	10/40/25/30/120	0.55	6×10^{-6}	[6]
Ti/Al/Mo/Au	10/50/20/30	0.429		[9]
Ti/Al/Ti/Au/Mo	10/50/20/30/140	0.416		[9]
Ti/Al/Mo/Au	10/50/40/30	0.420		[9]
Ti/Al/Ti/Au	10/50/20/30	0.502		[9]
Ti/Al/Ni/Au	10/50/20/30	0.371		[9]
Ti/Al/Pt/Au	10/50/20/30	0.410		[9]
Mo/Al/Mo/Au		0.22		[7]
V/Al/Mo/Au		0.35		[7]

of 0.2 Ω-mm (7.3×10^{-7} Ω-cm^{-2}) for Ti/Al/Ni/Au (300/1800/400/1500Å) contacts made on $Al_{0.25}Ga_{0.75}N$/GaN FET structures (R_{sh} = 550 Ω/sq).

6.2.2 Series Resistances

Figure 6.4 shows a typical geometric structure of a recessed gate FET. Each of the parasitic series resistances R_s and R_d are composed of the bulk resistance of the access region and a contact resistance component. As aforementioned, for widely used Au-Ge-Ni contacts on n^+-GaAs, contact resistivity per unit width is in the order of 0.1 Ω-mm (see Section 6.2.1). Thus, assuming a MESFET device with 300-μm gatewidth, this would mean a contact resistance of 0.3Ω.

The specific resistivity of the semiconductor layer is $\rho_{sh} = (qN_d\mu_0)^{-1}$. The bulk sheet resistance of the semiconductor layer R_{sh} is calculated as $R_{sh} = \rho_{sh}/(a - h)$ with the layer resistivity ρ_{sh}, layer thickness a, and depletion region height h. It comprises the bulk free carrier density multiplied by the thickness of the active channel. Then the bulk resistance of the source-gate access region $R_{s,bulk}$ yields

$$R_{s,bulk} = R_{sh}\frac{l_{gs}}{w} = \frac{\rho_{sh}}{a-h}\frac{l_{gs}}{w} = \frac{1}{(qN_D\mu_0)(a-h)}\frac{l_{gs}}{w} \tag{6.3}$$

with h from (4.64). A minimum resistance value is given for $V_{gs} = 0$. The bulk resistance per unit width reads

$$R'_{s,bulk} = R_{s,bulk}\cdot w \tag{6.4}$$

Example
Typical submicrometer-gate MESFET data: Gate-length is equal to 0.3 μm. The deposited epilayer with 1×10^{17} cm^{-3} doping concentration is 0.3 μm thick. The low-field mobility is assumed to be equal to 4,500 cm^2(Vs)$^{-1}$. Furthermore, we assume a gate-source spacing l_{gs} = 0.5 μm and a gate width w = 300 μm. V_d is assumed to be 0.8V and $V_{gs} = 0$.

For this data we find for the specific channel resistivity $\rho_{sh} = 13.88 \times 10^{-3}$ Ω-cm, for the sheet resistance 48 Ω/sq, and for the channel resistance 0.8Ω. Then, assuming a minimum contact resistance of 0.3Ω, the total value of R_s is estimated to be $R_s = R_c + R_{s,bulk} \approx 0.3\Omega + 0.8\Omega = 1.1\Omega$.

For MESFET devices the DC drain resistance R_D is calculated as

$$R_D \approx \frac{l_{gd} - X}{qN_D\mu_0 w(a - h(V_{gs}))} \tag{6.5}$$

where X denotes the space-charge layer extension [11],

$$X \leq \left[\frac{2\varepsilon}{qN_d(V_{gs} + V_d)}\right]^{1/2}(V_{dg} + V_d) \tag{6.6}$$

For typical GaAs MESFET data, $N_D = 1.5 \times 10^{17}$ cm^{-3}, $\varepsilon_r = 12.9$, $V_d = 0.7$V, $V_{gs} = 0$V and $V_{dg} = 5$V, the extension X is calculated to be 0.66 μm. This value agrees very well with that taken from Figure 1 in [12].

The length of the bulk region ($l_{gd} - X$) varies with the RF signal. The differential (small-signal) resistance R_d reads [11]

$$R_d \approx \frac{R_D}{1 - 2\dfrac{v_{sat}}{\mu_0}\dfrac{a - h}{a}\dfrac{\varepsilon}{qN_d h}} \tag{6.7}$$

Inserting typical numbers into the second term in the denominator provides a very small value compared with unity, so that the static and differential resistances are approximately comparable, $R_d \approx R_D$.

The recessed gate FET in Figure 6.4, a 0.4-μm gate MESFET of type MGFC 1423 manufactured by Mitsubishi, has the following geometry and semiconductor data: $N_D = 1.5 \times 10^{17}$ cm^{-3}, $a_2 = 0.2$ μm, $a_1 = 0.5$ μm, $L_{sg1} = L_{sg2} = 0.5$ μm, $L_{gd1} = L_{gd2} = 0.8$ μm, $w = 300$ μm, $R_{sc} = 0.3\Omega$. Then, with $V_{GS} = 0$V, the source resistance R_s is calculated as

$$\begin{aligned} R_s &= R_{sc} + R_{s1} + R_{s2} \\ &= 0.3\Omega + \frac{1}{qN_D\mu_0 w}\left(\frac{L_{sg1}}{a_1 - h} + \frac{L_{sg2}}{a_2 - h}\right) = 0.3\Omega + 1.4\Omega + 0.37\Omega = 2\Omega \end{aligned} \tag{6.8}$$

Similarly, the drain bulk resistance is calculated as [13]

$$R_d = R_{dc} + R_{d1} + R_{d2} = 0.3\Omega + 2.2\Omega + 0.6\Omega = 3.1\Omega \tag{6.9}$$

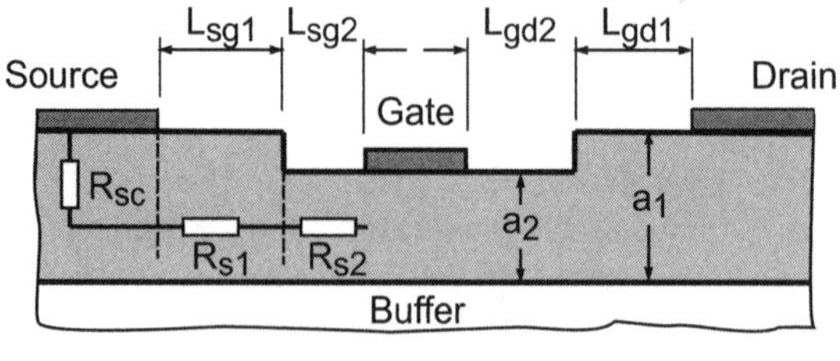

Figure 6.4 MESFET with recessed gate structure [13].

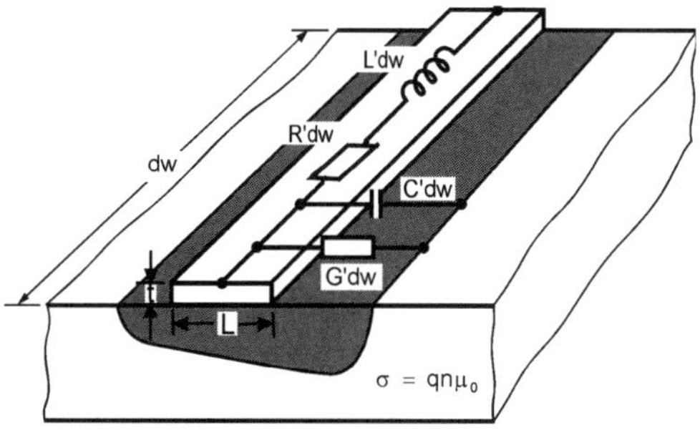

Figure 6.5 Gate finger section and its lumped element representation.

Regarding HEMT devices, the source resistance is calculated to be

$$R_s = \frac{1}{(qn_s\mu_{2DEG})d_{2DEG}}\frac{l_{sg}}{w} = \frac{\rho_{2DEG}}{d_{2DEG}}\frac{l_{sg}}{w} = R_{sh}\frac{l_{sg}}{w} \tag{6.10}$$

A similar relation is given for the drain resistance with l_{sg} replaced by the drain-gate distance l_{dg}. Typical sheet resistance values lie in the order of 500 Ω/sq. As reported by Chang et al. [14], very low sheet resistance values for AlGaN/GaN HEMTs could be realized. Increasing the sheet carrier concentration from 1×10^{13} cm^{-2} to 1.75×10^{13} cm^{-2}, the sheet resistance decreased from 425 to 250 Ω/sq. With a gate-source distance of $l_{gs} = 0.5$ μm, a gate width $w = 300$ μm, and sheet resistance $R_{sh} = 500$ Ω/sq, the source series resistance gives 0.8Ω.

6.2.3 Gate Resistance, Gate Inductance

As illustrated in Figure 6.5, the gate electrode of a MESFET with the conductive channel represents a two-electrode system with the space charge region as dielectric layer. Similar consideration can be undertaken with the HEMTconfiguration with the dielectric AlGaAs layer and the 2DEG channel. The characteristic impedance Z_0 of the lossy transmission line reads

$$Z_0 = [(R' + j\omega L')/(G' + j\omega C')]^{1/2} \tag{6.11}$$

and the propagation coefficient γ,

$$\gamma = [(R' + j\omega L')(G' + j\omega C')]^{1/2} \tag{6.12}$$

The input impedance Z of an open-ended (gate) line of length w, characteristic impedance Z_0 and load impedance $Z_L \to \infty$ is calculated as

$$Z(w) = \frac{Z_L + Z_0\tanh(\gamma w)}{1 + Z_L/Z_0\tanh(\gamma w)} = Z_0\coth(\gamma w) \tag{6.13}$$

With the series expansion of coth (γw),

$$\coth(\gamma w) = \frac{1}{\gamma w} + \frac{\gamma w}{3} - \frac{(\gamma w)^3}{45} + \frac{2(\gamma w)^5}{945} \pm \cdots \quad (0 < |\gamma w| < \pi) \tag{6.14}$$

and sufficiently small shunt conductance per unit length ($G' \ll \omega C'$),

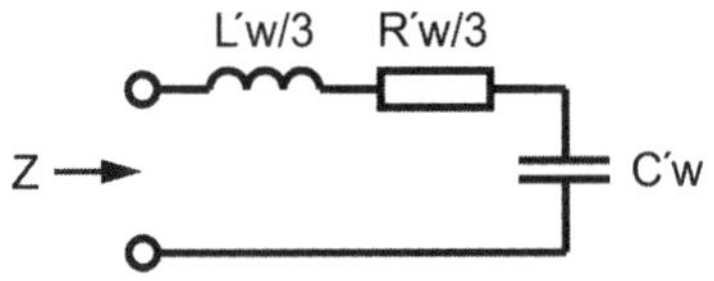

Figure 6.6 Approximated input impedance of the gate electrode.

$$\gamma = [j\omega C'(R' + j\omega L')]^{1/2} \tag{6.15}$$

the input impedance can be approximated for sufficiently low frequencies as

$$Z(w)/Z_0 \approx (\gamma w)^{-1/2} + \gamma w/3 \tag{6.16}$$

With

$$Z_0/\gamma = 1/j\omega C', \qquad Z_0\gamma = R' + j\omega L'$$

it follows

$$Z \approx \frac{1}{j\omega C'w} + \frac{1}{3}R'w + j\frac{1}{3}\omega L'w \tag{6.17}$$

Thus, the gate inductance L_g is approximated as

$$L_g = L'w/3 \tag{6.18}$$

and the gate resistance R_g as

$$R_g = R'w/3 \tag{6.19}$$

with $R' = \rho/(wt)$. t denotes the thickness of the gate metallization, and ρ the resistivity of the doped layer. Figure 6.6 illustrates the lumped element representation of a single gate finger.

The frequency limit of the lumped approximation of the gate transmission line can be estimated from (6.14) taking into account the third expansion term,

$$Z(w) \approx Z_0\left(\frac{1}{\gamma w} + \frac{\gamma w}{3}\right) - \frac{Z_0}{45}(\gamma w)^3 \tag{6.20}$$

The last term expressed by the transmission line parameters, reads

$$\begin{aligned} -\frac{Z_0}{45}(\gamma w)^3 &= -\frac{1}{45}(Z_0\gamma)\gamma^2 w^3 \\ &= \frac{1}{45}\left\{j\omega C'[R'^2 - (\omega L')^2] - 2\omega^2 R'L'C'\right\}w^3 \end{aligned} \tag{6.21}$$

Assuming a gate width $w = 100$ μm, gate length $L_g = 100$ μm, gate metallization thickness $t = 0.4$ μm, and a resistivity $\rho = 3.03\times10^{-6}$ Ωcm for aluminum, the resistance per unit length, R', is calculated as $R' = \rho/(wt) = 151.5 \times 10^3$ Ω/m (sheet resistance $= \rho/t = 151.1\Omega$). Furthermore, we suppose an inductance per unit length $L' = 10^{-6}$ H/m and a capacitance per unit length $C' = 1.5\times10^{-9}$ F/m. Then the lumped element values yield

$$R_g = R'w/3 = 5.05\Omega, \quad L_g = L'w/3 = 33.33\text{ pH}, \quad C_g = C_g'w = 0.15\text{ pF}$$

In this case the magnitude of the reflection coefficient for the lumped element equivalent circuit of a single gate finger turns out to be $|\Gamma| = 0.83$ and $|\Gamma| = 0.82$ at 50 GHz and 100 GHz, respectively.

We now estimate the contribution of the real and imaginary parts of the third expansion term in (6.21) for 50 GHz and 100 GHz. The resistive and inductive increase, ΔR_g and ΔL_g, are given as

$$\Delta R_g = \frac{2}{45}\omega^2 R'L'C'w^3, \quad \Delta L_g = \frac{1}{45}C'[(\omega L')^2 - R'^2]w^3 \tag{6.22}$$

ΔR_g is calculated as $\Delta R_g = 1.0\Omega$ for 50 GHz and 4.0Ω for 100 GHz, which means an increase of 19.8% and 79%, respectively. Similarly, the inductance rise ΔL_g is given as $\Delta L_g = 2.5$ pH for 50 GHz and 12.4 pH for 100 GHz, which implies an increase of 7.6% and 37.2%, respectively. The magnitudes of the respective reflection coefficients now read: $|\Gamma|_{50\mathrm{GHz}} =$ 0.79, $|\Gamma|_{100\,\mathrm{GHz}} = 0.74$. In comparison with the simple lumped element equivalent circuit, we find a change in the reflection coefficient of 4.6% and 10.1%, respectively.

A stronger deviation results from the skin effect, which is calculated from [1]

$$\delta/\mu\mathrm{m} = \sqrt{\frac{10^{16}}{4}\frac{\rho/\Omega\mathrm{cm}}{f/\mathrm{Hz}}} \tag{6.23}$$

With $\rho = 3.03 \times 10^{-6}$ Ωcm for aluminum and $f = 100$ GHz the skin depth becomes $\delta = 0.275$ μm and $R' = 220.36 \times 10^3$ Ω/m. In this case the skin effect contributes to a high resistive value of 5.8 Ω, so that the total gate resistance results in 10.8 Ω. Recalculation of the input reflection coefficient at 100 GHz now provides a rather reduced value of 63.1%. Related to the simple lumped element equivalent circuit this implies a change of 23.3%. These estimations agree well with the results reported by Heinrich [15] analyzing the frequency limitation of lumped element modeling approach of field-effect transistor devices, which are basically distributed networks at higher frequencies. It is stated that the distributed character of the gate electrode affects mostly the scattering coefficient S_{11} at higher frequencies.

Commonly, the rule of thumb is that a distributed network can be replaced by a lumped element circuit if the maximum dimension of the network is less than one tenth of the operating wavelength. Applying to the previous calculations and assuming an effective dielectric constant of 9.8 [1], an upper frequency limit of 96 GHz would have been estimated. Also, if the resistor becomes an appreciable electrical length (λ_g or so), it needs to be treated as a lossy transmission line section as opposed to a lumped element [16].

Due to (6.45) in Section 6.3.1 it is seen that the transconductance G_m is proportional to the gate width w ($G_m \sim w$). From this point of view the gate finger should be as long as possible. However, as we have just discussed, with longer gates, distributed effects would come in and undesired impedance transformation effects would occur.

Many authors adopt the approximate dimension $\lambda_g/10$ (λ_g being the electrical wavelength) as the limit above which distributed effects must be accounted for and as a rule of thumb to introduce the proper number of active slices along the gate metallization [17].

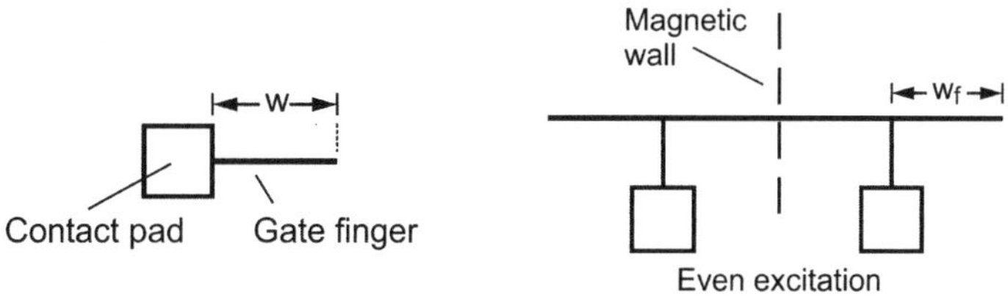

Figure 6.7 Single gate finger of large width replaced by short-width multifinger construction.

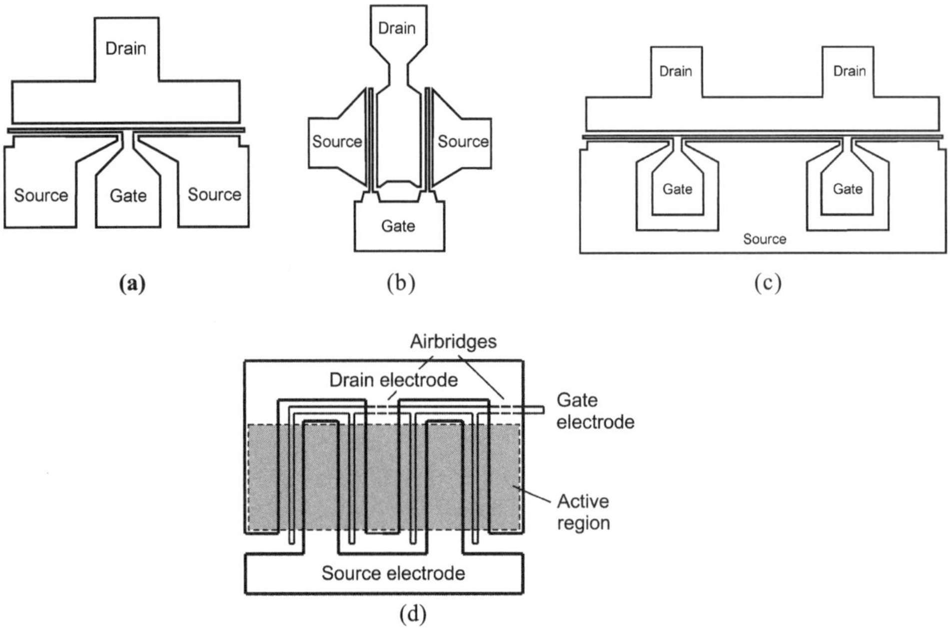

Figure 6.8 Various FET gate constructions: (a) and (b) T-gate, (c) pi-gate, and (d) interdigital gate with air bridges.

It is rather practiced, as shown in Figure 6.7, to divide the total width of the gate into N single fingers and interconnect them in a parallel fashion. Due to (6.19) the single gate finger exhibits a gate resistance of $R_{g,\text{single}} = R'w/3 = w/3 \times (\sigma L t)^{-1}$. With the multifinger solution, the total gate width is divided into N sections, so that the resistance of a single shortened gate width reads $R_g/N = (R'w/3)/N$. As the N fingers are connected in parallel, the resistance of the multifinger gate reads

$$R_g = \frac{1}{N}\left(\frac{R'w}{3\ N}\right) = \frac{1}{N^2}\frac{w}{3\sigma L t} \tag{6.24}$$

or

$$R_g = \frac{1}{N}\left(\frac{R'w}{3\ N}\right) = \frac{1}{3}\frac{1}{N^2}R_{sheet}\frac{w}{L}, \qquad R_{sheet} = \rho_{sh}/t = \frac{1}{\sigma t} \tag{6.25}$$

Various multifinger constructions are shown in Figure 6.8. A frequently used construction is the pi-gate FET, and typically for power FET devices, the interdigital gate periphery.

Figure 6.9 shows the top view of a pi-gate transistor chip of type NE70000. Relevant data of the MESFET: Gate length $L = 0.3$ µm, metallization thickness $t = 0.4$ µm, conductivity of Al-gate: $\sigma = 33.3 \times 10^4$ S/cm, total gate width $w = 280$ µm. The RF gate resistance for the pi-gate configuration is calculated as $R_{sh} = 1/\sigma t = 7.5 \times 10^{-2}\Omega$/sq. Then the gate resistance in (6.25) is found to be $R_g = 1.45\Omega$.

The contact resistance R_c may vary between 0.1 Ωmm and 0.5 Ωmm [1]. Assuming $R_c \approx 0.3$ Ωmm, the contact resistance contributes to 0.3 Ωmm/280 µm/16 = 1.15/16Ω = 0.07Ω.

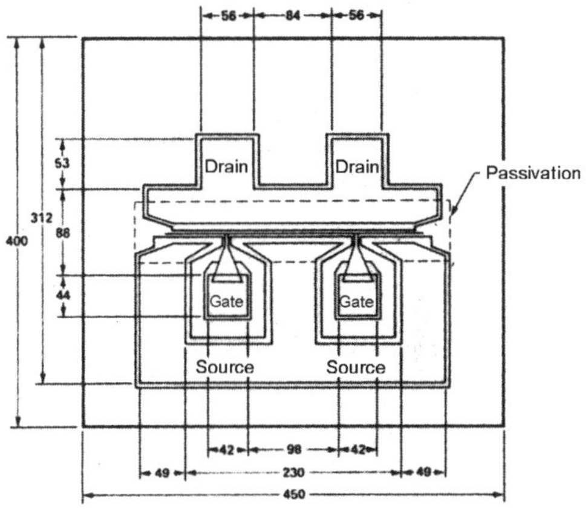

Figure 6.9 Top view of recessed 0.3-μm gate MESFET NE70000 chip with twofold driven gate (pi-gate); units in μm, chip thickness: 140 μm. [NE710 Data Sheet, NEC Corporation, 1983.]

Thus, the gate resistance is estimated as 1.52Ω, which agrees really very well with typical extraction results of 1.5Ω. The gate inductance L_g is calculated as

$$L_g = L'w/3 \tag{6.26}$$

Assuming a typical value $L' \approx 1$ nH/mm for the gate finger of a 0.5 μm-gate FET, then the inductance of a 70-μm gate width yields 23 pH. When contacting the gate of the device by wire-bonding, then the bondwire inductance will exhibit a much larger inductive value.

In addition to the distributed gate metallization resistance R_g, Rohdin et al. [18] reported about an additional interfacial gate resistive component R_{gi} that contributes to the gate resistance, so that the total gate resistance reads $R'_g = R_g + R_{gi}$. The interfacial gate resistance component is defined as

$$R_{gi} = \frac{r_{gi}}{w_g L_g} \tag{6.27}$$

with r_{gi} as the normalized interfacial gate resistance. For GaAs- and InP-based Schottky-barrier-gate field-effect transistors, r_{gi} is in the order of 10^{-7} Ω-cm^2. The resistance has been explained by electron tunneling between the three-dimensional (3D) gate metal and the 2D semiconductor surfaces states. Assuming a gate length $L = 0.25$ μm and a gate width $w = 100$ μm, the interfacial resistance reads 0.4Ω.

As pointed out in [1], for long channel devices a rectangular cross section of the gate electrode was commonly used. As the transit frequency f_t of a FET is inversely proportional to the gate length, extension of device operation to higher frequencies needed shortening of the gate length. However, in contrast to the high degree of improvement in f_t upon reducing the gate length, the maximum oscillation frequency f_{max} was improved only slightly because of the increase of the gate resistance R_g. This is due to the small sectional area of the gate finger due to the short gate length and thin gate metal. To improve the RF characteristics of the device, the gate cross-sectional area must be enlarged while keeping the gate length small. T- and Y-shaped gate electrodes have been utilized (Figure 2.32 in [1]).

Onodera et al. [19] have investigated GaAs MESFETs with a wide-head T-shaped gate.

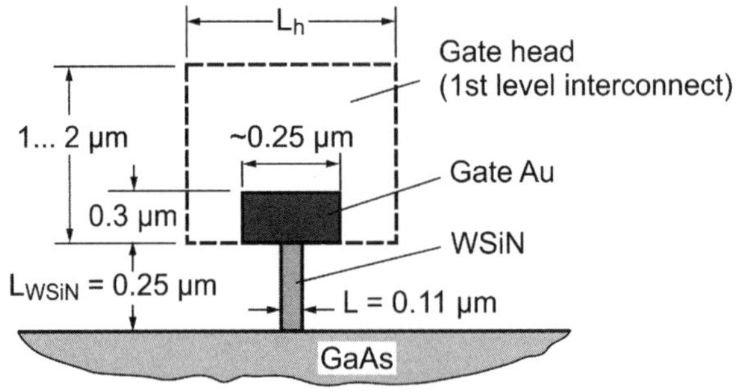

Figure 6.10 Wide-head T-shaped gate after Onodera et al. [19].

Table 6.2

L_h/µm	0	0.5	1.0	1.5
R_g/Ω	5.92	1.02	0.88	0.9

The Schottky gate is made of refractory metal WSiN, which is highly stable at elevated temperature and acts as a diffusion barrier to prevent intermetallic interactions mainly between As and GaAs. The resistivity is as high as 600 µΩcm. The gate resistance becomes extremely high if the gate length is shorted to a sub-half-micron range. Here, to reduce gate resistance, a Au-WSiN T-shaped gate structure was formed as shown in Figure 6.10.

Without first-level interconnect (denoted as $L_h = 0$), the area of the Au gate head is about 0.25 µm × 0.3 µm. The resistivity of the plated Au was measured as ~3.8×10^{-6} Ωcm. The gate-length of the investigated MESFET was ~0.11 µm and the total gate width $w = 2 \times 50$ µm = 100 µm. Then the gate sheet resistance reads $R_{sh} = \rho/L_{WSiN} \approx 3.8 \times 10^{-6}$ Ωcm/ 0.25 µm = 0.15Ω. The two-finger gate resistance is calculated as $R_g = R_{sh} \times w/L_{Au\text{-}gate}/12 \approx 5\Omega$. With increasing L_h, the value of the gate resistance R_g decreases dramatically as shown in Table 6.2 [19].

In comparison with purely Au gate heads the values given in Table 6.2 were about 0.5–0.6Ω higher. It has been analytically shown that the vertical gate resistance of the Schottky gate made of refractory metal WSiN has an increasing effect on the gate resistance according to

$$R_g = R_{g0}/3 + 1/G_{g0} \tag{6.28}$$

where R_{G0} is the metallurgical DC resistance of the Au gate and G_{G0} is the metallurgical conductance along the vertical direction; that is, the conductance of WSiN measured from top to bottom. In this case, the vertical resistance of the WsiN gate is calculated as $R_{Au/WSiN} = \rho_{WsiN} L_{WSiN}/L/w = 0.15\Omega$. As the difference is three or four times higher it is supposed that this gap may be mainly ascribed to the contact resistance of the Au gate head and WSiN gate.

6.2.4 Gate Charging Resistance

The gate (capacitance) charging resistance R_i, also termed intrinsic channel resistance, is in series with the gate-source capacitance C_{gs}. It controls the charging and discharging of the respective capacitance, and thus provides some input circuit delay between applied gate-

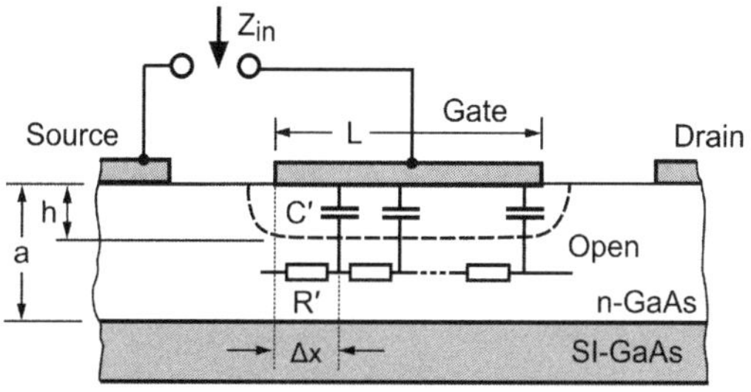

Figure 6.11 Planar Schottky contact with distributed gate capacitance and channel resistance as transmission line.

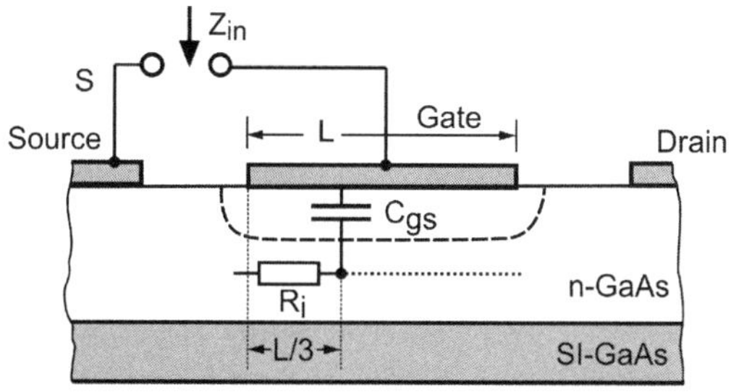

Figure 6.12 Cross section and lumped model with gate-source capacitance C_{gs} and gate charging resistance R_i for $|\gamma l| \leq 1$.

source voltage and inner effective gate potential. We first analyze the MESFET configuration. The results may then be used to also estimate the gate charging resistance of HEMT devices.

For the analysis we start with a simple consideration. For that we approximate the channel region by a planar Schottky contact exhibiting a lossy planar transmission line (see Figure 6.11). Due to the gate-to-source voltage, the distributed gate structure may be interpreted as an ideally open-ended line (i.e., open-end effects are not taken into account). The characteristic line parameters of the RC-line are the wave impedance Z_0, and the propagation coefficient γ,

$$Z_0 = (R'/j\omega C')^{1/2} \tag{6.29}$$

$$\gamma = (R' j\omega C')^{1/2} \tag{6.30}$$

where the resistance per unit length R' reads

$$R' = \frac{1}{\sigma A} = \frac{1}{qN_D \mu_0 w(a - h)} \tag{6.31}$$

The input impedance of the open-ended gate-channel transmission line structure is approximated (see Figure 6.12) as

$$Z_{in} = Z_0 \coth(\gamma l) \approx \frac{1}{j\omega C' w} + \frac{R' L}{3} \qquad \text{with} \quad \gamma l \ll 1 \tag{6.32}$$

Thus, with the input impedance of the lumped element circuit in Figure 6.6,

$$Z_{\text{in}} = \frac{1}{j\omega C_{\text{gs}}} + R_{\text{i}} \tag{6.33}$$

the charging resistance R_{i} is given as

$$R_i = (R'L)/3 \tag{6.34}$$

Substituting R' in (6.31) gives

$$R_{\text{i}} = \frac{L}{3qN_{\text{D}}\mu_0 wa(1 - h/a)} \tag{6.35}$$

With

$$R_{i0} = \frac{L}{3qN_D\mu_0 wa} \tag{6.36}$$

(6.35) reads

$$R_i = R_{i0}/(1 - h/a) \tag{6.37}$$

Taking typical values for a MESFET with 0.5-μm gate-length, $a = 0.1$ μm, $N_D = 1.5 \times 10^{17}$ cm^{-3}, $w = 300$ μm, and $\mu_o = 4500$ cm^2/Vs, R_{i0} is calculated to be 0.51Ω. Under the assumption that the short-channel saturation model can be applied (see Section 4.6), the ratio h/a can be approximated by $\sqrt{((-V_{\text{GS}} + V_{\text{d}})/V_{\text{pi}})}$,

$$R_i = \frac{R_{i0}}{1 - [(V_{GS} + V_d)/V_{pi}]^{1/2}} \tag{6.38}$$

For $V_{\text{GS}} = 0$ one obtains

$$R_i = \frac{R_{i0}}{1 - \sqrt{V_d/V_{pi}}} \tag{6.39}$$

The intrinsic pinch-off voltage V_{pi} (see Section 4.2) is calculated as $V_{\text{pi}} = (qN_{\text{d}}/2\varepsilon)a^2 = 1.05$V. Then, from (6.39) with $V_{\text{d}} = 0.8$V, the charging resistance R_{i} ($V_{\text{GS}} = 0$V) is obtained to be 4Ω.

With (4.63) (6.37) can be reformulated:

$$R_i \approx \frac{1}{3}\frac{v_s L_G}{\mu_0 I_{DS}} \tag{6.40}$$

Notice in [11] the factor 1/3 is missing because it is assumed that the low-field mobility μ_0 is degraded about one-third of its as-doped value due to transverse fields in the channel. From the literature, values from 2Ω to 10Ω are known.

The reliable extraction of R_{i} is seriously hampered by the fact that the input impedance of a field-effect transistor device is dependent on the sum of the gate, source, and gate charging resistance. Regarding the input impedance of the simplified FET equivalent circuit in Figure 6.13, and applying Kirchhoff's loop rule at the gate side, we obtain for the input impedance Z_{in}

$$Z_{in} = R_g + R_i + R_s + \frac{G_m L_s}{C_{gs}} + j\omega(L_g + L_s)(1 - (\omega_0/\omega)^2) + \frac{R_s G_m}{j\omega C_{gs}} \tag{6.41}$$

with $\omega_0 \approx C_{\text{gs}}(L_{\text{g}} + L_{\text{s}})$. Thus the real part of Z_{in} reads

$$\text{Re}(Z_{in}) = R_g + R_i + R_s + \frac{G_m L_s}{C_{gs}} \tag{6.42}$$

At pinch-off G_m is very small so that the last real term can be neglected. In this case we have

$$\mathrm{Re}(Z_{in}) \approx R_g + R_i + R_s \tag{6.43}$$

The input impedance is related to the measured eigen-reflection coefficient S_{11}. Notice, from experience, the sum of the resistances ($R_i + R_g + R_s$) is determined rather consistently. However, the distribution of the total resistance value among the individual resistances is by far more challenging.

The bulk resistance of a HEMT device beneath the gate electrode is given by

$$R = R_{sh}\frac{L}{w} = \frac{1}{q\mu_{ch}n_s}\frac{L}{w} \tag{6.44}$$

where R_{sh} is the sheet resistance in units of ohms per square, and w is the channel width. The sheet resistance is equal to $1/q\mu_{ch}n_s$, where q is the electron charge (is taken to be a positive number throughout this book), μ_{ch} is the channel mobility (cm^2/Vs), and n_s is the sheet charge density (cm^{-2}). For a heterostructure-based device such as a HEMT, n_s is simply the density of free carriers in the two-dimensional electron gas (2DEG).

With a channel mobility of $\mu_{ch} \geq 1400$ cm^2/Vs and sheet charge density $n_s \geq 8\times10^{12}$ cm^{-2} [20] we find $R_{sh} \leq 558$ Ω/sq. Assuming a gate length $L = 150$ nm and a gate width $w = 0.18$ mm, we get from (6.44) $R = 0.465\Omega$. The charging resistance R_i is one third of the bulk resistance (i.e., $R_i = 0.16\Omega$). It is clear that, going toward pinch-off, R_i increases with decreasing sheet carrier density.

Equation (6.40) can also be applied to HEMT devices considering the respective physical parameters. In this case μ_0 denotes the 2DEG carrier mobility, and v_s the carrier saturation velocity in the 2DEG channel. In [20] a 0.15-μm gate AlGaN/GaN HEMT with $w = 0.18$ mm was investigated. The measured drain current I_{DS} per unit width was measured to be 1100 mA/mm at $V_{GS} = 0$V. Then the current reads 198 mA. Using $v_s = 0.87 \times 10^7$ cm/s [21] and $\mu_{ch} = 1400$ cm^2/Vs, R_i is calculated as 0.157Ω, which agrees well with the result derived from (6.44).

6.3 CONDUCTANCES

Conductances comprise the transconductance and channel conductance. The trans-

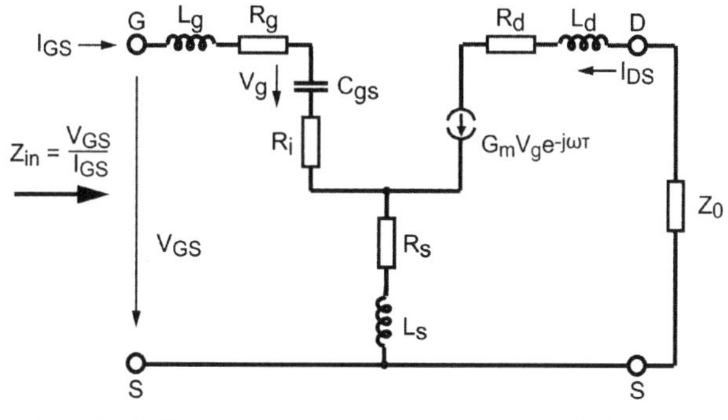

Figure 6.13 Approximate calculation of the input impedance of a FET device.

conductance is also termed transfer or mutual conductance and describes the gain capability of a transistor. The channel conductance describes the conductance between the source and drain electrodes. It is also termed drain or output conductance.

6.3.1 Transconductance

Regarding short channel MESFETs with gate widths L in the order of 0.35 μm and below, the saturation model can be applied. In Section 4.6 the intrinsic transconductance was found to be

$$G_m = v_s w (q N_D \varepsilon / 2)^{1/2} (-V_{gs} + V_d)^{-1/2} \tag{6.45}$$

With typical values for a GaAs-MESFET device, saturation velocity $v_s = 1.5 \times 10^7$ cm/s, doping density $N_D = 1.5 \times 10^{17}$ cm^{-3}, and diffusion voltage $V_d = 0.8$V, one obtains with $\varepsilon_r = 12.9$ (300K) and $w = 300$ μm under the assumption of $V_{gs} = 0$V, a maximum intrinsic transconductance $G_m = 60$ mS. Normalized to the gate width w we get for the transconductance per unit gate $G_m' = G_m/w = 200$ mS/mm.

In Section 4.7.1 the transconductance of a HEMT has been estimated as

$$G_m = \frac{\varepsilon_0 \varepsilon_{rA} v_{eff} w}{d_A + \Delta d} \tag{6.46}$$

In the following (6.46) will be verified against published data (Problems 1–4).

Problem 1

A 0.5-μm gate-length HEMT has the following data [22]: Gate width $w = 150$ μm, spacer layer thickness $\Delta d = 20$Å, doping layer thickness $d_A = 500$Å, AlGaAs doping density $N_D = 1.5 \times 10^{18}$ cm^{-3}. Values of 2×10^7 cm/s and 5×10^6 cm/s are assumed for the 2DEG and AlGaAs drift velocities, respectively. Source and drain resistances per unit width are assumed to be 0.9 Ω-mm. The intrinsic/extrinsic transconductance can be estimated as follows:

For $Al_xGa_{1-x}As$ ($x = 0.3$) the dielectric constant is 12.05 [23]. The sheet concentration n_{2D} is estimated to be $N_D \times d_A = 1.5 \times 10^{18}$ cm^{-3} $\times$ 500Å $= 7.5 \cdot 10^{12}$ cm^{-2}. Then Δd is approximated as 14Å using (4.78). With

$$\varepsilon_A w v_{eff} = 12.05 \times 8.85 \cdot 10^{-14} \text{As/(Vcm)} \times 150\ \mu\text{m} \times 2 \cdot 10^7 \text{cm/s} = 3.2 \cdot 10^{-7} \text{(A/V) cm}$$

and

$$d_A + \Delta d = 500\text{Å} + 20\text{Å} + 14\text{Å} = 534\text{Å}$$

the intrinsic transconductance G_m is calculated as 60 mS. The source resistance R_s is estimated as 0.9 Ω-mm/150 μm = 6Ω. Then the measured extrinsic transconductance $G_M = G_m/(1 + R_s \cdot G_m)$ is obtained as 44 mS, which agrees satisfactorily with the measured data of about 37 mS [22].

Problem 2

In [24] HEMT devices were grown by MBE on semi-insulating substrates with an undoped GaAs buffer followed by AlGaAs spacer and Si-doped AlGaAs layer ($N_D = 2.8 \times 10^{18}$ cm^{-3}).

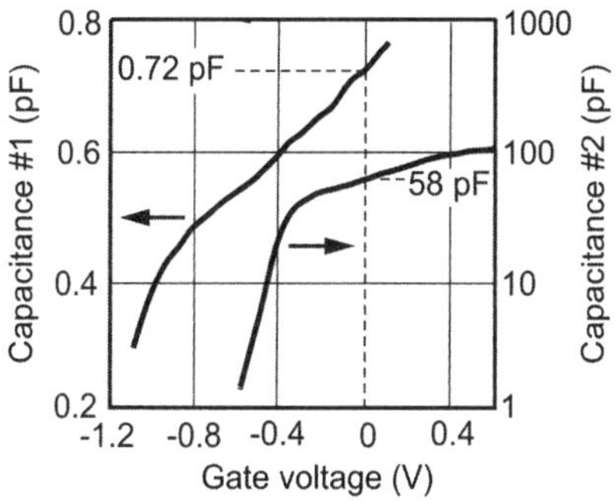

Figure 6.14 Experimental *C-V* characteristics. (After Park et al. [25].) #1: Result of Moloney et al. [26], #2: Result of Norris et al. [24].

Figure 6.14 shows the measured effective *C–V* response of a 400 × 50 μm² gate size MODFET (large size), measured at 1 MHz (sample #2). By accurate curve fitting of theoretical results to the measurement the following device data was obtained: Doped AlGaAs layer d = 217Å, undoped spacer thickness d_i = 150Å (i.e., total thickness of the AlGaAs layer $d_A = d + d_i$ = 376Å). We estimate the gate capacitance as follows:

Estimating the electron sheet concentration by $n_{2D} \approx N_D d = 6 \times 10^{12}$ cm^{-2}, and applying (4.78), the effective distance Δd becomes 16Å. Then, the total spacing between gate electrode and 2DEG is given as 392Å. Then from (4.87) the capacitance per unit area yields

$$C' = \varepsilon_A/d_A = 12.9 \times 8.85 \times 10^{-14}\text{As/(Vcm)/392Å} = 2.9 \times 10^{-3}\ \text{pF/μm}^2$$

Then, taking into account the gate size 400 × 50 μm², the gate capacitance is calculated as 58 pF, which agrees perfectly with the measured value of 58 pF at zero gate voltage in Figure 6.14.

Assuming an electron saturation velocity of 1.5 × 10⁷ cm/s, the normalized transconductance G_M' is calculated from (4.88) as 2.9 × 10^{-3} pF/μm² × 1.5 × 10^7 cm/s = 435 mS/mm. With w = 400 μm, the transconductance reads 174 mS.

Problem 3

In [26] a small-size 100 × 2 μm² HEMT is treated (sample #1 in Figure 6.14). The MBE grown structure consists of a 1-μm-thick undoped GaAs layer grown on a semiconducting substrate, followed by 30Å-thick undoped AlGaAs and then by 350Å-thick layer of this same composition, doped with Si to 2 × 10^{18} cm^{-3}. A 50Å n-GaAs cap layer completes the structure, which facilitates the formation of ohmic contacts. Gate capacitance measurement was performed with a LCR meter. We can estimate the transconductance and gate capacitance as follows.

Using the nominal device data, $(d_A + \Delta d)$ is calculated as 350Å + 30Å + 14Å = 394Å. Then, (4.86) provides a capacitance per unit area of $C' = 2.89 \times 10^{-3}$ pF/μm^{-2}. Then, for a gate area of 100 × 2 μm² the capacitance reads 0.58 pF in contrast to the measured value of 0.72 pF (V_{GS} = 0V) in Figure 6.14. For better agreement between measurement and theoretical analysis, a smaller doped AlGaAs of 300Å was taken for modeling [26]. In this case, a

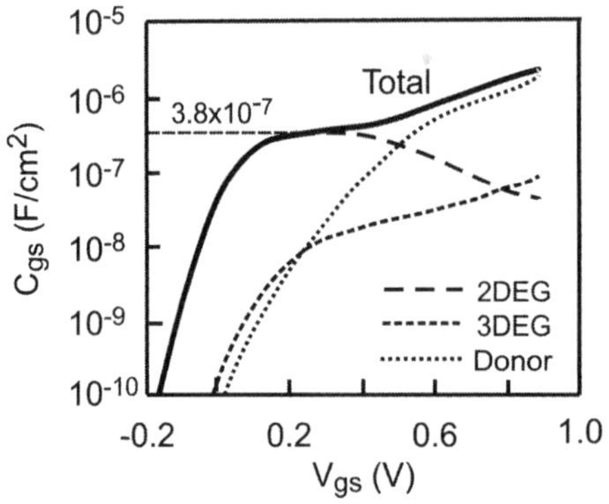

Figure 6.15 Gate capacitance versus voltage plots for conventional HEMTat 300K. 2DEG: capacitance due to 2DEG in the channel. 3DEG: 3DEG in the AlGaAs. Donor: Donors in the AlGaAs [27].

capacitance value of 66.4 pF was calculated that corresponds relatively well to the measured one. Thus, there is always some uncertainty between designed nominal model parameters and the realized ones originating from the final technological process. For the normalized transconductance we calculate G_M' to be 434 mS/mm. With w = 100 μm, the transconductance reads 43.4 mS.

Problem 4

In [27] a HEMT structure with a 3×10^{18} cm^{-3} doped and 200Å thick AlGaAs gate insulator over 200Å thick InGaAs channel was theoretically investigated. No spacer was introduced between the gate insulator and the channel. The gate capacitance is calculated by differentiating the charge density under the gate with respect to the applied voltage. The calculated C_{gs} per unit area is shown in Figure 6.15 as a function of V_{gs}.

The curve denoted by 2DEG is the capacitance due to electrons in the channel. 3DEG is due to electrons in the AlGaAs, and Donor is due to occupied donors in the AlGaAs. The solid line curve represents the total capacitance. Near threshold voltage, donors throughout the AlGaAs layer are fully ionized with no free electrons in the insulator, and considerable electrons are distributed deep into the GaAs buffer, resulting in a small capacitance. An increase in the gate voltage near threshold causes band bending at the channel and confines the electrons; thus, the capacitance increases rapidly. Once the 2DEG is formed, the *effective* position of the 2DEG from the heterointerface is no longer noticeably influenced. Thus, the capacitance remains almost constant. When the gate is further forward-biased, donor neutralization and parallel conduction become marked, with the conduction band edge lying near the Fermi level in the AlGaAs. As a result, the capacitance due to 2DEG in the channel per unit area is lower, while the total capacitance again increases rapidly. The intrinsic transconductance is estimated as follows.

Under the short-channel approximation, the intrinsic transconductance (G_m) per unit gate width is estimated by the following equation:

$$G_m' = C_{2DEG}' \cdot v_s \tag{6.47}$$

where v_{eff} is the saturation velocity for the electrons. In the calculation v_s was assumed to be

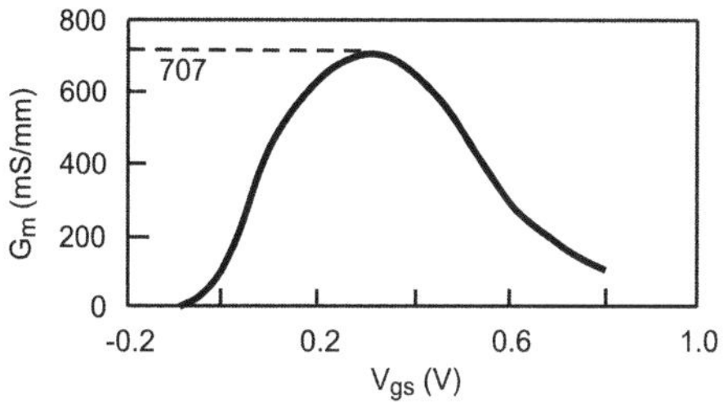

Figure 6.16 Numerically calculated transconductance versus gate voltage for analyzed HEMT. (After [27].)

2×10^7 cm/s. The dielectric constant of AlGaAs was assumed to be 12.2. The correction factor Δd is almost constant (~80Å). With this data the capacitance per unit area is calculated as 3.8×10^5 pF/cm^2, which corresponds perfectly to the value marking the transition from the nonconducting to the conducting AlGaAs layer in Figure 6.15. Then, with (6.47) the intrinsic transconductance G_m reads 3.8×10^5 pF/cm^2 $\times 2 \times 10^7$ cm/s = 760 mS/mm, which is about 7.5% higher than the calculated value based on an exact model for charge control employing a first principles theory, as shown in Figure 6.16 [27].

6.3.2 Channel Conductance

In short channel devices the most unpredictable parameter is G_{ds} [28]. We may distinguish between low-power (low-noise), medium-power and high-power devices. The first two groups might be fabricated on GaAs-based semiconductor material, whereas the last one mainly relates to GaN compound semiconductors.

Regarding the dependence of the drain-source conductance G_{ds} on the control voltages V_{gs} and V_{ds} the following observation can be made. From extracted 3D plots, there are three regions with strongly different behavior of the G_{ds} model element: (1) pinch-off, (2) saturation, and (3) ohmic region. As our experience has shown, under pinch-off operation the output conductance is very low. This was not always the case. However, it seems that meanwhile substrate current has become of minor importance due to continual improvement of device processing. Indeed, imperfect interfaces of channel-buffer layer and buffer layer-substrate can cause incomplete pinching off the channel so that residual channel current flows in spite of high negative gate-source voltages. Table 6.3 summarizes experimental results of extracted drain-source conductance values in the ohmic and saturation region for GaAs-based MESFET and HEMTs. It is seen that most of the devices exhibit a G_{ds} value between 2.5 mS and 5.5 mS, corresponding to drain-source resistances R_{ds} in the range of 180Ω and 400Ω. In the ohmic region, approaching low drain-source (V_{ds}) and high gate-source (V_{gs}) voltages, the output conductance (resistance) increases (decreases) strongly and reaches values >100 mS (<10Ω). Physically, G_{ds} models the variation of the drain current with the control voltages. Increasing the gate-source voltage (with V_{ds} small) decreases the depletion layer and increases the channel charge density, and thus, the channel current remarkably.

For AlGaN/GaN HEMTs similar distributions with respect to the control voltages V_{gs} and

Table 6.3
Drain-Source Conductances of Investigated GaAs FET Devices in the Ohmic and Saturation Region

Transistor			Saturation			Ohmic Region		
	Typ	$L_g \times w$ (μm^2)	G_{ds} (mS)	R_{ds} (Ω)	G_{ds}/w (mS/mm)	G_{ds} (mS)	R_{ds} (Ω)	$R_{ds} \times w$ (Ω-mm)
MESFET	MGFC1423	0.4 × 300	5.1	196	17.0	>10.6	<94	<28.1
	NE71000	0.3 ×260	5.6	178	21.5	133	7.5	1.95
	FSX03X		3.7	270		100	10.0	
HEMT	MGFC4401	0.4 × 200	2.37	422	11.8	133	7.5	1.5
	TGF2021-01	0.35 ×1000	5.32	188	5.32	1500	0.7	0.7
	JS8902-AS	0.3 × 200	3.8	263	19.0	100	10.0	2.0

Table 6.4
Drain-Source Conductances of Investigated GaN HEMT Devices in the Ohmic Region

w (mm)	G_{ds} (mS)	R_{ds} (Ω)	$R_{ds} \times w$ (Ω-mm)
1	130	7.7	0.77
10	1500	0.65	0.65
20	440	2.27	1.13
32	430	2.32	0.725

V_{ds} are given (i.e., approximately zero conductance in the pinch-off region). Regarding 1.0-, 2.0-, and 3.2-mm devices, typical values for the pinch-off region lie in the kiloohm region. Table 6.4 indicates typical extracted values of G_{ds} and R_{ds} in the ohmic region.

Averaging the results for the various gate widths of analyzed AlGaN/GaN HEMTs the resistance per unit width gives $R_{ds} \approx 0.81$ Ω-mm. Notice this does not correspond to the ON resistance, which also includes the series resistances R_s and R_d.

6.4 CAPACITANCES

In the following we consider estimations for gate-source, gate-drain, and drain-source capacitances.

6.4.1 Gate-Source Capacitance

As derived in Section 4.6, the intrinsic gate-source capacitance C_{gs} of a short-channel MESFET device can be estimated by the following relation

$$C_{gs} = wL(qN_D\varepsilon/2)^{1/2}(-V_{gs} + V_d)^{-1/2} \tag{6.48}$$

Frequently, the capacitance is normalized to the gate length L and gate width w. Then the capacitance per unit area reads

$$C'_{gs} = (qN_D\varepsilon/2)^{1/2}(-V_{gs} + V_d)^{-1/2} \tag{6.49}$$

As derived in section 4.2, normalized transconductance and capacitance are related to each other as

$$G'_m = C'_{gs}v_s \tag{6.50}$$

or in denormalized form,

$$G_m = C_{gs}v_s/L \tag{6.51}$$

Regarding short-channel field effect transistors the fringing effect of the gate electrode must be taken into account; that is, the effective gate length L_{eff} is simply approximated by

$$L_{eff} = L + 2d \tag{6.52}$$

with the extended gate length d, which is assumed to be equal to the height of the lateral depletion region of the zero-biased Schottky-gate. d is calculated as [1, 29]

$$d = [(2\varepsilon/qN_D)(-V_{gs} + V_d)]^{1/2} \tag{6.53}$$

With $\varepsilon_r = 12.9$ (300K), $N_D = 1.5 \times 10^{17}$ cm^{-3}, and $V_d = 0.7$V, d is calculated as $(2\varepsilon V_d/qN_D)^{1/2}$ = 0.082 μm for $V_{gs} = 0$V. The total gate extension then reads 0.164 μm.

For a 0.5-μm gate GaAs MESFET with a gate-width $w = 300$ μm, a capacitance C_{gs}' per unit area of 1.4×10^{-7} F/cm^2 is obtained ($V_{gs} = 0$V). With the given gate width and effective gate length of 0.5 μm + 0.164 μm = 0.664 μm, the intrinsic capacitance is calculated as 0.28 pF.

The following approximate relations for the 2DEG gate-source capacitance C_{gs} and transconductance G_m were found in Section 4.7.2:

$$C_{gs} = A\varepsilon_A(d_A + \Delta d)^{-1/2} \tag{6.54}$$

$$G_m/C_{gs} = wv_{eff}/A \tag{6.55}$$

$$G'_m = \frac{\varepsilon_A v_{eff}}{d_A} = C'_{gs}v_{eff} \tag{6.56}$$

G_m' is the transconductance per unit width and C_{gs}' the gate-source capacitance per unit area.

With the data of Problem 1 in section 6.3.1, $G_m = 60$ mS, $w = 150$ μm, $L = 0.5$ μm, $v_{\text{eff}} = 2 \times 10^7$ cm/s, we get for the normalized transconductance $G_m' = G_m/w = 400$ mS/mm. Then it is $C_{gs}' = G_m'/v_{\text{eff}} = 2 \times 10^{-9}$ F/mm^2. With $C_{gs} = C_{gs}' \times L \times w$ the gate-source capacitance reads $C_{gs} = 150$ fF, which agrees well with extracted values.

6.4.2 Gate-Drain Capacitance

In the saturation region the space charge depletion area is widely extended toward the drain electrode. In this case the gate-drain capacitance originates mainly from the stray field between the gate and drain electrode. As indicated in Figure 6.17, the electrode configuration can be interpreted as a very narrow and thick stripline of width t, which can be treated with Wheeler's theory [30]. The electrical stray fields on the side faces of the strip of height L

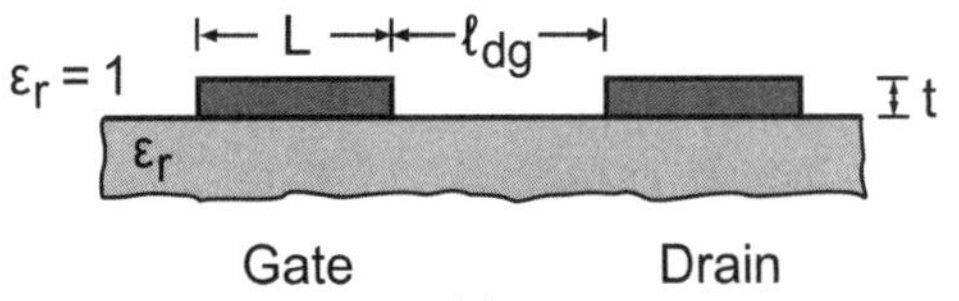

(a)

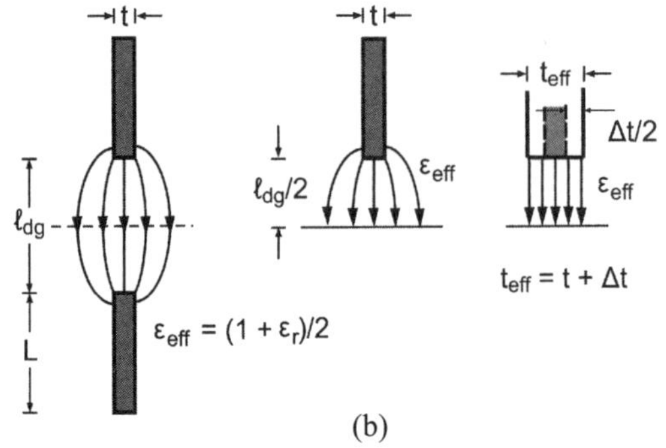

(b)

Figure 6.17 Approximate calculation of gate-drain capacitance C_{gd}. (a) Gate-drain electrode configuration, and (b) model of narrow thick stripline [30].

enlarge the physical width of the stripline by Δt; that is,

$$t' = t + \Delta t \tag{6.57}$$

where t' denotes the corrected strip with zero thickness. Δt is calculated from

$$\Delta t = \frac{t}{\pi}\left\{1 + \ln\left(\frac{l_{dg}}{L_G}\right)\right\}, \quad t/(l_{dg}/2) \geq 1/(2\pi) \tag{6.58}$$

Then, with the effective thickness of the electrodes t_{eff},

$$t_{eff} = t' + \frac{l_{dg}}{\pi}\left\{\ln\left[2\pi e\left(\frac{t'}{l_{dg}} + 0.92\right)\right]\right\} \tag{6.59}$$

the stray capacitance per unit length between the gate and drain electrode can be found by

$$C'_{gd} = \frac{\varepsilon_0 \varepsilon_{r,eff} t_{eff}}{l_{dg}/2}, \quad \varepsilon_{r,eff} = (1 + \varepsilon_r)/2 \tag{6.60}$$

We suppose a GaAs MESFET with the following data: $L = 0.3$ µm, $l_{dg} = 0.5$ µm, $w = 300$ µm, $\varepsilon_r\varepsilon_0 = 12.9 \times 8.854 \times 10^{-14}$ As/V/cm. The thickness t of the ohmic contact may be estimated to be 0.3 µm (Figure 2.44 in [34]). Then (6.58) delivers $\Delta t = 0.14$ µm, and t' reads $t' = t + \Delta t = 0.3$ µm + 0.14 µm = 0.44 µm.

Applying (6.59) gives $t_{eff} = 0.44$ µm + 0.54 µm = 0.98 µm. Then (6.60) delivers $C_{gd}' = 2.4\times10^{-12}$ Fcm, and the gate-drain capacitance is calculated to be $C_{gd} = C_{gd}' \times w = 70$ fF, which agrees well with the result in [13].

C_{gd} increases rapidly with increasing gate-source voltage under low drain-source voltage conditions. In this case the depletion region is diminished and uniformly distributed under the gate. Therefore, the total gate capacitance can be split into two approximately equal capacitances, C_{gd} and C_{gs}.Thus, (4.79) can also be applied to estimate C_{gd} in the ohmic region.

Regarding the pinch-off region (6.60) can also be applied to HEMT devices based on

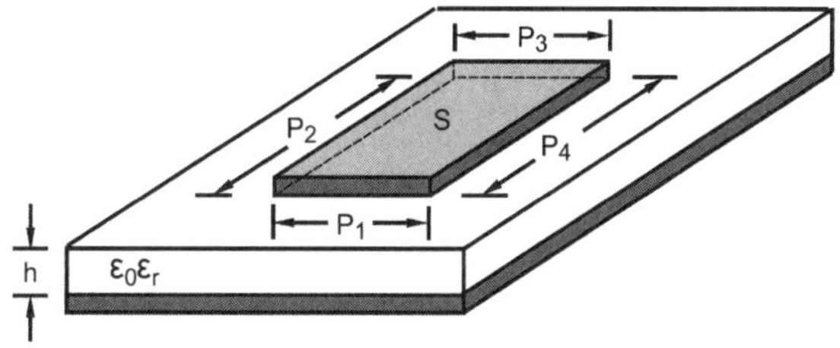

Figure 6.18 Estimation of the drain-source capacitance C_{ds}. The periphery P is composed of all lateral lengths P_1 ... P_4. It should not be regular.

GaAs and GaN technology basis. A GaN HEMT may have the following data: $L = 0.5$ μm, $l_{dg} = 0.5$ μm, $w = 100$ μm, $\varepsilon_{r,eff}\varepsilon_0 = (8.9 + 1)/2 \times 8.854 \times 10^{-14}$ As/V/cm. The thickness t of the ohmic contact may be estimated to be 0.18 μm (Figure 2.45 in [1]). Then (6.58) delivers $\Delta t = 0.05$ μm, and t' reads $t' = t + \Delta t = 0.18$ μm + 0.05 μm = 0.23 μm. Applying (6.59) gives $t_{eff} = 0.23$ μm + 0.50 μm = 0.73 μm. Then (6.60) delivers $C_{gd}' = 1.28\times10^{-12}$ Fcm, and the gate-drain capacitance is calculated as $C_{gd} = C_{gd}'\times w = 13$fF, which agrees well with the extraction result in [31].

6.4.3 Drain-Source Capacitance

C_{ds} is only slightly voltage-dependent because the capacitance mainly results from the whole drain metallization area over ground (source). In general it is $C_{ds} \ll C_{gs}$. Estimating the C_{ds} value, Higashisaka and Hasegawa [32] have suggested the following empirical formula for

squared and rectangular electrodes on semi-isolated GaAs with substrate thicknesses of 100 μm to 250 μm (see Figure 6.18):

$$C_{ds} \approx 1.3\,\varepsilon_0\varepsilon_r S/h + 0.92\,\varepsilon_0\varepsilon_r P/\log(10^4 h) \tag{6.61}$$

with drain electrode area S (in cm²), electrode periphery P (in cm), and substrate height h (in cm). When the geometric dimensions are given in cm, C_{ds} is calculated in farads.

Example

The dimensions of the drain electrode of the MESFET NE71000 in Figure 6.9 are estimated as follows: Total drain electrode area $S \approx 1.71\times10^{-4}$ cm², drain electrode periphery (contour) length $P \approx 664\times10^{-4}$ cm, chip thickness $h = 10^{-4}$ cm. Then the first term in (6.61) is calculated as

$$1.3\varepsilon S/h = 1.3\times8.85\cdot10^{-14}\text{As/V/cm}\times12.9\times1.71\cdot10^{-4}\text{cm}^2/140\cdot10^{-4}\text{cm} = 18.14\text{ fF}$$

The second term gives

$$0.92\varepsilon P/\log(10^4 h) = 0.92\times8.85\cdot10^{-14}\times12.9\times664\cdot10^{-4}/[\log(10^4\times140\cdot10^{-4})] = 32.5\text{ fF}$$

Then, in total, C_{ds} turns out to be 50 fF.

6.5 DELAY TIME

Regarding short-channel FETs, it can be assumed that the carriers pass the entire gate region with saturation drift velocity. Then the delay time is calculated as

$$\tau = L_G / v_s \tag{6.62}$$

Assuming a GaAs MESFET with $L_G = 0.5$ µm, $v_s = 1.8 \times 10^7$ cm/s, then the delay time is calculated to be $\tau = 2.8$ ps. Similar values have also been extracted for GaAs- and GaN-based HEMT devices. It is clear that carrier saturation velocity is dependent on the carrier density.

6.6 CONTACT AND INTERCONNECT STRUCTURES

Contacting pads are commonly connected to the gate, drain, and source electrodes. They provide electrical control of the transistor. Device contacting can be done with probe tips of a wafer prober, which allows flexible tip contact and release. In contrast, wire-bonding to a test fixture is a fixed connection that must be destroyed if the transistor die under measurement should be reused. This is avoided with the partitioning design approach proposed by Rui Ma [33]. Furthermore, air bridges and via holes are essential components of power device designs.

6.6.1 Device Contacting Pads

The physical origin of the capacitances C_{pg} and C_{pd} is simplified by Figure 6.19 showing a typical situation of a transistor chip where the gate (drain) electrode is wire-bonded to a microstrip transmission line. There are two effects that contribute to the parasitic external capacitances: the fringing electrical field at the open end of the microstrip and the offset L_b of the bonding pad.

Example

In-fixture measurements frequently use plastic substrate materials, such as low dielectric constant CuClad 250 LX ($\varepsilon_r = 2.5$, $h = 0.328$ mm) and high dielectric constant DiClad 810 ($\varepsilon_r = 10.2$, $h = 0.635$ mm) (e.g. [34]). The characteristic impedance of the microstrip transmission line should be 50Ω. We estimate the parasitic capacitances as follows.

Using simple Wheeler's theory, from diagram Figure 3.18 in [30], for 50Ω impedance we read $w/h \approx 1$ for DiClad and $w/h \approx 3$ for CuClad. Then from Figure 6.7 in [30] we found for the end effect capacitance $C/w \approx 0.056$ pF/mm (DiClad) and $C/w \approx 0.016$ pF/mm (CuClad), respectively. With the given microstrip widths we get $C \approx 36$ fF for high dielectric constant DiClad, and $C \approx 16$ fF for low dielectric constant CuClad.

The offset of the bonding pad L_b should be ideally zero. We estimate the contribution to the parasitic external capacitance for nonzero line length. The calculation can be done using (6.4) in [30],

$$C = \frac{\sqrt{\varepsilon_{eff}}\, L_b}{Z_0(\varepsilon_r)\, c_0} \tag{6.63}$$

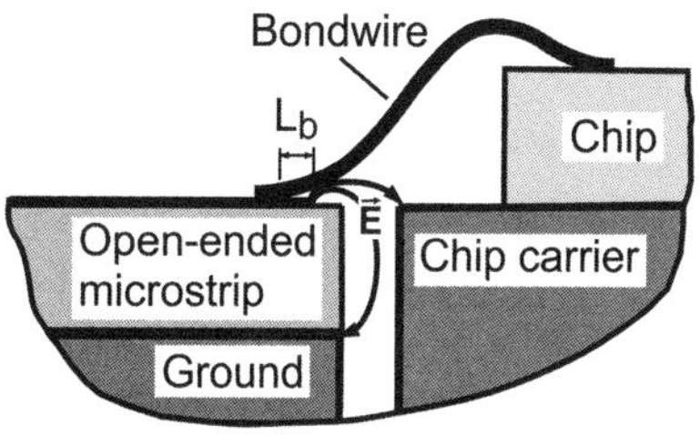

Figure 6.19 Illustration of physical origin of parasitic capacitances C_{pg} and C_{pd} of a wire-bonded transistor chip. L_b signifies the offset of the bond pad from the open end of microstrip line.

where ε_{eff} is the effective dielectric constant of the characteristic impedance $Z_0 = 50\Omega$. From diagram Figure 3.17 in [30] we read $\sqrt{\varepsilon_{eff}} \approx 2.6$ (for $\varepsilon_r = 10.2$) and 1.5 (for $\varepsilon_r = 2.5$). For a 50Ω microstrip line and an offset of 100 μm we get $C_{offset} \approx 17$ fF for high dielectric constant DiClad and $C_{offset} \approx 8$ fF for low dielectric constant CuClad. It is seen that it is very important to accurately set the bond as near as possible at the open end of the microstrip line. The parasitic stray capacitances C_{pg} and C_{pd} were extracted in the range of 0.02 pF to 0.04 pF.

6.6.2 Bondwire Inductance

Figure 6.20 illustrates the assembly situation of a wire-bonded transistor chip in a test fixture [34] with typical wire-bonding configuration for source, gate, and drain contacting. The inductance per unit length of a bondwire with diameter d in a distance D above a conducting plane with $D/d \gg 1$ is approximately calculated as

$$L' = \frac{\mu_0}{\pi} \ln\left(\frac{2D}{d}\right) = 0.4 \ln\left(\frac{2D}{d}\right) \frac{\text{nH}}{\text{mm}} \tag{6.64}$$

with $\mu_0 = 0.4\pi$ nH/mm. Assuming $D = 150$ μm and a bondwire diameter $d = 25$ μm, it follows $L' \approx 1$ nH/mm.

As is demonstrated Figure 6.20 the bondwire lengths must be as short as possible to minimize parasitic inductive effects. Analyzing the bond pattern, which is typically for transistor chips in a test fixture discussed in [34], an effective length of the gate bondwire $l_g = 400$ μm provides a single-wire gate-source inductance of $L_g \approx 0.4$ nH. For an effective length $l_d = 360$ μm for the drain bonding, a single-wire value $L_d \approx 0.36$ nH is estimated. As the gate and drain electrodes are commonly bonded twofold, the inductance values approximately halve. These effective values agree very well with typical extracted values: $L_g \approx 0.19$ nH, $L_d \approx 0.16$ nH, which include a mutual-inductance decrease of 5% and 25%, respectively.

The shortest bonding can be realized with the source-to-ground connection, leading the bondwire directly near the transistor chip side. In the example in Figure 6.20 the effective wire length is estimated as 330 μm, representing a single-wire inductance value $L_s \approx 0.33$ nH. Typically multiwirebonding is applied. With $N = 4$ wirebonds, L_s is reduced to $L_s/N \approx 0.08$ nH, if the mutual inductances are neglected. Extraction delivered values about 0.05 nH, which reflects a reduction of about 40% according to (6.67).

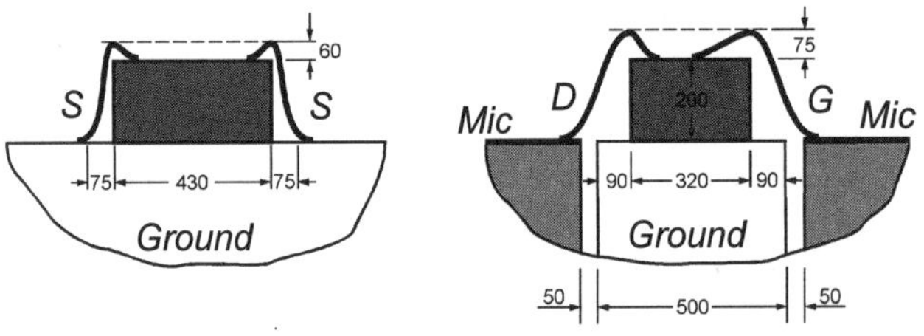

Figure 6.20 Assembling a wirebonded transistor chip: (left) source contacting (multiwirebonded), and (right) gate and drain contacting (dimensions in μm).

The inductance does grow faster than linearly with length because there is a mutual coupling between parts of the bondwire. As reported in [35], the DC inductance of a bondwire is given by

$$L = \left[\frac{\mu_0 l}{2\pi}\right]\left[\ln\left(\frac{2l}{r}\right) - 0.75\right] \approx 2\times10^{-7} l\left[\ln\left(\frac{2l}{r}\right) - 0.75\right] \tag{6.65}$$

From the logarithmic term, however, we see that this effect is minor. The inductance is similarly insensitive to the wire diameter, so even relatively large wire has an inductance on the order of 1 nH/mm.

The behavior of bond wire interconnections at millimeter-wave frequencies is determined by the fact that, as frequency is increased, the length of the wires reaches significant fractions of wavelength, and the wires exhibit transmission line properties [36, 37]. An interconnection with a return loss better than −10 dB requires bondwires not longer than $0.033\cdot\lambda_0$, which is only about 100 μm at 94 GHz. If the wire length exceeds this maximum value, due, for example, to different substrate thicknesses, an interconnection of the same quality can only be achieved at the cost of an additional bond wire compensation. A compensated bondwire interconnection has additional matching networks at either end of the wires. Symmetric bond wire compensation demands additional chip area and results in an output impedance that is different from the usual 50Ω reference. Therefore, in many cases an asymmetric compensation on only one side of the interconnection would be preferable, even though the matching is inferior and the bandwidth decreases by about 30%.

The effective resistance of the bond wire per length with the radius r of the wire can be estimated as

$$\frac{R}{l} = \frac{1}{2\pi r\delta\sigma} \tag{6.66}$$

where δ is the skin depth and σ is the conductivity. For example, at 1 GHz the skin depth δ is 2.5 μm. Then the resistance per length reads 125 mΩ/mm.

Mutual Inductance

Adjacent bond wires couple magnetically. A measure of this coupling is the mutual inductance between them. For two bondwires of equal length, this inductance is given by

$$M \approx \frac{\mu_0 l}{2\pi}\left[\ln\left(\frac{2l}{D}\right) - 1 + \frac{D}{l}\right] \tag{6.67}$$

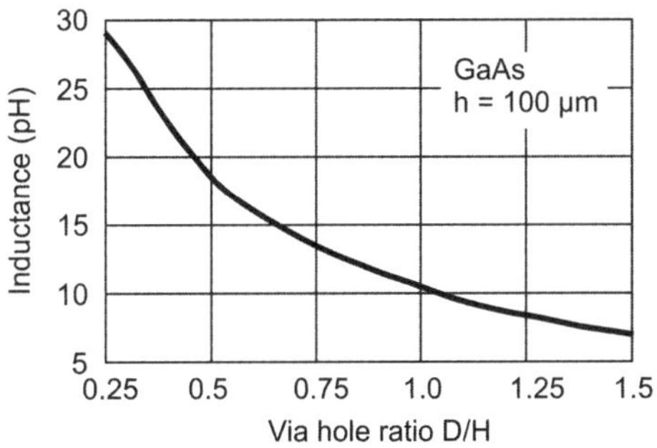

Figure 6.21 Inductance of a via as a function of ratio of via diameter to substrate height for GaAs (substrate height h = 100 μm).

where μ_0 (μ_0 = 0.4π nH/mm) is the permeability in free space, l the wire length of the bondwires, and D is the distance between them [35]. Mutual inductance has little frequency dependency.

Using bonding wires around several GHz, radiation is not a problem. However, to model higher-frequency bonding wires, capacitance should be taken into account [38]. RF designers usually use approximate analytical formulae for straight wires to estimate bonding wire inductance L and mutual inductance M as given by (6.65) – (6.67) [35].

6.6.3 Via Hole Inductance

The inductance of a single via from top metallization to ground can be calculated as proposed in [39]. This is an empirical solution that was found to fit data where the substrate height h was between 100 μm and 631 μm. r is the radius. It is supposed to be valid for heights less that 3% of a wavelength.

$$L_{via} = \frac{\mu_0}{2\pi}\left[h \times \ln\left(2h + \frac{\sqrt{r^2 + (2h)^2}}{r}\right) + \frac{3}{2}\left(r - \sqrt{r^2 - h^2}\right)\right] \tag{6.68}$$

If the units of all of the lengths are in meters, then the inductance is in henrys. In many cases it is desirable to keep r/h ratio to 0.25 to minimize the catch-pad area, so 20 pH is a good number to use as a guidance in circuit design. Figure 6.21 illustrates the inductance as function of normalized via hole diameter for a 100-μm thick GaAs substrate.

6.6.4 Air Bridge

For multifinger field-effect devices air bridges are required to connect the individual source fingers with the outer source contacts. Figure 6.22 indicates a typical air bridge in power devices between two neighbor source electrodes [40] overcrossing drain electrode with gate electrodes on both sides. With the drain electrode width w_D, bridge width w_B, and bridge height h, the crossover capacitance C_{dsc} is estimated as

$$C_{dsc} = \frac{\varepsilon_0 w_D w_B}{h} \tag{6.69}$$

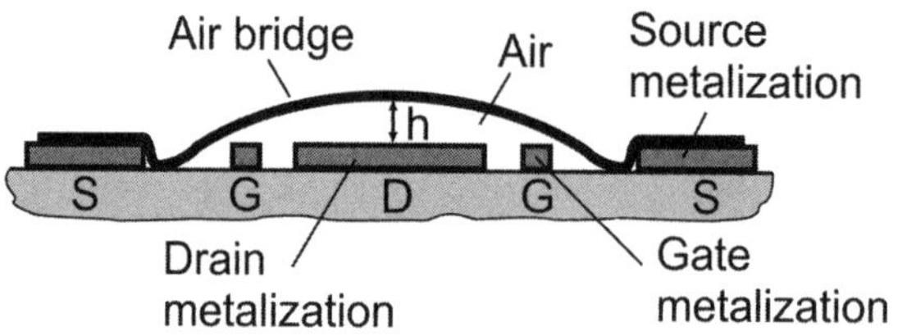

Figure 6.22 Typical air bridge configuration in multifinger FET devices.

Assuming $w_D = w_B = 25$ μm and $h = 3$ μm, (6.69) delivers 1.8 fF. Depending on the finger width, additional air bridges may be provided.

6.6.5 Field Plate

Field plates have been primarily introduced to modify the electric field profile at the drain-sided gate edge and to decrease its peak value, hence reducing the trapping effect and increasing breakdown voltages in high-power HEMT devices [1].

Generally, as shown in Figure 6.23, two different types of field plates are deployed: namely source-connected field plate and gate-connected field plate. Source-connected field plate is mainly used to minimize the drawback of gate to drain feedback capacitance introduced by gate-connected field plate that significantly reduces device gain. However, field plates introduce additional capacitances and reduce device transconductances [41]. Thus, the gate -source capacitance C_{gs} depends not only on the charge Q_g on the gate but also on the charge on the gate (and source) field plate [42].

After incorporating a field-plate (FP) structure in a HEMT, its breakdown voltage increases. Although FP-HEMTs exhibit lower unity current gain frequency (f_T) due to increased gate-to-drain (C_{gd}) and drain-to source (C_{ds}) capacitances due to incorporation of gate-connected FP (GC-FP) and source-connected FP, respectively, they show improved breakdown voltage, linearity, stability, reliability, and efficiency [43].

In [44] it is reported that the source-connected FP structure reduces C_{gd} but at the cost of a slight increase in C_{gs}, particularly for a long gate length. For a gate length of $L_G = 0.8$ μm and a source-connected FP length of $L_{FP} = 1.2$ μm an increase of about 40% of the gate-source capacitance C_{gs} was measured.

Concerning gate-connected FP, up to a certain value, the longer the FP, the more output

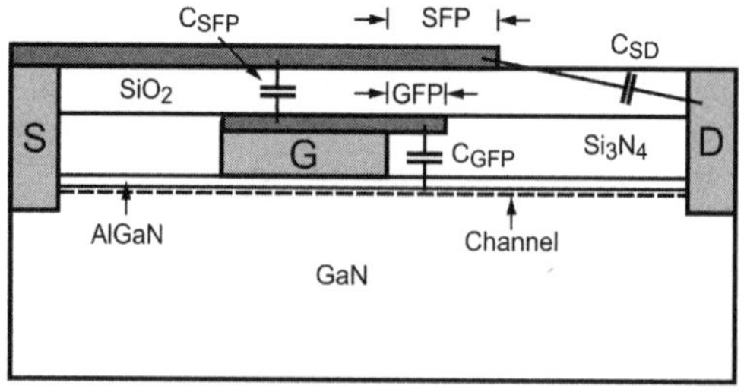

Figure 6.23 Basic source and gate field plate configurations and additional parasitic capacitances: C_{GFP} (gate field plate capacitance), C_{SFP} (source field plate capacitance), C_{SD} (source-drain field plate capacitance). GFP (SFP) is the length of gate (source) field plate.

power was achieved. However, in this configuration the capacitance between the FP and drain becomes gate-to-drain capacitance (C_{gd}), resulting in negative Miller feedback. This causes reduction in current gain and power-gain cutoff frequencies (f_t/f_{max}) [45].

In source-connected FP, the FP-to-channel capacitance becomes the drain-source capacitance, which could be absorbed in the output-tuning network. The drawback of additional C_{gd} by the FP is hence eliminated. Depending on the implementation, the source-connected field plate can add parasitic capacitance to the device input. However, this can also be absorbed into the input tuning circuit, at least for narrowband applications [45].

In [46], discussing the engineering of a source-FP, the influence of critical device parameters such as FP length, passivation dielectric, and dielectric thickness on the electric potential and E-field distribution in AlGaN/GaN HFETs are investigated. Interestingly, the following analytical formula indicates the parasitic capacitance increase due to the FP:

$$r_C = \frac{C_{eff}}{C_{gate}} = 1 + \frac{(t_{ox}/\varepsilon_{ox} + d_{AlGaN}/\varepsilon_{AlGaN})\, L_{FP}}{t_{ox}/\varepsilon_{ox} + d_{AlGaN}/\varepsilon_{AlGaN} + t_{pass}/\varepsilon_{pass})\, L_G} \tag{6.70}$$

where ε_{ox} is the equivalent relative dielectric constant of the whole passivation and gate dielectric layer and t_{ox} is the respective thickness. Correspondingly, ε_{AlGaN} and d_{AlGaN} are the dielectric constant and thickness of the AlGaN layer, and ε_{pass} and t_{pass} the passivation material dielectric constant and passivation layer thickness. Regarding typical passivation materials, such as SiO_2 ($\varepsilon_r = 4.5$) and SiN_x ($\varepsilon_r \approx 7$), the parasitic capacitance increase is about 16% for a field-plate length of 1.8 μm. The variation of FP lengths in the range of 1 μm to 2 μm provides a capacitance change of about 6%.

6.7 SUMMARY

Model parameter extraction procedures based on optimization rely on high-quality starting values for the model elements to obtain reliable extraction results. Based on simplified analytical description of the electrical properties of MESFETs and HEMTs, relations have been derived to estimate physical model parameters from technology and material data. The practical applicability was illustrated by several device data examples from the literature.

REFERENCES

[1] Kompa, G., *Basic Properties of III-V Devices–Understanding Mysterious Trapping Phenomena*, Kassel, Kassel University Press GmbH, 2014.

[2] Look, D. C., "Contact Resistance Measurements in GaAs MESFET's and MODFET's by the Magneto-TLM Technique," *J. Electrochem. Soc.: Solid-State Science and Technology*, Vol. 135(8), 1988, pp. 2054–2058.

[3] Soares, R. (Ed.), *GaAs MESFET Circuit Design*, Norwood: Artech House, Inc., 1988.

[4] Atherton, J. S., *Characterization of MESFETs Based on a Physical Model*, Doctoral Thesis, University of Leeds, 1993.

[5] Sai Saravanan, G., K. Mahadeva Bhat, K. Muraleedharan, H. P. Vyas, R. Muralidharan, and A. P. Pathak, "Ohmic Contacts to Pseudomorphic HEMTs With Low Contact Resistance Due to Enhanced Ge Penetration Through AlGaAs Layers," *Semicond. Sci. Technol.*, Vol. 23, 2008, 025019 (6pp).

[6] Nebauer, E., W. Österle, J. Hilsenbeck, J. Würfl, and A. Klein, "XTEM and TFXRD Investigations of Ohmic Ti/Al/Ti/Au/WSiN Contacts on AlGaN/GaN HFET Layer Systems," *Semicond. Sci. Technol.*, Vol. 17, 2002, pp. 249-254, 2004.

[7] Selvanathan, D., F. M. Mohammed, A. Tesfayesus, and I. Adesida, "Comparative Study of Ti/Al/Mo/Au, Mo/Al/Mo/Au, and V/Al/Mo/Au Ohmic Contacts to AlGaN/GaN Heterostructures," *J. Vac. Sci. Techno. B*, Vol. 22(5), 2004, pp. 2409-2415, 2004.

[8] Chaturvedi, N., U. Zeimer, J. Würfl, and G. Tränkle, "Mechanism of Ohmic Contact Formation in AlGaN/GaN High Electron Mobility Transistors," *Semicond. Sci. Technol.*, Vol. 21, 2006, pp. 175–179.

[9] Charturvedi, N., *Development and Study of AlGaN/GaN Microwave Transistors for High Power Operation*, Doctoral Thesis, Technical University Berlin, 2007.

[10] Jacobs, B., M. C. J. C. M. Kramer, E. J. Geluk, and F. Karouta, "Optimisation of the Ti/Al/Ni/Au Ohmic Contact on AlGaN/GaN FET Structures," *J. Cryst. Growth*, Vol. 241, 2002, pp. 15–18.

[11] Ladbrooke, P. H., *MMIC Design: GaAs FETs and HEMTs*, Artech House, Inc., 1989.

[12] Ladbrooke, P. H., A. J. Hill, and J. P. Bridge, "CAD for GaAs IC Manufacture-Ability - Thirteen Issues in FET Physics," *IEEE GaAs IC Symposium*, 1991, pp. 175–178.

[13] Schlechtweg, M., *Broadband Characterization and Modelling of GaAs-MESFETs and (AlGa)As/GaAs-MODFETs up to 40 GHz* (in German), Doctoral Thesis, Kassel University, 1989.

[14] Chang, C. Y., T. J. Anderson, F. Ren, S. J. Pearton, A. M. Dabiran, A. M. Wowchak, B. Cui, and P. P. Chow, "Very Low Sheet Resistance AlN/GaN High Electron Mobility Transistors," CS MANTECH, 2009 (4 pp.).

[15] Heinrich, W., "Limits of FET Modelling by Lumped Elements," *Electronics Lett.*, Vol. 22(12), 1986, pp. 630–632.

[16] Popović, R. S., "Metal-N-Type Semiconductor Ohmic Contact with a Shallow N^+ Surface Layer," *Solid-State Electronics*, Vol. 21, 1978, 1133–1138.

[17] Cidronali, A., G. Collodi, G. Vannini, A. Santarelli, and G. Manes, "A New Approach to FET Model Scaling and MMIC Design Based on Electromagnetic Analysis," *IEEE Trans.*, Vol. MTT-47, 1999, pp. 900–907.

[18] Rohdin, H., N. Moll, A. M. Bratkovsky, and C.-Y. Su, "Dispersion and Tunneling Analysis of the Interfacial Gate Resistance in Schottky Barriers," *Phys. Rev. B*, Vol.59(20), 1999-II, pp. 13102–13113.

[19] Onodera, K., K. Nishimura, S. Aoyama; S. Sugitani, Y. Yamane, and M. Hirano, "Extremely Low-Noise Performance of GaAs MESFET's with Wide-Head T-Shaped Gate," *IEEE Trans.*, Vol. ED-46(2), 1999, pp. 310–319.

[20] Quay, R., A. Tessmann, R. Kiefer, R. Weber, F. van Raay, M. Kuri, M. Riessle, H. Massler, S. Müller, M. Schlechtweg, and G. Weimann, "AlGaN/GaN HEMTs on SiC: Towards Power Operation at V-Band," *IEEE Int. Electron Device Meeting 2003, IEDM'03*, Technical Digest, pp. 567–570.

[21] Bajaj, S., O. F. Shoron, P. S. Park, S. Krishnamoorthy, F. Akyol, T.-H. Hung, S. Reza, E. M. Chumbes, J. Khurgin, and S. Rajan, "Density-Dependent Electron Transport and Precise Modeling of GaN High Electron Mobility Transistors," *Appl. Phys. Lett.*, Vol. 107, 2015, 153504/1–4.

[22] Hughes, W. A., and C. M. Snowden, "Nonlinear Charge Control in AlGaAs/GaAs Modulation-Doped FET's," *IEEE Trans.*, Vol. ED-34(8), 1987, pp. 1617–1625.

[23] Ioffe, *NSM Archive Aluminium Gallium Arsenide (AlGaAs) – Basic Parameters*, www.ioffe.ru/SVA/NSM/Semicond/AlGaAs/basic.html.

[24] Norris, G. B., D. C. Look, W. Kopp, J. Klem, and H. Morkoç, "Theoretical and Experimental Capacitance-Voltage Behavior of $Al_{0.3}Ga_{0.7}As$/GaAs Modulation-Doped Heterojunctions: Relation of Conduction-Band Discontinuity to Donor Energy," *Appl. Phys. Lett.*, Vol. 47(4), 1985, pp. 423–425.

[25] Park, K., H. B. Kim, and K. D. Kwack, "A Model for the Capacitance-Voltage Characteristics of MODFET′s," *IEEE Trans.*, Vol. ED-34, 1987, pp. 2422–2426.

[26] Moloney, M. J., F. Ponse, and H. Morkoç, "Gate Capacitance-Voltage Characteristics of MODFET's: Its Effect on Transconductance," *IEEE Trans.*, Vol. ED-32(9), 1985, pp. 1675–1985.

[27] Ando, Y., and T. Itoh, "Analysis of Charge Control in Pseudomorphic Two-Dimensional Electron Gas Field-Effect Transistors," *IEEE Trans.*, Vol. ED-35(12), 1988, pp. 2295–2301.

[28] Ahmed, A. A., N. Ahmed, K. S. Chaudhary, and M. F. Iqbal, "Estimation of Intrinsic Small Signal Parameters of a GaAs MESFET from DC Measurements," *Proc. IEEE INMIC Conf.*, 2001, pp. 97–103.

[29] Han, J., "Electron Transport Characteristics of Sub-100 Nanometer GaAs MESFETs," *J. Korean Phys. Soc.*, Vol. 42, 2003, pp. 297-301.

[30] Kompa, G., *Practical Microstrip Design and Applications*, Norwood: Artech House, Inc., 2005.

[31] Jarndal, A. H., *Large-Signal Modeling of GaN Device for High Power Amplifier Design*, Doctoral Thesis, University of Kassel, 2006.

[32] Higashisaka, A., and F. Hasegawa, "Estimation of Fringing Capacitance of Electrodes on S.I. GaAs Substrate," *Electronics Lett.*, Vol. 16(11), 1980, pp. 411–412.

[33] Rui Ma, *Reliable RF Power Amplifier DesignBased on a Partitioning Approach*, Doctoral Thesis, University of Kassel, 2009, Kassel: Kassel University Press GmbH, 2010.

[34] Kompa, G., M. Schlechtweg, and F. van Raay, "Precisely Calibrated Coaxial-to-Microstrip Transitions Yield Improved Performance in GaAs FET Characterization," *IEEE Trans.*, Vol. MTT-38, 1990, pp. 62–68.

[35] Lee, T. H., *The Design of CMOS Radio-Frequency Integrated Circuits*, Cambridge University Press, 2nd Edition, 2004, pp. 144–146.

[36] Krems, T., W. Haydl, H. Massler, and J. Rüdiger, "Millimeter-Wave Performance of Chip Interconnections Using Wire Bonding and Flip Chip," *1996 IEEE MTT-S Digest*, 1996, pp. 247–250.

[37] Alimenti, F., P. Mezzanotte, L. Roselli, and R. Sorrentino, "Modeling and Characterization of the Bonding-Wire Interconnection," *IEEE Trans.*, Vol. MTT-49(1), 2001, pp. 142–150.

[38] Qi, X., P. Yue, T. Arnborg, H. T. Soh, H. Sakai, Z. Yu, and R. W. Dutton, "A Fast 3-D Modeling Approach to Electrical Parameters Extraction of Bonding Wires for RF Circuits," *IEEE Trans. on Advanced Packaging*, Vol. 23(3), 2000, pp. 480–488.

[39] Goldfarb, M. E., and R. A. Pucel, "Modeling Via Hole Grounds in Microstrip," *IEEE Microwave and Guided Wave Lett.*, Vol. 1(6), 1991, pp. 135–137.

[40] Javorka, P., *Fabrication and Characterization of AlGaN/GaN High Electron Mobility Transistors*, Doctoral Thesis, RWTH Aachen, 2004.

[41] Islam, M. M., *Engineering Model of III-Nitride Power Heterostructure Field-Effect Transistor on Silicon Substrate*, Doctoral Thesis, University of South Carolina, 2016.

[42] Sharma, K., A. Dasgupta, S. Ghosh, S. A. Ahsan, S. Khandelwal, and Y. S. Chauhan, "Effect of Access Region and Field Plate on Capacitance Behavior of GaN HEMT," *2015 IEEE Int. Conf. On Electron Devices and Solid-State Circuits (EDSSC)*, 2015, pp. 499–502.

[43] Ahsan, S. A., S. Ghosh, K. Sharma, A. Dasgupta, S. Khandelwal, and Y. S. Chauhan, "Capacitance Modeling in Dual Field-Plate Power GaN HEMT for Accurate Switching Behavior," *IEEE Trans.*, Vol. ED-63(2), 2016, pp. 565–572.

[44] Xie, G., E. Xu, J. Lee, N. Hashemi, B. Zhang, F. Y. Fu, and W. Ng, "Breakdown-Voltage-Enhancement Technique for RF-Based AlGaN/GaN HEMTs With a Source-Connected Air-Bridge Field Plate," *IEEE Electron Device Lett.*, Vol. 33(5), 2012, pp. 670–672.

[45] Mishra, U. K., L. Shen, T. E. Kazior, and Y.-F. Wu, "GaN-Based RF Power Devices and Amplifiers," *Proceedings of the IEEE*, Vol. 96(2), 2008, pp. 287–305.

[46] Chen, T., Q. Zhou, D. Wei, C. Dong, W. Chen, and B. Zhang, "Physics-Based 2-D Analytical Model for Field-Plate Engineering of AlGaN/GaN Power HFET," *IEEE Trans.*, Vol. ED-66(1), 2019, pp. 116–125.

Chapter 7

Small-Signal Transistor Model Complexity

7.1 INTRODUCTION

The modeling of large-size gate periphery transistor structures with strong interelectrode capacitive effects is beyond the classical modeling approach that is typically represented by the widely used 15-element device model in Figure 7.1. Large-size transistors commonly comprise more than two gate fingers and include gate and/or source air bridges. Complex topologies may also be adopted for small-size transistors with one or two gate fingers at high frequencies (e.g., >80 GHz) due to stray capacitances [1]. Such extended transistor structures need more sophisticated model topologies as given, for example, in Figure 7.2 [2].

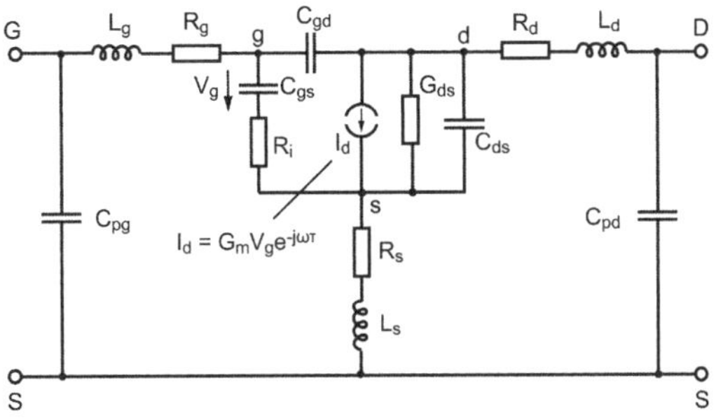

Figure 7.1 Widely used standard 15-element field-effect transistor model.

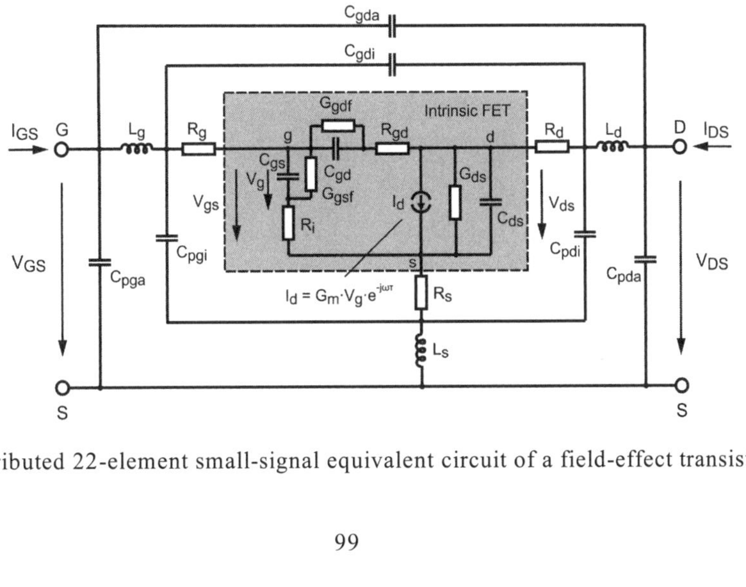

Figure 7.2 Distributed 22-element small-signal equivalent circuit of a field-effect transistor.

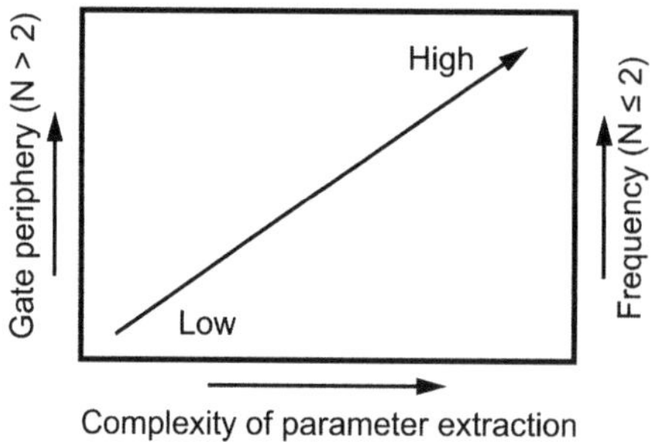

Figure 7.3 Increasing complexity of parameter extraction with increasing large-size gate peripheries (N > 2) or small-size transistors (N ≤ 2) with increasing frequency (*N* number of gate fingers).

Generally the complexity of parameter extraction depends both on the complexity of the device structure and on the frequency of device operation. Parasitic capacitance effects can become effective and noticeable with complex HEMT gate pattern designs at rather low frequencies. In contrast, parameter extraction of small-size gate periphery devices may become more complex with increasing frequency. This is schematically depicted in Figure 7.3, showing the dependence of complexity of parameter extraction with the complexity of the model: (1) multifinger devices with more than two gate fingers, and (2) single- or two-gate finger devices at higher frequencies. Direct model parameter extraction is in principle possible with the standard 15-element model. However, beyond optimization methods are commonly applied with the main disadvantage not being sure about the accuracy of extracted model parameter values.

Here we follow the strategy to find starting values for the optimization process, which are already near the extraction solution. In this case, global optimization is no longer needed. Instead, local optimizer can be employed avoiding the local minimum problem.

Therefore, in this chapter we try to find well-estimated model parameter values from DC, quasi-DC, and low and medium frequency measurements. The simplified equivalent circuits are presented for these frequency ranges. The potential for the determination of the respective model parameters are discussed.

7.2 SMALL-SIGNAL TRANSISTOR OPERATION

Figure 7.4 shows a field-effect transistor FET in common source operation with extrinsic terminals. We have already discussed the static (DC) behavior of a FET device in chapter 4. It has been shown that the drain current is nonlinearly dependent both on the drain-source and the gate-source voltage.

Using the notation for voltages and currents in Figure 7.4 the drain-source current can be written as $I_{DS} = f(V_{GS}, V_{DS})$. Similarly, the gate-source current reads $I_{GS} = f(V_{GS}, V_{DS})$. We assume that the FET has no dispersion (i.e., the small-signal transconductance and channel conductance are both frequency-independent). In this case the small-signal parameters can be obtained from the stationary nonlinear current-voltage characteristics on the basis of

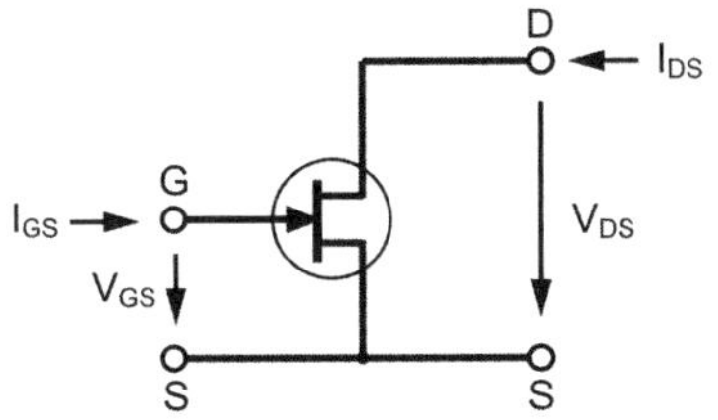

Figure 7.4 Common source field-effect transistor amplifier.

Taylor series expansion.

When we apply the Taylor series expansion to the nonlinear current function $I_{DS} = f(V_{GS}, V_{DS})$ in a given quiescent (bias) point (V_{GS0}, V_{DS0}), then the drain-source current follows incremental voltage changes ΔV_{GS} and ΔV_{DS} according to

$$\begin{aligned} I_{DS}(V_{GS}, V_{DS}) &= I_{DS}(V_{GS0}, V_{DS0}) \\ &+ \left.\frac{\partial I_{DS}(V_{GS}, V_{DS})}{\partial V_{GS}}\right|_{V_{GS0}, V_{DS0}} \Delta V_{GS} + \left.\frac{\partial I_{DS}(V_{GS}, V_{DS})}{\partial V_{DS}}\right|_{V_{GS0}, V_{DS0}} \Delta V_{DS} \\ &+ \frac{1}{2}\left.\frac{\partial^2 I_{DS}(V_{GS}, V_{DS})}{\partial V_{GS}^2}\right|_{V_{GS0}, V_{DS0}} (\Delta V_{GS})^2 + \frac{1}{2}\left.\frac{\partial^2 I_{DS}(V_{GS}, V_{DS})}{\partial V_{DS}^2}\right|_{V_{GS0}, V_{DS0}} (\Delta V_{DS})^2 \\ &+ \left.\frac{\partial I_{DS}(V_{GS}, V_{DS})}{\partial V_{GS}} \frac{\partial I_{DS}(V_{GS}, V_{DS})}{\partial V_{DS}}\right|_{V_{GS0}, V_{DS0}} \Delta V_{GS} \Delta V_{DS} + \ldots \end{aligned} \tag{7.1}$$

Assuming a sufficiently small step width (ΔV_{GS}, ΔV_{DS}), all terms of higher order with $(\Delta V_{GS})^k$, $(\Delta V_{DS})^k$ ($k > 1$) can be neglected with respect to the first three terms. Then the following linear relation is valid

$$\begin{aligned} I_{DS}(V_{GS}, V_{DS}) &= I_{DS}(V_{GS0}, V_{DS0}) \\ &+ \left.\frac{\partial I_{DS}(V_{GS}, V_{DS})}{\partial V_{GS}}\right|_{V_{GS0}, V_{DS0}} \Delta V_{GS} + \left.\frac{\partial I_{DS}(V_{GS}, V_{DS})}{\partial V_{DS}}\right|_{V_{GS0}, V_{DS0}} \Delta V_{DS} \end{aligned} \tag{7.2}$$

with

$$\Delta V_{GS} = V_{GS} - V_{GS0} \tag{7.3}$$

$$\Delta V_{DS} = V_{DS} - V_{DS0} \tag{7.4}$$

Furthermore,

$$\Delta I_{DS}(\Delta V_{GS}, \Delta V_{DS}) = I_{DS}(V_{GS}, V_{DS}) - I_{DS0}(V_{GS0}, V_{DS0}) \tag{7.5}$$

Regarding small stimulating alternating voltages (harmonic signals) v_{GS} and v_{DS}, the alternating drain-source current can be written as

$$i_{DS}(V_{GS}, V_{DS}) = \left.\frac{\partial I_{DS}(V_{GS}, V_{DS})}{\partial V_{GS}}\right|_{V_{GS0}, V_{DS0}} v_{GS} + \left.\frac{\partial I_{DS}(V_{GS}, V_{DS})}{\partial V_{DS}}\right|_{V_{GS0}, V_{DS0}} v_{DS} \tag{7.6}$$

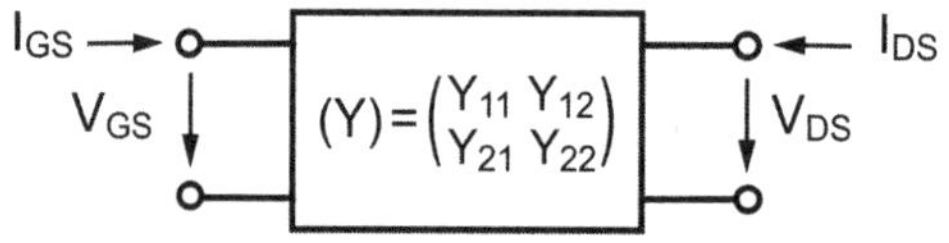

Figure 7.5 Small-signal two-port Y-matrix representation of a field-effect transistor.

The variation of the terminal drain current I_{DS} with terminal gate-source voltage V_{GS} is termed extrinsic (small-signal) transconductance G_M. The extrinsic (small-signal) output conductance G_{DS} describes the variation of the terminal drain current I_{DS} with terminal drain-source voltage V_{DS},

$$G_M = \left.\frac{\partial I_{DS}}{\partial V_{GS}}\right|_{V_{GS0},\,V_{DS0}} \tag{7.7}$$

$$G_{DS} = \left.\frac{\partial I_{DS}}{\partial V_{DS}}\right|_{V_{GS0},\,V_{DS0}} \tag{7.8}$$

For bipolar transistors the linearity region is well defined by the condition $V_{be} \ll kT/q$ ($\approx$ 25 mV at room temperature). A corresponding condition cannot be defined for field-effect transistors. For linear response of a FET device, it is roughly required that

$$|v_{GS}| \ll |-V_{GS0} + V_d| \tag{7.9}$$

$$|v_{DS}| \ll |V_{DS0}| \tag{7.10}$$

Note: In the following, for clarity, small-signal quantities will also be written with capital letters so that no confusion is possible with other quantities such like time-varying or large-signal quantities.

7.2.1 Two-Port Y-Matrix Transistor Model

We assume that the transistor is small-signal modulated in a fixed-bias point. In this case, the two-port can be characterized, for example, by the (extrinsic) small-signal Y-parameters as shown in Figure 7.5, which relate the extrinsic input and output currents to the input and extrinsic output voltages. Which kind of electrical circuit the black box exactly contains remains completely open. The terminal equations in terms of the terminal parameters are

$$\begin{aligned} I_{GS} &= Y_{11}V_{GS} + Y_{12}V_{DS} \\ I_{DS} &= Y_{21}V_{GS} + Y_{22}V_{DS} \end{aligned} \tag{7.11}$$

The Y-parameters are generic and independent of the details of the physical construction of the device. These model parameters are directly accessible by measurement. S-parameters are measured and converted into Y-parameters.

7.2.2 Generic Extrinsic Transistor Pi-Model

Figure 7.6 depicts the generic field-effect transistor pi-type equivalent model using the Y-matrix concept. The reference planes are the extrinsic gate-source and drain-source terminals.

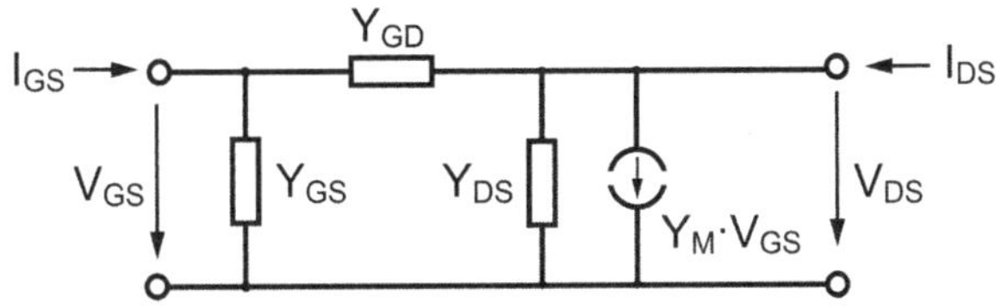

Figure 7.6 Generic FET pi-type equivalent network with extrinsic reference planes.

Y_{GS}, Y_{DS} and Y_{GD} are the extrinsic (generic) input, output, and feedback admittance parameters. Y_M denotes the extrinsic (generic) transconductance.

The extrinsic model parameters in Figure 7.6 are related to the two-port Y-parameters (Figure 7.5) as follows (see Appendix A):

$$\begin{aligned} Y_{GS} &= Y_{11} + Y_{12} \\ Y_{GD} &= -Y_{12} \\ Y_{DS} &= Y_{22} + Y_{12} \\ Y_M &= Y_{21} - Y_{12} \end{aligned} \tag{7.12}$$

7.3 TRANSISTOR MODEL COMPLEXITY

A variety of equivalent circuits have been published that describe the electrical behavior of microwave transistors (e.g. [1-7]). Frequently different device technologies, such as III-V semiconductor-based field-effect and heterobipolar as well as silicon-based lateral diffusion metal oxide semiconductor (LDMOS) transistors show strong similarities in their model structure. This, in particular, relates to the core of active part of the overall equivalent circuit.

Although the complexity of a model is not clearly defined, it may be identified with the number of model elements (parameters) in an equivalent circuit. In this way, the model complexity may be limited to a few (e.g., seven [1]) or may consists of a relatively large amount of model elements (e.g., more than 20 [2]).

The choice of model complexity depends on various conditions, such as layout of the device periphery and frequency of operation, and special technological features, such as the application of low-cost, large-size high-resistivity silicon (HRS) substrates.

At first, it is assumed that all model elements of the small-signal equivalent circuit are frequency-independent. However, the contribution of an element to the overall response may be very different under varying operation conditions. An inductive element is of minor importance if the operating frequency is very low. Its reactive impedance value is proportional to the frequency (i.e., at sufficiently low frequency inductances can be neglected). The opposite is true for capacitances. Their reactive impedance is inversely proportional to frequency, thus, at very low frequency, they behave like an open circuit. In contrast to inductive and capacitive model elements, resistances are purely real impedances and should be principally taken into account at any frequency. However, regarding the series connection with a capacitive element at low frequencies, the resistor can be mostly neglected as the resistive value of the series resistor is commonly much smaller than the reactive impedance of the capacitor at lower frequencies. Thus, although the device topology can be very complex, with respect to specific application it may be simplified under a given operation condition.

Another influencing factor is the biasing condition of the device; for example, under pinch-off operation the drain-current source vanishes. In forward-gate operation gate resistance and gate inductance predominate. Thus, with the appropriate choice of biasing and frequency, the complexity of circuit model topology can be reduced. This offers the possibility to determine model elements in a direct way from measurement without optimization (e.g., [4]). Of course, because all model parameters are coupled together, the individual determination of model elements based on different bias conditions, such as pinch-off, forward-gate, and saturation, may be in some way questionable. Nevertheless, this approach is very common and widely accepted. It is fast, simple, and traceable, and mostly useful for circuit simulation purposes. However, to what extent the determined parameter values are physically relevant remains unclear. Regardless, by focusing on the derivation of reliable physics-based equivalent circuits, this approach may be useful to generate good starting element values for an optimization-based model parameter extraction refinement process. Notice the small-signal equivalent circuits under discussion represent the fundamental building blocks for the development of nonlinear (large-signal) equivalent circuits.

It should be noted that with growing model complexity the model parameter extraction becomes more sophisticated and there is an increasing risk in the uncertainty of determined model parameter values. Therefore, for a given upper frequency of a limited measurement bandwidth only a limited number of model elements can be extracted consistently. This is discussed in the following Section 7.3, demonstrating the evolution of equivalent circuit complexity with increasing frequency (i.e., DC, quasi-DC, low, medium and high frequency). From a model parameter extraction point of view, it is needed to restrict the measurement bandwidth specifically for the determination of various model parameter (e.g., transconductance, channel conductance, capacitances, and inductances). Note that extracted values are not fixed but only regarded as estimated values for final overall circuit optimization.

Low-frequency determined well-estimated model element values may contribute to a more reliable optimization process. The search radius is reduced and the risk to be trapped in a local minimum is minimized. Therefore, the utmost goal should be to generate rather high-quality starting model element values for final optimization that already deliver good simulated results in agreement with measured scattering parameters. Even if this approach cannot completely eliminate the well-known local minimum problem, it makes the extraction procedure more reliable. More advanced model parameter extraction strategies generate high-quality starting element values that no longer need to be improved with regard to practical circuit design and simulation.

The transistor equivalent circuit should reflect the physical construction and physical behavior of the device under various operating conditions. For the general case, the topology of the equivalent circuit can become very complex. However, with lower operating frequency, the complexity decreases (i.e., the number of relevant model parameters to be taken into account is being reduced). In the following we will discuss the equivalent circuit complexity variation of a field-effect transistor with frequency for small- and large-periphery devices.

7.3.1 Small-Periphery Devices

Transistors can be grouped in small- and large-periphery devices. Although not clearly defined, we associate small-periphery devices with relatively simple gate layout structures. In the case of a maximum number of two transistor gate fingers, elaborate air-bridge interconnection technology is not necessary because it would introduce additional parasitic capacitive effects. Nevertheless, despite their structural differences, small-size devices may reveal similar complex model topologies at higher frequencies. In the following, model topologies with increasing frequency are presented and extraction of respective model parameters discussed.

7.3.1.1 Very Low-Frequency Topology

At very low frequency (quasi-DC, $f \approx 0$) the branch admittances Y_{GS} and Y_{GD} in Figure 7.6 are zero because of the capacitive input and capacitive feedback. When the device is not gate forward biased, which is normally given under small-signal operation, no gate current flows. As depicted in Figure 7.7(b), the circuit includes an extrinsic output conductance G_{DS} in shunt with the extrinsic transconductance G_M. The extrinsic (small-signal) output conductance G_{DS} describes the variation of the terminal drain current I_{DS} with terminal drain-source voltage V_{DS}, and the extrinsic (small-signal) transconductance G_M refers to the variation of the terminal drain current I_{DS} with terminal gate-source voltage V_{GS}.

Following the rules for calculation of the parameters, we obtain from (7.11)

$$Y_{11} = \left.\frac{I_{GS}}{V_{GS}}\right|_{V_{DS}=0} = 0 \qquad Y_{12} = \left.\frac{I_{GS}}{V_{DS}}\right|_{V_{GS}=0} = 0 \tag{7.13}$$

$$Y_{22} = \left.\frac{I_{DS}}{V_{DS}}\right|_{V_{GS}=0} = G_{DS} \qquad Y_{21} = \left.\frac{I_{DS}}{V_{GS}}\right|_{V_{DS}=0} = G_M$$

Therefore, (7.11) now reads

$$\begin{aligned} I_{GS}(V_{GS}, V_{DS}) &= 0 \\ I_{DS}(V_{GS}, V_{DS}) &= G_M V_{GS} + G_{DS} V_{DS} \end{aligned} \tag{7.14}$$

Referencing to the characteristic impedance Z_0, the normalized Y′-matrix reads

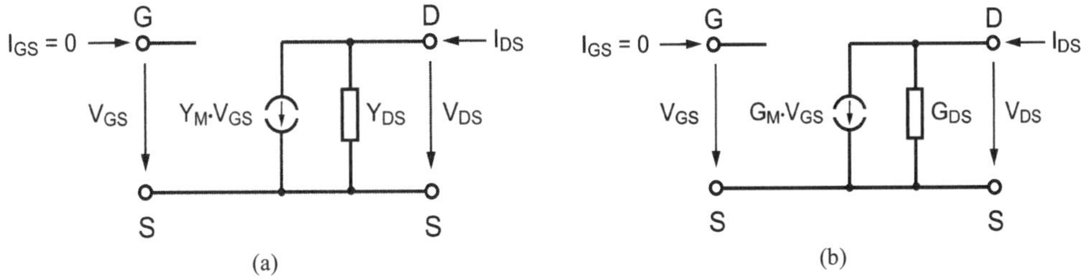

Figure 7.7 Quasi-DC field-effect transistor topology: (a) Generic Y-parameter and (b) generic model parameter representation (2-element model topology).

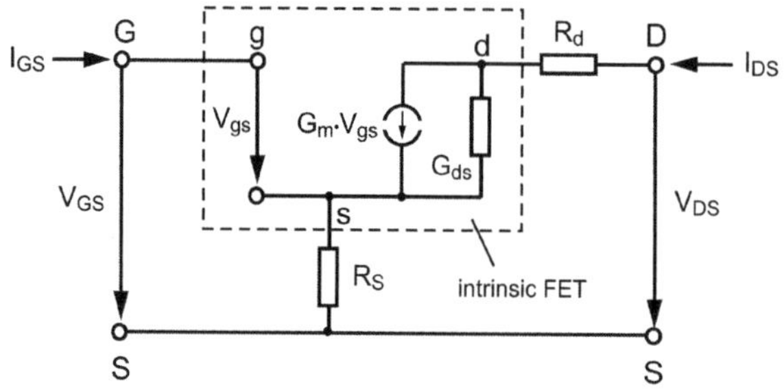

Figure 7.8 Physics-based equivalent circuit topology for quasi-DC operation with intrinsic transconductance G_m, intrinsic channel conductance G_{ds} and series resistances R_d and R_s (4-element model topology).

$$Y' = YZ_0 = \begin{pmatrix} 0 & 0 \\ G_M Z_0 & G_{DS} Z_0 \end{pmatrix} \tag{7.15}$$

Note the low-frequency extrinsic transconductance G_M and channel conductance G_{DS} are distinguished from their true DC counterpart due to trapping effects. G_M and G_{DS} are directly accessible by low-frequency measurement.

Under quasi-DC operation the FET device can be described by the reduced physics-based equivalent circuit in Figure 7.8. It comprises an intrinsic (active) part of the FET with intrinsic transconductance G_m and intrinsic channel conductance G_{ds}, and an extrinsic (passive) part with the parasitic series resistances R_d and R_s. Knowing the resistances (e.g., from physics-based estimation or separate DC measurements) the intrinsic elements can be approximated.

7.3.1.2 Low-Frequency Topology

For low frequencies the topology of the intrinsic equivalent circuit is extended by the gate-source capacitance C_{gs}, the gate-drain capacitance C_{gd}, and the drain-source capacitance C_{ds}, where C_{gs} predominates (Figure 7.9). The extrinsic part is expanded by the pad capacitances C_{pg} and C_{pd}. Inductive effects are not yet considered. R_i denotes the gate-source capacitance charging resistance.

The pad capacitances are commonly determined from cold pinch-off measurement.

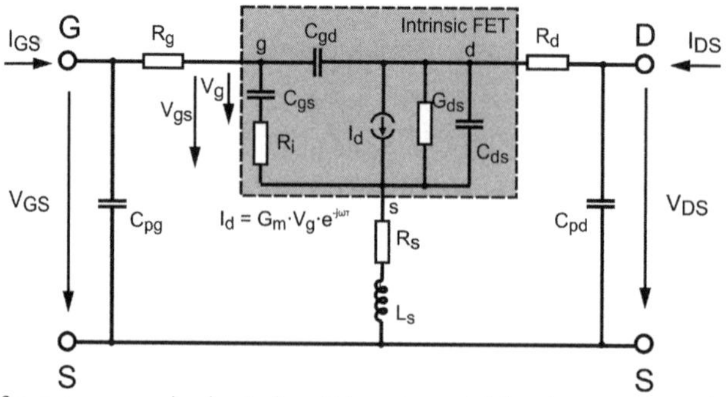

Figure 7.9 Low frequency equivalent circuit topology taking into account intrinsic capacitive effects (11-element model topology).

7.3.1.3 Medium-Frequency Topology

At medium frequencies the series gate, drain, and source inductances become effective as depicted in Figure 7.1. Larger inductance values occur when dies are electrically connected with bondwires. Lowering of the inductance value can be obtained by multibonding as is commonly done with the source electrode connection to ground.

Inductance values can be determined from cold pinch-off S-parameter measurements converted to Y-parameters. After de-embedding known pad capacitances the inductance values can be determined by a curve-fitting procedure.

7.3.1.4 High-Frequency Topology

Standard 15-element model (Figure 7.1) is applied over a wide range of frequencies. Its limitation will become noticeable at very high frequencies where additional parasitic effects become relevant; for example, distributed effects (nonuniform voltage distribution along gate width) or capacitive coupling and stray effects.

7.3.1.5 Very High-Frequency Topology

At very high frequencies additional capacitive effects must be taken into account as shown in Figure 7.2 [2]. Parasitic capacitances, which are independent of bias voltages, exist (C_{pgi}, C_{pdi}, C_{gdi}), and between an electrode and the ground plane (C_{pga}, C_{pda}), where C_{pgi}, C_{pdi}, C_{gdi}, C_{pga}, C_{pda} are gate-to-source, drain-to-source, gate-to-drain, gate-to-ground, and drain-to-ground parasitic capacitances. C_{gda} describes the capacitance between the extrinsic terminals of the gate and drain electrodes. It has been validated that the distributed model topology is appropriate to describe the device electrical behavior of small-periphery devices up to very high frequencies around 100 GHz and above. The parasitic capacitances C_{gda} and C_{gdi} are commonly very small and may be possibly neglected. Most of the published FET model variations can simply be derived from the proposed one by excluding appropriate elements.

As the total branch capacitances could be determined rather reliably from low-frequency measurements (see Section 7.3.1.2), it is a great challenge to distribute this value to the partial branch capacitances. This is a key topic in Chapter 21.

7.3.2 Large-Periphery Devices

The maximum output power of a field-effect transistor depends on the breakdown voltage and maximum channel current. Field-plate technology has strongly improved the breakdown behavior of power devices [8]. As already discussed in Chapter 4, the channel current of a FET device can be increased by increasing the width of the gate. However, due to wave propagation effects on a single-gate finger, the gate width is limited so that multifinger structures must be taken. When the number of fingers exceeds two, then the metallization pattern of the device becomes more complex. Air bridges and/or via holes have to be taken into account. From the author's experience such device structures cannot be reliably modeled on the basis of simplified models such as the standard 15-element equivalent circuit.

Thus, the distributed model in Figure 7.2 adequately describes the properties of large gate-periphery of power devices including interconnections of source fingers (source air bridges) and capacitive interelectrode effects.

The author and his group have studied 0.5-μm gate AlGaN/GaN HEMTs with large gate periphery for microwave power application (UMTS, WiMAX). Power devices with total gate widths up to 4 mm and multigates up to 16 fingers have been investigated, fabricated by the Fraunhofer Institute for Applied Solid-State Physics (IAF), Freiburg I. Br., and Ferdinand-Braun-Institute (FBH), Berlin. In all cases the distributed transistor model has provided excellent small-signal device models as reported in several research works (e.g., [9-12]) and publications, and based on that, proven large-signal HEMT device models.

7.3.3 High-Resistivity Silicon Substrates

In previous discussions it was assumed that the substrate is nearly perfect (i.e., the substrate is lossless). However, successful market introduction of microwave circuits depends heavily on the production cost. Materials that provide superior microwave performance for the realization of active devices and passive elements are typically very expensive. Instead, silicon has many advantages as a substrate material including low cost, stable material properties, high thermal conductivity, and is a proven mature technology. In addition, the wafer size of Si substrates is generally larger than that of a compound semiconductor. However, the lower resistivity of Si compared to compound semiconductor substrates for microwave applications is perceived as a major disadvantage. Nevertheless, silicon substrates are very interesting in view of future integration between compound semiconductor and silicon technologies.

7.3.3.1 Combining III-V and Silicon Technologies

The combination of III-V compound semiconductor and silicon materials encounters some severe technological difficulties. This particularly concerns the large lattice and thermal mismatch. Silicon has a lattice mismatch with GaAs (GaAs/Si substrate) of approximately +4%. GaN provides a much larger lattice mismatch of about −20%. GaAs active layers on silicon substrate, Si (100), have the advantage of dissipating higher power than those on GaAs substrates due to the three times higher thermal conductivity of silicon [8]. Regarding GaN material, the thermal mismatch between epilayer and substrate is rather high, estimated between +54% [8] and +77% [13].

Commonly, GaN-based devices are grown on sapphire (Al_2O_3) or silicon carbide (SiC) substrates. These are either thermally insulating or very expensive and not available in large diameter. Si is available from high *p*- and *n*-type conductive (0.01 Ω-cm) up to highly resistive material with around 10^4 Ω-cm. In comparison with the intrinsic resistivity of GaN, sapphire or SiC ($\gg 10^9$ Ω-cm) this is a rather low value. It can lead to high parasitic capacitances that are detrimental for high-frequency operation of devices. High dislocation density of the epitaxial nitride layer on silicon can also induce leakage current. It should be mentioned that continuous technological improvement of III-V-on-Si systems has been reported. This concerns, for example the insertion of thermal and lattice mismatch absorption transition layers or superlattice buffer between GaN epitaxial layer and Si substrate before

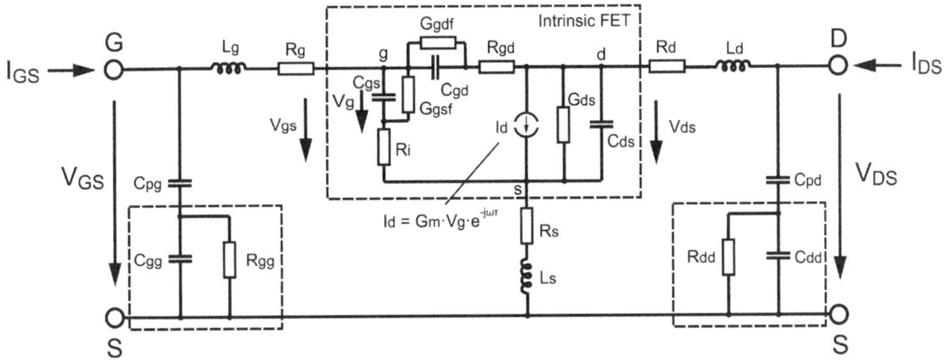

Figure 7.10 Small-signal equivalent circuit model for active GaN HEMT on Si including buffer-related parasitic conduction effect [22].

epitaxial growth of GaN layer [8]. Silicon on insulator (SOI) technology comprises a semiconductor structure that consists of a layer of single crystalline silicon separated from the bulk substrate by a thin layer of insulator, typically silicon dioxide (SiO_2).

7.3.3.2 HEMT-on-Si Equivalent Circuit Topology

Standard lumped element [14] and distributed [2, 15] equivalent circuit small-signal models were used successfully for simulating the RF small-signal measurements of GaN HEMT-on-SiC even for large devices [16]. With respect to this device, GaN HEMT-on-Si shows different RF characteristics; this applies in particular to the real part of the measured low-frequency Y-parameters under pinch-off bias condition [17]. This accordingly limits the capability of the standard model to accurately simulate *S*-parameter measurements for this device especially in the low-frequency range, and was also observed in [18] for the same device. This phenomenon was also previously reported for GaAs HEMT on Si substrate with respect
to a corresponding one fabricated on GaAs bulk wafers [19]. It can be related to a parasitic condition through the buffer (subbuffer) layer, the buffer-substrate interface, or the substrate [20–21]. This can be attributed to a defect (lattice match induced defect) assisted conduction mechanism in the buffer [21]. Figure 7.10 [22] indicates the extension of a common FET device equivalent circuit taking into account the additional parasitic effects mainly influenced by the pads. The capacitors C_{gg} and C_{dd} represent capacitances between the pads and the Si substrate, and the resistors R_{gg} and R_{dd} represent resistances due to the higher electrical conductivity of the Si substrate than the GaAs bulk [19]. However, these RC networks are not required if the conductivity of the silicon is higher than 10 kΩ-cm [20].

Small-signal equivalent circuit model including the buffer parasitic conduction was already proposed for GaN HEMT on sapphire substrate [17]. In this model, the additional extrinsic elements, which describe this parasitic effect, were determined from the measured S-parameters of an open dummy structure fabricated on the same wafer. However, as has been reported in [22], the elements of the parasitic transversal RC network can also be extracted from cold S-parameter measurements without the need for additional measurements or a separate test pattern.

7.4 SUMMARY

Small-signal device operation has been defined and characterized by a generic black-box two-port matrix model described by Y-parameters determined from measurement. A pi-type physics-based equivalent network with extrinsic terminals has been presented with network parameters directly calculated from two-port parameters.

The term model complexity was introduced denoting the number of model elements in an equivalent circuit. It was noted that transistors with large-size gate periphery with strong interelectrode capacitive effects comprise a large number of model elements, which is far beyond the usual scope. This also holds for small-size devices at very high-frequency operation.

Measurements were discussed to extract relevant extrinsic model parameters that can be used to constrict the search space of optimization-based extraction procedures.

Beyond the common substrates for AlGaN/GaN power devices, an alternative solution was discussed. Recent development of devices on silicon substrate was illuminated and their technological issues arising from thermal and lattice mismatch, were highlighted. It was pointed out, that high resistivity silicon used as substrate is relatively conductive in comparison with compound semiconductors. The additional parasitic effects provide an extension of the common transistor equivalent circuit in form of a transverse RC network. Measures were briefly addressed to improve the performance of III-V-on-Si technology.

REFERENCES

[1] Tasker, P. J., and J. Braunstein, "New MODFET Small Signal Circuit Model Required for Millimeter-Wave MMIC Design: Extraction and Validation to 120 GHz," *IEEE MTT-S Int. Microwave Symp. Digest*, 1995, pp. 611–614.

[2] Kompa, G., "Reliable Extraction of Small-Signal Elements of a Generalized Distributed FET Model," *IEEE MTT-S Int. Microwave Symp. Dig.*, Baltimore, USA, 1998, pp. 291–294.

[3] Müller, J.-E.,"Investigation of GaAs MESFET Small Signal Equivalent Circuits for Use in a Cell Library," *Proc. 19th European Microwave Conf.*, London, 1989, pp. 991-996.

[4] Dambrine, G., A. Cappy, F. Heliodore, and E. Playez, "A New Method for Determining the FET Small-Signal Equivalent Circuit," *IEEE Trans.*, Vol. MTT-36, July 1988, pp. 1151–1159.

[5] Hughes, B., and P. J. Tasker, "Bias Dependence of the MODFET Intrinsic Model Elements Values at Microwave Frequencies," *IEEE Trans.*, Vol. ED-36, 1989, pp. 2267–2273.

[6] Bandler, J. W., S. H. Chen, S. Ye, and S. Daijavad, "Integrated Model Parameter Extraction Using Large-Scale Optimization Concepts," *IEEE Trans.*, Vol. MTT-36, 1988, pp. 1629–1638.

[7] Berroth, M, and R. Bosch, "High-Frequency Equivalent Circuit of GaAs FET's for Large-Signal Applications," *IEEE Trans.*, Vol. MTT-39, 1991, pp. 224–229.

[8] Kompa, G., *Basic Properties of III-V Devices–Understanding Mysterious Trapping Phenomena*, Kassel, Kassel University Press GmbH, 2014.

[9] Mwema, W. N., *A Reliable Optimisation-Based Model Parameter Extraction*, Doctoral Thesis, Department of High Frequency Engineering, University of Kassel, 2002, Kassel: Kassel University Press GmbH, 2002.

[10] Jarndal, A. H., *Large-Signal Modeling of GaN Device for High Power Amplifier Design*, Doctoral Thesis, Department of High Frequency Engineering, University of Kassel, 2006, Kassel: Kassel University Press GmbH, 2007.

[11] Srinidhi, E. R., *Intermodulation Distortion Modelling and Measurement Techniques for GaN HEMT Characterization*, Doctoral Thesis, Department of High Frequency Engineering, University of Kassel, 2008, Kassel: Kassel University Press GmbH, 2009.

[12] Zamudio Flores, J. A., *Device Characterization and Modelling of Large-Size GaN HEMTs*, Doctoral Thesis, Department of High Frequency Engineering, University of Kassel, 2012, Kassel: Kassel University Press GmbH, 2012.

[13] Mion, C., *Investigation of the Thermal Properties of Gallium Nitride Using the Three Omega Technique*, Doctoral Thesis, North Carolina State University, 2005.

[14] Brady, R. G., C. H. Oxley, and T. J. Brazil, "An Improved Small-Signal Parameter-Extraction Algorithm for GaN HEMT Devices," *IEEE Trans.*, Vol. MTT-56(7), 2008, pp. 1535–1544.

[15] Kompa, G., and M. Novotny, "Frequency-Dependent Measurement Error Analysis and Refined FET Model Parameter Extraction Including Bias-Dependent Series Resistors," *IEEE Workshop on Experimentally Based FET Device Modelling & Related Nonlinear Circuit Design*, Kassel, Germany, July 17-18, 1997, pp. 6.1–6.16, https://archive.org/details/DTIC_ADA331468/page/n279.

[16] Jarndal, A., and G. Kompa, "An accurate Small-Signal Model for AlGaN-GaN HEMT Suitable for Scalable Large-Signal Model Construction," *IEEE Microwave and Wireless Components Lett.*, Vol. 16 (6), 2006, pp. 333–335.

[17] Crupi, G., D. Xiao, D. M. M.-P. Schreurs, E. Limiti, A. Caddemi, W. De Raedt, and M. Germain, "Accurate Multibias Equivalent-Circuit Extraction for GaN HEMTs," *IEEE Trans.*, Vol. MTT-54(10), 2006, pp. 3616–3622.

[18] Markos, A. Z., *Efficiency Enhancement of Linear GaN RF Power Amplifiers Using Doherty Technique*, Doctoral Thesis, University of Kassel, 2008.

[19] Goto, M., Y. Ohta, T. Aigo, and A. Moritani, "A Small-Signal Linear Equivalent Circuit of HEMT's Fabricated on GaAs-on-Si Wafers," *IEEE Trans.*, Vol. MTT-44(5), 1996, pp. 668–673.]

[20] Chumbes, E., A. Schremer, J. Smart, Y. Wang, N. MacDonald, D. Hogue, J. Komiak, S. Lichwalla, R. Leoni, and J. Shealy, "AlGaN/GaN High Electron Mobility Transistors on Si(111) Substrates," *IEEE Trans.*, Vol. ED-48, 2001, pp. 420–426.

[21] Kuzmik, J., M. Blaho, D. Pogany, E. Gornik, A. Alam, Y. Dikme, M. Heuken, P. Javorka, M. Marso, and P. Kordoš, "Backgating, High-Current and Breakdown Characterisation of AlGaN/GaN HEMTs on Silicon Substrates," *European Solid-State Research Conference (ESSRC)*, Estoril, Portugal, 2003, pp. 319–322.

[22] Jarndal, A., A. Z. Markos, and G. Kompa, "Improved Parameter Extraction Method for GaN HEMT on Si Substrate," *IEEE MTT-S Int. Microwave Symp. Dig.*, Jun. 2010, pp. 1668–1671.

Chapter 8

Reliable Parameter Estimates from Low-Frequency Measurements

8.1 INTRODUCTION

In the previous chapter discussed the variation of topology complexity of a generalized field-effect transistor model with frequency. Now we want to investigate which type of model parameter information can be obtained from low-frequency measurements and in which way it can support the model parameter extraction process. We will see that reliable estimates for some relevant elements can be derived, which can serve as very good starting values in a model parameter optimization process.

8.2 DETERMINATION OF GENERIC PI-MODEL PARAMETERS

From low-frequency *S*-parameter measurements generic pi-model parameters can be accurately determined; that is, the extrinsic transconductance G_M, extrinsic output conductance G_{DS}, extrinsic gate-source capacitance C_{GS}, extrinsic drain-source capacitance C_{DS}, and extrinsic gate-drain capacitance C_{GD} (Figure 8.1). Then, well-approximated physics-based model element values of the intrinsic transistor model can be derived from these generic model parameters introducing appropriate approximations.

8.2.1 Generic Transconductance and Output Conductance

We start with the simple quasi-DC model topology in Figure 7.8, which is valid at very low frequency. Applying mesh and node rules after Kirchhoff, we get

$$V_{gs} = V_{GS} - I_{DS}R_s \tag{8.1}$$

$$V_{ds} = -I_{DS}(R_d + R_s) + V_{DS} \tag{8.2}$$

$$V_{ds} = G_{ds}^{-1}(I_{DS} - G_m V_{gs}) \tag{8.3}$$

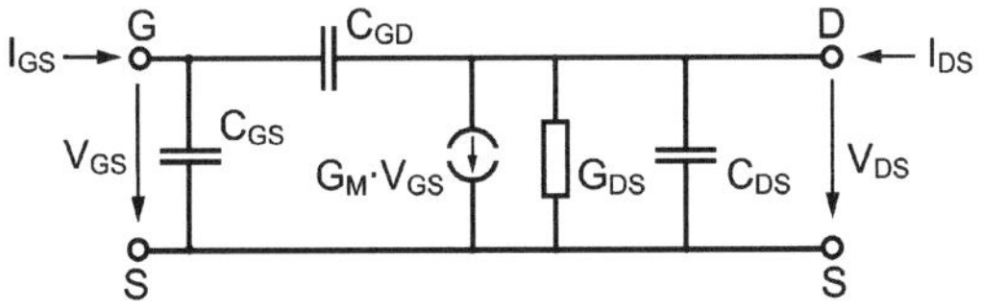

Figure 8.1 Low-frequency generic pi-model of a field-effect transistor.

Combining (8.2) and (8.3) yields

$$I_{DS} = \frac{G_m}{1 + G_m R_s + G_{ds}(R_s + R_d)} V_{GS} + \frac{G_{ds}}{1 + G_m R_s + G_{ds}(R_s + R_d)} V_{DS} \tag{8.4}$$

Comparing with (7.14), we identify

$$G_M = \frac{G_m}{(1 + D)} \tag{8.5}$$

$$G_{DS} = \frac{G_{ds}}{(1 + D)} \tag{8.6}$$

with

$$D = G_m R_s + G_{ds}(R_s + R_d) \tag{8.7}$$

It is seen that G_M gets degraded with high R_s and R_d and high G_{ds}, while G_{DS} gets improved with high R_s and R_d and high G_m. The resistances consist of ohmic contact resistance and active layer resistance between gate and source (for R_s) and between gate and drain (for R_d), which have been treated in Section 6.2.2 and Chapter 7. Taking the ratio of (8.5) and (8.6), we find that the ratio of the extrinsic transconductance and output conductance equals the intrinsic physics-based ones,

$$\frac{G_M}{G_{DS}} = \frac{G_m}{G_{ds}} \tag{8.8}$$

Similar to (7.15) we get for the normalized extrinsic Y'-parameters as function of intrinsic physics-based model parameters

$$Y' = YZ_0 = \begin{pmatrix} 0 & 0 \\ \dfrac{G_m Z_0}{1 + D} & \dfrac{G_{ds} Z_0}{1 + D} \end{pmatrix} \tag{8.9}$$

Conversion of the Y'-parameters to S-parameters results in

$$\begin{aligned} S_{11} &= (1 - Y'_{11} + Y'_{22} - \det Y')/N \\ S_{22} &= (1 + Y'_{11} - Y'_{22} - \det Y')/N \\ S_{12} &= -Y_{12}/N \\ S_{21} &= -Y_{21}/N \end{aligned} \tag{8.10}$$

where

$$N = 1 + Y'_{11} + Y'_{22} + \det Y' \tag{8.11}$$

With $\det(Y') = 0$ in (8.9) and $N = 1 + G_{DS}Z_0$ we obtain the following low-frequency S-matrix

$$S(f \to 0) = \begin{pmatrix} 1 & 0 \\ -\dfrac{2G_M Z_0}{1 + G_{DS}Z_0} & \dfrac{1 - G_{DS}Z_0}{1 + G_{DS}Z_0} \end{pmatrix} = \begin{pmatrix} S_{11} & S_{12} \\ S_{21} & S_{22} \end{pmatrix} \tag{8.12}$$

Then the extrinsic transconductance and extrinsic output conductance can be determined from measured scattering parameters as

$$S_{21} = -\frac{2G_M Z_0}{1 + G_{DS} Z_0} \tag{8.13}$$

$$S_{22} = \frac{1 - G_{DS} Z_0}{1 + G_{DS} Z_0} \tag{8.14}$$

Solving for the extrinsic parameters G_M and G_{DS}, we get

$$G_M = Z_0^{-1} \frac{|S_{21}|}{1 + |S_{22}|} \tag{8.15}$$

$$G_{DS} = Z_0^{-1} \frac{1 - |S_{22}|}{1 + |S_{22}|} \tag{8.16}$$

Taking the ratio of (8.15) and (8.16) we get with (8.8)

$$\frac{G_M}{G_{DS}} = \frac{G_m}{G_{ds}} = \frac{|S_{21}|}{1 - |S_{22}|} \tag{8.17}$$

Thus, the measured S-parameters at very low frequencies ($f \lesssim 200$ MHz, no appreciable trapping effects) deliver reliable information about the ratio of physics-based intrinsic transconductance and intrinsic output conductance.

8.2.2 Generic Capacitances

The extrinsic admittance parameters Y in (8.9) can be expressed in terms of the intrinsic physics-based parameters. At low frequencies, following the equivalent circuit topology in Figure 7.9, the admittance parameters of the physics-based equivalent circuit topology can be written in good approximation as follows [1]:

$$Y_{11} \approx j\omega\left[C_{gs}\left(1 - \frac{G_m R_s}{1 + D}\right) + C_{gd}\left(1 + \frac{G_m R_d}{1 + D}\right)\right] \tag{8.18}$$

$$Y_{12} \approx -j\omega\left[C_{gd}\left(1 - \frac{G_{ds} R_d}{1 + D}\right) + C_{gs}\left(1 + \frac{G_{ds} R_s}{1 + D}\right)\right] \tag{8.19}$$

$$Y_{21} \approx \frac{G_m}{1 + D} - j\frac{\omega}{1 + D}\left\{C_{gd}\left[1 + G_m(R_s + R_g) + G_{ds}R_s\right]\left(1 + \frac{G_m R_d}{1 + D}\right)\right. \tag{8.20}$$

$$\left. + C_{gs}\left[G_m(R_s + R_g) + G_{ds}R_s\right]\left(1 - \frac{G_m R_s}{1 + D}\right) + C_{ds}\frac{G_m(R_s + R_d)}{1 + D}\right\}$$

$$Y_{22} \approx \frac{G_d}{1 + D} + j\frac{\omega}{1 + D}\left\{C_{gd}\left[1 + G_m(R_s + R_g) + G_{ds}R_s\right]\left(1 - \frac{G_{ds} R_d}{1 + D}\right)\right. \tag{8.21}$$

$$\left. + C_{gs}R_s G_{ds}\frac{G_{ds}R_s + G_m(R_s + R_g)}{1 + D} + C_{ds}\left(1 - \frac{G_{ds}(R_s + R_d)}{1 + D}\right)\right\}$$

where

$$D = G_m R_s + G_{ds}(R_s + R_d) \tag{8.22}$$

It is seen that the model parameters cannot be directly extracted from measured Y- or S-parameters. As Y_{11} and Y_{12} are purely imaginary, only six equations are available for the determination of nine physics-based model elements. However, the physics-based equivalent circuit in Figure 7.9 with extrinsic terminals can be equivalently described by the generic pi-model topology shown in Figure 8.1. In this case, the extrinsic Y-parameters read

$$\begin{aligned} Y_{11} &= j\omega(C_{GS} + C_{GD}) \\ Y_{12} &= -j\omega C_{GD} \\ Y_{21} &= G_M - j\omega C_{GD} \\ Y_{22} &= G_{DS} + j\omega(C_{DS} + C_{GD}) \end{aligned} \tag{8.23}$$

Solving (8.23) for the extrinsic (generic) model elements yields

$$\begin{aligned} C_{GS} &= 1/\omega\cdot(\mathrm{Im}(Y_{11} + \mathrm{Im}(Y_{12})) \\ C_{GD} &= -1/\omega\cdot\mathrm{Im}(Y_{12}) \\ C_{DS} &= 1/\omega\cdot(\mathrm{Im}(Y_{22} + \mathrm{Im}(Y_{12})) \\ G_{DS} &= \mathrm{Re}(Y_{22}) + \mathrm{Re}(Y_{12}) \\ G_M &= \mathrm{Re}(Y_{21}) - \mathrm{Re}(Y_{12}) \end{aligned} \tag{8.24}$$

With (8.10) the following relations can be derived for the extrinsic model parameters:

$$C_{GD} = Z_0^{-1}\frac{|S_{12}|}{2\omega}(1 + G_{DS}Z_0) \tag{8.25}$$

$$C_{GS} = -\frac{\varphi_{11}}{2\omega Z_0} - C_{GD}\frac{1 + (G_M + G_{DS})Z_0}{1 + G_{DS}Z_0} \tag{8.26}$$

$$C_{DS} = -\frac{\varphi_{22}}{2\omega Z_0}\left[1 - (G_{DS}Z_0)^2\right] - C_{GD}(1 + G_M Z_0) \tag{8.27}$$

8.3 RELATIONS BETWEEN GENERIC AND PHYSICS-BASED PARAMETERS

The following relations between the physics-based model elements in Figure 7.9 and those of the generic two-port pi-model in Figure 8.1 can be derived:

$$G_M = \frac{G_m}{1 + D} \tag{8.28}$$

$$G_{DS} = \frac{G_{ds}}{1 + D} \tag{8.29}$$

$$C_{GS} = C_{gs}\frac{1 + R_d G_{ds}}{1 + D} + C_{gd}\frac{R_d(G_m + G_{ds})}{1 + D} \tag{8.30}$$

$$C_{GD} = C_{gs}\frac{R_s G_{ds}}{1 + D} + C_{gd}\left(1 - \frac{R_d G_{ds}}{1 + D}\right) \tag{8.31}$$

$$\begin{aligned} C_{DS} = {} & C_{gd}\left(\frac{G_m R_g}{1 + D} - R_d G_{ds}\frac{1 + G_m(R_s + R_g) + G_{ds}R_s}{(1 + D)^2}\right) \\ & + C_{gs}\left(R_s G_{ds}\frac{G_{ds}R_s + G_m(R_s + R_g + R_i)}{(1 + D)^2} - \frac{G_{ds}R_s}{1 + D}\right) \\ & + C_{ds}\left(\frac{1}{1 + D} - \frac{G_{ds}(R_s + R_d)}{(1 + D)^2}\right) \end{aligned} \tag{8.32}$$

with

$$D = G_m R_s + G_{ds}(R_s + R_d)$$

Equations (8.28) and (8.29) correspond to (8.5) and (8.6).

8.4 APPROXIMATE DETERMINATION OF PHYSICS-BASED INTRINSIC ELEMENTS FROM GENERIC MODEL PARAMETERS

In the following approximation equations concerning the calculation of physics-based model parameters from generic model parameters measured at low-frequency are derived [2].

8.4.1 Output Conductance G_{ds}

Commonly, the intrinsic output conductance is much smaller than the transconductance. Furthermore, the drain resistance is assumed to be of minor importance. Then, with

$$R_d \cdot G_{ds} \ll R_s(G_m + G_{ds}) \quad (\text{or } R_d = 0) \tag{8.33}$$

the extrinsic output conductance G_{DS} in (8.29) can be approximated as

$$G_{DS} \approx \frac{G_{ds}}{1 + R_s(G_m + G_{ds})} \tag{8.34}$$

Solving for intrinsic output conductance G_{ds} gives

$$G_{ds} \approx G_{DS}[1 + R_s(G_m + G_{ds})] \tag{8.35}$$

With (8.17),

$$G_m = G_{ds} \cdot G_M / G_{DS} \tag{8.36}$$

it follows

$$G_{ds} \approx \frac{G_{DS}}{1 - R_s G_{DS}\left(1 + G_M / G_{DS}\right)} \tag{8.37}$$

8.4.2 Transconductance G_m

Similarly, an approximated expression for the intrinsic transconductance G_m is found to be

$$G_m \approx \frac{G_M}{1 - R_s G_M \left(1 + G_{DS}/G_M\right)} \tag{8.38}$$

8.4.3 Gate-Source Capacitance C_{gs}

The intrinsic gate-source capacitance C_{gs} is related to the extrinsic model parameters as follows. Using the assumption in (8.33), we get from (8.29)

$$C_{GS} \approx \frac{C_{gs}}{1 + D} = C_{gs} \frac{G_{DS}}{G_{ds}} \tag{8.39}$$

Solved for C_{gs}, delivers

$$C_{gs} \approx G_{ds} \frac{C_{GS}}{G_{DS}} \tag{8.40}$$

8.4.4 Gate-Drain Capacitance C_{gd}

Starting with (8.31) and regarding (8.33) the extrinsic capacitance C_{GD} can be approximated as

$$C_{GD} \approx C_{gd} + C_{gs} \frac{G_{ds} R_s}{1 + D'} \tag{8.41}$$

Solving (8.41) for intrinsic gate-drain capacitance C_{gs} results in

$$C_{gd} \approx C_{GD} - C_{gs} \frac{G_{ds} R_s}{1 + D'} \tag{8.42}$$

with

$$D' \approx R_s (G_m + G_{ds}) \tag{8.43}$$

8.4.5 Drain-Source Capacitance C_{ds}

As the mathematical derivation of an approximation equation for C_{ds} as a function of extrinsic drain-source capacitance C_{DS} and further intrinsic model parameters is more extensive, only the final result for the approximated drain-source capacitance C_{ds} is given here:

$$C_{ds} \approx \left\{ C_{DS} - C_{gd} \frac{G_m R_g}{1 + D'} - C_{gs} \left[R_s G_{ds} \frac{R_s G_{ds} + G_m (R_s + R_g + R_i)}{(1 + D')^2} - \frac{G_{ds} R_s}{1 + D'} \right] \right\} \times \\ \times \left[\frac{1}{1 + D'} \left(1 - \frac{G_{ds} R_s}{1 + D'} \right) \right]^{-1} \tag{8.44}$$

with D' in (8.43).

It should be mentioned that the derived relations between the extrinsic and intrinsic model parameters are of general interest for the model parameter extraction process. Principally they can be applied to any of the discussed equivalent circuit topologies in Chapter 7. Regarding the highly complex network topology in Figure 7.2, at lower frequencies the respective distributed gate-source, drain-source, and gate-drain branch capacitances are merging into a single effective branch capacitances. Then, knowing these total branch capacitances, it will remain the great challenging task to determine the individual physics-based capacitive model elements. This will be thoroughly treated in Chapter 21.

8.5 ESTIMATION OF PHYSICS-BASED PARAMETERS FROM LOW-FREQUENCY S-PARAMETERS

The effectiveness of the low-frequency parameter extraction strategy will be demonstrated by the following example. Figure 8.2 shows low-frequency measured S-parameters of an AlGaN/GaN HEMT with the following data: Gate-length $L = 0.5$mm, gate-width $w = 1$ mm, gate-source spacing l_{gs} = 0.8 μm, gate-drain spacing l_{gd} = 1.6 μm (estimated from micrograph). The device was operated at $V_{GS} = -3$V and $V_{DS} = 19.8$V.

The extrinsic model parameters from the extrinsic transistor can be directly determined from measured scattering parameters. From the measuring diagrams the magnitudes of S_{21} and S_{22} read 9.868 and 0.778, respectively. Then, with (8.15) and (8.16) the extrinsic conductances are calculated as $G_M = 110$ mS and $G_{DS} = 2.5$ mS. For the estimation of C_{GD}, we determine first the slope of measured $S_{12}(f)$, which is 1.59×10^{-11} s. Then with (8.25) we find $C_{GD} = 0.18$ pF.

For C_{GS} computation, we first determine the slope of the phase of S_{11}, which reads $\varphi_{11}/\omega = 2.04 \times 10^{-10}$ s. Then according to (8.26) we calculate C_{GS} with all the previously derived parameters, $C_{GS} = 0.98$ pF. C_{DS} is derived from (9.27). With the measured slope of the phase of $S_{22}(f)$, $\varphi_{22}/\omega = 1.39 \times 10^{-10}$ s, we get $C_{DS} = 0.2$ pF.

Next step is the determination of intrinsic (physics-based) model parameters from extracted extrinsic (generic) model parameters. For this we make use of the presented approximation relations between generic and intrinsic model parameters. However, additional information must be given for the series resistances R_s and R_d and the gate resistance R_g. Estimates can be obtained from Section 6.2.2 on the basis of physical data of the transistor or on the basis of direct measurements (see Appendix B).

With typical contact resistance of 0.5Ω-mm per unit length for GaN HEMTs we estimate a contact resistance $R_{sc} = 0.5\Omega$ for the given gate-width (see Ssection 6.2.1). Assuming a typical bulk sheet resistance of 500Ω/sq we find with the given data for the bulk resistance $R_{s,bulk} = 0.4\Omega$ (see Section 6.2.2). Thus, in total, the source resistance is approximated to be $R_s = R_{sc} + R_{s,bulk} = 0.9\Omega$. Accordingly, the drain resistance is found to be $R_d = R_{dc} + R_{d,bulk} = 0.5\Omega + 0.8\Omega = 1.3\Omega$.

With the R_s value the intrinsic transconductance G_m is determined from (8.38) to be 123 mS. Using (8.8) the intrinsic channel conductance G_{ds} reads 2.8 mS. From (8.40) C_{gs} turns out to be 1.1 pF. Applying (8.42) we get $C_{gd} = 0.17$ pF.

Comparing with the results in [3] and [4] extracting for the same type of AlGaN/GaN

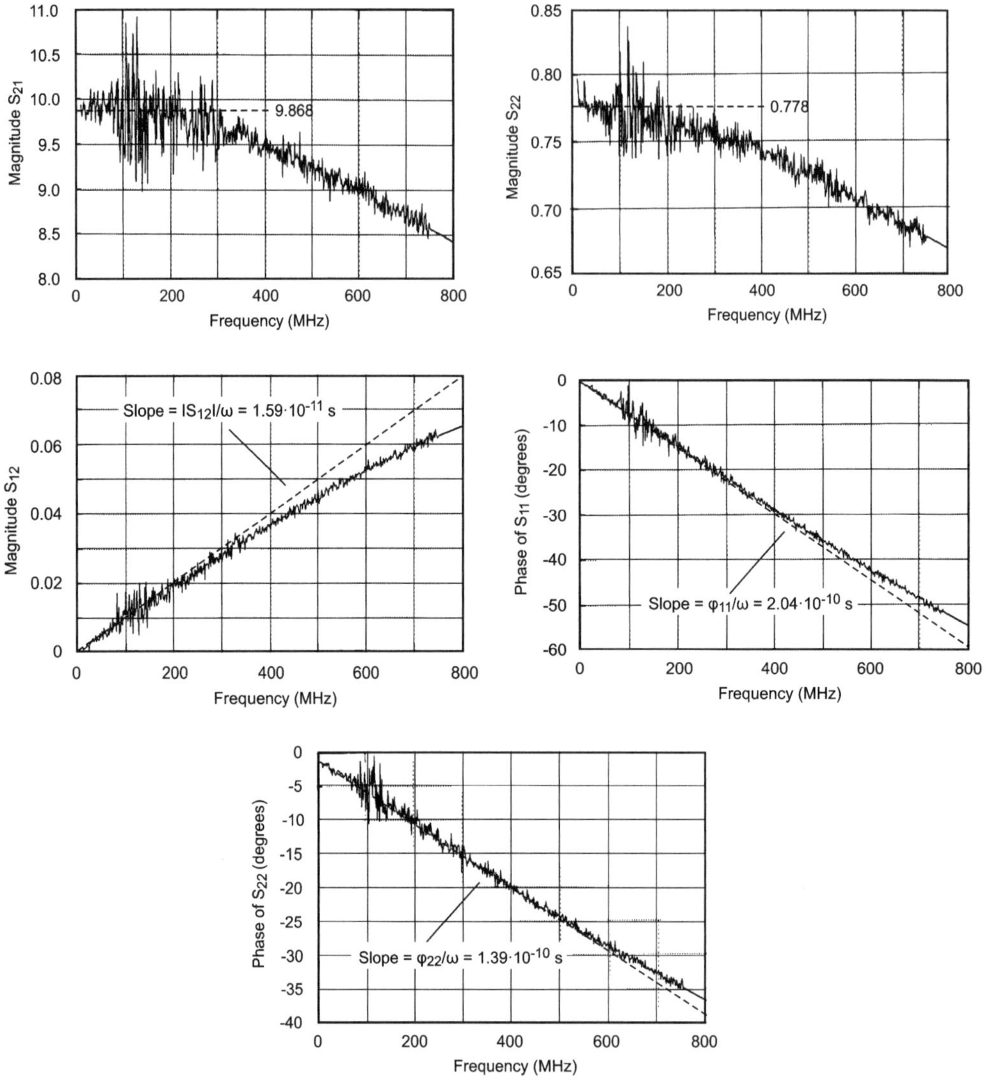

Figure 8.2 Determination of extrinsic model parameters of extrinsic FET device using low-frequency S-parameter measurements.

HEMT (but different devices) we may conclude that the estimated model parameter values are really in the same order of magnitude. Furthermore, it can be stated that with increasing gate width of high-power GaN devices, the adjustment effect brought about by the series resistances will be less relevant.

In [5] a 0.5 µm-gate GaAs MESFET of type MGFC1423 was operated at $V_{GS} = -0.35$V and $V_{DS} = 3$V ($I_{DS} = 30$ mA). The measured low-frequency scattering coefficients were $|S_{21}| = 4.64$, $|S_{22}| = 0.625$, $|S_{12}| = 0.032$ ($f = 2$ GHz), $\angle S_{11} = -27.7°$ ($f = 2$ GHz). The parameters of the extrinsic two-port FET model are then found as $G_M = 57$ mS, $G_{DS} = 5.3$ mS, $C_{GD} = 32.2$ fF.

Following Section 6.2.2 the series resistances R_s and R_d can be estimated as 2Ω and 3.1Ω, respectively. With the conductivity σ_{Al} of the Al-gate, $\sigma_{Al} = 33.3 \times 10^4$ S/cm, total gate width $w = 2 \times 150$ μm = 300 μm, and gate thickness $t = 0.1$ μm, the gate sheet resistance is found to be 0.3Ω/sq. Then the gate resistance is calculated as $R_g = 1/(3N^2) \times R_{sh} \times w/L = 3.75\Omega$.

From *S*-parameter measurement, $R_g = 4.11\Omega$ was obtained, which is very near the estimated value. Applying (6.43) with $v_s \approx 1\times 10^7$ cm/s, $\mu_o = 4500$ cm²/Vs, the gate charging resistance is determined to be $R_i \approx 1.23\Omega$ (extracted 2.42Ω).

The respective physics-based estimated and optimized model parameters (in brackets) were $G_m = 65$ mS (67 mS), $G_{ds} = 6.04$ mS (6.14 mS), $C_{gs} = 0.35$ pF(0.31 pF), $C_{gd} = 28.4$ fF (31.0 fF). With all these estimates, from (8.44), we get $C_{ds} = 57$ fF (76 fF), which is a rather good result, especially in view of the accumulation of model element uncertainties.

8.6 SUMMARY

For the low-frequency region approximate interrelations between extrinsic (generic) and intrinsic (physics-based) model parameters were derived. It has been shown that physics-based model parameters can be directly estimated from measured low-frequency *S*-parameters, which can serve as very good starting values in an optimizer-based model parameter extraction procedure. The usefulness of the proposed approach was validated by the evaluation of low-frequency measurement examples.

REFERENCES

[1] Novotny, M., *Investigations on the Consistency of Transistor Equivalent Circuits* (in German), Research Project, Department of High Frequency Engineering, University of Kassel, 1990.

[2] Kompa, G., "Small- and Large-Signal Characterization Methods," *Digest IEEE Workshop on Measurement Techniques for Microwave Device Characterization and Modelling*, Stuttgart, Germany, April 23, 1990, pp. 67–97.

[3] Srinidhi, E. R., *Intermodulation Distortion Modelling and Measurement Techniques for GaN HEMT Characterization*, Doctoral Thesis, Department of High Frequency Engineering, University of Kassel, 2008, Kassel: Kassel University Press GmbH, 2009.

[4] Mengistu, E. S., *Large-Signal Modeling of GaN HEMTs for Linear Power Amplifier Design*, Doctoral Thesis, Department of High Frequency Engineering, University of Kassel, 2007, Kassel: Kassel University Press GmbH, 2008.

[5] Novotny, M., *Extraction of Equivalent Circuit Parameters of GaAs FETs und HEMTs* (in German), Master Thesis, Department of High Frequency Engineering, University of Kassel, 1988.

Chapter 9

Static-/Pulsed-DC Measurements for the Analysis of Thermal and Trapping Effects

In this chapter we will treat static-DC (or just DC) and pulsed-DC measurements of transistors for microwave applications. DC IV measurements are restricted to the safe-operation region, whereas pulsed-DC measurements can overcome this operation limitation. Pulsed-DC IV measurements are of great importance for the development of nonlinear device models. This particularly applies to the large-signal modeling of the drain current in high-power field-effect transistors.

9.1 STATIC-DC MEASUREMENTS

Experimental device DC analysis is very important to identify its safe-operation region without exceeding its normal ratings such as maximum power and maximum drain current to prevent temperature and electrical field-induced breakdown. Standard large-signal FET models are essentially DC models. They are of limited applicability to highly nonlinear operation of modern high-power power devices (see Chapter 25). With respect to small-signal device modeling, scattering parameters have to be measured in many DC bias points to extract bias dependent model elements.

9.1.1 Principal DC IV Characteristics and Definitions

Static-DC IV measurements deliver the transfer and output characteristics of the device under analysis and describe the behavior of the DC drain current as a function of the terminal DC-voltage supplies. These characteristics are obtained by sweeping the gate and drain bias voltages with two power supply sources and measuring the DC drain current for each sweep step. In the output characteristics, the drain-source current I_{DS} is plotted versus the drain-source voltage V_{DS} for various constant gate-source voltages V_{GS}. In the transfer characteristics, I_{DS} is plotted versus V_{GS} with V_{DS} as the parameter.

Figure 9.1 illustrates principal DC output characteristics with various operation regions and specific voltage and current notations. The V_{DS} value known as the knee voltage V_{knee} generally characterizes the transition from the ohmic to the saturation region. The I_{DS} value denoted by I_{DSS} corresponds to V_{knee} with $V_{GS} = 0V$ and is commonly used as a reference for the maximum output current of the device. V_{th} is the gate-source voltage value known as threshold voltage, and it characterizes the transition from the saturation to the pinch-off

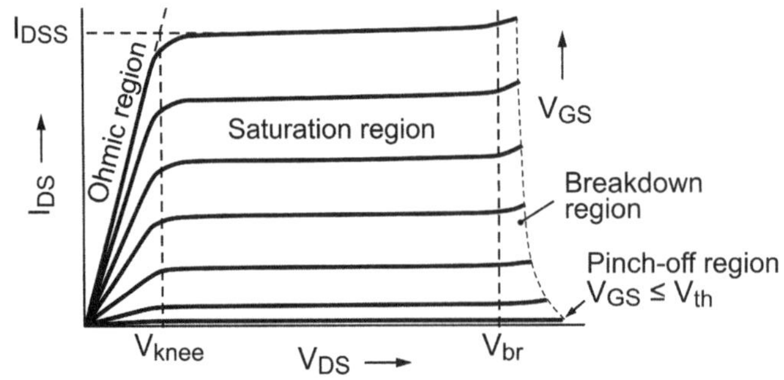

Figure 9.1 Operation regions and relevant values of an ideal DC IV output characteristics.

region, in which I_{DS} takes its minimum values. A precise definition of V_{th} and its actual values obtained from measured data are presented in Section 21.2.3. V_{br} denotes the onset of breakdown region. It is denoted as drain-source off-state breakdown voltage when the related V_{GS} value belongs to the pinch-off region ($V_{GS} \leq V_{th}$). When the V_{GS} value corresponds to the saturation region (e.g., open-channel conditions ($V_{GS} > V_{th}$)), the value that characterizes the transition to the breakdown region is called drain-source on-state breakdown voltage, $V_{br\text{-}on}$.

9.1.2 Measured DC IV Characteristics

Experiences in device characterization indicate that the most important aspect of DC measurement is the observation of the device limits in voltage, current, and power for the definition and realization of the bias voltage sweeps.

The DC voltage limits are denoted by BV_{GD} and V_{br}, respectively. BV_{GD} is the gate-drain breakdown voltage typically specified by the manufacturers as the voltage limit of FETs. BV_{GD} is related to a semiconductor material property called critical electric field, E_c, which defines the maximum permissible electric field in the device channel and thus, the maximum voltage applicable across a two-terminal semiconductor device before avalanche breakdown occurs [1]. If DC voltages above BV_{GD} (V_{br}) are applied, device failure is likely to occur. V_{br} is the V_{DS} value associated with BV_{GD} when the device is biased in the pinch-off region ($V_{GS} = V_{th}$); that is, V_{br} is the V_{DS} value of off-state breakdown, as defined for example in [2].

The DC power limit is related to the device safe-operation area (SOA). The SOA denotes the area of $I_{DS} - V_{DS}$ plane where operation without any risk of device failure is expected. The I_{DS} value I_{DSS} corresponds to the saturation current in open-channel conditions, $V_{GS} = 0$V, and is commonly used as a reference for the DC current limit. However, it does not represent a strict limit, but a reference of the highest current values that are safely generated in the device. Regarding GaN-based devices, DC measurements are often performed with positive V_{GS} values up to a few volts, and thus, with I_{DS} values above I_{DSS}. In any case, the maximum permissible value of DC current is included in the manufacturer definition of the SOA and generally complied as long as the actual DC power is observed within the SOA.

The setup of DC measurements may consist of an on-wafer probing station and DC voltage sources that also act as current sensors. When the power supplies provide only positive voltages, manual rearrangement of the connections is needed to produce negative V_{GS} values. Figure 9.2 illustrates typical DC IV output and transfer characteristics of measured large-size AlGaN/GaN HEMT devices [3]. Chosen maximum I_{DS} saturation current

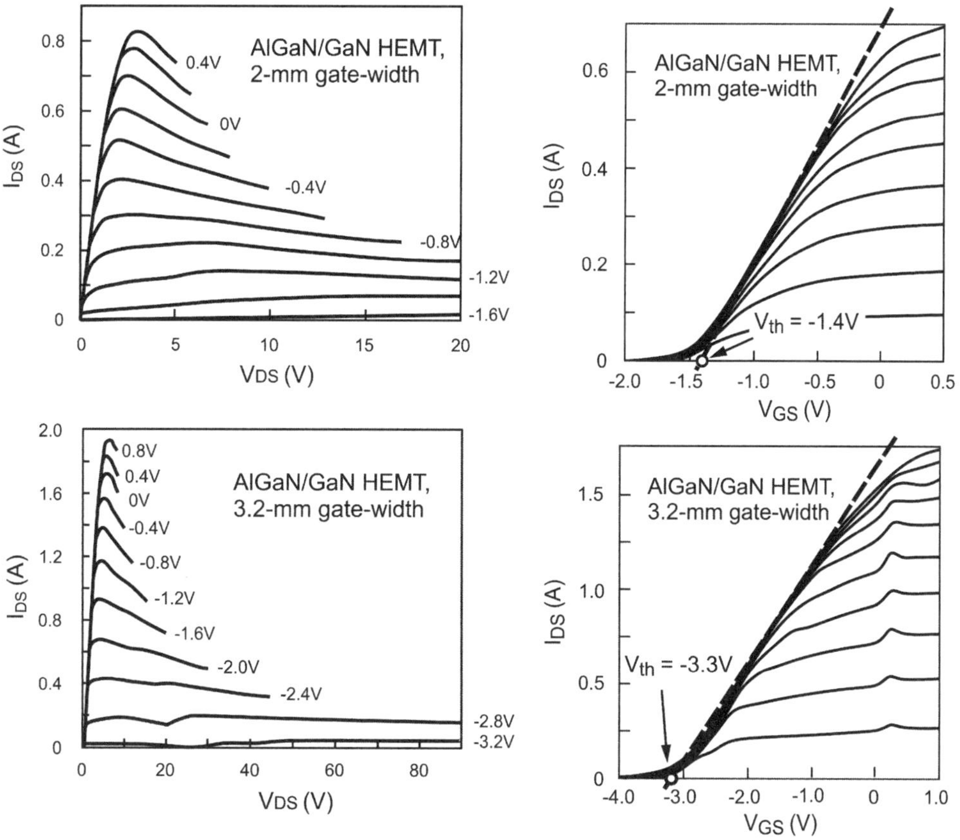

Figure 9.2 Measured static-DC IV output and transfer characteristics of large-size 2-mm (above) and 3.2-mm (below) AlGaN/GaN HEMTs (after [3]).

and maximum gate-drain voltage values were in compliance with the manufacturer's recommendations. The negative slope of the curves at higher DC drain current values is due to self-heating, which is attributed to the dissipated power in a given bias point.

Cold pinch-off S-parameters measured at the bias point (V_{DS} = 0V, $V_{GS} = V_{pinch\text{-}off}$) are a key database for the small-signal device modeling. The indicated V_{th} value represents a first approximation to the actual $V_{pinch\text{-}off}$ value required for cold *pinch-off* S-parameter measurements (see Section 21.2.3).

As already mentioned, V_{th} characterizes the transition into the pinch-off region. In classical textbook definitions, V_{th} is identified as the voltage value in the transfer characteristics at which linear extrapolation of I_{DS} intercepts the V_{GS}-axis. The V_{DS}-curve that must be extrapolated corresponds to the one with the largest slope, providing the peak transconductance, as indicated in Figure 9.2.

Besides the measurement in the cold *pinch-off* condition, S-parameters measured at bias points distributed over ohmic and saturation regions (active bias points) are also the database for device modeling. Sweeps of V_{DS} and V_{GS} values are selected for the active bias points according to the bias dependency of the current observed in the DC IV characteristics. The DC limits of devices and of the measurement equipment are also taken into account.

The selected V_{DS} sweeps should be dense enough in the low-V_{DS} value range to cover the nonlinearity of the transition from the ohmic to the saturation region. In the saturation region, this density is gradually decreased, reducing the number of bias points and overall duration of the measurement sessions.

9.1.3 Thermal Resistance from DC Measurements

Under assumption of negligible device trapping effects (e.g., [4, 5]) the dispersive current can be thought of as being composed of an isothermal (ideal) DC IV characteristics and a dispersive term describing the degradation due to temperature influence. Then we can express the thermal dispersion term as follows:

$$I_{DS}^{DC}(V_{GS}, V_{DS}) = I_{DS}^{ISO}(V_{GS}, V_{DS}) + f_{th}(V_{GS}, V_{DS})R_{th} \cdot P_{diss} \tag{9.1}$$

and solved for the dispersion term f_{th},

$$f_{th}(V_{GS}, V_{DS}) \cdot R_{th} = \frac{I_{DS}^{DC}(V_{GS}, V_{DS}) - I_{DS}^{ISO}(V_{GS}, V_{DS})}{P_{diss}} \tag{9.2}$$

The product $f_{\mathrm{th}}(V_{\mathrm{GS}}, V_{\mathrm{DS}}) \cdot R_{\mathrm{th}}$ is known; that is, the only unknown parameter is the thermal resistance, which can then be determined by optimization.

9.2 PULSED-DC MEASUREMENTS

Pulsed-DC measurements are very important to characterize the temperature and frequency dispersion of the drain current. Depending on the choice of the quiescent bias point, isothermal DC characteristics can be measured and the dynamic properties of surface and buffer traps can be studied. Moreover, pulsed operation can overcome the specified safe DC operation area, which is, for example, important for gate forward measurement to extract the gate resistance model element.

Measurements of the drain current are performed using pulsed-DC voltages as a bias condition instead of DC values. This was proposed to minimize thermal effects. Since such effects are delay processes with specific time constants, drain current is in fact measured without important thermal effects using widths of the voltage pulses significantly smaller than the time constants of thermal effects. If the pulse repetition time is much larger than the pulse width (i.e., the duty cycle is low), the heat produced by the current on each pulse cannot add up effectively to the next pulse. The DC bias point characterized by V_{GS} and V_{DS}, on which the pulses are superimposed, is referred to as the *quiescent* bias point. From the quiescent bias point, voltage pulses with varying amplitude produce any *pulsed-to* point desired; that is, any pair of v_{GS} and v_{DS}. Sweeping the v_{GS} and v_{DS} values of the pulsed-to points produces pulsed-DC IV characteristics that is specific to the quiescent bias point employed.

Figure 9.3 shows an example of pulsed-DC IV characteristic with the quiescent bias point marked by a cross ($V_{\mathrm{DS}} = 4.0\mathrm{V}$, $V_{\mathrm{GS}} = -2.0\mathrm{V}$, $I_{\mathrm{DS}} = 0.215\mathrm{A}$). Pulses are applied from this point to the pulsed-to points in the IV plane, indicated by a small circle ($v_{\mathrm{DS}} = 20.0\mathrm{V}$, $v_{\mathrm{GS}} = -1.0\mathrm{V}$, $i_{\mathrm{DS}} = 0.951\mathrm{A}$) [3].

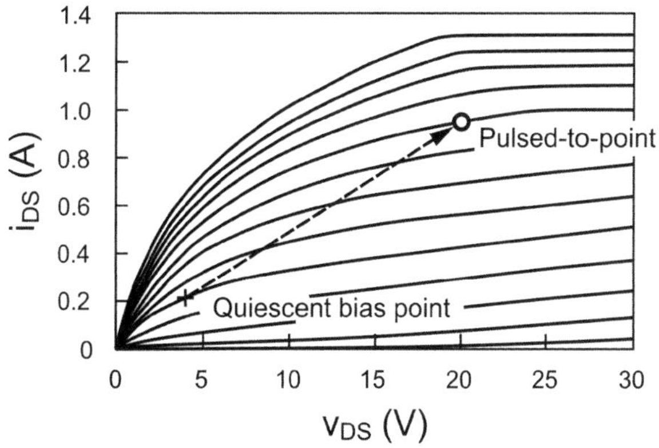

Figure 9.3 Example of measured pulsed-DC IV characteristic (solid lines). + quiescent point, ○ pulsed-to point (after [3]).

9.2.1 Implementation of Measurements

A dynamic IV analyzer (DiVA® D265EP by Accent®) has been used for pulsed-DC measurements [6]. It sets the quiescent bias point, generates the DC-voltage pulses with the desired duration and repetition rate and measures the dynamic drain current of the device. The analyzer has been connected to an on-wafer probing station. The reference temperature of the measurements is set with a thermochuck (prober surface of the on-wafer station) and a temperature control unit providing an accuracy of ±0.1 °C and maximum temperature of 150°C.

The power and current limits must be set in the controlling software of the DiVA® to assure that the measurements keep the SOA of the device. Thus, the DC power related to the quiescent bias points under consideration is always compared with $P_{DS,max}$ of the safe operating area (SOA) of the device (maximum permissible static-DC drain-source power). The DiVA® ratings are $I_{DSmax} = 2A$, $V_{DSmax} = 65V$, and $P_{DSmax} = 30W$.

Figure 9.4 shows the applied gate and drain voltage pulses and the measured current transient for a given pulsed-to point example in Figure 9.3. The arrow indicates the transition from the quiescent bias point (+ marker) to the pulsed-to point (○ marker). Generally, the measured current transient has four recognizable phases or regions in the time scale [6]:

(1) Region I is the turn-on transient. It usually includes ringing and the adjustment of the fast processes to the pulsed-to voltages.

(2) Region II represents a fast-process plateau of current where fast device processes have adjusted to the dynamic condition but slow device process have not yet changed from the quiescent condition.

(3) Region III contains slow-process adjustments to the pulsed-to voltages. It involves charge interaction with traps (typically in the order of μs to ms) and electrical changes due to the device heating/cooling (typically in the order of microseconds to seconds).

(4) Region IV corresponds to the steady-state (static-DC) current that appears when the adjustment to the dynamic condition is complete.

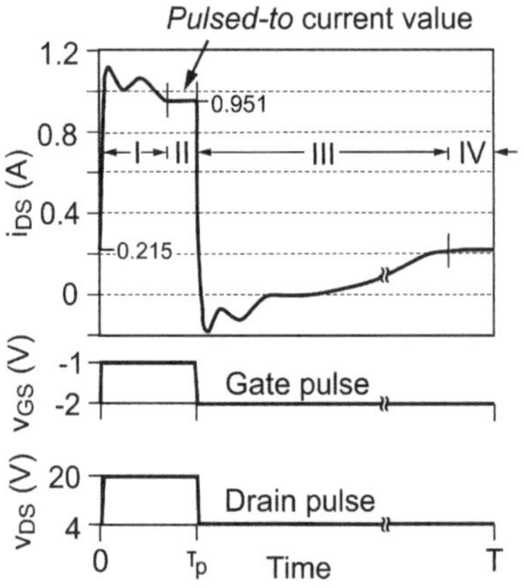

Figure 9.4 Stimulus and pulsed-DC measuring signal waveform. Quiescent bias point (V_{DS} = 4.0V, V_{GS} = - 2.0V, I_{DS} = 0.215A) and pulsed-to point (v_{DS} = 20.0V, v_{GS} = - 1.0V, i_{DS} = 0.951A) correspond to the example in Figure 9.3. (After [3].)

Each i_{DS} value in the pulsed-DC IV characteristics must be measured after the turn-on transient and before the onset of the slow-process adjustments. The dynamic IV analyzer automatically reports the current measured during the plateau of region II. In pulsed-DC IV measurements, the device temperature is defined by the case temperature (thermochuck) and by the device dissipated power at the quiescent bias point.

Note, meanwhile, after takeover of the DiVA system by Auriga in 2008, sales have ceased. Instead Auriga has offered the AU4550 Pulsed IV/RF system [7].

9.2.2 Instability Stabilization Techniques

As described in detail in [8], instability problems can occur in connection with pulsed-DC measurements. This is in particular given with devices with high gain. This effect is well known in microwaves. Instability can occur due to internal feedback of the device. However, in connection with the measurement instrument, unintended (electromagnetic) coupling between the two channels of the instrument (gate and drain pulser) may cause unwanted feedback paths. As the gain of the transistor and the instrumental feedback paths cannot be quantized, empirical approach to achieve device stability is recommended.

One technique used by microwave engineers to render a device stable is to load it with a relatively low resistance across the drain-source [8] to reduce the loop gain. As stated in [8], taking into account the value of the stabilizing resistor, the software-based correction in the instrument should ensure that the two sets of characteristics—the corrected and measured instable ones—should coincide. As the author's experiences have shown, correction may not be perfect.

As reported in [3], severe instability problems have been observed with the measurement of GaN-based high-power HEMTs displaying high gain (transconductance) areas in the output IV characteristics. In this case unstable operation (spurious oscillations) can occur as those shown in the measurements of Figure 9.5. Unstable measurements occur mainly in quiescent bias points located in the IV region associated with high transconductance (i.e., for V_{GS}- V_{DS} voltage combinations with the steepest I_{DS} slope in the static-DC IV transfer characteristic) and exclusively for large device sizes.

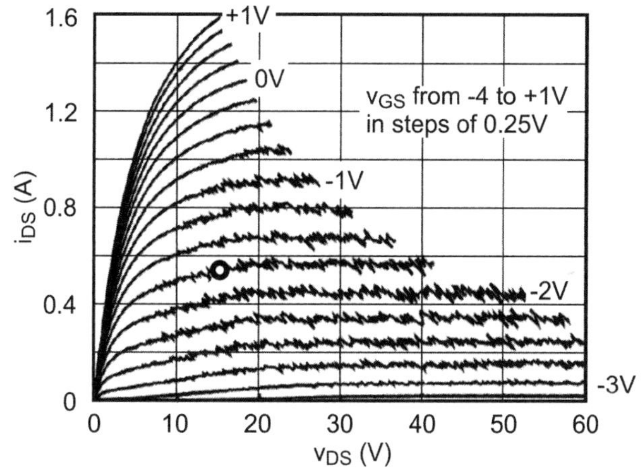

Figure 9.5 Unstable pulsed-DC IV measurements of a 3.2-mm AlGaN/GaN HEMT biased at the quiescent bias point $V_{GS} = -1.75V$, $V_{DS} = 15V$, $I_{DS} \approx 0.55A$, $P_{DS} \approx 8.3W$ (○-marker) [3].

Unstable pulsed-DC IV measurements are unsuitable to be part of the database for the calculation of the model. As well, strong instabilities generate significant oscillations of the current, which causes the DiVA® controlling software to stop responding and abrupt interruption of the measurement session. When that happens, there is an elevated risk of high-current damage for the device and the measurement equipment.

The appearance of instabilities in the measurements constrains the region where quiescent bias points can be selected and where reliable and safe measurements are achievable. Unstable operation has not emerged in quiescent bias points with zero or negligible drain power, but in points where the power is significant. For studied large-size devices power dissipation P_{DS} of quiescent bias points at which unstable operation appeared was as low as 1 W/mm. On the other hand, pulsed-DC IV measurements in such quiescent bias points with nonnegligible P_{DS} are needed to calculate key large-signal model parameters. Furthermore, unstable operation constrains achievable the pulsed-DC IV database and also the IV region where the model can be successfully verified, ultimately reducing the model validity region.

A stabilization technique can overcome these drawbacks. The manufacturer of the DiVA® system recommends using a stabilizing resistor connected in shunt to the output of unstable FET devices (between the drain terminal and ground) [6]. Moreover the controlling software of the DiVA® has a built-in feature that calibrates the effects on the measurements produced by the stabilization technique, where the user only must input the resistance value used as stabilizing resistor.

As illustrated in Figure 9.6 the stabilization technique is performed by inserting a stabilizing resistor in the drain path between the corresponding ports of the DiVA® and the

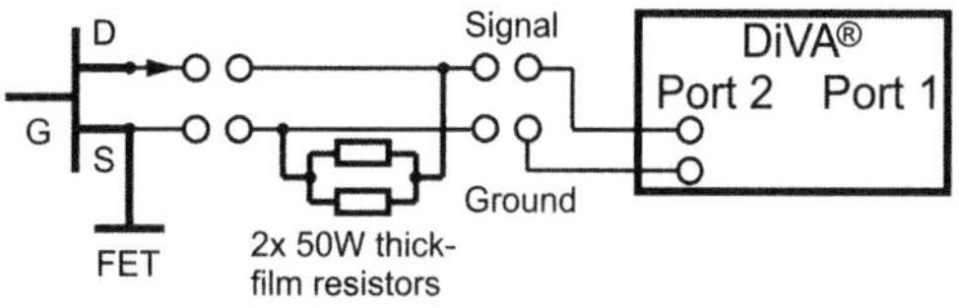

Figure 9.6 Stabilization of pulsed-DC IV measurements with shunted thick-film resistors between drain terminal and ground.

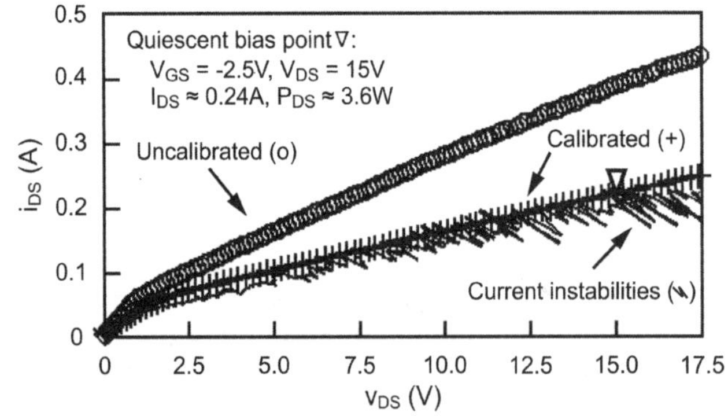

Figure 9.7 Stabilized pulsed-DC measurement of 3.2-mm AlGaN/GaN HEMT [3].

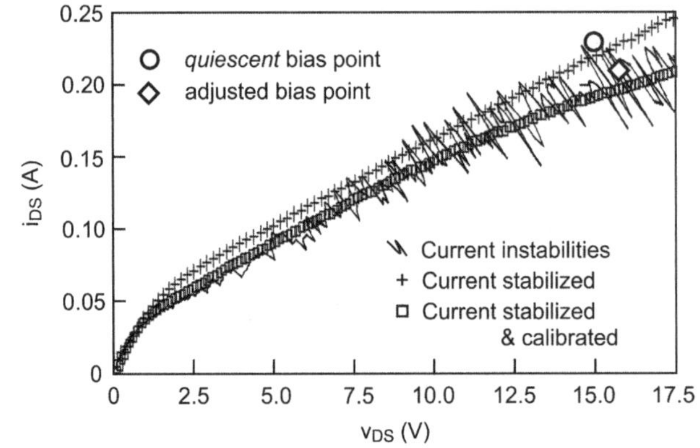

Figure 9.8 Bias-point adjustment to account for the residual error of the stabilizing element [3].

on-wafer probing station. The resistive element was composed by two shunted 50W thick-film resistors of type Vishay® model LTO 50 [3].

Starting with a high-impedance element, it has been shown that the stabilizing effect increases with decreasing resistance value. Figure 9.7 shows stable measurements with a stabilizing resistor of 110Ω (curve with circled symbols). As can be seen, this stabilized result differs on average from the originally measured data (uncalibrated stabilized data). Therefore, a final calibration, taking into account the effect of the stabilizing element, must be carried out, which is supported by the controlling software of the DiVA® (the calibration needs the input of the value of the stabilizing resistance).

After calibration (curve with +++ markers in Figure 9.7), a residual offset between calibrated and measured data remains, although arguably small. Nevertheless, this residual error is likely to impair the reliability of the large-signal modeling process. This, in particular, concerns the determination of the nonlinear drain current source, which is mainly based on pulsed-DC IV measurements.

As proposed in [3], the residual error can be taken into account with an adjustment of the operation bias point in the later large-signal simulations. The required adjustment is estimated to be lower than ±10% of the nominal V_{GS} and V_{DS} values. This was roughly tested with stabilized pulsed-DC IV measurements on several quiescent bias points of studied large-size devices. Figure 9.8 and Figure 9.9 show two examples of such tests.

Figure 9.8 depicts three different curves of pulsed-DC IV measurements at a quiescent

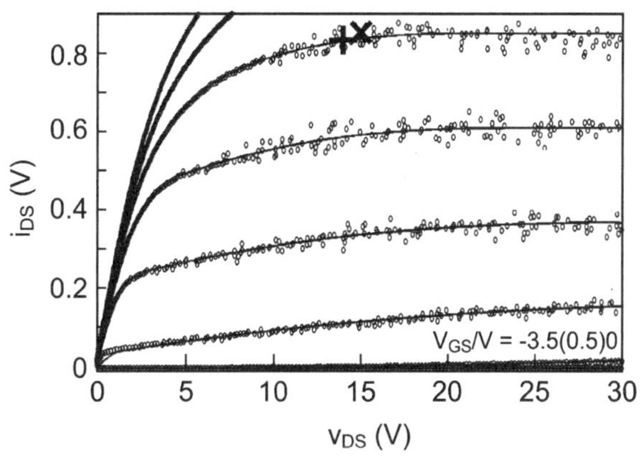

Figure 9.9 Perfect agreement of measured unstable "noisy" (circles) and error-corrected (solid lines) (stabilized and adjusted) pulsed-DC IV characteristics. Symbol × denotes the original quiescent bias point and symbol + the adjusted one [3].

bias point (V_{GS} = −2.5V, V_{DS} = 15V). The noisy solid line indicates the original measurement with current instabilities. The curve denoted by crosses indicates the measured data after calibration, and the curve with circle symbols includes additionally quiescent bias point adjustment (V_{GS} = −2.65V, V_{DS} = 16V).

Similarly, Figure 9.9 shows the good agreement of both the originally measured unstable pulsed-DC IV characteristics with quiescent bias point (V_{GS} = −1V, V_{DS} = 15V) and error-corrected (stabilized and adjusted) IV data with adjusted quiescent bias point (V_{GS} = −0.94V, V_{DS} = 14V).

For pulsed-DC IV measurements with stabilizing resistor, the maximum V_{DS} achievable is 54V in contrast to the DiVA® maximum rated value of 65V. Furthermore, in contrast to the specified nominally rated power of 30W of investigated devices, stable pulsed-DC IV measurements could only be performed up to approximately 17W (with V_{DS} = 30V).

9.2.3 Choice of Quiescent Bias Points

The dynamics of microwave devices is essentially dependent on the choice of a quiescent bias point. The following section will identify various device phenomena with particular bias conditions. Depending on the dissipated power of the device we distinguish between passive and active quiescent bias points.

9.2.3.1 Passive Quiescent Bias Points

At passive quiescent bias points the drain-source current I_{DS} is approximately zero and, thus, the power (I_{DS} ≈ 0A, P_{DS} ≈ 0W). At zero bias point (V_{GS} = 0V, V_{DS} = 0V) the pulsed-DC measurements usually deliver the maximum i_{DS} values, because charge-trapping and self-heating effects do not occur with adequate pulsing. Regarding large-size GaN-based power devices, the maximum drain current frequently exceeds the specified ratings of the pulsed-DC analyzer (maximum current of 2A in the case of DiVA® instrument).

Figure 9.10(a) illustrates the pulsed-DC IV characteristics of a 3.2-mm AlGaN/GaN HEMT at zero bias point (I_{DS} = 0A, P_{DS} = 0W). As seen, the curve for v_{GS} = −0.5V appears

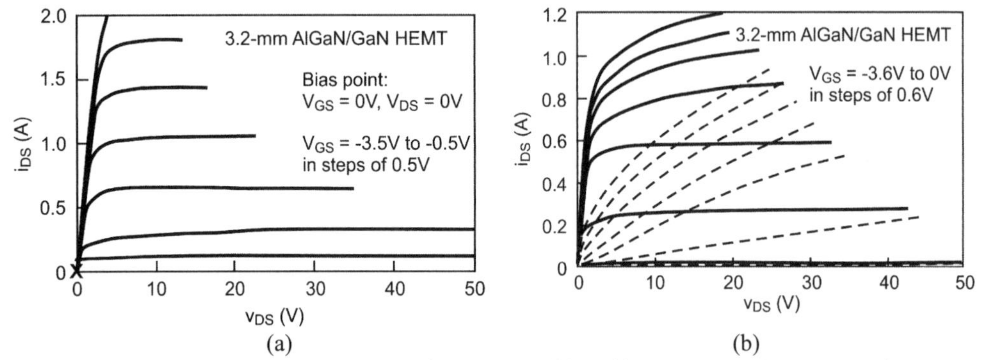

Figure 9.10 Pulsed-DC IV measurements of a 3.2-mm AlGaN/GaN power HEMT at passive *quiescent* bias points: (a) bias point: $V_{GS} = 0V$, $V_{DS} = 0V$, (b) bias points: $V_{GS} = -6V$, $V_{DS} = 0V$ (solid lines) and $V_{GS} = -6V$, $V_{DS} = 54V$ (dashed lines). (After [3].)

truncated before reaching the saturation current at $v_{DS} > 4V$. Measured i_{DS} curves for v_{GS} values $< -0.5V$ are complete and only constrained by the power limit. The limitation of pulsed-DC measurements toward high values of i_{DS} occurring in large-size power devices underlines one major reason why model scalability is of utmost importance in the empirical transistor modeling approach.

Figure 9.10(b) illustrates pulsed-DC IV measurement results for a 3.2-mm GaN device in two different quiescent bias points: (1) ($V_{GS} = -6V < V_{th}$, $V_{DS} = 0V$), and (2) ($V_{GS} = -6V$, $V_{DS} = 54V$). Since I_{DS} is approximately zero in both cases, the dissipation power is also zero ($P_{DS} \approx 0W$). Thus, both biasing situations describe a passive quiescent point. It is observed that strong trap-induced distortion is observed with high drain-source voltage V_{DS}.

9.2.3.2 Active Quiescent Bias Points

Studying the performance of a 3.2-mm GaN HEMT, Figure 9.11 depicts constant-P_{DS} contours in the I_{DS}-V_{DS} plane including discrete V_{GS} values (denoted by circles). Thus, each quiescent bias point is characterized by the control voltages (V_{GS}, V_{GS}) drain current I_{DS} and dissipation power P_{DS}. As seen, the selected active quiescent bias points cover a power range from approximately 1 to about 16W.

The distribution of quiescent bias points (V_{DS}, V_{GS}, P_{DS}) are essentially limited only by the device SOA and operation limits of the DiVA®. The selection of active quiescent bias points over broad V_{DS}, V_{GS}, and P_{DS} ranges is very important for the parameter extraction of the large-signal I_{ds} model, because the V_{DS}, V_{GS} and P_{DS} values control the basic equation for the low-frequency dispersive behavior of the drain current. The device characterization at manifold active *quiescent* bias points provides a broad database for the model parameter extraction and final model verification.

Figure 9.12 illustrates an example for pulsed-DC IV device characterization at active quiescent bias points for a large-size 3.2-mm device. The two sets of curves belong to different active quiescent bias points with the same V_{DS} but two different P_{DS} values: $V_{DS} = 40V$ with $P_{DS} \approx 1W$ and 12W. The two chosen bias points illustrate the influence of thermal effects related to quiescent power dissipation (self-heating effects) on the pulsed-DC IV measurements. It is seen that an increase of quiescent dissipated power produces a drop of the measured i_{DS} values due to the increased heat generation.

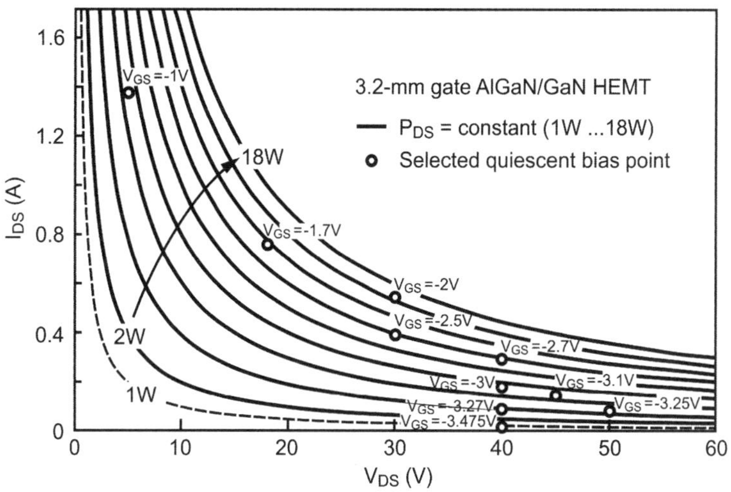

Figure 9.11 Set of active quiescent bias points (circles) selected for pulsed-DC IV characterization of a 3.2-mm AlGaN/GaN power HEMT. (After [3].)

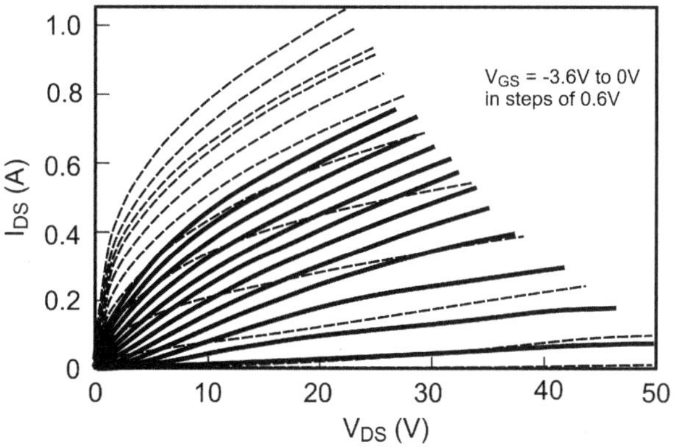

Figure 9.12 Pulsed-DC IV characteristics of a 3.2-mm AlGaN/GaN power HEMT at different active quiescent bias points. Solid lines: $V_{GS} = -2.7$V, $V_{DS} = 40$V, $P_{DS} \approx 12$W, dashed lines: $V_{GS} = -3.475$V, $V_{DS} = 40$V, $P_{DS} \approx 1$W. (After [3].)

9.2.4 Self-Heating Effects

Thermal effects are characterized by measuring pulsed-DC IV characteristics from a quiescent bias point ($V_{GS} = 0$V, $V_{DS} = 0$V) at different chuck temperatures. Measurements from this bias point have negligible trapping effects, thereby enabling to characterize self-heating effects only [9]. Figure 9.13 shows temperature-dependent pulsed-DC drain current characteristics at three chuck temperatures, 25°C, 50°C, and 75°C. Here, the effect of the change in channel temperature (ΔT_{ch}) on the drain current is measured directly. The increase in chuck temperature is assumed equivalent to the change in the channel temperature. As can be seen clearly from these IV characteristics, the effects of elevated channel temperature of the device are a decrease in the drain current and an increase in the knee voltage. In terms of small-signal parameters these imply a decrease in output conductance due to the flatted IV curves. These characteristics also imply a decrease in the transconductance due to the IV curves squeezed close together. All these are attributed to the decrease of both low-field

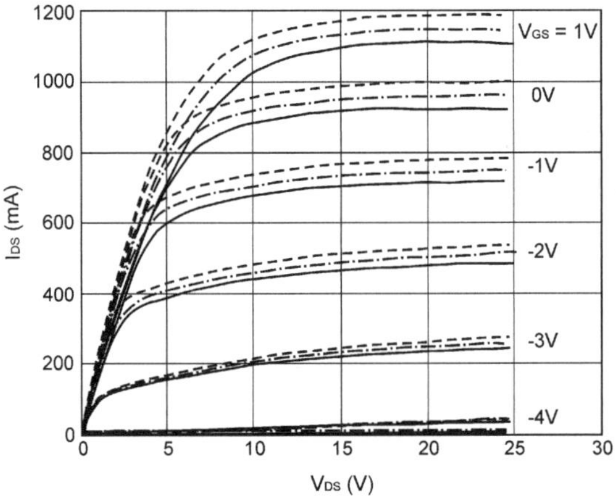

Figure 9.13 Pulsed-DC IV characteristics of a 1-mm gate-width GaN HEMT as function of chuck temperatures of 25 °C (dashed lines), 50 °C (dotdashed lines), and 75 °C (solid lines) from quiescent bias point (V_{GS} = 0V, V_{DS} = 0V) [9].

mobility and high-field saturation velocity of the carriers with increasing temperature [10].

9.3 THERMAL RESISTANCE EXTRACTION

The thermal resistance R_{th} of a FET structure is an important parameter in the device modeling process. It can be estimated analytically [11] or can be extracted from pulsed-DC measurements in different ways, as will be described next. As commented in [12], all thermal characterizations are performed for a given level of traps.

9.3.1 Pulsed-DC IV Isothermal Curves Overlapping

Studying a 3.2-mm AlGaN/GaN HEMT [13], pulsed-DC IV measurements have been performed from two different quiescent bias points for different controlled chuck-temperature values. The first pulsed-DC IV curve is pulsed from the reference quiescent bias point (V_{GS} = 0V, V_{DS} = 0V) with zero power dissipation value (P_{diss} = 0W). Since pulsed-DC IV current from this reference point delivers maximum drain current, the current level is reduced for comparison purpose; that is, the controlled case temperature is set to a higher value (T_{case1} = 75 °C). The second pulsed-DC IV curve is operated from an arbitrary quiescent point in the active region (V_{GS}, V_{DS}) with known nonzero power dissipation value (P_{diss2} = $I_{DS} \cdot V_{DS}$). The controlled case temperature (T_{case2}) of the second curve is varied until the drain curve approximately overlaps the first curve obtained from reference bias point ((V_{GS} = 0V, V_{DS} = 0V) [14]. As shown in Figure 9.14, the second pulsed-DC IV measurement *quiescent* bias point is (V_{GS} = -2V, V_{DS} = 5V) with I_{DS} = 927.4 mA and power dissipation P_{diss2} = 4.637W. The overlapping curves occur at case-controlled temperatures of (T_{case1} = 75 °C) and (T_{case2} = 10 °C) for the reference and the active region bias points, respectively. At this point the thermal resistance can be obtained from the equations of the equivalent channel temperatures as follows:

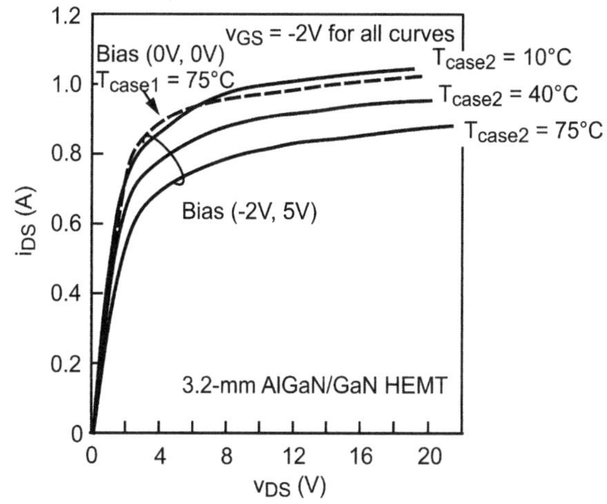

Figure 9.14 Thermal resistance extraction from overlapped pulsed IV curves. Dashed line: Pulsed IV measurement at case temperature T_{case1} = 75°C for bias point (V_{GS}, V_{DS}) = (0V, 0V) with P_{diss1} = 0W. Solid lines: Pulsed IV measurement at distinct case temperatures T_{case2} = 75°C, 40°C, 10°C for (V_{GS}, V_{DS}) = (−2V, 5V) with P_{diss2} = 4.638W. All curves are for v_{GS} = −1V. (After [13].)

(1) Pulsed from (0V, 0V): P_{diss1} = 0W. The channel temperature then reads

$$T_{ch} = R_{th} \cdot P_{diss1} + T_{case1} = 75°C$$

(2) Pulsed from (−2V, 5V): P_{diss2} = 4.638W. In this case we have

$$T_{ch} = R_{th} \cdot P_{diss2} + T_{case2} = R_{th} \cdot 4.638W + 10°C$$

Since the two drain current curves overlap, the channel temperature should be equivalent to T_{ch} = 75°C. Hence, the value of the thermal resistance R_{th} is equal to 14°C/W.

9.3.2 Pulsed-DC IV and Static-DC Crossing

Here, the thermal resistance is extracted from two drain current measurement curves (pulsed-DC and static-DC) for two different bias points with distinct case temperatures (Figure 9.15). The intersection point of the two measured drain current curves is the point at which equivalent channel temperature and drain current delivered are the same [15]. The equivalent channel temperature will be described for both curves as

$$T_{ch} = R_{th} \cdot P_{diss1} + T_{case1} \tag{9.3}$$

$$T_{ch} = R_{th} \cdot P_{diss2} + T_{case2} \tag{9.4}$$

where P_{diss1} and P_{diss2} are the dissipated power at bias point 1 and bias point 2 under controlled case temperature T_{case1} and T_{case2}, respectively. By equating the two expressions, the resulting thermal resistance can be extracted to be equal

$$R_{th} = \frac{\Delta T_{case}}{\Delta P_{diss}} \tag{9.5}$$

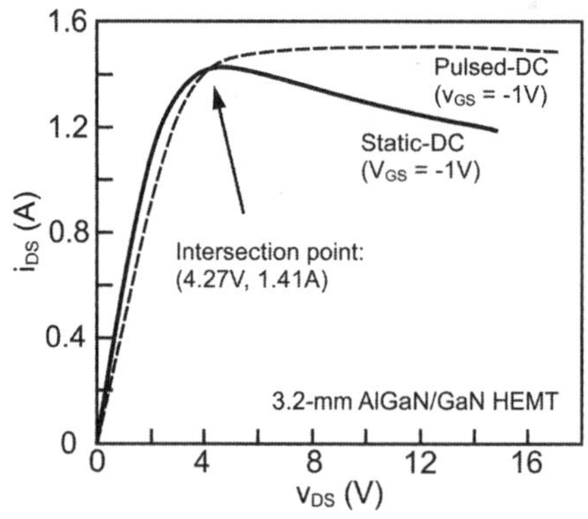

Figure 9.15 Thermal resistance extraction at the intersection point (V_{GS}, V_{DS}) = (- 1V, 4.27V) at drain current I_{DS} = 1.41A and power dissipation P_{diss} = 6.02W. Static-DC measurement at case temperature T_{amb} = 20°C (V_{GS} = - 1V), and pulsed-DC measurement at case temperature T_{case1} = 100°C (v_{GS} = - 1V). (After [13].)

To obtain this intersection, the first selected drain curve is based on pulsed-DC measurements from quiescent point (V_{GS} = 0V, V_{DS} = 0V) to a pulsed-to curve point (v_{GS}, v_{DS}). At this point the dissipated power is zero and the temperature is equal to the temperature of the case. The temperature of the case is varied until the intersection happens with a second curve obtained under static-DC measurement. The channel temperature will then be given by the relation

$$T_{ch} = R_{th} \cdot P_{diss1} + T_{case1} = T_{case1} \tag{9.6}$$

At the intersection point of the two curves, the power dissipation of the second curve is directly related to the product of the static-DC drain voltage and drain current value (P_{diss} = $I_{DS} \cdot V_{DS}$). With the temperature of the case (T_{case2}) equal to ambient temperature (T_{amb} = 20°C), the equivalent channel temperature will be given by

$$T_{ch} = R_{th} \cdot P_{diss} + T_{case2} \tag{9.7}$$

from which the thermal resistance can be found to be equal to

$$R_{th} = \frac{T_{case1} - T_{case2}}{P_{diss}} \tag{9.8}$$

Investigating a 3.2-mm (8 × 400 μm) AlGaN/GaN HEMT device, for pulsed-DC IV measurement a pulsed-to gate voltage v_{GS} = - 1V was chosen as indicated in Figure 9.15. The intersection point with the static-DC curve (V_{GS} = - 1V) is (I_{DS}, V_{DS}) = (1.41A, 4.27V) with a power dissipation of 6.02W (P_{diss} = $I_{DS} \cdot V_{DS}$). The obtained total thermal resistance is then obtained as R_{th} = 13.28°C/W.

9.4 TRANSIENTS

We have seen in previous sections that DC-pulsing using very short pulse widths enable measuring the drain current above the cutoff frequency of the dispersion effects.

Consequently, the model parameters derived from this data can predict the drain current behavior in the low-frequency range above thermal and trapping cutoff [9].

Next we focus on low-frequency excitations that cause changes in quiescent bias voltages and/or channel temperature. The changes in the average gate and drain bias voltage conditions affect the low-frequency charge exchange with surface and deep level [16]. Similarly, the channel temperature is function of low-frequency heating and cooling caused by low-frequency voltage and current variations (i.e., low-frequency dissipated power variations).

The DiVA® has been used to make the transient measurements presented below. The instrument enables to take transient data for 1-ms duration and at the following fixed time intervals (μs): 0.1, 0.2, 0.5, 1, 2, 5, 10, 20, 50, 100, 200, 500, and 1,000.

9.4.1 Thermal Time Constants

As discussed in earlier sections, pulsed-DC IV measurements from the quiescent bias point ($V_{GS}=0V$, $V_{DS}=0V$) involve negligible trapping effects. Hence transient measurements from this quiescent bias point can be used to obtain the thermal time constant(s) of a device. The transients shown in Figure 9.16, measured for a 2 × 250 μm AlGaN/GaN HEMT, are for $0 < t < 30$ μs duration. In these measurements, the pulsed-to gate voltage is kept constant to $v_{GS} = 0V$. Increased self-heating is clearly seen with increasing drain current magnitude due to the self-heating effects. The small transients in the first couple of microseconds at lower drain current levels also support the reasoning that trapping effects are negligible for pulsed-DC measurements from ($V_{GS} = 0V$, $V_{DS} = 0V$). With increasing drain current level, device self-heating becomes dominant and the thermal time constant can be extracted from such transients.

The drain current transient to the pulsed-to point ($v_{GS} = 0V$, $v_{DS} = 4V$) in Figure 9.16(a) is shown separately in Figure 9.16(b). This transient shows two different time constants, a fast initial fall then followed by a slow decrease. This may be explained by an initial fast heat transfer to the epilayer causing a fast temperature increase, which is then followed by a slower temperature increase due to a thicker and better heat-conducting SiC substrate.

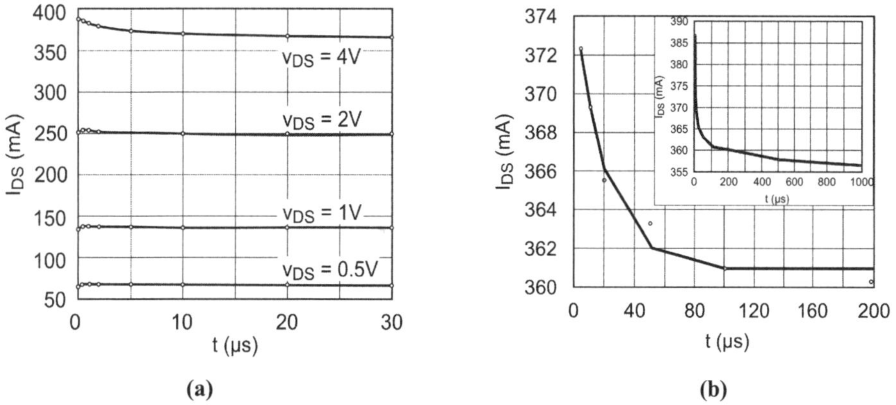

Figure 9.16 (a) Evolution of measured drain current transients for a 2 × 250 μm AlGaN/GaN HEMT from bias point ($V_{GS} = 0V$, $V_{DS} = 0V$). The pulsed-to points for drain voltage v_{DS} are indicated on the curves while $v_{GS} = 0V$ in all cases. (b) Drain current on enlarged scale for $v_{DS} = 4V$. Inset: Same drain current transient for $0 < t < 1$ ms. (After [9].)

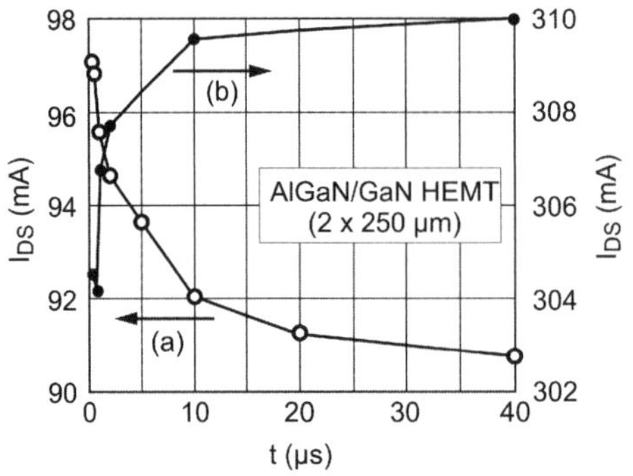

Figure 9.17 Measured drain current transients for a 2 × 250 μm GaN HEMT from bias points lying on constant power hyperbola. (a) Bias point: (V_{GS} = - 1V, V_{DS} = 4V), pulsed-to point: (v_{GS} = -3.5V, v_{DS} = 21V). (b) Bias point: (V_{GS} = -3.5V, V_{DS} = 21V), pulsed-to point: (v_{GS} = - 1V, v_{DS} = 4V). (After [9].)

Although multiple time constants are required to fit the transient, a first-order exponential curve fit gives a thermal time constant of 20.5 μs.

9.4.2 Trapping Time Constants

In order to obtain the trapping time constants, the thermal state of the device should remain unchanged at the quiescent bias and pulsed-to points. This may be accomplished by selecting these points along constant dissipated power curve [17]. Such drain current transient measurement results are shown in Figure 9.17 for a 2×250 μm AlGaN/GaN HEMT. The dissipated DC power at the two quiescent bias points (-1V, 4V) and (-3.5V, 21V) is approximately the same, which is equal to 1.4W.

The transient measurement from quiescent bias point (-1V, 4V) to the pulsed-to point (-3.5V, 21V) gives a drain current that decreases to its steady value as charge trapping progresses. Conversely, pulsing in the opposite direction from the quiescent bias (-3.5V, 21V) to the pulsed-to point (-1V, 4V) results in detrapping of charge and hence a gradual drain current increase to the steady-state value. Again, simple exponential curve fitting to these curves gives time constants of about 2 μs and 4.5 μs for Figure 9.17(a) and Figure 9.17(b), respectively.

However, these transient measurements require the alteration of both the gate and drain voltages from quiescent bias to the pulsed-to point in order to stay on the constant power curve. Hence, it is not possible to relate these time constants separately to the gate or drain voltages and thus to surface or buffer trapping. Therefore, this measurement technique is useful if either the gate lag or drain lag effect only is present. Nevertheless, the procedure provides a gross estimate of the trap time constants.

9.5 THERMAL MODEL

A major advantage of AlGaN/GaN HEMT technology is its relatively high-temperature operating level, which reduces the current flow in the channel, increases the knee voltage, and introduces self-heating memory effect in total drain-source current. To enhance the model capability to predict and simulate the response correctly in a CAD environment, thermal models are needed in large-signal model implementation for AlGaN/GaN HEMT devices.

The average dissipated power for a HEMT device is determined by its quiescent bias conditions (V_{GS0}, I_{DS0}),

$$P_{diss} = V_{DS0} \cdot I_{DS0} \tag{9.9}$$

The channel temperature T_{ch} depends on the ambient temperature (initial temperature reference) T_0 and the power dissipation P_{diss} of the device. Under static-DC operation condition this reads

$$T_{ch} = R_{th} \cdot P_{diss} + T_0 \tag{9.10}$$

or

$$P_{diss} = \frac{T_{ch} - T_0}{R_{th}} = \frac{\Delta T}{R_{th}} \tag{9.11}$$

where R_{th} (°C/W) is the equivalent thermal resistance of the device structure.

Similarly, a more general form of (9.11) can be obtained for a time-varying thermal analysis in semiconductor devices subjected to instantaneous time-dependent power dissipation $P_{diss}(t)$. The resulting expression for the above development in time domain can be reduced to [18]

$$P_{diss}(t) = \frac{\Delta T}{R_{th}} + C_{th} \frac{d}{dt}(\Delta T_{ch}) \tag{9.12}$$

The first term in (9.12) reflects the conductive heat generation as a function of channel temperature and thermal resistance, as previously explained. The second term in (9.12) reflects the dynamic heat flow due to the heat stored in the semiconductor material. Since this characteristic is directly related to the ability of the semiconductor material to store heat (defined as heat capacity in materials), the resulting coefficient C_{th} is defined as the thermal capacitance of the semiconductor material. Therefore, the total instantaneous power dissipation generated heat $P_{diss}(t)$ is the sum of heat flow term ($\Delta T_{ch}/R_{th}$) and heat storage term ($C_{th}(dT_{ch}/dt)$).

To predict the temperature change and its corresponding drain current variation in the device channel, an equivalent electrical circuit is used. By analogy to electrical circuits, replacing the instantaneous power dissipation $P_{diss}(t)$ and the channel temperature T_{ch} in (9.12) by a current source $I_{th}(t)$ and a potential difference $V_{th}(t)$, respectively, (9.12) reduces to

$$I_{th}(t) = \frac{\Delta V_{th}}{R_{th}} + C_{th} \frac{d}{dt}(\Delta V_{th}) \tag{9.13}$$

This equation clearly shows the sum of two currents under the same potential difference ΔV_{th}. The first term is the conductive current across the thermal resistance R_{th} and the second term is the displacement current across the thermal capacity C_{th}. Therefore, a parallel $R_{th} \cdot C_{th}$ filter is the electrical equivalent circuit that can be used to simulate the thermal response of the semiconductor device. The equivalent thermal capacitance is evaluated from the measured time constant τ_{th} of the drain current transient and previously determined thermal resistance R_{th} regarding the relation $\tau_{th} = R_{th} \cdot C_{th}$.

Figure 9.18 shows the first-order thermal equivalent model implementation in a CAD

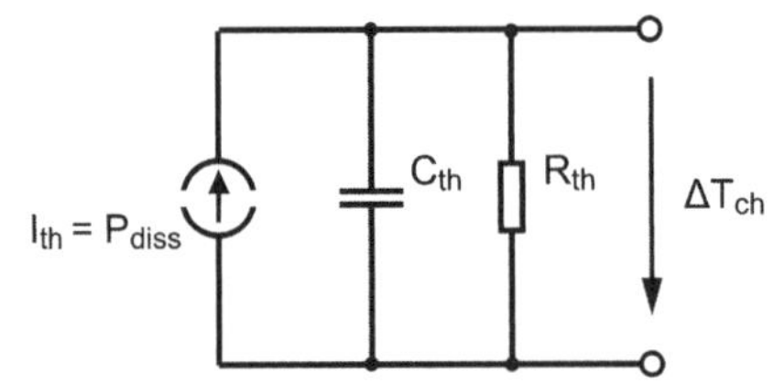

Figure 9.18 Thermal model equivalent subcircuit representation in a large-signal model of power devices to determine channel temperature (first-order low-pass filter approximation).

environment. A similar formulation can be extracted in complex notation. The instantaneous power dissipation is replaced by the complex current notation given by

$$I_{th}(\omega) = P_{diss}(\omega) = V_{ds}(\omega)\cdot I_{ds}(\omega) \tag{9.14}$$

The resulting channel temperature difference $\Delta T_{\mathrm{ch}} = T_{\mathrm{ch}} - T_0$ is equivalent to the simulated voltage across the thermal model terminals given by

$$\Delta T_{ch}(\omega) = \Delta V_{ch}(\omega) = H(\omega)\cdot R_{th}\cdot I_{th}(\omega) \tag{9.15}$$

where

$$H(\omega) = \frac{1}{1 + j\omega R_{th} C_{th}} \tag{9.16}$$

represents the first-order low-pass filter transfer function of the thermal model. Depending on the thermal model extraction method, a higher-order low-pass filter can be used [13].

9.6 TRAPPING SUBCIRCUIT

Trapped charges at RF frequencies are frozen (i.e., the time constants of trapping and detrapping are rather large so that trapped electrons cannot follow the gate and drain control voltage variations in a given bias point). Thus, the IV characteristics are not only dependent on (channel) temperature but also on the trap situation in the semiconductor interfaces and bulk materials. The trapping conditions under small-signal operation remain stable. This is even approximately valid for class-A operation and moderate stimuli. However, moving to class-AB, class-B or class-C operation, the quiescent bias point shifts under large-signal stimuli due to the asymmetric signal waveform. This leads to an effective bias point shifting that causes different trap conditions. Furthermore, regarding digital modulation, the modulation frequency band (baseband) covers just a lower frequency range where trapping and detrapping processes become relevant.

In Figure 9.19 illustrates schematically a large-signal intrinsic transistor model that has been extended to take care for low-frequency trapping dynamics. The large-signal intrinsic equivalent circuit includes RF nonlinear dynamic phenomena. The model was originally proposed by Filicoro et al. [19].

The gate (surface) trapping and drain (buffer or deep) trapping time constants τ_{GT} and τ_{DT}, respectively, are modeled, using a first-order RC high-pass circuit at the gate and drain sides, respectively. Resistances R_{GT} and R_{DT} in series with C_{GT} and C_{DT}, respectively, are used to model the small dispersive leakages currents in the gate and drain paths. The capacitors

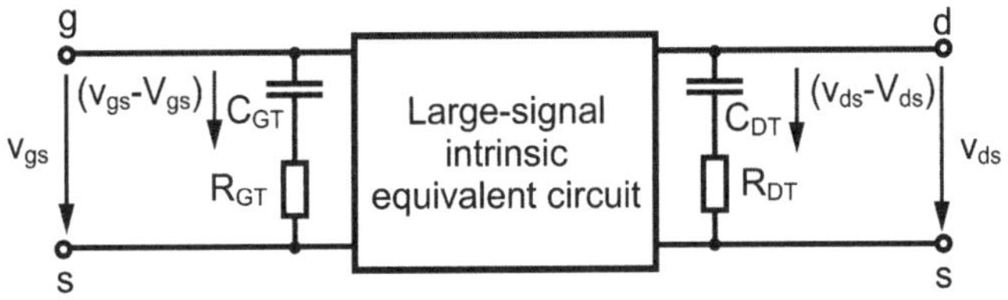

Figure 9.19 Large-signal intrinsic equivalent circuit of a HEMT device including buffer and surface trapping subcircuits. (After [19].)

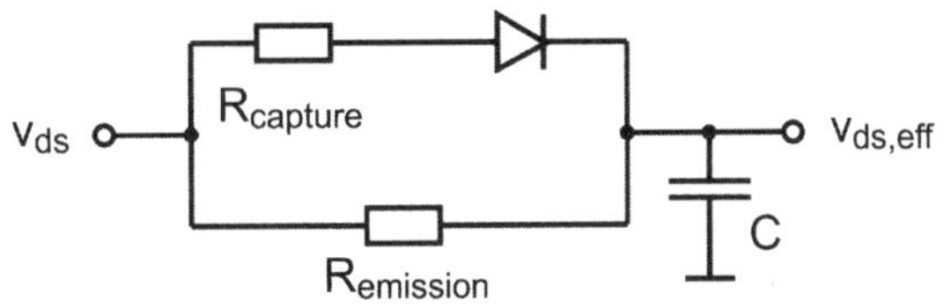

Figure 9.20 Drain-lag subcircuit. (After [23].)

values C_{GT} and C_{DT} are chosen in the order of 1 pF to model the small stored charges in the surface and buffer traps [20]. These charges are primarily related to the leakage current from the gate metal edge to the surface or from the channel into the buffer layer [21].

The values of R_{GT}, R_{DT}, C_{GT}, and C_{DT} are selected so that the time constants $R_{GT}C_{GT}$ and $R_{DT}C_{DT}$ products are equal to the trapping time constants τ_{GT} and τ_{DT}. In the case of lack of characterization means, approximate trapping time constants in the order of 10^{-5} to 10^{-4}s has been used with series resistances in the order of 1 MΩ [20]. This trapping implementation improves model accuracy for describing low-frequency channel transconductance, and output conductance, which is mainly related to the surface and buffer traps [22]. The discussed trapping subcircuit will be implemented in the nonlinear electrothermal model in Chapter 23.

Figure 9.20 illustrates a drain-lag trapping subcircuit that distinguishes between trap emission and capture processes [23]. The circuit comprises two RC branches: The upper one models the capture (filling of the traps) process, and the lower one describes the emission (emptying of the traps) process of the traps. As the emission time constant is commonly much larger than the capture time constant, the emission resistance $R_{emission}$ is much larger than the capture resistance $R_{capture}$ (i.e., $R_{emission} \gg R_{capture}$). Taking into account drain-lag effects, the drain current model is modified in that way that the dependence of the drain current drain on the drain voltage V_{ds} is now replaced by the effective drain-source voltage $V_{ds,eff}$ in Figure 9.20.

Similar trapping subcircuits can be defined for gate-lag phenomena. An overview of various published trapping subcircuits are given, for example, in [24].

9.7 SUMMARY

Static- and pulsed-DC measurements and measurement limitations with respect to maximum drain current, maximum power, and gate-drain breakdown were discussed. Device measurement instabilities of pulsed-DC characterized large-size AlGaN/GaN HEMTs were presented and remedies proposed. Temperature dependence of IV characteristics were analyzed. The thermal resistance was determined on the basis of static-DC and pulsed-DC measurements. Thermal and trapping time constants were provided by pulsed-DC measurements. Thermal and trapping subcircuits were presented.

REFERENCES

[1] Trew, R. J., "Wide Bandgap Semiconductor Transistors for Microwave Power Amplifiers," *IEEE Microwave Magazine*, Vol. 1, 2000, pp. 46–54.

[2] Tan, C. L., H. Wang, and K. Radhakrishnan, "Hot-Electron-Induced Degradation in BCB- and SiN-Passivated $Al_{0.25}Ga_{0.75}As/In_{0.2}Ga_{0.8}As$ PHEMTs," *IEEE Trans. on Device and Materials Reliability*, Vol. 7(3), 2007, pp. 438–446.

[3] Zamudio Flores, J. A., *Device Characterization and Modelling of Large-Size GaN HEMTs*, Doctoral Thesis, Kassel: Kassel University Press GmbH, 2012.

[4] Meneghesso, G., M. Meneghini, D. Bisi, I. Rossetto, A. Cester, U. Mishra, and E. Zanoni, "Trapping Phenomena in AlGaN/GaN HEMTs: A Study Based on Pulsed and Transient Measurements," *Semicond. Sci. Technol.*, Vol. 28, 2013, pp. 074021/15–8.

[5] Meneghini, M., D. Bisi, D. Marcon, S. Stoffels, M. Van Hove, T.-L. Wu, S. Decoutere, G. Meneghesso, and E. Zanoni, "Trapping and Reliability Assessment in D-Mode GaN-Based MIS-HEMTs for Power Applications," *IEEE Trans. on Power Electronics*, Vol. 29(5), 2014, pp. 2199–2207.

[6] Accent Optical Technologies Inc., *Dynamic I(V) Analyzer Applications Manual, Issue 1.0*, 131 NW Hawthorne Bend, Oregon 97701,2004.

[7] Auriga Measurement Systems, *AU4550 Pulsed IV/RF System*, Product Note, 2007.

[8] Ladbrooke, P. H., *Pulsed I(V) Measurement of Semiconductor Devices with Applications*, Accent Optical Technologies Inc., 2004.

[9] Mengistu, E. S., *Large-Signal Modeling of GaN HEMTs for Linear Power Amplifier Design*, Doctoral Thesis, University of Kassel, 2007.

[10] Anwar, A. F. M., S. Wu, and R. T. Webster, "Temperature Dependent Transport Properties in GaN, $Al_xGa_{1-x}N$, and $In_xGa_{1-x}N$ Semiconductors," *IEEE Trans.*, Vol. ED-48(3), 2001, pp. 567–572.

[11] Darwish, A. M., A. J. Bayba, and H. A. Hung, "Thermal Resistance Calculation of AlGaN–GaN Devices," *IEEE Trans.*, Vol. MTT-52, 2004, pp. 2611–2620.

[12] Teyssier, J. P., D. Barataud, A. Laloue, P. Bouysse, and R. Quéré, "Nonlinear Characterization of Microwave Transistors by the Means of Pulsed I(V) and Pulsed S-Parameter Measurements," *GAAS99*, Munich, 1999, pp. 393–398.

[13] Dahmani, S., *Large-Size AlGaN/GaN HEMT Large-Signal Electrothermal Characterization and Modeling for Wireless Digital Communications*, Doctoral Thesis, University of Kassel, 2006.

[14] Baylis II, C. P., L.P. Dunleavy, and J. E. Daniel, "Direct Measurement of Thermal Circuit Parameters Using Pulsed IV and Normalized Difference Unit," *IEEE MTT-S Int. Microwave Symp. Digest*, Vol. 6, 2004, pp. 1233–1236.

[15] Charbonniaud, C., S. De Meyer, R. Quere, and J. P. Teyssier, "Electrothermal and Trapping Effects Characterization of AlGaN/GaN HEMTs," *11th GaAs Symposium*, Munich 2003, pp. 201–204.

[16] Kompa, G., *Basic Properties of III-V Devices–Understanding Mysterious Trapping Phenomena*, Kassel, Kassel University Press GmbH, 2014.

[17] Siriex, D., D. Barataud, R. Sommer, O. Noblanc, Z. Ouarch, Ch. Brylinski, J. P. Teyssier, and R. Quéré, "Characterization and Modeling ofNonlinear Trapping Effetcs in Power SiC MESFETs," *2000 IEEE MTT-S Digest*, 2000, pp. 765–768.

[18] Schmale, I., and G. Kompa, "Integration of Thermal Effects into a Table-Based Large-Signal FET Model," *28th European Microwave Conference*, Amsterdam, 1998, pp. 102–107.

[19] Filicori, F., G. Vannini, A. Santarelli, A. Mediavilla Sánchez, and Y. Newport, "Empirical Modeling of Low-Frequency Dispersive Effects Due to Traps and Thermal Phenomena in III-V FET′ s," *IEEE Trans.*, Vol. MTT-43, 1995, pp. 2972–2981.

[20] Jarndal, A. H., *Large-Signal Modeling of GaN Device for High Power Amplifier Design*, Doctoral Thesis (2006), University of Kassel, published by the Kassel University Press GmbH, 2007.

[21] Kohn, E., I. Daumiller, M. Kunze, M. Neuburger, M. Seyboth, T. J. Jenkins, J. S. Sewell, J. Van Norstand, Y. Snorchkova, and U. K. Mishra, "Transient Characteristics of GaN-Based Heterostructure Field-Effect Transistors," *IEEE Trans.*, Vol. MTT-51, 2003, pp. 634–642.

[22] Meneghesso, G., G. Verzellesi, R. Pierobon, F. Rampazzo, A. Chini, U. K. Mishra, C. Canali, and E. Zanoni, "Surface-Related Drain Current Dispersion Effects in AlGaN-GaN HEMTS," *IEEE Trans.*, Vol. MTT-51, 2004, pp. 1554–1561.

[23] Jardel, O., F. De Groote, T. Reveyrand, J.-C. Jacquet, C. Charbonniaud, J.-P. Teyssier, D. Floriot, and R. Quéré, "An Electrothermal Model for AlGaN/GaN Power HEMTs Including Trapping Effects to Improve Large-Signal Simulation Results on High VSWR," *IEEE Trans.*, Vol. MTT-55(12), 2007, pp. 2660–2669.

[24] Rudolph, M., C. Fager, and D. Root (Eds.), *Nonlinear Transistor Model Parameter Extraction Techniques*, Cambridge University Press, 2012.

Chapter 10

Vector Network Analyzer: Operation Principle and Error Models

10.1 INTRODUCTION

Device models are truly nonlinear. However, if the device is operated in a fixed bias point and the modulation is low (small-signal operation), the model can be linearized. Small-signal models are the basis for the development of large-signal (nonlinear) device models. In common source configuration FET devices can be described by two-port network parameters, such as *Z*, *Y*, *H* (see Appendix A). The determination of these parameters rely on the measurement of voltages and currents with terminals being open or shorted. These impedances can be easily realized at low frequencies. However, at microwave frequencies, this is difficult to achieve as open and short circuits involve stray capacitances and lead inductances. Therefore, at higher frequencies S-parameters are commonly measured using a network analyzer to characterize the FET device under small-signal operation conditions. In this case, the device under test (DUT) is terminated with matched loads, which enables very broadband measurements.

As a huge amount of literature exists about vector network analyzers, this chapter gives only a brief overview about their development and state of the art, the basic construction, system measurement errors, and system calibration. The interested reader will find additional information, for example, in [1–4].

10.2 EVOLUTION OF THE VECTOR NETWORK ANALYZER

In 1967, the vector network analyzer HP 8410A, manufactured by Hewlett-Packard (now Keysight Technologies), was one of the earliest commercially available systems [5,6]. It was a pure analog system. Measurement error correction via an external computer was already included. Manual operation was both tedious and time-consuming because calibration of the system and measurement of the two-port under test had to be done at any desired measuring frequency in alternating steps always involving mechanical changes of the measurement setup. First a SHORT calibration standard was used for reflection measurement. Then, after replacing the device by the DUT, reflection measurement could be performed. Transmission measurement required a similar laborious approach (i.e., direct connection of the input and output ports using flexible coaxial arms, then insertion of the DUT and measurement). Afterward complete S-parameter measurement required turning of the device. The frequency range was from 110 MHz to 12.4 GHz or optionally to 18 GHz, depending on the system configuration. Frequency downconversion was done in two steps. The first IF was at 20 MHz, which ensured safe locking of the PLL. Magnitude and phase detection needed a second frequency conversion at 200 kHz. After amplification and signal processing the result was displayed on a screen. The system can be described as a dual channel receiver that performed the function of a ratiometer between two signals (one-path, two-port network analyzer). Complete two-port characterization required reversion of the device.

Today a full automatic operation is standard. In the calibration process the nonideal parasitic properties of the network analyzer are analyzed and quantified. Accurate calibration presumes the availability of calibration standards whose electrical characteristics are very accurately known (precision calibration standards). The influence of different error sources of the instrument on the measurement are structured in the form of error network models. The calibration-deduced knowledge of the error parameters then allows an error correction of the measured devices' scattering parameters.

The HP 8410A system was continuously improved until the manufacture was discontinued in 1984. The performance of the successor model HP 8510A (updated HP 8510B in 1988) had been substantially improved [7]. The microwave source was a synthesizer with 1-kHz resolution (optionally 1 Hz). The first mixer was a four-channel device; (i.e., all four S-parameters can be measured without turning the DUT). The PLL comprised its own processor. A powerful computer with permanently running control software was installed in the main unit. Signal processing, beginning with the second IF, included analog-digital converters and a digital signal processor DSP. Measurement and error correction were carried out in quasi-real-time. The system covered a frequency range from 45 MHz to 26.5 GHz, which was then extended to 40 GHz and 50 GHz, respectively. The system comprised the individual system component synthesizer, test set, and analyzer/display and was commonly mounted in a 19″-system rack. Using frequency extenders, the frequency range could be extended up to 110 GHz in waveguide technology.

In 1987, the competing coaxial Wiltron 360 vector network analyzer was brought to market, offering a frequency range up to 40 GHz. The system concept was very similar, as discussed. However, it used three intermediate frequencies (89 MHz, 2.25 MHz, and 83⅓ kHz). The analyzer comprised an RF source, test set and analyzer. The coaxial test set covered a frequency range up to 60 GHz (optionally 62.5 GHz) [8]. An upgrade for a 110-GHz version used waveguide technology. The measurement accuracy was comparable with the HP 8510A.

In 1997, the Agilent 8510XF vector network analyzer came onto the market, first presented on MTT-S in Denver, in May 1997 [9]. The system involved a broad frequency coverage, from 45 MHz to 110 GHz, in a single sweep. The 8510XF was therefore not a banded system and did not require the user to change any hardware in order to change measurement frequencies. Previous millimeter-wave systems (85106 and 85109) were implemented in waveguide and these systems required hardware changes when switching measurement frequencies across waveguide bands. The 8510XF systems are implemented in a 1.0- mm coaxial technology and are not limited by the banded nature of waveguide. The test ports were 1.0-mm male connectors. Cascade Microtech designed a wafer probe station specifically to accommodate the 8510XF test heads (internal port power level control, less than 1 dB over full frequency sweep). All frequency band switching was performed internally. There was no need to connect and disconnect banded coaxial test sets or waveguide modules as measurements move from one frequency band to the next.

Since 2014, Agilent has offered the N5251A vector network analyzer in 1.0-mm coaxial technology, which is considered as a direct replacement for the 8510XF with improved performance (higher measurement accuracy) [10]. The system can be fully integrated into a Cascade probe station and provides a single sweep from 10 MHz to 110 GHz using the Cascade 1.0-mm infinity probes. Regarding S-parameter measurements, the handling has become much simpler, so that broadband measurements can be performed from 10 MHz up to 110 GHz with fixed mounting of the DUT, which increases measurement accuracy. Former coaxial vector network analyzers with limited frequency coverage (e.g., 50 GHz (HP system) or 60 GHz (Wiltron) coaxial systems) needed additional waveguide extensions (two separate waveguide bands) to extend the frequency range. In the case of broadband measurements up to 110 GHz, the DUT had to be contacted at least twice. Regarding on-wafer measurements, probe heads had to be remounted, which made new calibration necessary unless a second wafer prober is in place. Earlier banded vector network analyzers delivered measured data with a discontinuous curve shape at the transition frequency of adjacent frequency bands (see Section 11.3).

Today, new trends in vector network analyzer (VNA) application have become visible. The classical VNA with a focus on S-parameter measurements has meanwhile mutated to a comfortable nonlinear measurement system with salient features supporting direct nonlinear device modeling [11]. Furthermore, the flexibility and complexity of VNAs have progressed. For example, in [12] the spectrum analyzer capability has been integrated into a VNA instrument.

Another trend focuses on the extension of frequency bands. The terahertz technology with challenging new applications, for example, in mobile communications, automotive, and security sectors, requires innovative electronic and optical devices, which need an appropriate measurement infrastructure that goes beyond traditional instrumental capabilities. With frequency extenders, the frequency of commercially available vector analyzers can currently be extended up to 500 GHz [13], to 1.1 THz [14], and to1.5 THz [12, 15, 16].

Network analyzer extenders for THz applications are offered, for example, by OML, Inc.© (formerly known as Oleson Microwave Labs) up to 500 GHz [17] and by Virginia Diodes Inc. up to 1.5 GHz, driving development to 3.1 THz in the near future [18–20]. Challenges in the fabrication of waveguides (mechanical tolerances, careful flange alignment) for S-parameter measurements well above 1 THz are discussed in [21].

10.3 VECTOR NETWORK ANALYZER CONSTRUCTION AND OPERATION

As already pointed out, a huge amount of literature exists concerning construction, functioning, and application of vector analyzers. Therefore, in this section only some basics are briefly replicated, which may be useful for later discussion.

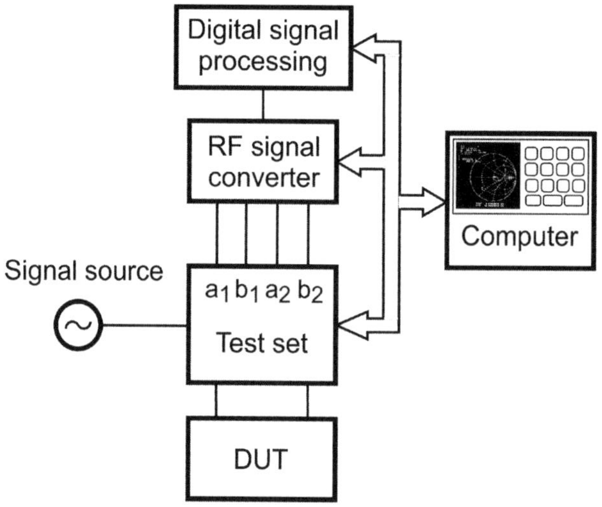

Figure 10.1 Major system components of a vector network analyzer.

Figure 10.1 depicts the major parts of a vector network analyzer system. The *signal source* supplies the stimulus for the system. Earlier network analyzers included a wobble generator. However, modern analyzers comprise programmable synthesizers so that the S-parameters can be automatically measured as function of frequency. Synthesizers are generally more stable and more accurate signal sources than wobble generators. In addition, synthesizers exhibit lower phase noise, which is advantageous for measuring low-noise devices. Synthesizers also permit higher frequency resolution.

The *test set* separates the incident and reflected/transmitted waves a_1, b_1, a_2, and b_2 at the test ports and delivers these to the receiver/detector unit (*RF signal converter*). Separation can be realized by resistive reflectometer bridges or dual-directional couplers, also termed reflectometer. In addition, power dividers (splitters) are needed to obtain the reference signal from the signal generator. We distinguish between a *one-path* two-port test set and a *two-path* two-port test set.

The one-path two-port test set permits reflection and transmission measurement of a two-port with only one signal flow direction. A full two-port measurement with such a system requires manually reversing the DUT between the test ports.

The two-path two-port test set enables all four S-parameters of the device-under-test DUT to be measured (after suitable calibration) with a single insertion between the two test ports. This is shown in Figure 10.2 illustrating the architecture of a four-channel vector network analyzer. The test set of the VNA comprises two reflectometers, which are made up of two directional couplers each (a dual directional coupler can also be used). The couplers sample the incident and reflected signals for the device under test and feed them to harmonic mixers that convert the signals directly to the 20-MHz IF of the network analyzer. Isolators placed ahead of the mixers prevent unwanted harmonics generated by the mixers from producing spurious responses that would interfere with the measurement.

The source can be switched to excite port 1 or port 2 of the device under test DUT. The switch must have excellent repeatability. The switch also provides a Z_0 termination (usually

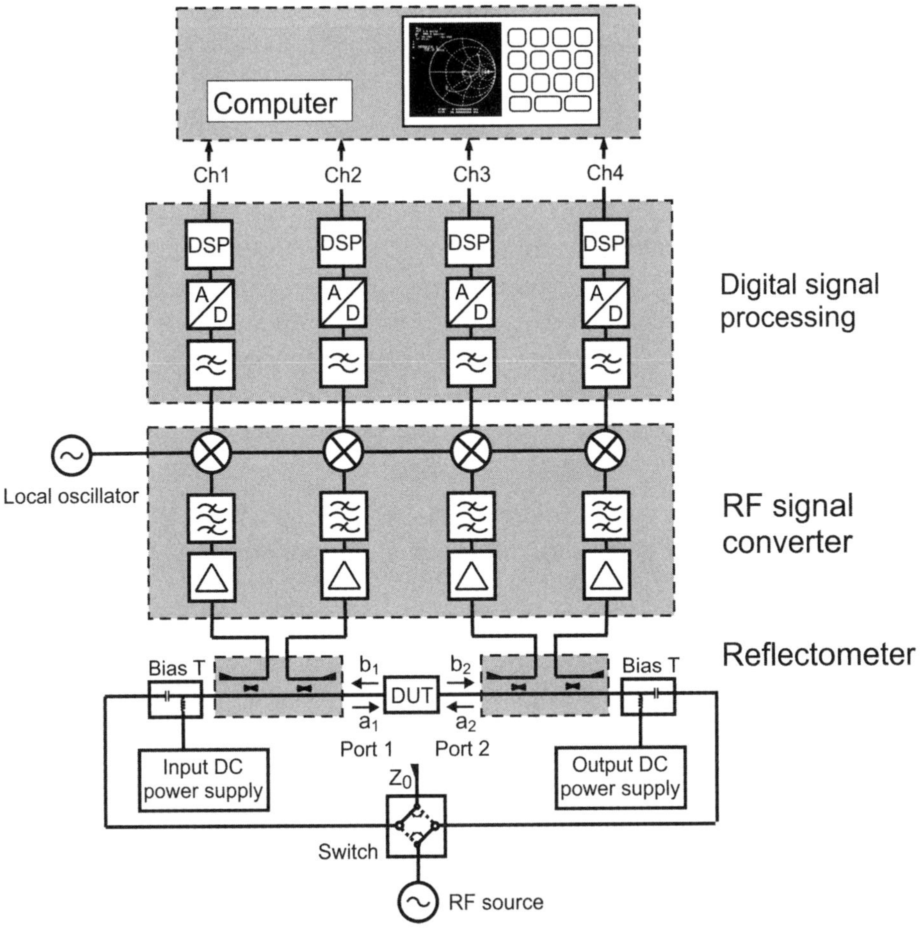

Figure 10.2 Simplified architecture of a four-channel vector network analyzer.

50Ω) for the output port in each direction.

The mixer in the *RF signal converter* unit is used to downconvert the detected RF signals to a fixed low intermediate frequency IF. The frequency conversion requires well-matched broadband mixers, which are controlled by two signals of different frequency: the measuring frequency f_S and the mixing oscillator frequency f_{LO}. At the output of the mixer an intermediate frequency signal (IF) with the difference frequency $f_{IF} = |f_S - f_{LO}|$ appears. Using a phase-locked loop (PLL) the difference frequency is kept constant [5]. For reason of moderate control velocity a first intermediate frequency is chosen to be relatively high (e.g., 20.278 MHz) so that final measurement of the amplitude and phase requires a further mixing stage. Finally, in the *digital signal processing* unit, the downconverted complex signals are digitized by an analog/digital converter (ADC).

Signal processing and control of the total system is commonly provided by an integrated *computer*. The obtained results can be displayed in different ways; for example, illustrative representation of frequency-dependent complex scattering parameters in the complex plane (Smith chart) or separate presentation of scattering parameters with respect to frequency-dependent magnitude and phase.

The S-parameters of the DUT are defined as follows: (1) $S_{11} = b_1/a_1$, switch in forward direction, (2) $S_{21} = b_2/a_1$, switch in forward direction, (3) $S_{12} = b_1/a_2$, switch in reverse direction, and (4) $S_{22} = b_2/a_2$, switch in reverse direction. Frequency-domain reflected or transmitted wave responses, $b_1(\omega)$ and $b_2(\omega)$, can be given as linear combinations of its frequency-domain incident wave excitations $a_1(\omega)$ and $a_2(\omega)$,

$$b_1(\omega) = S_{11}(\omega)a_1(\omega) + S_{12}(\omega)a_2(\omega) \tag{10.1}$$

$$b_2(\omega) = S_{21}(\omega)a_1(\omega) + S_{22}(\omega)a_2(\omega) \tag{10.2}$$

where these incident and reflected/transmitted waves can be related to the ports′ currents and voltages by

$$a_{1,2}(\omega) = (V_{1,2} + Z_0 \cdot I_{1,2})/2 \tag{10.3}$$

$$b_{1,2}(\omega) = (V_{1,2} - Z_0 \cdot I_{1,2})/2 \tag{10.4}$$

where Z_0 is the ports′ characteristic impedance, usually assumed to be 50Ω.

Because of nonidealities of the receiver and test sets, the readings of the *digital signal processing* and *display* unit are not the actual incident and reflected/transmitted waves in the DUT′s reference planes (port 1 and port 2), but raw (uncorrected) versions of them. Their correction requires a calibration model that is capable of describing the referred VNA nonidealities. Usually these are assumed to be linear (because appropriate controlled attenuators and variable gain amplifiers are used to prevent any possible VNA receiver saturation) and associated with the test set′s directional couplers in the following way.

10.4 ERROR-CORRECTED MEASUREMENTS

Here we will discuss the main error sources in connection with VNA measurements. The measurement capability relies on the basic component of a fourport directional coupler shown in Figure 10.3. The scattering matrix of the four-port device reads

$$(S) = \begin{pmatrix} S_{11} & S_{12} & S_{13} & S_{14} \\ S_{12} & S_{22} & S_{23} & S_{24} \\ S_{13} & S_{23} & S_{33} & S_{34} \\ S_{14} & S_{24} & S_{34} & S_{44} \end{pmatrix} = \begin{pmatrix} \rho & \tau & \kappa & \sigma \\ \tau & \rho & \sigma & \kappa \\ \kappa & \sigma & \rho & \tau \\ \sigma & \kappa & \tau & \rho \end{pmatrix} \tag{10.5}$$

with the transmission coefficient τ, coupling coefficient κ, eigen-reflection coefficient ρ, and isolation coefficient σ. Under ideal conditions, all ports are matched (ρ = 0), and ports 1 and 4, and ports 2 and 3 are decoupled (σ = 0). In this case the scattering matrix simplifies to

$$(S) = \begin{pmatrix} 0 & S_{12} & S_{13} & 0 \\ S_{12} & 0 & 0 & S_{24} \\ S_{13} & 0 & 0 & S_{34} \\ 0 & S_{24} & S_{34} & 0 \end{pmatrix} = \begin{pmatrix} 0 & \tau & \kappa & 0 \\ \tau & 0 & 0 & \kappa \\ \kappa & 0 & 0 & \tau \\ 0 & \kappa & \tau & 0 \end{pmatrix} \tag{10.6}$$

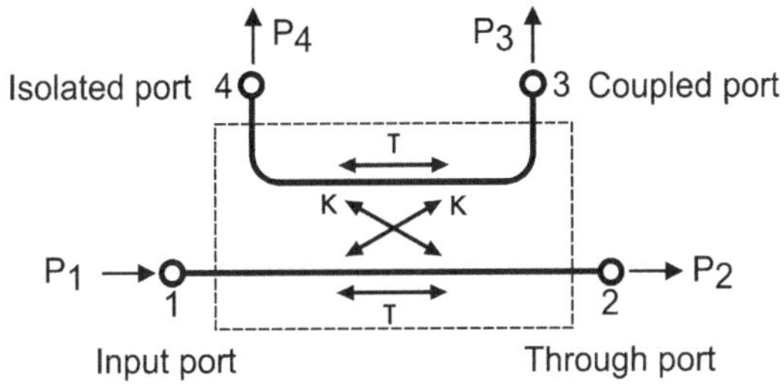

Figure 10.3 Ideal directional coupler with characteristic parameters.

The properties of the coupler are characterized by the following parameters:

Coupling C

C (dB) is the ratio of the incident power P_1 fed into the through transmission line at input port 1 to the power P_3 at coupled port,

$$C(\mathrm{dB}) \equiv 10\log\frac{P_1}{P_3} = 10\log\frac{1}{|S_{31}|^2} = 10\log\frac{1}{|\kappa|^2} \tag{10.7}$$

or

$$C = -20\log|\kappa| \ \mathrm{dB} \tag{10.8}$$

Frequency sensitivity (or flatness) characterizes the maximum peak-to-peak variation in coupling coefficient that may be expected over a specified frequency band.

Directivity D

D (dB) is defined as the ratio of the power P_3 at the coupled port 3 to the power P_3 transmitted to the isolated port 4,

$$D = 10\log\frac{P_3}{P_4} = 10\log\frac{P_3/P_1}{P_4/P_1} = 10\log\frac{|S_{31}|^2}{|S_{41}|^2} \quad (\mathrm{dB}) \tag{10.9}$$

or

$$D = 20\log\frac{|\kappa|}{|\sigma|} \quad (\mathrm{dB}) \tag{10.10}$$

If the directional coupler is ideal, then $D \to \infty$.

Isolation I

I (dB) describes the degree to which port 4 is isolated from the input port 1, and is defined as

$$I = 10\log\frac{P_1}{P_4} = 10\log\frac{1}{|S_{41}|^2} \tag{10.11}$$

or

$$I = -20\log|\sigma| \quad (\mathrm{dB}) \tag{10.12}$$

Similarly, if the directional coupler is ideal, then $I \to \infty$. These three quantities are related by

$$I = D + C \quad (\mathrm{dB}) \tag{10.13}$$

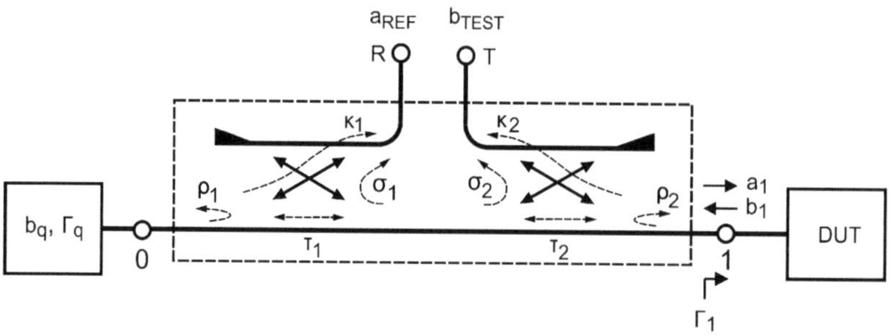

Figure 10.4 Real reflectometerr with characteristic parameters.

Insertion Loss L

L (dB) describes the power loss along the main transmission line from port 1 to port 2, and is defined as

$$L \equiv 10\log\frac{P_1}{P_2} = 10\log\frac{1}{|S_{21}|^2} \quad \text{(dB)} \tag{10.14}$$

or

$$L = -20\log|\tau| \quad \text{(dB)} \tag{10.15}$$

The reflectometer is a dual directional coupler, which comprises two directional couplers in cascaded connection as shown in Figure 10.4. It permits the measurement of the complex reflection coefficient of a one-port device.

Figure 10.4 shows the setup for the measurement of the reflection coefficient of a 1-port, using a reflectometer. Most of the power injected from the wave source (b_q, Γ_q) is transmitted to the DUT, and a smaller part is coupled out to the reference channel R. The wave reflected from the DUT is partly redirected to the source and is absorbed when the source is matched to the characteristic impedance of the lines. Another part of the reflected signal is coupled out to the test channel T. The reflection coefficient of the DUT is then calculated as the complex ratio of reflected wave b_{TEST} and incident wave a_{REF}. The connected signal generator does not influence the measurement of the reflection coefficient because its parameters b_q and Γ_q are cancelled with the ratio formation (i.e., the reflection measurement of the reflectometer is valid for arbitrary sources). It should be noted that indexed waves a and b are peak values and not effective values of the corresponding sine functions.

The nonperfect reflectometer can be modeled by a cascaded network comprising an ideal reflectometer and an error-two-port (Figure 10.5). The error-two-port covers all systematic errors of the component hardware.

Regarding an ideal reflectometer the wave quantities a_0 and b_0 at the reference and test ports correspond to the forward and reverse waves in the fictitious detector reference plane. The parameters of the two-port error-network are determined by VNA calibration. If the error-two-port is known, then the measured complex waves a_1 and b_1 can be mapped onto the inner fictitious detector plane providing the error-corrected waves a_0 and b_0.

The measurement of a two-port requires a second reflectometer (Figure 10.6). With a switch the output signal of the RF source can be transmitted either in forward direction to the port 1 or in reverse direction to port 2 of the device.

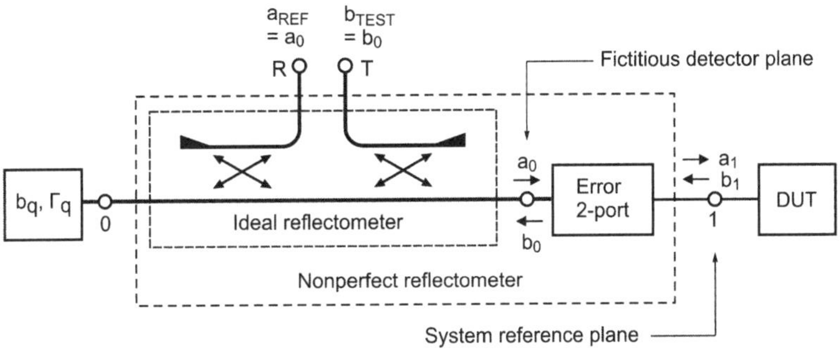

Figure 10.5 Modeling of a nonperfect reflectometer device.

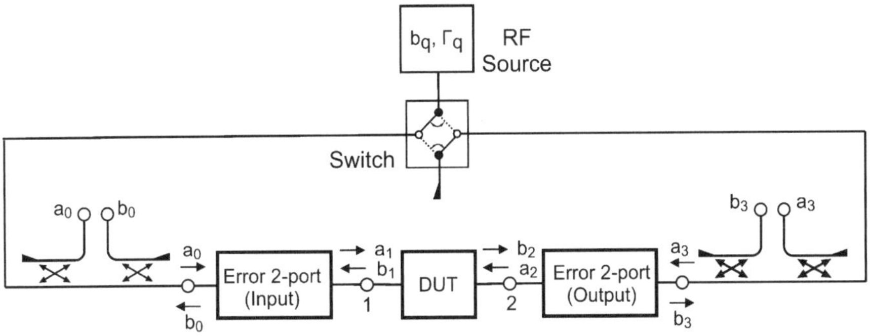

Figure 10.6 Two cascaded reflectometers used for two-port measurement.

10.4.1 Error Models

VNA error models and calibration methods have been widely discussed (e.g., [22, 23]) so that in the following, only a brief review of the various error sources originating in the nonperfect hardware of the network analyzer will be given.

A complete characterization of a two-port network needs a reflection and transmission measurement in forward and backward directions. Figure 10.7 shows the signal flow graph of the error model in forward direction. e_{00} and e_{11} are the eigen-reflection coefficients, and $e_{10}e_{01}$ the reflection transmission of the input reflectometer. e_{00} is also termed directivity error as it describes cross talk of the directional coupler, and e_{11} denotes the source match error, which characterizes multiple reflection due to non-Z_0 input and output impedance of the cables and connectors. The transmission product $e_{10}e_{01}$, also termed reflection tracking error, describes the frequency response of the couplers, the test cables, and the detectors. It is mainly a scaling factor. The eigen-reflection coefficient e_{22} of the output reflectometer is the load match error. Generally the DUT is not matched at the output. The product $e_{10}e_{32}$ denotes the transmission tracking error in the forward direction. The term e_{30} describes the cross talk (leakage) between port 1 and 2 of the DUT.

Similarly, for the backward measurement the same six-term error model can principally be used as shown in Figure 10.8. Corresponding error parameters are denoted as primed quantities.

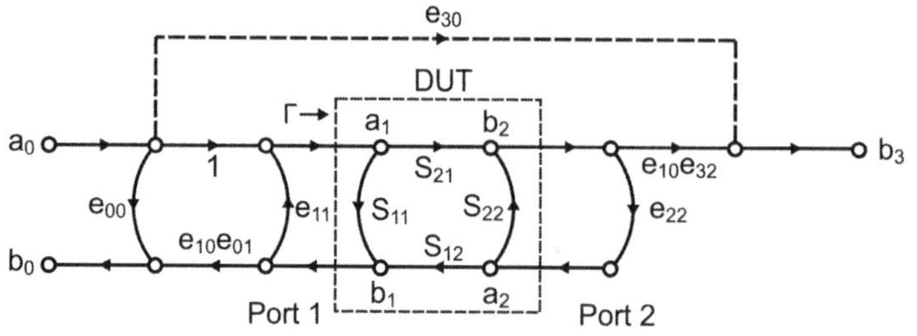

Figure 10.7 6-term error model in forward direction (e_{11} source match error).

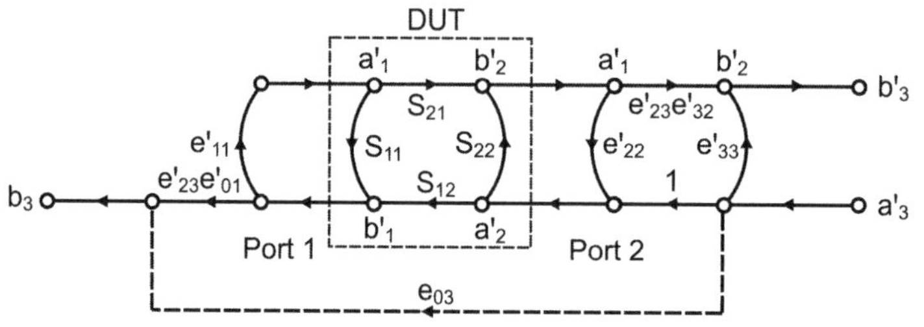

Figure 10.8 6-term error model in reverse direction.

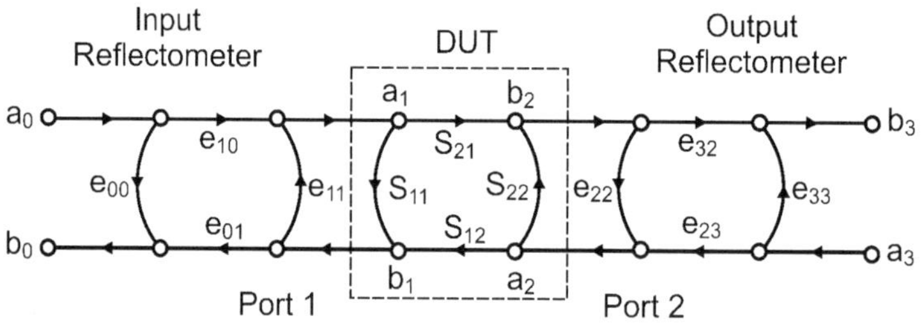

Figure 10.9 Signal flow diagram for the 8-term error model.

Thus, it has been shown that the complete error model of an automatic vector network analyzer is described by 12 error parameters (12-term error model). Under certain conditions the error model can be simplified. In this case we assume that the switch in Figure 10.6 is perfect. Then the following relations are valid: $e_{11} \approx e_{11}'$ and $e_{22} \approx e_{22}'$. In this case the error parameters are independent on the signal flow direction. If we, in addition, ignore the cross-talk, the error model reduces to an 8-term error model, which is mostly applied for vector network analyzer calibration purposes. Figure 10.9 shows the signal flow graph of the unique error model that can be used for measurement in both the forward and backward directions. Assuming that all four reflection and transmission coefficients in the forward and reverse directions were measured, and that the error parameters of the network analyzer are known, then the error-corrected two-port scattering parameters S_{11}, S_{12}, S_{21}, and S_{22} can be calculated.

10.4.2 Calibration

As has been pointed out in the previous section, at microwave frequencies, systematic effects such as leakage, test port mismatch, and frequency response of the vector network analyzer will affect measured data. However, in a stable measurement environment these effects are repeatable and can be measured by the network analyzer. This process is termed vector network analyzer calibration. During calibration, we connect a series of known devices called calibration standards. The systematic effects are determined as the difference between the measured and known responses of the standards. The process of mathematically removing these systematic effects is called error correction. The systematic errors of the measurement system are eliminated.

Frequently used calibration methods are SOLT, TRL, and LRM. SOLT calibration uses short, open, load, and through (shortly thru) standards. SOLT calibration method is often applied to the calibration of measurement systems in coaxial technique. Regarding planar transmission lines (e.g., the microstrip) the required precision standards are generally not known so that a more useful calibration is needed. TRL calibration uses through, reflect, and delay standards. Instead of direct through standard a line of finite length is also alternatively used. In this case the calibration method is termed LRL. Line standards are used as reference with low frequency limitation. LRM calibration employs line, reflect, and match standards. This method is easy to use and provides accurate calibration for all frequencies.

As was discussed, the systematic errors of a VNA can be entirely described with the 12- and 8-term error model, respectively. The quality of the determination of the error parameters during the calibration process depends decisively on the quality of the used standards. But beyond that, human intervention is a decisive factor for successful calibration. Indeed, calibration subject is not very attractive for young researchers, who nevertheless should become aware of the great importance of this starting step in device characterization so that the modeling task will have been in vain.

10.5 SUMMARY

The VNA is the basic instrument for the measurement of scattering parameters of microwave devices. Its rapid evolution with respect to accuracy and expansion of measurement frequency bandwidth was. VNA construction and operation was outlined. System error models and calibration methods were briefly discussed.

REFERENCES

[1] Hiebel, M., *Fundamentals of Vector Network Analysis*, 5th Ed., Rohde & Schwarz, 2011.

[2] Roblin, P., *Nonlinear RF Circuits and Nonlinear Vector Network Analyzers*, Cambridge, U.K.: Cambridge Univ. Press, 2011.

[3] Dunsmore, J. P., *Handbook of Microwave Component Measurements with Advanced VNA Techniques*, John Wiley & Sons, Ltd., 2012.

[4] Kirkby Microwave, *Useful Links and Other Resources—Application Notes (Keysight, Anritsu, Rohde & Schwarz), Books*, https://www.kirkbymicrowave.co.uk/Support/Links/.

[5] Hewlett Packard Co., *8410A Network Analyzer and S-Parameters*, Hewlett-Packard Journal, Total Issue February 1967 (24 pages).

[6] Hewlett Packard Co., *Microwave Network Analyzer Applications*, Application Note 117-1, June 1970.

[7] Hewlett Packard Co., *Millimeter-Wave Measurements Using the HP 8510B Network Analyzer*, Product Note 8510-12, November 1, 1987.

[8] Wiltron Co., *Model 360 Vector Network Analyzer*, Product Note, January 1991.

[9] Hewlett Packard Co., *HP 8510XF—Single-Connection, Single-Sweep Systems*, September 1997.

[10] Keysight Technologies, *Millimeter-Wave Network Analyzers 10 MHz to 110 GHz, with Extensions to 1.1 THz*, October 6, 2014.

[11] Root, D., "Future Device Modeling Trends," *IEEE Microwave Magazine*, November/December 2012, pp. 45-59.

[12] Keysight Technologies, *THz Spectrum Analysis with a Network Analyzer*, Microwave Journal, www.microwavejournal.com/articles/26933-thz-spectrum-analysis-with-a-vector-network-analyzer.

[13] Rohde & Schwarz, *R&S®ZNA Vector Network Analyzer—Masters the Most Challenging Measurement Tasks*, Product Brochure, Version 01.01, 2019.

[14] Anritsu Company, *VectorStar—High Performance, Broadband Network Analysis (Millimeter Waveguide System: 50 GHz to 1.1 THz)*, Technical Data Sheet, 2018.

[15] Keysight Technologies, *Paving the Way for Research and Innovation—Solution Brochure* (Device Characterization in TeraHertz, 2018.

[16] Keysight Technologies, *Banded Millimeter Network Analysis—Technical Overview*, 2018.

[17] OML, *VNA Extension Modules*, https://www.omlinc/products/vna-extension-moduls.

[18] Hesler, J. L., "From GHz to THz: The Essentials of Very High Frequency Signal Measurements," www.keysight.com/upload/cmc_upload/All/WS5_Monday_14.15_Jeffrey_Hesler.pdf.

[19] Hesler, J. L., Y. Duan, B. Foley, and T. Crowe, "THz Vector Network Analyzer Development & Measurements", 2010, Virginia Diodes, Inc., www.vadiodes.com.

[20] Koller, D., S. Durant, C. Rowland, E. Bryerton, and J. Hesler, "Initial Measurements with WM164 (1.1-1.5THz) VNA Extenders", *41st Int. IEEE Conf. on Infrared, Millimeter, and Terahertz Waves* (IRMMW-THz), 2016 (2 pages).

[21] Koller, D., S. Durant, E. Bryerton, J. Hesler, and A. Arsenovic, "Challenges of WM164 S-Parameter Measurements Due to Flange Dimensional Tolerances", *41st Int. IEEE Conf. on Infrared, Millimeter, and Terahertz Waves* (IRMMW-THz), 2016 (2 pages).

[22] Rytting, D., "Network Analyzer Error Models and Calibration Methods," *RF and Microwave Measurements for Wireless Applicants (ARFT/NIST Short Course Notes)*, 1996.

[23] Keysight Technologies, *Specifying Calibration Standards and Kits for Keysight Vector Network Analyzers*, Application Note, June 29, 2016.

Chapter 11

Uncertainties in the Device Modeling Process

11.1 Introduction

This chapter details imperfections inherent in the measurement process of device characterization illustrating their influence on the model parameter extraction and quality of device model finding. Commonly, measurement errors originate in measurement uncertainties of used measuring systems. Measurement results are also affected by the experimenter-dependent variances of measurement performance, including careful setup calibration and calibration verification, diligence in device contacting, as well as measurement data verification and confirmation by measurement repetition. Even relatively small measurement errors can cause large errors in the extraction of model parameter values, which applies, for example, to low-sensitive model elements such as the low-valued series resistances.

Using some relevant terms in the field of measurement uncertainty, some practical examples of frequently occurring measurement errors will be presented and discussed. At the end of this chapter, from experience, some recommendations are given that may help to perform the model parameter extraction successfully.

11.2 DEFINITION OF MEASUREMENT TERMS

There is a vast amount of literature concerning measurement uncertainty subjects such as accuracy, precision, repeatability, and reproducibility. Terminology and notation related to measurement uncertainty is not used consistently among experts. There are well-recognized standards organizations, such as the International Organization for Standardization (ISO) [1, 2] and the National Institute of Standards and Technology (NIST) ([3]). Each of these organizations promulgates standards covering a wide range of scientific and technological activities. In [4] the meaning of terms defined in the most often cited references [5-7] have been reviewed highlighting their different usage. Uncertainty is also discussed in connection with model parameter extraction, and is, therefore where appropriate, included in the following notations and remarks.

11.2.1 Uncertainty

A measurement result should never be accepted a priori as true. It should always be regarded with some doubt. Repeated measurements, for example, can enhance the confidence in the measured value. But that is not enough. All elements of the measuring chain (hardware and software) must be controlled and continuously subjected to critical inspections. This includes measurement system calibration as often as necessary, measurement system error analysis for error correction of measured data, and, in particular, permanent control of the (precision) calibration standards (wear, mechanical damage) that are being used. The last aspect is of great importance because the degree of degradation of the electrical performance of necessary calibration standards depends mainly on the user's skill and expertise. Note that the quality of a device commonly changes gradually with frequency of use.

Measurement uncertainty is also an important subject concerning model parameter extraction as will be discussed in the following chapters. The extraction results depend on the accuracy of the measured device data (depending on the systematic and random errors of the measuring instrument and the random errors commonly induced by the skill of the experimenter taking the measurements), but also, to a large extent, on the applied extraction algorithm aimed to find some global minimum in the model parameter space without being trapped in a local minimum.

Having taken care in reducing systematic errors by careful calibration of the measuring instrument, measurement (extraction) uncertainty can be analyzed. It characterizes the *dispersion* (variability, scatter, or spread) of the measured (extracted) values, and can be statistically analyzed, which is expressed as standard deviation. Most uncertainties (more or less) follow the Gaussian (normal) distribution. A normal distribution is described by the mean (μ) and variance (σ^2),

$$f(x) = \frac{1}{\sqrt{2\pi\sigma}} \exp\left[-\frac{(x-\mu)^2}{2\sigma^2}\right] \tag{11.1}$$

The standard deviation is defined as

$$\sigma = \sqrt{\frac{1}{N}\sum_{i=1}^{N}(x_i - \bar{x})^2} \tag{11.2}$$

where (x_1, x_2, ..., x_N) are the observed values of the measured (extracted) parameter, $\bar{x}$ is the mean value of these measurements (extractions), and N is the number of measurements (extractions). The arithmetic mean denoted by $\bar{x}$ is simply defined as the sum of x_i, $i = 1...n$, divided by n,

$$\bar{x} = \frac{1}{N}\sum_{i=1}^{N} x_i \tag{11.3}$$

Figure 11.1 shows a plot of the standard normal distribution (or bell-shaped curve) with the mean value $\mu = 0$ and standard deviation $\sigma = 1$.

The variance is the average squared distance of the standard deviation σ from the mean value μ,

$$\sigma^2 = \frac{1}{N}\sum_{i=1}^{N}(x_i - \bar{x})^2 \tag{11.4}$$

The standard deviation may be normalized, for example, to the mean value $\bar{x}$, thus becoming

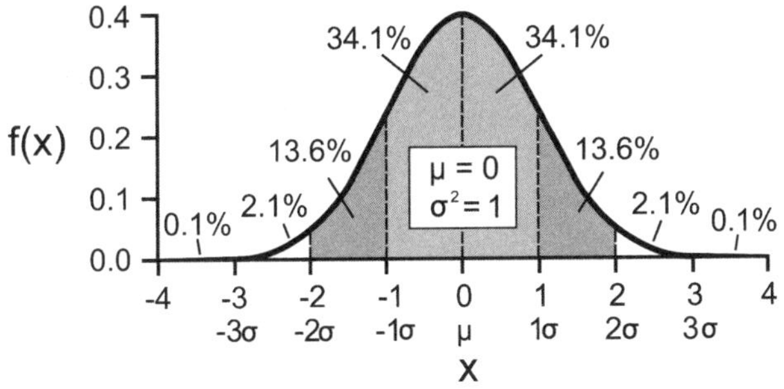

Figure 11.1 Normal distribution with the mean value $\mu = 0$ and standard deviation $\sigma = 1$. 68.27%, 95.45%, and 99.73% of the values lie within one, two, and three standard deviations of the mean value, respectively.

a dimensionless quantity expressed in percent

$$\sigma_n = \frac{\sigma}{|\bar{x}|} \tag{11.5}$$

σ_n is also denoted as the relative standard deviation or coefficient of variation. It indicates the extent of variability in relation to the mean value of distribution.

11.2.2 Measurement Errors

In general, measurements have imperfections that give rise to errors in the measurement results. Traditionally, an error is viewed as having two components: a random component and a systematic component.

11.2.2.1 Random Errors

Random errors presumably arise from unpredictable or stochastic temporal and spatial variations of influence quantities. The effects of such variations (random effects) give rise to variations in repeated measurements. Although it is not possible to compensate for the random error of a measurement result, it can usually be reduced by increasing the number of measurements (averaging). In a measurement setup, missing connector repeatability and insufficient connection cable stability can also contribute to random errors.

11.2.2.2 Systematic Errors

Systematic errors, like random errors, cannot be eliminated but can often be reduced. If a systematic error arises from a recognized effect of an influence quantity on a measurement result (systematic effect), the effect can be quantified, and if it is significant in size relative to the required accuracy of the measurement, a correction can be applied to compensate for the effect. It is assumed that, after correction, the expected value of the error arising from a systematic effect is zero. Systematic (or bias) errors of a vector network analyzer concern directivity, tracking, match, and dynamic accuracy (see Section 10.4). They cannot be removed by repeating measurements or averaging large numbers of results.

Regarding model parameter extraction, systematic errors can also result from model topology mismatch (see Sections 11.3.3 and 11.3.4), which indicates the discrepancy between the physical model and defined electrical equivalent circuit. The physical model based on physical simulation, allows the internal behavior of the device to be completely known, thereby leading to both better understanding of device operation and easier and more accurate identification of equivalent circuits [8]. An essential requirement for the acceptance of an equivalent model is the frequency independence of the extracted model parameter values.

Vaikus in his noteworthy contribution [9] defined a *minimal equivalent circuit* as one that matches a set of measured two-port S-parameters for a given set of discrete frequencies within a defined error tolerance with the smallest number of circuit parameters. A minimal equivalent circuit is the most economical representation of the measured data within the specified error tolerance. It also appears that a minimal equivalent circuit has the lowest element value uncertainty (the smallest ranges of acceptable element values). There can be more than one minimal equivalent circuit configuration. Furthermore, the author points out that there is a price to pay for the reduced element value uncertainty of minimal equivalent circuits. The larger the difference in element count between the physical and the minimal circuits, the more obscure are the connections between the elements of the minimal circuit and the physical model. Based on experimental investigations, Vaitkus concluded that, under assumption of measured S-parameters of limited accuracy and bandwidth, if the element count of the equivalent circuit is too high, the calculated ranges of acceptable element values are too broad to be useful. If, on the other hand, the element count of the equivalent circuit is too low, no satisfactory solution can be found at all. Because of the more limited number of elements in the minimal equivalent circuit, their values are combinations of the values of the physical equivalent circuit elements. Vaitkus suggested introducing additional measurements, for example, for parasitic resistances and capacitances to establish a closer tie between the electric circuit and the physical model without increasing the uncertainty of the extracted minimal-circuit-based parameters.

11.2.3 System Calibration

Measurement system calibration means identification and determination of measurement system error parameters to provide error-corrected reliable, precise, and accurate (repeatable) measured data. This can only be realized on the basis of well-calibrated (ideally *perfect*) calibration standards. Without doubt calibration is very important for successful model parameter extraction and device modeling. Nevertheless, calibration is often regarded as an annoying problem and not treated with the necessary care. Not being dilige in this preparatory work directly affects the quality of the final device modeling step.

A calibration procedure is applied to remove as many of the errors as possible. Regarding a vector network analyzer the loss and match error can be greatly reduced depending on the accuracy of the calibration standards used. However, the noise and linearity errors cannot be reduced using a simple calibration procedure [10].

Once the network analyzer is calibrated the drift, stability, and repeatability errors will degrade the system performance. This usually means that the system will need to be recalibrated at some interval depending on the system usage, environment and required accuracy [10].

It is now widely recognized that, when all of the known or suspected components of error have been evaluated and the appropriate corrections (calibrations) have been applied, there still remains an uncertainty about the correctness of the stated result; that is, a doubt about how well the result of the measurement represents the value of the quantity being measured [10].

11.2.4 Accuracy

Measurement accuracy or trueness is a determination of the performance of a calibrated measurement process run repeatedly under the entire range of measurement conditions. The intent of any calibration process is to remove all of the bias errors from the measurement process. Note that additional sources of error can be introduced by the calibration process itself, including random errors associated with the raw data for the calibration artifacts (from connector repeatability, noise effects, etc.), random errors associated with the calibration process, and systematic errors resulting from the imprecise knowledge about the calibration standards [11]. The accuracy of a measurement can be increased by taking a number of repeat readings to calculate the mean value of the readings [3].

11.2.5 Precision

Measurement precision is the ability of an instrument in measuring a quantity in a consistent manner with only small relative deviation between readings. It is the ability to replicate a given result. In other words, precision concerns the variability or scatter of the individual results of a set of measurements. Measurement precision is related to random measurement error and usually expressed numerically by measures of imprecision, such as standard deviation [3]. A measurement can be precise but wrong (inaccurate).

11.2.6 Repeatability

Measurement repeatability describes the variability in measurement results. It means that multiple measurements taken from a measurement system are consistent around the same value. It is the measurement precision under conditions that include the same measurement procedure, operators, measuring system, operation conditions, and location, and replicate measurements over a short period of time. Repeatability may be expressed quantitatively in terms of the dispersion characteristics of the results [1].

11.2.7 Reproducibility

In contrast to the measurement repeatability definition, reproducibility measures how the results of a measurement vary when the measurement is repeated under changed conditions and over long periods of time [11]. This may include *different* locations, operators, measuring systems, and replicate measurements on the object of interest. Thus, a measurement system must first be repeatable before it can be reproducible [3]. Reproducibility may be expressed quantitatively in terms of the dispersion characteristics of the results [1].

With respect to VNA measurements, it is stated in [11] that measurement reproducibility is a determination of the performance of a measurement process run repeatedly under typical measurement conditions. A measurement process usually requires some disassembly of the measurement system to change the DUT to another device or circuit, perhaps to insert a calibration artifact. Reproducibility should provide an indication of the total of all of the random errors such as connector repeatability or fixture repeatability.

11.2.8 Reliability

Measurement reliability refers to the ability to measure something consistently and the extent to which it is accurate and stable. Reliable measurement results are reproducible. A reliable measurement yields consistent (constant, stable) results when repeated measurements are taken. Measurement reliability refers to the instrument's consistency, stability, or precision. Reliability incorporates the meaning that the measurement is dependable and that it will give the same measurement result every time. This contributes to increased confidence in the measurement procedure.

The reliability of an instrument commonly decreases with time after calibration. Therefore, it should be calibrated periodically. Measurement robustness may include frequent calibration to properly compensate for drift errors and careful monitoring of the measurement uncertainty.

11.2.9 Validity

Measurement validity is a determination of the performance of a calibrated measurement process in gauging the actual (or true) value of some quantity. Since any a priori knowledge of this value must be derived from other measurements, it is extremely difficult if not impossible to know with absolute certainty the actual (or true) value of any quantity. Indeed, one can never be certain if the calibration process has removed all of the bias errors [11].

To get information on comparability of measurement results, independent interlaboratory comparisons (intercomparisons) can be performed to guarantee the validity of measurements. In this case two or more parties (laboratories) measure the same device and compare their results [12]. Such round-robin experiments and interlaboratory comparisons are good approaches to knowing about the measurement process (an easy way to have confidence in your measurements) [13]. A measured (calibrated) device starts out at a pivot lab, then travels to many other reference labs to be measured, returning to the pivot lab for closure of measured data. Such mobile comparisons are also carried out between national laboratories of various countries as a routine procedure to assure international measurements at the highest level. These include NIST in the United States, NPL in the United Kingdom, and the PTB in Germany. There are also numerous calibration service organizations [14].

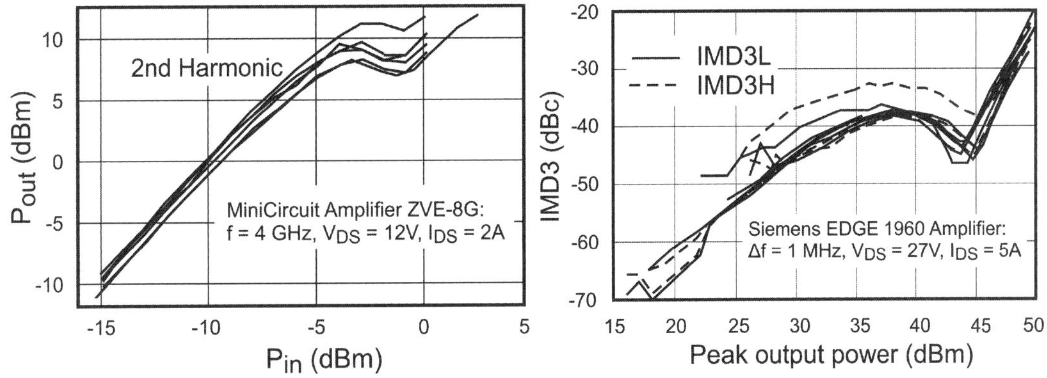

Figure 11.2 Exemplary round-robin measurement results from eight partners within TARGET activities: (a) comparison of measured second harmonic of MiniCircuit Amplifier ZVE-8G ($P_{out} \leq 30$ dBm), and (b) comparison of measured IMD3 of Siemens EDGE 1960 Amplifier ($P_{out} \leq 48$ dBm) [16].

Validity is also used in connection with measured data, which may verify the validity of a given analytical equation or expression (or approximation). Validity of a device model expresses the good agreement between simulation and experiment under various operating conditions.

Figure 11.2 shows a result of round-robin measurements carried out by several academic labs participating in the Sixth Framework Programme (FP6) of the European Union (EU) forming the consortium Top Amplifier Research Groups in a European Team (TARGET). The activity was created for the purpose of strengthening European research in the field of microwave power amplifiers for broadband wireless access [15, 16]. Eight partners participated in this task. Each partner had to measure two test amplifiers under several conditions: single-tone, two-tone, and wideband code division multiple access (WCDMA) signal measurements. The examples shown demonstrate mostly very good agreement in the measured results, indicating the validity of the performed measurements.

11.2.10 Blunders

Mistakes or blunders are procedural errors that should be avoided by paying careful attention [5]. Fortunately, these errors are usually apparent either as obviously incorrect data points or as results that are not reasonably close to expected values. They are classified as illegitimate errors and generally can be corrected by carefully repeating the operations [6].

11.3 ACCURATE MEASUREMENTS: A KEY CONDITION FOR SUCCESSFUL DEVICE MODELING

As illustrated in Figure 11.3, a successful circuit simulation is based on a reliable FET device model that can be attained from an experimental modeling approach. The device characterization has turned out to be the fundamental and most significant step in the whole

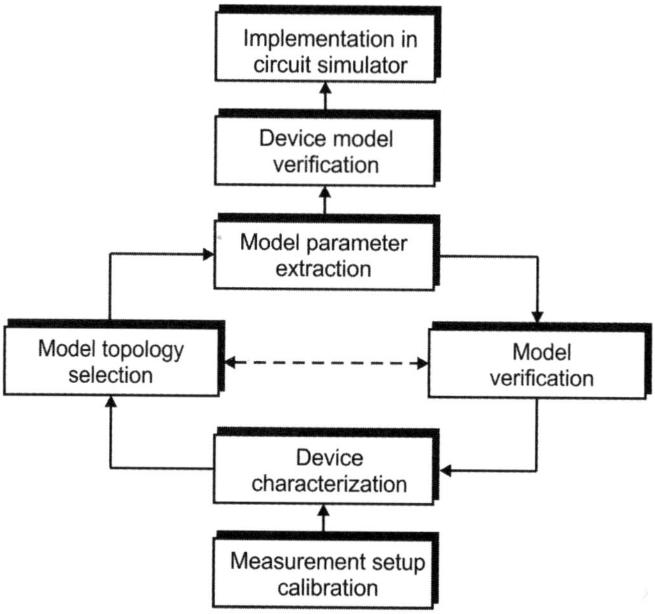

Figure 11.3 Measurement-based device modeling for reliable circuit simulation. (After [17].).

procedure. It can be based on DC data, scattering parameters, pulsed data, power spectra of harmonics, or signal waveform measurements. It is clear that the degree of accuracy in the measurement performance essentially influences the usability and reliability of the derived models. After introducing a physically relevant model topology, parameter extraction can be carried out on the basis of measured data. Usually model verification is performed by comparison of simulated and measured scattering parameters or power spectra of harmonics. Summarizing, experimental modeling is essentially focused on the following three areas, (1) creation of device-matched models, (2) development of precise measurement techniques, and (3) definition of a reliable model parameter extraction algorithm [17].

The different steps in the device modeling procedure is schematically shown in Figure 11.3 [17]. Commonly the nonlinear device model is derived from small-signal measurements at a large number of bias points. The nonlinear model elements are then attained by two-dimensional integration over the small-signal model elements (for some time, the drain current source is exceptionally derived from pulsed-DC measurements) . Therefore, the validity of the nonlinear model, and thus, the circuit simulation is critically dependent on the reliability of the derived small-signal model.

11.3.1 Importance of Diligent System Calibration

The attainable accuracy of measured device data and extracted results depend on the performance of the used VNA and to a large extent, on the quality of the applied calibration procedure to compensate for the systematic measurement errors. Although inexperienced device modelers are inclined to regard accurate and reliable calibration as a necessary evil that is time-wasting and tedious, everyone should be conscious about its great importance with regard to the final modeling result. Calibration is really the most important starting phase in the entire extraction and modeling process. Once the system calibration is completed, residual (postcalibration) errors become noticeable [18]. If you are not sure about the success of a calibration, repeat it once more or a second time until the calibration verification result can be accepted.

The vector network analyzer can be used to characterize devices directly having coaxial or waveguide terminations. Frequently, a transition connection network is needed to extend the measurement feasibility to planar devices. In this case wafer probe stations or commercial/in-house test fixtures can be used. Today, the interconnection of a vector network analyzer and a wafer prober is well developed so that calibration covers the whole combined measurement system, also termed *single-tier* calibration. This may be considered in a different way with test fixtures as discussed in [18].

11.3.1.1 Specified and Actual Vector Network Analyzer Performance

The potential sources of VNA measurement errors are well documented in [19]. The overall system performance specifications consider the different kinds of error sources such as systematic, random, and drift errors.

Systematic system errors result from stable imperfections and inaccuracies associated with the vector network analyzer hardware. For systematic errors magnitude and phase uncertainty specification is performed on a worst-case vector basis, which means that the total effect of such errors is summed up by conventional addition [20]. Experience shows that the actual measured errors are usually much lower than defined instrument specifications, as is illustrated in Table 11.1, which might be explained by a manufacturer's equipment statutory warranty claims. Hence, simple estimation of model parameter extraction uncertainty on the basis of overall measuring instrument performance specifications may not comply with the realistic capabilities in the model parameter extraction processing (see e.g., [21]).

Therefore, Vidkjaer [20] proposed a more distinct error analysis with respect to device modeling tasks. Instrument uncertainty is derived from a detailed analysis of the nature of all potential error sources, and in particular whether they are correlated or not rather than on systematic or random distinction. Random errors result from the normal variations that are

Table 11.1

Measured and Specified Total System Uncertainty of the HP 8510A Vector Network Analyzer*

	Measurement		Specification	
Total Uncertainty	Magnitude	Phase	Magnitude	Phase
Reflection (1.0 reflection)	0.15 dB	2.0°	0.6 dB	3.6°
Transmission (0-dB attenuation)	0.05 dB	0.5°	0.17 dB	1.0°

* Note that recent vector network analyzers may provide lower uncertainty data (see e.g., PNA type system in Chapter 21).

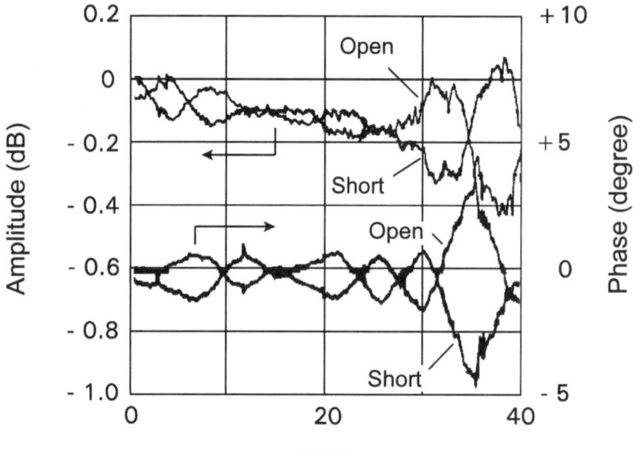

Figure 11.4 Calibration verification of the HP 8510 with a coaxial offset short and open (phase compensation for the additional offset length included).

characteristic of a given process. Drift errors are considered to be time- or temperature-dependent performance characteristics. Therefore, actual device measurements must be made within ±1 °C of the measurement calibration temperature, from 19°C to 27°C [22].

Residual errors result from the imperfections of calibration standards, the connector interface, the interconnecting cables, and the instrumentation. Finally, as is outlined in [19], the operator error focusing on improper connection technique, connector dimensions, and contact surfaces might be the largest error in the data acquisition. In the following some typical calibration and measurement errors are presented.

Figure 11.4 illustrates the residual errors measured [23] with a VNA of type HP 8510 [24] using coaxial SOLT calibration. Commonly, calibration verification is performed with a coaxial offset short or/and open standard. As discussed before, due to imperfection of the description of the calibration standards, and an assumed error model of the measuring system, residual systematic errors remain after the correction. These manifest themselves as ripples in the recorded frequency-dependent amplitude and phase of the measured offset calibration standards.The displayed ripple is the result of the interaction between the system error vector and the open or the short [25]. From the ripple distance a length of about 16 mm can be estimated.

The frequency-dependent decrease of the crossing points of *open* and *short* curves is caused by the offset transmission line loss. Due to this reference the residual systematic error of amplitude and phase is extremely low, up to about 30 GHz. At higher frequencies the amplitude and phase error increases to about 0.52 dB and 4 degrees, respectively.

Figure 11.5 illustrates a similar verification result for a Wiltron 40 GHz VNA system Model 360 [26]. It can be seen that the ripples representing the amplitude errors are close to the lower frequency data given in Figure 11.4 over the whole bandwidth. The residual systematic phase errors, however, are relatively large, and reach values up to 7 degrees at about 40 GHz.

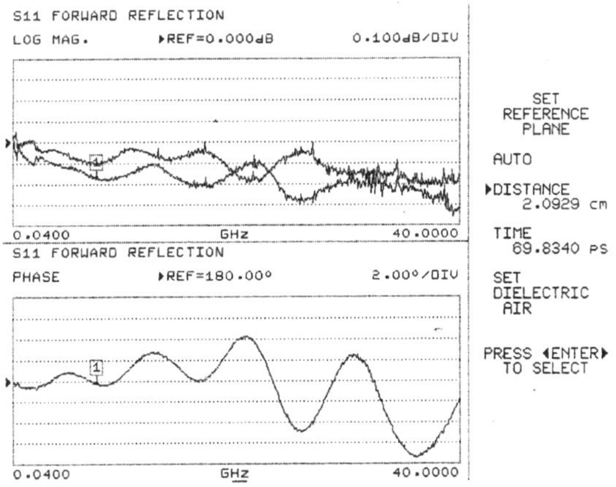

Figure 11.5 Amplitude and phase calibration verification of a 40-GHz Wiltron system [26]. Note that recent vector network analyzers may provide lower uncertainty data.

11.3.1.2 Multiband On-Wafer Measurement Issues

The VNA is commonly equipped with coaxial connectors applicable up to 100 GHz and above (banded frequency ranges are also available in waveguide configuration) so that mating with a (planar) DUT needs special adaptation for measurement. This can be established with wafer probe stations, or more conventionally, with commercial or in-house test-fixtures incorporating coaxial-to- planar transmission line transitions.

As already discussed, a combined measurement setup of a vector network analyzer and wafer probe station is commonly calibrated in a single-tier calibration procedure; that is, the complete measurement setup with the planar input and output reference planes of the DUT is considered for calibration. In this case a conclusion about measurement uncertainty can only be drawn with respect to the entire measurement system. Separate control of measurement uncertainty with respect to the individual parts of the setup is not possible.

Multiband on-wafer measurement setups are in use to perform S-parameter measurements over a broad frequency range from 45 MHz up to about 118 GHz [27]. In this case calibration is performed in three different frequency bands, 45MHz - 50 GHz, 50 - 75 GHz, and 75 - 118 GHz. The lower band uses a coaxial technique, whereas the latter ones are realized in a waveguide technique.

Figure 11.6 shows the measured response of two samples of a 0.15-μm gate-length HEMT fabricated at the Fraunhofer Institute of Applied Solid State Physics (IAF). As can be seen the recorded data for sample 1 shows a sharp inconsistency in the trace at the transition of the middle to the upper frequency band, which is totally unreasonable and unexpected from the device performance. Such illegitimate error [6], which may have resulted in this case from faulty calibration of the equipment used, can be dealt with immediately. Generally, it is recommended to repeat and to improve the experiment before extraction if obviously incorrect data points have been measured. Calibration should be successfully refined until

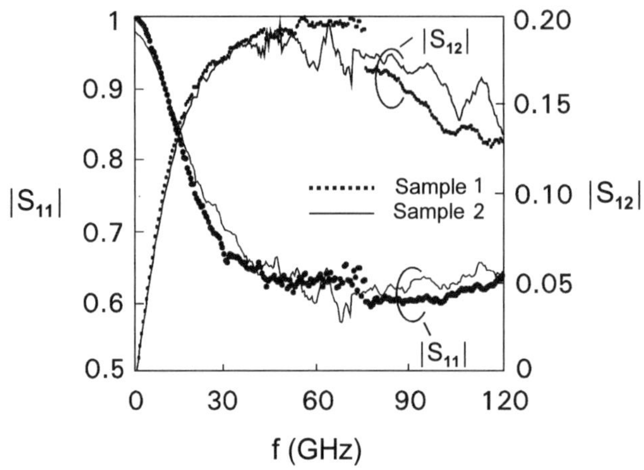

Figure 11.6 Measured S-parameters of a 0.15-μm T-gate HEMT (IAF) with a multiband vector network analyzer in the frequency range from 45 MHz up to 120 GHz revealing calibration errors at the frequency band transitions.

the measured data can be accepted by visual inspection with some confidence to be a reliable response of the device. This has successfully been attained with a repeated measurement (performed by a different measuring person). Measured data of sample 2 show a smooth transition between the different calibration regimes. As will be shown, the fast oscillating ripples, although undesired, may be controlled to a large extent by a proper extraction strategy. Meanwhile, an available broadband single-connection vector network analyzer, for example, HP 8510XF [28], which covers 45 MHz to 110 GHz in one sweep and single connection (i.e., with no band breaks), eliminates the discussed calibration errors.

It is assumed that the ripples in the trace of S_{11} in Figure 11.6 mainly originate in the incomplete calibration of the probe-tip-to-transmission-line transition and the transition from the coplanar line to the shielded line on the probe, for which a distance less than 1 mm could be roughly estimated by visual inspection. This is consistent with the measurement-based estimation of about 0.7 mm.

11.3.1.3 Test-Fixture Measurement Uncertainty

Regarding in-fixture measurement, additional errors are contributed by the test fixture. Commercially available test fixtures have been available for many years, and have attained a high-quality performance. Nevertheless, repeatability of measurements is insistently doubted. In this case, separate investigation of the measurement uncertainty of the given test fixture may be advisable as well as performing a two-tier (two-step) calibration of the entire measurement system [18]. Applying a single-tier (one-step) calibration, the complete measurement setup comprising the VNA and the connected test fixture is subjected to calibration. In this case overall uncertainty can be determined, but neither VNA nor test fixture uncertainty is available.

Following a two-tier calibration, the network analyzer is first calibrated as discussed in

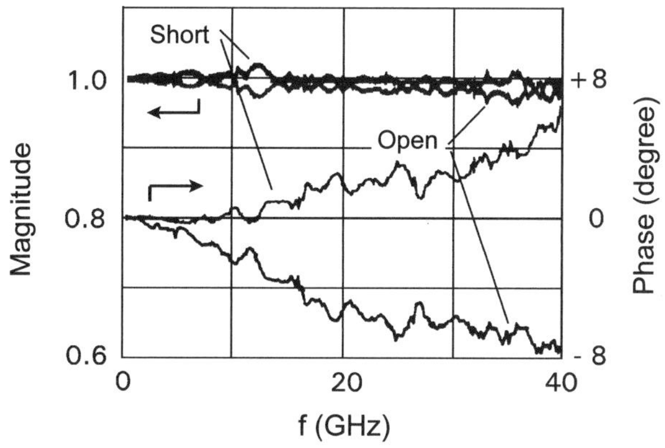

Figure 11.7 Microstrip test-fixture verification measurement with offset open and offset short.

Section 10.4.2; calibration data is stored. In a second step, after having the test fixture connected, the entire measuring system (i.e., VNA and the test fixture) is calibrated. Applying a deembedding procedure, the uncertainty of the test-fixture can be determined. In this case, full control is given over the residual calibration errors of both the VNA and test fixture.

In [18] a microstrip test fixture has been presented that uses (ideal) through, high, and low reflection standards (THLR). "Through" characterizes a continuous microstrip transition of both halves of the test fixture with zero length. "High reflection" means nonperfect short and open, and "low reflection" denotes a matched load; i.e., fixed load at low frequencies and sliding load at higher frequencies. As discussed, two-tier calibration is well suited to distinguish between the residual calibration errors due to the VNA and the errors resulting from the test fixture. Figure 11.7 illustrates the calibration verification result including the system errors of the used VNA HP 8510 (see Figure 11.4). The ripples are attributed to the length of the microstrip lines being 1 inch. It can be seen that a maximum amplitude error of 0.02 (0.08 dB) is observed for frequencies less than 30 GHz. Above that, errors up to 0.06 (0.26 dB) have been measured. The phase error with respect to the mean straight lines is within about 2 degrees up to 30 GHz. Above that, error values up to about 4 degrees have been attained. In comparison with the uncertainty of the VNA in Table 11.1 only slight degradation of the reflection uncertainty is observed up to 40 GHz. As seen before, the amplitude and phase errors of the short are in antiphase of those for the open circuit.

11.3.2 Device Measurement Analysis Issues

In the previous sections, measurement errors originating due to incomplete calibration of the vector network analyzer and test fixture were described. In this section, measurement uncertainty induced limitation of reliable model parameter extraction is discussed.

It is of fundamental importance that, in general, measurements should be performed with utmost care. Nevertheless, residual errors will remain and must be taken into account. The evaluation of measured data with respect to the utility for parameter extraction assumes some

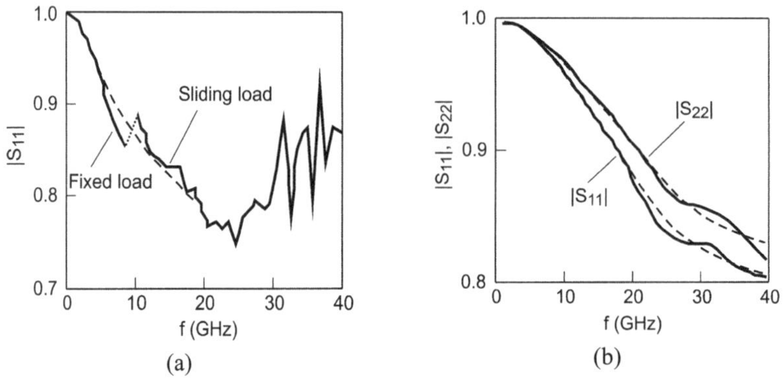

Figure 11.8 Measured response of MESFET devices in-fixture (a) and on-wafer (b) measurements.

experience and expert knowledge in this field. It is sometimes advisable to repeat the experiment and to refine the measurement technique and extraction method instead of working with obviously faulty S-parameter data, and become frustrated over disappointing results after intensive extraction effort. To preempt this frustrating experience, it is very important to analyze the potential measurement errors in advance before starting with the final extraction process. Model parameter extraction can only be performed on a database that can be trusted to be a reliable description of the device properties and is not dominated by instrument errors or faulty operator manipulation.

Figure 11.8(a) shows a typical in-fixture measurement result for a wire-bonded MESFET MGFC1423 ($V_{GS} = -2.0$V, $V_{DS} = 0$V) using a 40-GHz HP 8510 system [22]. A first obviously invalid measurement result can be seen around 10 GHz that originates in the frequency-dependent properties of the microstrip calibration standards used. At lower frequencies a 50-ohm chip resistor is used as a fixed load, showing increasing load mismatch with growing frequency. The electrical length of the surface ferrite absorber used as a sliding load [29] at frequencies above 10 GHz determines the matching properties. Improved load matching is given with growing frequency. Looking at the scale of the diagram, such measurement errors would hardly be noticeable in Smith chart representation, and hence, seem to be without any serious effect on the extraction process. Reasonable match between calculated and measured device response [21] may be convenient for the extraction for most of the highly sensitive (with respect to device response) model parameters; however, this is not true for the less sensitive resistive elements.

On-wafer measurements exhibit internal multireflections with slow-moving undulations due to shorter length of transmission lines. If in this case the device response is measured in a limited frequency interval, there is a temptation to accept the faulty data partly as valid values. Figure 11.8(b) shows the measured magnitude of S_{11} and S_{22} for a MESFET fabricated at the Plessey foundry (2 × 100 μm, $V_{GS} = -2.0$V, $V_{DS} = 0$V). Up to about 25 GHz the traces are very smooth, which one may accept with some confidence to be reliable results. However, as is indicated by the mean-valued dotted line of the slowly varying measured data, extraction based on a limited frequency range would yield model parameter values having no physical relevance at all.

A major problem encountered when making network measurements in microstrip or other noncoaxial media is the need to separate the effects of the transmission medium (in which the device is embedded for testing) from the device characteristics. While it is desired to predict how a device will behave in the environment of its final application, it is difficult to measure this way. The accuracy of this measurement depends on the availability of quality calibration standards. Unlike coaxial measurements, the three discrete impedance standards typically used are often impossible to produce for noncoaxial transmission media.

11.3.3 Choice of Model Topology Complexity

All observed relevant device features should be adequately reflected by the model. Model topology definition may be supported by physics-based device modeling. The higher the operation frequency of small-size devices as well as the higher the level of complexity of multifinger power devices, the larger the number of model elements to be taken into account because of the increasing distributive nature of the device (see Chapter 7).

The motivation to develop a device model is twofold. It should serve as a reliable tool in computer-aided design (CAD)-based circuit design. In this connection there is often a debate about model accuracy and computational efficiency. It is said that circuit designer wishes nothing better than reliable accurate models. On the other hand, it is often argued that the model should be as simple as possible to spare computation time during circuit performance optimization. Model topology simplification *can* be done if an accurate device model is known at all. Otherwise, uncontrolled model restrictions might lead to unreliable simulation results. Knowing accurate (maybe complex) models, controlled simplification for special applications may be acceptable.

On the other hand, device modeling can be used to understand the device performance in a better way on the basis of technology and semiconductor physics. This includes the capability to improve device processing. Intermodulation is a good example. To fulfill intermodulation distortion requirements many linearization methods have been developed to compensate for the linearity deficiencies of AlGaN/GaN power devices. It seems that localizing the main intrinsic intermodulation sources and reducing these by proper device redesign and optimization will be more successful in the long run.

11.3.4 Challenges in the Parameter Extraction Process

The block diagram in Figure 11.9 illustrates the manifold extraction problems occuring in the modeling process and illustrates the interaction of the constitutive process functions [30]. M^D signifies the true (i.e., matched device topology), and S^D the true device response. M^D and S^D are, of course, not known. Device measurements, although error-corrected, always exhibit residual measurement errors; i.e., testing of the device can only deliver approximated device data $\tilde{S}^D$.

Based on physical considerations an approximated model topology $\tilde{M}^D$ with the model parameter vector $\tilde{P}^M$ is adopted for the device. Then $\tilde{\varphi}^M$ is an approximated model parameter *value* vector for the measured device data and for the supposed topology.

It is generally accepted that even in a simple case of a linear equivalent device network

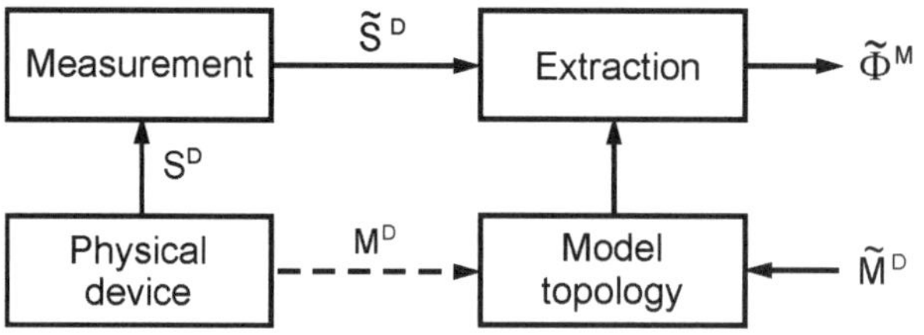

Figure 11.9 Block diagram illustrating the interrelation and dependence of various quantities in the extraction process [30] .

the consistent model parameter extraction in a given bias point is a challenging task. This is mainly attributed to the large number of unknowns (typically more than 12). Measurement errors have significant influence on the determination of the model element values. Relative small measurement errors can also lead to larger distribution of the extracted model parameters.

The sharpness of the extraction solution depends on different factors, mostly on the degree of topology mismatch and on measurement uncertainties. The validity of a lumped element device model with a limited number of elements is generally restricted to a lower frequency range. The higher the operation frequency of the device (or the larger the size of devices for high-power application), the larger the number of circuit elements is needed to simulate the increasingly distributive character of the device. The experience has shown that the distributed model topology in Figure 7.2 with 22 model elements works well at least up to 120 GHz.

11.3.5 Challenges of Consistency in Device Modeling

A valid bias-dependent small-signal or *linear* model should accurately simulate small-signal operation in various regions in the IV plane (multibias device characterization), as illustrated in Figure 11.10.

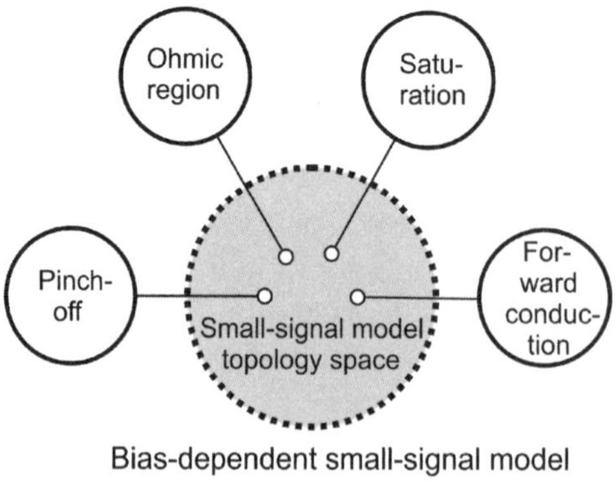

Figure 11.10 A valid bias-dependent linear device model should accurately simulate small-signal operation in any bias point in the IV plane.

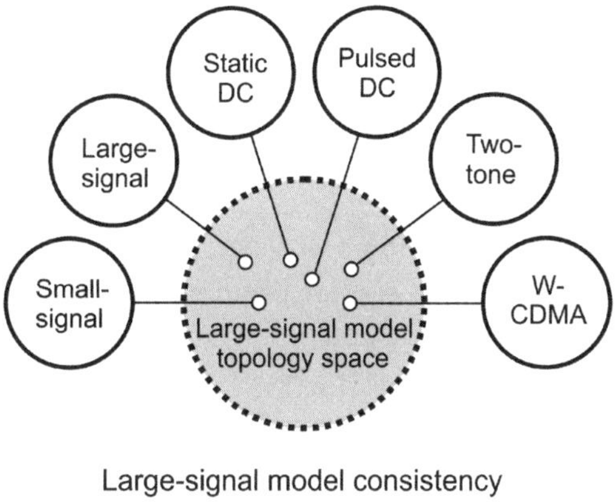

Figure 11.11 A good consistent nonlinear device model should accurately simulate large-signal operation of any kind in any bias point in the IV plane, including small-signal behavior as linearized case.

Model verification is performed by comparison of simulated and measured scattering parameters under such operation conditions that are different from those used in the model parameter extraction process.

In contrast to small-signal models, large-signal (*nonlinear*) models should ideally predict device behavior under any operating condition and stimulation (bias point selection, pulsed (DC, RF) or CW, single-/multitone, and complex WCDMA modulation) as is illustrated in Figure 11.11. Again, model validation comprises test signals that are unknown to the device response with respect to the previous modeling approach.

11.3.6 Model Implementation into Circuit Simulator

Useful simulation results could only be achieved with valid device models derived from reliably and accurately determined model parameters. Reliable simulation results are absolutely dependent on having good device models. Measurement-based device models are mostly table-based ones, which can be easily implemented, for example, in the ADS software of Keysight, including SDD (symbolically-defined device) block for the nonlinear inner part of the device under consideration. Note that nonlinear models are commonly formulated in the Verilog-A language. Verilog-A has emerged as the de facto standard language for defining and distributing compact models [31]. The SDD interface is similar to Verilog-A. The SDD describes the model in terms of charges and currents and automatically generates the derivatives and load information, as does Verilg-A. However, SDDs automatically determine the relationship between equations whereas Verilog-A is a procedural language where the order of the equations is important [32].

11.4 EXTRACTION STRATEGY RECOMMENDATIONS

The device modeling process is a very complex, tedious, and time-consuming process, which demands utmost care and effort. From experience, the following recommendations may be helpful to accomplish device modeling successfully.

11.4.1 Check of Measured Data

Once measurement is completed primary attention is paid to the assessment of the quality of the measurement results critically addressing, for example, the reliability of the measured data, sufficiency of broadness of measured frequency range, and expected curve trace of the measured parameter. If obvious measurement errors are discovered, extraction should not be started at all; but rather, measurements should be repeated. Check the measurements a second or third time before you proceed and do not accept them without further reflection. Look at the measured results with a healthy level of mistrust.

There may be general agreement on the principal curve profile of a measured parameter. However, there are significant outlying data points. Eliminating them by a smoothing function can be reasonable, but then control is lost over the validity of the extracted results. This applies in particular to the less sensitive model parameters such as the drain and source resistances. Measurement repetition is then a better option than to remain in doubt about extracted results. Simulation results using fictitious device data based on experience or estimated physical device parameters (see Chapter 6) can be useful to increase the confidence in the validity of measured results.

11.4.2 Choice of Topology

Regarding efficiency of circuit simulation, device model topology should comprise as small a number of equivalent circuit elements as possible. However, note a too-small number of equivalent circuit elements may hamper good fitting, and the extraction may deliver unphysical element values. In this case, topology-dependent element values may be extracted. To avoid such shortage a distributed equivalent circuit topology is recommended. As experience has shown, a three-shell model represents a suitable device topology both for high-frequency and high-power applications. With a reliable model parameter extraction minor important parameters may become evident and thus negligible. With respect to specific device application, simplifications may be justified. In case of considerable deviation of simulation from the measurements (i.e., principal fitting with the chosen model topology is not recognizable), the chosen topology must be modified.

11.4.3 Choice of Extraction Algorithm

In Chapter 12, both local and global optimization techniques are discussed, always confronted with the fundamental issue of trapping into a local minimum. In recent years much progress has been observed regarding the applicability of global optimization algorithms. Nevertheless, finding the absolute minimum not only depends on the algorithm but specifically from the quality of measured device data; i.e., whether they are noisy, extremely flat around the minimum, or have multiple minima in tight neighborhood or strong minima in immediate proximity to the global minimum.

The classical simplex algorithm is classified as a local search method. However, as will be discussed in Section 12.7.5, appropriate modifications have expanded its capabilities for global search. Based on much experience the modified simplex algorithm has proven a reliable optimization algorithm for the model parameter extraction process.

The development of a reliable optimization algorithm should always follow this trivial principle: The nearer the starting value vector at the global minimum, the smaller the trapping probability. A single optimization result may not be meaningful, so a better solution is the probing of the properties of the minimum of an error function using different starting value vectors and applying statistical evaluation. The definition of starting values is extremely important for the optimization run. Initial values of model parameters may be obtained from physics- and low-frequency-based estimation of device model elements as discussed in Chapters 6 and 8.

11.5 SUMMARY

Measurement uncertainties were classified and described. It was concluded that accurate measurements are the basis of successful model parameter extraction and the premise for establishing good device models. It was noted that before starting the model parameter extraction process very careful analysis of the measured device data should be performed in order to ensure extraction results of a high confidence level. By means of typical measurement plots, errors should be identified, described, and assessed with respect to reliable model parameter extraction.

REFERENCES

[1] JCGM 100: 2008, GUM 1995 with Minor Corrections, Corrected Version 2010, *Evaluation of Measurement Data—Guide to the Expression of Uncertainty in Measurement.*

[2] Brinkmann, B., Internationales Wörterbuch der Metrologie: *Grundlegende und Allgemeine Begriffe und Zugeordnete Benennungen (VIM), Deutsch-englische Fassung ISO/IEC-Leitfaden 99:2007*, Beuth Verlag GmbH, Berlin, 2012.

[3] NIST, *Technical Note 1297, Appendix D1–Terminology*, https://www.nist.gov/pml/ nist-tn-1297-appendix-d-clarification-and-additional-guidance.

[4] Deardorff, D. L., *Introductory Physics Students' Treatment of Measurement Uncertainty*, Dissertation, North Carolina State University, 2001.

[5] Taylor, J., *An Introduction to Error Analysis: A Study of Uncertainties in Physical Measurements*, 2nd Edition, University Science Books: Sausalito, CA, 1997.

[6] Bevington, P. R., and D. K. Robinson, *Data Reduction and Error Analysis for the Physical Sciences*, 3rd Edition, New York: McGraw-Hill, 2003.

[7] Baird, D. C., *Experimentation: An Introduction to Measurement Theory and Experiment Design*, 3rd Edition, Prentice Hall: Englewood Cliffs, NJ, 1995.

[8] Ghione, G., C. U. Naldi, and F. Filicori, "Physical Modeling of GaAs MESFET's in an Integrated CAD Environment: From Device Technology to Microwave Circuit Performance," *IEEE Trans.*, Vol. MTT-37(3), 1989, pp. 457–468.

[9] Vaitkus, R. L., "Uncertainty in the Values of GaAs MESFET Equivalent Circuit Elements Extracted from Measured Two-Port Scattering Parameters," *Proc. 1983 IEEE Conf. High Speed Semiconductor Devices Circuits*, Cornell Univ, Ithaca, NY, pp. 301–308.

[10] Rytting, D. K., "Network Analyzer Accuracy Overview," *58th ARFTG Conf. Digest*, San Diego, CA, 2001 (13 pages).

[11] Wilker, C., and D. Schreurs, "Guest Editorial on Measurement," *IEEE Microwave Magazine*, August 2007, pp. 6–13.

[12] Raggio, J., "Results of a Multi-Site Round-Robin to Examine Probed Measurements of an Impedance Standard Substrate," *IEEE Trans.*, Vol. IM-39(6), 1990, pp. 962–963.

[13] Ginley, R. A., "Confidence in VNA Measurements," *IEEE Microwave Magazine*, August 2007, pp. 54–58.

[14] Keysight Technologies, *Fundamentals of RF and Microwave Power Measurements–Introduction to Power, History, Definitions, International Standards & Traceability*, Application Note, July 31, 2014.

[15] Magerl, G., "Introduction —TARGET Special Issue," *Int. Journal of Microwave and Wireless Technologies*, Vol. 1(2), 2009, p. 89.

[16] Devlin, C., T. Brazil, and G. Magerl, "TARGET Network: Microwave Power Amplifiers," *IEEE Microwave Magazine*, April 2006, pp. 34–48.

[17] Kompa, G., "Modeling of Dispersive Microwave FET Devices Using a Quasi-Static Approach," *Int. J. of Microwave and Millimeter-Wave Computer-Aided Engineering* (MMWCAE), Vol. 5, No. 3, 1995, pp. 173–194.

[18] Kompa, G., M. Schlechtweg, and F. van Raay, "Precisely Calibrated Coaxial-to-Microstrip Transitions Yield Improved Performance in GaAs FET Characterization," *IEEE Trans.*, Vol. MTT-38, pp. 62-68, Jan. 1990.

[19] Donecker, B., "Determining the Measurement Accuracy of the HP 8510 Microwave Network Analyzer," *Hewlett Packard, RF & Microwave Symposium and Exhibition*, March 1985, pp. 1–71.

[20] Vidkjaer, J., *Network Analyzer Uncertainty Computations for Small-Signal Model Extractions*, Lab. Report R 549 February 1994, Electromagnetics Institute, Tech. Univ. of Denmark, Lyngby.

[21] Müller, J.-E., "Investigation of GaAs MESFET Small Signal Equivalent Circuits for Use in a Cell Library," *Proc. 19th European Microwave Conf.*, London, 1989, pp. 991-996.

[22] Hewlett Packard, *HP 8510B Network Analyzer – System Manual*, April 1988.

[23] Schlechtweg, M., *Broadband Characterization and Modelling of GaAs-MESFETs and (AlGa)As/GaAs-MODFETs up to 40 GHz*, Doctoral Thesis (in German), University of Kassel, 1989.

[24] Hewlett Packard, *HP 8510 Network Analyzer, Operating and Service Manual*, Vol. 1, Part No. 08510-90001, March 1984, Section I. (General Information).

[25] Oldfield, B., "VNA S11 Uncertainty Measurement - A Comparison of Three Techniques," *ARFTG Conf. Dig.*, June 1992, pp. 86-105.

[26] Wiltron, *Wiltron Network Analyzer, Model 360*, Wiltron Product Catalog, 1989.

[27] Tasker, P. J., and J. Braunstein," New MODFET Small Signal Circuit Model Required for Millimeter-Wave MMIC Design: Extraction and Validation to 120 GHz," *IEEE MTT-S Int. Microwave Symp. Digest*, 1995, pp. 611–614.

[28] Hewlett Packard, *HP 8510XF (110 GHz Coaxial Single Sweep NVA System)*, Hewlett Packard, 1997.

[29] Kompa, G., *Practical Microstrip Design and Applications*, Norwood: Artech House, Inc., 2005.

[30] Kompa, G., "Fundamentals in Successful Experimental Active Device Modelling," *Proc. of 'Workshop on Advanced Microwave Devices, Characterisation and Modelling'* (during 23th European Microwave Conf.), Madrid, Spain, Sept. 10, 1993, pp. 22–25.

[31] McAndrew, C. C., G. J. Coram, K. K. Gullapalli, J. R. Jones, L. W. Nagel, A. S. Roy, J. Roychowdhury, A. J. Scholten, G. D. J. Smit, X. Wang, and S. Yoshitomi, "Best Practices for Compact Modeling in Verilog-A," *IEEE Electron Devices Society*, Vol. 3(5), 2015, 383-396.

[32] Keysight Technologies, *Using Verilog-A in Advanced Design System*, http://literature.cdn.keysight.com/litweb/pdf/ads2005a/veriloga/index.html, 2005.

Chapter 12

Optimization Methods for Model Parameter Extraction

12.1 INTRODUCTION

Optimization plays an essential role in the model parameter extraction process. During the last five decades, a vast amount of optimization methods have been developed and applied in the most diverse fields. Extensive literature exists that include a number of excellent text (e.g., [1-11]). This chapter provides an overview about various optimization techniques, their specific features and issues, with a focus on the application of model parameter extraction and device modeling. This should support the device modeler in the selection of an appropriate optimization method for any given modeling task.

Generally, optimization indicates a process of finding the *best* solution to a problem. In the case of device modeling, best solution may imply a high coincidence of measurement and simulation response of an equivalent device network. In addition, optimization should select the best decision that reflects the underlying physical reality.

Typically, a physically substantiated network topology is chosen and tentative values are assigned to the network elements. Then, in each iteration, these values are altered in such a way that a preassigned error function is reduced. This iterative process is continued until the error function is minimized. A major disadvantage of optimization algorithms is that generally there is no guarantee that the iterations will converge to an optimal (in the best case even physics-based) network.

An optimizer needs *good* input data. In the case of small-signal operation, these are mostly error-corrected frequency-dependent scattering parameters. Nevertheless, such device data still contains measurement noise and structural (topology-related) mismatch (see section 11.3.4). Residual calibration errors originating from nonperfect error models of the used measurement setup and from worn or insufficiently characterized calibration standards may contribute to the uncertainty of the input data. Further influencing parameters result from random errors, operator manipulation, and thermal drift of the measuring instrument.

The optimization run is under the control of various conditions, such as choice of object function, starting values, termination criteria, limitation of searching space of variables, suppression of unphysical (negative) model parameter values, and the provision to escape from local minima.

The user's expectations regarding the optimizer are high. The extraction algorithm should be fast and robust. For example, extraction with different starting parameter vectors should deliver (in the ideal case) the same or at least very similar extraction results. Robustness indicates the radius of stability against uncertainties. This can be studied by sensitivity analysis. The problem is first solved with fixed values of the parameters (perhaps the most likely values) defining the optimal solution. Then it is observed in which way this solution is affected by small perturbations.

The optimizer should work well with noisy data. The optimized results should be reliable and accurate. The model parameter extraction should provide high-quality fitting of measured scattering parameters, and it should deliver nonnegative model element values. In particular, the extracted data should physically reflect meaningful element values.

Preferably the optimizer should not require information about the derivatives of the error function to be minimized. From the user's point of view, the working principle of the optimization algorithm should simply be easy to understand, easy to program, and easy to use.

12.2 LOCAL MINIMUM PROBLEM

In the mathematical usage, an n-dimensional optimization problem is to find for a given n-dimensional function $f(\boldsymbol{x})$ a parameter vector $\boldsymbol{x} = [x_1, x_2, ..., x_n]^T$ such that $f(\boldsymbol{x})$ is minimized. If the variable parameter vector $\boldsymbol{x}$ has to satisfy certain conditions, the problem is called a constrained optimization problem; otherwise, it is called unconstrained.

Regarding small-signal model parameter extraction the device is characterized by its scattering matrix $\boldsymbol{S} = (S_{11}, S_{12}, S_{21}, S_{22})$. At a frequency point f_k let us denote the response function $\boldsymbol{S}_c^{fk}$. With the measured response at the frequency point f_k, characterized by $\boldsymbol{S}_m^{fk}$, the error E_{ij}^{fk} $(i, j = 1, 2)$ between measured and calculated response of a scattering parameter can be written as $E_{ij}^{fk} = \sum_{f_k} |w^{fk}(S_{c,ij}^{f} - S_{m,ij}^{fk})|$ with a weighting factor w^{fk}, which may be a complex quantity. Taking into account all four parameters the error function could be defined as

$$E(\boldsymbol{p}) = \sum_{k=1}^{K} \sum_{i,j=1}^{2} |w^{f_k}(S_{m,ij}^{f_k} - S_{c,ij}^{f_k}(\boldsymbol{p}))| \tag{12.1}$$

In this case, the absolute differences of measured and calculated frequency-dependent (complex) values of all four scattering parameters at the considered frequency points are summed up. The *objective* (also *target* or *cost)* function in (12.1) is a real-valued function. Thus, the optimization problem is to minimize the error function $E(\boldsymbol{p})$ with $\boldsymbol{p} = (p_1, p_2, \cdots, p_n)$. Commonly the vector $\boldsymbol{p}$ represents a set of independent parameters. This might be true for a huge complex vector with thousands of components where a single parameter may have only an insignificant effect on the response of the error function to be optimized. In contrast, regarding transistor model parameter extraction with a limited number of variables, we will recognize throughout the subsequent chapters that all model elements are closely related and interact with each other.

In general, there will be constraints that must be satisfied either during optimization or by the optimum. Each parameter might be constrained explicitly by an upper and lower bound as follows:

$$p_{li} \le p_i \le p_{ui} \qquad (i = 1, 2, \cdots, n) \tag{12.2}$$

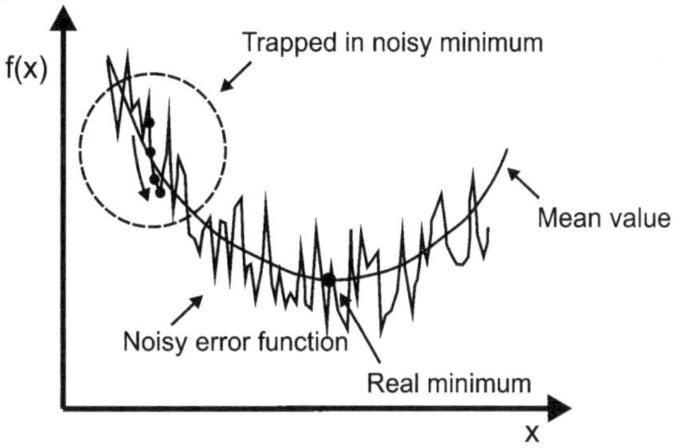

Figure 12.1 Illustration of trapping in a local minimum of a noisy error function.

where p_l and p_u are lower and upper bounds, respectively.

We distinguish between *global* and *local* optimization. Global optimization relies on the explicit use of a computer to find one set of model parameters that will best fit the available experimental (measured) data. This methodology may give the minimum average error between measured and simulated (calculated) data points, but it also treats each parameter as just a mathematical fitting parameter. Parameters extracted in such a manner might yield values that are not consistent with their expected physical meaning. Local optimization describes an optimization process near a minimum of an error function. This may be a local or global one. In any case, there is no intention and no means of the optimization algorithm to escape from the given minimum.

Experience shows that most of optimization methods published in the literature converge into a local minimum. Generally there is no effort to overcome local minima. If there is only one minimum, then the local minimum problem does not exist. Confidence in having found the global minimum may be inspired by comparison of other existing minima. With the conglomeration of local minima, the definition of a global minimum can become questionable. Noisy error functions hamper the optimization process, as is illustrated in Figure 12.1.

12.3 OPTIMIZATION STRATEGIES

The definition of an effective strategy for model parameter extraction is a key problem. Many techniques have been described in the literature. Optimization algorithms can be broadly classified into two categories: gradient-free (or direct) and gradient-based optimization methods, as is illustrated in Figure 12.2. In direct search methods, the search process is based only on function values at different points in the search space, whereas in gradient-based methods, derivative information is used to contrive a search. Gradient-based methods use first and/or second derivatives, while direct search methods do not use any derivative information.

Regarding direct search methods one may distinguish between deterministic and stochastic gradient-free optimization algorithms. Deterministic algorithms refer to methods whose

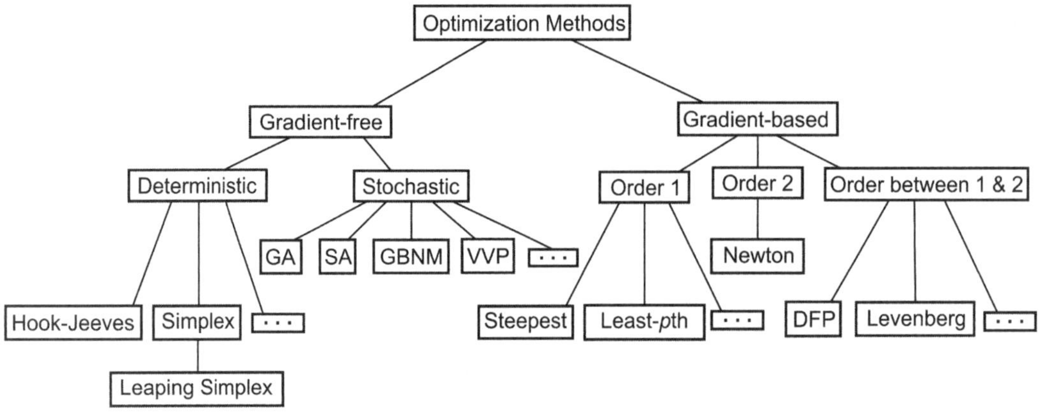

Figure 12.2 Overview of n-dimensional optimization methods investigated so far.

output depends on the input parameters (i.e., the initialization parameter vector). It is stated that they converge rather slowly in the neighborhood of a minimum, but can be rather tolerant to the presence of noise in the object function. However, their main weakness is to become trapped at local minima if the initial guess is far from the optimum solution. Therefore they are also characterized as local search methods. Among the deterministic group are, for example, the Hook-Jeeves optimization technique, described in Section 12.6.1, and the simplex algorithm, discussed in Section 12.6.2.

The direct search methods require only the computation of objective function values. These algorithms neither require nor estimate derivative information to determine a direction of descent. These algorithms have remained popular for practical reasons. Often, particularly in experimental settings, derivatives are simply unavailable. In addition, function values based on experimental data are often noisy (i.e., they can only be trusted to a few digits of accuracy) so that finite-difference approximations may prove unreliable. Finally, most of the direct search methods are easy to understand, easy to program, and easy to use.

Stochastic optimization methods, on the other hand, rely on randomness and retrials to better sample the parameter space in searching for an optimum solution. They are classified as global search methods. They do not rely on only one starting parameter vector instead the parameters are randomly generated within a defined value interval. Thus, the search is performed over the entire search space in a parallel manner. The results are sorted, weighted, and selected for the next iteration cycles. The decision for the actual search direction is based on the experience of past trials. The classical global optimization methods, the genetic algorithm (GA), and the simulating annealing (SA), are discussed in Section 12.7.2 and Section 12.7.3, respectively.

The nonlinear simplex algorithm is classically considered as a local search method. However, recently many contributions have been published to overcome this limited performance. The research group of the author has thoroughly studied and tested, for many years, the *Nelder-Mead simplex* method for model parameter extraction. It has been found that, with some modifications included, detrapping from local minima is really well tackled. Because of its leaping ability, this simplex variant may be termed *leaping simplex*, which is discussed in Section 12.7.5.

The group of gradient-based optimization techniques utilize gradient information of the objective function to find the direction and size of the step towards the optimum solution. Algorithms that use only the gradient are called first-order algorithms. These include the *steepest descent method*, described in Section 12.4.1, and the *least-squares optimization* with *Gauss-Newton* method, discussed in Section 12.5.

An algorithm that uses the second derivatives—or Hessian—is referred to as second-order algorithm to which Newton's method belongs (Section 12.4.2). The Hessian greatly speeds up the convergence as the algorithm comes close to the solution. The gradient reveals the search direction but gives no information about the distance. In contrast, the use of the Hessian allows the algorithm to precisely localize the minimum if the quadratic approximation of the object function is correct. In general, the object function is not quadratic, but this expression becomes valid near the solution.

The Hessian computation is often time-consuming so the second-order method is hardly ever used. Instead, approximations of the Hessian have been proposed that leads to *quasi-Newton* methods. They allow determining the Hessian without using second-order derivatives. Thus, the Hessian matrix may be computed using gradient evaluation of previous iterations. This applies, for example, to the *Davidon-Fletcher-Powell* (DFP) algorithm treated in Section 12.4.3.

The Levenberg-Marquardt algorithm (Section 12.5.2), focuses on least-squares error function optimization and combines the gradient descent method and the Gauss-Newton method. The combination of two or more optimization techniques is also termed *hybrid methods*.

In the following these optimization methods are outlined and their special properties are examined more closely with respect to device modeling. However, note that a detailed analysis of the optimization algorithms is beyond the scope of this chapter and the interested reader is referred to the extensive literature about this topic (e.g., [1-11]) for more comprehensive treatment.

12.4 DESCENT METHODS

Decent methods are based on the gradient of the objective function. The parameter vector $\boldsymbol{p}$ is determined such that the error function $E(\boldsymbol{p})$ is minimized. The search methods generate trial points recursively. In particular, if a trial point $\boldsymbol{p}^i$ is known, a new trial point $\boldsymbol{p}^{i+1}$ is generated according to

$$\boldsymbol{p}^{i+1} = \boldsymbol{p}^i + \delta\boldsymbol{p}^i \tag{12.3}$$

with

$$\delta\boldsymbol{p}^i = \alpha^i \boldsymbol{d}^i \tag{12.4}$$

where $\delta\boldsymbol{p}^i$ denotes the parameter vector change, $\boldsymbol{d}^i$ the search direction, and α^i the step-size parameter in the ith iteration step. Thus, the optimization process can be viewed as a combination of two separate processes: (1) a decision is made to which direction $\boldsymbol{d}$ to proceed in the parameter space $\boldsymbol{p}$, and then (2) a decision is made as to how far (proportionality constant α) to proceed in that direction. The search direction $\boldsymbol{d}$ will be related to the gradient of the error function $E(\boldsymbol{p})$. When the direction of the parameter change $\boldsymbol{d}$ has been determined, then quadratic interpolation can be used to find the proportionality constant α [4].

12.4.1 Steepest Descent Method

Deepest (or *steepest*) *descent* method is a first-order iterative optimization algorithm and is also known as the method of *gradient descent*. This application needs the calculation of the first derivative of the objective function. The error function $E(\boldsymbol{p})$ is a function of many parameters. That is, $\boldsymbol{p}$ is a vector that may be written as

$$\boldsymbol{p} = (p_1, p_2, ..., p_n) \tag{12.5}$$

For this multidimensional case, the gradient ∇E is written as

$$\nabla E = \left(\frac{\partial E}{\partial p_1}, \frac{\partial E}{\partial p_2}, \dots, \frac{\partial E}{\partial p_n} \right) = \left[\frac{\partial E}{\partial p_1} \quad \frac{\partial E}{\partial p_2} \quad \cdots \quad \frac{\partial E}{\partial p_n} \right]^T \tag{12.6}$$

The gradient is an n-dimensional vector, the ith component of which is obtained by finding the partial derivative of the object function with respect to p_i. We are interested in the parameter vector change $\delta\boldsymbol{p}$, as to minimize the error function $E(\boldsymbol{p})$. If the error function is a function of only one variable, then it can be expanded in terms of a Taylor series as

$$E(p + \delta p) = E(p) + \frac{dE}{dp}\delta p + \frac{d^2E}{dp^2}\frac{\delta p^2}{2!} + \dots \tag{12.7}$$

If $\delta\boldsymbol{p}$ is small, then this can be approximated as

$$E(p + \delta p) \approx E(p) + \frac{dE}{dp}\delta p \tag{12.8}$$

The approximation in (12.8) can be generalized to n dimensions. For this case, if the n parameter changes $\delta p_1, \delta p_2, ..., \delta p_n$ are sufficiently small, then it follows that

$$E(\boldsymbol{p} + \delta\boldsymbol{p}) \approx E(\boldsymbol{p}) + \frac{\partial E}{\partial p_1}\delta p_1 + \frac{\partial E}{\partial p_2}\delta p_2 + \dots + \frac{\partial E}{\partial p_2}\delta p_n \tag{12.9}$$

With (12.6), it follows

$$\nabla E^T \delta\boldsymbol{p} = \left[\frac{\partial E}{\partial p_1} \quad \frac{\partial E}{\partial p_2} \quad \cdots \quad \frac{\partial E}{\partial p_n} \right] \begin{bmatrix} \delta p_1 \\ \delta p_2 \\ \vdots \\ \delta p_n \end{bmatrix} \tag{12.10}$$

$$= \frac{\partial E}{\partial p_1}\delta p_1 + \frac{\partial E}{\partial p_2}\delta p_2 + \dots + \frac{\partial E}{\partial p_n}\delta p_n$$

Comparing this with (12.9), it follows

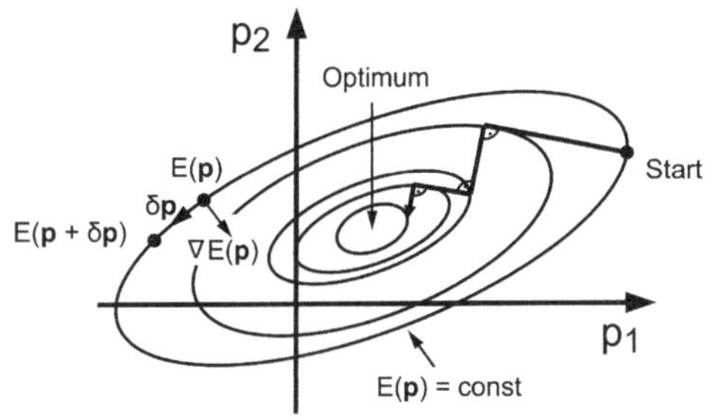

Figure 12.3 The method of deepest descent approaches the optimum (minimum) in a zigzag line, where the new search direction is orthogonal to the previous.

$$E(\boldsymbol{p} + \delta\boldsymbol{p}) \approx E(\boldsymbol{p}) + \nabla E^{T}\delta\boldsymbol{p} \tag{12.11}$$

where $\nabla E^{\mathrm{T}}\ \delta\boldsymbol{p}$ is equivalent to the scalar product of the gradient ∇E and parameter change $\delta\boldsymbol{p}$.

In the steepest-descent optimization technique one first chooses an initial set of parameters $\boldsymbol{p}$, which results in an error $E(\boldsymbol{p})$. Next the gradient ∇E is calculated at point $\boldsymbol{p}$. The parameter changes are then chosen to be in the direction of the negative gradient; that is

$$\delta\boldsymbol{p} = -\alpha\nabla E \tag{12.12}$$

where α is a positive proportionality constant. α can then again be determined by quadratic interpolation. In the steepest-descent optimization technique the direction is that of

$$\boldsymbol{d} = -\nabla E \tag{12.13}$$

Finally, the new parameters can be calculated from the old parameters using relations (12.3) and (12.4).

Choosing the parameter changes to be in the direction of the gradient results in a path that is orthogonal to contours of constant error. Recalling (12.11), if $\delta\boldsymbol{p}$ is chosen to be along a contour of constant error, then $E(\boldsymbol{p} + \delta\boldsymbol{p}) = E(\boldsymbol{p})$, so that $\nabla E^{\mathrm{T}}\ \delta\boldsymbol{p} = 0$. This implies that ∇E is orthogonal to $\delta\boldsymbol{p}$, (i.e., the gradient is orthogonal to contours of constant error [4]). Figure 12.3 illustrates the generated successive search directions, which are orthogonal to each other, and the minimum is approached in a kind of zigzag line.

Discussion

Gradient search is a typical local optimizer, which was one of the frequently used optimization tools in microwave laboratories in the eighties and nineties. It had been found that the method is fast at larger distance from the minimum. However, near the minimum it converges very slowly, as also commented in [12]. A main drawback is that this optimization method is susceptible to getting trapped at local minima in a multimodal search space.

Model parameter extraction indicates an optimization problem with usually many local minima. A satisfactory minimum can be found by conventional gradient-based techniques only with considerable expert guidance. But even when these algorithms are restarted from several randomly chosen initial points, results are seldom satisfactory. Expert consideration of the equivalent circuit at a local minimum with respect to the actual physics of the device can be used to estimate a superior starting point for another gradient-based optimization. This process is painstaking but can result in realistic equivalent circuits for microwave devices such as the HBT transistor [13].

The performance of gradient-based methods strongly depends on the initial values supplied. Several optimization runs with different initial guesses might be necessary if no a priori knowledge is available. There is a tremendous risk that the optimizer is stopped at a local minimum. The gradient-based optimizer does not yield a good result, although its initial progress is fast [14].

The local gradient-based method is the fastest if the initial guess is chosen appropriately but stops in a local minimum or even fails to converge. In this case, the whole optimization must be restarted with a different initial guess. This requires user interaction whenever a local optimum is encountered [15, 16].

The major cost of gradient-based algorithms is the engineering time required to examine the extracted circuit, decide on its merits, and to choose physical starting points [13].

12.4.2 Newton´s Method

Newton´s method is an extension of the steepest descent method insofar that is considers not only the first-order gradient (slope) of the object function but additionally the second-order gradient (curvature), and thus should be more accurate.

The extended Taylor approximation, including the second quadratic terms in (12.7), reads

$$E(\boldsymbol{p} + \delta\boldsymbol{p}) \approx E(\boldsymbol{p}) + \sum_{i=1}^{n} \frac{\partial E}{\partial p_i}\delta p_i + \frac{1}{2}\sum_{i=1}^{n}\sum_{j=1}^{n} \frac{\partial^2 E}{\partial p_i \partial p_j}\delta p_i \delta p_j \tag{12.14}$$

or in vector notation

$$E(\boldsymbol{p} + \delta\boldsymbol{p}) \approx E(\boldsymbol{p}) + \nabla E^T \delta\boldsymbol{p} + \frac{1}{2}\delta\boldsymbol{p}^T \boldsymbol{H} \delta\boldsymbol{p} \tag{12.15}$$

where $\boldsymbol{H}$ is the Hessian matrix, which is defined as

$$\boldsymbol{H}(\boldsymbol{p}) = [\boldsymbol{H}(\boldsymbol{p})]_{ij} = \begin{bmatrix} \frac{\partial^2 E}{\partial p_1^2} & \frac{\partial^2 E}{\partial p_1 \partial p_2} & \cdots & \frac{\partial^2 E}{\partial p_1 \partial p_n} \\ \frac{\partial^2 E}{\partial p_2 \partial p_1} & \ddots & & \vdots \\ \vdots & & \ddots & \vdots \\ \frac{\partial^2 E}{\partial p_n \partial p_1} & \cdots & \cdots & \frac{\partial^2 E}{\partial p_n^2} \end{bmatrix} \tag{12.16}$$

The gradient of E; that is, $\partial E/\partial(\delta\boldsymbol{p})$, is easily calculated from (12.15) as

$$\frac{\partial E}{\partial(\delta\boldsymbol{p})} = \nabla E + \boldsymbol{H}\delta\boldsymbol{p} \tag{12.17}$$

For a minimum we require $\partial E/\partial(\delta\boldsymbol{p}) = 0$, which delivers the solution

$$\delta\boldsymbol{p} = -\boldsymbol{H}^{-1}\nabla E \tag{12.18}$$

This gives the iterative update

$$p^{i+1} = p^i + \delta p^i = p^i - H^{-1}\nabla E \tag{12.19}$$

Line minimization is proposed to ensure global convergence. With (12.4), (12.19) modifies to

$$p^{i+1} = p^i - \alpha^i d = p^i - \alpha^i H^{-1}\nabla E \tag{12.20}$$

As seen by (12.20) the inverse of the Hessian matrix $\boldsymbol{H}$ needs to be computed at each iteration. This is not always feasible or is too expensive to compute at every iteration. The *classical Newton* method uses the Hessian at the initial point $\boldsymbol{p}^0$ throughout the optimization process. The effectiveness depends on how fast the Hessian is changing.

Recall that the Hessian is a matrix that contains second-order partial derivatives. We know that the evaluation of second derivatives by numerical methods is an inaccurate process. Therefore, direct evaluation of $\boldsymbol{H}^{-1}$ is commonly avoided. Instead, approximations are used. Any method that replaces the exact Hessian (or Jacobian) with an approximation is termed a *quasi-Newton* method. The most famous approximation is used in the Fletcher-Powell optimization technique described in the following section.

12.4.3 Davidon-Fletcher-Powell Method

The DFP method was initially proposed by Davidon in 1959 [17] and then modified by Fletcher and Powell in 1963 [18]. The DFP algorithm approximates the inverse of the Hessian by another matrix $\boldsymbol{G}$, which iteratively approaches $\boldsymbol{H}^{-1}$. Initially, the approximation is chosen to be the identity matrix $\boldsymbol{I}$, so that $\delta \boldsymbol{d}$ in (12.18) points in the direction of the negative gradient. Thus, the first Fletcher-Powell iteration is equivalent to the steepest descent, but the following iterations use better and better approximations for $\boldsymbol{H}^{-1}$ to improve the search direction.

The approximation $\boldsymbol{G}$ is obtained as follows. We assume that we are at a point $\boldsymbol{p}$ and have a value for $\boldsymbol{G}$ (initially, this would be the identity matrix). We then go in the direction $\boldsymbol{d} = -\boldsymbol{G} \cdot \text{GRAD}(\boldsymbol{p})$ for a distance $\Delta \boldsymbol{p}$ (the distance can be found by quadratic interpolation). The notation GRAD ($\boldsymbol{p}$) is used to indicate that the gradient is evaluated at point $\boldsymbol{p}$. Thus, the next point is $\boldsymbol{p}' = \boldsymbol{p} + \Delta \boldsymbol{p}$. At this point, a better approximation $\boldsymbol{G}'$ is found by using the parameter change $\Delta \boldsymbol{p}$ and also the gradient change

$$y = \text{GRAD}(p') - \text{GRAD}(p) \tag{12.21}$$

These changes are used to adjust $\boldsymbol{G}$ as follows:

$$G' = G + \frac{\Delta p \Delta p^T}{d_1} - \frac{(Gy)(Gy)^T}{d_2} \tag{12.22}$$

where the scale factors d_1 and d_2 are given by

$$d_1 = y^T \Delta p, \qquad d_2 = y^T G y \tag{12.23}$$

For a derivation of (12.22), the interested reader is referred to [3].

Discussion on the DFP Method:
The author's experiences have shown that concerning the extraction of model parameters, the parameter values have been changed only in small steps after each optimization cycle. For this reason the method runs relatively slowly. However, near the optimum the convergence was rapid. Therefore it seemed reasonable to combine the DFP method with the steepest descent method. The overall convergence of the solution could be definitely improved by starting the search first with the deepest descent method, and then, after some iterations, switching over to the DFP method. Nevertheless, the risk was relatively high that the algorithm moved into a local minimum.

12.5 NONLINEAR LEAST-SQUARES DATA FITTING

The least-*p*th optimization technique uses an error function as follows:

$$E(\boldsymbol{p}) = \frac{1}{2}\sum_{i=1}^{m} w_i [e_i(\boldsymbol{p})]^p \tag{12.24}$$

The individual error $e_i(\boldsymbol{p})$ is first raised to an even positive power p to guarantee that each contribution to the total error E is positive, and then multiplied by a weight w_i before the summation process. Please note that the vector $\boldsymbol{p}$ indicates the parameter vector and the scalar p the least-p exponent. The factor 1/2 in (12.24) is introduced to simplify an annoying factor of 2 in a lot of expressions. It has no effect on the minimum search.

In the following, for simplification, we assume the weighting factor included in the error function. Then the error function can be written in the form

$$E(\boldsymbol{p}) = \frac{1}{2}\sum_{i=1}^{m} [e_i(\boldsymbol{p})]^p \tag{12.25}$$

When the error function is raised to the power $p = 2$, the least *p*th optimization technique is commonly referred to as the *least-squares* optimization technique. In this case the error function can be written in the form

$$E(\boldsymbol{p}) = \frac{1}{2}\sum_{i=1}^{m} [e_i(\boldsymbol{p})]^2 = \frac{1}{2}\sum_{i=1}^{m} e_i^2 \tag{12.26}$$

12.5.1 Gauss-Newton Method

We know that the gradient of the error function must be zero at a minimum. It will be made zero by proceeding in an iterative manner. Parameter changes for $\boldsymbol{p}$ will be chosen. These increments will usually not reduce ΔE to zero at the first step because a linear approximation must be made. However, by iterating, the gradient may be made as close as desired.

To learn how to choose the parameter change $\delta\boldsymbol{p}$, it is necessary to calculate the gradient of the error function. For the special form that was assumed in (12.26) it follows that

$$\nabla E(\boldsymbol{p}) = \begin{bmatrix} \dfrac{\partial E}{\partial p_1} \\ \dfrac{\partial E}{\partial p_2} \\ \vdots \\ \dfrac{\partial E}{\partial p_n} \end{bmatrix} = \begin{bmatrix} \sum_{i=1}^{m} \left(e_i \dfrac{\partial e_i}{\partial p_1} \right) \\ \sum_{i=1}^{m} \left(e_i \dfrac{\partial e_i}{\partial p_2} \right) \\ \vdots \\ \sum_{i=1}^{m} \left(e_i \dfrac{\partial e_i}{\partial p_n} \right) \end{bmatrix} \tag{12.27}$$

In this equation it has been assumed that the error function depends on n parameters, so that the gradient vector has n components. Any of these n components can be found by evaluating a summation of m terms, where m represents the number of individual error functions.

The result in (12.27) can be written in a simpler form if we introduce the $m \times n$ Jacobian matrix, which is defined as

$$\boldsymbol{J}(\boldsymbol{p}) = \begin{bmatrix} \dfrac{\partial e_1}{\partial p_1} & \dfrac{\partial e_1}{\partial p_2} & \cdots & \dfrac{\partial e_1}{\partial p_n} \\ \dfrac{\partial e_2}{\partial p_1} & \dfrac{\partial e_2}{\partial p_2} & \cdots & \dfrac{\partial e_2}{\partial p_n} \\ \vdots & \vdots & & \vdots \\ \dfrac{\partial e_m}{\partial p_1} & \dfrac{\partial e_m}{\partial p_2} & \cdots & \dfrac{\partial e_m}{\partial p_n} \end{bmatrix} = \begin{bmatrix} \dfrac{\partial e_i}{\partial p_k} \end{bmatrix} \tag{12.28}$$

That is, the element in the ith row and kth column is $\partial e_i/\partial p_k$.

Then, for a least-squares error formulation, the gradient can be written as

$$\nabla E(\boldsymbol{p}) = \boldsymbol{J}^T(\boldsymbol{p})\boldsymbol{e}(\boldsymbol{p}) \tag{12.29}$$

where $\boldsymbol{e}(\boldsymbol{p})$ is a vector that has the individual errors e_i as components. That is, $\boldsymbol{e}(\boldsymbol{p}) = (e_1, e_2, ..., e_n)$.

Recall that at a minimum of the error function E, the gradient should be zero. The initial parameter vector $\boldsymbol{p}$ will usually not be at the minimum; thus $\nabla E(\boldsymbol{p})$ will not be zero. However, (12.29) can be used to obtain a change $\delta\boldsymbol{p}$ such that the magnitude of the gradient is reduced. The increment $\delta\boldsymbol{p}$ can be found by using (12.29) to express the gradient at $\boldsymbol{p} + \delta\boldsymbol{p}$ as

$$\nabla E(\boldsymbol{p} + \delta\boldsymbol{p}) = \boldsymbol{J}^T(\boldsymbol{p} + \delta\boldsymbol{p})\boldsymbol{e}(\boldsymbol{p} + \delta\boldsymbol{p}) \tag{12.30}$$

It often happens that the Jacobian does not vary very much as the parameters are changed. Even if this is not true at the start of an iterative search for the minimum, it will be true as the minimum is approached. Thus we can use the approximation

$$\nabla E(\boldsymbol{p} + \delta\boldsymbol{p}) \approx \boldsymbol{J}^T(\boldsymbol{p})\boldsymbol{e}(\boldsymbol{p} + \delta\boldsymbol{p}) \tag{12.31}$$

In order to choose the parameter change $\delta\boldsymbol{p}$ so that the gradient becomes zero, we must find an approximation for $\boldsymbol{e}(\boldsymbol{p} + \delta\boldsymbol{p})$. Expanding this in a Taylor series and keeping the first-order terms yields

$$\boldsymbol{e}(\boldsymbol{p} + \delta\boldsymbol{p}) \approx \boldsymbol{e}(\boldsymbol{p}) + \sum_{k=1}^{n} \frac{\partial \boldsymbol{e}}{\partial p_k} \delta p_k \tag{12.32}$$

This approximation can be written in the following simpler form by using the Jacobian defined in (12.28):

$$\boldsymbol{e}(\boldsymbol{p} + \delta\boldsymbol{p}) \approx \boldsymbol{e}(\boldsymbol{p}) + \boldsymbol{J}(\boldsymbol{p})\delta\boldsymbol{p} \tag{12.33}$$

The approximations (12.31) and (12.33) can be used to estimate the parameter changes $\delta\boldsymbol{p}$ that will produce a zero gradient at $\boldsymbol{p} + \delta\boldsymbol{p}$. Substituting (12.33) into (12.31), setting $\nabla E(\boldsymbol{p} + \delta\boldsymbol{p})$ equal to zero, and rearranging yields

$$\boldsymbol{J}^T\boldsymbol{J}\delta\boldsymbol{p} \approx -\boldsymbol{J}^T\boldsymbol{e}(\boldsymbol{p}) \tag{12.34}$$

12.5.2 Levenberg-Marquardt Method

The Levenberg-Marquardt algorithm was developed in the early 1960s to solve nonlinear least-squares problems. It is an iterative technique that locates the minimum of a multivariate function that is expressed as the sum of squares of nonlinear real-valued functions. It has become a standard technique for nonlinear least-squares problems and can be thought of as a combination of steepest descent and the Gauss-Newton method. When the current solution is far from the correct one, the algorithm behaves like a steepest descent method: slow, but guaranteed to converge. When the current solution is close to the correct solution, it becomes a Gauss-Newton method [19].

Levenberg (1944) and later Marquardt (1963) suggested using a *damped Gauss-Newton method* [20, 21]. The step $\delta\boldsymbol{p}$ is defined by the following modification to (12.34):

$$(\boldsymbol{J}^T\boldsymbol{J} + \lambda\boldsymbol{I})\delta\boldsymbol{p} \approx -\boldsymbol{J}^T\boldsymbol{e}(\boldsymbol{p}) \tag{12.35}$$

with $\boldsymbol{J} = \boldsymbol{J}(\boldsymbol{p})$. $\boldsymbol{I}$ denotes the identity matrix. The damping factor λ has several effects [12]:

(1) For $\lambda > 0$, $\delta\boldsymbol{p}$ is a descent direction.

(2) For large values of λ we get

$$\delta\boldsymbol{p} \approx -\frac{1}{\lambda}\boldsymbol{J}^T(\boldsymbol{p})\boldsymbol{e}(\boldsymbol{p}) \tag{12.36}$$

that is, a short step in the steepest descent direction (see (12.29)). This is good if the current iterate is far from the solution.

(3) If λ is very small, then $\delta\boldsymbol{p}$ approaches to the parameter vector change of Gauss-Newton method (see (12.34)), which is a good step in the final stages of the iteration, when $\boldsymbol{p}$ is close to $\boldsymbol{p}^*$. If $E(\boldsymbol{p}^*) = 0$ (or very small), then we get (almost) quadratic final convergence.

The Levenberg-Marquardt algorithm is adaptive because it controls its own damping. It raises the damping if a step fails to reduce the error. Otherwise it reduces the damping. In this way the LM algorithm is capable of alternating between a slow descent approach when being far from the minimum and a fast convergence when being at the minimum's neighborhood.

The damping parameter λ is initialized to be large so that first updates are small steps in the steepest-descent direction. If any iteration happens to result in a worse approximation ($E(\boldsymbol{p} + \delta\boldsymbol{p}) > E(\boldsymbol{p})$), then λ is increased. Otherwise, as the solution improves, λ is decreased, the Levenberg-Marquardt method approaches the Gauss-Newton method, and the solution typically accelerates to the local minimum.

Discussion on Levenberg-Marquardt
Application of the Levenberg-Marquardt (LM) method considerably improves the performance of gradient-based parameter extractors although entrapment in local minima can still be a problem [13, 22]. To improve the search, it is suggested to combine the LM method with a global search algorithm, as it needs a sufficiently close initial approximation [22].

12.6 DIRECT SEARCH METHODS

Direct search methods need not rely explicitly on evaluation or estimation of partial derivatives of the objective function at any point [23]. Frequently direct search methods are preferred to gradient-based optimization techniques for very practical reasons. Often, particularly in experimental settings, analytic derivatives are simply unavailable. Finite-difference approximations to the gradient could be used, but in an experimental setting it is not at all unusual to have function values that can only be trusted to a few significant digits so that finite-difference approximations to the gradient may prove unreliable [24].

In the following we will describe two direct search methods: the *Hook-Jeeves* pattern search and *Nelder-Mead simplex* algorithm. In particular, the latter optimization method has proven very successful in the scope of model parameter extraction of microwave devices.

12.6.1 Hook-Jeeves

Hooke and Jeeves presented the following pattern search strategy in 1961 [25]. The method involves two types of moves:

- *Exploratory move*: This is a very local search that looks for an improving direction in

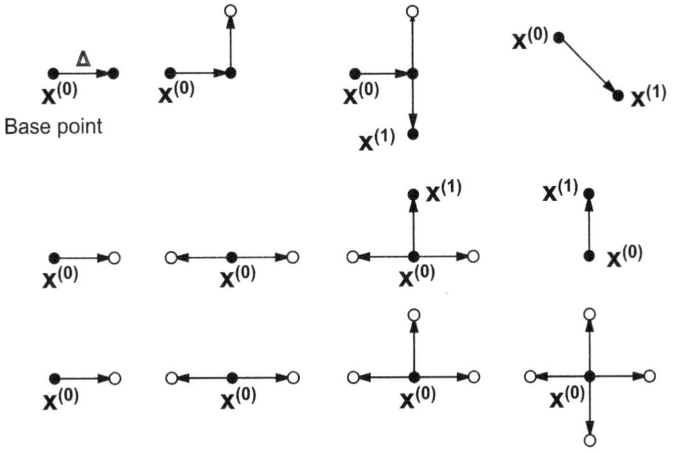

Figure 12.4 Hooke-Jeeves pattern search: Examples of exploratory searches. The last example denotes a failure, which entails reduction of the increment vector **Δ**.

which to move. In some senses it is a crude search for the gradient direction.

- *Pattern move*: This is a larger search in the improving direction. Larger and larger moves are made as long as the improvement continues.

Figure 12.4 illustrates examples of exploratory searches. $x^{(0)}$ signifies a current base point (starting point) from where to find a better solution of the objective junction. The increment vector $\mathbf{\Delta}$ denotes the variable incremental change, which may vary during the optimization process. In the exploratory move, the current point $x^{(0)}$ is perturbed in positive and negative directions along each variable one at a time and the best point ($x^{(1)}$) is recorded (Figure 12.4). If the exploratory search is not successful at all, then the increment vector $\mathbf{\Delta}$ is reduced by a reduction factor of 2.

After finalized exploratory search, pattern move starts. The original point $x^{(0)}$ and found best point ($x^{(1)}$) indicate an auspicious direction to move in. A new point $x^{(2)}$ is generated by moving from $x^{(0)}$ through $x^{(1)}$ as follows:

$$\boldsymbol{x}^{(2)} = \boldsymbol{x}^{(0)} + a(\boldsymbol{x}^{(1)} - \boldsymbol{x}^{(0)}) \tag{12.37}$$

where a is the *acceleration factor* that just multiplies the length of the improving direction vector by $\boldsymbol{x}^{(1)} - \boldsymbol{x}^{(0)}$. A common choice is $a = 2$, in which the equation reduces to

$$\boldsymbol{x}^{(2)} = 2\boldsymbol{x}^{(1)} - \boldsymbol{x}^{(0)} \tag{12.38}$$

Note that pattern moves are repeated as long as they are successful, and usually become longer and longer. But as soon as a pattern move fails by producing an $f(\boldsymbol{x}^{(2)})$ that is worse than the previous best value of the object function $f(\boldsymbol{x}^{(1)})$, then the pattern move is retracted and we go back to an exploratory search around the best point seen so far.

The algorithm stops when an exploratory search around the best point seen so far fails at all sizes (compare the last example in Figure 12.4) of perturbation, even the smallest as specified $\boldsymbol{x}^{(0)}$ by the perturbation tolerances. This means that the algorithm cannot find an improving direction away from the best point; hence it must be the optimum point. An alternative direct search method is the so-called simplex algorithm, which is discussed in the following section.

12.6.2 Simplex

The nonlinear simplex optimization algorithm was proposed by Nelder and Mead in 1965 [26]. It is widely used and has often been described in the literature. The book written by Walters et al. [6] is devoted entirely to the Nelder-Mead method and its variations and contains a bibliography with thousands of references. It is also treated in the handbook *Numerical Recipes* [10], and in MATLAB [27]. Of all the direct search methods, the Nelder-Mead algorithm is the one most often found in numerical software packages [28]. For MATLAB users, the simplex algorithm is implemented in the function fminsearch [29].

The term *simplex* is the topological generalization of a triangle, which is shaped by three points on a two-dimensional surface. A nondegenerate simplex in an n-dimensional space (encloses a finite inner n-dimensional volume) is formed by a set of $(n + 1)$ vertices. In a three-dimensional space the vertices of the simplex form a tetrahedron. The points of a *regular* simplex are all in equal distance (equilateral triangle).

The search with regular simplexes, originally proposed by Spendley, Hext, and Himsworth [30], exhibited certain practical difficulties in the regular simplex procedure; namely, that it did not provide for acceleration of the search and encountered difficulty in carrying on the search in curving valleys or on curving ridges. In the Nelder and Mead technique the simplex is permitted to alter in shape and thus no longer stays a simplex; hence the use of the more descriptive nameflexible polyhedron search. The FPS method, as opposed to the rigid simplex, is designed to adapt itself to the topography of the objective function, elongating down long inclined planes, changing directions on curving valleys, and contracting in the neighborhood of a minimum [2].

In the author's research group the features of the simplex optimization method were thoroughly analyzed over several years. It has been proven as a reliable optimization tool in the area of model parameter extraction of microwave devices. It is clear that the classical simplex algorithm is primarily applicable for local minimum search. However, as will be discussed in Section 12.7.5, some modifications have been introduced that essentially widens the search capability; that is, after having been trapped in a local minimum, the algorithm enables to escape to look for the next minimum, possibly a global one. In Section 12.8, various means are discussed in which way the classical localized simplex can be globalized. The global search capability of an optimization method is a primary requirement for the minimization of a multidimensional highly nonlinear error function comprising about 15 optimization variables or more.

In the following the simplex optimization strategy is briefly described. We suppose that an error function $\varepsilon(\boldsymbol{P})$ should be minimized. $\varepsilon(\boldsymbol{P})$ is a function of n variables $p_1, ..., p_n$. $\boldsymbol{P}$ describes an n-dimensional vector, and therefore it comprises an n-dimensional optimization concept.

The nonlinear simplex optimization procedure starts with the definition of a parameter vector with $(n + 1)$ components that spans an n-dimensional space. For each optimization run, values for $p_1, ..., p_n$ must be given (an initial guess must be made for the variables $p_1, ..., p_n$). The simplex optimization method varies these starting values in such a way that $\varepsilon(\boldsymbol{P})$ is minimized. The arbitrarily chosen starting points are stored in an n-dimensional vector field as a vector $\varepsilon(\boldsymbol{P}_0)$,

$$\boldsymbol{P}_0 = (p_1, p_2, ..., p_n) \tag{12.39}$$

This starting vector is used to generate further vectors $\boldsymbol{P}_1, \boldsymbol{P}_2, ..., \boldsymbol{P}_n$. The vector $\boldsymbol{P}_1$ is derived from $\boldsymbol{P}_0$ as follows:

$$\boldsymbol{P}_1 = (k{\cdot}p_1, p_2, ..., p_n) \tag{12.40}$$

The vector $\boldsymbol{P}_i$ is given as

$$\boldsymbol{P}_i = (p_1, p_2, ..., k{\cdot}p_i, ..., p_n) \tag{12.41}$$

where k denotes the initialization factor and is frequently chosen as 0.9 and 1.1, respectively. The iterative part of the simplex algorithm initially determines the error values $\varepsilon(\boldsymbol{P}_0)$, $\varepsilon(\boldsymbol{P}_1)$, ..., $\varepsilon(\boldsymbol{P}_n)$ of the vectors $\boldsymbol{P}_1, \boldsymbol{P}_2, ..., \boldsymbol{P}_n$. The vectors that result in the lowest, the highest, and the second highest error are kept as P^L, P^H, P^S. The simplex algorithm replaces always P^H, the vector with highest error, during the procedure. This results is a movement toward the error minimum, which is performed by a series of reflections, contractions, and extensions, as illustrated in Figure 12.5. The execution of the simplex algorithm can be described by the following three basic steps:

Step 1: Generation of an initial simplex with $(n + 1)$ vertices (initialization).

Step 2: Determination of all error function values, in particular P^H vertices of the simplex with highest error $\varepsilon(P)$, P^S vertices with second highest error $\varepsilon(P)$, P^L vertices with smallest error $\varepsilon(P)$, P^0 centroid of value vectors P^i with the exception of P^L, defined as

$$P^0 = \frac{1}{n}\sum_{\substack{i=1 \\ i \neq H}}^{n+1} P_i = \frac{1}{n}\left(\sum_{i=1}^{n+1} P_i - P^H\right) \tag{12.42}$$

Step 3: Replacing the point with largest error by the reflection point provides the expansion point or contraction point. If these trials fail, shrinking of the simplex may be necessary.

We regard a two-dimensional model parameter vector $P\,(p_1, p_2)$. The initial simplex with an initialization factor k (starting vector) reads

$$\begin{pmatrix} p_1 & p_2 \\ kp_1 & p_2 \\ p_1 & kp_2 \end{pmatrix} \tag{12.43}$$

In the first row the starting vector is placed; in the following rows, one parameter value of the starting vector is always multiplied with the initialization factor k. Thus, the simplex comprises $n + 1 = 3$ vertices: $\mathbf{P}_0$, P_1, and P_2, as illustrated in Figure 12.6.

Reflection Test: The first trial to replace P^H is performed by reflection at the centroid P^0: $P^R = (1 + \alpha)P^0 - \alpha P^H$; that is, P^H is reflected at P^0 and yields P^R. When $\alpha = 1$ then the

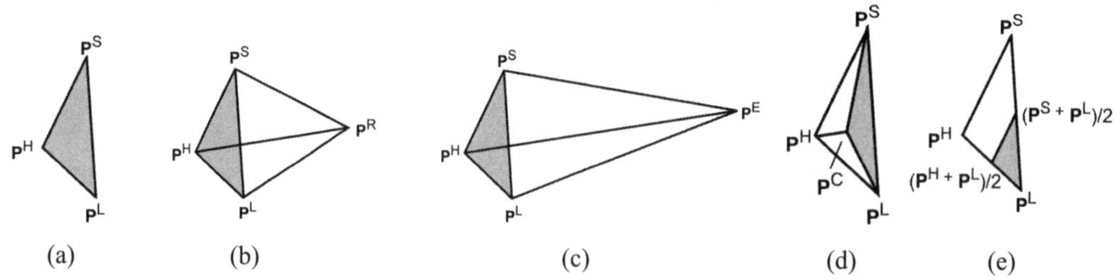

Figure 12.5 Possible simplex mutations. (a) Initial simplex with highest, second highest, and lowest error points P^H, P^S, and P^L, respectively. The point with highest error P^H should be replaced by a point of lower value. For this, the following trials are stepwise carried out until the desired result is obtained: (b) reflection away from the point of highest error P^H, (c) reflection and expansion away from point P^H, (d) contraction along one dimension from point P^H, or (e) contraction along all dimensions toward the point with lowest error value P^L (shrinking). An appropriate sequence of such steps will always converge to a minimum of the function, which is not necessarily the global one.

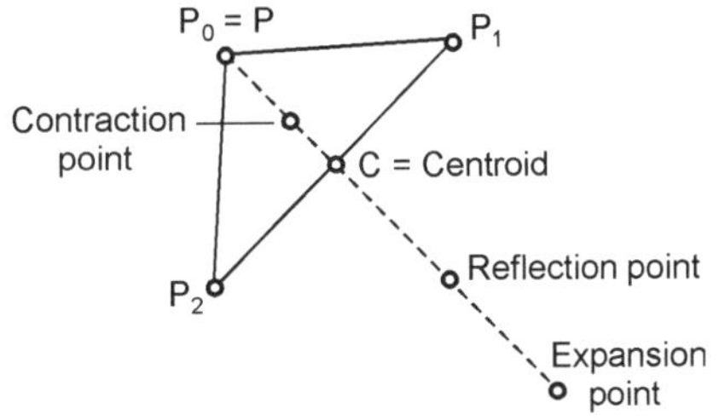

Figure 12.6 Replacing point $P_0 = P^H$ with the highest error by a lower error point found by contraction, reflection, or expansion.

reflection point has the same distance to the centroid as $\boldsymbol{P}^H$. The reflection coefficient α controls the length of the reflection. It is $\alpha \approx 1$. In order to avoid zero results, α is defined as 0.9985 in the computer program.

If $\varepsilon(\boldsymbol{P}^R) \le \varepsilon(\boldsymbol{P}^S)$ and $\varepsilon(\boldsymbol{P}^R) \ge \varepsilon(\boldsymbol{P}^L)$, then $\boldsymbol{P}^H$ can be replaced by $\boldsymbol{P}^R$. After that it can be restarted with this new simplex. If $\varepsilon(\boldsymbol{P}^R) < \varepsilon(\boldsymbol{P}^L)$, then the error function value is smaller than all other points of the simplex. In this case it will be expanded in the same direction.

If $\varepsilon(\boldsymbol{P}^R) > \varepsilon(\boldsymbol{P}^H)$, then no improvement was obtained by the reflection point. This implies that the minimum lies probably within the simplex. Therefore, the simplex is contracted with respect to the value vector with smallest value $\boldsymbol{P}^L$ according to $\boldsymbol{P}_i = (\boldsymbol{P}_i + \boldsymbol{P}^L)/2$.

Expansion Test: When the reflection point generates a new minimum so that $\varepsilon(\boldsymbol{P}^R) < < \varepsilon(\boldsymbol{P}^H)$ and $\varepsilon(\boldsymbol{P}^R) < \varepsilon(\boldsymbol{P}^S)$, then it seems to be in all probability that we iterate in a promising direction so that the simplex continues to search in this direction according to the expression $\boldsymbol{P}^E = \beta \boldsymbol{P}^R + (1 - \beta)\boldsymbol{P}^0$. β is the expansion coefficient. If $\beta = 2$ is chosen the point $\boldsymbol{P}^E$ found by expansion removes itself from the centroid $\boldsymbol{P}^0$ by twice the distance to the reflection point. In the computer program $\beta = 1.95$ is chosen to prevent potential instabilities during optimization. $\boldsymbol{P}^H$ is replaced by $\boldsymbol{P}^R$ or by $\boldsymbol{P}^E$ with the lowest error.

If $\varepsilon(\boldsymbol{P}^R) < \varepsilon(\boldsymbol{P}^H)$ and $\varepsilon(\boldsymbol{P}^R) > \varepsilon(\boldsymbol{P}^S)$, then only a slight improvement has been obtained, and $\boldsymbol{P}^H$ is replaced by $\boldsymbol{P}^R$. The contraction test follows.

Contraction Test: When the reflection operation does not succeed; that is, $\varepsilon(\boldsymbol{P}^R) < \varepsilon(\boldsymbol{P}^H)$, then it is expected that it was searched on the wrong side of the centroid. Now, a contraction is started, according to the instruction $\boldsymbol{P}^C = \gamma \boldsymbol{P}^H + (1 - \gamma)\boldsymbol{P}^0$. γ is the contraction coefficient. If $\gamma = 0.5$, then the contraction point is in the middle of $\boldsymbol{P}^H$ and the centroid. From the practical point of view $\gamma = 0.4985$ is chosen. If $\varepsilon(\boldsymbol{P}^C) < \varepsilon(\boldsymbol{P}^H)$, the contraction was successful, and we set $\boldsymbol{P}^H = \boldsymbol{P}^C$. In this case the result is the same as the reflection process would have been successful, and it is—as described above—again continued with the reflection process.

Shrinking (scaling): If reflection, expansion, and contraction fail, we resort to the shrinking operation. This operation retains the point of lowest error $\boldsymbol{P}^L$ and shrinks the simplex; that is, for all points of the simplex except the best one, we compute a new position $\boldsymbol{P}_i + k(\boldsymbol{P}^L - \boldsymbol{P}_i)$ for all $\boldsymbol{P}_i$, $i = 0, 1, \ldots, n$. If $k = 0.5$, then the distance is halved between the vector $\boldsymbol{P}^L$, which delivers the smallest error, and the other vectors $\boldsymbol{P}_i$. Commonly the simplex becomes smaller with $k > 0$ around point $\boldsymbol{P}^{\mathrm{L}}$. If $k = -1$, then the distance is doubled between $\boldsymbol{P}^{\mathrm{L}}$ and $\boldsymbol{P}_i$.

The search continues until the simplex size is less than a specified amount (i.e., $\boldsymbol{P}^{\mathrm{H}} - \boldsymbol{P}^{\mathrm{L}} < \varepsilon_{\mathrm{min}}$). This indicates that either a local or global a minimum point has been reached. Implementing the Nelder-Mead method is relatively simple due to the fact that the algorithm is already implemented as the MATLAB function *fminsearch*. *fminsearch* requires a cost function $f(x)$ where x is a vector. The algorithm needs a starting point and returns the minimum cost function value [29]. *fminsearch* does not admit bound constants. *fminscon* is used exactly like *fminsearch*, except that bounds are applied to the variables. The bounds are inclusive inequalities, which admit the boundary values themselves. The Nelder-Mead method is also implemented as *Nminimize* in the software package *Mathematica* [31].

Discussion on the Nelder-Mead Simplex

Direct search methods have the significant advantage of requiring no gradient information of the objective function. Working previously with the standard nonlinear Nelder-Mead simplex, we observed that the individual model parameter values were strongly changed after each optimization cycle. This lowered the risk to get stuck in local minima, but nevertheless the optimizer also delivered extracted results that were obviously unphysical (e.g., negative resistance values). Repeated model parameter extraction of a common 15-element equivalent network, the nonlinear Simplex algorithm, using different starting vectors delivered solutions with a residual error of a comparable order of magnitude. However, the solutions for the individual model elements were different. Improvement could be obtained by reducing the search space based on the approximation of relevant model elements from a low-frequency measurement (see Chapter 8) [32]. Dramatic improvement of the simplex algorithm could be obtained by implementation of a reinitialization function, as described in Section 12.7.6.

12.7 GLOBAL OPTIMIZATION

A global search can be performed by repeatedly restarting a local optimization. To void finding the same local minima, the new initial points should be different and preferably far from previous initial points and already known local solutions [33].

Global optimization methods play, in particular, an important role in physical device modeling. Physics-based models require extraction of numerous parameters [34]. For example, the well-known Hefner model for an insulated gate bipolar transistor (IGBT) [35] includes 20 physical and structural parameters. As reported in [13], measurement errors coupled with the large number of elements of a physically-based equivalent circuit lead to an error function with many nonphysical local minima in addition to the global minimum. The number of additional minima grows rapidly as the number of elements of the equivalent circuit is increased, and gradient-based algorithms generally terminate in one of these.

Nevertheless, global optimization methods are commonly needed where numerous variables of a very complex model have to be determined from measurements. This is the case, for example, in the field of *technology computer-aided design* (TCAD), which is concerned with the simulation of semiconductor device performance under the consideration of manufacturing process data [36].

In this section we briefly review, in addition to *grid* and *random search* methods, two frequently used global optimization techniques: *genetic algorithm* (GA) and *simulated annealing* (SA). These stochastic techniques have been widely used for years in complex optimization problems. Application to microwave CAD can be found in [13, 37–39].

12.7.1 Multistart Methods

The simplest method for global optimization are perhaps the *multistart* methods. Initial starting parameter values can be generated in a systematic way (grid search method) or can be randomly generated (random search method).

12.7.1.1 Grid Search Method

The grid search method starts with setting up an equally spaced grid in the parameter space. Then objective function values are evaluated at the grid points to obtain a set of local minima (by using any local optimization method). The point corresponding to the best objective function value is considered as the optimum solution. Improved solutions may be found by a restart with a finer grid. Furthermore, nonequal or adaptive grid search approaches can be chosen.

Indeed, if a sufficient number of initial points are generated and if the local minima are also relatively evenly distributed in the search space, the chance of finding a global minimum is rather high. The more initial points generated, the higher the chance of finding the global minimum. However, with the number of optimization variables, the number of grid points increases exponentially, and thus the cost of computation.

12.7.1.2 Random Search Method

A random optimization consists of a number of trials in the parameter space. In each trial every optimized variable is perturbed by a random amount. The modification is such that the search tends to go in the direction of improvement and to avoid directions of failure. Solution quality depends on the number of tested points. Since there is no reason to favor areas of the search space, uniform density distributions are used to generate parameter starting values.

The advantage of a random search is its ease of programming and the freedom of a priori selecting the number of iterations.

Discussion of Multistart Methods

In [13] the extraction of modified Materka model of a GaAs MESFET is reported applying different optimization algorithms. Starting model element values for a multistart gradient optimization were obtained from previously optimized element values of the equivalent circuit of a similar device. Typically three to seven restarts were required to obtain a quality solution. Between each restart the physical merit of the extracted equivalent circuit was examined to determine the starting point for subsequent gradient optimization. With more restarts the likelihood is increased to find some global minimum but the high cost of human intervention practically limits the number of restarts that can be tried. All minima encountered during optimization are recorded, thus providing alternatives in case of a nonphysical final solution. Nevertheless, this approach is easy to implement and it has proved to be the most successful of the commercially available gradient-descent-based approaches.

12.7.2 Genetic Algorithm

A GA is an evolutionary optimization technique using Darwin′s principle of survival of the fittest to improve a population of solutions. If the population size is suitably large, GA is not at risk of being stuck in a local minimum.

The evolution strategies (ES) algorithm is an optimization method developed in the 1970s by Rechenberg and Schwefel [41, 42]. It is a population-based, natureinspired, stochastic, derivativefree, and global algorithm. ES is based on competition among individuals: Only strong individuals survive; that is, individuals that are associated with a good fitness value [40].

Genetic algorithms represent an intelligent exploitation of a random search used to solve optimization problems.While randomized, genetic algorithms are no simple random walk. They efficiently exploit previous information to speculate on new search points with expected improved performance. Genetic algorithms are robust. They do not break easily in the presence of reasonable noise. GAs are suited for searching a large state-space, multimodal state-space, or n-dimensional surface. The genetic algorithm is an iteration process.

Figure 12.7 illustrates the operation principle of the genetic algorithm. It manipulates a population of individuals that compete with each other through biological inspired mechanisms called operators: *selection*, *crossover*, *mutation,* and *replacement*. The fittest

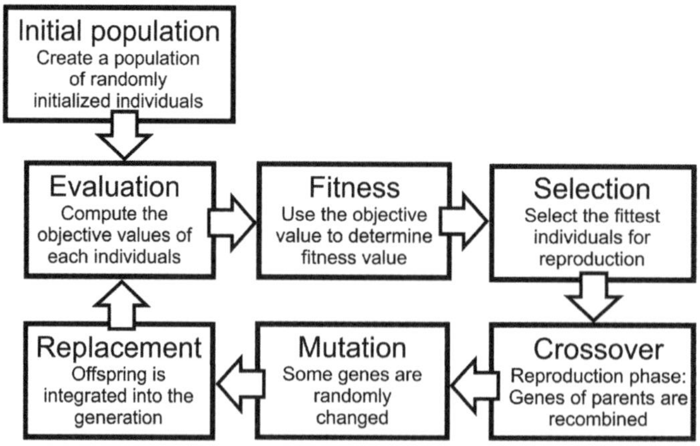

Figure 12.7 Generic structure indicating random generation of starting parameter values (individuals) and following an iterative optimization loop with stochastically changing parameter values (genes) according to crossover and mutation selection rules [40].

individuals produce more offspring. Generation after generation, the population improves and converges to a solution. In other words, the selection operator defines the procedure that selects parents that will be used by the crossover. Selection gives the evolution a direction, whereas mutation and recombination maintain genetic diversity among the population. Some operator, sometimes considered as an elitist, will make sure that the best individual out of all parents and all children is not replaced by a worse one.

The mutation operator consists of applying a random perturbation to individuals′ genes (parameters) of the population. It is used to maintain genetic diversity from one generation to the next. The use of this operator on an individual′s parameter is controlled by the probability of mutation (for example 0.05 or 5%). In each generation, a random number is drawn in [0; 1] for each parameter of each individual. If this number is lower than the probability of mutation, the mutation operator is applied. It works either by replacing the parameter by a randomly drawn value, by adding a random perturbation, or by switching two different parameters.

It should be mentioned that the search process operates on encoded parameter values and not on the real values. The binary code is most commonly used. For example, a problem with two parameters p_1 and p_2 may be mapped onto the gene in the following way:

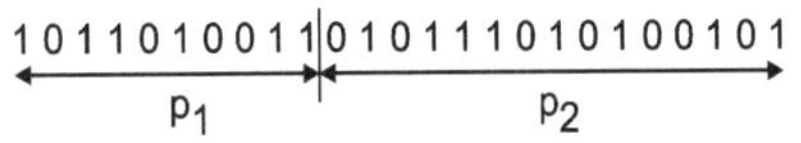

where the parameter value p_1 is encoded with 10 bits and p_2 with 15 bits [43, 44]. After having undergone the various genetic operations, the genes are decoded to commence the selection of individuals on the basis of objective function evaluation.

The initial population (set of initial potential solutions) comprises individuals that are real-valued vectors representing vector points in the model parameter space. A gene is simply one component of this vector. It has the same meaning as parameter. A generation is a set of individuals (i.e., population) at a specified iteration. Let us assume that the object function, here called *fitness*, is *n*-dimensional. The main utility of the fitness function is to generate a population of optimized parameters, specified by the user, in order to obtain the minimum error and solve the optimization problem. For that purpose, a population of points at each iteration is generated and evaluated by the genetic algorithm; different options are available for the user, allowing then the possibility to set constraints to this generation and decide the population size and generation size, among others. At the end the best parameters generated are saved as a variable. However, one never knows if the current best solution is the optimal one.

Discussion on the Genetic Algorithm

Experience with the genetic algorithm *GAlib*, freely available from lancet.mit.edu/ga, has been reported by Binder [15, 16]. GAlib includes different types of genetic algorithms. Three basic types differ in the way that they create new individuals and replace old individuals during the course of an evolution. Several experiments with different crossover and mutation probabilities were carried out. It is noted that the best individual within a population sometimes occurs at a lower evaluation number. Binder concluded: For the case of the genetic optimizer, the crossover and mutation probabilities are very crucial to the performance. It turned out to be rather time-consuming to find acceptable values for those parameters. As pointed out in [33], one of the main problems associated with GA is the high computational intensity and the premature convergence, which may happen if the initial population is not appropriately selected.

12.7.3 Simulated Annealing

SA is a stochastic optimization method introduced by Metropolis et al. [45], which mimics the annealing process in metallurgy. It was first proposed by Kirkpatrick et al. in 1983 [46] for optimization of combinatorial problems (in which the objective function is defined in a discrete domain) and was successfully employed for objective functions involving a very large number of variables (even tens of thousands). Corona et al. [47] were among the many researchers who extended the method for objective functions involving continuous variables.

The method closely mimics the physical annealing process. At high temperatures, atoms move freely in molten solid. It must be gradually cooled down in order to reach the crystalline state. If the cooling occurs too quickly, the atoms lose their thermal mobility and never reach a well-ordered low-energy crystalline state (this is called quenching). Therefore, annealing is the process of slow cooling of a material, which allows the atoms to distribute themselves in a minimum energy state.

In thermal equilibrium at temperature T the energy of the atoms obey the Boltzmann probability distribution [10],

$$P(E) \sim e^{-E/kT} \tag{12.44}$$

where k is Boltzmann's constant. That is, the system has its energy probabilistically distributed among all different energy states E. Even at low temperature, there is a chance, albeit a very small one, of the atoms of a system being in a high energy state. Therefore, there is a corresponding chance for the system to get out of a local energy minimum in favor of finding a better, more global one.

In 1953, Metropolis et al. [45] first incorporated these kinds of principles into numerical calculations. Offered a succession of options (random changes in the configuration), a simulated thermodynamic system was assumed to change its configuration from energy E_1 to energy E_2 with probability $P = \exp[-(E_2 - E_1)/kT]$. Notice that if $E_2 < E_1$, this probability is greater than unity; in such cases the change is arbitrarily assigned a probability $P = 1$ (i.e., the system always took such an option). This general scheme of always taking a downhill step while sometimes taking an uphill step has come to be known as the Metropolis algorithm [10].

A simulated annealing algorithm considers the objective function for global optimization as the energy function of an imagined physical system and mimics the physical annealing process so that, in the end, the global minimum of the objective function can be located by reaching the global energy minimum of the imaged physical system.

In the following section, the SIMANN SA algorithm developed by Goffe et al. [48] is described, which is based on the methodology proposed by Corana et al. [47] for objective functions involving continuous variables. This was one of the earliest public domain implementation of the *classical* simulated annealing, written in FORTRAN (and GAUSS) code [49, 50]. The file is well documented in [51]. The value assignment of the input parameters was essentially adopted from [47].

12.7.3.1 Classical Simulated Annealing Algorithm

The simulated annealing algorithm is schematically shown in Figure 12.8 [47]. Let $\boldsymbol{p}$ be an n-dimensional parameter vector and $(p_1, p_2, ..., p_n)$ its components. Let $E(\boldsymbol{p})$ be the object (error) function with the parameter vector $\boldsymbol{p}$ $(p_1, p_2, ..., p_n)$ to be minimized. The range of parameter values are defined within lower bounds (l_b) and upper bounds (u_b), i.e., $l_{b,1} < p_1 < u_{b,1}, ..., l_{b,n} < p_n < u_{b,n}$, each ranging in a finite, continuous interval. Before starting, different parameter settings have to be done. If the algorithm chooses occasionally undesired variable values $p_i < l_{b,i}$ or $p_i \geq u_{b,i}$ $(i = 1, n)$, a point from inside is randomly selected. The starting temperature T_0 must be chosen sufficiently high so that at the beginning both downhill and uphill moves are possible.

As shown in Figure 12.8, the algorithm starts with the initialization of the n parameter (i.e., each parameter must be assigned a starting value). This can be found by a previous random search (Section 12.7.1.2) or it is simply chosen as the mean value of the specified value interval of each parameter. The function evaluation is made at the starting point $\boldsymbol{p}$ and its value E is recorded. Then the global minimum is approached iteratively by the following steps:

1. Generation of trial points: A new point $\boldsymbol{p}'$ is generated around the current $\boldsymbol{p}$ by varying element i of $\boldsymbol{p}$ (index i refers to the control variable being optimized $(1 \leq i \leq n)$),

$$p_i^{/} = p_i + \Delta p_i = p_i + r v_i \tag{12.45}$$

where r is a uniformly distributed random number from $[-1, 1]$ and v_i is element of $\boldsymbol{v}$. The step vector $\boldsymbol{v}$ denotes the maximum increments possible along each direction. For element p_i, the next trial point selected is from $p_i - v_i$ to $p_i + v_i$. If the point falls outside the definition domain of E, a new point is randomly generated until a point belonging to the definition domain is found. Regarding the starting step, v_i would be estimated from the specified lower and upper bounds, $v_i = u_{b,i} - p_i = p_i - l_{b,i}$. Since $\boldsymbol{v}$ is adjusted after each cycle of random moves so that only half of all points are accepted, the input value is not very important.

2. Selection criteria: The object function $E' = E(p')$ is then computed. If E' is smaller than E, $\boldsymbol{p}'$ is accepted, $\boldsymbol{p}$ is set to $\boldsymbol{p}'$, and the algorithm moves downhill. If this is the lowest E, it and $\boldsymbol{p}$ are recorded since this is the best current value of the optimum. If E' is greater than or equal to E, then accept or reject the point with the acceptance probability P (Metropolis criterion):

$$P = \exp(-\Delta E / T(k)) \tag{12.46}$$

where $\Delta E = E' - E$ and T is the annealing temperature (index k refers to the external annealing loop). In practice, a pseudo-random number P' is generated in the range $[0, 1]$ and

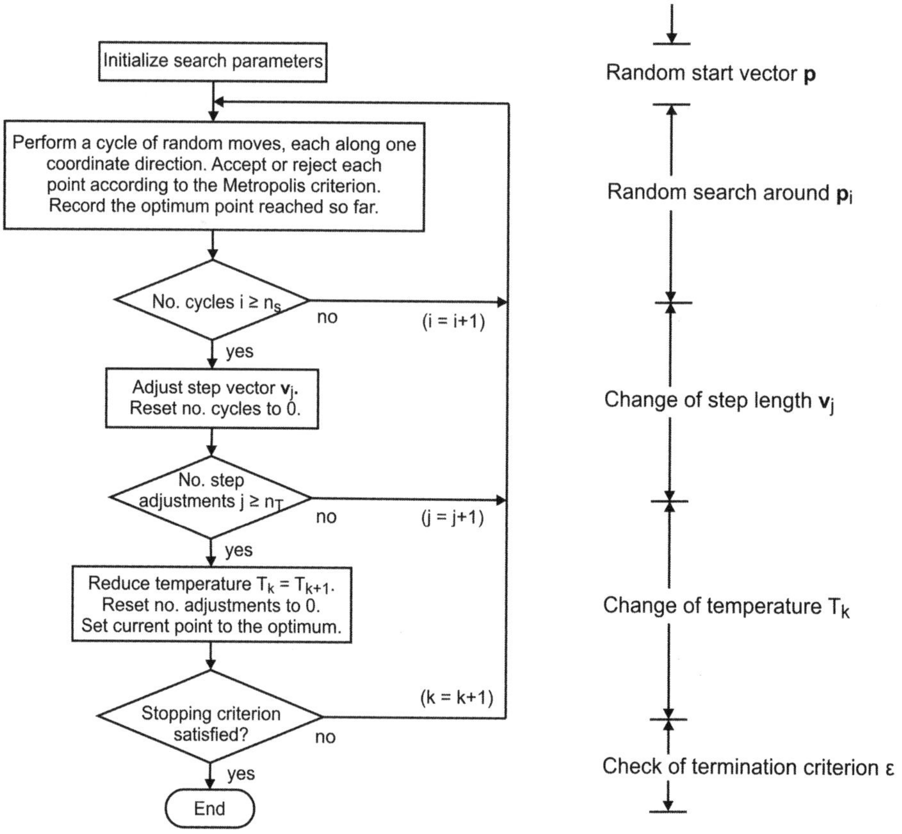

Figure 12.8 The SA minimization algorithm. (After [47].)

is compared with P. If $P > P'$, the point is accepted; otherwise it is rejected. Two factors decrease the probability of an uphill move: lower temperatures and larger differences in the function's value.

3. Step length adjustment: After all parameters of $\boldsymbol{p}$ have been randomly changed after n_S steps (i.e. this first optimization loop comprises $n \cdot n_S$ function evaluations. n_S is suggested to be 20) and the corresponding object function values evaluated, the step vector $\boldsymbol{v}$ is updated as follows. For each direction i the new step vector component v'_i is defined by [47]

$$
\begin{aligned}
v_i' &= v_i\left(1 + c_i\frac{n_i/n_s - 0.6}{0.4}\right) && \text{if } n_i > 0.6n_s \\
v_i' &= \frac{v_i}{1 + c_i\dfrac{0.4 - n_i/n_s}{0.4}} && \text{if } n_i < 0.4n_s \qquad (12.47)\\
v_i' &= v_i && \text{otherwise}
\end{aligned}
$$

where the vector $\boldsymbol{c}$ controls the step length adjustment. The suggested value for all elements is 2.0. Then the previous $\boldsymbol{v}$ is replaced by the new $\boldsymbol{v}'$. The aim of these variations in step length in (12.47) is to maintain the average percentage of accepted moves at about one-half of the total number of moves.

4. *Annealing temperature reduction*: n_T is the number of iterations before temperature reduction. This includes $n \cdot n_S \cdot n_T$ function evaluations. After n_T (suggested value for n_T: maximum of (100, $5 \cdot n$)) times through the above loops, the temperature, T, is gradually reduced. The new temperature is given by (exponential temperature function)

$$T_{k+1} = r_T T_k = T_0 \cdot (r_T)^k \tag{12.48}$$

where r_T is the temperature reduction coefficient (suggested value is 0.85). A lower temperature makes a given uphill move less likely, so the number of rejections increases and the step length declines. In addition, the first point tried at the new temperature is the current optimum.

5. *Convergence criteria*: The iterations are repeated until the reduction of the objective function value in n_ε successive cycles (suggested value is 4) is less than a small number ε, or if the total number of function evaluations exceeds the user-specified value.

It has been found that the algorithm requires a large amount of iterations, depending sensitively on the starting parameter values. With reference to the standard parameter setting in [47] it was shown that by elaborate parameter tuning, computing time could be decreased by 99.6%.

12.7.3.2 Refined Simulated Annealing Algorithms

A fixed simulated annealing schedule that is characterized by a fixed initial temperature, a constant temperature decrement, and a constant number of steps at each temperature is usually not practically applicable to general optimization problems since it requires several tests with different parameter values [15, 16].

Many variants of the simulated annealing algorithm have been published during the last decade, differing mainly in the schedule for temperature changes and method for obtaining a neighboring solution. Adaptive algorithms, which change their parameters during the evolution of the cost function, are more appropriate. These can be grouped in *standard* [52–54], *fast* [53–55], and *very fast* [53, 56, 57] simulated annealing.

In MATLAB [57], two types of annealing function are available: fast annealing function (annealingfast), and Boltzmann annealing function (annealingboltz). Bounds must be specified using the function simulannealbnd. The public domain software has been published by Ingber [56, 58].

The right choice of the critical parameters (variable bounds, initial temperature, function limit, and number of iterations performed at a particular T, i.e., cooling schedule) is absolutely essential for successful working of the SA method. Hence, some amount of trial and error is needed for finalizing these parameters [59].

Discussion on Simulated Annealing

Simulated annealing optimization has been applied in various fields (e.g., [60–62]). Chibante et al. [63] reported about the model parameter extraction of an insulated-gate bipolar transistor IGBT. The physical model requires in total 16 parameters, including geometrical, physical, and electrical parameters. Because of the large number of parameters a two-step strategy is used by dividing parameters into two groups: a first set of parameters is extracted using DC characteristics, while a second set is extracted using transient time waveforms with optimum parameters from DC extraction. For an initial satisfying guess of the simulated annealing algorithm, it is recommended to extract some parameters from measurements, exploit datasheet information, and consider equations relating some parameters.

It has been proved that the simulated annealing algorithm is efficient for the extraction of a large number of parameters needed for physics-based IGBT models. Good agreement has been attained between simulation and measurement.

It is noted that one disadvantage of simulated annealing is that definition of some control parameters (initial temperature, cooling rate, etc.) is somewhat subjective and must be defined from an empirical basis. The algorithm should be initially tested with several sets of control parameters to evaluate how they influence an algorithm's performance. This agrees with the experience reported by Shojaee et al. [64]. Although simulated annealing algorithms are conceptually simple, finding optimal parameters such as initial temperature, the annealing schedule, and the acceptance function parameters is by no means simple and straightforward. First of all, setting parameters for simulated annealing is problem dependent, and it is best accomplished through trial and error. Furthermore, many studies have demonstrated that simulated annealing algorithms are very sensitive to parameters, and their performances are largely dependent on fine tuning of the parameters.

In [65], the simulated annealing approach was used for the parameter extraction of an intrinsic seven-element equivalent circuit for a MODFET. It was noted that simulated and measured data of unilateral power gain were in good agreement. It is not really clear why the authors preferred simulated annealing optimization in this case. Nevertheless, it was stated that the great advantage of the simulated annealing algorithm is that no human intervention is required. To carry out model parameter extraction only the upper and lower limits of the element values have to be specified by the user.

12.7.4 Tree Annealing

Tree annealing is a variant of the simulated annealing algorithm and was proposed by Bilbro et al. [13, 66]. We have seen in the previous section that regarding the simulated annealing optimization method, a new candidate $\boldsymbol{p}'$ is generated by perturbation of a current solution $\boldsymbol{p}$; that is, $\boldsymbol{p}' = \boldsymbol{p} + \Delta\boldsymbol{p}$, where $\Delta\boldsymbol{p} = r\boldsymbol{v}$. Initially the step vector $\boldsymbol{v}$ equals the specified parameter value intervals but is reduced with progressive optimization steps. r indicates a random number generated uniformly between 0 and 1. Comparison of the values of the current and new objective function decides about acceptance or rejection of the sampled point in the parameter space. The Metropolis criterion also ensures uphill move that is likely to escape from a local minimum. Thus, simulated annealing may be classified as a local search optimizer not getting stuck in local minima with a certain probability, which depends on the current annealing temperature.

The principle of tree annealing is illustrated in Figure 12.9 for a two-dimensional

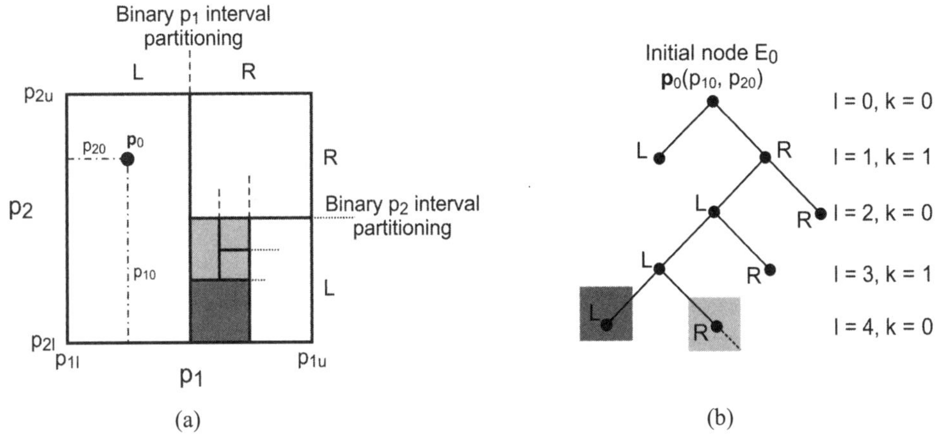

Figure 12.9 A two-dimensional search space and corresponding binary partition tree.

optimization problem with search parameters p_1 and p_2. Note in [13] the operation for a simple one-dimensional problem is detailed. The entire search space is limited by the specified upper and lower bounds (index u and l) of the parameters (Figure 12.9(a)), and the search space is structured, as depicted in Figure 12.9(b) as a k-dimensional (k-d) tree (or more precisely: a *k-d tree* [67], which by binary partitioning of the parameter space at a given tree level always provides subspaces of the same size).

The procedure starts choosing a point $\boldsymbol{p}(p_{10}, p_{20})$ at random with uniform probability from the entire search space. This point with initial objective function value E_0 becomes the first accepted point, and indicates the *root* node (tree level $l = 0$). The number k of node-level l is generally calculated as $k = 1 \bmod d$ (that is, $k = 0, 1$ in the two-dimensional case) and indicates the partition with respect to the kth degree of freedom (DOF). Regarding the initial node with $k = 0$, partition of the entire search space is done by dividing the p_1 interval (marked by the dashed line). In the case of $k = 1$, the actual value interval of p_2 would be halved (marked by the dotted lines).

After the root at zero level has been initialized with randomly chosen parameter vector $\boldsymbol{p}(p_{10}, p_{20})$ and the initial temperature T has been specified, two daughter nodes ($l = 1$) are created corresponding to a division of the search space in half L (left) and half R (right) along the first DOF, which is parameter p_1.

The binary tree records previous moves and uses this information to optimally generate subsequent candidates. At each node, two numbers are stored, n_L and n_R, representing how many times in the past an acceptable point has been found in the left and right subtrees, respectively. Then the left/right decision depends on the probability of $n_L/(n_L + n_R)$ or $n_R/(n_L + n_R)$. The n_L and n_R are both initialized to 1 for the root node.

A new point $\boldsymbol{p}'$ is generated uniformly at random over the subspace defined by the node and compared with $\boldsymbol{p}$ as the current sample; if $\boldsymbol{p}'$ is rejected, $\boldsymbol{p}$ remains the current sample. If $\boldsymbol{p}'$ was accepted, the current node $\boldsymbol{p}'$ is split, and two new daughter nodes are created, thus making more resolution available at this node. At this point, the number of right and left nodes at each node will have changed and the tree will need to be updated. Figure 12.9 indicates the growing of the tree structure with increasing levels of node associated with the gradual reduction of the global search space.

As can be seen, the data structure of the tree becomes very complex with the increasing degree of the search space. Assuming, for example, fifteen parameters of an equivalent circuit, k would take the numbers 0, 1, 2, ...,14. Thus the tree becomes very large. Suppose the domain of DOF 8 is $0 \le p_8 \le 1$, then a node at level 50 represents a partition whose moderate width is $1/(2^{[50/15]}) = 0.0625$.

As illustrated in Figure 12.9(a), tree annealing initially samples the entire search space in a coarse grid, then gradually increases the resolution when moving in the vicinity of a local minimum. Regarding the acceptance/rejection rule of a new point, the tree annealing algorithm leads to a modified Metropolis criteria. Instead of directly minimizing the objective function $E(\boldsymbol{p})$, simulated annealing algorithm considers maximizing the related probability density,

$$p(\boldsymbol{p}) \propto \exp(-E(\boldsymbol{p})/T) \tag{12.49}$$

Since the exponential function is monotonic, this does not change the problem and would not facilitate its solution except that p is now amenable to the powerful Metropolis sampling procedure. The modes of p coincide with the minima of E at all T, and at low enough values of T, $p(\boldsymbol{p})$ is so sharply peaked at the lowest values of $E(\boldsymbol{p})$ that virtually no other points contribute to the distribution. In simulated annealing a current state $\boldsymbol{p}$ is randomly perturbed into a candidate state $\boldsymbol{p}'$. The candidate is either accepted or rejected by the Metropolis criterion. The new point is accepted if

$$\frac{p(\boldsymbol{p}')}{p(\boldsymbol{p})} = \exp[-(E(\boldsymbol{p}') - E(\boldsymbol{p}))/T] > r \tag{12.50}$$

where r is a random number generated uniformly between 0 and 1. If the candidate is accepted, it replaces the current state in the next time step; otherwise, the current state persists and the candidate is discarded.

In tree annealing the generation of new candidates is provided by global approximation of $p(\boldsymbol{p})$, $g(\boldsymbol{p})$, rather than by perturbation of the current state. $g(\boldsymbol{p})$ is a uniform distribution in the subregion related to the respective node, and thus approximates the correct $p(\boldsymbol{p})$. For example, in the one-dimensional case, assuming an initial parameter interval L, the uniform distribution $g(\boldsymbol{p}) = g(\boldsymbol{p}')$ is simply calculated as $1/L$. Correspondingly, for the two-dimensional case with L_1 and L_2 parameter intervals, uniformly distributed $g(\boldsymbol{p})$ is calculated as $1/(L_1 \cdot L_2)$. In general, $g(\boldsymbol{p}')$ is computed from the path of the descent down the tree by

$$g(\boldsymbol{p}') = \frac{1}{V_{p'}} \prod_l \hat{p}_l \tag{12.51}$$

where $\hat{P}_l = a_l/(a_l + b_l)$, l represents each node visited, a_l represents n_L or n_R, according to which direction was chosen at each node. Similarly, b_l represents the n of the direction not chosen. Final division by $V_{p'}$, the hypervolume containing $\boldsymbol{p}'$, ensures that g represents a proper probability density. The generation of candidates from approximated $g(\boldsymbol{p})$ instead of correct $p(\boldsymbol{p})$ requires a modification of the Metropolis criterion,

$$P = \frac{g(\mathbf{p})\,p(\mathbf{p}')}{g(\mathbf{p}')\,p(\mathbf{p})} \tag{12.52}$$

where $p(\lambda) \propto \exp(-E(\lambda)/T)$ with $\lambda = (\mathbf{p}, \mathbf{p}')$. The annealing schedule is very similar to the simulated annealing algorithm (see (12.48)),

$$T \leftarrow rT \tag{12.53}$$

The temperature may be reduced with each node addition to the tree, which tightens the acceptance criterion. Stopping criteria may be specified on the basis of various metrics, such as maximum CPU limit, maximum number of calculations of the objective function, or maximum number of temperature changes.

Discussion on Tree Annealing
With respect to the consistent model parameter extraction, it is reported that the tree annealing procedure also reveals that various combinations of the elements of the transistor equivalent circuit are closely correlated so that the associated elements cannot be adequately resolved. This is in particular valid for series connections of a resistor R with a RC parallel circuit. Also most of the capacitors are tightly correlated. Therefore, if it is important to distinguish them, additional or independent measurements are required [13].

Focusing on model parameter extraction of microwave devices, Bilbro et al. [13, 68] have reported about the extraction of the parameters of a 14-element physically-based equivalent circuit model of a heterobipolar transistor using deembedded S-parameter measurements from 45 MHz to 26.5 GHz [68]. Simulated and measured stability factor k and maximum available gain G_{max} show very good agreement. A maximum deviation of about 7% is found for G_{max} at 10 GHz, and about 16% for k at 20 GHz.

As reported in [69], tree annealing is generally more reliable than the simulated annealing for global optimization over large dimensional search domains although it may be subject to a noticeable overhead cost of CPU time and memory space.

12.7.5 Leaping Simplex

The original nonlinear simplex algorithm is prone to converging to local minima. To overcome the local minimum problem we have modified the used simplex algorithm in such a way that after 10 successful optimization steps (i.e. decreasing of the error-function value), the simplex is forced to become enlarged to take the opportunity to escape from any instantaneous local minimum (modified simplex algorithm with reinitialization) [70].

There are two breaking criteria of the used simplex algorithm. The first breaking criterion depends on the error in the various points of the simplex. If the difference

$$\max(E(P_0), \ldots, E(P_n)) - \min(E(P_0), \ldots, E(P_n)) < W_{\min_1} \tag{12.54}$$

falls below a previously indicated value, it is assumed that the object function has reached a minimum. The search can be terminated.

The second termination condition depends on the size of the simplex. One takes the difference between the point of smallest error and all other points of the simplex,

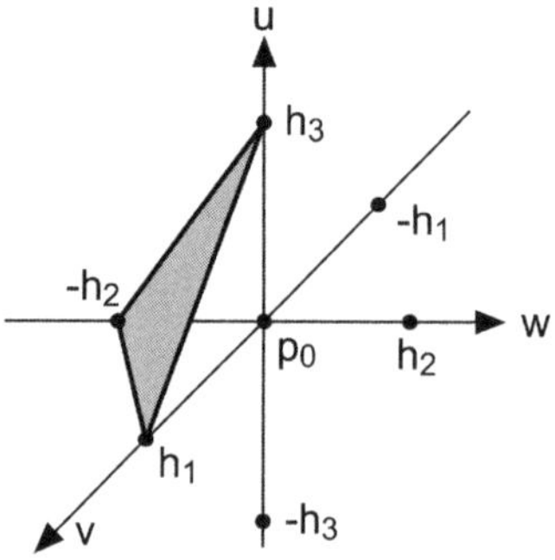

Figure 12.10 Determination of adapted simplex (p_0, $p_0 + h_1$, $p_0 - h_2$, $p_0 + h_3$) on the basis of actual minimum (p_0) and last measuring quantity ($h_i = h_i'/1.16$) [72].

$$E_{\min} \rightarrow P_k = (x_{1_{\min}}, ..., x_{n_{\min}})$$

$$Dif_i = P_i - P_k = (x_{1_i} - x_{1_{\min}}, ..., x_{n_i} - x_{n_{\min}}) \qquad (12.55)$$

If none of the parameter of the resulting vectors is larger than a previously defined value,

$$\max(Dif_1, ..., Dif_n) < W_{\min_2} \qquad (12.56)$$

then this is construed that all points are sufficiently near the point with the lowest error, so that the search can be terminated.

In order to actually stop the optimizer, both conditions must be fulfilled simultaneously. The first termination condition determines the region within the n-dimensional space where the object function has is a minimum (medium diameter of simplex). The second termination condition aims at finding the location of the minimum with utmost accuracy. The required minimal difference can be defined with the second termination condition.

In order to avoid trapping in local minima a new parameter was implemented that controls a reinitialization of the simplex algorithm. The actual simplex is then adjusted as follows. The actual minimum point x_0 represents the new starting point. Finding the new simplex, from this reference point further points are generated by addition or subtraction of a defined quantity. This quantity (h_i) results from the value of previous (suboptimal) iteration process (h_i') divided by a factor 1.16 (that is, $h_i = h_i'/1.16$ (see [71]). The error function values are calculated in the determined points. The presentation in Figure 12.10 reflects the results: $E(p_0 + h_1) < E(p_0 - h_1)$, $E(p_0 - h_2) < E(p_0 + h_2)$, and $E(p_0 + h_3) < E(p_0 - h_3)$. Thus, the gray shaded simplex is chosen for the next iteration cycle. It brings the vertices out of the local minimum and forces new optimization, which could find a better solution point.

We have investigated different breaking schemes for reinitialization. Three modifications of the classical simplex algorithm are exemplarily illustrated in Figure 12.11: (1) staticreinitialization is always performed after 10 iterations of the simplex algorithm, (2) dynamic reinitialization depends on the size dynamic (shrinkage) of the Simplex during the iteration process. When conditions (12.54) and (12.56) are fulfilled within a specified re-initialization limit, restart of the simplex is carried out. Figure 12.11 shows an example of the iteration sequence of a simplex optimization concerning the extraction of the gate resistance of a 15-element FET model from measured scattering coefficients, using different

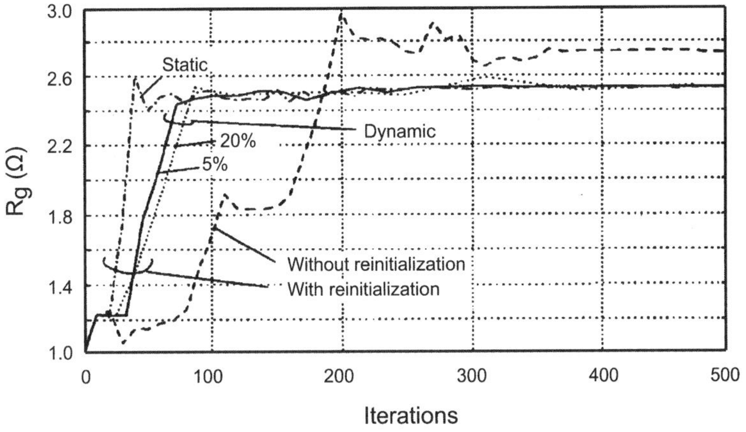

Figure 12.11 Simplex optimization of gate resistance of a 15-element FET model using different reinitializing schemes.

reinitializing scheme: static and dynamic reinitialization with 5% and 20% simplex shrinkage.

It is seen that, regarding the starting situation, all versions are entrapped in a local minimum after about eight iterations. The static reinitialization scheme with forced reinitialization after 10 iterations (leading to simplex expansion) is capable escaping from the local minimum and rapidly converging to the global minimum after about 40 iterations. In the dynamic case of only 5% simplex shrinkage, the simplex first remains in the local minimum, reducing its size. Then reinitialization is activated after the specified limit of the simplex size has been reached, successfully approaching the global minimum. For the 20% simplex shrinkage limit, the simplex is able to escape from the local minimum earlier than the aforementioned case. In contrast to the three discussed reinitialization schemes, which converge differently, but to the same resistance value of 2.53Ω, the classical simplex algorithm jumps from one local minimum to another if it is not finding the final global minimum. The solution is stabilized at 2.75Ω. This example demonstrates the superiority of a modified simplex algorithm based on the specified static reinitialization scheme. The extended simplex algorithm overcomes the local minimum problem. The new algorithm has

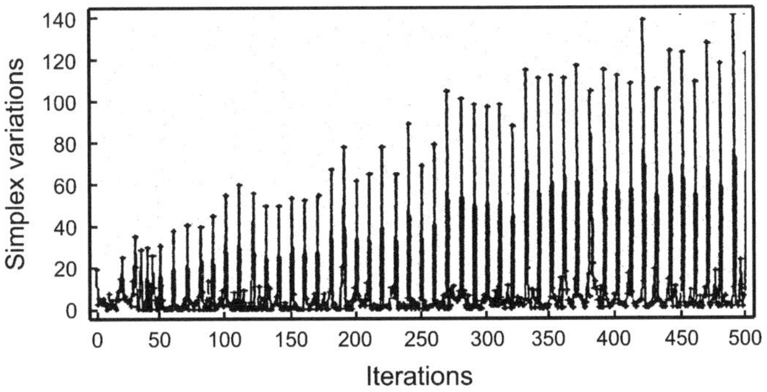

Figure 12.12 Simplex reinitialization after 10 successful iterations.

proven to be an extremely robust extraction method. The example has also demonstrated the great disadvantage of the classical simplex algorithm in a multidimensional search domain (i.e., the high risk of getting trapped in local minima).

The reinitialization of the simplex algorithm after having converged provides an reenlargement of the shrunken simplex, thus enabling an escape from a local minimum and to search in the neighborhood for a better solution. Figure 12.12 illustrates the periodical evolution of the simplex (simplex variations) after each reinitialization. It is seen that at the beginning of the optimization process, only a few simplex variations are needed to receive an improvement (see Figure 12.11). Up to about the 50th iteration, the evolution is irregular, indicating strong local minima issues. Up to about 260 iterations, the course becomes more regular but simplex variations are still increasing. Above that, the curve is relatively periodical, showing that the simplex algorithm cannot find a better solution. The observations agree well with the static curvature in Figure 12.11. Typically the number of iterations is limited to 500, which is generally sufficient to settle the optimization.

Discussion on Leaping Simplex

The leaping simplex algorithm denotes the classical Nelder-Mead simplex with an additionally implemented reinitialization function. Reinitialization is always performed after 10 iterations, leading to a larger search space radius. Reinitialization is based on a sequential parameter change of 10% (or 16%). Dependent on three parameter results with the lowest, medium, and second-highest error, the algorithm is starting again.

Reinitialization ensures that the risk of getting caught in a local minimum is nearly eliminated. In contrast to gradient-based methods, leaping simplex offers rather high rate of convergence. The optimization algorithm has proven to be a robust and reliable method for the optimization of equivalent circuits. The initial parameter values to start the optimizer is very important and guarantees reliable performance. These values may be derived from physics-based device analysis (Chapter 6) and/or low-frequency measurements (Chapter 8).

12.8 HYBRID OPTIMIZER

Local optimizers commonly suffer from getting trapped in local minima when the object function includes a rather large number of optimization variables. In contrast, global optimizers are testing the parameter space in a larger scale. Although they may be captured by a local minimum they have the ability to escape. This can be done by restarting the algorithm in a systematic (e.g., grid search method) or randomized way (e.g., random search). Other global optimizers such as genetic, and simulated and tree annealing possess the property that the probability getting entrapped is greatly reduced.

Thus, it is fairly obvious to embed an efficient local algorithm with good convergence into a robust global one forming a hybrid search method, which will speed up the global search and reduce computational cost.

The Nelder-Mead optimization method seems to be the most popular direct method for local optimization of unconstrained problems. It has been shown to be effective in practice by producing a rapid initial decrease in function values [73]. The popular algorithm has been used in various combinations with global search algorithms such as random search [74–78], genetic algorithm [79], Tabu search [80], and simulated annealing [81–85].

In [35, 69], tree annealing and gradient descent optimization were combined to perform parameter extraction of a GaAs MESFET.

Discussion on the Classical Simplex and Fletcher-Powell Combination

With both algorithms the risk is that a local minimum of the object function is reached. Is the starting vector far away from the minimum of the object function, the simplex algorithm converges faster. However, near the minimum, Fletcher-Powell converges faster. Thus, both algorithms are complementing each other [1]. Testing the combination for model parameter extraction of a 15-element equivalent network of MESFET and HEMT devices, the following search strategy was applied. Optimization was started with the simplex algorithm with rapid convergence. Then, after this preoptimization, it was continued with the Fletcher-Powell method. When observing only small changes of the object function value, the simplex was again initiated. The final optimization runs were performed with the Fletcher-Powell method [86]. The following problems occurred: (1) The object function converged to a functional value that is far away from the specified minimum target value, (2) resistances were calculated incorrectly with negative sign, and (3) capacitances converted to inductances and vice versa.

Discussion on the Tree Annealing and Gradient Descent Combination

Skaggs et al. [38] reported about model parameter extraction of a GaAs MESFET using a hybrid optimization procedure that combines tree annealing optimization with gradient descent (quasi-Newton). The modified Materka-Kacprzak MESFET model included a dispersive RC branch across the drain current source to account for frequency dispersion. The circuit encompasses a total number of about 30 variables in the nonlinear I–V and C–V expressions and for the linear elements . However, as pointed out by the authors, only 15 parameters were optimized. The parameters that were not optimized were established by previous parameter extraction. Measured DC data and scattering parameters were fitted simultaneously.

The procedure was as follows. First, an initial starting point is chosen (this can be chosen at random or by the user), and a gradient optimization is performed. Optimization using tree annealing then proceeds. Whenever the error decreases below the previous best error obtained by tree annealing, a gradient-descent search is initiated. At any time, if the residual error is below a user-specified tolerance, parameter extraction is halted. In this way several local minima can be detected before the error is suitably low. All minima encountered throughout the optimization process are recorded along with their errors.

Rather good agreement with respect to the measured and simulated S-parameter phases could be observed. Many local minima with comparable residual error of the objective function were recorded, possibly related to measurement errors. It is concluded that any further study of the extracted model element values, notably with respect to their physical meaning, is not possible.

As pointed out by [69], tree annealing gradually narrows the width of the subspace to close in around the global minimum. However, this method does not find the global minimum but rather an arbitrary small interval where the minimum lies. Therefore, local search algorithms, providing stronger local search power, may also be combined with the tree annealing approach.

12.9 SUMMARY

In this chapter various optimization algorithms such as gradient-based and direct search methods, and global and hybrid optimization algorithms were presented and evaluated particularly with regard to model parameter extraction application. Based on several extraction examples, advantages and disadvantages of the discussed optimization procedures become obvious. Gradient search works only reliably very near a local minimum, which may occasionally be the global one. Direct search algorithms such as the classical Nelder-Mead-simplex do not need first/second derivatives of the object function. In far distance they iterate fast. However, near the optimum they move slowly. The modified leaping simplex algorithm has turned out, over the years, to be a very robust and reliable optimization method for model parameter extraction purposes of microwave device equivalent networks. The genetic algorithm and simulated and tree annealing overcome the main disadvantage of the aforementioned algorithms merely moving downhill. Probabilistic rules of decision permit escape from local minima (hillclimbing). The convergence decreases with rapprochement to the optimum. Hybrid optimizers combine the advantage of global search of a large multimodal domain and fast local exploration in the neighborhood of a minimum.

REFERENCES

[1] Cooper, L., and D. Steinberg, *Introduction to Methods of Optimization*, Saunders Company, 1970.

[2] Himmelblau, D. M., *Applied Nonlinear Programming*, New York: McGraw-Hill, 1972.

[3] Gottfried, B. S., and J. Weisman, *Introduction to Optimization Theory*, Englewood Cliffs, N.J.: Prentice-Hall, 1973.

[4] Daniels, R. W., *An Introduction to Numerical Methods and Optimization Techniques*, New York: North-Holland, 1978.

[5] Fidler, J. K., and C. Nightingale, *Computer Aided Circuit Design*, Nelson, 1978.

[6] Walters, F. H., L. R. Parker, S. L. Morgan, and S. N. Deming, *Sequential Simplex Optimization*, CRC Press, Boca Raton, FL, 1991.

[7] Melanie, M., *An Introduction to Genetic Algorithms*, Cambridge, Mass.: MIT Press, 1996.

[8] Nocedal, J., and S. J. Wright, *Numerical Optimization*, Springer-Verlag New York, Inc., 1999.

[9] Arora, J. S., *Introduction to Optimum Design*, Second Edition, Elsevier Academic Press, 2004.

[10] Press, W. H., S. A. Teukolsky, W. T. Vettering, and B. P. Flannery, *Numerical Recipes*, 3rd Edition, Cambridge University Press, Cambridge, UK, 2007.

[11] Deb, K., *Optimization for Engineering Design—Algorithms and Examples*, 2nd Edition, PHI Learning Private Ltd., New Delhi, November 2012.

[12] Madsen, K, H. B. Nielsen, and O. Tingleff, *Methods for Non-Linear Least Squares Problems*, 2nd Edition, Technical University of Denmark, April 2004. Lecture notes, available at http://www.imm.dtu.dk/courses/02611/nllsq.pdf (56 pages).

[13] Bilbro, G. L., M. B. Steer, R. T. Trew, C.-R.Chang, and S. G. Skaggs, "Extraction of the Parameters of Equivalent Circuits of Microwave Transistors Using Tree Annealing," *IEEE Trans.*, Vol. MTT-38(11), 1990, pp. 1711–1718.

[14] Heitzinger, C., T. Binder, and S. Selberherr, "Parallel TCAD Optimization and Parameter Extraction for Comptationally Expensive Objective Functions," *Proc. 15th European Simulation Multiconference: Modelling and Simulation (ESM 2001)*, Prague, Czech Republic, June 2001, pp. 534–538.

[15] Binder, T., *Rigorous Integration of Semiconductor Process and Device Simulators*, Dissertation, TU Wien, 2002.

[16] Binder, T., C. Heitzinger, and S. Selberherr, "A Study on Global and Local Optimization Techniques for TCAD Analysis Tasks," *IEEE Trans. on Computer-Aided Design of Integrated Circuits and Systems*, Vol. 23(6), 2004, pp. 814–822.

[17] Davidon, W. C., *Variable Metric Method for Minimization*, AEC Research and Development Report, ANL-5990 (Rev.), Nov. 1959 (21 pages).

[18] Fletcher, R., and M. J. D. Powell, "A Rapidly Convergent Descent Method for Minimization," *Computer J.*, Vol. 6, pp. 163–168.

[19] Lourakis, M. I. A., "A Brief Description of the Levenberg-Marquardt Algorithm Implemented by 'levmar'", *Foundation of Research and Technology*, Vol. 4, 2005, pp. 1-6.

[20] Levenberg, K., "A Method for the Solution of Certain Non-Linear Problems in Least Squares," *Quaterly of Applied Mathematics*, Vol. II, No. 2, 1944, pp. 164-168.

[21] Marquardt, D. W., "An Algorithm for Least-Squares Estimation of Nonlinear Parameters," *J. Soc. Indust. Appl. Math.*, Vol. 11, No. 2, June, 1963, pp. 431-441.

[22] Zahradnický, T., *MOSFET Parameter Extracton Optimization*, Doctoral Thesis, Czech Technical University in Prague, 2010.

[23] Bandler, J. W., "Optimization Methods for Computer-Aided Design," *IEEE Trans.*, Vol. MTT-17, 1969, pp. 533–552.

[24] Torczon, V., "On the Convergence of the Multidirectional Search Algorithm," *Siam J. Optimization*, Vol. 1(1), 1991, pp. 123–145.

[25] Hooke, R., and T. A. Jeeves, "Direct Search' Solution of Numerical and Statistical Problems," *Journal of the ACM (JACM)*, Vol. 8(2), 1961, pp. 212–229.

[26] Nelder, J. A., and R. Mead, "A Simplex Method for Function Minimization," *Computer Journal*, Vol. 7, 1965, pp. 308–313.

[27] MATLAB, *The Math Works*, Natick, MA, 2016.

[28] Lewis, R. M., V. Torczon, and M. W. Trosset, "Direct Search Methods: Then and Now," *Journal of Computational and Applied Mathematics*, Vol. 124, 2000, pp. 191–207.

[29] MathWorks, MATLAB program *fminsearch,* https://de.mathworks.com/ help/matlab/ref/-fminsearch.html.

[30] Spendley, W., G. R. Hext, and F. R. Himsworth, "Sequential Application of Simplex Designs in Optimisation and Evolutionary Operation," *Technometrics*, Vol. 4(4), 1962, pp. 441–461.

[31] Wolfram Research Inc.: *Global Optimization*, http://www.wolfram.com/ products/ applications/globalopt/, 2016.

[32] Novotny, M., *Extraction of Equivalent Circuit Parameters of GaAs FETs und HEMTs* (in German), Master Thesis, Department of High Frequency Engineering, University of Kassel, 1988.

[33] Ghiasi, M. H., *Multi-Objective Optimization of Composite Structures Made by Resin Transfer Moulding*, Ph.D. thesis, McGill University, Montreal, 2009.

[34] Chibante, R., A. Araújo, and A. Carvalho, "Modeling Buffer Layer IGBTs with an Efficient Parameter Extraction Method," *Proc. 36th IEEE Power Electronics Specialists Conference (PESC '05)*, 2005, pp. 2194–2200.

[35] Hefner, A., and S. Bouché, "Automated Parameter Extraction Software for Advanced IGBT Modeling," *7th IEEE Workshop on Computers in Power Electronics*, 2000, pp. 10–18.

[36] Strasser, R., *Rigorous TCAD Investigations on Semiconductor Fabrication Technology*, Dissertation, Technical University Wien, 1999.

[37] Vai, M.-K., J.-S. Lin, and S. Prasad, "Acceleration of Simulated Annealing and its Application to Microwave Device and Circuit Optimization," *IEEE MTT-S Digest*, 1991, pp. 1213-1216.

[38] Skaggs, S. G., J. Gerber, G. Bilbro, and M. B. Steer, "Parameter Extraction of Microwave Transistors using a Hybrid Gradient Descent and Tree Annealing Approach," *IEEE Trans.*, Vol. MTT-41, 1993, pp. 726–729.

[39] Menozzi, R., A. Piazzi, and F. Contini, "Small-Signal Modeling for Microwave FET Linear Circuits Based on a Genetic Algorithm," *IEEE Trans. Circuits Syst. I: Fundamental Theory and Applications*, Vol. 43(10), 1996, pp. 839- 847.

[40] Ruffio, E., D. Saury, D. Petit, and M. Girault, "Tutorial 2: Zero-Order Optimization Algorithms," http://www.sft.asso.fr/Local/sft/dir/ user-3775/ documents/ actes/Metti5_ School/Lectures& Tutorials-Texts/ Text-T2-Ruffio.pdf (2011).

[41] Rechenberg, I., *Evolutionsstrategie: Optimierung technischer Systeme nach Prinzipien der biologischen Evolution*, Doctoral Thesis, Technical University of Berlin, 1971.

[42] Schwefel, H.-P., *Evolutionsstrategie und numerische Optimierung*, Technical University of Berlin, 1975.

[43] Chipperfield, A., P. Fleming, H. Pohlheim, and C. Fonseca, *Genetic Algorithm Toolbox for Use with Matlab®*, University of Sheffield, Dept. Automatic Control and Systems Engineering, 1994.

[44] Chipperfield, A. J., and P. J. Fleming, "The MATLAB Genetic Algorithm Toolbox," *The Institution of Electrical Engineers* (IEE), 1995, pp. 10/1–10/4.

[45] Metropolis, N., A. Rosenbluth, M. Rosenbluth, A. Teller, and E. Teller, "Equation of State Calculations by Fast Computing Machines," *J. Chem. Phys.*, Vol. 21, 1953, pp. 1087–1090.

[46] Kirkpatrick, S., C. D. Gelatt, M. P. Vecchi, "Optimization by Simulated Annealing," *Science*, Vol. 220, No. 4598, 1983, pp. 671–680.

[47] Corana, A., M. Marchesi, C. Martini, and S. Ridella, "Minimizing Multimodal Functions of Continuous Variables with the Simulated Annealing Algorithm," *ACM Trans. on Mathematical Software*, Vol. 13(3), 1987, pp. 262– 280.

[48] Goffe, W. L., G. D. Ferrier, and J. Rogers, "Global Optimization of Statistical Functions with Simulated Annealing," *Journal of Econometrics*, Vol. 60, 1994, pp. 65–99.

[49] Goffe, W. L., G. D. Ferrier, and J. Rogers, "Simulated Annealing: An Initial Application in Econometrics," *Computer Science in Economics and Management*, Vol. 5, 1992, pp. 133–146.

[50] Goffe, W. L., "SIMANN: A Global Optimization Algorithm Using Simulated Annealing," *Studies in Nonlinear Dynamics & Econometrics*, Vol. 1(3), 1996, pp. 169–176.

[51] Simulated Annealing FORTRAN Code Provided on [sepwww.stanford.edu/sep/claudio/ Thesis/ Prst_ExpRefl/Tese/src/ areal/ERM/optrnd/simann1.f90].

[52] Geman, S., and D. Geman, "Stochastic Relaxation, Gibbs Diostributions, and the Bayesian Restoration of Images," *IEEE Trans.*, Vol. PAMI-6(6), 1984, pp. 721–741.

[53] Ingber, L., "Very Fast Simulated Re-Annealing," *Math. Comput. Modelling*, Vol. 12(8), 1989, pp. 967–973 (http://www.ingber.com/asa89_vfsr.ps.gz).

[54] Rosen, B. E., "Function Optimization Based on Advanced Simulated Annealing," *Proc. IEEE Workshop on Physics and Computation (PhysComp '92)*, 1993, pp. 289–293.

[55] Szu, H., and R. Hartley, "Fast Simulated Annealing," *Phys. Lett. A*, Vol. 122 (3,4), 1987, pp. 157–162.

[56] Ingber, L., "Adaptive Simulated Annealing (ASA): Lessons Learned," *Control and Cybernetics*, Vol. 25(1), 1996, pp. 33–54.

[57] Matlab 2012a: *Optimization Toolbox – Product Documentation.*

[58] Ingber, L., *Software Package Documentation*, 1996 (http://www.ingber.com/ asa96_lessons.ps.gz).

[59] Shukla, R., and T. Marwala, "Modeling of Software Development Effort," *Int. J. Advances in Engineering Science*, Vol. 3(3), 2013, pp. 52–57.

[60] Han, Y.-S., W. E. Snyder, and G. L. Bilbro, "Pose Determination using Tree Annealing," *Proc. IEEE Intern. Conf. Robotics and Automat.*, Cincinnati, 1990.

[61] Snyder, W., G. Bilbro, A. Logenthiran, and S. Rajala, "Optimal Thresholding—A New Approach," *Pattern Recognition Lett.*, Vol. 11, 1990, pp. 803–810.

[62] Snyder, K. I., and W. E. Snyder, "Determination of Exponential Parameters,“ *Report TR-91/11*, August 1991, Center for Communications and Signal Processing, Dept. Electrical and Computer Engineering, North Carolina State University.

[63] Chibante, R., A. Araújo, and A. Carvalho, "Parameter Identification of Power Semiconductor Device Models Using Metaheuristics," in: *Simulated Annealing Theory with Applications* (Ed. R. Chibante), 2010 Sciyo, pp. 1–16 (www.intechopen.com).

[64] Shojaee, K., H. Shakouri, and M. Behnam Taghadosi, "Importance of the Initial Conditions and the Time Schedule in the Simulated Annealing," in: Simulated Annealing, Theory with Applications (Ed. Rui Chibante), 2010, pp. 217–233 (http://www.intechopen.com/books/simulated-annealing-theory-with-applications).

[65] Vai, M., S. Prasad, and N. C. Li, "Modeling of Microwave Semiconductor Devices Using Simulated Annealing Optimization," *IEEE Trans.*, Vol. ED-36, No. 4; April 1989, pp. 761-762.

[66] Bilbro, G. L., and W. E. Snyder, "Optimization of Functions with Many Minima," *IEEE Trans. on Systems, Man, and Cybernetics*, Vol. 21(4), 1991, 840–849.

[67] Omohundro, S. M., "Efficient Algorithms with Neural Network Behavior," *Complex Systems*, Vol. 1, 1987, pp. 273–347.

[68] Steer, M. B., G. L. Bilbro, R. J. Trew, and S. G. Skaggs, "Parameter Extraction of Microwave Transistors Using Tree Annealing," *Proc. IEEE Cornell Conf. On Advanced Concepts in High Speed Semiconductor Devices and Circuits*, 1989, pp. 195–200.

[69] Sun, M., "Tree Annealing for Constrained Optimization," *Proc. 34th Southeastern Symposium on System Theory*, 2002, pp. 412–416.

[70] Kompa, G, and M. Novotny, "Highly Consistent FET Model Parameter Extraction Based on Broadband S-Parameter Measurements," *IEEE MTT-S Int. Microwave Symp. Dig.*, 1992, pp. 293–296.

[71] Rhea, R. W., "The Effectiveness of Four Direct Search Optimization Algorithms," *IEEE MTT-S Digest*, 1987, pp. 697–702.

[72] Lin, F., *A Method for Reliable Experimental Modeling of Microwave FETs*, Doctoral Thesis (in German), University of Kassel, 1993; published by VDI Verlag, Düsseldorf, Fortschritt-Berichte VDI, Reihe 9 (Elektronik), Nr. 164, 1993.

[73] Lagarias, J., J. A. Reeds, M. H. Wright, and P. E. Wright, "Convergence Properties of the Nelder-Mead Simplex Method in Low Dimensions," *SIAM J. OPTIM.* (Society for Industrial and Applied Mathematics), Vol. 9, 1998, pp. 112–147.

[74] Kolondzija, B. M., and D. I. Olcan, "Antenna Optimization Using Combination of Random and Nelder-Mead Simplex Algorithms," *IEEE Antennas and Propagation Society, AP-S International Symposium (Digest)*, Vol. 1, 2003, pp. 185–188.

[75] Luersen, M. A., and R. L. Riche, "Globalized Nelder-Mead Method for Engineering Optimization," *Computers and Structures*, 2004, pp. 2251–2260.

[76] Luersen, M. A., R. L. Riche, and F. Guyon, "A Constrained, Globalized, and Bounded Nelder-Mead Method for Engineering Optimization," *Struct. Multidisc. Optim.*, Vol. 27(1), May 2004, pp. 43–53. Also published online as Research Paper, 2003, DOI 10.1007/s00158-003-0320-9.

[77] Ghiasi, M. H., D. Pasini, and L. Lessard, "Improved Globalized Nelder-Mead Method for Optimization of a Composite Bracket," *16th International Conference on Composite Materials (ICCM-16)*, July 3-8, Kyoto, Japan, 2007 (9 pages).

[78] Ghiasi, M. H., D. Pasini, and L. Lessard, "Constrained Globalized Nelder-Mead Method for Simultaneous Structural and Manufacturing Optimization of a Composite Bracket," *Journal of Composite Materials*, Vol. 42(7), 2008, pp. 717–736.

[79] Chelouah, R., and P. Siarry, "Genetic and Nelder-Mead Algorithms Hybridized for a More Accurate Global Optimization of Continuous Multiminima Functions," *European Journal of Operational Research*, Vol. 148(2), 2003, pp. 335–348.

[80] Chelouah, R., and P. Siarry, "A Hybrid Method Combining Continuous Tabu Search and Nelder-Mead Simplex Algorithms for the Global Optimization of Multiminima Functions," *European Journal of Operational Research*, Vol. 161(3), 2005, pp. 636–654.

[81] Kvasnička, V., and J. Pospíchal, "A Hybrid of Simplex Method and Simulated Annealing," *Chemometrics and Intelligent Laboratory Systems*, Vol. 39, 1997, pp. 161–173.

[82] Efstratiadis, A., and D. Koutsoyiannis, "An Evolutionary Annealing-Simplex Algorithm for Global Optimisation of Water Resource Systems," *Hydroinformatics*, 2002, pp. 1423–1428.

[83] Tsoukalas, I., P. Kossieris, A. Efstratiadis, and C. Makropoulos, "Surrogate-Enhanced Evolutionary Annealing Simplex Algorithm for Effective and Efficient Optimization of Water Resources Problems on a Budget," *Environmental Modelling & Software*, Vol. 77, 2016, pp. 122–142.

[84] Behzadi, B., C. Ghotbi, and A. Galindo, "Application of the Simplex Simulated Annealing Technique to Nonlinear Parameter Optimization for the SAFT-VR Equation of State," *Chemical Engineering Science*, Vol. 60, 2005, pp. 6607–6621.

[85] Salazar, R., and R. Toral, "Simulated Annealing Using Hybrid Monte Carlo," Journal of Statistical Physics, Vol. 89, Issue 5-6, 1997, pp. 1047–1060.

[86] Kersten, B., *Investigation of Appropriate Equivalent Network Topologies for GaAs MESFETs on the Basis of Measured Frequency-Dependent Scattering Parameters*. Master Thesis (in German), University of Kassel, 1986.

Chapter 13

Extraction Methods: An Overview

This chapter gives an overview about model parameter extraction methods discussed so far, including both classical approaches and the recently developed methods particularly in relation to large gate-periphery GaN power HEMTs as well to FET devices with one or two gate fingers operated at very high frequencies such as 80 GHz and above. As Figure 13.1 indicates, we have a first group detailing all-at-once model parameter extraction. In this case all model elements of an equivalent network, mostly a 15-element FET model, are simultaneously optimized. Because of the large number of optimization variables, the search space domain contains numerous local minima. Applying local optimizers, as was frequently done in the eighties and nineties, a single optimized result had only limited significance. Therefore, various starting parameter value vectors were tested until some consistency in the extraction results could be observed. In order to reduce the risk of getting trapped in a local minimum, increased efforts were made to mitigate this challenge by various modifications to minimize the number of variables.

Optimization with reduced variables tries to introduce additional conditions for the model elements (e.g., from low-frequency analysis). Similar strategy is given with the *decomposition* extraction method, which is based on a sensitivity analysis of the individual model elements with respect to the scattering parameters. Optimization is performed iteratively with differently grouped model elements.

Commonly the object function is referred to the extrinsic terminals of the device comprising all model parameter variables. The *bidirectional search* performs the optimization with respect to an inner reference plane, dividing the whole search space into an extrinsic and intrinsic part, thus reducing the radius of the search space dramatically.

Extraction methods were developed that tried to solve the extraction problem purely analytically to circumvent the optimizer-inherent local minimum problem. It was found that *pure analytical extraction* approaches react very sensitively to measurement errors. Optimization methods based on curve fitting of rational functions have been proposed including both extrinsic parasitic and intrinsic bias-dependent model parameters.

The model parameter optimization can also be performed sequentially. That is, first the extrinsic linear part of the equivalent network is regarded under passive operation such as cold pinch-off or gate-forward biasing. If the extrinsic equivalent circuit is known, the residual inner model elements of the *intrinsic equivalent circuit* can be determined in different ways (i.e., by *overall optimization*, *branch-related optimization,* or *analytical extraction*).

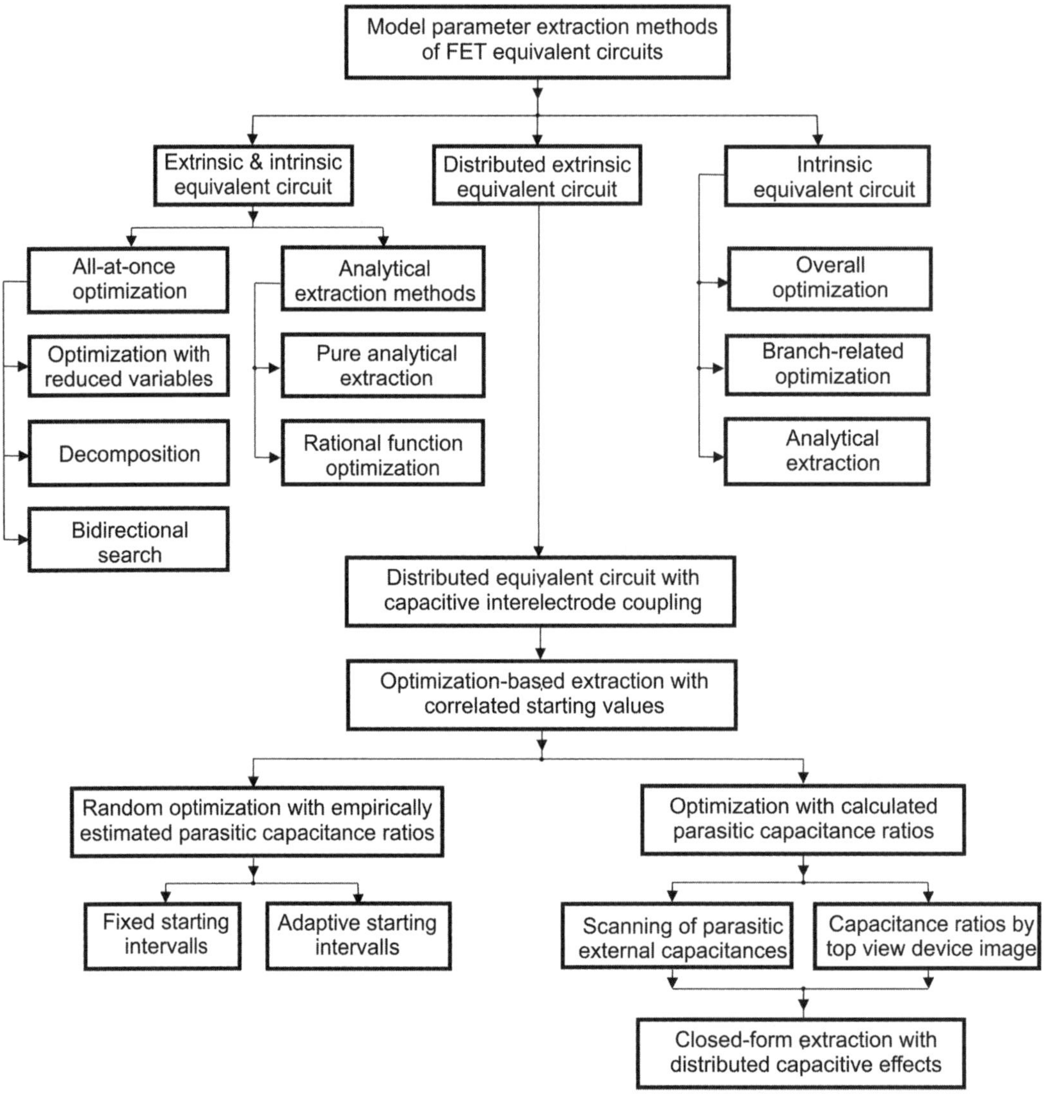

Figure 13.1 Overview of various model parameter extraction methods.

It is well known that the intrinsic model elements depend on the extracted results of the extrinsic equivalent circuit. Therefore, the most important part of the discussed optimization procedures is related to the model parameter extraction of these extrinsic model elements. It thoroughly treats recently developed advanced extraction procedures, which basically taking into account the *correlation of model elements* in the extraction process.

Classical equivalent circuits for microwave transistors comprising about 15 model elements (see e.g., Figure 8.1) are of increasingly limited validity with increasing operation frequency because of growing distributed capacitive effects. This particularly applies to multifinger power transistors already at low and moderate frequencies. Large-size gate peripheries suffer from parasitic capacitive interelectrode coupling.

Thus, for the general case, it is necessary to extend the classical equivalent circuit analysis. Figure 8.2 shows the proposed distributed equivalent circuit that considers the intradevice capacitive interactions. The model parameter extraction starts with the determination of effective branch capacitances (gate-source, gate-drain, and drain-source) at low frequencies. This recalls the classical extraction approaches. Frequently the extrinsic model elements are determined from a set of scattering parameters at different bias points. However, as the model parameters of the equivalent circuit are not independent, they should be extracted from S-parameters in a single bias point, commonly under pinch-off operation. In this way mutual correlation of the model parameters is assured. The term *measurement-correlated* parameter extraction should express that the S-parameter measurement in a single-bias point meets a priori this correlation requirement.

The partitioning of the effective capacitances into distributed branch capacitive networks is a really sophisticated undertaking. Several approaches will be discussed, such as random optimization with estimated parasitic capacitances and extraction with parasitic capacitances determined by a scanning method or from top view inspection of fabricated device. Knowing the distributed capacitive device network, the series inductances are found from S-parameters at moderate frequencies after deembedding the respective capacitances by the curve-fitting procedure.

It should be emphasized that the determined model parameter values represent only estimated values, which, however, already indicate a good extraction solution as is apparent from the comparison of measured and calculated S-parameters. Final optimization (refinement) on the basis of estimated parameters only cause minor enhancement at higher frequencies.

Finally, with the distributed capacitive network known, a novel closed-form inductance extraction is described, which leads to a closed-form extraction strategy of physically meaningful parameters of small-signal HEMT models with distributed parasitic capacitive effects.

Chapter 14

All-at-Once Model Parameter Extraction

14.1 INTRODUCTION

This chapter deals with a very general approach to extract the model parameter values from measured scattering coefficients of a microwave transistor. Initially, no distinction is made with respect to extrinsic and intrinsic model elements of a classical 15-element equivalent circuit of a field-effect transistor. The optimization plane is the outer reference plane of the device. Optimization is aimed to minimize the error between the measured and simulated device data.

We will discuss the starting value vector finding, optimization algorithm, and the distribution of extracted model element values. Because of many optimization variables, the extraction process is strongly disturbed by numerous local minima of the highly nonlinear object function. This means that the extracted model parameter values determined from a single optimization cycle may not be meaningful. Therefore, it is necessary to probe the parameter space in a region where the global minimum is expected. Estimation of the parameter values have been discussed in Chapter 6 and Chapter 8. As described in Section 12.7.1, initial starting parameter vectors can be generated in a systematic way (grid search method) or can be randomly generated (random search method).

In a second step we want to address the question of whether the interrelations between some model parameters are useful to be taken into account to simplify the optimization procedure. Since this approach reduces the number of variables, the radius of search space is reduced, which should minimize the uncertainty of extracted results.

14.2 RANDOM SEARCH COMBINED WITH LOCAL OPTIMIZER

In this section we discuss simultaneous extraction of model parameters using a hybrid optimizer, which combines a global random search with a local optimizer such as gradient-based (12.4) and, alternatively, the leaping simplex (Section 12.7.4) method. As described in Section 12.7.1.2, random search is considered as one of the simplest model parameter extraction algorithms. It is a global, stochastic, and direct optimization algorithm. This method is commonly available in commercial optimization software packages today, and is also still used frequently. References [1, 2] provide some insight into the beginning and evolution of microwave circuit optimization in the eighties and nineties. At that time researchers were frequently compelled to develop their own optimization software for a given

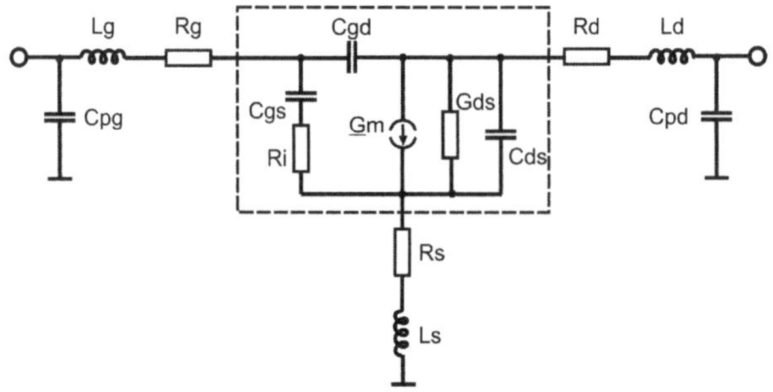

Figure 14.1 Widely used 15-element microwave transistor model.

application. In our case, we tested the combination of random search with two different local optimizers, changing optionally between Davidon-Fletcher-Powell (section 12.4.3) and modified simplex (section 12.7.5) algorithm. It has been found by many practical model parameter extraction trials that the modified simplex optimization method is superior regarding fast and reliable convergence. The developed statistically working global optimizer combined with the modified Simplex algorithm has proved to be a robust and efficient model parameter extraction method concerning low-noise and medium and high power GaAs- and GaN-based microwave devices. "Robust" indicates fast iteration and convergence for any chosen initial value of optimization variable within reasonably defined bounds, and the method is "efficient" due to moderate computing time and capacity.

The following results were derived from in-fixture measurements as described in [3]. Wire bonding chip connection technology entails increased parasitic inductance effects in contrast to the on-wafer measurement technique. The device under investigation was a Ku-band GaAs MESFET NE70000 chip with gate geometry of 0.5 μm × 300 μm, and gate and channel passivated with a thin layer of SiO_2 for mechanical protection (see Section 6.2.3, Figure 6.9). All bonding pads use a Ti-Pt-Au metallization system. Typical output power at 1-dB compression point (V_{DS} = 3V, I_{DS} = 20 mA, f = 12 GHz) was specified as 14.5 dB (28 mW). Note that the presented extraction results may be considered also representative for more advanced field-effect transistors such as GaAs and GaN HEMTs.

Model parameter extraction using the 15-element equivalent network (Figure 14.1) was based on measured S-parameter fitting in the frequency range from 2 to 26.5 GHz (saturation region). As is known, results of model parameters extracted by an optimization procedure for the best fit between measured and modeled S-parameters are affected considerably by the chosen starting values [4, 5]. To overcome this drawback, a statistical optimization procedure has been developed [6].

As illustrated in Figure 14.2 the model parameter values of the starting vector $\boldsymbol{p} = (p_1, p_{2, \ldots,} p_K)^T$ are chosen at random uniformly within the interval $[p_{il}, p_{iu}]$, which is predefined for each model element. The interval limits in Table 14.1 for the elements of the starting vectors have been chosen much larger than would be expected from analytical estimations or previously extracted results (see Chapters 6 and 8). Then the optimizer was started from each initially defined vector $\boldsymbol{p}_i$ for same scattering parameter data. Finally, the set of extracted model parameter value vectors were statistically evaluated, with regard to mean value $\bar{p}_i$ and

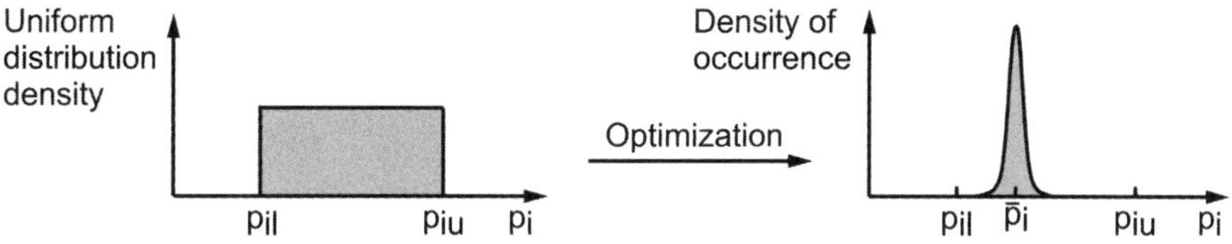

Figure 14.2 Uniformly distributed starting values of model parameter p_i within the interval $[p_{il}, p_{iu}]$ and distribution of extracted results with mean value $\bar{p}_l$ after optimization [7].

relative standard deviation σ_n on the basis of 100 single optimizations. The object function $\varepsilon(\boldsymbol{p})$ has been defined as

$$\varepsilon(\boldsymbol{p}) = \sum_{f=1}^{N} \sum_{i,j=1}^{2} \frac{\left| S_{c,ij}^{f}(\boldsymbol{p}) - S_{m,ij}^{f} \right|^2}{\left| S_{m,ij}^{f} \right|^2} \tag{14.1}$$

where indices c and m refer to calculated and measured scattering coefficient (indexed ij), respectively, and index f denotes the selected frequencies. In (14.1) relative squared errors between measured and simulated scattering parameters at considered discrete frequency points are summed up.

In a first experiment, the robustness of the optimizer was tested. In this case a preliminary extraction was performed based on measured S-parameters. Then from the extracted results a new (fictitious) S-parameter data set was calculated, which was indeed totally matched to the 15-element device model under discussion [8]. Starting the optimizer with 100 random parameter vectors, the following observations could be made: (1) the statistical distribution of the extracted results approximates a Gaussian distribution, (2) the mean value of the distributions agree very well with the fictitiously assumed model elements, and (3) the extracted model parameter values approximate the fictitious scattering parameters with high accuracy. It should be recalled that the distribution of the extracted individual parameter vectors $\boldsymbol{p}_i$ ($i = 1,2,\ldots K$) can also be interpreted as the distribution of local minima where the optimizer has stuck in. With an increasing number of starting vectors the distribution approaches the perfect Gaussian distribution function. But the experience also shows that in a case of limited density of parameter space sampling, standard distribution is approximately obtained.

Table 14.1

Typical Interval Boundaries of Element Values for Statistical Consistency Analysis

C_{gs} / pF	0.25 - 0.60	R_i / Ω	1 - 10
G_m / mS	40 - 100	G_{ds} / mS	3 - 10
C_{ds} / pF	0.05 - 0.15	C_{gd} / pF	0.01 - 0.06
R_g / Ω	0.5 - 5.0	L_g / nH	0.1 - 0.2
R_d / Ω	0.5 - 5.0	L_d / nH	0.1 - 0.2
R_s / Ω	0.5 - 5.0	L_s / nH	0.02 - 0.06
C_{pg} / pF	0.01 - 0.04	C_{pd} / pF	0.01 - 0.04
τ / ps	1 - 3		

Table 14.2
Classification of Extraction Uncertainty of Model Elements Based on Randomly Chosen Starting Parameter Values

Standard Deviation σ_n	Model Elements
$\leq 3\ \%$	C_{gs}, C_{gd}, G_m, τ, C_{ds}, L_g, L_d, L_s, C_{pg}, C_{pd}, R_Σ
$\leq 10\ \%$	R_s
$> 10\ \%$	R_g, R_i, R_d

From: [8].

In view of the statistical distributions based on fictitious (i.e., model-matched and accurate device data), the model elements can be grouped in three classes with respect to their extraction uncertainty (Table 14.2). Classification uses the relative standard deviation σ_n, which is the standard deviation divided by the mean parameter value $\bar{p}_i$ ($\sigma/\bar{p}_i$).

It is seen that the extraction uncertainty of the resistances R_g, R_i, and R_d is very large. In contrast, the sum of the three model elements in the gate-source branch, $R_\Sigma = R_g + R_i + R_s$, can be extracted with high consistency of $\sigma_n = 2\%$ (see Section 6.2.4). Note that the resistances of R_g and R_i are only decoupled through the low-value gate-drain capacitance C_{gd}.

The model parameters are not extracted with equal sensitivity. This has been studied as follows. The model parameter values of a MESFET of type NE70000 at a bias point (V_{GS} = 0.05V, V_{DS} = 3.0V) were extracted from S-parameters measured within a frequency range from 2 to 26.5 GHz and fixed. Then, the model element value of a single model element was successively changed over a deviation range of ±50%. Figure 14.3 shows the characteristic curve of the error function with increasing deviation from the optimum model element value for the inductance L_g.

Figure 14.4 shows a bar diagram indicating the error function value for all model elements at ±50% deviation from the optimum value. The specified width of the sensitivity function indicates the parameter influence on the optimized object function. The larger the error function limit value, the higher the extraction sensitivity of the considered element. It can be clearly seen that the highest sensitivities are provided by the model elements of the inner field-effect transistor (exceptionally the resistance R_i). With lower sensitivity the parasitic inductances L_g, L_d, and L_s follow. Sensitivity decreases further for τ, R_s, C_{pd}, and C_{pg}, and is

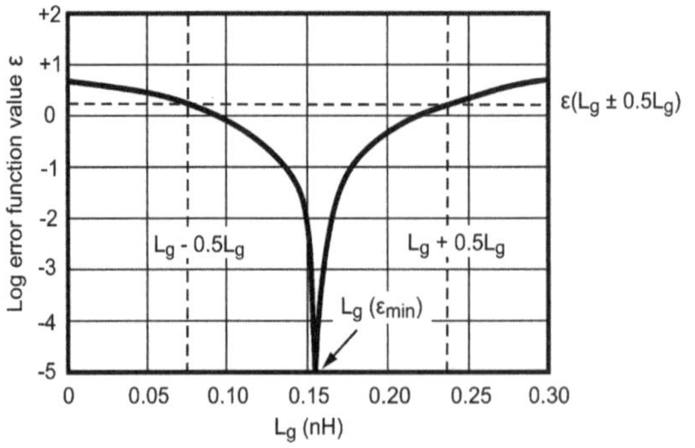

Figure 14.3 Sensitivity function of the extracted gate inductance L_g [9].

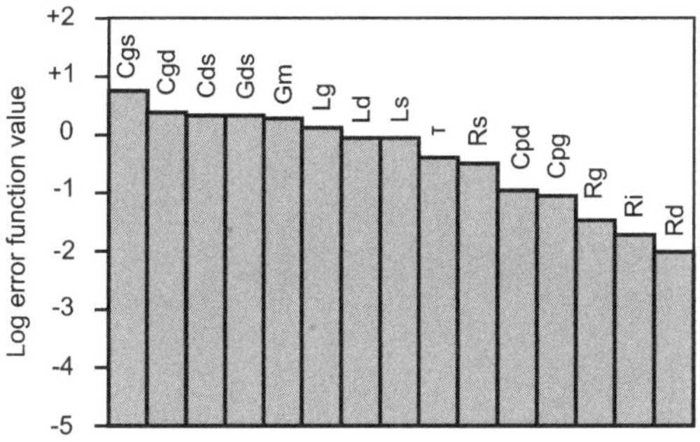

Figure 14.4 Sensitivity ranking of the extracted model parameters of a 15-element equivalent network.

lowest for the parasitic resistances R_g, R_i, and R_d. The sensitivity ranking of the model elements also illustrates how robust the object function is to small changes in the values of optimized model parameters.

Remember that the parasitics are really not desired because they are worsening the performance of the device. For model parameter extraction, however, these are, at first, only formally regarded as further variables in the optimization process. Nevertheless, they increase the number of optimization variables, and thus expand the search space with higher risk of local minimum trapping.

Similar results could be obtained on the basis of measured scattering parameters as illustrated in Figure 14.5. Inductance L_g is extracted with the lowest uncertainty, followed by extrinsic inductances L_s and L_d and intrinsic model elements. The determination of pad capacitances and series resistances is rather uncertain.

14.2.1 Measurement Bandwidth Variation

Using extracted model elements from the fictitious S-parameters in Section 14.2, model parameters were optimized on the basis of varying limited upper frequency (i.e., 10 GHz, 20 GHz, 30 GHz, and 40 GHz [7]).

It turns out that most determined model elements exhibit higher extraction certainty with

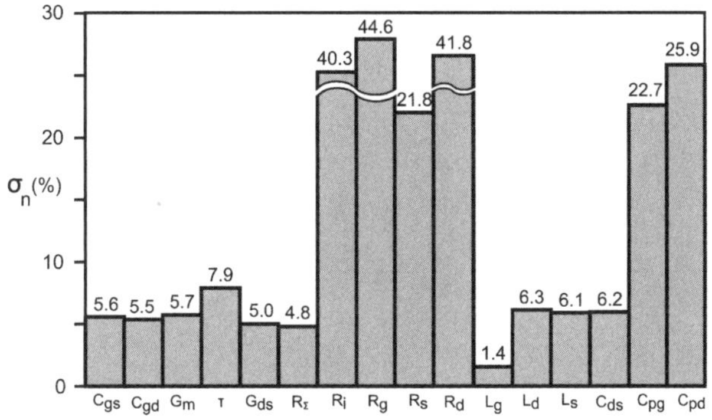

Figure 14.5 Relative standard deviation σ_n of extracted model element values based on 100 randomly distributed starting vectors (45 MHz $\leq$ f $\leq$ 26.5 GHz).

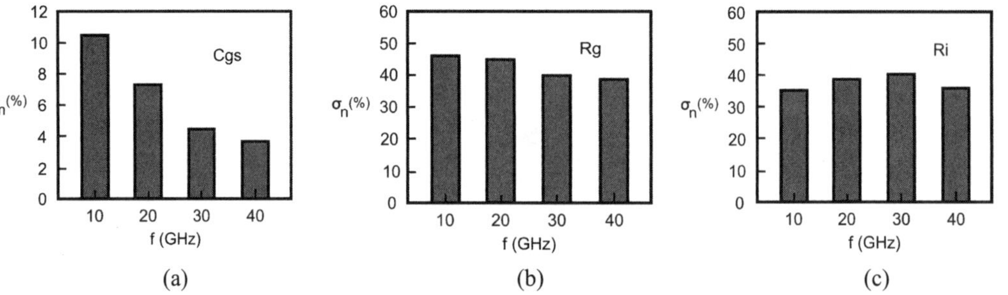

Figure 14.6 Relative standard deviation σ_n of some extracted model element values from fictitious S-parameter data as a function of the frequency bandwidth (saturation bias point).

increasing bandwidth [5], as is illustrated for C_{gs} in Figure 14.6(a). Nevertheless, C_{pg} needs the full bandwidth of 40 GHz to reach a moderate σ_n value of 5.6, whereas for C_{pd} a worse value of 12.5 is obtained (not illustrated). Again, L_g is optimally extracted for a limited frequency bandwidth of 30 GHz ($\sigma_n = 0.88$). The uncertainty of the resistance R_g decreases only marginally as shown in Figure 14.6(b). Slightly better uncertainty evolution is found for R_d, and in particular, for R_s, although the results for all series resistances are not really acceptable. With R_i no discernible trend could be observed (Figure 14.6(c)).

14.2.2 Measurement Error Simulation

The influence of measurement errors on extraction accuracy was investigated by modifying the measured scattering coefficients $S_{ij}(k)' = S_{ij}(k)\,(1 + \varepsilon_k)$ with $\varepsilon_k = |\varepsilon| \exp(j\varphi_k)$, $\varphi_k \in [0, 2\pi]$, $|\varepsilon| = 0.01, 0.02, ..., 0.05$. The phase of the error term φ_k has been generated by a randomizer over the 2π range. The element group G_m, C_{gd}, C_{gs}, L_g, L_d, L_s, C_{ds}, G_{ds} is relatively constant for small simulated errors, whereas all parasitic series resistances R_g, R_i, R_s, R_d, parasitic capacitances C_{pg}, C_{pd} and τ, also behave more inconsistently for relatively accurate data with errors less than 3%. The results emphasize the requirement for high accuracy of measured data in particular for the determination of the parasitic resistances.

The results show that with increasing error term $|\varepsilon|$ the uncertainty of extracted model elements grows significantly. This is illustrated by selected model elements in Figure 14.7.

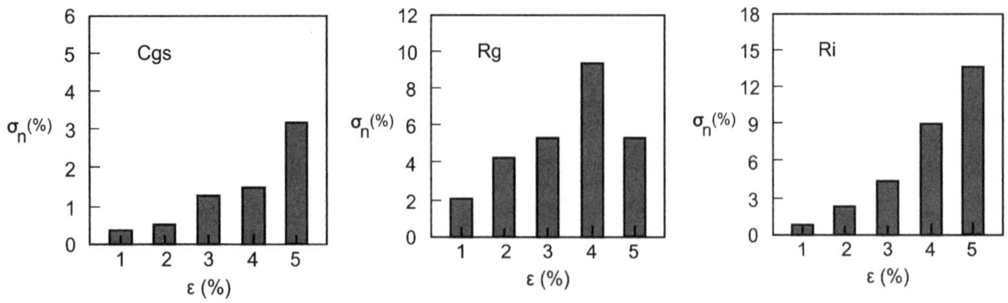

Figure 14.7 Relative standard deviation σ_n of some extracted model element values as a function of additional simulated measurement error ε.

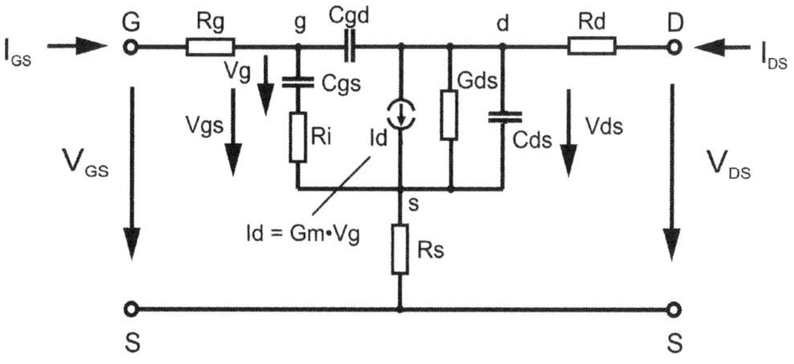

Figure 14.8 Simplified low-frequency equivalent circuit.

14.3 SEARCH SPACE MINIMIZATION

As discussed in Section 8.5, estimated model parameter values can be obtained from low-frequency device characteristics analysis. This opens up the possibility to decrease the bounded value range of optimization variables, which will deliver (hopefully) more consistent results. Figure 14.8 shows the low-frequency equivalent network under assumption $\omega L_k \ll R_k$ $(k = s, g, d)$ and $\omega\tau \ll 1$.

Recalling the approximation of model elements from low-frequency measurements in Chapter 8, the initially defined model parameter intervals in Table 14.1 for starting the optimization process could be considerably restricted as is evident from Table 14.3. As the extraction of R_Σ has proved rather consistent, the assumption $R_g \approx R_i \approx R_s \approx R_\Sigma/3$ was made. Then with the extracted R_Σ distribution, a minimum and maximum starting value for the elements G_{ds}, G_m, C_{gs}, C_{gd}, and C_{ds} can be estimated from (8.37), (8.38), (8.40), and (8.42), and (8.44), respectively.

Figure 14.9 illustrates the extraction results for the limited bounds of search intervals for the model elements. A significant improvement in the consistency of extracted model parameter values can be observed. However, the extraction of pad capacitances C_{pg} and C_{pd} and drain resistance R_d remains fairly uncertain.

Table 14.3
Restricted Interval Boundaries of Element Values Using Low-Frequency Model Parameter Estimation

C_{gs} / pF	0.42 - 0.43	R_i / Ω	2.76 - 3.0
G_m / mS	77.46 - 79.00	G_{ds} / mS	6.22 - 6.34
C_{ds} / pF	0.13 - 0.14	C_{gd} / pF	0.028 - 0.029
R_g / Ω	2.76 - 3.0	L_g / nH	0.1 - 0.2
R_d / Ω	0.5 - 5.0	L_d / nH	0.1 - 0.2
R_s / Ω	2.76 - 3.0	L_s / nH	0.02 - 0.06
C_{pg} / pF	0.01 - 0.04	C_{pd} / pF	0.01 - 0.04
τ / ps	1 - 3		

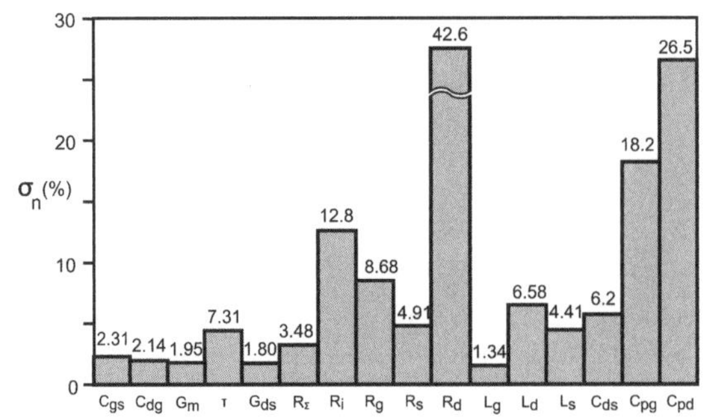

Figure 14.9 Relative standard deviation σ_n of the extracted model element values based on 100 randomly distributed starting vectors (45 MHz $\leq f \leq$ 26.5 GHz) with limited starting value ranges.

14.4 REDUCTION OF OPTIMIZATION VARIABLES

The 15-element model in Figure 14.1 comprises eight extrinsic and seven intrinsic model elements. Assuming that the extrinsic parameters are known, these can be easily deembedded by Y- and Z-matrix transformation, respectively. Then the intrinsic elements can be analytically determined from the deembedded equivalent circuit topology.

14.4.1 Interrelation of Intrinsic and Extrinsic Model Elements

Using the analytical relations between the Y-parameters and the model elements of the intrinsic device [7],

$$G_{ds} = \mathrm{Re}(Y_{22}) \tag{14.2}$$

$$C_{gd} = \frac{-\mathrm{Im}(Y_{12})}{\omega} \tag{14.3}$$

$$C_{ds} = \frac{\mathrm{Im}(Y_{22}) - \omega C_{gd}}{\omega} \tag{14.4}$$

$$C_{gs} = \frac{\mathrm{Im}(Y_{11}) - \omega C_{gd}}{\omega}\left\{1 + \frac{[\mathrm{Re}(Y_{11})]^2}{[\mathrm{Im}(Y_{11}) - \omega C_{gd}]^2}\right\} \tag{14.5}$$

$$R_i = \left\{\frac{\omega C_{gs} - [\mathrm{Im}(Y_{11}) - \omega C_{gd}]}{[\mathrm{Im}(Y_{11}) - \omega C_{gd}]\,[\omega C_{gs}]^2}\right\}^{1/2} \tag{14.6}$$

$$\tau = \frac{1}{\omega}\arctan\left\{\frac{\omega C_{gs}R_i + [\mathrm{Im}(Y_{21}) + \omega C_{gd}]/\mathrm{Re}(Y_{21})}{\omega C_{gs}R_i\,[\mathrm{Im}(Y_{21}) + \omega C_{gd}]/\mathrm{Re}(Y_{21}) - 1}\right\} \tag{14.7}$$

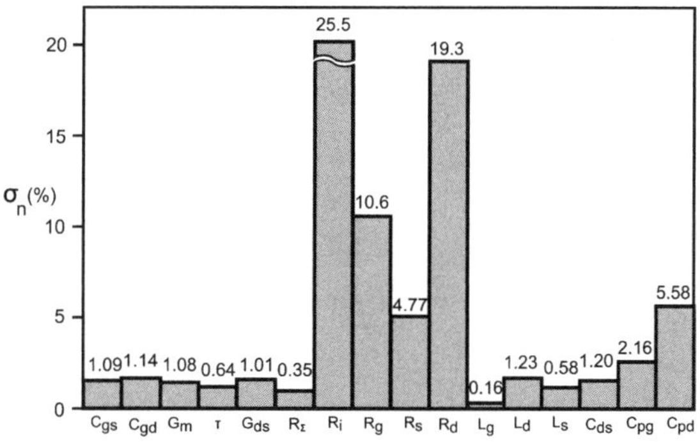

Figure 14.10 Relative standard deviation σ_n of the extracted model element values based on 100 randomly distributed starting vectors (45 MHz ≤ f ≤ 26.5 GHz) with limited starting value ranges and reduced number of optimization variables [7].

$$G_m = \frac{\mathrm{Re}(Y_{21})\,[1 + (\omega C_{gs} R_i)^2]}{\cos(\omega\tau) - \omega C_{gs} R_i \sin(\omega\tau)} \tag{14.8}$$

only R_s, R_d, R_g, L_s, L_d, L_g, C_{pg}, and C_{pd} remain as residual variables for the optimization process. As is shown in Figure 14.10, the inclusion of restricted starting conditions and reduction of variables highly improve the calculated standard deviation of the extracted element values in comparison to the statistical approach with low-frequency related limited starting value intervals (see Figure 14.9).

Figure 14.11 depicts the extracted results for the commonly less accessible model element R_s. Further results can be found in [7]. Without any additional constraint for the optimizer, a widespread distribution of the extracted element values between 0.82 Ω and 3.3 Ω can be noticed (Figure 14.11(a)). Closer inspection of the frequency of occurrence of extracted resistance values of R_s reveals that the optimization process suffers from local minimum

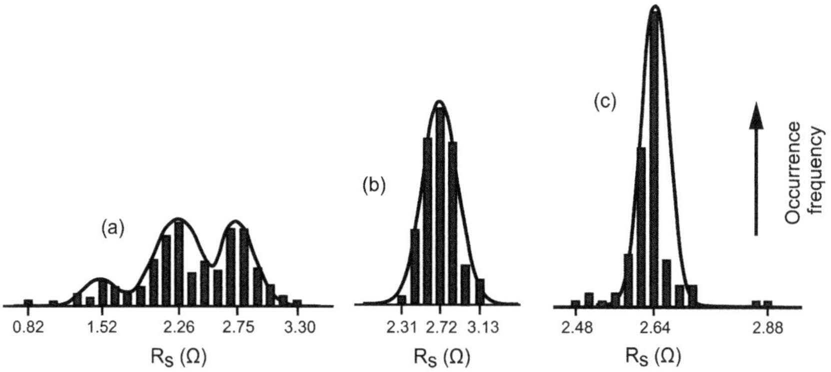

Figure 14.11 Optimizer-based R_s extraction from in-fixture measured S-parameter data: (a) without any additional constraints for the optimizer, (b) with confined starting values and implementation of analytical expressions for the intrinsic elements, and (c) the same as (b) with an additional controlled mean value of R_g.

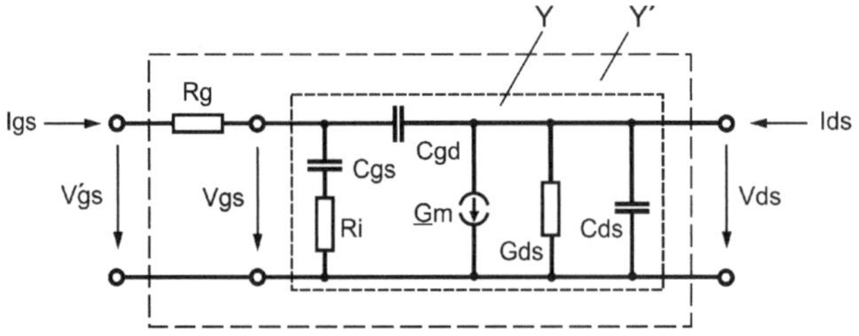

Figure 14.12 R_g is incorporated into the intrinsic model network denoted by the extended Y'- matrix. *Y* describes the intrinsic admittance matrix.

problem. It is clearly seen that the optimizer has been entrapped in three distinct local minima placed about 0.5 Ω and 0.75 Ω apart from each other. The results indicate the high density of local minima of a multivariable error function as given with the discussed equivalent circuits of microwave transistors. It is also confirmed that by fixing one series resistance, in this case R_g, increased extraction consistency is achieved.

14.4.2 Gate Resistance Allocation to the Intrinsic Elements

Under the assumption that that the real and imaginary parts of the Y-parameters of a two-port network are nonzero, then model networks up to eight parameters can be identified. This suggests to assign the gate resistance R_g to the intrinsic set of model parameters. When the common intrinsic model is characterized by the admittance matrix *Y*, and the extended network (including R_g) by Y', the following relations can be found between the intrinsic and extended parts [9]:

$$\mathrm{Im}(Y_{12}) = \frac{\mathrm{Re}(Y'_{12})}{\mathrm{Im}(Y'_{11})} \cdot \mathrm{Re}(Y'_{11}) + \mathrm{Im}(Y'_{12}) \tag{14.9}$$

$$R_g = \frac{\mathrm{Re}(Y'_{12})}{\mathrm{Im}(Y'_{11}) \cdot \mathrm{Im}(Y'_{12})} \tag{14.10}$$

$$\underline{Y}_{11} = \frac{\underline{Y}'_{11}}{\underline{Y}'_{12}} \cdot \mathrm{Im}(Y_{12}) \tag{14.11}$$

$$\underline{Y}_{21} = \frac{\underline{Y}'_{21}}{1 - \dfrac{\underline{Y}_{11} \cdot R_g}{1 + \underline{Y}_{11} \cdot R_g}} \tag{14.12}$$

$$\underline{Y}_{22} = \underline{Y}'_{22} + \frac{\underline{Y}_{21} \cdot \mathrm{Im}(Y_{12}) \cdot R_g}{1 + \underline{Y}_{11} \cdot R_g} \tag{14.13}$$

The Y'-parameter are obtained by conventional matrix transformation from the measured S-parameters. Then the unprimed Y-parameters can be deduced from the known primed admittance matrix (Y').

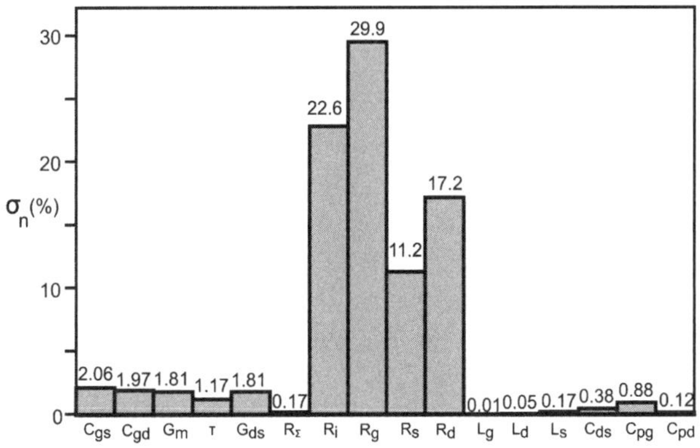

Figure 14.13 Relative standard deviation σ_n of the extracted model element values based on 100 randomly distributed starting vectors (45 MHz $\leq f \leq$ 26.5 GHz) with R_g incorporated into the intrinsic model network [9].

The further extraction procedure is provided as before with the only exception that the gate resistance R_g is now determined analytically, which means that the degree of freedom has further been reduced (only seven optimization variables remained). The extraction results are presented in Figure 14.13, which are to a certain extent unexpected. The relative standard deviation σ_n of the bondwire inductances L_g, L_d, and L_s and capacitances C_{pg}, C_{pd}, and C_{ds} have become very small and are much lower as in the previous case (Figure 14.10). However, it is not entirely clear why the standard deviation of R_g, which has been analytically determined after deembedding, has become considerably worse. Consequently, with uncertain R_g extraction, also all other intrinsic model parameters C_{gs}, C_{gd}, G_m, and G_{ds} show poorer extraction certainty.

14.5 LINEAR RELATIONSHIP AMONG SERIES RESISTANCES

In contrast to the large uncertainties of extracted resistance values, it is remarkable that the sum of the resistances $R_\Sigma = R_g + R_i + R_s$ can be determined with high consistency (constant sum extraction stability) (see Section 6.2.3); that is, each single resistance is dependent of the others. Note that the measurement frequency should be chosen sufficiently high so that R_g and R_i would become decoupled by the capacitance C_{gd}. Fixing the extracted mean value of $R_d = 0.8436\ \Omega$, further extractions have been performed, changing the R_d value between 0.5 and 2.0 Ω. Figure 14.14 illustrates the extraction results as a function of the drain resistance R_d. As can be seen, there is a linear relation between the individual resistances in the interval of $R_d = 0.6\ \Omega$ and $R_d = 1.5\ \Omega$. Beyond, the functions become nonlinear. This also results in a strong increase of the residual error function value.Therefore, a physically meaningful solution is assumed within the mentioned value interval.

Regarding the uncertainties of the resistances, we can estimate the uncertainty ranges of other model parameters. This is illustrated in Table 14.4. For most model parameters these uncertainties are well accepted (e.g., all capacitances and inductances as well as τ). Conversely all parasitic resistances, transconductance, and channel conductance suffer from

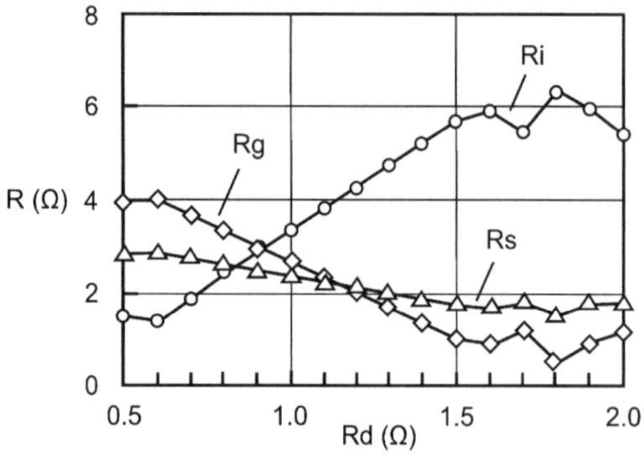

Figure 14.14 Extracted results for R_g, R_i, and R_s as a function of R_d [9].

Table 14.4
Final Extracted Uncertainty Ranges of Model Element Values

Model Element	$R_{d,min} = 0.6\ \Omega$	$R_{d,max} = 1.5\ \Omega$	Model Element	$R_{d,min} = 0.6\ \Omega$	$R_{d,max} = 1.5\ \Omega$
C_{gs} / pF	0.42575	0.38853	R_i / Ω	1.5205	5.5806
G_m / mS	77.873	71.844	G_{ds} / mS	0.0063	0.0058
C_{ds} / pF	0.99506	0.99350	C_{gd} / pF	0.02949	0.03225
R_g / Ω	3.3329	1.0564	L_g / nH	0.15620	0.15702
R_d / Ω	0.60882	1.4823	L_d / nH	0.15582	0.15545
R_s / Ω	2.7835	1.7202	L_s / nH	0.03799	0.03827
C_{pg} / pF	0.02742	0.02753	C_{pd} / pF	0.01523	0.01519
τ / ps	2.0481	1.9448			

larger uncertainties. Even though these extractions may not be acceptable as a final result, they can be used for further refinement with alternative optimization procedures.

The experiments have shown that reliable model parameter extraction is not possible without further knowledge of the resistances. However, if one of the four parasitic resistances R_g, R_i, R_s, or R_d is known, stable extraction consistency can be expected. Because of bias independence, R_g was selected as the element to be predetermined. Figure 14.15 demonstrates the dependence of the extracted element values on the chosen R_g value. Beyond the plotted R_g interval no physically meaningful solution could be found. Taking the mean value of R_g derived from the statistical approach, a rather good consistent result could be obtained, as shown in Figure 14.11.

14.6 SUMMARY

A statistical extraction approach combining the random search and modified simplex optimization methods, was applied to the model parameter extraction of a widely used 15-element equivalent circuit of field-effect transistors in the measurement range from 45 MHz

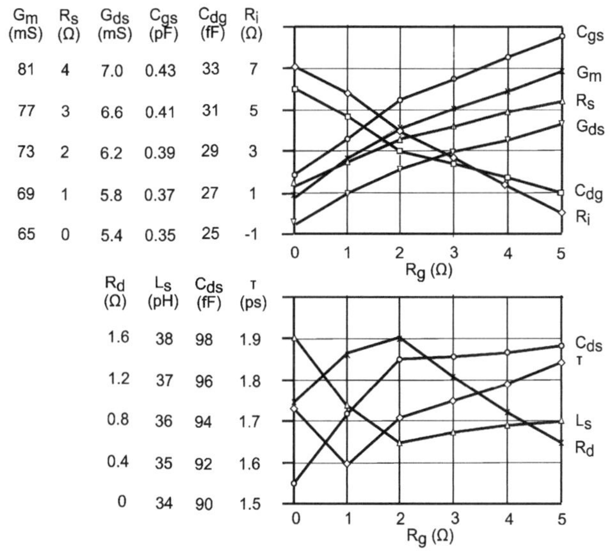

Figure 14.15 Extracted model element values from measured scattering parameters in the frequency range from 45 MHz to 40 GHz as a function of R_g. The ordinate axis is scaled differently for each model element [10].

to 26.5 GHz and 40 GHz, respectively.

Based on fictitious scattering parameters, the sensitivity and consistency of extracted model parameter values were analyzed. It has been found that extraction consistency increases with increasing measurement frequency bandwidth and measurement accuracy. The model elements are extracted with different certainty. For example, the intrinsic elements (except for delay τ) exhibit high consistency, whereas the determination of the series resistances is rather uncertain.

The scattering of extracted model parameter values largely obey normal distribution. This result contradicts previous observations [11].

It has been shown that extraction consistency can be improved in different ways; for example, by downsizing the search domain and number of optimization variables. From low-frequency measured data, several model parameter values can be well estimated, providing high-quality starting parameter values for the optimization procedure. This also implies the capability to drastically reduce the search interval size of optimization variables.

A further improvement of the extraction approach can be attained if the intrinsic model elements are analytically determined from the intrinsic Y-matrix, which is obtained by an deembedding approach from measured extrinsic Y'-matrix comprising extrinsic model parameter optimization variables.

Commonly, the gate resistance is not regarded as an intrinsic element. However, because of seven intrinsic elements in the 15-element model, it can be additionally included and analytically determined. The extraction uncertainty of the residual extrinsic parameters was drastically reduced as expected.

It was found that the sum of resistances $R_\Sigma = R_g + R_i + R_s$ is consistently extracted within a limited value range. In this interval most model parameters depend linearly on each other. Thus, it is necessary to specify the value of one resistor. An approximated resistance value may be obtained from the statistically evaluated mean value of normal model parameter distribution. It may be appropriate to choose the gate resistance because of its bias independence.

REFERENCES

[1] Bandler, J., "Have You Ever Wondered About the Engineer's Mysterious 'Feel' for a Problem?" *IEEE Canadian Review*, Summer 2013, pp. 51–60.

[2] Microwaves101.com, *History of Microwave Software*, https://www.microwaves101.com/encyclopedias/history-of-microwave-software.

[3] Kompa, G., M. Schlechtweg, and F. van Raay, "Precisely Calibrated Coaxial-to-Microstrip Transitions Yield Improved Performance in GaAs FET Characterization," *IEEE Trans.*, Vol. MTT-38, 1990, pp. 62–68.

[4] Kondoh, H., "An Accurate FET Modelling from Measured S-Parameters," *IEEE MTT-S Int. Microwave Symp. Digest*, 1986, pp. 377–380.

[5] Vaitkus, R. L., "Realistic Small-Signal Parameter Extraction Methods for Microwave Transistors," *GaAs Microwave Monolithic Integrated Circuits (MMICs) and Radar Cross-Section (RCS)*, Workshop Dig., Arizona State University, April 11, 1988.

[6] Kompa, G., and M. Schlechtweg, "Generalized Modelling of GaAs MESFETs and MODFETS Based on Highly Accurate Broadband Measurement," *Proc. of the 19th European Microwave Conf.*, London, 1989, pp. 179–186.

[7] Kompa, G., "Small- and Large-Signal Characterization Methods," *Digest IEEE Workshop on Measurement Techniques for Microwave Device Characterization and Modelling*, Stuttgart, Germany, April 23, 1990, pp. 67–97.

[8] Schlechtweg, M., *Broadband Characterization and Modelling of GaAs-MESFETs and (AlGa)As/GaAs-MODFETs up to 40 GHz* (in German), Doctoral Thesis, Kassel University, 1989.

[9] Novotny, M., *Investigation on the Consistency of Transistor Equivalent Circuits* (in German), Research Project, University of Kassel, 1990.

[10] Kompa, G., "Experimental Modelling of Microwave FETs Based on ′Hot′ S-Parameters," *Archiv für Elektrotechnik* (AEÜ), Vol. 74, 1991, pp. 219–224.

[11] van Niekerk, C., and P. Meyer, "Performance and Limitations of Decomposition-Based Parameter-Extraction Procedures for FET Small-Signal Models," *IEEE Trans.*, Vol. MTT-46, 1998, pp. 1620–1627.

Chapter 15

Decomposition-Based Extraction Methods

15.1 INTRODUCTION

Decomposition (or partitioning) is a general approach to solving a problem by breaking it up into smaller ones (subproblems) and solving each of the smaller ones separately. Regarding model parameter extraction of a microwave device network with a large number of parameters we may sort the model elements into different groups with stronger or weaker influence on the device response. As the model parameters interact with each other, the problems cannot be solved independently but rather iteratively.

It is often argued that decomposition provides less complex functions to be optimized. This reduces the amount of local minima in the decomposed subfunctions leading to robust extraction with less uncertainty results. Figure 15.1 illustrates the principal decomposition procedure. Starting from a group of optimizable variables, a sensitivity analysis of model parameters is performed to recognize their different degrees of influence on the individual scattering parameters. Thus, as an example, the model elements may be divided into high-, medium-, and low-sensitive subgroups, where each subgroup may be assigned a specific error function to be optimized.

15.2 EMPIRICAL DECOMPOSITION

Model partitioning was first proposed by Kondoh in 1986 [1] and applied to a 13-element FET model without external parasitic capacitances C_{pg} and C_{pd} (see Figure 14.1, Section 14.2). The optimizer used was based on a conventional gradient search algorithm [2]. To estimate the absolute accuracy of the new algorithm, optimization was performed on fictitious measured data. The data was generated for a frequency range of 1.5 GHz to 25 GHz by calculating S-parameters from a FET equivalent circuit that simulates a 0.5-μm gate-length and 300 μm gate-width microwave FET device.

The decomposition method after Kondoh comprises eight different error functions. In each of the error functions certain model parameters are optimized with respect to only one particular S-parameter coefficient. The eight error functions are repeatedly optimized in a specific order until all the model parameters have converged to their final results. Table 15.1 summarizes the combinations of selected elements optimized with a single S-parameter coefficient and a frequency range fitted in each step.

First, only G_{ds}^{-1} and C_{ds} are optimized to fit S_{22} over the entire frequency range. The other

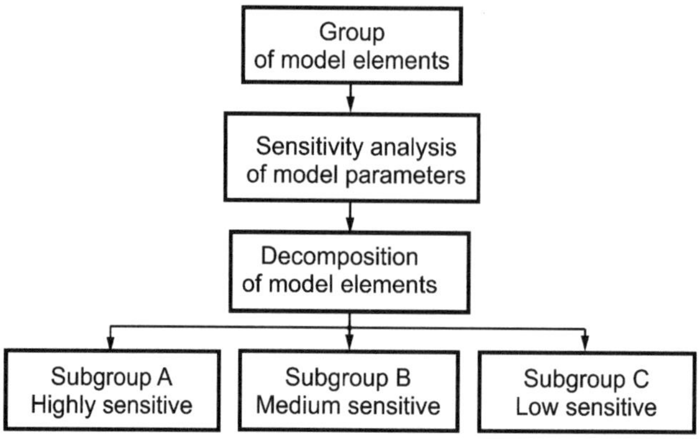

Figure 15.1 Decomposition of model elements into subgroups of high-, medium- and low-sensitive elements based on model element sensitivity analysis.

model elements are fixed in value to those obtained in their last optimization cycle. In the second step only C_{gs} is optimized to fit the calculated S_{11} in the entire frequency range and the subsequent optimization steps follow as shown in the table. It is pointed out that the order of the eight steps is important to enhance the numerical stability and the rate of convergence of the modeling algorithm. The combinations listed in Table 15.1 followed from intensive experimental investigation of S-parameter sensitivity performed on practical FETs. After all elements have been optimized once with respect to measured scattering parameters, the next iteration cycle starts again as described. The optimization ends when the error functions could be minimized successfully after a few iteration cycles.

It is pointed out that, with respect to the conventional optimization method, the new modeling method provides better accuracy. Based on fictitious S-parameter data for a frequency range of 1.5 GHz to 25 GHz, the modeling error was less than 3% for all the element values except for R_d and L_d with modeling errors of −17% and 15%, respectively.

Table 15.2 summarizes the extracted model parameters of a 0.5-µm gate width and 600 µm gate length (10-finger) GaAs FET measured at frequencies from 1.5 GHz to 25 GHz

Table 15.1
Eight Steps of One Iteration Cycle after Kondoh

Step	Elements Optimized	S-Parameter Fitted	Frequency Range
1	G_{ds}^{-1}, C_{ds}	S_{22}	All
2	C_{gs}	S_{11}	All
3	C_{gd}, R_s	S_{12}	All
4	G_m	S_{21}	All
5	R_d, L_d	S_{22}	Upper half band
6	R_g, R_i, L_g	S_{11}	Upper half band
7	L_s	S_{12}	Upper half band
8	τ	S_{21}	Upper half band

From: [1].

using an on-wafer prober. Extraction was first performed with a conventional optimizer and then with the decomposition-based optimization algorithm for comparison. S1 and S2 indicate two different starting values for the model parameters. In addition the new optimization routine was tested with and without R_i, as the extraction algorithm delivered consistent values for the sum of $(R_i + R_g)$, but the determination of the individual resistances was very uncertain. The two starting vectors read [3]:

S1: $G_m = 200$ mS, $\tau = 5.6692$ ps, $C_{gs} = 1.2$ pF, $C_{ds} = 0.25$ pF, $C_{gd} = 0.1$ pF, $R_g = 5\Omega$, $R_i = 2\Omega$, $R_s = 5\Omega$, $R_d = 3\Omega$, $G_{ds} = 15$ mS, $L_g = L_d = L_s = 10$ pH.

S2: $G_m = 61$ mS, $\tau = 3.45$ ps, $C_{gs} = 0.45$ pF, $C_{ds} = 0.15$ pF, $C_{gd} = 0.068$ pF, $R_g = 1.1\Omega$, $R_i = .7\Omega$, $R_s = 2.3\Omega$, $R_d = 2.3\Omega$, $G_{ds} = 7.2$ mS, $L_g = 50$ pH, $L_d = 2.0$ pH, $L_s = 2.0$ pH.

Bandler and Zhang [4] developed an automated approach for assigning model elements to the different subfunctions using a Monte Carlo analysis, which confirmed the assignment of model elements to subfunctions proposed by Kondoh.

Table 15.3 shows a comparison of optimized model parameter values based on decomposition algorithm after Kondoh and leaping simplex algorithm as described in Section 12.7.5. The same initial model element values were used in both cases [3]. As can be seen, the extraction results after Kondoh are pretty good, although there are some elements with larger inaccuracy such as R_d and L_d. In contrast the modified simplex algorithm provides, in general, model element values with much higher accuracy. This is in particular

Table 15.2

Extraction Results of Investigated GaAs FET*

Element	Conventional Routine/ Gradient Search			Decomposition-Based Routine with R_i			Decomposition-Based Routine without R_i		
	S1	S2	Inconsistency (%)	S1	S2	Inconsistency (%)	S1	S2	Inconsistency (%)
G_m (mS)	100.	98.9	1.44	88.7	87.9	0.9	88.8	87.8	1.2
τ (ps)	1.9	1.67	12.7	1.87	1.84	1.6	1.88	1.88	0
C_{gs} (pF)	0.631	0.6	5.0	0.554	0.546	1.4	0.557	0.552	0.8
C_{ds} (pF)	0.111	0.118	6.0	0.105	0.105	0	0.105	0.104	0.8
C_{gd} (pF)	0.0482	0.0465	3.6	0.0524	0.0533	1.7	0.0525	0.0527	0.4
R_g (Ω)	2.35	1.78	27.4	2.3	2.04	**12.2**	2.48	2.48	0
R_i (Ω)	0.263	0.641	84.0	0.268	0.664	**85.0**	—	—	—
$R_g + R_i$	2.61	2.43	7.4	2.57	2.70	4.9	2.48	2.48	0
R_s (Ω)	3.0	3.06	2.0	2.4	2.27	5.7	2.44	2.4	1.4
R_d (Ω)	1.88	3.13	50.0	1.35	1.58	**15.7**	1.18	1.18	0
G_{ds}^{-1} (Ω)	111.	105.	5.5	119.	121.	1.3	120.	121.	0.9
L_g (pH)	10.9	29.7	92.7	29.3	29.2	0.6	28.8	29.2	1.3
L_d (pH)	14.3	3.72	117.0	11.2	10.2	**9.2**	11.8	12.9	8.9
L_s (pH)	0.81	1.06	27.0	6.86	6.97	1.6	7.28	7.61	4.4

After: [1].

* S1 and S2 indicate the optimization results for respective two different sets of initial parameter values (inconsistency = (S1 − S2)/ [(S1+ S2)/2]×100%).

Table 15.3
Comparison between Kondoh and Leaping Simplex Algorithm

			Kondoh [1]		Simplex Algorithm	
Element	True Value	Initial Value [3]	Value Obtained	Modeling Error (%)	Value Obtained	Modeling Error (%)
G_m (mS)	50	100	49.5	-0.94	49.98	-0.22
τ (ps)	3.0	6.0	3.02	+0.5	3.002	+0.06
C_{gs} (pF)	0.25	0.5	0.25	-0.1	0.25	-0.01
C_{ds} (pF)	0.08	0.2	0.078	-1.4	0.08	-0.03
C_{gd} (pF)	0.025	0.2	0.025	±0.0	0.025	±0.0
R_g (Ω)	4.0	8.0	4.03	+0.6	4.0	+0.01
R_s (Ω)	4.0	8.0	3.94	-1.4	3.99	-0.05
R_d (Ω)	3.0	8.0	2.49	-17.0	2.98	-0.64
G_{ds}^{-1} (Ω)	250	125	251	+0.5	249.9	-0.03
L_g (pH)	60.0	150	60.0	-0.1	60.0	±0.0
L_d (pH)	25.0	100	28.8	+15.0	25.02	+0.1
L_s (pH)	15.0	100	15.5	+3.0	15.004	+0.03

After: [5].

also valid for the aforementioned low-sensitive elements.

As pointed out by Bandler [4], the different suboptimization problems are not linearly independent. In order to ensure convergence, they have to be solved in a specific sequence, which is repeated until all the model elements have converged to their final values. Kondoh determined his optimization sequence through experimentation, while Bandler demonstrated an adaptive algorithm that determines the optimization sequence as a function of the subfunction error and dimensions.

15.3 AUTOMATIC DECOMPOSITION BASED ON SENSITIVITY ANALYSIS

Parameter sensitivity analysis is frequently aimed at reducing the dimensionality of an optimization problem. In this way parameter optimization is simplified with the reduced number of necessary parameters to consider. The sensitivity analysis of model parameters evaluates the rank of sensitivity in parameters and thereafter removing parameters with trivial impact on model output. This aspect may be interesting to find out which and how many model elements are really necessary to simulate a measured device response with specified uncertainty. In this case, the sensitivity analysis will identify those elements that have less noticeable influence on the object function.

In the following, the sensitivity analysis is applied to automatically generate a model parameter vector, which ranks the individual model elements (or combination of elements) according to their degree of sensitivity concerning the object function. High-sensitive model parameters have great impact on the error function so that highly consistent extraction values are expected. On the other hand, low-sensitive elements have only a small effect on the optimization process, and thus extraction exhibit model parameter values with high uncertainty.

As will be discussed in the following, the sensitivity rank for parameters can be determined by the analysis of eigenvalues of the Hessian matrix $\boldsymbol{H}$.

15.3.1 Principal-Component Sensitivity Analysis

As noted previously, the sensitivity of the error function is poor in some model parameters. This, for example, applies to all series resistances. In addition, model topology mismatching gives further rise to ill-conditioning of the optimization problem. This in turn causes inaccurate parameter estimates and slow convergence when finite optimization is employed. Patterson et al. [6, 7] proposed *principal-component sensitivity analysis* to calculate the sensitivity of the object function with respect to each model parameter of a device equivalent network. The sensitivity analysis reveals ill-conditioning sensitivity properties of model parameters and provides estimation on the reliability of optimized model elements. It is stated that the proposed method improves the model parameter extraction, in particular, with respect to obtain accurate estimates of the insensitive model parameters such as the parasitic resistances [7].

A sensitivity analysis of the error function $E(\boldsymbol{p})$ to variations only along the directions of the model parameter axes, as was discussed in the previous section, gives an inadequate description of the model sensitivity characteristics. This is so because the parameters are correlated and the effect of combinations of parameters may be more important than the effect of individual parameters.

Contours of constant error function value are shown in Figure 15.2 for an arbitrary ill-conditioned least-squares problem having two parameters p_1 and p_2 [7]. For this problem the parameters p_1 and p_2 are equally sensitive since the contours of constant $E(\boldsymbol{p})$ are equally spaced along these two directions. As will be discussed, the problem is ill-conditioned even though the optimizable model parameters are equally sensitive.

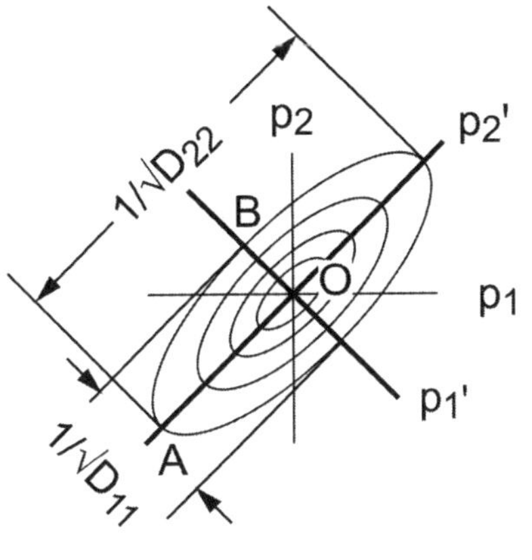

Figure 15.2 Contours of two-dimensional error function $E(\boldsymbol{p})$ with sensitivity directions. The most sensitive direction is OB since the highest rate of change is along this direction (the contours are closest together along this direction). Conversely, OA is the least sensitive direction. (After [7].)

Under application of rotation transformation, the search directions can be optimized as indicated in Figure 15.2. In contrast to the less optimal individual parameter directions p_1 and p_2, the transformed new search variables p_1' and p_2' point in the direction of maximum (OB) and minimum (OA) sensitivity of the error function. These sensitivity directions are called the principal components. In the direction of maximum sensitivity, the position of the minimum of $E(\boldsymbol{p})$ is likely to be more consistently determined. However, as should be recognized, optimization of the object function is no longer performed with respect to individual model parameters p_1 and p_2. The primed optimization variables p_1' and p_2' indicate linear combinations of individual parameters. Ill-conditioning in the model parameter extraction process is indicated by a large condition number that is the ratio of the sensitivities of the error function to changes in the directions comprising the most and least sensitive combinations of parameters. The condition number provides a measure of the spread of sensitivities of parameter combinations and is defined as the ratio of largest eigenvalue to the smallest eigenvalue (i.e., D_{11}/D_{22}) in Figure 15.2. The number of principal components equals the number of model parameters.

15.3.2 Sensitivity Analysis by Hessian Matrix Diagonalization

The sensitivity analysis proposed by Patterson et al. [6, 7] exploits the particular properties of the Hessian matrix. The Hessian matrix is symmetric, and in this case it can be diagonalized by appropriate matrix transformation. Recalling (12.26), least-squares error function is defined as

$$E(\boldsymbol{p}) = \frac{1}{2}\sum_{i=1}^{m}[e_i(\boldsymbol{p})]^2 = \frac{1}{2}\sum_{i=1}^{m}e_i^2 \tag{15.1}$$

In the neighborhood of a solution $\boldsymbol{p}$, $E(\mathbf{p})$ can be expressed as the Taylor series (12.15):

$$E(\boldsymbol{p} + \delta\boldsymbol{p}) \approx E(\boldsymbol{p}) + \nabla E^T\delta\boldsymbol{p} + \frac{1}{2}\delta\boldsymbol{p}^T\boldsymbol{H}\delta\boldsymbol{p} \tag{15.2}$$

where $\boldsymbol{H}$ is the Hessian matrix. At a local minimum the gradient should be zero so that the second term in (12.15) is less meaningful. The Hessian matrix for the particular case of least-squares error function is calculated as follows. With the Jacobian $J(\boldsymbol{p})$,

$$J(\boldsymbol{p}) = \left[\frac{\partial e_i}{\partial p_k}\right]_{i=1,\dots,m;\ k=1,\dots,n} \tag{15.3}$$

the first derivative of the least-squares error function $\nabla E(\boldsymbol{p})$ is calculated as

$$\nabla E(\boldsymbol{p}) = \sum_{i=1}^{m} e_i(\boldsymbol{p})\nabla e_i(\boldsymbol{p}) = J(\boldsymbol{p})^T \tag{15.4}$$

The second derivative denotes the Hessian,

$$H(\boldsymbol{p}) = \nabla^2 E(\boldsymbol{p}) = \sum_{i=1}^{m} \nabla e_i(\boldsymbol{p}) \nabla e_i(\boldsymbol{p})^T + \sum_{i=1}^{m} e_i(\boldsymbol{p}) \nabla^2 e_i(\boldsymbol{p})$$
$$= \left[\boldsymbol{J}(\boldsymbol{p})^T \boldsymbol{J}(\boldsymbol{p}) + \sum_{i=1}^{m} e_i(\boldsymbol{p}) \nabla^2 e_i(\boldsymbol{p}) \right] \quad (15.5)$$

When the residuals e_i themselves are small, or when the second derivative of the residuals is small, we can approximate the Hessian of the least squares error function $E(\boldsymbol{p})$ by

$$\boldsymbol{H}(\boldsymbol{p}) \approx \boldsymbol{J}(\boldsymbol{p})^T \boldsymbol{J}(\boldsymbol{p}) \quad (15.6)$$

As the Hessian matrix is symmetric, it is orthogonally diagonalizable [8],

$$\boldsymbol{H}(\boldsymbol{p}) = \boldsymbol{Q}\boldsymbol{D}\boldsymbol{Q}^T \quad (15.7)$$

where $\boldsymbol{Q}$ is the orthonormal matrix of eigenvectors, and $\boldsymbol{D}$ is a diagonal matrix with D_{ii} the ith eigenvalue of $\boldsymbol{H}(\boldsymbol{p})$. Therefore, writing $\boldsymbol{p}' = \boldsymbol{Q}^T \delta \boldsymbol{p}$, we have

$$E(\boldsymbol{p}') = E(\boldsymbol{p}) + \boldsymbol{p}'^T \boldsymbol{Q}^T \nabla E(\boldsymbol{p}) + \frac{1}{2} \boldsymbol{p}'^T \boldsymbol{D} \boldsymbol{p}' \quad (15.8)$$

The contours of $E(\boldsymbol{p}')$ describe a family of ellipsoids whose axis lengths are inversely proportional to $\sqrt{D_{ii}}$, and whose axes lie along the p_i' axes, as indicated in Figure 15.2.

The eigenvectors of $\boldsymbol{H}(\boldsymbol{p})$ (i.e., the columns of the matrix $\boldsymbol{Q}$) are unit vectors which define the p_i' axes. These eigenvectors (i.e., the p_i' axes) are the principal component directions. The eigenvector or principal component corresponding to the direction of maximum eigenvalue D_{ii} is the direction of maximum sensitivity to changes in the value of $\boldsymbol{p}'$. Conversely, the principal component corresponding to the direction of minimum eigenvalue D_{jj} is the direction of minimum sensitivity to changes in the value $\boldsymbol{p}'$. Information concerning these two principal component directions is most useful since they describe the extremes of the model sensitivity characteristics [7].

Regarding model parameter extraction, the condition number provides a measure of the spread of sensitivities exhibited by the device model and is defined as the ratio of largest eigenvalue to the smallest eigenvalue, D_{ii}/D_{jj}. The number of principal components equals the number of model parameters.

Table 15.4 depicts an extract from Table I in [7], showing a summary of results for a typical principal components sensitivity analysis for a MESFET device. The optimizable parameters and their final values are shown in the first two columns to the left. The remaining columns give the principal components directions (eigenvectors) expressed as unit vectors in terms of the model parameters. For example, the first principal component with the highest eigenvalue of 3×10^4 (sensitivity coefficient) comprises a unit vector with a component of 0.9 in the direction of the gate to the drain capacitance (C_{dg}) and 0.1 in the direction of the drain to source capacitance (C_{ds}); that is, $0.9C_{dg} + 0.1C_{ds}$. Further components of this principal component direction are much smaller and have been omitted for the purpose of presentation. This first principal component is the most sensitive direction (i.e., of all possible directions we could choose to move along from this base point, a step in the direction $0.9C_{dg} + 0.1C_{ds}$ will result in the largest change of error function value). The

Table 15.4

Sensibility Analysis on the Investigated Small-Signal MESFET Equivalent Circuit Model

Optimizable Parameter	Final Value	Principal Components Directions (Eigenvectors)					
Cgs (pF)	0.226	—	0.1	0.7	−0.6	−0.3	...
Ri (Ω)	8.803	—	—	0.1	—	—	...
Lg (nH)	0.013	—	—	0.1	−0.3	0.9	...
Ls (nH)	0.019	—	—	0.7	0.7	0.2	...
Cds (pF)	0.058	0.1	0.9	−0.1	—	—	...
Cdg (pF)	0.030	0.9	−0.1	0.1	0.1	—	...
Rds (Ω)					0.1	0.1	
Ld (nH)						0.2	
⋮	⋮	⋮	⋮	⋮	⋮		⋮
Sensitivity Coefficient (Eigenvalue)		30000	2000	1000	500	150	...

After: [7].

eigenvalue corresponding to each principal component direction is shown at the bottom of each column. A step from the base point along a given principal component direction will result in a change in error function value that is proportional to the corresponding eigenvalue. Therefore the eigenvalues can be considered to be sensitivity coefficients that indicate the sensitivity of the error function to move along the principal component directions. The principal component directions are arranged in order from the most sensitive direction to least sensitive directions from left to right.

From the sensitivity analysis given in Table 15.4 we can now estimate how reliable the component values in the least squares solution are. The most sensitive direction almost coincides with the direction of the gate to drain capacitance (C_{dg}) with just a small component in the direction of the drain to source capacitance (C_{ds}). Therefore C_{dg} will be precisely defined in the optimized solution since it is a sensitive component and a very small change in its value results in a large change in the error function value. The other sensitive principal component directions—say for example those with eigenvalues greater than 100, have large components in the directions of the elements C_{ds}, C_{gs}, L_s, and L_g. Since the error function is sensitive to small changes in these component values, these components will also be precisely defined in the optimized solution.

On the other hand, components such as G_m, τ, R_i, and the parasitic resistances R_s, R_d, and R_g form the major components of the principal component directions of low sensitivity—those with eigenvalues less than 1 (not shown in Table 15.4) [7]. These components are less accurately defined since large changes in their values have a very small effect on the error function value. Since the sensitivity coefficients vary over such a wide range from 3×10^4 to 9×10^{-6} it is clear that the equivalent circuit element values are resolved with widely varying degrees of uncertainty.

15.3.3 Scaling of Variables for Condition Number Adjustment

The condition number, which usually used to measure the degree of the ill-conditioning, is typically very large since the error function most often used is sensitive to changes in some combinations of model parameters but extremely insensitive to changes in other combinations. Therefore, Patterson et al. proposed to scale the p_i' variables inversely proportional to the square root of their corresponding eigenvalues D_{ii}. In this case a unit change in any of the p_i' variables will have equal effect on the sum of squares function for all i. The error function is now locally spherical, and the model condition number is 1. Since the error function is now equally sensitive to moves in each of the principal component directions, the model parameters are all optimized with similar accuracy. Furthermore, the spherically shaped error function provides the fastest convergence rate possible. The transformation used is $\boldsymbol{p}'' = \boldsymbol{Q}\boldsymbol{D}^{-1/2}\boldsymbol{p}'$, where $\boldsymbol{p}''$ is the new set of optimization variables.

15.3.4 Decomposition Optimization in Transformed Model Space

However, as is pointed out, when the condition number is very large, rounding and truncation errors can be involved in computations, which leads to increased uncertainty of the extracted parameters and causes slow convergence. In order to improve the performance of the numerical optimization, the sensitivity analysis results can be partitioned on the basis of the distribution of eigenvalues (see Table 15.4). Because the transformed variables are uncorrelated, they can be systematically divided into groups of similarly sensitive variables, which is denoted as subspace reduction. Table 15.4, for example, could be divided into two sub-spaces bounded by sensitivity coefficients 3×10^4 - 0.1 and 0.06 - 9×10^{-6} having condition numbers of 3×10^5 and 7×10^3, respectively.

15.3.4.1 Parameter Extraction Example

The proposed optimization method was tested on packaged and unpackaged chip MESFET devices measured over frequencies from 2 to 18 GHz. The number of parameters was 18 and 23 parameters, respectively. In the latter case, technological device parameters, such as the channel thickness, diffusion voltage, gate width, doping density, and saturation drift velocity were included in the set of optimized variables. Optimization (i.e., minimization of the sum of squares of the errors between the measured data and the model responses) was based on the *Gauss-Newton* algorithm (Section 12.5.1). Starting values were estimated for the package components in the electrical equivalent circuit model by trial and error.

It is stated that mostly convergence improvement has been emphasized with respect to conventional full-space optimization using the same starting points. Insensitive parameters such as the parasitic resistances were determined more uniquely since the sensitivity of the error function to these parameters was enhanced by automatic scaling. Nevertheless, it is pointed out that the largest variance was seen in the values of the least sensitive components (e.g., the parasitic resistances).

15.3.4.2 Robustness Analysis

It is unfortunate that Patterson et al. [6, 7] have not tested the sensitivity analysis approach on the basis of known (fictitious) S-parameters. This would have given more information about the robustness and accuracy of the proposed method for model parameter estimation and final numerical (gradient-based) optimization. From mere comparison of simulated and measured scattering parameters over a rather limited frequency range up to 18 GHz a final conclusion about extraction quality might not yet be drawn. This would be particularly valid with the focus on the extraction of physically meaningful model parameter values.

Summarizing, the method proposed by Patterson et al. [7] takes advantage of the properties of the symmetric Hessian matrix, which can be diagonalized (eigen decomposition). The eigenvectors comprise combinations of individual model parameters and the eigenvalues represent sensitivities of the error function to model elements. Using the principal components sensitivity analysis, parameter estimation can be made on how reliable the optimized component values are in the final solution. By performing a coordinate transformation of the model variables to the principal component axes that are uncorrelated, the new principal component variables can be automatically scaled providing uniform sensitivity for all variables (i.e., a condition number of unity).

Gu and Liu [9] pointed out that the proposed sensitivity analysis relies on a linear combination of several model elements and the analysis results cannot give accurate sensitivity values for each model element.

15.3.5 Decomposed Optimization of Individual Parameters

As described before, Patterson et al. [7] employed principle-component sensitivity analysis to identify the sensitivity of individual model parameters (more precisely, combinations of model parameters). Scaling is introduced to adjust the large differences in the evaluated condition numbers. Because of numerical issues in handling widespread condition numbers, decomposition has been performed in the model parameter space. Since the transformed variables are uncorrelated they can be systematically grouped according to their sensitivities so that two or more well-conditioned subspaces are formed [6]. As pointed out in [10], the optimization approach after Patterson et al. still suffers from large amounts of local minima, because it uses only one global error function.

In 1996, van Niekerk and Meyer [10] applied and extended the principal component sensitivity analysis suggested by Patterson et al. In contrast to only one global error function, now every model element is optimized on its own, leading to a number of one-dimensional suboptimization problems equal to the number of model elements. The optimization order of the error function is determined by the principal component sensitivity analysis. The model parameters are optimized in this specific order starting with the most sensitive and proceeding toward the least sensitive. The damped Gauss-Newton optimization algorithm was used for minimization of the error functions (Section 12.5.2).

15.3.5.1 Robustness Tests

The proposed parameter extraction procedure was tested on fictitious and measured S-parameter data [10]. The fictitious S-parameters were generated from a 13-element transistor equivalent circuit with the use of typical model parameter values from the Fujitsu 1993 GaAs FET data book (FLR016XV (#1), FSX52X (#2), FLK052XV (#3)). The starting values of the model parameters were chosen to be twice their correct value. The results of the proposed extraction method were compared to those obtained from hybrid optimizer combining Gauss-Newton algorithm [11] and simulated annealing (Section 12.7.3). Indeed, for most model parameters the extracted accuracy was better than 1%. Lower accuracy was obtained as follows: sample #1: $R_g = 9.7\%$; sample #2: $R_g = 23.7\%$, $R_i = 3.8\%$, $R_s = 3.0\%$; sample #3: $R_g = 12.1\%$, $R_i = 2.6\%$. In general, it could be observed that the extraction results were by far more accurate than those obtained with the hybrid optimizer proposed in [11].

Extractions were performed on fictitious S-parameter data sets in the frequency range from 1 to 44 GHz. Using a reduced span of 1 to 22 GHz resulted in larger errors. Therefore, the authors suggested that for model parameter extraction, the S-parameters should be measured over as wide frequency range as possible with the use of as many data points as possible (see Section 14.2.1).

Further robustness tests were published in 1998 [12]. The algorithm was tested with fictitious (ideal) data to provide an absolute measure for determining accuracy. The S-parameters were generated at 40 frequency points, from 1 to 40 GHz. In a first test, 100 random starting values were chosen in the search space using a uniform distribution. The search space ranged from 0.1 times to 5 times the nominal parameter values. An extraction was performed using each set of starting values.

A basic pattern emerged: Capacitors were always dominant, followed by τ and the extrinsic inductances. The intrinsic and extrinsic resistors and the transconductance were always the least sensitive elements in the model with the following maximum errors: $R_g = 398\%$, $R_i = 2.92\%$, $R_s = 3.93\%$. A second transistor delivered no better results: $R_g = 104.8\%$, $R_i = 89.91\%$, $R_s = 61.62\%$, $R_d = 95.51\%$, $\tau = 7.48\%$; all others > 3% exceptionally L_g and L_s < 1%.

The optimization was then repeated using a larger search space. Model element starting values were chosen in the ranges 0.05 – 10 and, respectively, 0.033 – 20 times the nominal element values. Both ranges caused some of the extractions in the experiments not to converge to the correct answer. These extractions were easy to identify since most of the model elements would take on values close to the optimization boundaries, and in some cases, the final global error was larger than the global error at the start of the search. However, better than 90% of the extractions converged to the correct element values.

Looking at such extracted results it is pointed out that the decomposed optimization algorithms are by no means globally convergent and can still be caught in local minima if the search is started too far from the global minimum. Nevertheless, it is stated that the developed algorithms are far more robust than most conventional search algorithms, and are capable of covering a far larger search area.

For a second test, the fictitious S-parameters were contaminated with noise. The noise was defined to have a Gaussian magnitude distribution and a uniform phase distribution between 0 and 2π radians. The magnitude distribution has a standard deviation of 2% of magnitude of the measured S-parameters. The second test used 500 random starting values. It has been found that the uncertainty in the determination of the less dominant parameters greatly increases in the presence of imperfections of data. This is especially true for the parasitic resistances R_g, R_d, R_s, and the channel resistance R_i.

15.3.5.2 Measurement-Based Parameter Extraction

Extraction on measured S-parameters could only be performed with a limited number of data points. Therefore, in this case, the starting value of R_d was fixed. For a given starting vector (doubling of the circuit parameter values supplied by Fujitsu [13]), extracted results are only presented by comparison of simulated and measured S-parameters so that any discussion about physical meaningfulness remained open.

15.3.5.3 Multibias Decomposition-Based Parameter Extraction

Based on wide experience with decomposition-based optimization strategies, van Niekerk et al. [14] commented about the extraction problems arising at a single bias point: R_g values are sensitive to the starting values used in the extraction. The values and sensitivity toward optimization starting values of the less sensitive elements, such as the parasitic resistances, are very much dependent on the bias point used in the extraction. This leads to high uncertainty in the extracted values of these elements.

Therefore, an integrated multibias parameter-extraction method was proposed in [14]. Modeling were performed with a 13-element model topology. The parasitic elements C_{pg} and C_{pd} were not included in the model because due to the small values of the series inductances L_g and L_d, the highest measurement frequency was still too low to reliably separate the effect of these elements from the intrinsic capacitors C_{gs} and C_{ds}.

Based on S-parameter measurements over a frequency range of 1 – 26.5-GHz for a 0.25 μm × 300-μm low-noise MESFET and 1 – 25 GHz for a 0.2-μm x 100-μm GaAs pHEMT at five bias points, it turned out that the multibias technique has proven to be independent of starting values and produced more consistent results for the nondominant model elements. The analysis was performed on the basis of 100 randomly generated starting model parameter value vectors. It is stated that most of the model elements consistently converge to a very small range of values. However, regarding the MESFET extraction results, R_i and R_g still exhibit an uncertainty of 32.9% and 34.6%, respectively. In particular, it is emphasized that the extracted resistances R_s and R_d always tended to be the lowest optimization boundry regardless of the starting values used (see also Chapter 20, error function curve shape in the R_s–R_d plane). In the MESFET case, R_d had strong tendency to move to negative values. The authors concluded that the multibias algorithm including sensitivity-induced model parameter decomposition cannot solve the nonuniqueness problem of parasitic resistances.

The 13- and 15-element pHEMT models provided comparable modeling errors, but the inclusion of elements C_{pg} and C_{pd} in the model destroyed the physically sound bias-dependent behavior of C_{gs} and C_{ds} and made it impossible to consistently converge to a unique solution. The pHEMT model has also been investigated with the addition of resistor R_{gd} in series with C_{gd}. Experiments were performed with such a model, but it was once again not possible to converge to a unique solution for all the model elements. This is not unexpected since R_{gd} will only have a pronounced effect at very high frequencies.

15.3.5.4 Hybrid Decomposition-Based Parameter Extraction

In 2003, van Niekerk et al. [15] proposed a *hybrid multibias analytical/decomposition-based* parameter extraction procedure to improve earlier published results. Additional analytical calculations of the intrinsic model elements as proposed by [16] were implemented into the multibias decomposition-based optimizer to further increase the robustness of the algorithm. Furthermore, a diversity-driven point selection algorithm had been developed which chooses a set of bias points that differ as much as possible from each other.

A robustness test was performed on a lattice-matched 0.2-μm × 100-μm GaAs pHEMT device with measured S-parameters in a frequency range of 1 GHz to 45 GHz. First a S-parameter data set of 13 bias points were used for extraction based both on the multibias method discussed in Section 15.3.5.3 and the new hybrid multibias extraction method, including analytical calculation of the inner model elements. The robustness test consists of repeating a particular extraction 20 times while using randomly chosen optimization.

Table 15.5 represents a summary of the extrinsic elements obtained with the hybrid multibias search algorithms. The table provides the range for each element in which the optimization starting values were chosen, as well as the best value, the mean, and the variation Δ of the extracted extrinsic element values. The best value is defined as the parameter corresponding to the extraction with the lowest modeling error, while the variation Δ is defined as the difference between the maximum and minimum extracted element value, thus indicating the range of solutions in absolute terms. The last column indicates the maximum deviation (Δ/2) from the mean value and relative standard deviation σ_n (%), respectively.

Table 15.5

Exemplary Extracted Results with the Hybrid Search Algorithm

	Starting Interval		Multibias Hybrid Search Algorithm #1Without Cold Bias Points				Multibias Hybrid Search Algorithm #2With Cold Bias Points			
Parameter	Min.	Max.	Best	Mean	Δ	Δ/2 (%)	Best	Mean	Δ	σ_n
L_g (pH)	0.0047	235	63.66	69.86	27.18	19.4	61.66	60.11	437	2.30
L_d (pH)	0.0030	150	50.24	48.68	67.01	68.8	51.54	52.16	4.34	2.71
L_s (pH)	0.0004	20	9.63	11.56	4.11	17.7	10.53	9.15	2.45	7.15
R_g (Ω)	0.0003	17.5	0.70	0.33	4.85	734.8	3.56	3.69	1.78	12.44
R_d (Ω)	0.0006	30	12.2	12.32	3.90	15.8	2.73	2.55	1.00	10.14
R_s (Ω)	0.0002	10	0.261	2.45	1.53	31.2	1.68	1.85	1.11	16.13
C_{pg} (fF)	0.0020	100	20.07	26.97	29.65	54.9	18.81	17.69	5.03	7.72
C_{pd} (fF)	0.0020	100	24.14	23.66	27.48	58.0	17.43	16.85	2.32	5.24

After: [15].

Case #1 denotes extraction results without separate determination of some parasitics from cold pinch-off measurements. It seems that the extension option concerning the analytical variation Δ of the resistances R_g, R_s, and R_d is still large. It is a pity that the relative standard deviation was not provided but the total spreading of extracted results was. Taking the maximum deviation to the mean value; that is, $(\Delta/2)/p_i$), rather large uncertainty values occur. For example, $L_d \approx 6.7\%$, $R_g \approx 17\%$, $R_d \approx 16\%$, $R_s \approx 17.5\%$.

These tests were repeated with the additional inclusion of cold S-parameter data at five bias points with $V_{GS} \leq V_{th}$. Cold pinch-off measurements were used to extract R_d using only measurements made at frequencies above 15 GHz, whereas C_{pg} and C_{pd} as well as inductors and resistors are only determined from data measured at frequencies below 10 GHz. It is noted that the inclusion of the cold S-parameter data has dramatically improved the range of solutions obtained for the intrinsic elements. Indeed, taking additionally into account cold pinch-off bias points, some improvement in the model parameter extraction should be expected. However, also in this case, between hand selected and diversity driven selected active bias points, no really significant progress can be observed, except for giving different weights of value spreading to different model parameters.

Viewing the results, there seems to be no significant difference between the multibias hybrid extraction algorithms with and without [14] analytical determination of the inner model elements.The partially widely spread extracted uncertainties are similar in both cases but change among the extracted model parameters. It is observed that the extracted values of the gate resistor R_g varies significantly with a clear tendency to the lower optimization boundary. In contrast, R_d is less varying. However, the value of R_d is far larger than what is realistically expected from such HEMT device. With proposed diversity driven selection of active bias points the standard deviation of R_s, R_g, and R_d was calculated to be 16.13%, 12.44%, and 10.14%, respectively.

A special remark is necessary regarding the best value, which is defined as the parameter recognized to provide the lowest modeling error. In the author´s own experience, the term "best value" may be misunderstood. As will be discussed in Chapter 19, one should distinguish between a mathematically and physically optimal extraction solution. For the first, optimization is a purely mathematical process to find the best fitting solution for a set of variables to some (simulated/measured) data points. In this respect the optimized variables are only formal fitting parameters without any deeper physical significance. The perspective changes when the variables are identified by model elements of a defined equivalent network that is expected to model the electrical performance of the device as closely as possible. However, as is known, all models are only approximations of the real behavior of the physical device. Thus, in this respect, the "best result" refers, first of all, to the chosen equivalent network topology, which may not be fully matched to the physical characteristics of the device under test. As will be shown in Chapter 19, discussing adaptive random optimization, the minimum residual error function value does not correspond to the best solution with respect to physically acceptable model parameter values. In contrast, this is frequently observed with residual error function values being still less than optimal. Limited device response data can also provide large uncertainty in the extraction of model parameter values.

15.4 SUMMARY

In this chapter decomposition-based model parameter extraction methods were investigated. Based on a parameter sensitivity analysis (empirically or automatically) the model parameters (or a combination of them) were ranked in descending order with respect to their influence on the object function. Optimization was then performed in sequential order, starting with the parameter group of highest sensitivity, followed by less sensitive groups in ranking order. The sensitivity ranking was realized by the determination of the eigenvectors/eigenvalues of the Hessian matrix. With respect to the large spread of condition numbers appearing in least-squares model parameter optimization problems, scaling for equalization was performed. Parameter groups with similar scale could then be optimized in sequential order. It was stated that insensitive model parameters were determined more uniquely. Nevertheless, large variances were found with the least sensitive components, such as the parasitic resistances.

Ongoing research on the decomposition-based method focused on further improvement of the consistency of extracted results. This, in particular, concerned the low resistance parameters. The following extensions were proposed: (1) the data basis for optimization was expanded by consideration of multibias S-parameter sets and (2) intrinsic model parameters were analytically determined.

The concept idea of decomposition breaking down the complex extraction problem into smaller subproblems looks very attractive at first view. However, as has been shown, the practical implementation and application concerning transistor modeling is challenging. Although considerable and notable efforts have been undertaken, the results obtained from the decomposition concept are not really conclusive. Classified as a global optimizer, the large spreading of extracted results indicate a high risk of getting caught in local minima. The question arises whether it is practical to test model parameter optimization algorithms with starting values that are physically by far less relevant.

REFERENCES

[1] Kondoh, H., "An Accurate FET Modelling from Measured S-Parameters," *IEEE MTT-S Int. Microwave Symp. Digest*, 1986, pp. 377–380.

[2] Microwaves101.com, *History of Microwave Software*, https://www. microwaves101. com/encyclopedias/history-of-microwave-software.

[3] Kondoh, H., *private communication*, 1988.

[4] Bandler, J. W., and Q. J. Zhang, "An Automatic Decomposition Technique for Device Modelling and Large Circuit Design," in *IEEE MTT-S Int. Microwave Symp. Dig.*, 1987, pp. 709–712.

[5] Novotny, M., *Investigation on the Consistency of Transistor Equivalent Circuits* (in German), Research Project, University of Kassel, 1990.

[6] Patterson, A. D., V. F. Fusco, J. J. McKeown, and J. A. C. Stewart, "Microwave Device Modelling Using Systematic Optimization Techniques," *IEE Colloquium on Computer Based Tools for Microwave Engineers Dig.*, Oct. 1991, pp. 3/1–3/4.

[7] Patterson, A. D., V. F. Fusco, J. J. McKeown, and J. A. C. Stewart, "A Systematic Optimization Strategy for Microwave Device Modelling," *IEEE Trans.*, Vol. MTT-41, 1993, pp. 395–405.

[8] Press, W. H., S. A. Teukolsky, W. T. Vettering, and B. P. Flannery, *Numerical Recipes*, 3rd Edition, Cambridge University Press, Cambridge, UK, 2007.

[9] Gu, C., and Y. Liu, "A New Systematic Optimization Strategy for Microwave Power Device Modeling," *Microwave and Optical Technology Lett.*, Vol. 27(4), 2000, pp. 269–273.

[10] van Niekerk, C., and P. Meyer, "A New Approach for the Extraction of an FET Equivalent Circuit from Measured S Parameters," *Microwave and Optical Technology Lett.*, Vol. 11(5), 1996, pp. 281–284.

[11] Leong, M. S., P. S. Kooi, T. S. Yeo, and B. L. Ooi, "A Hybrid Gauss-Newton–Simulated Annealing Optimizer for Extraction of MESFET Equivalent Circuit Elements," *Microwave and Optical Technology Lett.*, Vol. 6(8), 1993, pp. 461–466.

[12] van Niekerk, C., and P. Meyer, "Performance and Limitations of Decomposition-Based Parameter-Extraction Procedures for FET Small-Signal Models," *IEEE Trans.*, Vol. MTT-46(11), 1998, pp. 1620–1627.

[13] Fujitsu, *Microwave Semiconductor Databook*, 1993, pp. 282–284.

[14] van Niekerk, C., and P. Meyer, D. M. M.-P. Schreurs, and P. B. Winson, "A Robust Integrated Multibias Parameter-Extraction Method for MESFET and HEMT Models," *IEEE Trans.*, Vol. MTT-48(5), 2000, pp. 777–786.

[15] van Niekerk, C., J. A. du Preez, and D. M. M.-P. Schreurs, "A New Hybrid Multibias Analytical/ Decomposition-Based FET Parameter Extraction Algorithm with Intelligent Bias Point Selection," *IEEE Trans.*, Vol. MTT-51(3), 2003, pp. 893–902.

[16] Lin, F., and G. Kompa, "FET Model Parameter Extraction Based on Optimization with Multiplane Data-Fitting and Bidirectional Search—A New Concept," *IEEE Trans.*, Vol. MTT-42, 1994, pp. 1114–1121.

Chapter 16

Bidirectional Search Method

16.1 INTRODUCTION

The model parameter extraction method discussed in this chapter exhibits some salient features. It incorporates the *bidirectional search*, which is a well-understood area in artificial intelligence (e.g., [1-3]). *Unidirectional search* designates a search method which, after having started in an initial (starting) point, approaches the target (or goal, solution) in the *forward* direction. This procedure can be compared with traditional model parameter extraction algorithms. Based on an error function, which considers all model elements as variables, optimization is performed, for example, by the least-squares method with the goal that measurement and simulation with optimizer-based model parameter values are in preferably accurate matching. Widely used deembedding extraction procedures also fall in the same category. They all have in common that the solution search is basically restricted to the extrinsic reference planes, although the deembedding strategy proceeds in separate forward-directed search steps.

Conversely, the bidirectional search algorithm tackles the solution search of a problem not only in a unidirectional way (forward search) as just described, but simultaneously also in a backward or reverse direction from the solution point. The search stops when the forward and backward search meet in the middle. This search strategy reduces the search space dramatically and it makes the search more efficient and faster.

Regarding model parameter extraction we may divide the entire search space (model equivalent network topology) into the extrinsic and intrinsic transistor model part. This is similar to a *topology decomposition*. With the topology partitioning we obtain two sets of optimization variables. The extrinsic model elements comprise a first set of optimization parameters whereas the intrinsic elements comprise the second set. The objective function is minimized by a bidirectional search. Based on defined starting parameter values, the extrinsic elements are optimized with respect to measured S-parameters; i.e., the optimization is performed with respect to the extrinsic reference planes. In the next step, knowing the extrinsic parameters, intrinsically related S-parameters (Y-parameters, respectively) can be determined in the inner reference planes of the equivalent circuit by applying the deembedding technique. At this point analytical calculation of the inner model elements would provide frequency-dependent solutions because of the initially mismatched extrinsic model elements. Therefore the inner elements are quasi-analytically determined by a simple weighted polynomial curve fitting. With the intrinsic parameters known, the extrinsic elements are updated. The iterative process ends upon falling under a certain limit of the

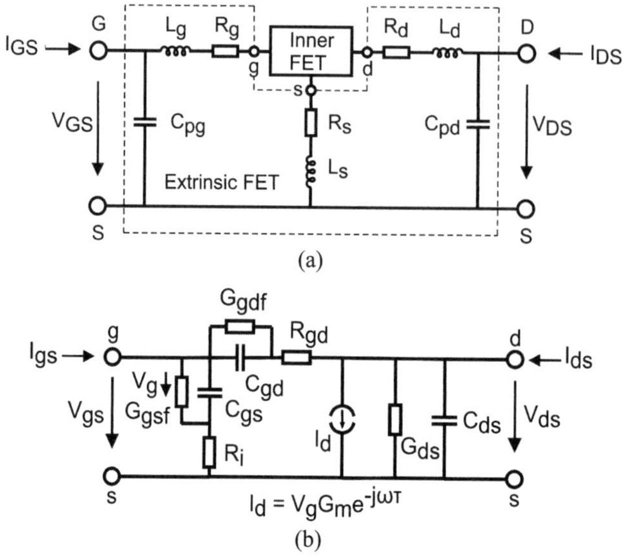

Figure 16.1 FET equivalent network (a) and inner FET topology (b).

error function. Because optimization is simultaneously and iteratively performed both in the extrinsic and intrinsic reference planes the search is more specifically termed *multiplane data fitting and bidirectional search* [4].

The bidirectional search technique has proven a robust and efficient model parameter extraction procedure originally developed for wire-bonded devices in a testfixture [1]. However, the experience has shown that on-wafer measured devices can also be treated likewise highly successfully.

The bidirectional search method has been applied to the FET model shown in Figure 16.1. The method originally proposed by Lin [4] combines two different extraction approaches: optimizer-based and analytical methods. The method has been published in several papers (e.g. [5, 6]).

16.2 BIDIRECTIONAL SEARCH STRATEGY

Traditionally the determination of the FET equivalent circuit elements with an optimization-based approach is carried out by minimizing an error function in such a way that starting from initial values all elements are changed independently and simultaneously by the optimizer until the error function reaches a minimum (see Chapter 14). Only one error criterion is examined for all circuit elements in the extrinsic measurement reference planes G–S and D–S. With respect to the increasing topology complexity of current large-size high-power GaN devices and ever-frequency-increasing operation of HEMT devices, equivalent circuit models commonly comprise a large number of network elements up to 22 circuit elements including gateforward operation (see e.g., [7–10]) for broadband operation.

With an increasing number of optimization variables, local minima are likely to occur

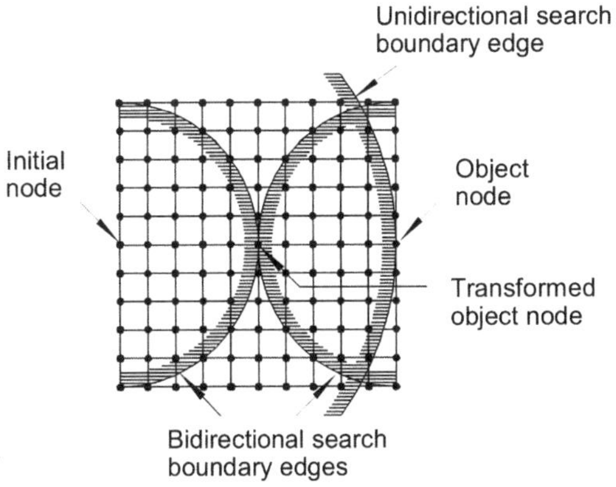

Figure 16.2 Illustration of unidirectional and bidirectional search approaches.

more frequently. This reveals the ultimate problem of optimizer based extraction methods in defining parameter starting vectors that locate the object function value sufficiently near the global minimum (if existing for example, in case of noisy and faulty measurement data). Indeed, this is a great challenge, and the experience shows that the extraction results are commonly dependent on the chosen initial parameter values.

The bidirectional search procedure was originally inspired by the artificial intelligence knowledge treating the efficiency of search strategies [9]. Figure 16.2 shows as an example a simple grid search graph [3, 6]. In grid search, a uniform grid is defined in the parameter space (see Section 12.7.1.1). Then points in each of the grid are evaluated and the global minimum is found in this space (assuming the grid is sufficiently dense).

Grid search is a straightforward and popular parameter search method and is regarded as the most reliable approach. The only problem is its high computational demands. Even moderately high-resolution searches can result in a large number of evaluations and unacceptably long run times. This is symbolized by the very large search space with a large number of nodes (parameter settings) in Figure 16.2.

A special grid search is given when the object node is already known (e.g., transport problems). Then the solution can be approached in different ways. Following the unidirectional search the state space graph is traversed in one direction from the initial to the object node (forward search). An alternative approach is the bidirectional search, where the search is performed simultaneously forward from the start node and backward from the goal node. The search stops when both meet in the middle. Note that the searching space can be reduced significantly with a simultaneous bidirectional search, as shown schematically in Figure 16.2. The search process comprises much less nodes (model parameter settings) than that of a unidirectional search. In addition to the reduction of the search space, bidirectional search establishes a sharp bend of the search boundary close to the object node, which yields an increased possibility in finding the global minimum.

Regarding conventional optimization, approaching to the objective is performed only in one directional search. The common optimization algorithms begin with an initial value vector for all variables and approach to the object of data-fitting (forward search).

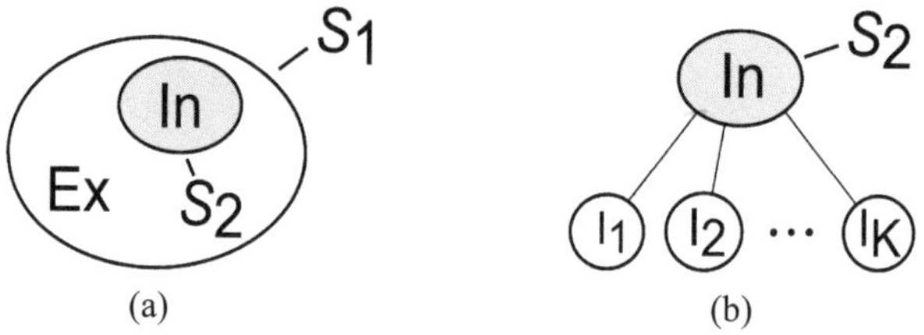

Figure 16.3 Bidirectional search: (a) illustration of data-fitting carried out in two reference planes S_1 and S_2, and (b) decomposition of the intrinsic search space *In* into primitive subproblems I_n ($i = 1 \ldots k$) [6].

With the bidirectional search, an additional internal data-fitting plane (device intrinsic reference planes $g-s$ and $d-s$) is introduced so that the total search space is divided into an outer search space *Ex* and inner search space *In*. The parameter extraction of model elements in the outer region is still based on optimization, however, with a much less number of optimization variables. It will be shown that with respect to the pinch-off condition of the common 15-element model (Figure 14.1) only two elements need to be optimized: the extrinsic capacitances C_{pg} and C_{pg}. As the respective two-dimensional error function contour is well-shaped, the starting value problem can be eliminated. Figure 16.3(a) illustrates the two search regions and indicates the common external data-fitting plane S_1 and the additional internal data-fitting plane S_2.

In contrast to the optimizer-based parameter search in the outer region in the forward direction, the search in the inner region is done in reverse search. In this case the model parameter values are (quasi-) analytically determined from the deembedded data in the inner data-fitting plane. To realize this, a further novel idea was adopted from pattern recognition theory [11], where complex scenes are decomposed into easily comprehensible subproblems (*primitive* subproblems). The complex problem of parameter extraction in the inner search space is decomposed into branch-related subproblems of the pi-network, as is illustrated in Figure 16.3(b).

Bidirectional search is an iterative optimization algorithm. Initial values of extrinsic elements are estimated from cold pinch-off operation. The determination of intrinsic elements (being coefficients of rational functions) uses the least-squares optimization technique, whereas extrinsic elements are optimized with l_1-norm.

16.3 DEFINITION OF OBJECT FUNCTION

Optimizer-based model parameter extraction methods calculate the model element values by minimizing an error function (object function), which contains the difference between measured and modeled scattering parameters. Commonly the object function $\boldsymbol{\varepsilon}$ is defined by the general l_p-norm (see Section 12.5),

$$\boldsymbol{\varepsilon} = \left(\sum_{k=1}^{K} |\varepsilon_k|^p \right)^{1/p} \tag{16.1}$$

The error term ε_k is the weighted difference between the calculated and measured response in the form:

$$\varepsilon_k = w_k[S_{ij}^c(f_k,\vec{p}) - S_{ij}^m(f_k)] \qquad (i,j = 1,2;\ k = 1,2,\dots,K) \tag{16.2}$$

$S_{ij}^m(f_k)$ is the measured response at frequency point k. $S_{ij}^c(\vec{p},f_k)$ is the calculated response from the model with the vector $\vec{p} = [C_{pg}, C_{pd}, L_g, L_d, L_s, R_g, R_d, R_s, \dots]^T$. The weighting factor w_k generally covers two functions. It is defined to eliminate the common denominator of a rational branch response expression with the aim of solving the primitive problems analytically (see Appendix C). It is used furthermore to control the error term ε_k with respect to frequency. Different values of p are used in the inner and outer reference planes of the data-fitting procedure. The objective function of l_2-norm ($p = 2$) is applied to the inner reference plane because of the necessity of calculating the derivatives.The l_1-norm ($p = 1$) is used in the outer reference plane because of its known tolerance of large errors in microwave device modeling [12].

16.3.1 Extrinsic Data Fitting

Regarding FET devices, the characterization is usually based on S-parameter measurements. We use a normalized l_1-norm in the outer measurement reference plane with the objective function

$$\varepsilon = \frac{1}{4K}\sum_{i,j=1}^{2}\sum_{k=1}^{K}\frac{\left|\mathrm{Re}\left\{S_{ij}^m(f_k) - S_{ij}^c(\boldsymbol{p},f_k)\right\}\right| + \left|\mathrm{Im}\left\{S_{ij}^m(f_k) - S_{ij}^c(\boldsymbol{p},f_k)\right\}\right|}{m_{ij}} \qquad (i,j = 1,2) \tag{16.3}$$

with

$$m_{ij} = \max_k\left\{|S_{ij}^m(f_k)|\right\}$$

where $S_{ij}^m(f_k)$ is the measured S-parameter at frequency f_k, $S_{ij}^c(\vec{p}, f_k)$ is the calculated corresponding S-parameter coefficient derived from extracted values of the model parameters, and $\vec{p}$ the vector of model parameters. K is the number of considered frequency points and m_{ij} the largest magnitude of the measured S-parameters S_{ij}^m.

As can be seen, the defined objective function includes the separate calculation of the real and imaginary parts of the scattering parameters in contrast to the commonly used definition, where only the amplitude is considered. Our experience shows that the convergence is faster using this definition. The searching space is reduced and the error function becomes more sensitive.

16.3.2 Intrinsic Data Fitting

Inner branch admittances as a function of the intrinsic Y-parameters read

$$\begin{aligned} Y_{gs} &= Y_{11} + Y_{12}, & Y_{gd} &= -Y_{12} \\ Y_{ds} &= Y_{22} + Y_{12}, & Y_{gm} &= Y_{21} - Y_{12} \end{aligned} \tag{16.4}$$

The branch admittances comprise the following branch model elements (see Figure 16.1):

$$\begin{aligned} Y_{gs}^{-1} &= R_{gs} + (G_{gsf} + j\omega C_{gs})^{-1}, & Y_{gd}^{-1} &= R_{gd} + (G_{gdf} + j\omega C_{gd})^{-1} \\ Y_{ds} &= G_{ds} + j\omega C_{ds}, & Y_{gm} &= G_m e^{-j\omega\tau}/[1 + R_{gs}(G_{gs} + j\omega C_{gs})] \end{aligned} \tag{16.5}$$

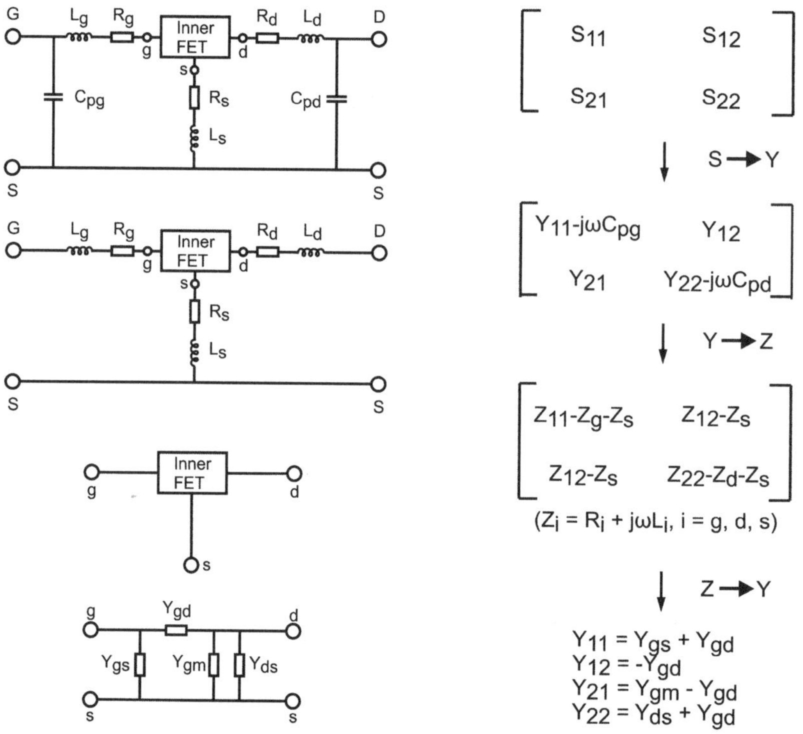

Figure 16.4 Deembedding of the 15-element model to obtain the topology of the inner FET.

The intrinsic Y-parameters are obtained by deembedding, as illustrated in Figure 16.4. The S-parameters are characteristic data of a device measured in a selected bias point within a defined frequency range. After converting S-matrix to Y-parameters, the pad capacitances are subtracted from Y-matrix since the pad capacitances C_{pg} and C_{pd} are connected in parallel. The pad capacitances matrix Y can be expressed as

$$Y_p = \begin{bmatrix} j\omega C_{pg} & 0 \\ 0 & j\omega C_{pd} \end{bmatrix} \tag{16.6}$$

The obtained Y-parameter matrix $(Y - Y_p)$ is then converted into a Z-parameter matrix where the Z-matrix is the inverse of the Y-matrix,

$$Z = (Y - Y_p)^{-1} \tag{16.7}$$

Since the parasitic resistances and inductances are connected in series with the intrinsic device, the parasitic elements matrix Z_{ext} is subtracted from Z resulting in the intrinsic Z-parameter matrix Z_{int}. Z_{ext} denotes the Z-parameters of the parasitic elements, which can be expressed as

$$Z_{ext} = \begin{bmatrix} Z_g + Z_s & Z_s \\ Z_s & Z_d + Z_s \end{bmatrix} = \begin{bmatrix} (R_g + R_s) + j\omega(L_g + L_s) & R_s + j\omega L_s \\ R_s + j\omega L_s & (R_d + R_s) + j\omega(L_d + L_s) \end{bmatrix} \tag{16.8}$$

The last step in calculating Y-parameters of the intrinsic device Y_{int} is to convert Z_{int} to Y_{int},

$$Y_{int} = Z_{int}^{-1} \tag{16.9}$$

Regarding the deembedding process, it is assumed that all the parasitic elements and pad capacitances are known either from other mathematical processes at different bias points or from optimized parameters. Equation (16.9) provides the deembedded Y-parameters from the measured S-parameters as illustrated in Figure 16.4.

16.4 BIDIRECTIONAL SEARCH OPERATION

As an example, the extracted results for a carrier-mounted 0.5-μm MESFET of type FSX 03X from Fujitsu Ltd. are presented and discussed. In-fixture measured S-parameters from two different operating bias points have been chosen: *Data Set A* with the FET biased in the saturation region ($V_{GS} = -0.2$V, $V_{DS} = 3$V, $I_D = 30$ mA), and *Data Set B* with the FET biased in the passive (cold) pinch-off region ($V_{GS} = -7$V, $V_{DS} = 0$V, $I_D = 0$ mA). The S-parameters were measured in the frequency range from 1 to 40 GHz in 1 GHz steps.

The small-signal FET equivalent circuit model considered is the widely used 15-element model shown in Figure 16.5.

It is well known, that there is no unique solution [13] if all elements are assumed as variables and only the measured extrinsic terminal S-parameters are attempted to be matched. Different element sets may result in a comparable quality of data-fitting; in particular the resistances R_g and R_i cannot be clearly distinguished. The results depend heavily on the starting condition. If the data-fitting is performed by the proposed technique with respect to both the external and the internal reference planes (Figure 16.5), this uncertainty can be eliminated.

In this example, only eight extrinsic parasitic elements are assigned as ordinary optimization variables in a forward search. All intrinsic elements (variables in a reverse search) are quasi-analytically calculated incorporating a least squares data-fitting formulation in this way: Measured terminal S-parameters are first de-embedded with respect to the extrinsic elements yielding the Y-parameters of the intrinsic FET; branch admittances of the intrinsic pi structure are then obtained and fitted to each branch element by means of l_2 data-

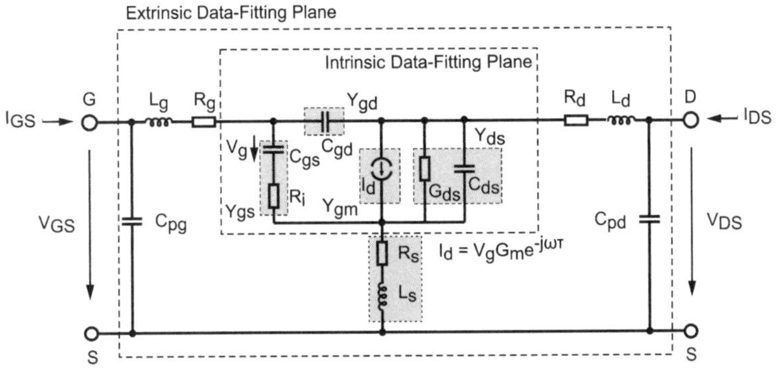

Figure 16.5 Common 15-element FET model with outer and inner reference planes and intrinsic branch admittances.

fitting with a reasonably chosen frequency-dependent weighting factor. They read (see Appendix C):

$$R_i = \frac{\sum_k \mathrm{Re}(Y^m_{gs,k})}{\sum_k |Y^m_{gs,k}|^2}, \qquad G_{ds} = \frac{\sum_k \mathrm{Re}(Y^m_{ds,k})/f_k}{\sum_k 1/f_k} \tag{16.10}$$

$$C_{ds} = \frac{1}{2\pi}\frac{\sum_k \mathrm{Im}(Y^m_{ds,k})}{\sum_k f_k}, \qquad G_m = \frac{\sum_k |Y^m_{gm,k}(1 + j2\pi f_k C_{gs} R_i)|/f_k}{\sum_k 1/f_k} \tag{16.11}$$

$$C_{gl} = \frac{1}{2\pi}\frac{\sum_k \left(|Y^m_{gl,k}|/f_k\right)^2}{\sum_k \mathrm{Im}(Y^m_{gl,k})/f_k}, \quad (l = s, d) \tag{16.12}$$

The summations are carried out over all K supporting frequency points f_k (e.g. 1 – 40 GHz).

16.4.1 Starting Vector Analysis at Pinch-Off

The problem of a starting value vector can be easily eliminated with the bidirectional search technique. In our example only initial values for the extrinsic elements are required for the optimization procedure. They can be generated, for example, by using the measured data of a FET operating in a cold pinch-off bias point ($V_{DS} = 0$, $V_{GS} < -V_p$) [14]. In this case the intrinsic FET model can be simplified to a pi structure of only three capacitances, which can then be transformed into a T structure (Figure 16.6). Together with the gate, drain, and source parasitic resistive and inductive model elements, simple series R-L-C branches are established that can be analytically calculated in terms of their branch impedance (see Appendix C):

$$R_l = Z_0\frac{x_2}{x_1}, \quad L_l = \frac{Z_0}{2\pi f_0}\frac{x_3}{x_1}, \quad C_l = \frac{1}{2\pi f_0 Z_0}x_1 \quad (l = g,d,s) \tag{16.13}$$

where x_1, x_2, and x_3 are the solution to the following matrix equation:

$$\begin{bmatrix} \sum_k \tilde{f}_k^2|z_k|^2 & -\sum_k \tilde{f}_k^2\mathrm{Re}(z_k) & -\sum_k \tilde{f}_k^3\mathrm{Im}(z_k) \\ -\sum_k \tilde{f}_k^2\mathrm{Re}(z_k) & \sum_k \tilde{f}_k^2 & 0 \\ -\sum_k \tilde{f}_k^3\mathrm{Im}(z_k) & 0 & \sum_k \tilde{f}_k^4 \end{bmatrix} \begin{bmatrix} x_1 \\ x_2 \\ x_3 \end{bmatrix} = \begin{bmatrix} -\sum_k \tilde{f}_k\mathrm{Im}(z_k) \\ 0 \\ \sum_k \tilde{f}_k^2 \end{bmatrix} \tag{16.14}$$

with

$$\tilde{f}_k = f_k/f_0, \quad z_k = Z^m_{l,k}/Z_0, \quad (l = g,d,s)$$

and f_0 and Z_0 being the normalization frequency and impedance.

The topology partitioning in an extrinsic and intrinsic region is done in such a way that only the parasitic capacitances C_{pg} and C_{pd} appear as extrinsic optimization variables, as shown in Figure 16.6.

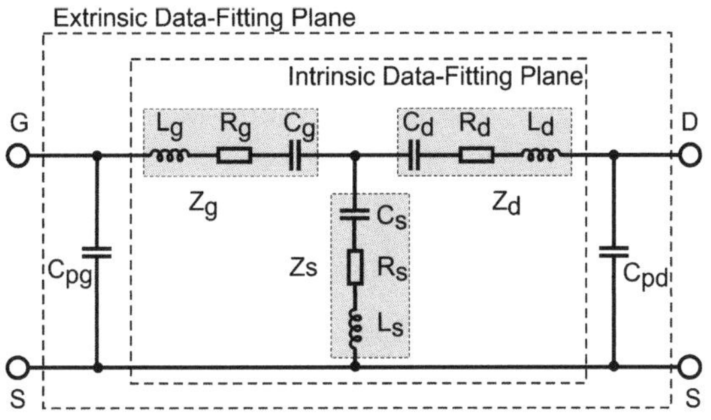

Figure16.6 FET model under cold pinch-off condition used for generating starting values of extrinsic elements.

16.4.2 Default Starting Vector

As the extrinsic model elements, including the series resistances R_g, R_i, R_s and R_d, should be bias independent, the starting values for the model parameter extraction can be obtained from the measured scattering parameters in the pinch-off region. In this case the intrinsic gate resistance R_i, the transconductance G_m, and the channel conductance G_{ds} can be neglected so that the equivalent circuit simplifies. The topology partitioning in an outer and inner region is done in such a way that only the parasitic capacitances C_{pg} and C_{pd} appear as extrinsic optimization variables.

The local minima problem was systematically analyzed starting the optimization process with different combinations of C_{pg} and C_{pd} values. The result is depicted in Figure 16.7. As shown, a global minimum of the error function can be found for a large search radius around the coordinate zero point. For larger capacitance values, terraced planes with increased residual error appear, indicating convergence areas of local minima.

Figure 16.8 presents an enlarged excerpt from the region around the coordinate zero point

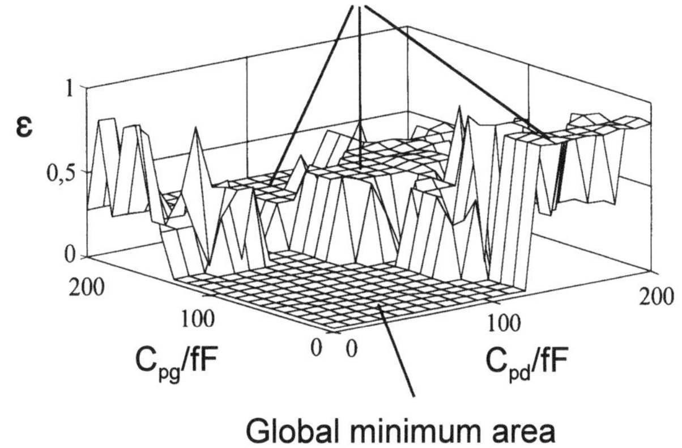

Figure 16.7 Visualization of the local minimum problem in the pinch-off region of in-fixture measured HEMT JS8902-AS at bias point $V_{GS} = -4.0V$, $V_{DS} = 0V$.

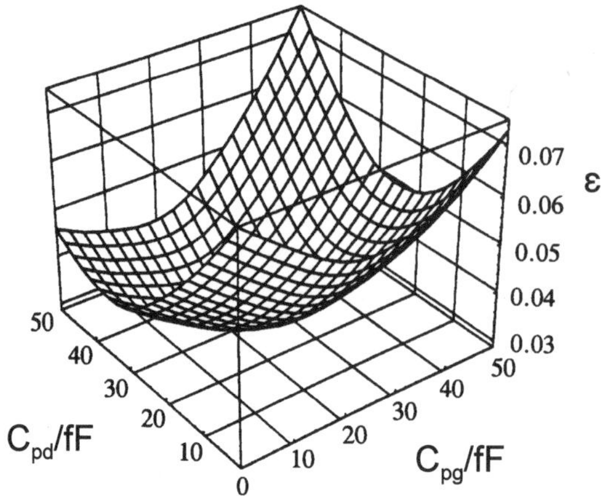

Figure 16.8 Error function versus C_{pg} and C_{pd} of the investigated MESFET (FSX 03X) model under pinch-off with single local (i.e., global) minimum (sectional view of Figure 16.6).

Table 16.1
Initial Values of Extrinsic Elements Obtained from Cold FET

C_{pg}/fF	C_{pd}/fF	L_g/pH	L_d/pH	L_s/pH	R_g/Ω	R_d/Ω	R_s/Ω
31.9	25.9	193.8	199.6	31.0	2.18	1.26	0.56

indicating a single (global) minimum. As the capacitance values lie in the order of 20 fF for wire-bonded FET chips, the starting values for both C_{pg} and C_{pd} can be set to zero. It follows that with C_{pg} and C_{pd} precisely determined, series inductances and resistances can also be found with high quality. Regarding on-wafer measurements, the pad capacitances are commonly less than 10 fF so that the minimum error function is less pronounced.

Table 16.1 shows the initial values for the extrinsic elements obtained from the cold pinch-off model.

16.5 MULTIBIAS EXTRACTION

In the bidirectional search method the device topology is partitioned in an outer and inner region (Figure 16.3). Thereby the number of optimization variables is reduced so that only the extrinsic bias-independent model parameters are varied during the optimization process. The frequency-independent but bias-dependent intrinsic model elements are (quasi-) analytically determined with the method of least squares. A particular property of the method is that the number of optimization variables is not increased when optimization is performed simulataneously in several bias points as indicated in Figure 16.9.

The multibias model parameter extraction runs as follows: Suppose, Ex_0 is the starting vector for the extrinsic elements. Then it follows from the S-parameter data set S_1^m an intrinsic Y-matrix after deembedding the extrinsic elements, from which the intrinsic elements in the given bias point can be analytically determined. Using the extracted element values, the

$$
\begin{array}{lccccc}
S_1^m: & Ex_0 & \rightarrow & In_1^{(1)} & \rightarrow & S_1^c \\
S_2^m: & Ex_0 & \rightarrow & In_1^{(2)} & \rightarrow & S_2^c \\
 & & & \vdots & & \\
S_N^m: & Ex_0 & \rightarrow & In_1^{(N)} & \rightarrow & S_N^c
\end{array}
$$

$$
Ex_1: \quad \frac{1}{N}\sum_{\mu=1}^{N} \left| S_\mu^m - S_\mu^c \right|^2 = \min
$$

Figure 16.9 Simultaneous optimization in a multiple of N bias points (Ex_0: extrinsic starting value vector).

simulated scattering parameters S_1^c can be obtained. In a similar way, the data set in a further bias point is treated. Then an improved extrinsic parameter vector Ex_1 is obtained by minimizing the deviation of the simulated and respective measured S-parameters in all regarded bias points.

In Table 16.2 [19] extracted results of extrinsic model elements of a FET device derived from various extraction methods are compared; that is, a bidirectional search procedure after Lin [6], a direct extraction method after Dambrine et al. [15], and an analytical method after Yanagawa et al. [16] (see Chapter 18). The results based on Lin′s method were obtained from an iteratively optimized extraction process based on the complete extrinsic equivalent circuit and using both cold pinch-off and gate-forward biases. Following the classical method after Dambrine, the parasitic inductances and resistances are extracted from gate-forward measurements, whereas the capacitances are determined in the pinch-off operation region. This extraction algorithm overestimates the resistances because part of the channel resistance occurring during gate-forward operation is included. Similarly, the pad capacitance C_{pd} is overestimated because the drain-source capacitance C_{ds} is not considered [17]. The extraction after Yanagawa et al. delivers negative source inductance. The pad capacitances are not considered because of the minor impact on the device response.

Table 16.2

Comparison of Extracted Results of Extrinsic Model Parameters Using Various Extraction Methods

	Lin		Dambrine	Yanagawa
	Starting Values	Multibias Results		
Inductances	Pinch-Off	Multiple Bias	Forward-Bias	Pinch-Off
L_g	97 pH	102 pH	101 pH	72 pH
L_s	0.5 pH	7.0 pH	4.5 pH	-9.2 pH
L_d	82 pH	88 pH	102 pH	61 pH
Resistances	Pinch-Off	Multiple Bias	Forward-Bias	Pinch-Off
R_g	2.5Ω	1.67Ω	1.4Ω	2.1Ω
R_s	0.8Ω	1.99Ω	4.7Ω	0.53Ω
R_d	2.6Ω	2.06Ω	5.2Ω	2.5Ω
Capacitances	Pinch-Off	Multiple Bias	Pinch-Off	
C_{pg}	38 fF	44 fF	88 fF	Omitted
C_{pd}	29 fF	40 fF	95 fF	

From: [18].

Table 16.3 represents an example of model parameter extraction based on the bidirectional search method using single and multiple bias points. The extraction results of mean values P_0 and standard deviations σ using only one set of measured hot S-parameters in a normal bias point are compared with an extraction based on multibias data-fitting. In the latter case an additional cold S-parameter set (pinch-off bias point) has been included.

A comparison has been performed by carrying out 100 individual extractions with randomly generated starting value vectors. One hundred starting vectors have been varied within ±50% on the basis of the initial values in Table 16.1 for the inductances and capacitances, and within ±100% for the resistances. These starting vectors are then optimized using the modified simplex algorithm (see Section 12.7.5). The following conclusions can be drawn from Table 16.3:

(1) Combined hot and cold parameter data set extraction (two-bias data-fitting) shows the smallest deviation for all elements. Uncertainty is below the termination condition (0.5%) of the modified simplex algorithm.

(2) Applying a bidirectional search technique to only one hot data set, extraction yields rather satisfying results for most of the model elements. However, that is not true for the resistances R_g, R_d, and R_s. This is clearly improved with a second bias point at pinch-off.

(3) The calculated residual error of multibias extraction is larger than the other because the measured cold S-parameters have been additionally taken into account for the multi-bias data-fitting process.

Table 16.3

Mean Value P_0 and Standard Deviation σ of Extracted Model Parameters in the Saturation Region*

Circuit element	One-Bias Data-Fitting		Two-Bias Data-Fitting	
	P_0	σ	P_0	σ
C_{pg} / fF	38.6	.038	40.0	.020
C_{pd} / fF	32.6	.178	29.8	.011
L_g / pH	206.8	.122	202.2	.065
L_d / pH	202.8	.417	197.7	.008
L_s / pH	38.4	.109	36.8	.009
R_g / Ω	0.30	.222	0.72	.005
R_d / Ω	.002	.004	0.99	.002
R_s / Ω	1.21	.124	0.92	.002
C_{gs} / fF	344.3	2.21	346.2	.127
C_{gd} / fF	32.5	.168	34.1	.014
C_{ds} / fF	77.4	.380	78.1	.006
G_{ds} / mS	3.777	.020	3.613	.001
G_m / mS	47.51	.316	47.72	.008
τ / ps	2.71	.007	2.71	.001
R_i / Ω	6.84	.339	6.49	.010
Residual error*	0.031		0.041	

* 0.5-μm Gate MESFET FSX 03X.

16.5.1 Three-Bias Measurement and Simulation

In Figure 16.9 and Figure 16.10, measured and simulated scattering parameters of in-fixture and on-wafer measured FET devices are plotted in the Smith chart over a frequency range of 1 to 40 GHz. The model parameter extraction was performed with the bidirectional search method using three bias points simultaneously, and those in the saturation, pinch-off, and ohmic regions. In all three cases an excellent agreement of measured and simulated data could be observed.

A GaAs MESFET of type MGFC1423 manufactured by Mitsubishi was in-fixture measured with the THLR measurement method described in [19]. The chosen bias points are ohmic region (V_{GS} = -0.1V, V_{DS} = 0.2V), pinch-off region (V_{GS} = -6.0V, V_{DS} = 0.0V), saturation region (V_{GS} = -0.5V, V_{DS} = 3.0V). Figure 16.9 shows measured and simulated scattering parameters in the Smith chart.

A further example of an on-wafer measured InGaAs HEMT (0.25 μm × 100 μm), manufactured by IAF, Freiburg, was measured and modeled using the bidirectional search and multibias parameter extraction. The chosen bias points are given as ohmic region: V_{GS} = -0.1V, V_{DS} = 0.2V; pinch-off region: V_{GS} = -1.5V, V_{DS} = 0.0V; saturation region: V_{GS} = -0.1V, V_{DS} = 3.0V.

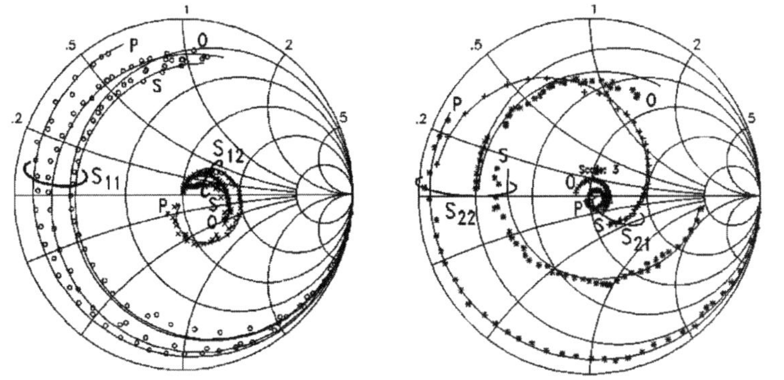

Figure 16.9 In-fixture measured and simulated S-parameters of a GaAs MESFET MGFC 1423 in the saturation (S), cold pinch-off (P), and ohmic (O) region in the frequency range of 1-40 GHz.

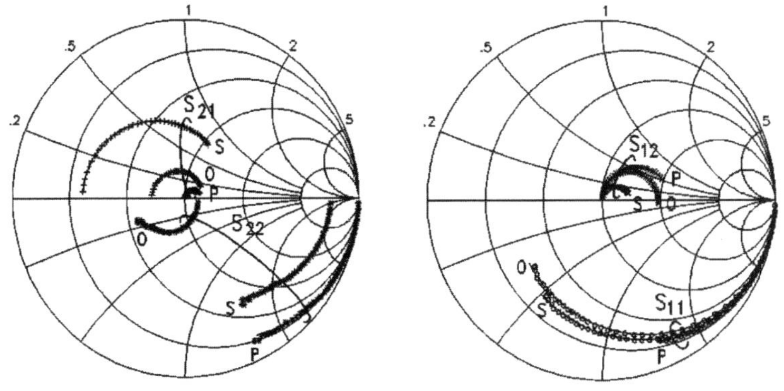

Figure 16.10 On-wafer measured and simulated S-parameters of a InGaAs HEMT (IAF) in the saturation (S), cold pinch-off (P) and ohmic (O) region in the frequency range of 1-40 GHz.

16.5.2 Extracted Model Parameters

The following diagrams in Figure 16.11 show typical 3D plots of the extracted bias-point dependent intrinsic model parameters of the InGaAs HEMT based on S-parameter sets of the considered three bias points. Note that each mesh point was separately extracted. The surfaces show very smooth curvature. A smoothing algorithm such as spline-interpolation was not used. The extracted bias-dependent model elements show very smooth three-dimensional contours and are physically consistent, which is briefly discussed below.

Transconductance G_m

As discussed in Section 6.3, the sheet channel concentration under the gate n_{2D} is proportionally dependent on the gate-source voltage V_{gs} and approximately on the barrier thickness d_A. The transconductance G_m is defined as the differential quotient of drain-source current I_{ds} over gate-source voltage V_{gs}. In the linear (ohmic) region the drain current is proportional to the drain-source voltage V_{ds} and charge density n_{2D}. The latter is a function of gate-source voltage. Thus, the drain current reads $I_{ds} \sim n_{2D}(V_{gs}) \cdot V_{ds}$. Differentiation with respect to V_{gs} delivers simply $I_{ds} \sim V_{ds}$; that is, the drain current increase linearly in the ohmic region with increasing drain-source voltage.

In the saturation region, the carrier velocity has become saturated and thereby the current is independent on the drain voltage; the drain current is then only dependent on the channel carrier density under control of the gate voltage, $I_{ds} \sim n_{2D}(V_{gs})$. Because of the linear relation between sheet carrier density and the gate-source voltage, $n_{2D} \sim V_{gs}$, the transconductance is constant in the saturation region.

Regarding the variation of G_m with the gate-source voltage, $G_m \approx 0$ at gate-source pinch-off voltage and below. Then, with increasing V_{gs}, the sheet carrier density n_{2D} increases strongly and thus, also the drain current. This implies a growing transconductance. However, as is known, the 2DEG density then saturates so that the transconductance becomes a maximum. Further increase of V_{gs} causes bypass conduction in the supply layer so that the mean carrier velocity drastically reduces and thereby the transconductance G_m starts to decrease.

Channel Conductance G_{ds}

The channel conductance G_{ds}, also called the output conductance, describes the variation of the drain current I_{ds} with the drain voltage V_{ds}. In the saturation region the drain current is expected to be nearly constant with growing drain voltage and therefore G_{ds} should be very small.

However, the drain current in practical devices consists of both the current through the active channel and the parallel currents in the substrate and supply layers, which generally increase with the applied drain voltage. Hence, G_{ds} should increase slightly with the drain voltage even in saturation.

With regard to the variation with the gate voltage V_{gs}, the increase in G_{ds} can be explained in terms of increasing channel current since the charge density in the 2DEG channel increases with the gate voltage at a given drain voltage. The increase is particularly

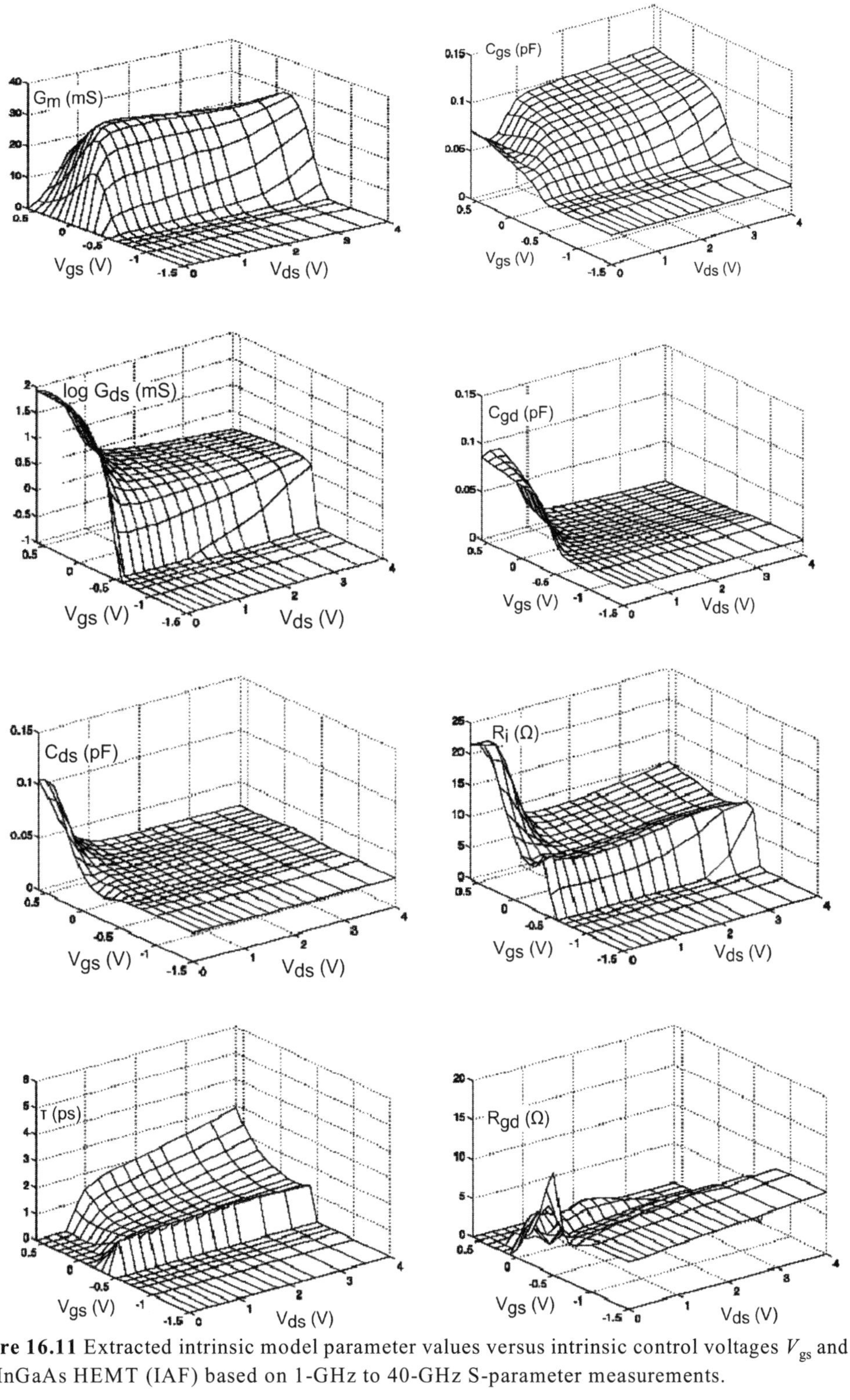

Figure 16.11 Extracted intrinsic model parameter values versus intrinsic control voltages V_{gs} and V_{ds} of a InGaAs HEMT (IAF) based on 1-GHz to 40-GHz S-parameter measurements.

remarkable at low drain voltages due to the reduced depletion of the channel by the drain voltage.

G_{ds} is expected to present its lowest value outside the ohmic region; there, it is often considered constant with respect to V_{ds}. In the ohmic region, G_{ds} is expected to increase quickly with increasing V_{gs} and decreasing V_{ds} values.

Gate-Source Capacitance C_{gs}

The gate-source capacitance C_{gs} appears as a parallel-plate capacitance formed by the gate metal and the 2DEG channel charge as a movable capacitor plate. The effective separation of this capacitor is determined by the depletion layer depth under the gate.

At pinch-off and below the gate-source capacitance is zero with $n_{2D} \approx 0$. When the gate-source voltage rises in the saturation region, the 2DEG charge density grows rapidly and thereby the capacitance. As the 2DEG carrier density saturates with further increase of V_{gs}, C_{gs} increase will slow down. With the formation of a 3DEG due to electrons in the supply layer at higher gate-source voltages, a renewed moderate rise of C_{gs} is observed.

Furthermore, C_{gs} slightly increases as the drain voltage V_{ds} increases due to the lateral electric field, which accelerates charge carriers in the channel to scatter into the barrier layer. This results in a reduction of the depletion layer depth, and therefore gradual increase for C_{gs} with increasing the drain voltage.

Gate-Drain Capacitance C_{gd}

The gate-drain capacitance C_{gd} represents charge storage effects of the gate-drain region between the gate and the 2DEG as was similarly discussed with the gate-source capacitance. At pinch-off, the charge storage in the gate-drain region is negligibly small, and thereby the value of C_{gd} is low. However, C_{gd} increases rapidly with increasing gate voltage under low drain voltage conditions. Under these conditions the depletion region is diminished and uniformly distributed under the gate, so that the displacement current equally divides roughly between the source and the drain. Therefore, the gate capacitance can be split into two approximately equal capacitances, C_{gd} and C_{gs}.

The variation with the drain voltage is somewhat gradual. Physically, C_{gd} is formed by the extension of the high field region of the channel into the gate-drain space. This extension increases with increasing drain voltage so that C_{gd} can be expected to become smaller as the drain voltage increases. However, in recessed-gate transistor structures, the extension of the depletion layer is reduced due to the nondepleted cap layer. Hence C_{gd} variation can be expected to be gradual.

Drain-Source Capacitance C_{ds}

The drain-source capacitance C_{ds} represents the geometrical capacitance between drain and source. With $V_{ds} = 0$V, V_{gs} supports or suppresses the formation of the 2DEG. Near the pinch-off region and with more negative V_{gs}, the 2DEG formation does not occur, setting the maximum electrostatic separation between drain and source and therefore small values of C_{ds}. C_{ds} value increases with increasing V_{gs} and thereby allows 2DEG formation.

Large V_{ds} values produce high electric fields along the channel that diminish the geometrical capacitance between drain and source, especially with V_{gs} near the pinch-off region or at more negative values. Thus, C_{ds} is expected to show a decrease with respect to the drain voltage C_{ds}, particularly in the pinch-off region.

Charging Gate Resistance R_i
The charging gate resistance R_i, also called intrinsic channel resistance, is series connected with the gate-source capacitance C_{gs}. As discussed in Section 6.2.4, the resistance value is inversely proportional to the 2DEG density. That means that R_i tends to become infinite with $n_{2D} \rightarrow 0$ at pinch-off. Indeed, viewing the 3D plot of R_i in Figure 16.11a slight increase is recognizable toward pinch-off region.

R_i models the undepleted part (low field region) of the channel under the gate, through which the gate-source capacitance is charged. The value of R_i is equal to the ratio of the potential drop in this channel part and the channel current, which is approximately equal to the drain current. Therefore, it is expected that the value of R_i should be high in the low drain current region.

Charging Resistance R_{gd}
The charging resistance R_{gd} is included in the model mainly to simulate the symmetrical distribution of the depletion region under the gate in the ohmic-forward region ($V_{gs} > 0V$). It is expected that R_{gd} should have a similar behavior as R_i in this region. Interpreted generously, with R_{gd} a strong increase of the resistance value is also observed in a similar way as described for R_i.

Transit Time τ
The transit time is a measure of the channel electron to pass the depletion region under the gate electrode. In the case of HEMT devices, τ represents the time delay due to the drift of electrons along the 2DEG.

The delay time τ is strongly dependent on the control voltages V_{gs} and V_{ds}. In the pinch-off region the magnitude of the complex transconductance $\underline{G}_m = G_m \cdot \exp(-j\omega\tau)$ is very small so that the phase of $\underline{G}_m$, and thus τ, is not well defined. Therefore, when G_m drops below a certain level (e.g., 1 mS), then τ may be set to zero in the graphic representation.

τ reaches a minimum at $V_{gs} = 0V$, where the transconductance G_m takes a maximum, and increases strongly toward the gate-forward direction. The latter effect originates in the formation of a parasitic MESFET structure in HEMT devices at sufficiently large positive gate voltages (see Section 6.3.1), at which additional current flow occurs in the charge supply layer. As the carrier velocity is much lower in the bypass channel, the effective delay time τ is strongly increased.

In the ohmic region, again the transconductance is very small, and therefore, the parameter τ cannot be extracted consistently.

16.5.3 Confidence Limits on Extracted Parameters

The uncertainty of extracted results has been analyzed as function of the overall measurement inaccuracy. The reliability of extracted model elements is estimated by calculating the confidence limits on the simulated results (Table 16.3) on the basis of Monte Carlo

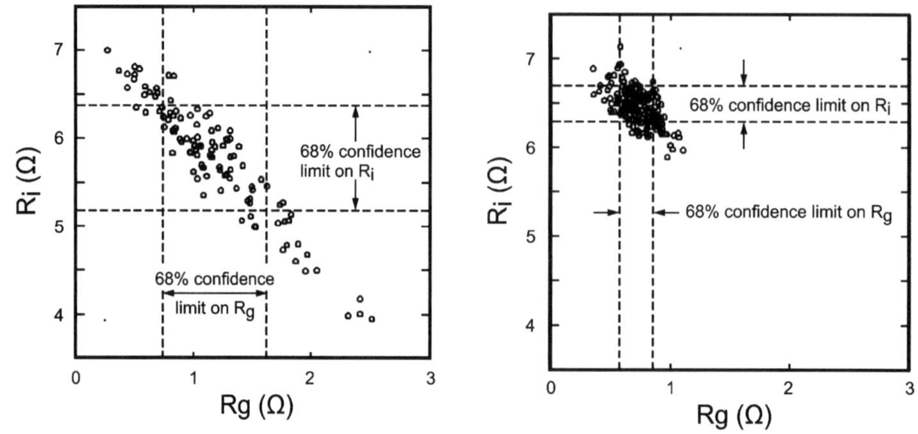

(a) (b)

Figure 16.12 Confidence limits on the model elements R_g and R_i calculated on the basis of 200 Monte Carlo simulations using (a) only one bias data set, and using (b) two bias data sets [4–6].

Table 16.4
Measurement Uncertainty of In-Fixture Measured S-Parameters*

f / GHz	ΔS	ΔS / dB	$\angle S$ /degree
0.04–28.0	0.02	0.087	1.2
28.0–30.0	0.04	0.177	4.0
30.0–40.0	0.06	0.268	5.0

* From: [4, 21, 22].

simulations [20] with 200 synthetic data sets. Such synthetic data sets are computer-generated experiments derived from a set of model parameters that have been well fitted to real measured data in a given operating bias point.

The following expressions were used to simulate measurement uncertainties taking into account actual S-parameter measurement errors according to Table 16.4:

$$\begin{aligned} |S_{ij}^{s}(f_k)| &= |S_{ij}^{c}(\vec{p},f_k)|[1+(2\alpha-1)\Delta S(f_k)] \\ \angle S_{ij}^{s}(f_k) &= \angle S_{ij}^{c}(\vec{p},f_k)+(2\beta-1)\Delta S_p(f_k) \quad (i,j=1,2) \end{aligned} \tag{16.15}$$

where S_{ij}^{c} $(i, j = 1, 2)$ are the calculated S-parameters with the fitted parameter vector $\vec{p}$ of values given in Table 16.3. α and β are equally distributed pseudorandom numbers ($0 \le \alpha, \beta \le 1$). ΔS and $\angle S$ are estimated overall measurement uncertainties with the values given in Table 16.4 and interpolation for other frequencies.

From the obtained simulation results [4] it can be followed that the elements R_i and R_g can not be reliably extracted anyhow using only one hot data set, as seen in Figure 16.12(a). Figure 16.12(b) indicates that a unique and reliable model parameter extraction for the extrinsic bias independent elements, under consideration of measurement uncertainties, is ensured with a multibias data-fitting procedure.

16.6 REPEATIBILITY AND REPRODUCIBILITY CONFIRMATION TEST

In his doctoral thesis. Khalaf studied the bidirectional search method after Lin. The method is well described and tested on the basis of self-written programming codes [23]. His main conclusions are as follows:

Using hypothetical (ideal) S-parameters the simulation results show good agreement between extracted and predefined model parameter values. With recommended initial zero values of the pad capacitances C_{pg} and C_{pd} the solution set for ideal data was typically identical to the correct solution.

However, introducing noise, he observed a serious shift of the minimum point. In this respect it is noted that the author used the standard simplex algorithm, denoted as FMIN() function in MATLAB [24]. In Figure 16.13 the principal characteristic of an ideal error function $\varepsilon(\boldsymbol{p})$ with the model parameter vector $\boldsymbol{p}$ showing a clear minimum is plotted. In addition this curve is disturbed by noise. Note that in the ideal case a local optimizer will certainly find the single (global) minimum. However, this is by far not the case for the noisy function. The local search has now mutated to a global search task with numerous noisy local minima. Thus, it is not surprising that the search result may be different in both cases. Repeatedly restarting the simplex after a few optimization steps should eliminate this issue (see leaping simplex in Section 12.7.5). The author himself stated that the minimum shift may result from the noise "leading to multiple solutions depending on the noise."

It is clear that the extracted model parameter values from ideal and noisy S-parameter data will be different. The author expressed the deviations in percent and critisized the scale of the less accurately determined elements (i.e., $\Delta R_d \approx -8\%$ and $\Delta R_i \approx -30\%$). In absolute values this does not sound very dramatic. The extraction delivered for the ideal and noisy S-parameters 4.95Ω and 5.3Ω, respectively, for R_d, and 0.52Ω and 0.65Ω, respectively, for R_i.

A further test confirmed that the starting values should be taken from the passive pinch-off bias point. Having started the optimization with zero initial values for the model elements delivered wrong results.

The author also tested the error function value around the obtained solution to discover some points at which the error function value is less than its value at the obtained solution. This would indicate that the obtained solution is a local minimum and not a global minimum.

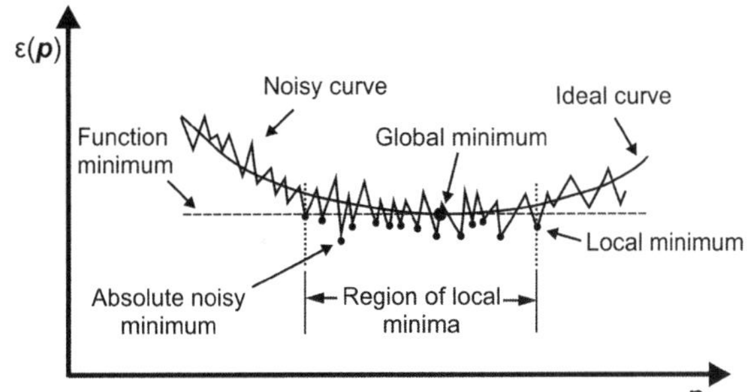

Figure 16.13 Illustration of ideal and noisy error function near the global minimum. Referencing to the error function minimum indicated by the dashed line, the noisy error function may provide a significant extraction uncertainty region with numerous local minima.

The author states that the philosophy of Lin's method is to find an initial set of values for the optimization parameters as close as possible to the correct solution, and then the optimizer should look for the correct solution in the neighborhood of the initial values set. The correct solution supposedly lies at a local minimum. Furthermore, the author points out that this observation was confirmed by also applying Lin's method to actual (measured) device S-parameter data.

It is concluded that it is important for this method to work with relatively broadband measurements (1–40 GHz) for accurate results. It also needs initial values as close as possible to the expected minimum.

The technique shows good results to avoid the local minimum problem and to desensitize the extracted values to the measurement errors.

16.7 SUMMARY

The bidirectional search method was described and illuminated. Model parameter extraction procedure was exemplified by both in-fixture and on-wafer measured FET devices. The main features of the bidirectional search method are found to be:

- Partitioning of the device topology in an extrinsic and intrinsic region (model partitioning).
- Optimization variables are only the extrinsic bias-independent model parameters.
- No starting value problem exists. A self-consistent generation of starting values is obtained from cold pinch-off.
- Analytical determination of the intrinsic frequency-independent but bias-dependent model elements using least-squares method.
- No increase of the number of optimization variables when optimization includes several bias points simultaneously.
- Partitioning of the intrinsic FET topology in *primitive* pi branches improves the fitting to the *Y*-parameters and, respectively, to the S-parameters.

REFERENCES

[1] Sturtevant, N. R., and A. Felner, "A Brief History and Recent Achievements in Bidirectional Search," *The Thirty AAAI Conference on Artificial Intelligence (AAAI-18)*, 2018, pp. 8000–8006, https://aaai.org> AAAT>AAAI18>paper>download.

[2] Russell, S., and P. Norvig, *Artificial Intelligence—A Modern Approach*, Third Edition, Prentice Hall, 2010.

[3] Paluszynski, W., *Artificial Intelligence and Machine Learning*, Lecture Notes, No. 2, Search Methods, Slide 36, 2014.

[4] Lin, F., *A Method for Reliable Experimental Modeling of Microwave FETs* (in German), Doctoral Thesis, University of Kassel, 1993; published by VDI Verlag, Düsseldorf, Fortschritt-Berichte VDI, Reihe 9 (Elektronik), Nr. 164, 1993.

[5] Lin, F., and G. Kompa, "Efficient FET Model Parameter Extraction Using Multi-Plane Data-Fitting and Bidirectional Search Technique," *IEEE MTT-S Int. Microwave Symp. Dig.* (Atlanta, USA), 1993, pp. 1021-1024.

[6] Lin, F., and G. Kompa, "FET Model Parameter Extraction Based on Optimization with Multiplane Data-Fitting and Bidirectional Search—A New Concept," *IEEE Trans.*, Vol. MTT-42, 1994, pp. 1114–1121.

[7] Kompa, G., "Reliable Extraction of Small-Signal Elements of a Generalized Distributed FET Model," *IEEE MTT-S Int. Microwave Symp. Dig.*, Baltimore, USA, 1998, Paper TUA-8, pp. 291–294.

[8] Jarndal, A., and G. Kompa, "A New Small-Signal Modeling Approach Applied to GaN Devices," *IEEE Trans.*, Vol. MTT-53, 2005, pp. 3440–3447.

[9] Jarndal, A., and G. Kompa, "A New Small-Signal Model Parameter Extraction Method Applied to GaN Devices," *IEEE MTT-S Int. Microwave Symp. Dig.*, Long Beach, CA, 2005, pp. 1423–1426.

[10] Jarndal, A., A. Z. Markos, and G. Kompa, "Improved Parameter Extraction Method for GaN HEMT on Si Substrate," *IEEE MTT-S Int. Microwave Symp. Dig.*, Jun. 2010, pp. 1668–1671.

[11] Parekh, R. G., and V. Honavar, "Efficient Learning of Regular Languages Using Teacher- Supplied Positive Samples and Learner-Generated Queries," *Computer Science Technical Reports*, Iowa State University, 10-19-1993.

[12] Bandler, J.W., S.H. Chen, and S. Daijavad, "Microwave Device Modeling Using Efficient l_1 Optimization: A Novel Approach," *IEEE Trans. Microwave Theory Tech.*, Vol. MTT-36, Dec. 1986, pp. 1282-1293.

[13] Vaitkus, R.L., "Uncertainty in the Values of GaAs MESFET Equivalent Circuit Elements Extracted from Measured Two-Port Scattering Parameters," in *Proc. of 1983 IEEE Cornell Conference on High Speed Semiconductor Devices and Circuits*, (Cornell U., USA), 1983, pp. 301-308.

[14] Kompa, G, and M. Novotny, "Highly Consistent FET Model Parameter Extraction Based on Broadband S-Parameter Measurements," *IEEE MTT-S Int. Microwave Symp. Dig.*, 1992, pp. 293–296.

[15] Dambrine, G., A. Cappy, F. Heliodore, and E. Playez, "A New Method for Determining the FET Small-Signal Equivalent Circuit," *IEEE Trans.*, Vol. MTT-36, 1988, pp. 1151–1159.

[16] Yanagava, S., H. Ishihara, and M. Ohtomo, "Analytical Method for Determining Equivalent Circuit Parameters of GaAs FET's," *IEEE Trans.*, Vol. MTT-44(10), 1996, pp. 1637–1641.

[17] Schmale, I., *Development of a Consistent Electrothermal Large Signal Model for Field-Effect Transistors Used for the Reliable Design of Monolithic Frequency Doublers* (in German), Doctoral Thesis, Kassel University, 1999.

[18] Schmale, I., "Extraction of Extrinsic Model Elements," *Seminar*, Dpt. of High-Frequency Engineering, University of Kassel, July 1997.

[19] Kompa, G., M. Schlechtweg, and F. van Raay, "Precisely Calibrated Coaxial-to-Microstrip Transitions Yield Improved Performance in GaAs FET Characterization," *IEEE Trans.*, Vol. MTT-38, 1990, pp. 62–68.

[20] Press, W. H., B. P. Flannery, S. A. Teukolsky, and W. T. Vetterling, *Numerical Recipes in C—the Art of Scientific Computing*. Cambridge, UK: Cambridge University Press, 1988.

[21] Kompa, G., "Frequency-Dependent Measurement Error Analysis and Refined FET Model Parameter Extraction Including Bias-Dependent Series Resistors," *IEEE Workshop on Experimentally Based FET Device Modelling & Related Nonlinear Circuit Design*, Kassel, Germany, July 17-18, 1997, pp. 6.1–6.16; https://archive.org/details/DTIC_ADA331468/page/n279.

[22] Schlechtweg, M., *Broadband Characterization and Modeling of GaAs-MESFETs and (AlGa)As/GaAs-MODFETs up to 40 GHz* (in German), Doctoral Thesis, Kassel University, 1989.

[23] Khalaf, Y. A., *Systematic Optimization Technique for MESFET Modeling*, Doctoral Thesis, Virginia Polytechnic and State University, July 2000.

[24] Grace, A., *Optimization Toolbox User's Guide for Use with MATLAB*, Natick, Mass.:The MathWorks, Inc., December 1992.

Chapter 17

Pure Analytical Model Parameter Extraction

In the previous chapters we discussed various model parameter extraction procedures. What they have in common is that they rely totally or partly on optimization. In principle, optimization procedures have the following features: The risk of the local minimum problem is always present. This is in particular valid for noisy (faulty) measurements. The starting parameter value vector must be carefully chosen because it influences the efficiency of the optimizer. The nearer the initial values are located at the solution minimum, the faster the algorithm works. Optimizer-based results should always be critically scrutinized and validated by additional considerations or estimations. Optimizer procedures operate more reliably with a minor number of variables. Some parameter values might be determined by using test patterns or additional measurements.

In this chapter we investigate an extraction method that is totally analytic. We expect no starting problem, no need for the definition of a starting vector, and no local minimum problem, and a fast straightforward and exact solution for our model parameter extraction problem in case that proper data is available. The model parameter extraction strategy relies entirely on the analytical determination of model parameters from measured scattering coefficients. The widely used 15-element equivalent circuit of a microwave device is supposed in the following. This would principally require four data sets of S-parameters at four frequencies assuming that all relations are linear-independent. From a pure analytical approach the following advantages are expected: The algorithm is very fast, it includes no starting value problem, there is no local minimum problem, and extraction should be performed for a single bias condition.

We will see that the treatment of a pure analytical extraction algorithm will provide very useful general information regarding capabilities and limitations of reliable and consistent model parameter extraction with respect to model topology, measurement bandwidth, and measurement uncertainty. However, with respect to practical application, the required high measurement accuracy and potential model topology mismatch present a major challenge. This problem may be mitigated by statistical evaluation approaches.

17.1 INTRODUCTION

The analytical extraction algorithm presented here was originally developed by Lin [1-3] and applied to a 15-element FET model shown in Figure 17.1. The question arises whether all model elements can be determined on the basis of a pure analytical approach, avoiding

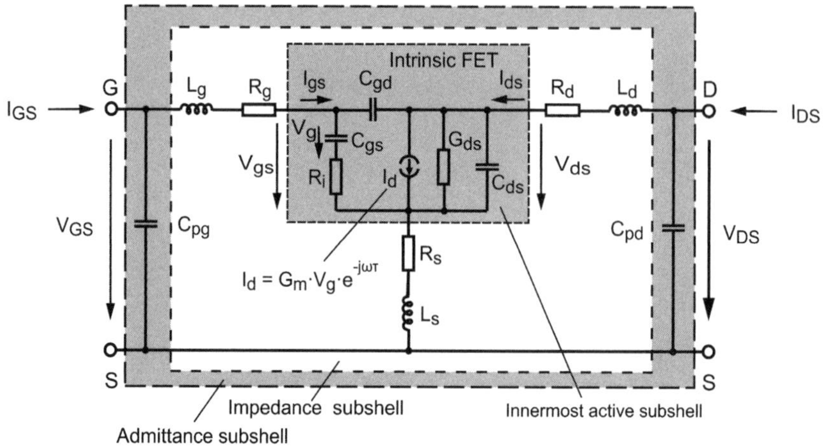

Figure 17.1 Widely used 15-element equivalent circuit used for the analysis of the analytical model parameter extraction algorithm.

the necessity of defining model parameter starting values and preventing the local minimum problem. It is assumed that the model topology is appropriate for frequencies up to 40 GHz or higher. Furthermore, it is expected that all elements in the model are frequency-independent. We will illuminate the practicability of the analytical approach, the dependence of parameter element value calculation on measurement accuracy and measurement bandwidth, and will draw general conclusions on the extraction sensitivity of each single-model element.

17.2 THEORETICAL ANALYSIS

Regarding the intrinsic field-effect transistor in Figure 17.1, the constitutive relations between the intrinsic currents I_{gs} and I_{ds} and the intrinsic voltages V_{gs} and V_{ds} are

$$\begin{bmatrix} I_{gs} \\ I_{ds} \end{bmatrix} = \begin{bmatrix} Y_{11} & Y_{12} \\ Y_{21} & Y_{22} \end{bmatrix} \begin{bmatrix} V_{gs} \\ V_{ds} \end{bmatrix} \tag{17.1}$$

Note that the control voltage V_g is related to the intrinsic gate-source voltage V_{gs} as

$$V_g = V_{gs}/(1 + j\omega C_{gs} R_i) = V_{gs}/\underline{D} \tag{17.2}$$

with

$$\underline{D} = 1 + j\omega C_{gs} R_i = 1 + j\omega\tau_i$$

where τ_i denotes the input time delay. With the branch admittances

$$\begin{aligned} Y_1 &= 1/R_i + j\omega C_{gs} \\ Y_2 &= \underline{G}_m + G_{ds} + j\omega C_{ds} \\ Y_3 &= j\omega C_{gs} \end{aligned} \tag{17.3}$$

the Y-parameter of the intrinsic FET are calculated as (see Appendix A, (A.9)):

$$
\begin{aligned}
Y_{11} &= (R_i + 1/j\omega C_{gs})^{-1} + j\omega C_{gd} = j\omega C_{gd} + j\omega C_{gs}/\underline{D} \\
Y_{12} &= -j\omega C_{gd} \\
Y_{21} &= G_m e^{-j\omega\tau}/\underline{D} - j\omega C_{gd} = \underline{G}_m/\underline{D} - j\omega C_{gd} \\
Y_{22} &= G_{ds} + j\omega(C_{ds} + C_{gd})
\end{aligned} \tag{17.4}
$$

with

$$\underline{G}_m = G_m e^{-j\omega\tau} \tag{17.5}$$

Substituting the Y-parameters in (17.1) leads to

$$\begin{bmatrix} I_{gs} \\ I_{ds} \end{bmatrix} = \begin{bmatrix} j\omega C_{gd} + j\omega C_{gs}/\underline{D} & -j\omega C_{gd} \\ \underline{G}_m/\underline{D} - j\omega C_{gd} & G_{ds} + j\omega(C_{ds} + C_{gd}) \end{bmatrix} \begin{bmatrix} V_{gs} \\ V_{ds} \end{bmatrix} \tag{17.6}$$

As discussed by Lin [1], taking into account the complex embedding parasitic network, this kind of two-port representation with one coefficient matrix (Y-matrix in (17.6)) is not useful for the development of a pure analytical model parameter extraction approach; that is, the extraction of the model parameters directly from the matrix coefficients has proven to be impractical. Therefore, the following two-port representation is employed, comprising two coefficient matrices denoted by [*A*] and [*B*]. This is found by eliminating the denominator in (17.6) and multiplying both sides with $\underline{D}$,

$$\begin{bmatrix} \underline{D} & 0 \\ 0 & \underline{D} \end{bmatrix} \begin{bmatrix} I_{gs} \\ I_{ds} \end{bmatrix} = \begin{bmatrix} j\omega C_{gd}\underline{D} + j\omega C_{gs} & -j\omega C_{gd}\underline{D} \\ \underline{G}_m - j\omega C_{gd}\underline{D} & \left(G_{ds} + j\omega(C_{ds} + C_{gd})\right)\underline{D} \end{bmatrix} \begin{bmatrix} V_{gs} \\ V_{ds} \end{bmatrix} \tag{17.7}$$

or more compactly,

$$[A]\boldsymbol{I} = [B]\boldsymbol{V} \tag{17.8}$$

which may be called extended constitutive (current-voltage) matrix-vector relation of the two-port network [1]. The embedding of the intrinsic FET network (innermost active subshell) into the impedance subshell (13-element model), and finally into the admittance subshell (15-element model) (see Figure 17.1) is treated in Appendix D. The relations between the terminal currents I_{GS} and I_{DS} and the terminal voltages V_{GS} and V_{DS} are found to be (note *a*, *b* quantities should not be mixed up with forward and backward power waves)

$$a_{11}I_{\mathrm{GS}} + a_{12}I_{\mathrm{DS}} = b_{11}V_{\mathrm{GS}} + b_{12}V_{\mathrm{DS}} \tag{17.9}$$

$$a_{21}I_{\mathrm{GS}} + a_{22}I_{\mathrm{DS}} = b_{21}V_{\mathrm{GS}} + b_{22}V_{\mathrm{DS}} \tag{17.10}$$

or in expanded matrix notation

$$\begin{bmatrix} a_{11} & a_{12} \\ a_{21} & a_{22} \end{bmatrix} \begin{bmatrix} I_{GS} \\ I_{DS} \end{bmatrix} = \begin{bmatrix} b_{11} & b_{12} \\ b_{21} & b_{22} \end{bmatrix} \begin{bmatrix} V_{GS} \\ V_{DS} \end{bmatrix} \tag{17.11}$$

with $a_{\mu\nu}$, $b_{\mu\nu}$ ($\mu, \nu = 1, 2$) being functions of model elements and frequency:

$$
\begin{aligned}
a_{11} &= \{1 + j\omega C_{gd}(R_g + j\omega L_g)\}\underline{D} + j\omega C_{gs}\{R_g + R_s + j\omega(L_g + L_s)\} \\
a_{12} &= j\omega C_{gs}(R_s + j\omega L_s) - j\omega C_{gd}(R_d + j\omega L_d)\underline{D}
\end{aligned} \tag{17.12}
$$

$$a_{21} = \underline{G}_m\{R_g + R_s + j\omega(L_g + L_s)\} + \underline{D}\{(G_{ds} + j\omega C_{ds})(R_s + j\omega L_s) - j\omega C_{gd}(R_g + j\omega L_g)\}$$

$$a_{22} = \underline{G}_m(R_s + j\omega L_s) + \underline{D}\{1 + (G_{ds} + j\omega C_{ds})(R_d + R_s + j\omega(L_d + L_s)) + j\omega C_{gd}(R_d + j\omega L_d)\}$$

and

$$\begin{aligned} b_{11} &= j\omega C_{gs} + j\omega C_{gd}\underline{D} + j\omega C_{pg}a_{11} \\ b_{12} &= -j\omega C_{gd}\underline{D} + j\omega C_{pd}a_{12} \\ b_{21} &= \underline{G}_m - j\omega C_{gd}\underline{D} + j\omega C_{pg}a_{21} \\ b_{22} &= \{G_{ds} + j\omega(C_{ds} + C_{gd})\}\underline{D} + j\omega C_{pd}a_{22} \end{aligned} \qquad (17.13)$$

with

$$\underline{D} = 1 + j\omega C_{gs}R_i \qquad (17.14)$$

$$\underline{G}_m = G_m e^{-j\omega\tau} \qquad (17.15)$$

Note that $\underline{G}_{\rm m}$ appears only in (17.10) with a_{21}, a_{22}, and b_{21}. Therefore we concentrate first on (17.9) with the coefficients a_{11}, a_{12}, b_{11}, and b_{12}.

17.2.1 Determination of Extrinsic Capacitances and Inductances

Calculating the real and imaginary parts of the coefficients in (17.9), we get

$$a_{11} = 1 - \omega^2[C_{gs}(L_g + L_s) + C_{gd}(L_g + C_{gs}R_iR_g)] + j\omega[C_{gs}(R_i + R_g + R_s) + C_{gd}R_g - \omega^2 C_{gs}C_{gd}R_iL_g] \qquad (17.16)$$

$$a_{12} = -\omega^2[C_{gs}L_s - C_{gd}(L_d + C_{gs}R_iR_d)] + j\omega(C_{gs}R_s - C_{gd}R_d + \omega^2 C_{gs}C_{gd}R_iL_d) \qquad (17.17)$$

$$b_{11} = -\omega^2\{C_{gd}C_{gs}R_i + C_{pg}[C_{gs}(R_i + R_g + R_s) + C_{gd}R_g]\} + \omega^4 C_{pg}C_{gs}C_{gd}R_iL_g + j\omega\{C_{gs} + C_{gd} + C_{pg} - \omega^2 C_{pg}[C_{gs}(L_g + L_s) + C_{gd}L_g + C_{gd}C_{gs}R_iR_g]\} \qquad (17.18)$$

$$b_{12} = -\omega^2[C_{pd}(C_{gs}R_s - C_{gd}R_d) - C_{gd}C_{gs}R_i] + \omega^4 C_{pd}L_dC_{gd}C_{gs}R_i + j\omega[-C_{gd} - \omega^2 C_{pd}(C_{gs}L_s - C_{gd}L_d - R_dC_{gd}C_{gs}R_i)] \qquad (17.19)$$

With the abbreviations

$$\begin{aligned} T_{11} &= C_{gs}(R_g + R_s + R_i) + C_{gd}R_g \\ T_{21} &= C_{gs}R_s - C_{gd}R_d \\ T_{31} &= C_{gs} + C_{gd} + C_{pg} \\ T_{41} &= -C_{gd} \end{aligned} \qquad (17.20]$$

$$\begin{aligned} T &= C_{gd}C_{gs}R_i \\ T_{12} &= C_{gs}(L_s + L_g) + C_{gd}L_g + R_gT \\ T_{22} &= C_{gs}L_s - C_{gd}L_d - R_dT \end{aligned} \qquad (17.21)$$

Equation (17.9) reads

$$\begin{aligned}
&\{[1 - \omega^2 T_{12}] + j\omega[T_{11} - \omega^2 L_g T]\} I_{GS} \\
&+ \{[\quad -\omega^2 T_{22}] + j\omega[T_{21} + \omega^2 L_d T]\} I_{DS} \\
&= \{[-\omega^2(C_{pg}T_{11} + T) + \omega^4 C_{pg}L_g T] + j\omega[T_{31} - \omega^2 C_{pg}T_{12}]\} V_{GS} \\
&+ \{[-\omega^2(C_{pd}T_{21} - T) - \omega^4 C_{pd}L_d T] + j\omega[T_{41} - \omega^2 C_{pd}T_{22}]\} V_{DS}
\end{aligned} \tag{17.22}$$

Denoting the real parts with a single prime and the imaginary parts with a double prime, (17.22) reads

$$\begin{bmatrix} a'_{11} + ja''_{11} & a'_{12} + ja''_{12} \\ a'_{21} + ja''_{21} & a'_{22} + ja''_{22} \end{bmatrix} \begin{bmatrix} I_{GS} \\ I_{DS} \end{bmatrix} = \begin{bmatrix} b'_{11} + jb''_{11} & b'_{12} + jb''_{12} \\ b'_{21} + jb''_{21} & b'_{22} + jb''_{22} \end{bmatrix} \begin{bmatrix} V_{GS} \\ V_{DS} \end{bmatrix} \tag{17.23}$$

Next we introduce scattering parameters. Generally, the total terminal currents I are the difference of the incident currents I^+ and the reverse currents I^- (i.e., $I = I^+ - I^-$). For the terminal voltages it reads $V = Z_0 \cdot I = Z_0(I^+ + I^-)$. The complex scattering coefficients are defined as $S_{ij} = (I_i^-/I_j^+)|_{Ii+=0}$ (i, j = 1, 2) and can be split into real and imaginary parts S_{ij}' and S_{ij}'', respectively. Then the following four real-valued equations are obtained:

$$\begin{aligned}
&a'_{11}(S'_{11} - 1) - \omega a''_{11}S''_{11} + a'_{12}S'_{21} - \omega a''_{12}S''_{21} \\
&\quad + Z_{01}[b'_{11}(S'_{11} + 1) - \omega b''_{12}S^{i}_{11}] + Z_{02}[b'_{12}S'_{21} - \omega b''_{12}S''_{21}] = 0
\end{aligned} \tag{17.24}$$

$$\begin{aligned}
&a'_{11}S'_{12} - \omega a''_{11}S''_{12} + a'_{12}(S'_{22} - 1) - \omega a''_{12}S''_{22} \\
&\quad + Z_{01}[b'_{11}S'_{12} - \omega b''_{11}S''_{12}] + Z_{02}[b'_{12}(S'_{22} + 1) - \omega b''_{12}S''_{22}] = 0
\end{aligned} \tag{17.25}$$

$$\begin{aligned}
&a'_{11}(S''_{11} + \omega a''_{12}(S'_{11} - 1) + a'_{12}S''_{21} + \omega a''_{12}S'_{21} \\
&\quad + Z_{01}[b'_{11}(S''_{11} + \omega b''_{12}(S'_{11} + 1)] + Z_{02}[b'_{12}S''_{21} + \omega b''_{12}S'_{21}] = 0
\end{aligned} \tag{17.26}$$

$$\begin{aligned}
&a'_{11}S''_{12} + \omega a''_{11}S'_{12} + a'_{12}S''_{22} + \omega a''_{12}(S'_{22} - 1) \\
&\quad + Z_{01}[b'_{11}S''_{12} + \omega b''_{11}S'_{12}] + Z_{02}[b'_{12}(S''_{22} + \omega b''_{12}(S'_{22} + 1)] = 0
\end{aligned} \tag{17.27}$$

In matrix form,

$$\begin{bmatrix} S'_{11}-1 & S'_{21} & S'_{11}+1 & S'_{21} \\ S'_{12} & S'_{22}-1 & S'_{12} & S'_{22}+1 \\ S''_{11} & S''_{21} & S''_{11} & S''_{21} \\ S''_{12} & S''_{22} & S''_{12} & S''_{22} \end{bmatrix} \begin{bmatrix} a'_{11} \\ a'_{12} \\ Z_{01}b'_{11} \\ Z_{02}b'_{12} \end{bmatrix} + \omega \begin{bmatrix} -S''_{11} & -S''_{21} & -S''_{11} & -S''_{21} \\ -S''_{12} & -S''_{22} & -S''_{12} & -S''_{22} \\ S'_{11}-1 & S'_{21} & S'_{11}+1 & S'_{21} \\ S'_{12} & S'_{22}-1 & S'_{12} & S'_{22}+1 \end{bmatrix} \begin{bmatrix} a''_{11} \\ a''_{12} \\ Z_{01}b''_{11} \\ Z_{02}b''_{12} \end{bmatrix} = 0 \tag{17.28}$$

Equation (17.28) reads with (17.22) and (17.23)

$$[S^{(1)}]\begin{bmatrix} 1 - \omega^2 T_{12} \\ \omega^2 T_{22} \\ Z_{01}[-\omega^2(C_{pg}T_{11} + T) + \omega^4 C_{pg}L_g T] \\ Z_{02}[-\omega^2(C_{pd}T_{21} - T) - \omega^4 C_{pd}L_d T] \end{bmatrix} + \omega[S^{(2)}]\begin{bmatrix} [T_{11} - \omega^2 L_g T \\ T_{21} + \omega^2 L_d T \\ Z_{01}(T_{31} - \omega^2 C_{pg}T_{12}) \\ Z_{02}(T_{41} - \omega^2 C_{pd}T_{22}) \end{bmatrix} = 0 \qquad (17.29)$$

and after reorganizing,

$$[S^{(1)}]\left(\begin{bmatrix}1\\0\\0\\0\end{bmatrix} - \omega^2\begin{bmatrix} T_{12} \\ T_{22} \\ Z_{01}(C_{pg}T_{11} + T) \\ Z_{02}(C_{pd}T_{21} - T) \end{bmatrix} + \omega^4 \begin{bmatrix} 0 \\ 0 \\ Z_{01}C_{pg}L_g T \\ Z_{02}C_{pd}L_d T \end{bmatrix}\right) + \omega[S^{(2)}]\left(\begin{bmatrix} T_{11} \\ T_{21} \\ Z_{01}T_{31} \\ Z_{02}T_{41} \end{bmatrix} - \omega^2 \begin{bmatrix} L_g T \\ L_d T \\ Z_{01}C_{pg}T_{12} \\ Z_{02}C_{pd}T_{22} \end{bmatrix}\right) = 0 \qquad (17.30)$$

or

$$\left[\omega[S^{(2)}] \quad -\omega^2[S^{(1)}] \quad -\omega^3[S^{(2)}] \quad \omega^4[S^{(1)}]\right]\begin{bmatrix} \boldsymbol{X}_1 \\ \boldsymbol{X}_2 \\ \boldsymbol{X}_3 \\ \boldsymbol{X}_4 \end{bmatrix} = -[S^{(1)}]\boldsymbol{X}_0 \qquad (17.31)$$

with

$$\begin{aligned} \boldsymbol{X}_0 &= (X_{01}\ X_{02}\ X_{03}\ X_{04})^T = (1 \quad 0 \quad 0 \quad 0)^T \\ \boldsymbol{X}_1 &= (X_{11}\ X_{12}\ X_{13}\ X_{14})^T = (T_{11} \quad T_{21} \quad Z_{01}T_{31} \quad Z_{02}T_{41})^T \\ \boldsymbol{X}_2 &= (X_{21}\ X_{22}\ X_{23}\ X_{24})^T = [T_{12} \quad T_{22} \quad Z_{01}(C_{pg}T_{11} + T) \quad Z_{02}(C_{pd}T_{21} - T)]^T \\ \boldsymbol{X}_3 &= (X_{31}\ X_{32}\ X_{33}\ X_{34})^T = [L_g T \quad L_d T \quad Z_{01}C_{pg}T_{12} \quad Z_{02}(C_{pd}T_{22})]^T \\ \boldsymbol{X}_4 &= (X_{41}\ X_{42}\ X_{43}\ X_{44})^T = [0 \quad 0 \quad Z_{01}C_{pg}L_g T \quad Z_{02}C_{pd}L_d T]^T \end{aligned} \qquad (17.32)$$

where the T-terms are given in (17.20) and (17.21).

The coefficients of the set of equations (17.31) are all frequency-dependent, but not the vectors. Therefore it is possible to set up any number of equations by choosing different frequencies, but under the restrictive condition that the equations are linear-independent (i.e., the space between selected frequencies must be sufficiently large).

The linear set of (17.31) comprises four equations ($m = 4$) and 14 unknown variables of the model elements ($n = 14$) for one frequency. This is an underdetermined linear equation system that cannot be solved without additional information.

As the coefficient matrices of the set of equations are nonlinear functions of frequency, further information can be obtained by scattering parameters at additional frequencies. As each new frequency provides four further equations, the set of equations is principally solvable with four different frequencies. Because of overdetermination the set of equations provides redundant solutions.

It is shown that 11 model elements, which appear in the vectors $\boldsymbol{X}_k$ ($k = 1, 4$) in (17.32), can be directly extracted. The remaining four model elements can be easily calculated with (17.10) based on one frequency. Figure 17.2 illustrates the sequence of analytical extraction.

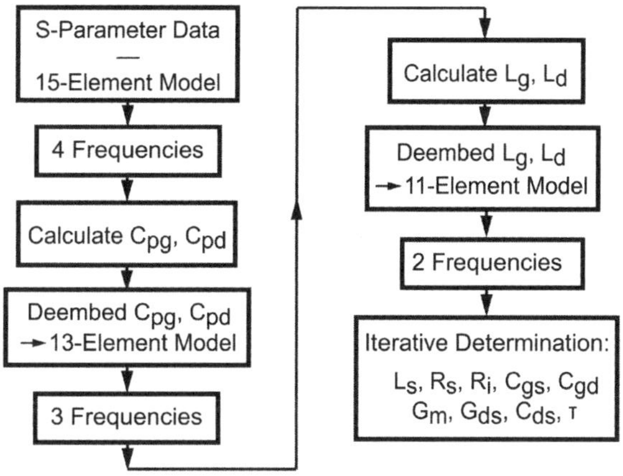

Figure 17.2 Schematic flow diagram of the analytical extraction algorithm.

Calculation of C_{pg} and C_{pd}

First, the extrinsic parasitic capacitances C_{pg} and C_{pd} are determined from scattering parameters at four different frequencies. Because the set of equations is overdetermined the solution is not unique. Therefore, the elements C_{pg} and C_{pd} can be determined in different ways:

$$C_{pg} = X_{4,3}/X_{3,1}/Z_{01} \quad \text{or} \quad X_{3,3}/X_{2,1}/Z_{01} \tag{17.33}$$
$$C_{pd} = -X_{4,4}/X_{3,2}/Z_{02} \quad \text{or} \quad X_{3,4}/X_{2,2}/Z_{02}$$

The observed multiple solutions of the parasitic capacitances can be used to control the stability of the extraction algorithm. Studying solutions of the set of equations it was found that the vectors $\boldsymbol{X}_k$ (k = 1, 4) are very sensitively dependent on the given scattering coefficients. However, the ratios of the vector components in (17.33); that is, $X_{4,3}/X_{3,1}$, $X_{3,3}/X_{2,1}$, $X_{4,4}/X_{3,2}$, $X_{3,4}/X_{2,2}$, were proven to be very stable.

Calculation of L_g and L_d

In a second step, with C_{pg} and C_{pd} known, the model elements L_g and L_d are extracted from the deembedded 13-element network using three frequencies. The procedure can be described by the following symbolic deembedding relation,

$$[S] \Rightarrow [Y] \Rightarrow [Y'] = [Y] - [C_p] \Rightarrow [S'] \tag{17.34}$$

From (12.22) we get

$$[(1 - \omega^2 T_{12}) + j\omega(T_{11} - \omega^2 L_g T)]\, I'_{GS} + [-\omega^2 T_{22} + j\omega(T_{21} + \omega^2 L_d T)]\, I'_{DS}$$
$$= (-\omega^2 T + j\omega T'_{31})\, V'_{GS} + (\omega^2 T + j\omega T_{41})\, V'_{DS} \tag{17.35}$$

Then, (12.29) reads

$$[S'^{(1)}] \begin{bmatrix} 1 - \omega^2 T_{12} \\ -\omega^2 T_{22} \\ -\omega^2 T Z_{01} \\ \omega^2 T Z_{02}) \end{bmatrix} + \omega[S'^{(2)}] \begin{bmatrix} T_{11} - \omega^2 L_g T \\ T_{21} + \omega^2 L_d T \\ T'_{31} Z_{01} \\ T_{41} Z_{02} \end{bmatrix} = 0 \tag{17.36}$$

Reorganized,

$$[S'^{(1)}]\left(\begin{bmatrix}1\\0\\0\\0\end{bmatrix}-\omega^2\begin{bmatrix}T_{12}\\T_{22}\\TZ_{01}\\-TZ_{02}\end{bmatrix}\right)+\omega[S'^{(2)}]\left(\begin{bmatrix}T_{11}\\T_{21}\\T'_{31}Z_{01}\\T_{41}Z_{02}\end{bmatrix}-\omega^2\begin{bmatrix}-L_gT\\L_dT\\0\\0\end{bmatrix}\right)=0 \qquad (17.37)$$

or

$$\begin{bmatrix}\omega[S^{(2)}] & -\omega^2[S^{(1)}] & -\omega^3[S^{(2)}]\end{bmatrix}\begin{bmatrix}X_1\\X_2\\X_3\end{bmatrix} = -[S^{(1)}]X_0 \qquad (17.38)$$

with

$$\begin{aligned} X_0 &= [1 \quad 0 \quad 0 \quad 0]^T \\ X_1 &= [T_{11} \quad T_{21} \quad Z_{01}T'_{31} \quad Z_{02}T_{41}]^T \\ X_2 &= [T_{12} \quad T_{22} \quad Z_{01}T \quad -Z_{02}T\,]^T \\ X_3 &= [L_gT \quad -L_dT \quad 0 \quad 0\;]^T \end{aligned} \qquad (17.39)$$

where the unprimed T-terms are given in (17.20) and (17.21). T'_{31} corresponds to T_{31} with $C_{pg} = 0$, i.e. $T'_{31} = C_{gs} + C_{gd}$. Three frequencies are needed to solve the equation. L_g and L_d are determined from the ratio of two vector elements; that is,

$$\begin{aligned} L_g &= X_{3,1}/T = Z_{01}X_{3,1}/X_{2,3} \quad \text{or} \quad -Z_{02}X_{3,1}/X_{2,4} \\ L_d &= X_{3,2}/T = Z_{02}X_{3,2}/X_{2,4} \quad \text{or} \quad -Z_{01}X_{3,2}/X_{2,3} \end{aligned} \qquad (17.40)$$

Calculation of L_s

The deembedding of L_g and L_d from the 13-element model (S'-matrix) can be symbolically described as

$$[S'] \Rightarrow [Z] \Rightarrow [Z'] = [Z] - [j\omega L] \Rightarrow [S''] \qquad (17.41)$$

Figure 17.3 depicts the remaining 11-element model. From (17.22) we obtain

$$\begin{aligned} &(1-\omega^2T'_{12}+j\omega T'_{11})\,I''_{GS} + (-\omega^2T'_{22}+j\omega T'_{21})\,I''_{DS} \\ &\quad = (-\omega^2T + j\omega T'_{31})\,V''_{GS} + (\omega^2T + j\omega T_{41})\,V''_{DS} \end{aligned} \qquad (17.42)$$

with

$$\begin{aligned} T &= C_{gd}C_{gs}R_i & T'_{11} &= C_{gs}(R_g + R_s + R_i) + C_{gd}R_g \\ T'_{12} &= C_{gs}L_s + R_gT & T'_{21} &= C_{gs}R_s - C_{gd}R_d \\ T'_{22} &= C_{gs}L_s - R_dT & T'_{31} &= C_{gs} + C_{gd} \\ & & T_{41} &= -C_{gd} \end{aligned} \qquad (17.43)$$

Then (17.42) reads

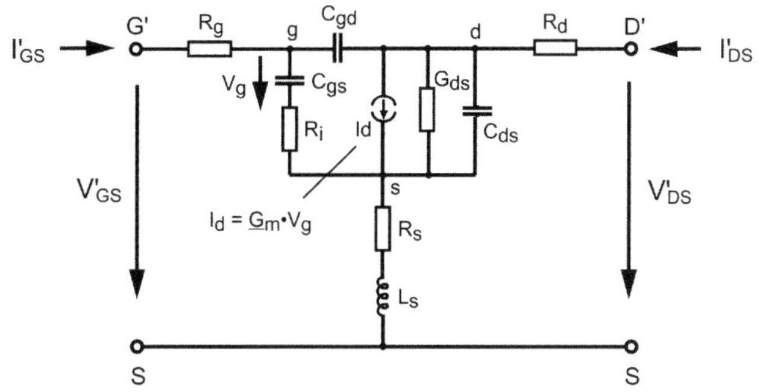

Figure 17.3 The 11-element small-signal model after deembedding parasitic capacitances C_{pg} and C_{pd} and inductances L_g and L_d from the 15-element model.

$$[S^{//(1)}]\begin{bmatrix} 1 - \omega^2 T'_{12} \\ -\omega^2 T'_{22} \\ -\omega^2 T Z_{01} \\ \omega^2 T Z_{02} \end{bmatrix} + \omega[S^{//(2)}]\begin{bmatrix} T'_{11} \\ T'_{21} \\ T'_{31} Z_{01} \\ T'_{41} Z_{02} \end{bmatrix} = 0 \tag{17.44}$$

Reorganized,

$$[S^{//(1)}]\left(\begin{bmatrix} 1 \\ 0 \\ 0 \\ 0 \end{bmatrix} - \omega^2 \begin{bmatrix} T'_{12} \\ T'_{22} \\ T Z_{01} \\ -T Z_{02} \end{bmatrix}\right) + \omega[S^{//(2)}]\begin{bmatrix} T'_{11} \\ T'_{21} \\ T'_{31} Z_{01} \\ T'_{41} Z_{02} \end{bmatrix} = 0 \tag{17.45}$$

or

$$\begin{bmatrix} \omega[S^{//(2)}] & -\omega^2 [S^{//(1)}] \end{bmatrix}\begin{bmatrix} \boldsymbol{X}_1 \\ \boldsymbol{X}_2 \end{bmatrix} = -[S^{//(1)}]\boldsymbol{X}_0 \tag{17.46}$$

with

$$\begin{aligned} \boldsymbol{X}_0 &= [1 \quad 0 \quad 0 \quad 0]^T \\ \boldsymbol{X}_1 &= [T'_{11} \quad T'_{21} \quad Z_{01}T'_{31} \quad Z_{02}T'_{41}]^T \\ \boldsymbol{X}_2 &= [T'_{12} \quad T'_{22} \quad Z_{01}T \quad -Z_{02}T\,]^T \end{aligned} \tag{17.47}$$

Two frequencies are needed to solve the equation for the seven unknowns. From (17.39) the following model parameters can be directly determined:

$$C_{gd} = -X_{1,4}/Z_{02}, \quad C_{gs} = X_{1,3}/Z_{01} - C_{gd}, \quad R_i = T/(C_{gd}C_{gs}) \text{ with } T = X_{2,3}/Z_{01} = -X_{2,4}/Z_{02} \tag{17.48}$$

Note that (T′) provides five equation for the determination of the series resistances. Only three of them are well-conditioned: that are the expressions of T, T'_{11}, and T'_{21}. The first term T is suited to calculate R_i. However, T'_{12} and T'_{22} are ill-conditioned because these parameters comprise quadruple products of model parameters.

For the source inductance L_s and the resistances R_s, R_g, and R_d the following matrix equation is valid:

$$\begin{bmatrix} C_{gs} & 0 & T & 0 \\ C_{gs} & 0 & 0 & -T \\ 0 & C_{gs} & 0 & -C_{gd} \\ 0 & C_{gs} & (C_{gs} + C_{gd}) & 0 \end{bmatrix} \begin{bmatrix} L_s \\ R_s \\ R_g \\ R_d \end{bmatrix} = \begin{bmatrix} T'_{12} \\ T'_{22} \\ T'_{21} \\ T'_{11} - C_{gs}R_i \end{bmatrix} \tag{17.49}$$

This method cannot give accurate values for R_i, R_s, R_g, and R_d because $T \ll T'_{12}$, $T'_{22} \ll (T'_{11}$, T'_{21}, $Z_{01}T'_{31}$, $Z_{021}T'_{41})$, for example, $T \sim 10^{-7}$, $(T'_{12}, T'_{22}) \sim 10^{-4}$, $(T'_{11}, T'_{21}, Z_{01}T'_{31}, Z_{021}T'_{41}) \sim 10^{-2}$. However good approximation can be found for L_s:

$$\begin{aligned} T'_{12} + T'_{22} &= 2C_{gs}L_s + T(R_g - R_d) \approx 2C_{gs}L_s \quad \text{with} \quad T(R_g - R_d) \ll C_{gs}L_s \\ T'_{31} + T'_{41} &= C_{gs} \end{aligned} \tag{17.50}$$

Then, from (17.50), L_s is determined as

$$L_s \approx \frac{1}{2}\frac{T'_{12} + T'_{22}}{T'_{31} + T'_{41}} \tag{17.51}$$

The approximated inductance L'_s is deembedded from the 11-element model:

$$[S''] \Rightarrow [Z] \Rightarrow [Z'] = [Z] - [j\omega L_s] \Rightarrow [S'''] \tag{17.52}$$

and results in approximated parameters C_{gs}', C_{ds}', R_g', R_d', R_s', R_i'. This equation is the same as before (see (17.42)) but now, with L_s approximately known, the T-parameter simplifies to $T'_{12} = R_gT$ and $T'_{22} = -R_dT$. The approximated capacitances C_{gd}' and C_{gs}' are calculated according to (17.48). The approximated resistances are given as

$$\begin{aligned} R'_g &= T'_{12}/T \\ R'_d &= -T'_{22}/T \\ R'_s &= (T'_{21} + C'_{gd}R'_d)/C'_{gs} \\ R'_i &= [T'_{11} - C'_{gs}R'_s - (C_{gs} + C'_{gd})R'_g]/C'_{gs} \end{aligned} \tag{17.53}$$

As will be shown in the following accurate model elements can be found by an iterative extraction process (Section 17.2.3) also including the residual model parameters G_m, G_{ds}, C_{ds} and τ in the 10-element model shown in Figure 17.4.

17.2.2 Conditioning of Equation System for the 10-Element Model

Recalling (17.10),

$$a_{21}I_{GS} + a_{22}I_{DS} = b_{21}V_{GS} + b_{22}V_{DS} \tag{17.54}$$

with

$$a_{2j} = a'_{2j} + j\omega a''_{2j} \quad \text{and} \quad b_{2j} = b'_{2j} + j\omega b''_{2j} \quad (j = 1, 2) \tag{17.55}$$

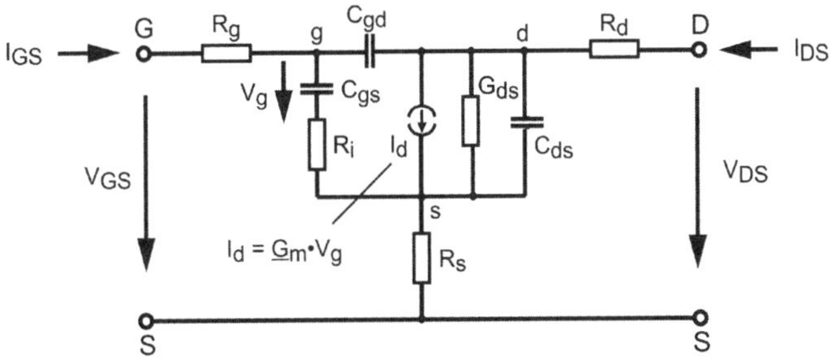

Figure 17.4 The 10-element small-signal model after deembedding parasitic capacitances and inductances.

and using the S-parameter concept, the current-voltage relation can be expressed by the following nonlinear homogeneous set of equations:

$$[S^{(1)}]\boldsymbol{A}_2 + \omega[S^{(2)}]\boldsymbol{B}_2 = 0 \tag{17.56}$$

where

$$[S^{(1)}] = \begin{bmatrix} S'_{11}-1 & S'_{21} & S'_{11}+1 & S'_{21} \\ S'_{12} & S'_{22}-1 & S'_{12} & S'_{22}+1 \\ S''_{11} & S''_{21} & S''_{11} & S''_{21} \\ S''_{12} & S''_{22} & S''_{12} & S''_{22} \end{bmatrix}, \quad [S^{(2)}] = \begin{bmatrix} -S''_{11} & -S''_{21} & -S''_{11} & -S''_{21} \\ -S''_{12} & -S''_{22} & -S''_{12} & -S''_{22} \\ S'_{11}-1 & S'_{21} & S'_{11}+1 & S'_{21} \\ S'_{12} & S'_{22}-1 & S'_{12} & S'_{22}+1 \end{bmatrix} \tag{17.57}$$

and

$$\begin{aligned} \boldsymbol{A}_2 &= (a'_{21} \;\; a'_{22} \;\; Z_{01}b'_{21} \;\; Z_{02}b'_{22})^T \\ \boldsymbol{B}_2 &= (a''_{21} \;\; a''_{22} \;\; Z_{01}b''_{21} \;\; Z_{02}b''_{22})^T \end{aligned} \tag{17.58}$$

The features of the vectors $\boldsymbol{A}_2$ and $\boldsymbol{B}_2$ depend on the given topology of the equivalent network. In the case of the considered 10-element model topology in Figure 17.4 the components of the vectors read as follows:

$$\boldsymbol{A}_2 = \begin{bmatrix} G_m\cos(\omega\tau)(R_g + R_s) + R_sG_{ds} + \omega^2(TR_g - C_{gs}C_{ds}R_gR_s) \\ 1 + G_m\cos(\omega\tau)R_s + (R_d + R_s)G_{ds} - \omega^2C_{gs}C_{ds}R_i(R_d + R_s) - \omega^2TR_d \\ Z_{01}[\omega^2T + G_m\cos(\omega\tau)] \\ Z_{02}(G_{ds} - \omega^2C_{gs}C_{ds}R_i - \omega^2T \end{bmatrix} \tag{17.59}$$

$$\boldsymbol{B}_2 = \begin{bmatrix} -G_m\sin(\omega\tau)(R_g + R_s) + \omega[C_{gs}R_iR_sG_{ds} - C_{gd}R_g + C_{ds}R_s] \\ -G_m\sin(\omega\tau)R_s + \omega\left\{C_{ds}(R_d + R_s) + C_{gd}R_d + C_{gs}R_i[1 + (R_d + R_s)G_{ds}]\right\} \\ -Z_{01}[\omega C_{gd} + G_m\sin(\omega\tau)] \\ Z_{02}(\omega C_{gs}R_iG_{ds} + \omega C_{gd} + \omega C_{ds}) \end{bmatrix} \tag{17.60}$$

with $T = C_{gs}\cdot C_{gd}\cdot R_i$.

The nonlinear set of equations (17.56) is underdetermined for one frequency and needs additional information. The question arises whether (17.56) can be consistently solved with two different frequencies. The equations should be sufficiently nonlinear in the regarded frequency band. Experience has shown that the frequency difference should be at least 10 GHz.

If small variations in the S-parameters cause large variations in the solution, one usually speaks of an ill-conditioned system. The condition numbers are practical ways to measure the quality of the solution of a given linear system $\boldsymbol{Ax} = \boldsymbol{b}$. In the mathematical literature a variety of condition numbers have been discussed. The *condition number of Hadamard* for the matrix $\boldsymbol{A}$ is defined as [4]

$$K_H(A) = \frac{|\det(A)|}{a_1 a_2 \cdots a_n} \quad \text{with} \quad a_i = \sqrt{a_{i1}^2 + a_{i2}^2 + \cdots + a_{in}^2} \tag{17.61}$$

where $a_{11}, a_{12}, \ldots, a_{1n}$ are elements of the ith row of matrix $\boldsymbol{A}$. For this number, the following empirical rules hold:

$K_H(\boldsymbol{A}) < 0.01$	Ill-conditioned
$K_H(\boldsymbol{A}) > 0.1$	Well-conditioned
$0.01 \leq K_H(\boldsymbol{A}) \leq 0.1$	Undefined

It follows that the condition number of set of equations (17.9) is less than 2.73×10^{-7} for all combinations of two frequencies up to 40 GHz. In contrast, for the second system of equations (17.10) the condition number is calculated at a low frequency of 2 GHz as 0.093, but increases with frequency, and reaches $K_H = 0.203$ at 35 GHz. Thus, system 1 is ill-conditioned, whereas system 2 is well-conditioned. It should be noted that the extraction quality of system 2 depends totally on the determined model parameter values of system 1 [5].

Until now we have investigated the analytical extraction algorithm under the assumption of ideal (model-matched) scattering parameters. The question arises how stable the extracted results will behave under additional measurement errors.

Extraction with noisy S-parameters confirms the high susceptibility of system 1 to data corruption. For a normalized standard deviation $\leq 3\%$ for C_{gs} and C_{gd} the simulated measurement error must be $< 0.006\%$ and for G_{ds}, G_m, C_{ds}, and $R_s < 0.002\%$. The smallest error of 0.0003% is required for the parameters R_d, R_i, R_g, and τ. This is indeed far from practical applicability.

In the following we discuss measures to improve the extraction conditioning. This could include reduction of variables; for example, the gate resistance R_g could be predetermined.

17.2.3 Iterative Determination of Intrinsic Model Elements

From (17.10) the following relation with G_m can be obtained:

$$\left\{[S^{(1)}][R_1] + [S^{(2)}][R_2]\right\} \boldsymbol{X}_3 = [S^{(1)}]\boldsymbol{C}_1 + [S^{(2)}]\boldsymbol{C}_2 \tag{17.62}$$

where

$$\boldsymbol{X}_3 = [G_m\cos(\omega\tau),\ G_m\sin(\omega\tau),\ G_{ds},\ \omega C_{ds}]^T \tag{17.63}$$

$$[R_1] = \begin{bmatrix} R_g + R_s & 0 & R_s & -\omega C_{gs} R_g R_s \\ R_s & 0 & R_d + R_s & -\omega C_{gs} R_i (R_d + R_s) \\ Z_0 & 0 & 0 & 0 \\ 0 & 0 & Z_0 & -\omega C_{gs} R_g Z_0 \end{bmatrix} \tag{17.64}$$

$$[R_2] = \begin{bmatrix} 0 & -R_g + R_s & \omega C_{gs} R_i R_s & R_s \\ 0 & -R_s & \omega C_{gs} R_i (R_d + R_s) & R_d + R_s \\ 0 & -Z_0 & 0 & 0 \\ 0 & 0 & \omega C_{gs} R_g Z_0 & Z_0 \end{bmatrix} \tag{17.65}$$

$$\begin{aligned} C_1 &= (-1, -1, 0, 0)^T \\ C_2 &= \left(-\omega C_{gs}(R_i + R_g + R_s),\ -\omega C_{gs}(R_i + R_s),\ -\omega C_{gs} Z_0,\ 0\right)^T \end{aligned} \tag{17.66}$$

The set of equations (17.62) are solvable for one frequency if the model elements C_{gs}, C_{gd}, R_i, R_d, R_s, and R_g are already known. That means that (17.62) can be solved with one frequency if (17.9) has already been solved. After deembedding C_{gs}', C_{ds}', R_g', R_d', R_s', R_i' from the 10-element model, G_m', G_{ds}', C_{ds}', and τ' can be approximately determined.

At very low frequencies ($f \leq 2$ GHz) (Figure 17.5) the admittance coefficients read after $S \rightarrow Y$ matrix conversion ($S_{LF} \rightarrow Y_{LF}$):

$$\begin{aligned} Y_{21}^{LF} &= \frac{G_m}{1 + R_s(G_m + G_{ds}) + R_d G_{ds}} \\ Y_{22}^{LF} &= \frac{G_{ds}}{1 + R_s(G_m + G_{ds}) + R_d G_{ds}} \end{aligned} \tag{17.67}$$

$$Y_{21}^{LH} / Y_{22}^{LF} = G_m / G_{ds}$$

With Y_{21} and Y_{22} known from low-frequency measurement, a new approximate value R_d'' can be calculated as

$$R_d'' = 1/Y_{22}^{LF} - 1/G_{ds}' - R_s'(1 + Y_{21}^{LF}/Y_{22}^{LF}) \tag{17.68}$$

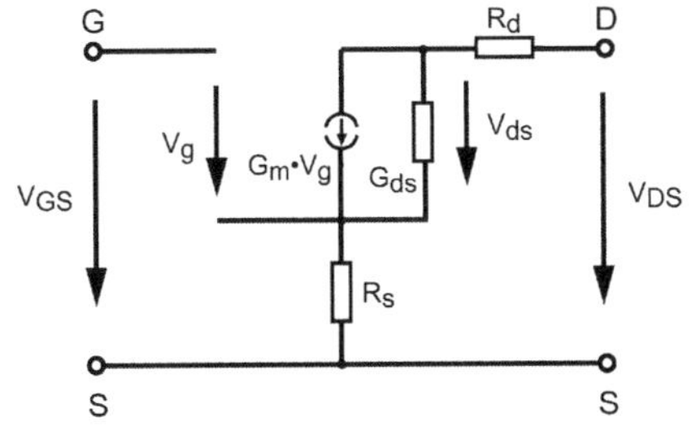

Figure 17.5 The 4-element small-signal model for very low frequencies.

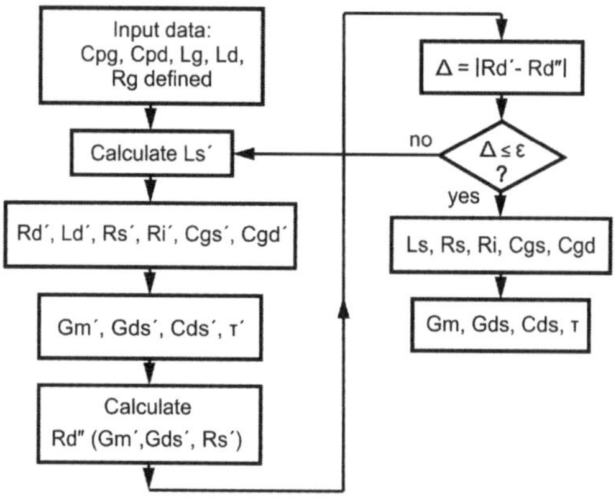

Figure 17.6 Illustration of iterative extraction of L_s, C_{gs}, C_{gd}, R_d, R_s, R_i and G_m, G_{ds}, C_{ds}, and τ.

Setting $R_d' = R_d''$ a new improved value of the drain resistance is calculated. The iteration cycle stops when the change in the drain resistance value becomes marginal (i.e., less than a specified value ε: $|R_d' - R_d''| = \Delta \le \varepsilon$). The iterative extraction procedure is illustrated in Figure 17.6.

17.3 FREQUENCY BANDWIDTH DEPENDENT EXTRACTION ACCURACY

The frequency requirements for the scattering coefficients have been analyzed by generating four fictitious scattering parameter sets within a defined frequency range. The analytical extraction algorithm should then deliver the element values used for the S-parameter generation process.

The investigation has shown that the data sets should cover a frequency range as large as possible (i.e., about 30 GHz) for the 15-element model. In the case of reduced frequency range <15 GHz or even more, below the X-band (e.g., < 5 GHz, see [6]), the original element values can no longer be retrieved. The reason is that the equations at frequencies within a narrow frequency bandwidth comprise only minor numerical deviations.

Table 17.1 illustrates the difference of extracted model elements from rated values for two frequency combinations. The chosen frequency sampling points should include the available upper limit frequency and should have maximum distance. It can be seen that a frequency spacing of 6 GHz over a total bandwidth of 20 GHz already provided accurate results, whereas a smaller extraction bandwidth of 10 GHz and closer frequency sampling point distribution of 3 GHz encloses higher extraction inaccuracy. Regarding the 15-element model, all model elements can be regained from the fictitious S-parameters with an accuracy of four significant digits using a frequency combination of 10(10)40 GHz.

17.4 SIMULATED MEASUREMENT ERROR DEPENDENT EXTRACTION ACCURACY

To study the stability of the analytical extraction algorithm, fictitious S-parameters with

Table 17.1
Extraction on the Basis of Fictitious S-Parameters Delivers Only Unique Results for Sufficiently Broad Frequency Ranges

Frequencies (GHz):		1, 7, 13, 20	1, 4, 7, 10
	Nominal	Extracted: Deviation from	
C_{pg}/pF	0.027447	0.00	0.00
C_{pd}/pF	0.015227	-0.00	-0.00
L_g/nH	0.156375	0.00	0.02
L_d/nH	0.155735	-0.00	-0.01
L_s/nH	0.038049	-0.00	-0.13
R_g/Ω	3.16958	0.03	1.99
R_d/Ω	0.843553	-1.05	-74.08
R_s/Ω	2.50817	-0.02	-1.51
R_{gs}/Ω	2.63133	-0.02	-1.08
C_{gs}/pF	0.415500	0.00	0.06
C_{gd}/pF	0.030252	-0.00	-0.06
C_{ds}/pF	0.094641	-0.02	-1.43
G_{ds}/mS	6.12404	0.00	-0.45
G_m/mS	76.2202	-0.01	-0.69
τ/ps	2.02067	0.05	3.72

From: [1].

different random variations have been used to simulate an erroneous device response $\tilde{S}^D$ [3]. The permitted limited absolute accuracy of scattering parameters with respect to sufficient extraction stability and reliability has been found between 10^{-2} ... 10^{-3} (10^{-0} ... 10^{-1}) for the magnitudes (phases). This is above all a highly impractical requirement. But if only one of the four parasitic resistances R_s, R_g, R_d and R_i is predetermined, extraction stability increases dramatically. Because of bias-independence, the element R_g is selected as predetermined element.

In order to analyze the extraction stability on the basis of noisy data, the fictitious S-parameter data S^D were corrupted by a random variation ΔS^D,

$$\tilde{S}^D = S^D + \Delta S^D \tag{17.69}$$

where

$$|\Delta S^D| = \alpha \cdot 10^{-N}$$
$$\angle(\Delta S^D) = \beta \cdot 10^{-N+2} \quad \text{with} \quad (0 \le \alpha, \beta \le 1) \tag{17.70}$$

α and β are random numbers between zero and one ($0 \le \alpha, \beta \le 1$). The simulation accuracy parameter N ($1 \le N \le 8$) denotes the absolute accuracy or the number of significant digits of magnitude and phase of the scattering parameters. $\tilde{S}^D$ denotes the noisy device data for the study of stability and reliability of extracted results. Assuming, for example, a fictitious S_{21} parameter, S_{21} = 3.123...∠82.123...°, then the notation N = 2 denotes the following uncertainty in magnitude and phase: $\tilde{S}_{21}$ = 3.1<u>1</u>3...- 3.1<u>3</u>3..., $\angle\tilde{S}_{21}$ = (8<u>1</u>.123... - 8<u>3</u>.123...)°.

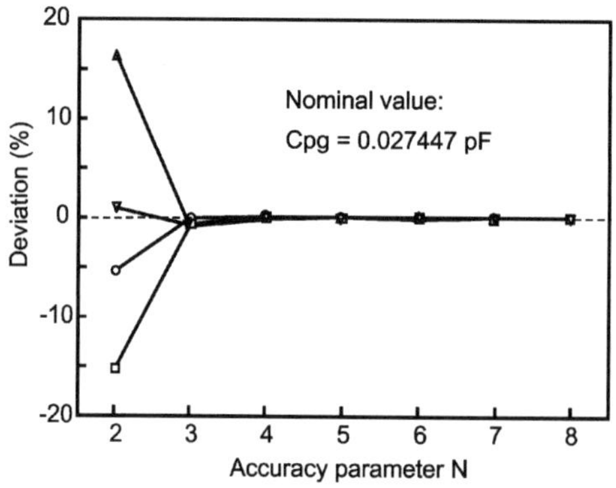

Figure 17.7 Deviation of extracted C_{pg} from the nominal value as a function of accuracy parameter N based on different frequency combinations (○ = 10, 20, 30, 40; □ = 5, 16, 25, 40; Δ = 5, 16, 32, 40; ∇ = 14, 21, 30, 36).

As an example, Figure 17.7 illustrates the simulated deviation of the nominal value of the extracted model element C_{pg} as a function of the absolute accuracy of scattering parameters for four combinations of four frequencies. It turns out that the extraction is unique with an accuracy parameter $N \geq 3$. Greater variances are observed with $N \leq 2$. However, for some combinations useful solutions could also be found in this region.

The followed limit of $N \geq 3$ for the extraction from noisy data means that there is a very high requirement regarding the accuracy of scattering parameters (measurement accuracy requirement for unique extraction):

$$\Delta|S_{ij}| \approx 0.001, \qquad \measuredangle S_{ij} \approx 0.10°, \qquad (i,j = 1,2) \tag{17.71}$$

Further studies have shown that extraction consistency will be increased if one of the four parasitic resistances R_g, R_i, R_s, or R_d is known. Because of bias independence, R_g was selected as the element to be predetermined. Appropriate value may be estimated from the device's physical data (Section 6.2.3) or measurement (Appendix B) .

17.5 MEASUREMENT-BASED ANALYTICAL MODEL PARAMETER EXTRACTION

S-parameter measurements in the frequency range of 1 GHz to 40 GHz were used to apply the analytical extraction method to the model parameter extraction of the model elements of 0.5-μm gate MESFET NE70000 (V_{GS} = 0.05V, V_{DS} = 3.0V, I_{DS} = 30 mA). In contrast to synthetic S-parameters, real data includes measurement uncertainties, and in particular, systematic errors due to model mismatching. We have seen that the analytic algorithm shows highly sensible response to only slight model mismatching. In order to properly meet this weakness, the analytic extraction method was applied to four four-frequency sets instead of a minimum single four-frequency set.

The frequency combinations in Table 17.2 were selected for extraction of the pad capacitances . Note in Table 17.4, for the extraction of R_d and R_s, lower frequency group f(3)

Table 17.2
Combinations of Chosen Lower, Medium, and Upper Frequencies for the Extraction of Parasitic Capacitances C_{pg} and C_{pd} Using 256 Frequency Combinations

Frequency Group	Selected Frequencies			
	f_1 / GHz	f_2 / GHz	f_3 / GHz	f_4 / GHz
f(1)	4.0	4.8	5.6	6.4
f(2)	15.2	16.0	18.4	19.2
f(3)	25.6	26.4	27.2	28.0
f(4)	35.2	37.6	39.2	40.0

Table 17.3
Combinations of Chosen Lower, Medium, and Upper Frequencies for the Extraction of Parasitic Inductances L_g and L_d Using 128 Frequency Combinations

Frequency Group	Selected Frequencies			
	f_1 / GHz	f_2 / GHz	f_3 / GHz	f_4 / GHz
f(2)	15.2	16.0	18.4	19.2
f(3)	25.6	26.4	27.2	28.0
f(4)	35.2	37.6	39.2	40.0

Table 17.4
Two Used Frequencies for the Extraction of Parasitic Resistances R_d and R_s Using 100 Frequency Combinations

Frequency Group	Selected Frequencies			
	f_1 / GHz	f_2 / GHz	f_3 / GHz	f_4 / GHz
f(1)	4.0	4.8	5.6	6.4
f(3)	25.6	26.4	27.2	28.0

was taken instead of higher frequency group f(4) because of lower measurement uncertainty of the used vector network analyzer (see Section 11.3.1.3).

For the determination of the inductances L_g and L_d scattering parameters are required at least at three frequencies. The chosen set of frequencies are given in Table 17.3.

Figures 17.7 to 17.9 represent the calculated results. It can be seen that all bar graphs have a clear maximum so that a mean value can be determined consistently. The shape of the graphs approximate normal distributions. In particular, the scattering results of the resistance values are very moderate.

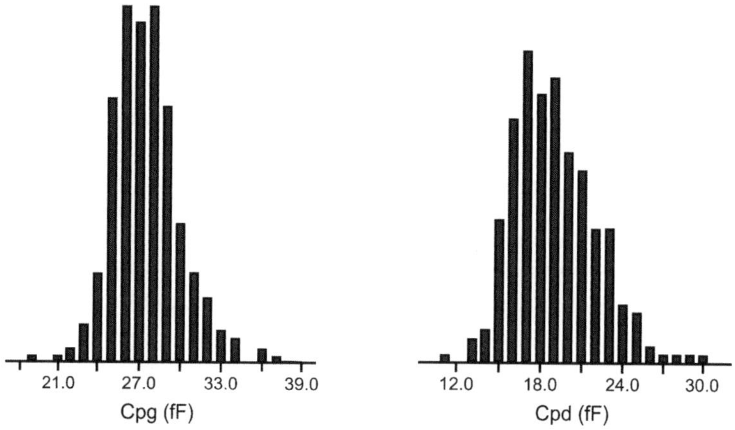

Figure 17.7 Distribution of extracted values for C_{pg} and C_{pd} using 256 frequency combinations of measured scattering parameters in the frequency range from 45 MHz to 40 GHz (see Table 17.2). The mean values of C_{pg} and C_{pd} are given as 27.0 fF and 18.0 fF, respectively.

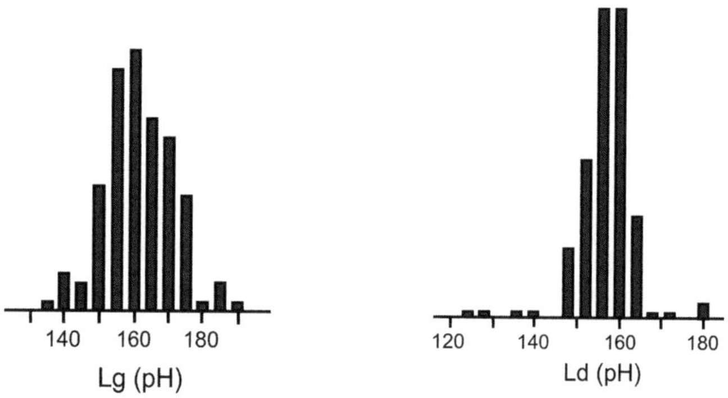

Figure 17.8 Distribution of extracted values for L_g and L_d using 128 frequency combinations of measured scattering parameters in the frequency range from 45 MHz to 40 GHz (see Table 17.3). The mean values of L_g and L_d are given as 161 pH and 157 pH, respectively.

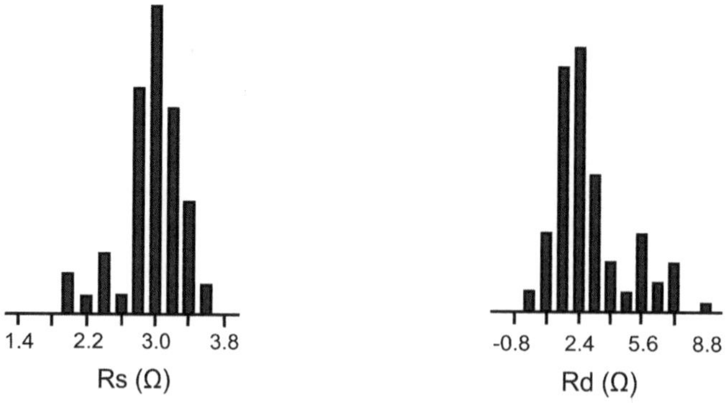

Figure 17.9 Distribution of extracted values for R_s and R_d using 100 frequency combinations of measured scattering parameters in the frequency range from 45 MHz to 40 GHz (see Table 17.4). The mean values of R_s and R_d are given as 3Ω and 2.4Ω, respectively.

17.6 SUMMARY

A fully analytical method for the extraction of all model parameters of a 15-element FET equivalent network was presented. The extraction method, using only *hot* scattering parameters in a single bias point, is extremely fast and eliminates the starting vector problem.

Extraction accuracy was first investigated on the basis of fictitious *S*-parameters. It has been shown that the extraction accuracy depends strongly on the available bandwidth of S-parameter data. The used S-parameter data sets must be sufficiently linear-independent.

On the basis of four data sets of fictitious S-parameters, each comprising scattering coefficients at four widely spaced frequencies, the model element values of a 15-element model can be uniquely determined. However, introducing model mismatching, which is simulated by the ongoing truncation of significant digits of the fictitious scattering coefficients, reveals some weakness of the method. It is shown that the method is extremely susceptible to input data uncertainties. The influence of limited accuracy in the S-parameter data on the extraction of the model parameters was discussed. It was found that one resistance value of R_s, R_g, R_d, or R_i should be predefined to keep the stability of all other extracted parameters. It is reasonable to choose R_g because this parameter is bias independent.

Improved calculated results could be obtained by using a multiple set of four-frequency combinations. This is demonstrated in particular when applying the analytical method onto measured data including model mismatching and measurement uncertainties. It was shown that with four four-frequency sets consistent results for a 15-element model under active bias operation could be obtained.

As the analytic approach is extremely fast, the method may be further refined by extending the number of four-frequency sets. The extension of frequency range could also minimize the uncertainty of the calculated results. This would be appropriate for transistors operated at higher frequencies.

The selection of the four-frequency sets could be automated. As well, relaxing the pure analytical claim, the method may be combined with optimization features. This has not yet been tested, for example, to solve (17.31) by the least-squares method.

REFERENCES

[1] Lin, F., *A Method for Reliable Experimental Modeling of Microwave FETs* (in German), Doctoral Thesis, University of Kassel, 1993; published by VDI Verlag, Düsseldorf, Fortschritt-Berichte VDI, Reihe 9 (Elektronik), Nr. 164, 1993.

[2] Kompa, G., and F. Lin, "New Algorithm Approach for Consistent Model Parameter Extraction Method of GaAs MESFET Equivalent Circuits," *Proc. of Microwaves and Optronics Conf.*, Stuttgart, Germany, 1990, pp. 257–262

[3] Kompa, G., and F. Lin, "FET Modelling Using an Analytic Extraction Method Based on Broadband S-Parameter Measurement," *European Microwave Conf.*, 1990, pp. 778–783.

[4] Engeln-Müllges, and F. Uhlig, *Numerical Algorithms with C*, Springer-Verlag, 1996.

[5] Haller, X., *Theoretical Investigation on Consistent Model Parameter Extraction of Microwave Field-Effect Transistors* (in German), Master Thesis, University of Kassel, 1991.

[6] Dambrine, G., A. Cappy, F. Heliodore, and E. Playez, "A New Method for Determining the FET Small-Signal Equivalent Circuit," *IEEE Trans.*, Vol. MTT-36, 1988, pp. 1151–1159.

Chapter 18

Analytical Model Parameter Extraction Using Rational Functions

18.1 INTRODUCTION

Mostly optimizer-based algorithms have been discussed so far, such as the *All-at-Once* model parameter extraction (Chapter 14), decomposition-based model parameter-extraction (Chapter 15), and the bidirectional search method (Chapter 16).

A pure analytical model parameter extraction was discussed in Chapter 17, which is based on the solution of an analytical $(m \times n)$ matrix equation with $m = 4$ (number of equations) and $n = 14$ unknown variables of the model elements from measured (or hypothetical) S-parameters at a single frequency point for a given bias point. As the linear set of equations is underdetermined, at least S-parameters at four frequencies are needed for the calculation of the 14-model elements. It was concluded that the method is sensitive with respect to measurement errors and model mismatching.

The model parameter extraction method discussed in the following differs significantly from the previously described extraction approaches. The application includes two steps. First, the frequency-dependent response of a small-signal model is simulated by a rational function. The fitting of this function onto measured S-parameters can be done with the highest accuracy. The more complex the equivalent network is, the higher the degree of the polynomials. In reality, there should be a compromise between reasonable complexity of topology and mathematical effort. Indeed, the coefficients of the polynomials can be determined with high accuracy. But in the second step, the model parameters have to be determined from the polynomial coefficients. As these frequently denote combinations of single-model elements, this task may become sophisticated. If this might be successfully resolved, then the agreement between simulated and measured S-parameters are commonly striking.

In the following, a detailed description of the analytical model parameter extraction method using rational functions will be given applied to GaAs-based FET devices. In a second step, the application to HBT devices is briefly addressed.

18.2 RATIONAL FUNCTIONS-BASED PARAMETER EXTRACTION

Wasige et al. reported about an analytical model parameter extraction procedure using rational functions [1, 2]. All small-signal equivalent circuit elements of the 17-element FET model depicted in Figure 18.1, are determined under active bias conditions. The extraction problem is first considered as a fitting problem that is analytically solved using the method of least squares by modeling the measured data using rational functions. The equivalent circuit elements are then directly calculated from coefficients of the rational functions. The extraction results show very good agreement between measurement and simulation.

Extraction on the basis of rational function approximation of the measured S-parameters using the method of least squares was, to the best of the authors knowledge, first reported by Nagatomo et al. [3] analyzing a 13-element model excluding C_{pgs} and C_{pds}. The order of numerator and denominator of the rational functions was empirically determined even though this can be determined automatically by an appropriate software package. With an increased number of model elements, the order of rational functions increases. Nagatomo et al. pointed out that since higher-order coefficients of ω are smaller, these contain relatively large errors. As a result, it is difficult to extract extrinsic elements that correspond to higher-order coefficients. Consequently, the six extrinsic parasitic model parameters were determined separately using an iterative optimization approach. This means that the analytical direct parameter extraction was restricted only to the intrinsic part of the equivalent model.

In the following the analytical extraction procedure proposed in [1] is described and extraction results are discussed. This approach redefines the model parameter extraction problem by first using the method of least squares to fit the chosen equivalent circuit topology, defined explicitly by its S-, Y-, or Z-parameters as rational functions of frequency, to the measured data.

18.2.1 Least-Squares Approximation by Rational Functions

We suppose that the S-, Y-, or Z-parameters of a two-port can be generally expressed by a rational function $F(\omega)$,

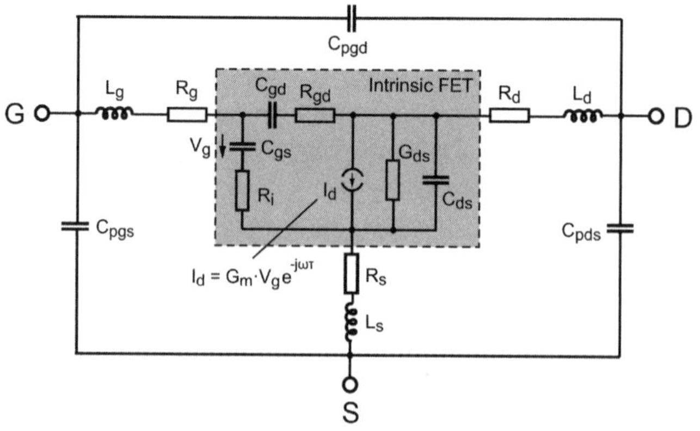

Figure 18.1 Analyzed 17-element small-signal FET equivalent circuit.

$$F(j\omega) = \frac{A(j\omega)}{B(j\omega)} \tag{18.1}$$

where

$$A(j\omega) = \sum_{i=0}^{M} a_i (j\omega)^i \quad \text{and} \quad B(j\omega) = 1 + \sum_{j=1}^{N} b_j (j\omega)^j \tag{18.2}$$

are polynomials in angular frequency ω and a_i and b_i are the coefficients of the polynomials. The task now is to determine first the coefficients a_i and b_j from the measured F-parameters, which denote S-, Y- or Z-parameters. Then, knowing the coefficients, the model elements can be calculated.

At a discrete sampled frequency point ω_k the rational function $F(j\omega)$ reads

$$F(j\omega_k) = \frac{\sum_{i=0}^{M} a_i (j\omega_k)^i}{1 + \sum_{j=1}^{N} b_j (j\omega_k)^j} \tag{18.3}$$

Deviation from the corresponding measured value $F_m(\omega_k)$ is

$$\varepsilon_k = F_m(\omega_k) - \frac{\sum_{i=0}^{M} a_i (j\omega_k)^i}{1 + \sum_{j=1}^{N} b_j (j\omega_k)^j} \tag{18.4}$$

where index k denotes the measured frequency point. Multiplying the denominator and taking the sum of the squared absolute value through all data points, the following error function ε is defined:

$$\varepsilon = \sum_k \left| \left\{ 1 + \sum_{j=1}^{N} b_j (j\omega_k)^j \right\} F_m(\omega_k) - \sum_{i=0}^{M} a_i (j\omega_k)^i \right|^2 \tag{18.5}$$

With $|A|^2 = A \cdot A^*$ for a complex variable A, (18.5) can be rewritten as

$$\varepsilon = \sum_k \left[\sum_{i=0}^{M} a_i (j\omega_k)^i - \left(1 + \sum_{j=1}^{N} b_j (j\omega_k)^j \right) F_m(\omega_k) \right] \left[\sum_{i=0}^{M} a_i (j\omega_k)^i - \left(1 + \sum_{j=1}^{N} b_j (j\omega_k)^j \right) F_m(\omega_k) \right]^* \tag{18.6}$$

The coefficients a_i and b_j, which minimize the error function value ε, can be determined by solving the following simultaneous equations:

$$\frac{\partial \varepsilon}{\partial a_m} = 0, \quad m = 0, 1, \ldots, M \tag{18.7}$$

$$\frac{\partial \varepsilon}{\partial b_n} = 0, \quad n = 0, 1, \ldots, N \tag{18.8}$$

Note that $a_i(j\omega)$ and $b_j(j\omega)$ denote multiplication of a_i and b_j with $(j\omega)$, whereas $F_m(\omega)$ is a function of ω. Substituting (18.6) into (18.7) and (18.8) yields

$$\begin{aligned}\frac{\partial\varepsilon}{\partial a_m} = \sum_k & \left[\sum_{i=0}^{M} a_i (j\omega_k)^i - \left(1 + \sum_{j=1}^{N} b_j (j\omega_k)^j\right) F_m(\omega_k) \right] (-j\omega)^m \\ & + \left[\sum_{i=0}^{M} a_i (j\omega_k)^i - \left(1 + \sum_{j=1}^{N} b_j (j\omega_k)^j\right) F_m(\omega_k) \right]^* (j\omega)^m = 0\end{aligned} \tag{18.9}$$

$$\begin{aligned}\frac{\partial\varepsilon}{\partial b_n} = \sum_k & \left[\sum_{i=0}^{M} a_i (j\omega_k)^i - \left(1 + \sum_{j=1}^{N} b_j (j\omega_k)^j\right) F_m(\omega_k) \right] (-j\omega)^n F_m^*(\omega_k) \\ & + \left[\sum_{i=0}^{M} a_i (j\omega_k)^i - \left(1 + \sum_{j=1}^{N} b_j (j\omega_k)^j\right) F_m(\omega_k) \right]^* (j\omega)^n F_m(\omega_k) = 0\end{aligned} \tag{18.10}$$

Expanding (18.9) gives

$$\begin{aligned}\frac{\partial\varepsilon}{\partial a_m} = & \sum_k \sum_{i=0}^{M} a_i \left[(j\omega_k)^i (-j\omega_k)^m + (-j\omega_k)^i (j\omega_k)^m \right] - \sum_k \left[(-j\omega_k)^m F_m(\omega_k) + (j\omega)^m F_m^*(\omega_k) \right] \\ & - \sum_k \sum_{j=1}^{N} b_j \left[(j\omega_k)^j (-j\omega_k)^m F_m(\omega_k) + (-j\omega_k)^j (j\omega_k)^m F_m^*(\omega_k) \right] = 0\end{aligned} \tag{18.11}$$

With $X + X^* = 2\mathrm{Re}(X)$ for complex variables, (18.11) reads

$$\begin{aligned}\sum_k \left[\sum_{i=0}^{M} a_i \mathrm{Re}\left\{ (j\omega_k)^i (-j\omega_k)^m \right\} - \sum_{j=1}^{N} b_j \mathrm{Re}\left\{ (j\omega_k)^j (-j\omega_k)^m F_m(\omega_k) \right\} \right] \\ = \sum_k \mathrm{Re}\left\{ (j\omega_k)^m F_m(\omega_k) \right\}\end{aligned} \tag{18.12}$$

Similarly, from (18.10) we obtain

$$\begin{aligned}\sum_k \left[\sum_{i=0}^{M} a_i \mathrm{Re}\left\{ (j\omega_k)^i (-j\omega_k)^n \right\} - \sum_{j=1}^{N} b_j \mathrm{Re}\left\{ (j\omega_k)^j (-j\omega_k)^n |F_m(\omega_k)|^2 \right\} \right] \\ = \sum_k \mathrm{Re}\left\{ (-j\omega_k)^n |F_m(\omega_k)|^2 \right\}\end{aligned} \tag{18.13}$$

Summarizing, (18.12) and (18.13) provide a system of equations whose order is dependent on M and N [1],

$$\sum_k \begin{bmatrix} \mathrm{Re} \sum_{i=0}^{M} [(j\omega)^i (-j\omega)^m] & -\mathrm{Re} \sum_{j=1}^{N} [(j\omega)^j (-j\omega)^m F_m(\omega_k)] \\ \mathrm{Re} \sum_{i=0}^{M} [(j\omega)^i (-j\omega)^n F_m^*(\omega_k)] & -\mathrm{Re} \sum_{j=1}^{N} [(j\omega)^j (-j\omega)^n |F_m(\omega_k)|^2] \end{bmatrix} \begin{bmatrix} [a_i] \\ [b_j] \end{bmatrix} = \sum_k \begin{bmatrix} \mathrm{Re}\, F_m(\omega_k)(-j\omega)^m \\ \mathrm{Re}\, |F_m(\omega_k)|^2 (-j\omega)^n \end{bmatrix} \tag{18.14}$$

or more explicitly,

$$\begin{aligned}\sum_k \sum_{p,q} & \begin{bmatrix} \mathrm{Re} \sum_{i=0}^{M} [(j\omega)^i (-j\omega)^m] & -\mathrm{Re} \sum_{j=1}^{N} [(j\omega)^j (-j\omega)^m F_{m,pq}(\omega_k)] \\ \mathrm{Re} \sum_{i=0}^{M} [(j\omega)^i (-j\omega)^n F_{m,pq}^*(\omega_k)] & -\mathrm{Re} \sum_{j=1}^{N} [(j\omega)^j (-j\omega)^n |F_{m,pq}(\omega_k)|^2] \end{bmatrix} \begin{bmatrix} \begin{bmatrix} a_{0,pq} \\ \vdots \\ a_{M,pq} \end{bmatrix} \\ \begin{bmatrix} b_1 \\ \vdots \\ b_N \end{bmatrix} \end{bmatrix} \\ & = \sum_k \sum_{p,q=1} \begin{bmatrix} \mathrm{Re}\, F_{m,pq}(\omega_k)(-j\omega)^m \\ \mathrm{Re}\, |F_{m,pq}(\omega_k)|^2 (-j\omega)^n \end{bmatrix}\end{aligned} \tag{18.15}$$

Equation (18.15) is a system of linear equations that can be analytically solved to yield the required coefficients a_i and b_j. By carefully choosing the form of the measured parameter, for FET devices either S-, Y- or Z-parameters, and/or appropriately scaling the frequency, a system of well-conditioned equations can be obtained and a precise determination of the coefficients is possible [1]. It can also be noted that this system of equations consists of only real parts of different expressions involving the measured parameter and frequency. This can be utilized to minimize numerical noise, thereby improving the accuracy in determining the coefficients.

18.2.2 Extrinsic and Intrinsic Y-Parameters

We consider the 17-element FET model in Figure 18.1. The determination of the model parameter elements is done sequentially. First, the parallel parasitic capacitances are calculated and de-embedded from the measured data. Next, the series parasitics are calculated and also de-embedded. In a final step, the intrinsic elements are directly calculated. An optimization algorithm is not required. The measured or de-embedded (measured) data is represented by a set of calculated coefficients from which the model elements are evaluated.

Starting with the Y-parameters of the intrinsic FET, we obtain the Y-matrix of the entire model topology by an embedding procedure as follows. The Y-matrix of the intrinsic FET device (Y_{int}) reads

$$Y_{int} = \begin{bmatrix} Y_{11} & Y_{12} \\ Y_{21} & Y_{22} \end{bmatrix} = \begin{bmatrix} Y_{gs} + Y_{gd} & -Y_{gd} \\ Y_{gm} - Y_{gd} & Y_{ds} + Y_{gd} \end{bmatrix} \tag{18.16}$$

with

$$Y_{gs} = \frac{j\omega C_{gs}}{1 + j\omega R_i C_{gs}} \tag{18.17}$$

$$Y_{gd} = \frac{j\omega C_{gd}}{1 + j\omega R_{gd} C_{gd}} \tag{18.18}$$

$$Y_{gm} = \frac{G_m e^{-j\omega\tau}}{1 + j\omega R_i C_{gs}} \tag{18.19}$$

$$Y_{ds} = G_{ds} + j\omega C_{ds} \tag{18.20}$$

To describe all parameters as rational functions of frequency, the exponential term of the transconductance is approximated by the first two terms of its power series expansion; that is,

$$G_m e^{j\omega\tau} \approx 1 - j\omega\tau \tag{18.21}$$

Conversion into Z-parameters of intrinsic FET (Z_{int}) gives

$$Y_{int} = \begin{bmatrix} Y_{11} & Y_{12} \\ Y_{21} & Y_{22} \end{bmatrix} \rightarrow Z_{int} = \begin{bmatrix} Z_{11} & Z_{12} \\ Z_{21} & Z_{22} \end{bmatrix} = \frac{1}{\Delta}\begin{bmatrix} Y_{22} & -Y_{12} \\ -Y_{21} & Y_{11} \end{bmatrix} \tag{18.22}$$

with $\Delta = Y_{11}Y_{22} - Y_{12}Y_{21}$.

The parasitic series Z-parameters (Z_{ser}) comprising the series elements L_g, L_d, L_s, R_g, R_d, R_s are then added to the intrinsic Z-parameters (Z_{int}):

$$Z_T = Z_{int} + Z_{ser} = \begin{bmatrix} Z_{11} & Z_{12} \\ Z_{21} & Z_{22} \end{bmatrix} + \begin{bmatrix} Z_g + Z_s & Z_s \\ Z_s & Z_d + Z_s \end{bmatrix} \tag{18.23}$$

$$Z_T = \begin{bmatrix} \dfrac{Y_{ds} + Y_{gd}}{\Delta} + Z_g + Z_s & \dfrac{Y_{gd}}{\Delta} + Z_s \\ \dfrac{-Y_{gm} + Y_{gd}}{\Delta} + Z_s & \dfrac{Y_{gs} + Y_{gd}}{\Delta} + Z_d + Z_s \end{bmatrix} \tag{18.24}$$

Next, matrix Z_T is converted into Y_T-matrix:

$$Z_T = \begin{bmatrix} Z_{T11} & Z_{T12} \\ Z_{T21} & Z_{T22} \end{bmatrix} \rightarrow Y_T = \begin{bmatrix} Y_{T11} & Y_{T12} \\ Y_{T21} & Y_{T22} \end{bmatrix} \tag{18.25}$$

Adding the parasitic Y_{pi}-matrix yields the Y-parameters of the complete equivalent circuit:

$$Y_{ec} = Y_T + Y_{pi} = \begin{bmatrix} Y_{T11} & Y_{T12} \\ Y_{T21} & Y_{T22} \end{bmatrix} + \begin{bmatrix} Y_{pgs} + Y_{pgd} & -Y_{pgd} \\ -Y_{pgd} & Y_{pds} + Y_{pgd} \end{bmatrix} \tag{18.26}$$

Using mathematical software tools (e.g., Mathcad® or Mathematica® [4]), which provide analytical simplification of mathematical equations involving only variables or symbols, it is easier to show that the Y-parameters of (18.26) can be written as polynomials of order 8 and 7 for the numerator and denominator, respectively. These polynomials, however, used to determine the parasitic capacitances have many and very long expressions of the coefficients and are not suitable for illustrating the concept. For example, as shown in [2], the presentation of the rational function of matrix element Y_{12} requires about 500 lines. Therefore, it would distract from the simple message to describe the direct determination of the extrinsic parasitic capacitances in detail.

18.2.3 Direct Model Parameter Extraction

The measured S-parameters of the device are converted into Y-parameters first: $S_m \rightarrow Y_m$. With determined $N = 7$ and $M = 8$ and measured response $F_m(\omega_k) = Y_m(\omega_k)$, the coefficients a_i and b_j in (18.15) can be calculated. As $M > N$ the rational function $f(j\omega)$ can be expanded as

$$f(j\omega) = \frac{h(j\omega)}{g(j\omega)} = p(j\omega) + \frac{q(j\omega)}{g(j\omega)} \tag{18.27}$$

with $p(j\omega)$ of order $(M - N)$. With $Y = f(j\omega)$; that is,

$$Y_{ec} = Y_{pi} + Y_T = \frac{h(j\omega)}{g(j\omega)} = p(j\omega) + \frac{q(j\omega)}{g(j\omega)} \tag{18.28}$$

$p(j\omega) = Y_{pi}$, which, regarding the power of the fractional components, presents the extrinsic parasitic Y-components. In this way, the extrinsic parasitic capacitances can be determined from the measured parameters. Then the Y_T-matrix can be calculated by de-embedding the parasitic components (Y_{pi}) from the measured parameters (Y_m):

$$Y_{T,m} = Y_m - Y_{pi} = Y_m - \begin{bmatrix} Y_{pgs} + Y_{pgd} & -Y_{pgd} \\ -Y_{pgd} & Y_{pds} + Y_{pgd} \end{bmatrix} \tag{18.29}$$

After the parasitic capacitances have been de-embedded from the measured Y-parameters, conversion into Z-parameters is done: $Y_{T,m} \rightarrow Z_{T,m}$. Recalling (18.24), $Z_{T,12} = Y_{gd}/\Delta + Z_s$ occurs in all components. To simplify the following calculation, $Z_{T,12}$ is subtracted:

$$Z_T' = Z_T - \begin{bmatrix} Z_{T12} & 0 \\ Z_{T12} & Z_{T12} \end{bmatrix} = \begin{bmatrix} \frac{Y_{ds}}{\Delta} + Z_g & \frac{Y_{gd}}{\Delta} + Z_s \\ \frac{-Y_{gm}}{\Delta} & \frac{Y_{gs}}{\Delta} + Z_d \end{bmatrix} \tag{18.30}$$

As before, the Z-parameters can be explicitly represented as rational functions using software tools like Mathematica®:

$$Z_T' = f(j\omega) = \frac{h(j\omega)}{g(j\omega)} \tag{18.31}$$

In this case the order of numerator and denominator is $M = 3$ and $N = 2$, respectively. As Z_{12} has been subtracted from three components of the Z-matrix of the model parameters, the same is done with the measured ones:

$$Z_T' = Z_T - Z_{T,m,12} \tag{18.32}$$

With $F_m(\omega) = Z_{T,m}'$ and $M = 3$ and $N = 2$, the coefficients a_i and b_j, which describe the function $f(j\omega) = Z_T'$, can be determined from (18.14). Setting $Z_T' = Z_{T,m}' = f(j\omega)$, then from (18.27) it follows

$$Z_g, Z_d, Z_s = p(j\omega) \tag{18.33}$$

All the steps can be evaluated symbolically with the help of a mathematical software package such as Mathematica [4] to explicitly determine the coefficients of the resulting rational functions.

The determination of the parasitic and intrinsic elements of the device can be provided by coefficient comparison. In this case the coefficients obtained from measured parameters are compared with the analytically obtained coefficients $a_{i,pq}$ and b_j (using Mathematica®). The parasitic components are calculated first and subtracted from the equivalent circuit Y-parameters [Y_{ec}]. Then the resulted Y-parameters are converted into Z-parameters. After the parallel parasitic capacitances have been determined and de-embedded, the measured data is converted to Z-parameters. Then the following rational functions for the components of the Z_T'matrix can be found (x = jω):

Z_{11} = (1 + x*Cgd*Rgd + x*Cgs*Ri + x*Cds*Rds + x^2*Cds*Rds*Cgd*Rgd + x^2*Cgs*Ri*Cds*Rds
+ Rg*x*Cgs + Rg*x*Cgd + x^2*Lg*Cgs + x^2*Lg*Cgd + Rg*x^3*Cgs*Cds*Rds*Cgd*Rgd
+ Rg*x^3*Cgd*Cgs*Ri*Cds*Rds + Rg*x^2*Cgs*Cgd*Rgd + Rg*x^2*Cgs*Cgd*Rds
- Rg*x^2*Cgd*Rds*Gm*Tau + Rg*x^2*Cgd*Cds*Rds + Rg*x^2*Cgd*Cgs*Ri + Rg*x^2*Cgs*Cds*Rds

+ Rg*x*Cgd*Rds*Gm + x^4*Lg*Cgs*Cds*Rds*Cgd*Rgd + x^4*Lg*Cgd*Cgs*Ri*Cds*Rgd
+ x^3*Lg*Cgs*Cgd*Rgd + x^3*Lg*Cgs*Cgd*Rds - x^3*Lg*Cgd*Rds*Gm*Tau
+ x^3*Lg*Cgd*Cds*Rds + x^3*Lg*Cgd*Cgs*Ri + x^3*Lg*Cgs*Cds*Rds + x^2*Lg*Cgd*Rds*Gm
+ x^2*Cgs*Ri*Cgd*Rgd + x^3*Cgs*Ri *Cds*Rds*Cgd*Rgd)/ D (18.34)

Z_{12} = (Cgd*Rds + Cgd*Rs + Cgs*Rs + Cgd*Gm*Rds*Rs + Cgd*Ls*x + Cgs*Ls*x + Cgd*Gm*Ls*Rds*x
+ Cgd*Cgs*Rds*Ri*x + Cds*Cgd*Rds*Rs*x + Cds*Cgs*Rds*Rs*x + Cgd*Cgs*Rds*Rs*x
+ Cgd*Cgs*Rgd*Rs*x + Cgd*Cgs*Ri*Rs*x - Cgd*Gm*Rds*Rs*Tau*x + Cds*Cgd*Ls*Rds*x^2
+ Cds*Cgs*Ls*Rds*x^2 + Cgd*Cgs*Ls*Rds*x^2 + Cgd*Cgs*Ls*Rgd*x^2 + Cgd*Cgs*Ls*Ri*x^2
+ Cds*Cgd*Cgs*Rds*Rgd*Rs*x^2 + Cds*Cgd*Cgs*Rds*Ri*Rs*x^2 + Cgd*Gm*Ls*Rds*Tau*x^2
+ Cds*Cgd*Cgs*Ls*Rds*Rgd*x^3 + Cds*Cgd*Cgs*Ls*Rds*Ri*x^3) / D (18.35)

Z_{21} = Gm*(1 - xTau)*Rds*(1 + x*Cgd*Rgd) / D (18.36)

Z_{22} = (Rd*Cgs*x^2*Cds*Cgd*Rgd + Rd*Cgd*x^2*Cgs*Ri*Cds*Rds + Rd*Cgs*x*Cgd*Rgd
+ Rd*Cgs*x*Cgd*Rds - Rd*Cgd*Rds*x*Gm*Tau + Rd*Cgd*x*Cds*Rds + Rd*Cgd*x*Cgs*Ri
+ Rd*Cgs*x*Cds*Rds + Rd*Cgs + Rd*Cgd*Rds*Gm + Rd*Cgd + x^3*Ld*Cgs*Cds*Rds*Cgd*Rgd
+ x^3*Ld*Cgd*Cgs*Ri*Cds*Rds + x^2*Ld*Cgs*Cgd*Rgd + x^2*Ld*Cgs*Cgd*Rds
- x^2*Ld*Cgd*Rds*Gm*Tau + x^2*Ld*Cgd*Cds*Rds + x^2*Ld*Cgd*Cgs*Ri + x^2*Ld*Cgs*Cds*Rds
+ x*Ld*Cgs + x*Ld*Cgd*Rds*Gm + x*Ld*Cgd + Cgs*Rds + Cgs*Rds*x*Cgd*Rgd) / D (18.37)

with

D = Cgd + Cgs + Cgd*Gm*Rds + Cds*Cgd*Rds*x + Cds*Cgs*Rds*x + Cgd*Cgs*Rds*x +
Cgd*Cgs*Rgd*x + Cgd*Cgs*Ri*x - Cgd*Gm*Rds*Tau*x + Cds*Cgd*Cgs*Rds*Rgd*x^2
+ Cds*Cgd*Cgs*Rds*Ri*x^2

Note that in this case the degree of the numerator and denominator polynomials is 3 and 2, respectively, and the polynomial coefficients are explicitly expressed in terms of the equivalent circuit elements. Therefore, by setting $M = 3$ and $N = 2$ and applying (18.11) on the measured Z_{12}, coefficients having a one-to-one correspondence to products of equivalent circuit elements can be calculated. All the measured four parameters in (18.30) are usually simultaneously considered using a correspondingly expanded version of (18.11).

Equation (18.30) is treated with Mathematica® and delivers, for example, for Z_{12}:

$$Z_{12} = \frac{a_{0,12} + a_{1,12}(j\omega) + a_{2,12}(j\omega)^2 + a_{3,12}(j\omega)^3}{1 + b_1(j\omega) + b_2(j\omega)^2} \tag{18.38}$$

The numerator coefficients are identified as

$$\begin{aligned} a_{0,12} &= R_s + \frac{C_{gd}R_{ds}}{C_{gd} + C_{gs} + C_{gd}G_mR_{ds}} \\ a_{1,12} &= L_s + b_1R_s + \frac{C_{gd}R_{ds}C_{gs}R_i}{C_{gd} + C_{gs} + C_{gd}G_mR_{ds}} \\ a_{2,12} &= b_1L_s + b_2R_s \\ a_{3,12} &= b_2L_s \end{aligned} \tag{18.39}$$

The denominator polynom $[1 + b_1(j\omega) + b_2(j\omega)^2]$ with the coefficients

$$b_1 = \frac{(C_{ds}C_{gd}R_{ds} + C_{ds}C_{gs}R_{ds} + C_{gd}C_{gs}R_{ds} + C_{gd}C_{gs}R_{gd} + C_{gd}C_{gs}R_i -.C_{gd}G_mR_{ds}\tau)}{(C_{gd} + C_{gs} + C_{gd}G_mR_{ds})}$$
$$b_2 = \frac{(C_{ds}C_{gd}C_{gs}R_{ds}R_{gd} + C_{ds}C_{gd}C_{gs}R_{ds}R_i)}{(C_{gd} + C_{gs} + C_{gd}G_mR_{ds})} \quad (18.40)$$

is equal for all Z-components. Analog to Z_{12}, the coefficients of the numerator of Z_{11} are obtained as

Z_{11}:

$$a_{0,11} = \frac{1}{C_{gd} + C_{gs} + C_{gd}G_mR_{ds}}$$
$$a_{1,11} = R_g + \frac{C_{ds}R_{ds} + C_{gd}R_{gd} + C_{gs}R_i}{C_{gd} + C_{gs} + C_{gd}G_mR_{ds}}$$
$$a_{2,11} = L_g + b_1R_g + \frac{C_{ds}R_{ds}C_{gd}R_{gd} + C_{ds}R_{ds}C_{gs}R_i + C_{gd}R_{gd}C_{gs}R_i}{C_{gd} + C_{gs} + C_{gd}G_mR_{ds}} \quad (18.41)$$
$$a_{3,11} = b_1L_g + b_2R_g + \frac{C_{ds}R_{ds}C_{gd}R_{gd}C_{gs}R_i}{C_{gd} + C_{gs} + C_{gd}G_mR_{ds}}$$
$$a_{4,11} = b_2L_g$$

Z_{21}:

$$a_{0,21} = \frac{G_mR_{ds}}{C_{gd} + C_{gs} + C_{gd}G_mR_{ds}}$$
$$a_{1,21} = \frac{C_{gd}G_mR_{ds}R_{gd} + C_{gs}G_mR_{ds}R_{gi} - G_mR_{ds}\tau}{C_{gd} + C_{gs} + C_{gd}G_mR_{ds}}$$
$$a_{2,21} = \frac{C_{gd}C_{gs}G_mR_{ds}R_{gd}R_i - C_{gd}G_mR_{ds}R_{gd}\tau - C_{gs}G_mR_{ds}R_i\tau}{C_{gd} + C_{gs} + C_{gd}G_mR_{ds}} \quad (18.42)$$
$$a_{3,21} = \frac{C_{gd}C_{gs}G_mR_{ds}R_{gd}R_i\tau}{C_{gd} + C_{gs} + C_{gd}G_mR_{ds}}$$

Z_{22}:

$$a_{0,22} = R_d + \frac{C_{gs}R_{ds}}{C_{gd} + C_{gs} + C_{gd}G_mR_{ds}}$$
$$a_{1,22} = L_d + b_1R_d + \frac{C_{gd}R_{gd}C_{gs}R_{ds}}{C_{gd} + C_{gs} + C_{gd}G_mR_{ds}} \quad (18.43)$$
$$a_{2,22} = b_1L_d + b_2R_d$$
$$a_{3,22} = b_2L_d$$

With these coefficients the model elements of the transistor can be expressed as

$$L_g = a_{4,11}/b_2 \quad (18.44)$$

$$L_s = a_{3,12}/b_2 \tag{18.45}$$

$$L_d = a_{3,22}/b_2 \tag{18.46}$$

$$R_d = (a_{2,22} - b_1 L_d)/b_2 \tag{18.47}$$

$$R_s = (a_{2,12} - b_1 L_s)/b_2 \tag{18.48}$$

$$R_g = \frac{a_{3,11} - b_1 L_g - [a_{1,11} - a_{0,11}(X_{gd} + X_i)]X_{gd}X_i}{b_2 - X_{gd}X_i} \tag{18.49}$$

$$R_{ds} = (a_{0,12} - R_s)(1 + A + a_{0,21}/a_{0,11}) \tag{18.50}$$

$$G_m = \frac{a_{0,21}}{a_{0,11}}\frac{1}{R_{ds}} \tag{18.51}$$

$$C_{gs} = \frac{1}{R_{ds}}\frac{a_{0,22} - R_d}{a_{0,11}} \tag{18.52}$$

$$C_{gd} = \frac{1}{R_{ds}}\frac{a_{0,12} - R_s}{a_{0,11}} \tag{18.53}$$

$$C_{ds} = \frac{1}{R_{ds}}\frac{a_{1,11} - R_g - a_{0,11}(X_{gd} + X_i)}{a_{0,11}} \tag{18.54}$$

$$R_{gd} = X_{gd}/C_{gd} \tag{18.55}$$

$$R_i = X_i/C_{gs} \tag{18.56}$$

$$\tau = X_{gd} + X_i - a_{1,21}/a_{0,21} \tag{18.57}$$

with

$$X_{gd} = \frac{a_{1,22} - L_d - b_1 R_d}{a_{0,22} - R_d}, \quad X_i = \frac{a_{1,12} - L_s - b_1 R_s}{a_{0,12} - R_s}, \quad A = \frac{C_{gs}}{C_{gd}} = \frac{a_{0,22} - R_d}{a_{0,12} - R_s} \tag{18.58}$$

$$\Delta Y = (Y_{gs} + Y_{gd})(Y_{ds} + Y_{gd}) + Y_{gd}(Y_{gm} - Y_{gd}) \tag{18.59}$$

18.2.4 Results

The proposed analytical extraction procedure was first validated using fictitious S-parameter data. The extracted model parameters could perfectly simulate the original fictitious S-parameters. This agrees very well with the results given by Ooi et al., presenting a similar analytical approach in 2009 [5].

Figure 18.2(a) shows excellent agreement between measured and modeled S-parameters (1–40 GHz) for a wire-bonded 0.5×2×150 μm GaAs MESFET. The modeled parameters have been calculated from coefficients of the rational functions. This is an alternative to using measured data directly in circuit design since it requires only a few coefficients (33 coefficients for the Y-parameter representation) to completely characterize the microwave

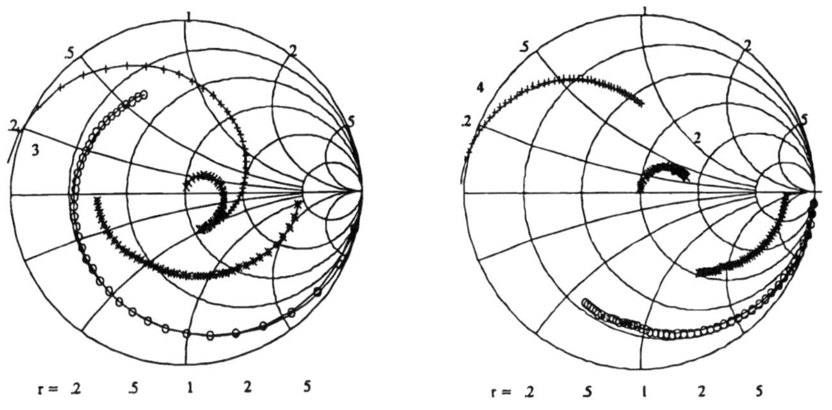

Figure 18.2 Measured and modeled S-parameters: (a) 0.5×2×150 μm GaAs MESFET (Plessey), V_{GS} = -0.4V, V_{DS} = 3.0V, f = 1 - 40 GHz, and (b) 0.3×2×50 μm GaAs HEMT (Eurochip), V_{GS} = 0V, V_{DS} = 2.0V, f = 1 - 26 GHz. Crosses/circles = measured values; — = simulation using the calculated coefficients.

Table 18.1
Extracted FET Equivalent Circuit Elements Based on Least-Squares Approximation by Rational Functions

Element	MESFET*	HEMT†	Element	MESFET*	HEMT†
C_{pds} (fF)	21.481	—	C_{gs} (fF)	282.61	145.29
C_{pgd} (fF)	0.9945	—	C_{gd} (fF)	23.631	19.463
C_{pgs} (fF)	24.144	—	C_{ds} (fF)	72.552	30.107
R_g (Ω)	7.8357	6.7612	R_i (Ω)	1.6711	1.1294
R_d (Ω)	3.0294	3.3751	R_{gd} (Ω)	0	27.412
R_s (Ω)	1.8506	2.8005	G_{ds} (mS)	3.7711	2.0723
L_g (pH)	157.2	32.217	G_m (mS)	40.12	51.398
L_d (pH)	176.4	6.8118	τ (ps)	1.9307	0.9043
L_s (pH)	2.4	0			

* Wire-bonded 0.5×2×150 μm GaAs MESFET (-0.4V, 3.0V).
† On-wafer measured 0.3×2×50 μm GaAs HEMT (0V, 2.0V).

behavior of a device. In addition, the least squares formulation results in smoothed curves as compared to discrete measured data.

Figure 18.2(b) shows, for example, the measured and simulated S-parameters of an on-wafer measured 0.3×2×50 μm HEMT (high electron mobility transistor) in the frequency range from 1 to 26 GHz). The agreement is nearly perfect with respect to measured data. Extraction results for the measured FET devices are given in Table 18.1.

Looking at the results in Table 18.1, note that the values of the extracted elements, in particular the parasitic resistances, are well acceptable regarding physical properties of the devices.

18.3 DISTRIBUTED SMALL-SIGNAL MODEL OF HBT DEVICE

Sun et al. have addressed the analytical model parameter extraction using rational functions to the distributed small-signal equivalent circuit of a SiGe HBT under an off-state condition [6–9]. Distributed effects in the intrinsic transistor, the link base region under the spacer, and extrinsic base-collector junction were taken into account. Equivalent circuits were derived for the different regions, then assembled together to a complete complex distributed HBT model shown in Figure 18.3. Distributed networks are shown in grey. The substrate network adopts a conventional parallel RC block. Pad capacitances are not shown. They were determined previously by measuring test patterns.

The distributed HBT model exhibits as many as 14 unknown elements to be extracted. Therefore, as the author pointed out, it is necessary to simplify the equivalent circuit in order to extract the model parameters accurately. On the basis of parameter estimations, the network in Figure 18.3 could be simplified (notice deleted resistances). The approximated network is shown in Figure 18.4, which comprises nine unknowns. The resistances R_{EU} and R_{CU} and the capacitance C_{TX} represent combinations from other elements.

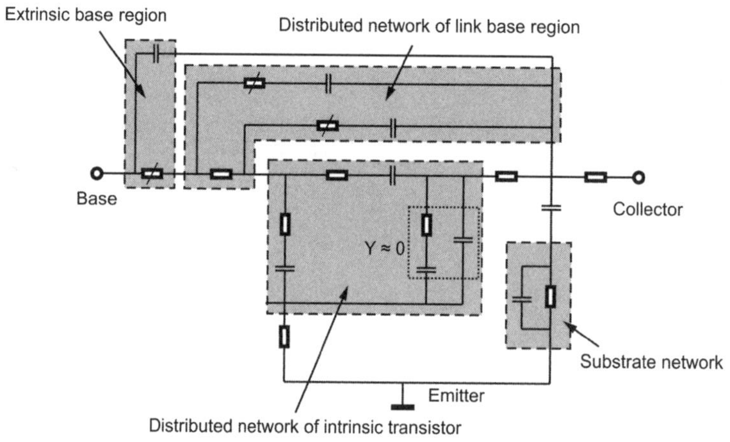

Figure 18.3 Distributed small-signal equivalent circuit for SiGe HBTs under common emitter configuration. Denotation of model elements can be found in [8]. Crossed-out resistors were neglected for simplification of the complex network.

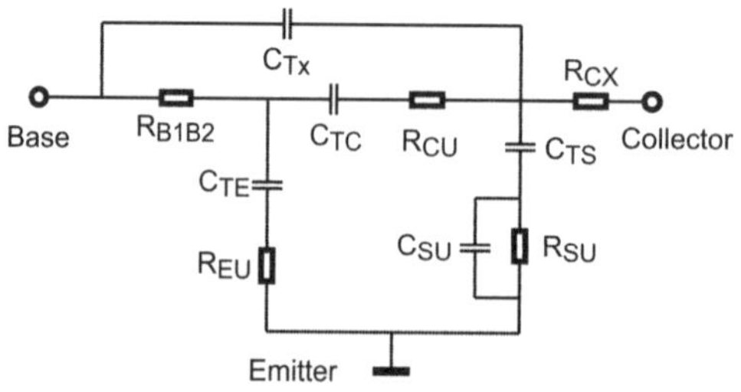

Figure 18.4 Simplified small-signal equivalent circuit of a HBT device. Collector resistance R_{CX} is already known from conventional R_C flyback method [10]. Notation of the model elements can be found in [8].

The admittance parameters [Y] of the pi network are then expressed according to (18.1) by rational functions [8]:

$$Im(Y_{11} + Y_{12}) = \frac{N_{10}\omega + N_{11}\omega^3}{1 + M_{11}\omega^2 + M_{12}\omega^4} \tag{18.60}$$

$$Re(Y_{11} + Y_{12}) = \frac{N_{20}\omega^2 + N_{21}\omega^4}{1 + M_{11}\omega^2 + M_{12}\omega^4} \tag{18.61}$$

$$-Im(Y_{12}) = \frac{N_{30}\omega + N_{31}\omega^3 + N_{32}\omega^5}{1 + M_{11}\omega^2 + M_{12}\omega^4} \tag{18.62}$$

$$-Re(Y_{12}) = \frac{N_{40}\omega^2 + N_{41}\omega^4}{1 + M_{11}\omega^2 + M_{12}\omega^4} \tag{18.63}$$

$$Im(Y_{12} + Y_{22}) = \frac{N_{50}\omega + N_{51}\omega^3 + N_{52}\omega^5 + N_{53}\omega^7}{1 + M_{21}\omega^2 + M_{22}\omega^4 + M_{23}\omega^6} \tag{18.64}$$

$$Re(Y_{12} + Y_{22}) = \frac{N_{60}\omega^2 + N_{61}\omega^4 + N_{62}\omega^6 + N_{63}\omega^8}{1 + M_{21}\omega^2 + M_{22}\omega^4 + M_{23}\omega^6} \tag{18.65}$$

where

$$N_{10} = C_{TE} \tag{18.66}$$

$$N_{20} = C_{TE}(R_{EU}C_{TE} + R_{B1B2}C_{TE} + R_{B1B2}C_{TC}) \tag{18.67}$$

$$N_{30} = C_{TC} + C_{TX} \tag{18.68}$$

$$N_{40} = C_{TC}(R_{B1B2}C_{TE} + R_{B1B2}C_{TC} + R_{CU}C_{TC}) \tag{18.69}$$

$$N_{50} = C_{TS} \tag{18.70}$$

$$N_{60} = R_{SU}C_{TS}^2 - R_{B1B2}C_{TE}C_{TC} \tag{18.71}$$

Higher-order terms of the polynomials N_{ij} and M_{ij} are neglected. In [8, 10] the excellent agreement of measured admittance parameters and fitted rational functions is demonstrated. The final step is to calculate the model elements from the fitted coefficients, as has already been demonstrated in the case of field-effect transistor in Section 18.2.3.

18.4 SUMMARY

In this chapter the analytical model parameter extraction using rational functions was discussed and applied to GaAs FETs and to SiGe HBTs.

The field-effect transistors comprising 17 model parameters operated under active bias conditions. The extraction problem is, first of all, considered as a fitting problem that is analytically solved using the method of least squares by modeling the measured data using rational functions. The equivalent circuit elements are then directly calculated from coefficients of the rational functions. Extraction results show that this approach results in excellent fitting of the measured parameters. The procedure gives physically meaningful equivalent circuit elements.

The modeling with rational functions has the great advantage that noisy (corrupted) measured data are smoothed out. On the other hand, it can cause mismatching between measurement and simulation if the model topology (i.e., the number of model parameters corresponding to the order of rational functions) is less valid to describe the device parasitics at very high frequencies or the complex model topology of power devices at moderate frequencies.

Frequently, the extraction problem has been mitigated by focusing previously on the parasitic network of the equivalent circuit using additional measurements or applying the analytical direct parameter extraction based on reduced order of rational functions (e.g., [6, 7]).

The method has also been applied to a SiGe HBT device. The complex distributed equivalent network under the off-state was described and analytical model parameter determination was discussed.

REFERENCES

[1] Wasige, E., H. Ntchalhatou, and G. Kompa, "An Analytical Model Parameter Extraction Procedure for GaAs FETs Based on Least Squares Approximation," *IEEE Workshop on Experimentally Based FET Device Modelling & Related Nonlinear Circuit Design*, Kassel, Germany, July 17-18, 1997, pp. 40.1–40.7.

[2] Ntchalhatou, H., *An Analytical Model Parameter Extraction Procedure for GaAs FETs Based on Least-Squares (in German)*, Master Thesis, University of Kassel, 1997.

[3] Nagatomo, K., Y. Daido, M. Shimizu, and N. Okubo., "GaAs MESFET Characterization Using Least Squares Approximation by Rational Functions," *IEEE Trans.*, Vol. MTT-41, 1993, pp. 199–205.

[4] S. Wolfram, *MATHEMATICA: A System of Doing Mathematics by Computer*, 2nd Ed., Addison Wesley, 1991.

[5] Ooi, B.-L., Z. Zhong, and M.-S. Leong, "Analytical Extraction of Extrinsic and Intrinsic FET Parameters," *IEEE Trans.*, Vol. MTT-57(2), 2009, pp. 254–261.

[6] Qin, M., Y. Sun, X. Li, and Y. Shi, "Analytical Parameter Extraction for Small-Signal Equivalent Circuit of 3D FinFet into Sub-THz Range, *IEEE Access*, Vol. 6, 2018, pp. 19752–19761.

[7] Olvera-Cervantes, J.-L., J. D. Cressler, J.-L. Medina-Monroy, T. Thrivikraman, B. Banerjee, and J. Laskar, "A New Analytical Method for Robust Extraction of the Small-Signal Equivalent Circuit for the SiGe HBTs Operating at Cryogenic Temperatures," *IEEE Trans.*, Vol. MTT-56(3), 2008, pp. 568–574.

[8] Sun,Y., J. Fu, J. Yang, J. Xu, Y. Wang, W. Zhou, W. Zhang, J. Jui, G. Li, and Z. Liu, "Novel Analytical Parameter Extraction for SiGe HBTs Based on the Rational Function Fitting," *Superlattices and Microstructures*, Vol. 80, 2015, pp. 11–19.

[9] Sun,Y., J. Fu, J. Yang, J. Xu, Y. Wang, J. Cui, W. Zhou, Z. Wei, and Z. Liu, "An Improved Small-Signal Model for SiGe HBT Under OFF-State, Derived from Distributed Network and Corresponding Model Parameter Extraction," *IEEE Trans.*, Vol. MTT-63(10), 2015, pp. 3131–3141.

[10] Sun, Y., Z. Liu, X. Li, and Y. Shi, "Distributed Small-Signal Equivalent Circuit Model and Parameter Extraction for SiGe HBT," *IEEE Access*, Vol. 7, 2019, pp. 5865–5873.

[11] Sun, Y., *Research on the Radiation Effects and Compact Model of SiGe HBT*, Doctoral Thesis, Tsinghua University, Beijing, China. Published by Springer, 2017.

Chapter 19

Repetitive Random Optimization and Adaptive Search Space

19.1 INTRODUCTION

Recalling previously discussed extraction procedures, we may keep in mind the following: (1) conventional optimization procedures such as gradient-based algorithms are local search methods. Because of the starting value problem, the found minima remain questionable. (2) the global optimization with multistart requires the definition of search intervals for each model parameter. In this case the starting values of the model parameters are arbitrarily chosen or statistically obtained using a random generator. As the model parameters are highly correlated with each other, the arbitrary choice of starting values enlarges the search space unnecessarily and thereby increases the local minima density dramatically.

This is avoided with analytical extraction algorithms. We have seen that pure analytical extraction method (Chapter 17) requires appropriate matching of the equivalent network topology to the device physics and high-quality measurement data. The described analytical extraction approach in Chapter 18 using rational functions are capable of determining the model parameters from a rational function with polynomials fitted to measured scattering parameters. It is observed that simulation commonly matches the measurement very well. As the coefficients of the polynomials comprise mostly combinations of model elements, the calculation of single-model parameters may prove challenging. We have also seen that the method should be applied to equivalent networks with lower complexity to limit the degree of polynomials. For that purpose the physics-based distributed equivalent circuit of a HBT presented in Chapter 18 was first simplified prior to analytical calculation of the model elements.

Independent of previously discussed extraction methods it must be noted that the model network topology under consideration was similar to the widely used 15-element equivalent circuit with possibly slight modifications. In the following chapters we will focus on more complex topology structures preferably for field-effect transistors that for example comprise e.g. 22 elements, as discussed in Chapter 7. Indeed the modeling concept may also be applied to complex linear models of HBTs as treated in the previous chapter without introducing estimations and approximations for model simplification.

The complex equivalent circuit topologies under discussion comprises three capacitive shells, as shown in Figure 19.1. In Chapter 7 the complexity was justified for large-size gate peripheries of field-effect transistors because of additional capacitive interelectrode coupling as well as small-size gate peripheries (one or two gate fingers, without source air-bridges) at high frequencies due to distributed effects and stray capacitances. The strategy is discussed in the following (and strengthened and refined in Chapter 21) how to approach the extraction problem with high confidence in the sense that extracted model elements are trustful and physically meaningful.

Based on an extraction example the evolution of the extraction (optimization) process is thoroughly pursued. As the optimization is basically a mathematical operation, the question therefore arises whether, beyond the mathematical optimum, the extracted parameters are also physically relevant.

Random search was already discussed in Section 12.7.1.2 and practically applied in Chapter 14. The extraction method of repetitive random optimization to be discussed now, is an extension in two ways. The parameter space is not only sampled once on the basis of a defined starting vector, but after sorting the residual error values of the object function in ascending order, optimization is repeated with a limited number of model parameter value vectors with lower residual error of the error function. It is expected that the new iteration cycle samples the parameter space nearer the global minimum. Thus, after a first initial sampling of the whole search space, this is then continuously reduced by focusing on search areas with lower error function values.

The second point concerns the definition of the starting model parameter value vector, as already mentioned. Instead of using statistically distributed initial model element values or values derived from separate measurement of test patterns or under different bias conditions, the starting value vector results from correlated model parameter values derived from a single measurement data set. In this regard, the term *measurement-correlated starting values* has been introduced by the author [1,2]. Since in this case all model parameter values are directly related to the device under actual operation in a fixed bias point, the extracted model parameter values may be assumed to be physically meaningful (under assumption that the device physics fits to the chosen equivalent circuit topology).Then, using these as starting values in an optimization algorithm, it is expected that the search space has already shrunk closely to the optimum solution. Then, based on such high-quality estimated model parameters only some further optimization step may be necessary for final extraction refinement. In the most favorable case, the previously found starting element values are of

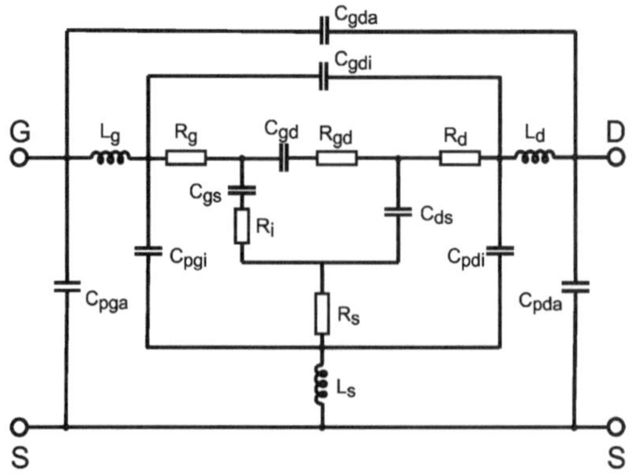

Figure 19.1 Broadband pinch-off field-effect transistor equivalent circuit with distributed topology.

such high quality that optimization is no longer necessary, in particular, taking into account extraction uncertainties such as model mismatch and measurement errors. Indeed, as we will see later, the series resistances require a special treatment. The common experience is that the extracted parameters mostly reveal very small values (zero in the worst case). A possible reason for this extraction issue is discussed in Section 20.3, and in Section 21.8 a well-proven approach for reliable determination of the series resistances is described.

The model parameter extraction concept described in this chapter can be briefly characterized as follows. Accurate effective correlating capacitances and inductances are determined from pinch-off S-parameter data. Regarding the distributed nature of the broadband equivalent circuit, well-defined initial values for the gate-, drain-, and source-inductance are derived in dependence of estimated distribution of the effective capacitances on the gate and drain side. Optimization of model parameters is based on a repetitive random search with adaptive starting interval sizes of model parameter values. The leaping simplex algorithm discussed in section 12.7.5 is used for optimization. The new extraction concept converges very fast. Human intervention for guessing suitable initial element values is omitted.

Nearly all model elements are consistently extracted. However, due to S-parameter measurement uncertainties and the unfavorable shape of the error function near its minimum, stable but not always physically meaningful values for the resistances R_s and R_d are obtained. This problem is addressed in more detail in Chapter 20.

19.2 STARTING VALUES

The most important point for the optimization process is the definition of an *appropriate* starting value vector for the model elements. Appropriate means that the starting vector should be located near the global (optimum) minimum. (Note later on in Section 19.3, we make a finer distinction between an optimum solution in the mathematical sense and a solution that reflects the device physics.) This can be obtained as follows: (1) all bias-independent elements are derived under cold pinch-off condition, (2) the composed distributed capacitances (effective capacitances) in the gate-source, drain-source, and gate-drain branches are determined at low frequencies from the measured admittance coefficients, (3) with known effective branch capacitances, inductances are determined from measurements at medium frequencies, and (4) splitting of the effective capacitances according to the distributed equivalent network in Figure 19.1. This last point is a great challenge, and has initiated various realization concepts over the course of time (see Sections 21.4 - 21.6).

19.2.1 Effective Capacitances

As pointed out, for lower frequencies, an effective gate-source capacitance C_{gs}', gate-drain capacitance C_{gd}', and drain-source capacitance C_{gs}' can be evaluated from measured data. C_{gs}' has three components as indicated in Figure 19.1: (1) C_{pga} due to the extrinsic contact pads,

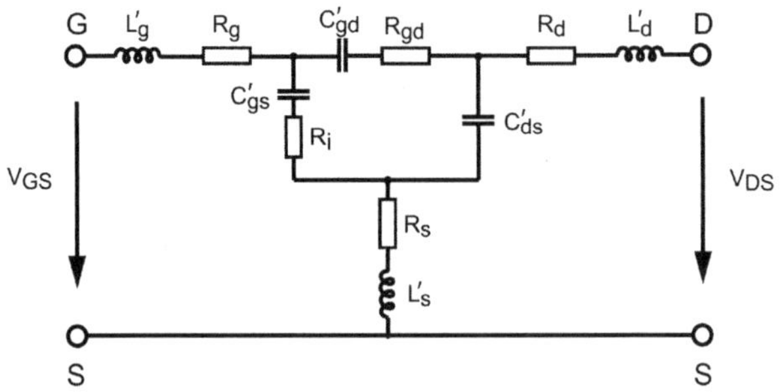

Figure 19.2 Equivalent circuit of a field-effect transistor under pinch-off with effective capacitances and inductances.

(2) C_{pgi} due to the fringe capacitance of the inner electrode metallization pattern, and (3) intrinsic capacitance C_{gs} that is bias-dependent. Similar splitting of the effective values can be considered for C_{gd}' and C_{ds}' as well. The effective capacitances can be written as follows:

$$C_{gs}^{/} = C_{gs} + C_{pgi} + C_{pga} \tag{19.1}$$

$$C_{ds}^{/} = C_{ds} + C_{pdi} + C_{pda} \tag{19.2}$$

$$C_{gd}^{/} = C_{gd} + C_{gdi} + C_{gda} \tag{19.3}$$

Figure 19.2 shows the equivalent circuit of a field-effect transistor under pinch-off with effective capacitances as defined in (19.1) - (19.3) and effective inductances. It should be noted that the primed effective gate-, drain-, and source inductances are not equal to L_g, L_d, and L_s in Figure 19.1. The inductance values in the distributed model depend strongly on the distribution of the effective capacitances.

Experience has shown that for the various investigated devices (wire-bonded, probe tip contacted) the generalized topology in Figure 19.1, which is of a distributed nature at the input and output ports, describes the electrical properties for all investigated devices very well. Moreover, it includes many topology variations proposed in the literature. Indeed, the quantitative splitting of the effective capacitances into three separate capacitances, which are generally required for broadband data fitting, depends on the given device technology and construction. Simplification is possible in particular cases.

As the probability of trapping into a local minimum is increased with an increased number of optimization parameters, complex structures as given in Figure 19.1 are commonly avoided. Nevertheless, the extractor should be able to output element values within confidential limits (i.e., values which can be accepted by physical inspection of model elements). This should also be true with the more insensitive resistors; (i.e.. R_g, R_i, R_{gd}, and R_d). Because the pad capacitances associated with C_{gd} are generally small we may first assume

$$C_{gdi} = C_{gda} = 0 \tag{19.4}$$

and with (19.3) it follows

$$C_{gd}^{/} = C_{gd} \tag{19.5}$$

Under pinch-off condition, a FET device with symmetric gate provides a gate-source

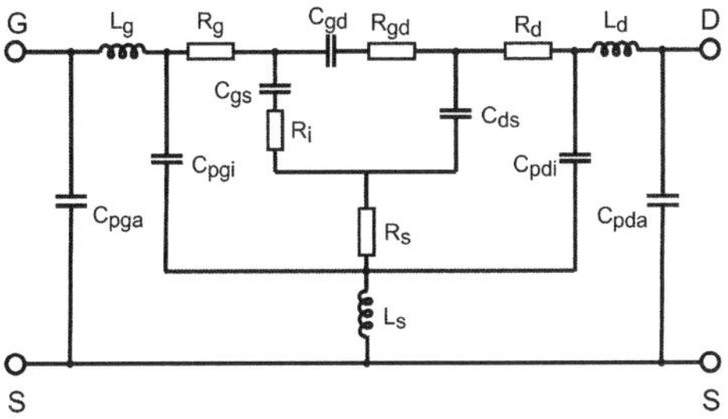

Figure 19.3 Used distributed pinch-off model for the FET device without C_{gdi} and C_{gda}.

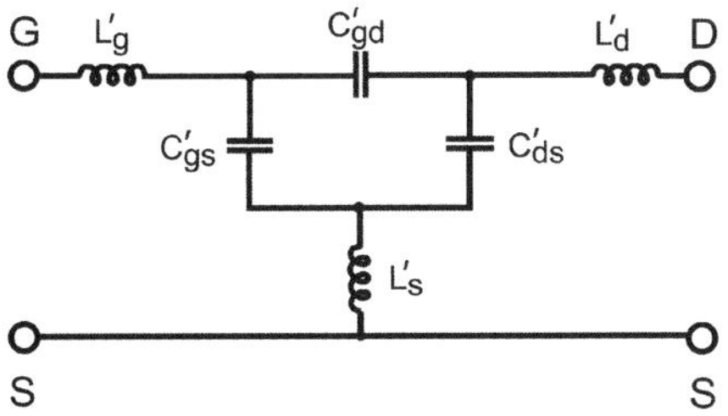

Figure 19.4 Pinch-off model with effective capacitances and inductances (resistances neglected).

capacitance C_{gs} and a gate-drain capacitance C_{gd} of equal size. Indeed, this is not valid in the general case. Nevertheless, as we are only searching for meaningful starting model element values for an optimization process, we may approximately write

$$C_{gs} \approx C_{gd} \tag{19.6}$$

Figure 19.3 shows the pinch-off model finally used. For very low frequencies ($f \rightarrow 0$) the equivalent network of the device behaves like a capacitive network (Figure 19.4). The effects of the series resistances and inductances become negligible compared to the capacitances.

Calculating the Y-parameters from the scattering coefficients the asymptotic approximation for $f \rightarrow 0$ results in

$$Y_{11} = j\omega\,(C_{gs}' + C_{gd}') \tag{19.7}$$

$$Y_{22} = j\omega\,(C_{ds}' + C_{gd}') \tag{19.8}$$

$$Y_{12} = Y_{21} = -j\omega\, C_{gd}' \tag{19.9}$$

Experience shows that the effective capacitances can be determined from measured data in the lower frequency range with an upper frequency less than 10 GHz [3]. From (19.7) - (19.9) the effective capacitances can be calculated as

$$C_{gd}' = -\frac{Im\{Y_{12}\}}{\omega} \tag{19.10}$$

$$C_{gs}' = \frac{Im\{Y_{11}\} + Im\{Y_{12}\}}{\omega} \tag{19.11}$$

$$C_{ds}' = \frac{Im\{Y_{22}\} + Im\{Y_{12}\}}{\omega} \tag{19.12}$$

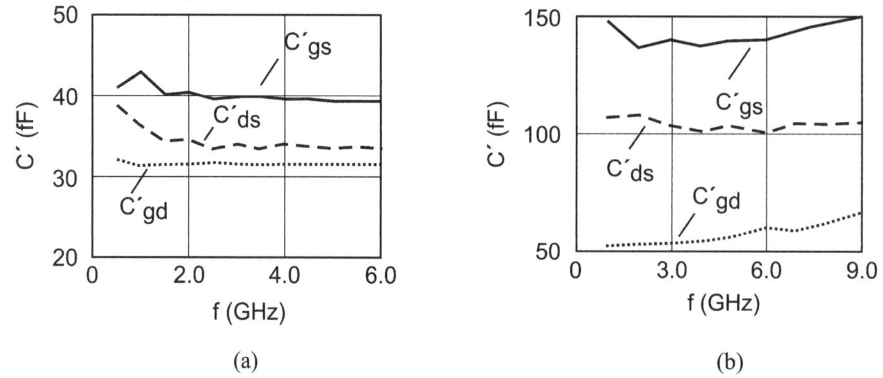

Figure 19.5 Extracted effective capacitances: (a) on-wafer measured 0.15-μm gate HEMT (IAF), and (b) in-fixture-measured wire-bonded 0.4-μm gate MESFET (MGFC1423).

Figure 19.5(a) shows the determined effective capacitances versus frequency for a 0.15-μm GaAs-HEMT fabricated by the Fraunhofer Institute for Applied Solid State Physics (FhG-IAF), Freiburg i. Br.. Regarding the high resolution of the diagram scale, the effective capacitances can be evaluated very precisely, neglecting values below about 1 GHz. Figure 19.5(b) illustrates the result for a wire-bonded FET chip of type MGFC1423. Also in this case the capacitances can be derived accurately from low-frequency measurements. The increase of C_{gd}' and C_{gs}' with frequency can be explained by the growing inductive influence.

19.2.2 Effective Inductances

As the effective capacitances are known, the effective inductances can be derived as follows. Considering Figure 19.4, the capacitive pi-network is transformed into a T-network with the elements

$$C_g = C_{gs}' + C_{gd}' + C_{gs}' C_{gd}' / C_{ds}' \tag{19.13}$$

$$C_d = C_{ds}' + C_{gd}' + C_{ds}' C_{gd}' / C_{gs}' \tag{19.14}$$

$$C_s = C_{gs}' + C_{ds}' + C_{gs}' C_{ds}' / C_{gd}' \tag{19.15}$$

Knowing the effective capacitances we can then derive effective inductances. We transform the capacitive pi-structure into a T-network (Figure 19.6) and calculate the effective inductances from the impedance coefficients. Then the primed effective series inductances can be derived from the Z-parameters as follows:

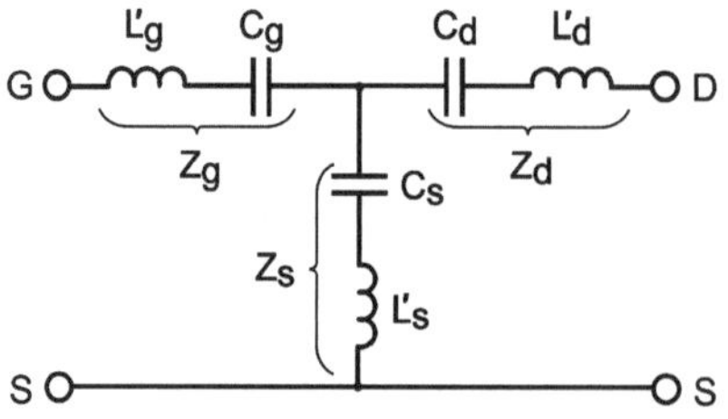

Figure 19.6 T-type pinch-off model with effective parameters.

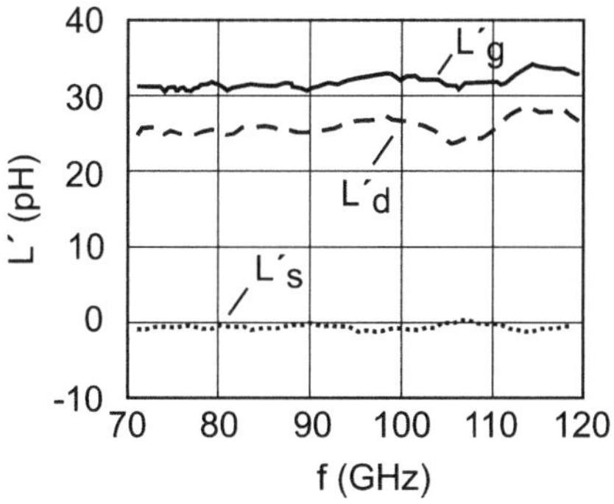

Figure 19.7 Effective inductances extracted for the 0.15-μm gate HEMT (IAF).

$$L'_{\mu} = \frac{Im\{Z_{\mu}\}}{\omega} + \frac{1}{\omega^2 C_{\mu}}, \quad \mu = (g, d, s) \tag{19.16}$$

Figure 19.7 shows the derived effective inductances versus frequency for the 0.15-μm IAF-HEMT. The asymptotic values at lower frequencies can be well determined. The effective source inductance is evaluated to be about 0 pH.

A problem may arise with wire-bonded devices that comprise relatively large external parasitic capacitances C_{pga} and C_{pda}. These can lead to parallel resonance effects (see Figure 19.8) within the measured frequency range that manifest themselves in a monotonous increase of the L'_{μ} values with frequency. Estimation of the parallel resonance frequency can be easily estimated.

Considering Figure 19.3, the circuit in Figure 19.9(a) can be deduced under the following assumptions: (1) the resistances are neglected, (2) the extrinsic capacitances are much smaller than the others, and (3) known effective inductances are taken instead of the unprimed quantities for resonance frequency estimation.

The inductive T-structure can be transformed into a pi-structure (Figure 19.9(b)) with

$$L_{gd} = L'_g + L'_d + L'_g L'_d / L'_s \tag{19.17}$$

$$L_{gs} = L'_g + L'_s + L'_g L'_s / L'_d \tag{19.18}$$

$$L_{ds} = L'_d + L'_s + L'_d L'_s / L'_g \tag{19.19}$$

L_{gd} is found to be very large so that the parallel resonance frequencies at the input and output

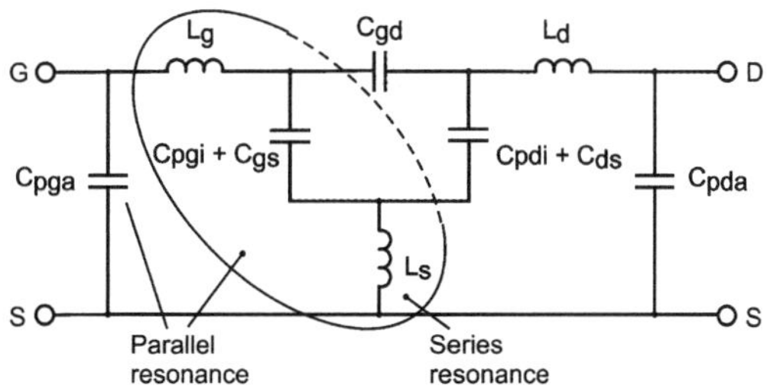

Figure 19.8 Existence of parallel resonance at higher frequencies due to the external capacitances.

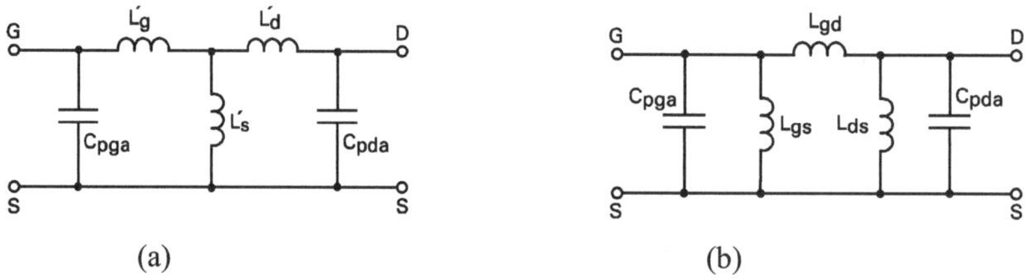

Figure 19.9 Circuit for parallel resonance frequency estimation: Initial (a) and pi-transformed (b) inductance T-structure.

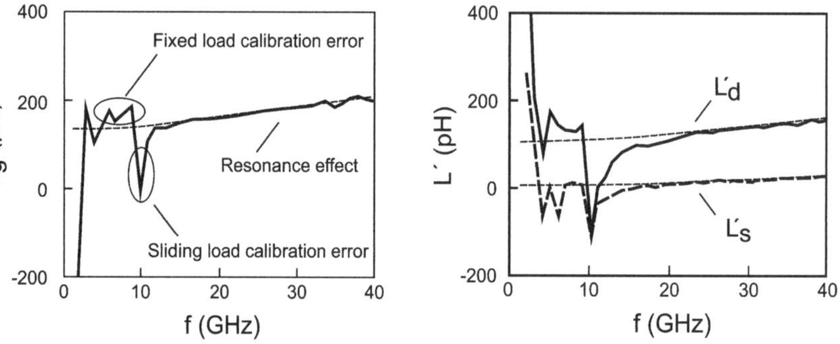

Figure 19.10 Extracted effective inductance values for in-fixture measured MESFET MGFC1423.

re estimated separately. It turns out that the lowest parallel resonance existing with the 0.15-μm gate HEMT lies far beyond the measured frequency range of 118 GHz. However, regarding the wire-bonded MDFC1423, parallel resonance phenomena are already expected to be around 79 GHz.

Figure 19.10 shows the curvature for the determination of effective inductances for the MGFC1423 device. The increase of the inductance with frequency can be clearly seen. At lower frequencies some irregularities arise that originate in the nonperfect calibration standards used in the THLR calibration procedure [4, 5]. As we will utilize the derived inductance values only as initial starting parameter values for subsequent optimization, the approximation indicated by the thin solid line is entirely acceptable.

19.2.3 Distributed Model Elements

For the final optimization step the derived effective capacitive and inductive model parameters must be reasonably split according to the distributed equivalent transistor network in Figure 19.1. The starting element values are defined as follows.

With the assumptions (19.4) made before, C_{gd} and C_{gs} are known from measurement. Extrinsic capacitances C_{pga} and C_{pgi} are set equal half of the difference of primed C_{gs}' and C_{gs}. C_{ds} is equal to C_{pdi} (i.e., primed C_{ds}' minus C_{pda}). The unprimed inductances distinguish from the effective values and are a function of the distribution of the capacitances.

At this point we have found a basic starting vector for optimization with model elements strongly correlated to each other and fully matched to the given model topology at low and medium frequencies. To circumvent the local minimum problem we randomly generate up to 100 initial starting vectors from the basic vector within a value interval of ± 40%. By using this local testing of the global minimum, relevant information about its properties can be gained. The extracted results after initial optimization are then used to refine the value intervals for the next random optimization step. It has been shown that after only a few re-optimization cycles the extraction process stabilizes (e.g., three reruns for investigated MESFET, and five re-runs for HEMT devices).

To start optimization we assume that the total input and output capacitances C_{gs}' and C_{ds}' are distributed as follows (see (19.1) - (19.3)):

$$C_{pga} = C_{pgi} = (C_{gs}' - C_{gs})/2 = C_{pda} \tag{19.20}$$

$$C_{ds} = C_{pdi} = (C_{ds}' - C_{pda})/2 \tag{19.21}$$

Equation (19.20) means that the residual gate-source capacitance (i.e., effective gate-source capacitance minus gate-source capacitance) is equally subdivided into the extrinsic and intrinsic parasitic gate capacitances. Equation (19.21) is similar to the first one concerning the distribution of the effective drain-source capacitance. As has been shown, the effective gate-source and drain-source capacitances can be determined very accurately from the Y-parameters with $f \to 0$. With the effective capacitance values known, the inductance values of L_g, L_d, and L_s can approximately be derived from the effective inductive quantities taking into account the defined capacitance distribution (i.e., the model inductances L_g, L_d, and L_s depend both on the related primed inductances and the distributed capacitive gate-source and drain-source network). As has been shown the model inductances L_g, L_d, and L_s can be estimated as

$$L_\mu \approx L_\mu' q_\mu \tag{19.22}$$

with

$$q_\mu \approx (1 + q_{c,\mu})/[1 - (q_{c,\mu}/(1 + q_{c,\mu})]$$

where $q_{c,\mu}$ [$\mu = (g,d)$] defines the ratio of the extrinsic capacitance C_{pga} and the sum of the capacitances $(C_{pgi} + C_{gs})$, and the ratio of C_{pda} and $(C_{pdi} + C_{ds})$, respectively. For $q_{c,\mu} \ll 1$, (19.22) gives

$$L_\mu \approx (1 + 2q_{c,\mu}) L_\mu' \tag{19.23}$$

The estimation of L_s from known effective values L_s' is based on the following relation

$$L_s \approx L_s'(q_g q_d)^2 \tag{19.24}$$

If we approximate $C_{ds} \approx C_{pdi} \approx C_{pda}$, the last squared term becomes 3.16 and

$$L_s \approx 3.16 q_g^2 L_s' \tag{19.25}$$

Note in Sections 26.6 and 26.7 the inductances are directly determined from the known distribution of the capacitances.

For the 0.15-μm gate HEMT the values for $q_{c,g}$ and $q_{c,d}$ are 0.10, and 0.13 respectively. Hence, from (19.22) - (19.25) the inductances can be derived: $L_g \approx 1.20\,L_g'$, $L_d \approx 1.26\,L_d'$, and $L_s \approx 4.40\,L_s'$. These are only small corrections because of the low values of the extrinsic parasitic capacitance. For the MGFC1423 MESFET the capacitance distribution factors have been calculated as $q_{c,g} \approx 0.22$ and $q_{c,d} \approx 0.30$, leading to the following inductance values: $L_g \approx 1.49\,L_g'$, $L_d \approx 1.69\,L_d'$, and $L_s \approx 7.00\,L_s'$.

An alternative approach using estimated C_{pga} and C_{pda} has been discussed in [1].

Converting measured S-parameters to Y-parameters ($S \rightarrow Y$) and de-embedding C_{pga} and C_{pda}, a capacitive pi-network comprising $C_{gs}' = C_{gs} + C_{pgi}$, $C_{ds}' = C_{ds} + C_{pdi}$ and $C_{gd}' = C_{gd}$ is obtained that can be converted to a T-network (C_μ, $\mu = (g, d, s)$). Then the inductances can be determined by curve fitting,

$$L_\mu = \frac{\mathrm{Im}(Z_\mu^{de})}{\omega} + \frac{1}{\omega^2 C_\mu}, \quad \mu = (g, d, s) \tag{19.26}$$

19.3 ADAPTIVE SEARCH SPACE ALGORITHM

Trapping into a local minimum is a fundamental problem in optimization. It is expected that the nearer the starting values of the model parameters are located at the global minimum, the lower the trapping probability. Because of measurement errors the environment of the global minimum may be noisy for getting lost among other local minima. In this case extraction includes some degree of uncertainty. Furthermore, systematic errors can predominate so that any extraction effort may become unsuccessful although the extracted result for some parasitics may be reasonable.

The idea of the proposed extraction procedure discussed in the following is to probe the object function in the neighborhood of the expected global minimum of the object function with a variety of carefully defined starting value vectors for the model elements. It is expected that the statistical evaluation of the extracted results will give essential information about the usefulness of the defined starting parameter values (i.e., measurement-based correlated effective model parameters) and the properties of the searched global minimum.

Extraction has been carried out as follows. One hundred starting vectors were based on uniformly distributed random model parameters within ± 20% of the derived starting values. The series resistances R_g, R_s, and R_d varied from 1Ω to 5Ω. R_i and R_{gd} were chosen within 10Ω and 30Ω. Optimization was started from each random vector using the leaping simplex algorithm with reinitialization to overcome the local minimum problem (see Section 12.7.5). The extracted parameter values are then sorted upward with a residual error of the object function. Reoptimization is performed using reduced starting intervals according to the extracted value distribution of each parameter. It has been observed that already after a few adaption cycles of interval size, the extraction values of the model elements become stationary. The bidirectional search algorithm ([6] (see Chapter 16) has been applied for optimization.

19.4 RESULTS: MATHEMATICAL VERSUS PHYSICS-BASED SOLUTION

Figures 19.11–19.17 show the extraction results for the wire-bonded 0.4-μm gate MGFC1423 device. Figure 19.11 shows the residual errors of objective function versus randomly chosen starting vectors sorted in ascending residual error value. As is shown, the initial extractions denoted as *initial*, indicate virtually a linear function whereas the curve denoted as *after iteration* provides the following characteristic features: First, a low-valued error-region is observed characterized by a strong increase of the error function value. It

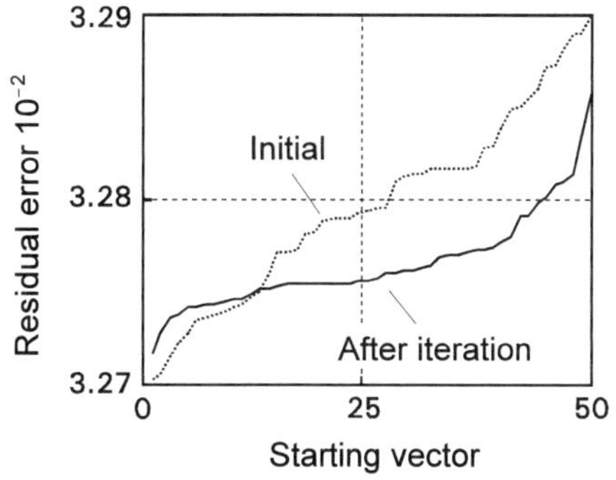

Figure 19.11 Residual errors of objective function in ascending order versus corresponding randomly defined starting vectors (MGFC1423).

follows a broad gradually increased region. Then the curve grows over-proportionally. As will be discussed later, the three periods can be identified, respectively as, the mathematically best solution, the physically meaningful solution, and the local minima solution in the neighborhood of the optimum one.

It can be seen that the extraction based on the initial starting vector delivers highly consistent values for the individual gate-source capacitances (Figure 19.12(a)). Iterative optimization based on adaptive reduction of parameter interval sizes does not essentially improve the extracted values. The obtained traces of the inductances (Figure 19.12(b)) can be interpreted in a similar way.

As can be seen in Figure 19.12(c), the determination of the extrinsic capacitance C_{pda} is rather stable. Conversely, the residual part of the effective drain-source capacitance is initially not clearly related to C_{ds} and C_{pdi}. After some iteration cycles, however, the distribution adjustment has been stabilized.

Regarding the resistances (Figures 19.12(d) - 19.12(f)), the values initially scatter strongly as expected. Note the extracted value of R_g equal to 0.18 ohm for the best fit with a residual error value of 3.27×10^{-2} for the initial starting vectors is truly unphysical. From the information of the manufacturer [7] the aluminum gate of the discussed device has a gate length of 0.4 μm, a gate width of 260 μm, and a gate thickness of 0.4 μm. Assuming a conductivity for aluminum between 33×10^6 S/cm (bulk material) and 19.6×10^6 S/cm (for typical MESFET gate structures [8]), the gate metallization resistance value R_g is expected to be higher than 1.0 and at the highest 1.7 ohm, which is approximately found along the weakly increasing part of the error function in Figure 19.11. This indicates that the best fit (minimum residual error function value) of the final extraction run may not deliver the most physically meaningful parameter values. *Best fit* reflects the optimum solution in the mathematical sense. The discrepancy between a physical and mathematical solution may partially be reasoned with measurement uncertainties, but the main cause is expected to be attributed to model topology mismatching.

Analyzing the recessed gate structure of the MGFC1423 device [7], and assuming that only the thinned channel under the gate will be depleted under pinch-off, a minimum value of 0.3Ω (contact resistance) and maximum value of 0.67Ω may be expected for the source resistance R_s. The extracted value of about 1.5 ohm (Figure 19.12(e)) might not really be

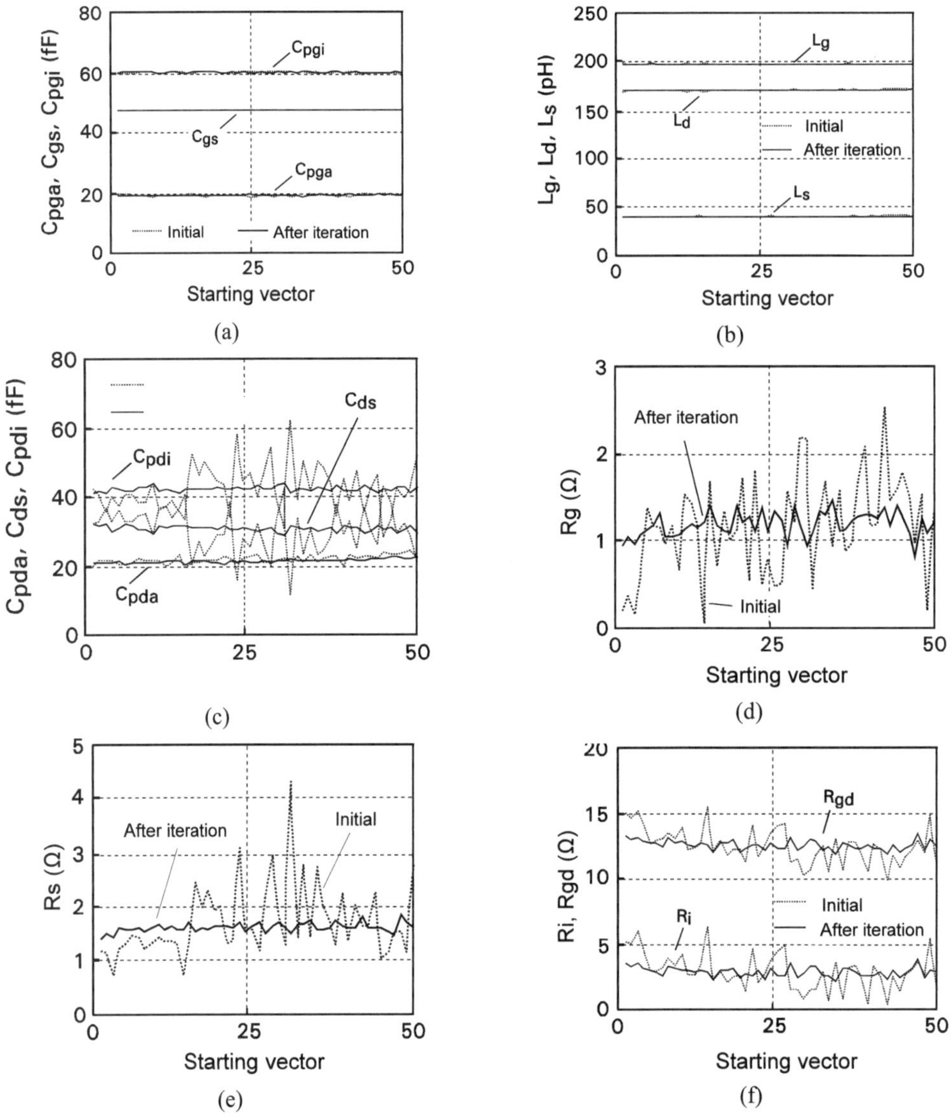

Fgure 19.12 Extracted model parameter values of in-fixture measured MESFET MGFC1423: (a) gate capacitances, (b) inductances, (c) drain capacitances, (d) gate resistance R_g, (e) source resistance, and (f) gate-source R_i and gate-drain R_{gd} charging resistances.

justified from a physical point of view.

Experience has shown that the best overall fit will not automatically imply the best fit with respect to physical meaningful parameter values. This is in particular evidenced by the extracted values of series resistances R_s and R_d. As can be seen in Figure 19.11, the minimum of the sorted residual error values has increased after iteration with the effect that the initially totally unphysical R_g value in Figure 19.12(d) has been raised to a more realistic value.

Figures 19.13(a) - 19.13(f) show the corresponding extraction for the IAF-

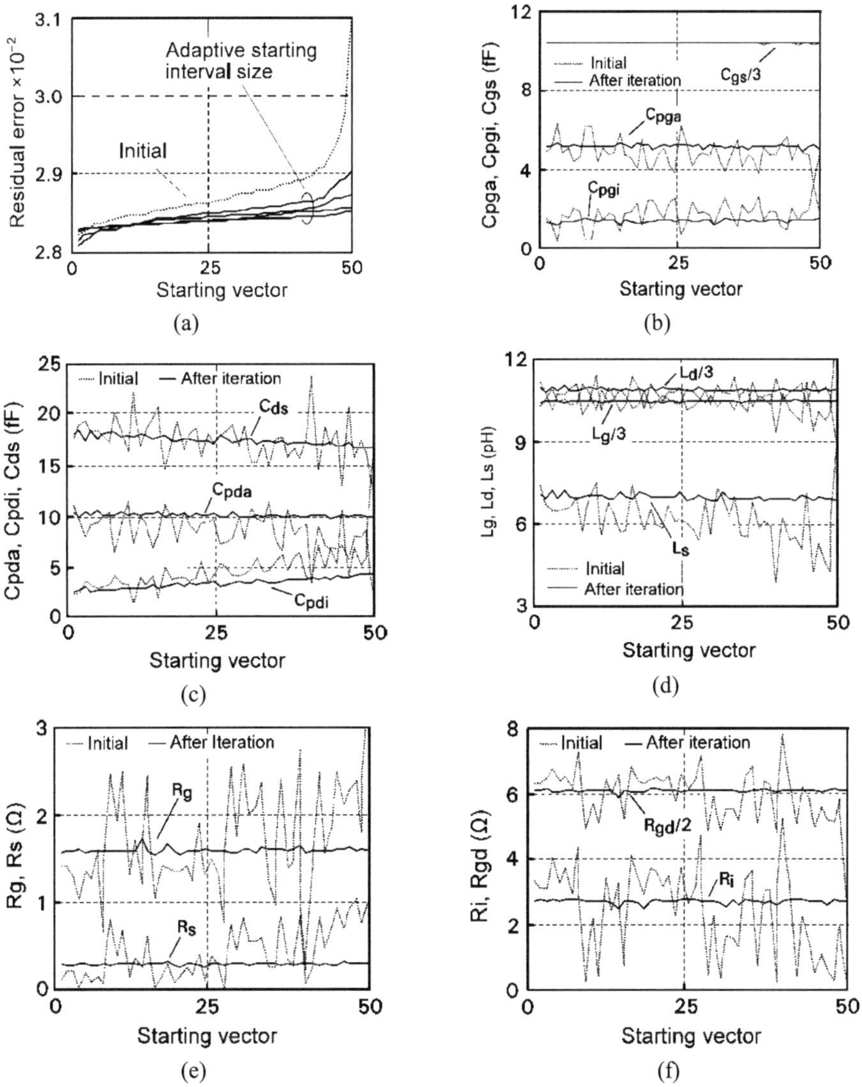

Figure 19.13 Extracted model parameter values of on-wafer measured 0.15-μm gate HEMT: (a) residual error of error function sorted according to size versus respective starting vector, (b) extracted distributed gate capacitances, (c) extracted distributed drain capacitances, (d) extracted inductances, (e) extracted gate and source resistances, and (f) gate- and drain-charging resistances.

HEMT. It is seen that the residual error in Figure 19.13(a) is very low over a wide range of randomly chosen starting vectors in comparison to Figure 19.11 for the in-fixture measured device. Opposite interchanging of model element values is observed mainly between L_g and L_d, and C_{pga} and C_{pgi}. There is also a vital interchange recognized between C_{pda} and C_{ds}. After four reoptimization cycles the extracted values become stabilized. From device geometry and material R_g and R_s could be roughly estimated (see Section 6.2). It turns out that R_g can be accepted as physics-related; however, R_s was extracted much too low. This issue is treated separately more closely in the following chapter.

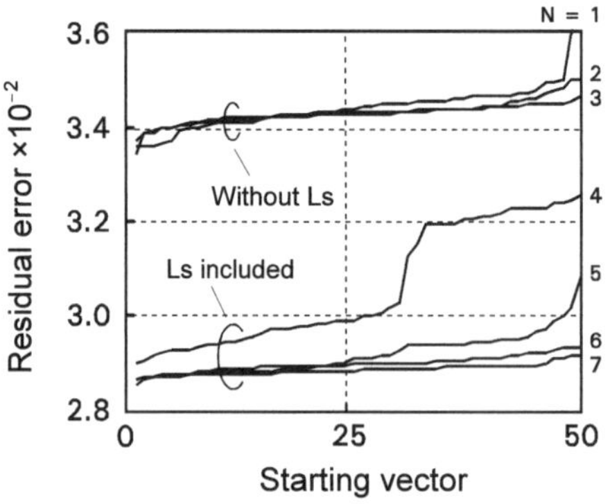

Figure 19.14 Residual error function value sorted according to size versus respective starting vector using arbitrarily chosen (uncorrelated) starting intervals (0.15-μm gate HEMT). N = 1 indicates the initial randomly chosen model parameter value vector. N = 2, 3... are reruns with adaptive search intervals.

The extracted value for R_s being about 0.5 ohm seems to lie outside of the confidence limits. The contact resistance measured under a gate-forward condition is expected to be 0.15 Ω-mm; that is, 1.5 ohm for a gate-length of 2 × 50 μm [9].

No exact data of the cross section of the three-layer (Ti-Pt-Au) T-gate was available. Nevertheless, geometry could be estimated from an e-beam microscope photo. Assuming a gold conductivity of 41×10^4 S/cm and a gate cross-section area of about 0.25 μm^2, a gate resistance of 0.8Ω has been calculated. Assuming a conductivity reduction roughly in the order of 20% for evaporated material, an R_g value about 1 ohm is expected. Frequency dependent increase due to the skin-effect should also be discussed within the measured frequency range up to 118 GHz. Estimation results in an increase less than 10%, so that this effect is not further considered. In conclusion, the value of $R_g \approx 1.6\Omega$ might be accepted, whereas the extracted R_s values are not acceptable.

Considering the individual gate-source capacitances the initially extracted intrinsic gate-source capacitance C_{gs} remains stable (practically the same) after optimization. Conversely, there is an interchange of value contributions between C_{pga} and C_{pgi}, which disappears with continuous iteration. Analyzing the traces for the drain-side capacitances, the extraction shows some interaction between C_{ds} and C_{pda}. After some cycles of iteration C_{pda} turns out to be very stable, and there is some slight value interchange between C_{ds} and C_{pdi}.

Figure 19.13(d) indicates correlation between L_g and L_s. A value interchange is observed between both of these and L_d.

Figure 19.14 and Figure 19.15 illustrate a result for seven iteration cycles based on the usual definition of uncorrelated and very large starting value intervals for the capacitive and inductive elements [10]. The source inductance was first neglected. The residual error stabilized around 0.034. The inclusion of L_s (1.0 – 5.0 pH) drastically reduced the residual errors. A very pronounced step-like rise can be observed in the residual error trace. Figures 19.15(a) – 19.15(d) show the corresponding intermediate ($N = 4$) results. The abrupt change in the residual error function manifests itself in strong parameter value shifts among the capacitances and inductances.This has not been observed with extraction using well-defined

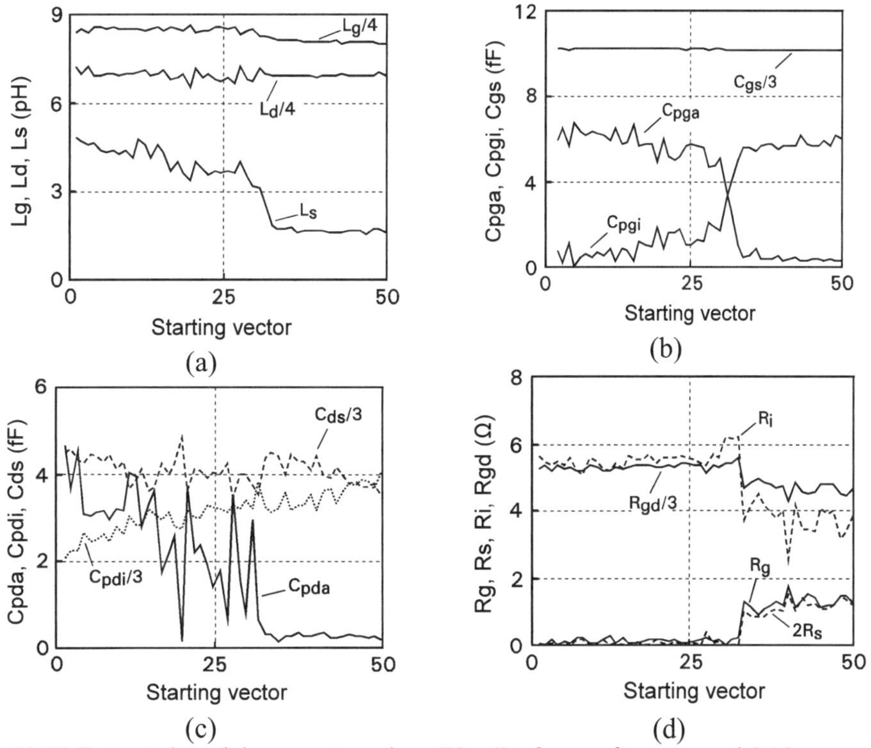

Figure 19.15 Extracted model parameter values ($N = 4$) of on-wafer measured 0.15-μm gate HEMT based on uncorrelated starting values of model parameters: (a) extracted inductances, (b) extracted distributed gate capacitances, (c) extracted distributed drain capacitances, (d) extracted resistances.

starting vectors as described before. From Figure 19.15(d) it is obvious that following minimum residual error as best fit would lead to a zero-value R_g and R_s. Therefore, to restart the optimization procedure, the definition of new starting vectors should rely on the upper part of sorted starting vectors.

Nearly all model elements are consistently extracted. However, due to S-parameter measurement uncertainties, stable but not always physically meaningful values for the resistances R_s and R_d are obtained. As already mentioned, this problem will be discussed in the following chapter.

19.5 SUMMARY

Problems well-known in measurement-based device modeling were highlighted. It was concluded that before starting the final model parameter extraction process very careful analysis of the measured device data should be performed in order to ensure extraction with confidential results. Common random optimization using arbitrarily defined starting model parameter value intervals may give rise to very strong changes of the parameters during the extraction process. This enlarges the risk to terminate in a local minimum with partly unphysical model parameter elements. This is avoided by the proposed model parameter extraction concept based on well-defined starting values both for the capacitive and inductive parameters derived from single-bias (pinch-off) measurements. The inherent correlative properties of the starting parameter values and gradual adaptation of individual search intervals provide very fast and stable convergence of the parameter values.

REFERENCES

[1] Kompa, G., and M. Novotny, "Frequency-Dependent Measurement Error Analysis and Refined FET Model Parameter Extraction Including Bias-Dependent Series Resistors," *IEEE Workshop on Experimentally Based FET Device Modelling & Related Nonlinear Circuit Design*, Kassel, Germany, July 17–18, 1997, pp. 6.1–6.16, https://archive.org/details/DTIC_ADA331468/page/n279.

[2] Kompa, G., "Reliable Extraction of Small-Signal Elements of a Generalized Distributed FET Model," *IEEE MTT-S Int. Microwave Symp. Dig.*, Baltimore, USA, 1998, Paper TUA-8, pp. 291–294.

[3] G. Dambrine, A. Cappy, F. Heliodore, and E. Playez, "A New Method for Determining the FET Small-Signal Equivalent Circuit," *IEEE Trans.*, Vol. MTT-36, 1988, pp. 1151-1159.

[4] Kompa, G., M. Schlechtweg, and F. van Raay, "Precisely Calibrated Coaxial-to-Microstrip Transitions Yield Improved Performance in GaAs FET Characterization," *IEEE Trans.*, Vol. MTT-38, 1990, pp. 62–68.

[5] Kompa, G., "Small- and Large-Signal Characterization Methods, Modelling and Verification," *Digest IEEE Workshop on Measurement Techniques for Microwave Device Characterization and Modelling*, Stuttgart, Germany, April 23, 1990, pp. 67–97.

[6] Lin, F., and G. Kompa, "FET Model Parameter Extraction Based on Optimization with Multiplane Data-Fitting and Bidirectional Search—A New Concept," *IEEE Trans.*, Vol. MTT-42, 1994, pp. 1114–1121.

[7] Schlechtweg, M., *Broadband Characterization and Modelling of GaAs-MESFETs and (AlGa)As/ GaAs-MODFETs up to 40 GHz* (in German), Doctoral Thesis, University of Kassel, 1989.

[8] Pengally, R.S., *Microwave Field-Effect Transistors – Theory, Design and Applications*, New York: Wiley & Sons, reprinted April 1984, p. 37.

[9] Schlechtweg, M., private communications, 1997.

[10] Novotny, M., and G. Kompa, "Unique and Physically Meaningful Extraction of the Bias-Dependent Series Resistors of a 0.15 μm PHEMT Demands Extremely Broadband and Highly Accurate Measurements," *IEEE MTT-S Int. Microwave Symp. Dig.*, San Francisco, California,1996, pp. 1715-1718.

Chapter 20

Bias Dependence of Source and Drain Resistances

20.1 INTRODUCTION

In Section 6.2.2 approximate relations for the calculation of series resistances were derived to be used as starting element values in an optimization process. Simple closed-form expressions were found by investigating the device performance under limit operation in the saturation region.

In this chapter we are interested in the general bias dependence of series resistances of GaAs- and GaN-based field-effect transistors. In particular, we want to find out in which way the resistance value of the model elements are influenced by the control voltages V_{gs} and V_{ds}, and how to implement the results in a device modeling concept.

Bias-dependence of series resistances has received little attention in the past. Various extraction techniques were applied. Mostly pinch-off and gate-forward measurements were used even though it was known that these values were not representative of the device under normal (active) operation. In Section 21.7 these methods will be briefly reviewed, and an alternative procedure is presented.

We may differentiate between devices based on GaAs and GaN technology and related III-V semiconductor compounds. GaN stands for high-power application and offers high-temperature operation. Thus, though multifinger GaN HEMT devices may provide relatively low series resistance values, the high channel current may cause remarkable voltage drop that must be taken into account for device operation. Above all, high current value and high voltage drop provide power dissipation, which may result in temperature increase and increase of the resistance value. Thus, the bias dependence of access resistances in GaN HEMTs is currently under discussion.

20.2 BIAS-DEPENDENT VERSUS BIAS-INDEPENDENT SERIES RESISTANCES

This section provides a brief review about various proposals that have been published in recent years concerning the determination of the parasitic resistances in a field-effect transistor equivalent network, both analytically and experimentally. For example, DC measurements have been suggested to extract the series resistances R_s, R_d, and R_g under a gate-forward bias condition (e.g., [1 - 7]). This approach leads to high gate currents that could destroy the Schottky barrier diode. Furthermore the determined resistance values commonly differ from those valid under normal operating conditions. This also holds for an approach reported by Baier et al. [8] using a modified transmission-line model (TLM) structure, where Schottky metal is deposited between each of the ohmic contacts (gated TLM structure). Under assumption that the gate lengths are large enough depletion at the edges can be neglected. The total ohmic-to-ohmic resistance R_{GD} (i.e., resistance between two adjacent ohmic contacts gate G and drain D) is then measured as a function of gate lengths for several gate voltages. For $L \to 0$, the minimum R_{GD} value corresponds to $2(R_c + R_p)$; that is, twice the sum of contact and bulk resistances.

As already pointed out in [9], series resistance values at microwave frequencies are expected to differ from those obtained from DC measurements. Thus, extraction methods have been proposed based on S-parameter measurement under cold FET biasing (e.g., [10, 11]). In this mode of operation, the equivalent model can be simplified so that the model elements of the remaining circuit can be analytically determined from measured S-parameters. Both cold pinch-off and cold gate-forward device operation have been investigated. However, as already pointed out, large forward gate current may degrade the Schottky contact, and therefore, this method has only limited applicability. This problem particularly arises with wide bandgap devices such as AlGaN/GaN power HEMTs (see Section 21.8.2). Cold pinch-off operation provides only low current flows so that the device is not stressed in any way (e.g., [12]).

Summing up, it can be said that extracted resistance values differ strongly with respect to various measurement conditions. Both cold gate-forward and pinch-off bias conditions refer to some specific device testing situation that frequently differs from the final application.

Based on various definitions of the limited charge control region, bias-dependent source and drain resistances were calculated in different ways. In Section 6.2.2 charge control is based on a geometric definition, whereby the resistance is defined for the region from the source to the edge of the gate for R_s, as is indicated in Figure 20.1 for the gate-source region, and from the drain to the edge of the gate for R_d. However, looking more closely, the bulk source resistance can be divided into the resistance R_{s1} from the source to the edge of the lateral gate depletion region, and the resistance R_{s2} from the edge of the lateral depletion region to the edge of the gate [13]. As indicated in the figure, the depletion region of a FET device depends strongly on the gate bias. As the forward bias increases ($V_{GS1} \to V_{GS2} > V_{GS1}$), the depletion region becomes shallower. From the figure it is clearly seen that the bulk

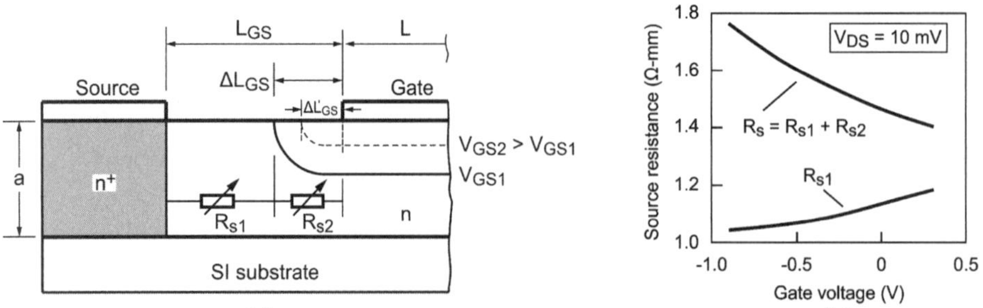

Figure 20.1 (a) MESFET structure (with thickness a of active layer) showing the lateral depletion region ΔL_{GS}. Definition of the source resistance: $R_s = R_{s1} + R_{s2}$. (b) Simulated bias-dependent total source resistance R_s and nondepleted resistive part R_{s1}. (After [13].)

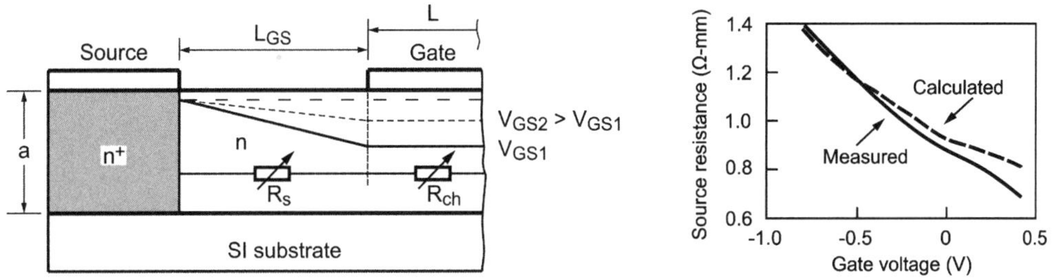

(a) (b)

Figure 20.2 (a) MESFET structure (with thickness a of active layer) showing linear change of the surface potential in the ungated portion (gate-to-source) of the channel. (b) Simulated and measured bias-dependent source resistance R_s. (After [15].)

resistance R_{s1} increases with increasing gate voltage, whereas R_{s2} decreases. The edge of the gate's lateral depletion region, denoted by the edge depletion region length ΔL_{GS}, has been located where the surface charge concentration decreased to 90 percent of the nondepleted value [13]. Similar conditions exist on the drain side concerning partial resistances R_{d1} and R_{d2}. As pointed out in [13], in contrast to R_{s2}, the corresponding gate-sided drain resistance R_{d2} becomes part of the gate charge control region under the saturation bias condition so that only R_{d1} remains as effective bulk drain resistance.

As is known, the crystal lattice at the semiconductor surface is perturbed and therefore a high density of surface states with acceptor character appears in the surface region between the ohmic contacts. Under condition of thermal equilibrium the negative charge of the surface states is neutralized by a positive charge of ionized donors of equal value [14]. States on the free surface of GaAs are known to produce a surface band-bending of about 0.8 eV (i.e., close to that of a zero-biased metallic Schottky barrier). At V_{GS} = 0V, the width of the depletion region is, therefore, approximately constant from source to drain [14].

Normally it is tacitly assumed that the population of these free surface states is unchanged when the gate is reverse-biased so that (apart from fringing close to the gate) the depletion depth is independent of V_{GS}. This is equivalent to assuming that the surface potential is held at the potential of the source and drain electrodes right up to the edges of the gate [14].

A physically more reasonable assumption is that the surface potential varies along the access regions so that the free surface depletion depth varies along the length of the access regions, as illustrated schematically in Figure 20.2(a) [14].

Byun et al. [15] have analytically calculated the source (and drain) resistance value as function of the gate-source voltage. Looking at Figure 20.2(a), gradual channel approximation (see Chapter 4) has been applied to calculate the channel current I_{ch} in the gate-to-source spacing. Then with the potential $V(L_{GS})$ at the edge of the gate electrode, the source resistance reads $R_S(V_{GS}) = V(L_{GS})/I_{ch}$. Figure 20.2(b) confirms the expected result: With increasing V_{GS} the depth of the fringing charge in the ungated portion decreases, and therefore the source resistance diminishes.

Dutta [16] applied a similar charge distribution model as depicted in Figure 20.2(a), and his numerical calculations, based on gradual channel approximation, delivered bias dependence of both source and drain resistance. As previously found in [15], the source

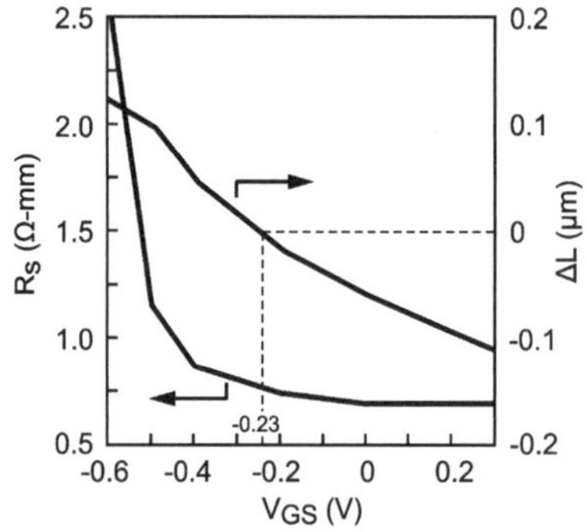

Figure 20.3 Simulated gate-length variation $\Delta L(V_{GS})$ and parasitic source resistance R_s of a 0.3-μm MESFET as function of gate-source voltage V_{GS}. (After [17].)

resistance decreases with increasing gate voltage and is almost constant over drain voltage. In contrast, the drain resistance is enhanced both with increasing drain and decreasing gate voltage.

Menozzi et al. [17–19] reported on the gate voltage dependence ($V_{DS} = 0$) of the series resistances of short-channel MESFETs. The intrinsic device is defined as the region where the gate control of the channel charge at low drain bias is described by the gradual channel approximation within a given tolerance of, for example, 100%. The channel charge per unit area in the central section at $x = x_0$ is calculated as $Q^*(V_{GS}) = q\int_0^\infty n(x_0, y, V_{GS})\,dy$ (y denotes the vertical and x the lateral direction). Recalling (4.15) of gradual channel approximation in Chapter 4, the width of the depletion region is calculated as

$$h(V_{GS}, V_{DS} = 0) = \sqrt{2\varepsilon/(qN_D)}\sqrt{-V_{GS} + V_d} \tag{20.1}$$

The minimum space charge extension $h(V_{GS} = 0)$ reads

$$h(V_{GS} = 0, V_{DS} = 0) = \sqrt{2\varepsilon/(qN_D)}\sqrt{V_d} \tag{20.2}$$

Then $Q^*(V_{GS})$ is given as $qN_D\,[h(V_{GS} = 0) - h(V_{GS})]$, which yields

$$Q*(V_{GS}) = \sqrt{2\varepsilon/(qN_D)}\sqrt{V_d}(1 - \sqrt{-V_{GS} + V_d / V_d} \tag{20.3}$$

Two-dimensional simulations confirm the channel charge $Q^*(x_0, V_{GS})$ dependence on the gate bias as predicted by the gradual channel approximation, which is partly a consequence of the simulated structure. However, for x closer to the gate edges, the agreement becomes progressively worse. Based on the defined tolerated channel charge change for the intrinsic device, an effective length, $L_{eff}(V_{GS}) = L - \Delta L(V_{GS})$, can be determined. As illustrated in Figure 20.3, $\Delta L(V_{GS})$ is decreasing with increasing gate-source voltage. More specifically, it is seen that the effective length is shorter than the physical gate length for gate voltage values < -0.23V and larger for > -0.23V. Thus, correspondingly, the source resistance increases strongly as soon as the effective length falls to below the physical gate length L.

At higher V_{GS} values the effective length exceeds continuously, reducing the access region and therefore, the resistance decreases. The validity of the simulated results was supported by measurements [18].

20.3 PRACTICAL EXPERIENCES IN THE EXTRACTION OF SERIES RESISTANCES OF GaAs FETs

Investigations on the reliability of model parameter extraction have shown [20] that most of the parameters can be extracted with satisfying uncertainty. However, extracted resistance values of the series resistances R_s and R_d commonly cannot be verified by physical inspection. Uncertainty in the extracted values originates in the limited measurement accuracy and possibly in the limited frequency bandwidth (see e.g. Section 14.2). Therefore, in the following, a measurement-error sensitive frequency scanning method is described [21, 22], which permits the extraction of the bulk series resistances together with all other bias-dependent parameters.

An optimization method based on a random search technique with adaptive starting interval sizes was described in Chapter 19. Nearly all model elements are consistently extracted. However, due to S-parameter measurement uncertainties, stable but not always physically meaningful values for the resistances R_s and R_d have been obtained. This special extraction problem is solved by using only two single-frequency S-parameter data sets, one at a fixed low frequency controlling the effective capacitances and inductances and the other is scanned over the upper measured frequency range mainly controlling the parameter distribution and the relatively insensitive resistances. An upper frequency is chosen within a measurement interval that reveals confidential (i.e., physically meaningful) parameter values. As will be shown, the extracted bias-dependence of the source resistance R_s for a wire-bonded 0.4-μm recessed-gate MESFET and a 0.15-μm gate HEMT reflects the expected physical properties.

Standard extraction methods are aimed at reaching the best fit between simulated and measured scattering coefficients in a limited frequency range mostly determined by the available measurement feasibilities. We noted previously that various extraction methods indeed allow the consistent extraction of most of the model elements with physical significance. However, regarding parasitic resistances such as R_s, R_d, R_g, and R_{gd}, these model parameters are commonly extracted with large uncertainty or, if consistently determined, their value cannot be physically interpreted. Often zero-values for R_s and R_d (or even negative element values if not principally excluded by the specified boundary conditions of the optimizer) are derived from pinch-off device data, which is entirely unphysical. The extracted value should at least reach a minimum value given by the contact resistance.

Figure 20.4 illustrates this situation. It shows the residual error of measurement-based object function in dependence of R_s and R_d resistance values. In this case extraction has been performed in a saturated bias point taking into account all values of the parasitics extracted from pinch-off except for R_s and R_d. Intrinsic elements were then quasi-analytically

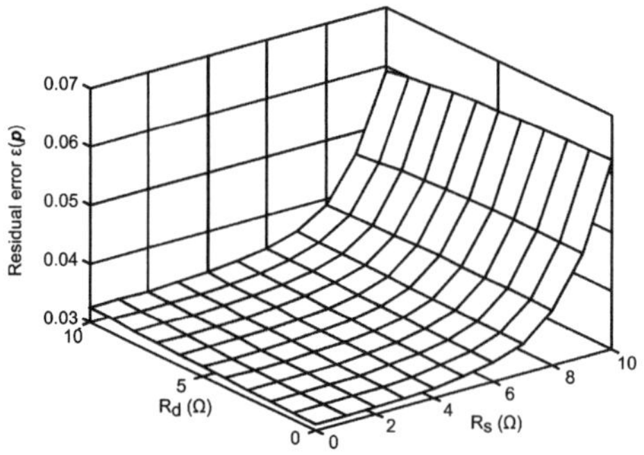

Figure 20.4 Residual error versus R_s and R_d for measured S-parameters in the saturated region (0.15-μm gate HEMT).

determined for different resistance values (see Chapter 16). As can be seen, no local (global) minimum with R_s and $R_d > 0$ can be recognized. In this case the extractor would deliver zero-values for both resistances. If no lower bound resistance values were defined (unconstrained optimization), negative values would have been extracted. It can be shown that due to the present measurement errors, commonly the individual suberror-functions ε_{ij},

$$\varepsilon_{ij}(\boldsymbol{p}) = \sum_{f=1}^{N} \left| S^{f}_{c,ij}(\boldsymbol{p}) - S^{f}_{m,ij} \right|^2 \Big/ \left| S^{f}_{m,ij} \right|^2, \quad (i,j) = (1,2) \tag{20.4}$$

do not indicate any absolute minimum, and if it does happen, then their minima are located at different frequencies so that the total error function

$$\varepsilon(\boldsymbol{p}) = \sum_{i,j=1}^{2} \varepsilon_{ij}(\boldsymbol{p}) \tag{20.5}$$

does not reveal any local (global) minimum. It can be concluded that the common extraction method taking into account S-parameter data at numerous frequencies over a broad frequency range is, due to present measurement errors, not very practical to extract reasonable values for R_s and R_d.

To get closer to the extraction problem we have first used synthesized S-parameters. We performed the extraction for different pairs of R_s and R_d. Figure 20.5 shows a result based on synthetic device data. As can be seen from this figure a clear minimum is identified. In this case a minimum of the residual error function is clearly seen for the chosen R_s = R_d

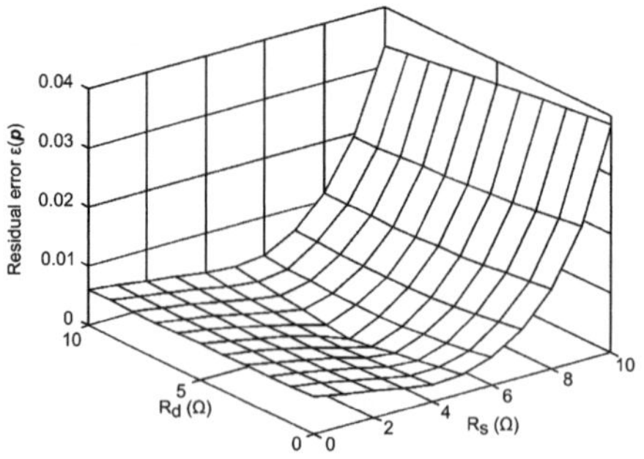

Figure 20.5 Residual error versus R_s and R_d using synthetic S-parameters ($R_{s0} = R_{d0} = 5\Omega$).

= 5Ω. Regarding the R_s dependence, the residual error increases rapidly in the direction of increasing R_s values, but changes weakly with decreasing resistance value. Variation of R_d has only a marginal effect on the residual error of the object function so that R_d was primarily chosen equal to R_s. It is clear that in view of potential measurement errors, the physically meaningful extraction of less sensitive resistances can prove itself to be a great challenge.

20.3.1 Frequency Scanning Extraction Method

If fixed values are chosen for R_s and R_d, the internal elements can be extracted from the S-parameter set at any measured frequency point. We used two frequencies, a fixed one in the lower frequency range (e.g., 0.5 or 1.0 GHz), where the intrinsic capacitances strongly influence the device response. As can be shown, the determination of R_s and R_d needs very high frequencies. Therefore, a second upper frequency is scanned over the higher frequency range. The quasi-analytical determination in dependence of assumed R_s and R_d values have been performed as described in [23].

Figure 20.6 shows the scanning result for the investigated recessed-gate MESFET MGFC 1423 operating in the bias point (V_{GS} = -0.5V, V_{DS} = 3.0V). The residual error of the object function depends on the chosen R_s and scanning frequency value. Because of low influence of R_d on the device response, R_d was primarily set equal R_s.

As can be seen from this figure, above 30 GHz minimum R_s values are spreading and are less confidential due to the increased measurement uncertainty of the VNA (see Section 11.3.1.3). At low frequencies (approximately until about 10 GHz) the device response is mainly determined from capacitive influences that let extraction of the resistances appear hopeless. There is a medium frequency range where the scattering of R_s minimum values is rather low. Analysis of the individual residual error functions of S_{11}, S_{12}, $S_{21,}$ and S_{22} shows that in this region their minima appear approximately at the same R_s value. For the frequency scanning model parameter extraction we used a higher frequency where the minima of the four error-functions totally overlapped. This was a frequency of 16 GHz for the MESFET device.

Figures 20.7 illustrates the result for the investigated 0.15μm-gate HEMT. In this case the individual suberror functions versus the scanned value of R_s from 0Ω to 6Ω show a very

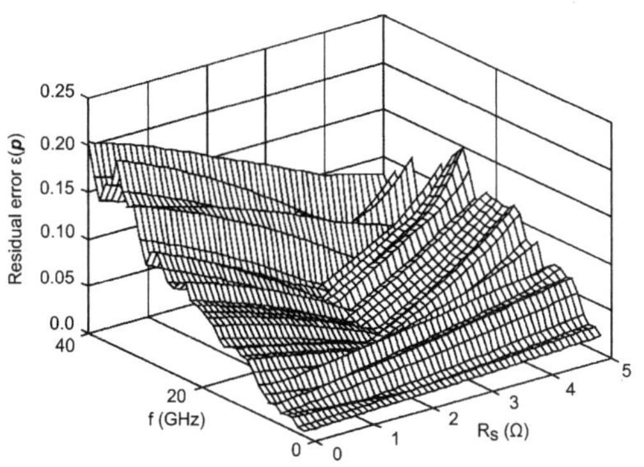

Figure 20.6 R_s-dependent residual error for scanned upper frequency (MGFC1423).

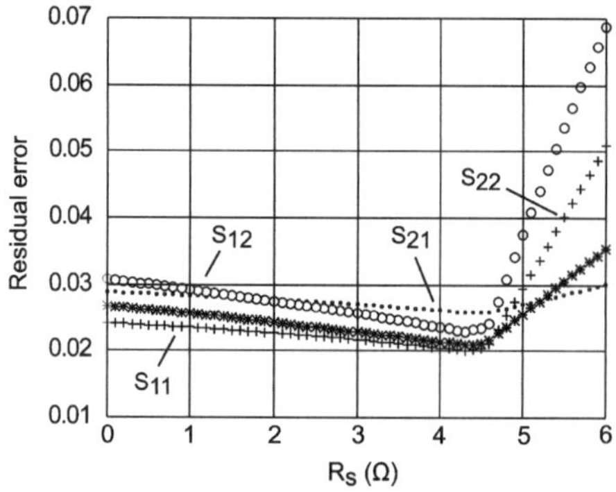

Figure 20.7 Residual value of the suberror functions versus R_s for f = 117 GHz (0.15-μm gate HEMT: V_{GS} = -0.5V, V_{DS} = 3.0V). Note the similarity of asymmetric curve shape in Figure 20.5.

good agreement at a frequency of 117 GHz. A clear minimum exists for all four scattering parameter curves at the same R_s value of 4.25Ω. Such concurrent conditions imply small measurement uncertainties. In other cases the minima would have different values.

20.3.2 Extracted Bias-Dependent Source Resistances

Figure 20.8(a) shows the extraction result for the bias-dependent source resistance R_s of the recessed-gate MESFET MGFC1423. It is obvious that the extracted bias-dependent source resistance differs considerably from the results discussed previously, which may be explained by the construction of the investigated device.

Figure 20.9 shows the principle structure of the measured gate recessed MESFET. The source resistance in this case comprises four parts; i.e. the source contact resistance R_{sc}, the access resistance R_{s1} with constant channel cross section, the transition access resistance R_{s2}, and the recess resistance R_{s3}.

At pinch-off and low drain-source voltage V_{DS}, the source resistance R_s is very small. The extracted value of about 0.4Ω agrees reasonably with the estimated contact resistance of 0.3Ω

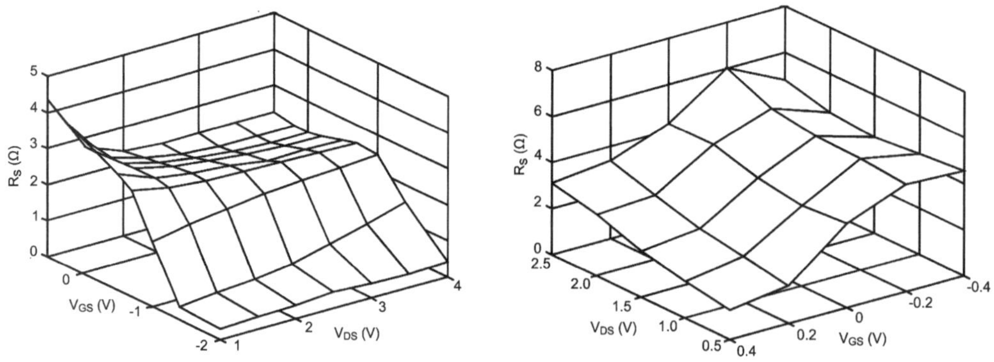

(a) (b)

Figure 20.8 Extracted bias dependency of R_s. (a) MESFET MGFC1423 , fixed frequency 16 GHz, and (b) 0.15-μm gate HEMT (IAF), fixed frequency 117 GHz.

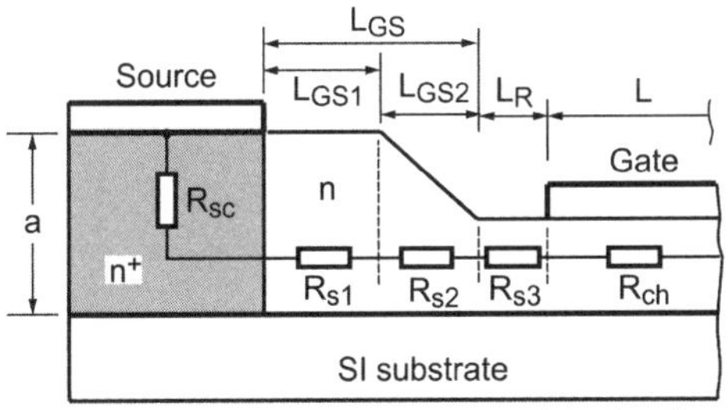

Figure 20.9 Schematic view of a recessed-gate MESFET with distributed source resistance elements: contact resistance R_{sc}, bulk resistance R_{s1} with constant cross-sectional area, resistance R_{s2} of the transition section, and recess resistance R_{s3}.

(Section 6.2.1). Toward higher V_{GS}, the resistance rises up to about 2.5Ω, which may correspond to the transition access region in Figure 20.9. Approaching the gate-forward operation mode, a further strong rise is observed, reaching a maximum value of about 4.2Ω. This growth may be attributed to the continuous lateral extension of the channel conduction in the ungated recess region. The bias-dependence of the drain-source voltage V_{DS} is moderate. Only at high gate-source voltage $V_{GS} \gtrsim 0$V, a considerable decrease is observed resulting from the extension of charge depletion region with increasing V_{DS}. Regarding the extraction results, it should be noted that the source resistance in the saturation region ($V_{GS} > -1.0$V) is rather bias-independent, approximately 2.5Ω.

Very similar bias-dependence of R_s was presented in [24] for a 0.3-µm recessed-gate MESFET NE71000, applying all-at-once extraction approach as described in Chapter 14. In this case the measurement covered a frequency range of 26.5 GHz. In [25] a 0.2-mm GaAs MESFET (2×100 µm) was measured in the frequency range of 45 MHz to 26 GHz. Parameter extraction was based on a 15-element model with nine bias-dependent (inclusively R_g, R_s, $\tau_m = R_i C_{gs}$) and seven bias-independent (inclusively R_d, τ) model elements. Extraction was based on the bidirectional search technique implemented in IC-CAP software tool [26]. Interestingly, the two-dimensional surface of the extracted bias-dependent source resistance agrees very well with Figure 20.8(a), such as the nearly constant R_s value at pinch-off ($V_{GS} \approx V_{pa}$, V_{DS}). Nevertheless, the extracted source contact resistance of 1.7Ω seems to be rather high. It is also recognized that the extraction results in the linear region are not fully reliable (occasionally negative values for source and gate resistance were extracted), although, as stated by the authors, good agreement between measured and simulated S-parameters could be observed.

Figure 20.10 presents an entire set of plots showing the sensitivity of model parameters (MGFC14423) with respect to the residual error of object function under pinch-off (see Figure 20.6). Although extraction has only be applied to the standard 15-element model in this case, the results are a positive demonstration that R_s and R_d can be extracted separately [27]. Furthermore, the plot of R_g confirms that the gate metallization resistance can be reliably determined by overall data fitting as was stated in Section 19.3.

Figures 20.8(b) shows a similar result for a 0.15-µm gate HEMT from IAF. With the frequency fixed at 117 GHz, the bias-dependent device model elements have been determined. Comparing with the previously discussed result in Figure 20.8(a) for the

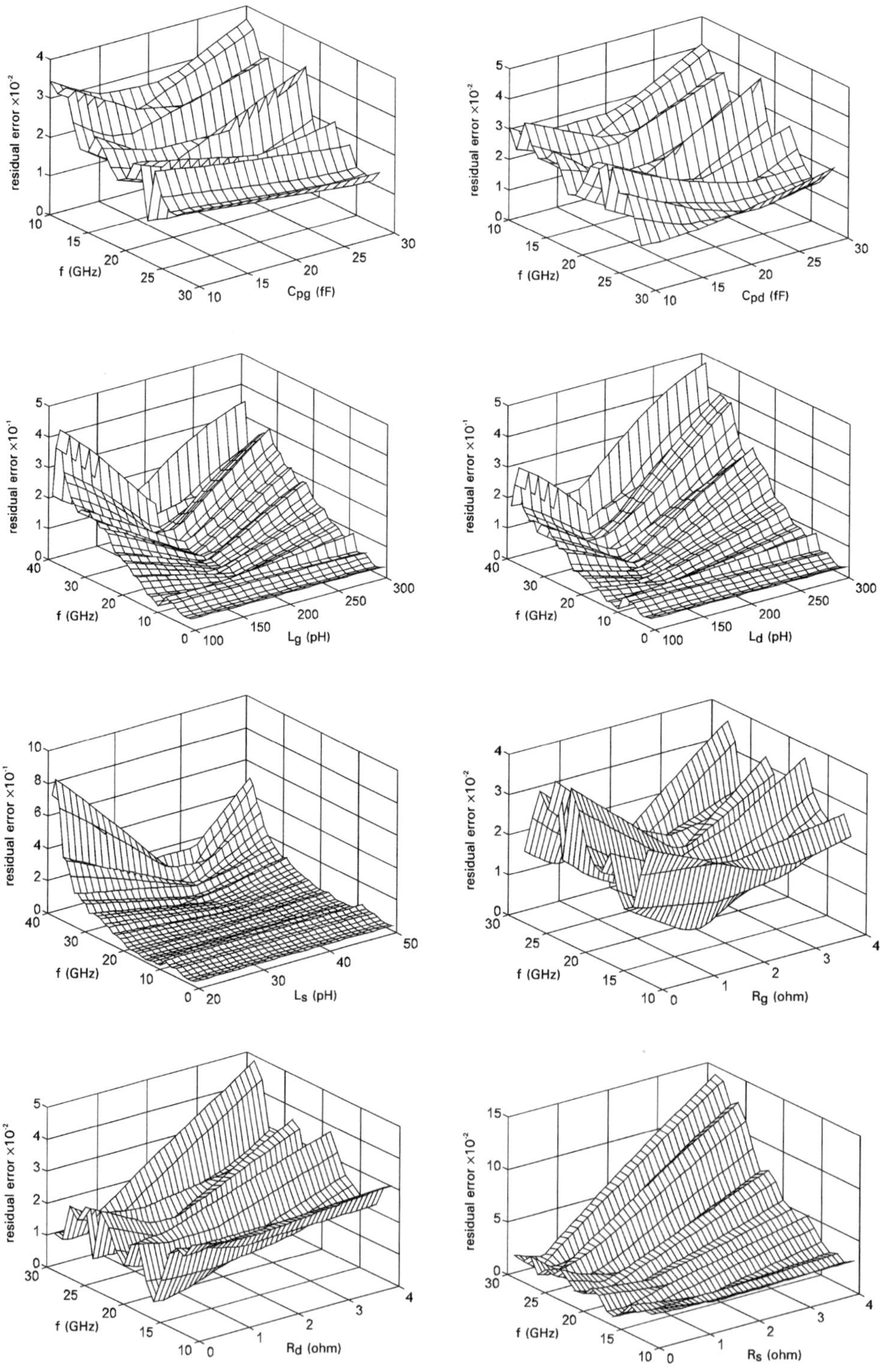

Figure 20.10(a) Extracted model parameters C_{pg}, C_{pd}, L_g, L_d, L_s, R_g, R_d, and R_s of MESFET MGFC1423 based on the two-frequency scanning method at pinch-off bias.

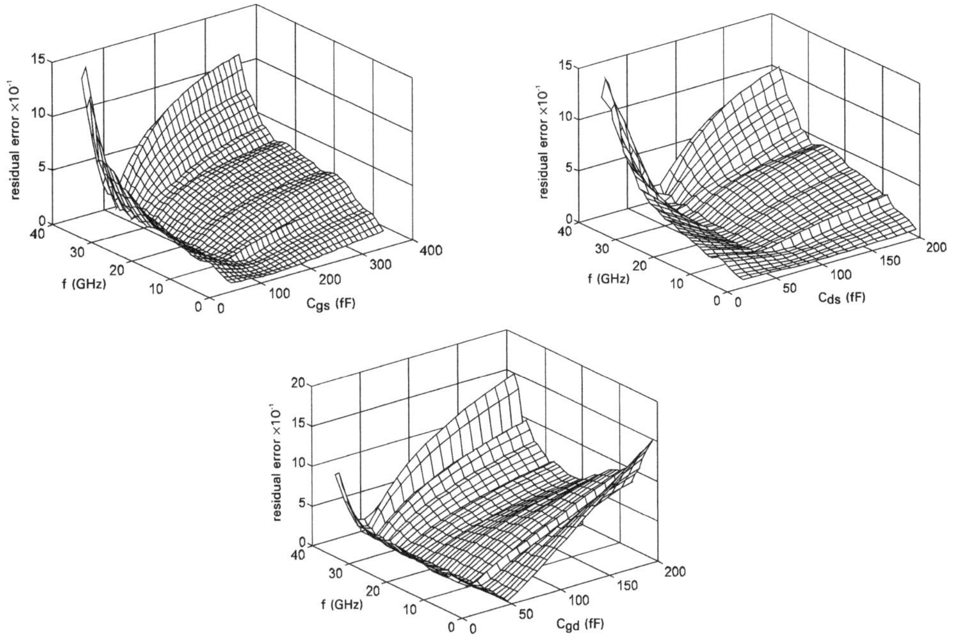

Figure 20.10(b) Extracted model parameters C_{gs}, C_{ds}, and C_{gd} of MESFET MGFC1423 based on the two-frequency scanning method at pinch-off bias.

MESFET device, the source resistance R_s behaves quite differently. Now the resistance increases toward pinch-off. At a typical bias point in the saturated region ($V_{GS} = 0V$, $V_{DS} = 2.5V$) an R_s value of 4.2 ohm has been extracted that correlates well with the expected value in the order of 3.5 ohm [28]. It can be seen that at pinch-off ($V_{GS} = -0.4V$) an R_s value of 4.4Ω is extracted, including contact and residual 2D channel resistance. The estimation of the series resistances of HEMTs is not trivial due to the multilayer structure of the source and drain regions [29], which leads to an interaction between the cap layer conductivity, the channel conductivity, and the tunneling conductance [30]. With respect to available data, a minimum value of 1.4Ω has been evaluated for ($V_{GS} = 0.4V$, $V_{DS} = 0.5V$).

Feuer [29] has pointed out that the high-bandgap AlGaAs layer in a HEMT structure provides a barrier to electron flow from the conductive cap layer to the 2DEG channel and vice versa (see Figure 20.11). As the two conductive layers are relatively thin, one can use an analytical model in which each layer is considered two-dimensional. The AlGaAs layer is essentially nonconductive. Even under the gate, where positive bias can induce charge in

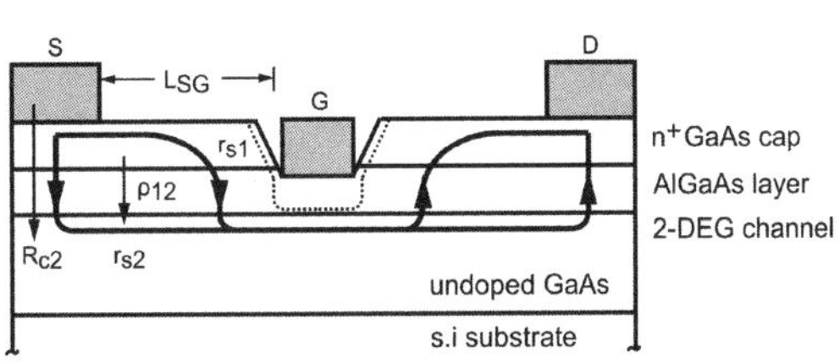

Figure 20.11 Schematic representation of a AlGaAs/GaAs HEMT device with cap layer and gate recess. The dotted line indicates the gate depletion zone and the arrows describe paths of electron flow.

the AlGaAs (parasitic MESFET effect), its very low mobility ensures a small conductivity. Furthermore, for the low-voltage part of the IV curve, the barrier can be characterized by a specific barrier resistivity ρ_{12}. The gate recess forces all the current into the channel (layer 2).

The calculation of current flow in the two layers (i.e., cap layer and 2DEG channel) results in coupled differential equations. The solution reveals an interesting result for small values of L_{SG}. In this case

$$R_s = R_{c2} + \left(r_{s2} - \frac{R_{c2}^2}{\rho_{12}} \right) L_{SG} \tag{20.6}$$

where R_{c1} is the metal-semiconductor contact resistance to cap layer (layer 1), R_{c2} the metal-semiconductor contact resistance to layer 2, r_{s1} the sheet resistance of the cap layer, r_{s2} the sheet resistance of the channel layer, and ρ_{12} the barrier resistance in the HEMT structure. Equation (20.6) shows the surprising possibility that R_s may decrease with increasing L_{SG}. This occurs when the increase in L_{SG} decreases the resistance of the cross-barrier path more rapidly than it increases the current due to current along the layers.

A very simple explanation for the decrease of the R_s resistance with increasing gate-source voltages has been proposed by Oxley investigating bias-dependent source resistance in GaN HEMTs [31]. R_s was found to decrease as V_{GS} is increased, implying that the conduction channel becomes wider, which indicates a parallel conduction path to the 2D electron gas.

20.4 BIAS-DEPENDENT ACCESS RESISTANCES IN GaN HEMTs

Oxley investigated the bias dependence of parasitic resistances in a 0.25-μm gate AlGaN/GaN HEMT with a total gate width of 100 μm, consisting of two unit gate-widths of 50 μm in Pi configuration [31]. S-parameter measurements were performed from 0.25 to 50 GHz. Extraction of the model parameters used de-embedding techniques. The results for the GaN HEMTs agree principally with those discussed in section 20.3.2 for the GaAs HEMT. A strong decrease of R_s has been observed toward forward gate operation. This is in total contradiction to other publications reporting about a strong increase of access resistances with increasing channel current flow (i.e., with growing drain- or gate-voltage).

We have previously seen that the bias-dependence of the parasitic bulk resistances of a field-effect transistor depends on the definition of the resistive areas between gate-source and gate-drain electrodes. Traditionally, the resistances describe the region outside the space charge beneath the gate. The location of the transition between the more or less depleted channel zone, which is under control of the gate potential, and the adjacent conductive regions on both sides of the gate, is not fixed but varies with the gate and drain potentials (source commonly grounded). With this in mind, measured (extracted) bias-dependent resistance values were interpreted.

However, the situation with high-power GaN devices may be different. It seems that a defined access resistance covering the fixed-region between source and gate electrode is favored. In this case a dramatic increase of resistance value has been reported with growing channel current.

Based on DC measurements Palacios et al. investigated the source access resistance of a 0.25-μm gate-length AlGaN/GaN HEMT with a gate width of 150 μm. The source-to-drain and source-to-gate distance were 1.9 μm and 0.5 μm, respectively [32].

To measure the dynamic source access resistance, Palacios et al. forward-biased the gate contact with respect to the channel by introducing a 0.5-mA current through the gate of the transistor into the channel, while the source contact was grounded. The drain voltage was varied from 0 to 3V in order to measure the differential source resistance under different drain current levels. In this measurement, it was assumed that the voltage drop through the gate Schottky barrier is constant. Therefore, the gate voltage is equal to the channel voltage at the source edge of the gate plus the turn-on voltage of the Schottky gate diode. The derivative of the gate voltage with respect to the drain current gives the differential source access resistance, R_s. It is shown that R_s is at high current levels more than three times larger (about 5.5Ω) than under low current conditions (about 1.5Ω) [32].

The origin of the strong increase of R_s is explained as follows: Based on two-dimensional device simulation, the longitudinal electric field in the source access region turned out to be higher than 15 kV/cm for gate voltages higher than −6V. At this electric field, the electron velocity has already started its quasi-saturation. This quasi-saturation that occurs in nitride-based structures for electric fields higher than 10 kV/cm produces a decrease in the electron mobility, which translates into an increase in the differential access resistance. The origin for the quasi-saturation in the electron velocity profile is generally associated with the high phonon energy in nitride-based semiconductors[32].

Figure 20.12 shows the cross section of an AlGaN/GaN HEMT as discussed in [33]. The device has a gate of length L_G and width w_G, flanked by ungated lengths L_{SG} and L_{DG} to the source and drain contacts, respectively. In GaN HEMTs, traps may convert portions Δ of the ungated regions in the vicinity of the gate into virtual gates having the same potential as the gate [14]. When V_{DS} is small, Δ is approximately same on either side of the gate. The 2-DEG beneath the length $(L_G + 2\Delta)$, separated from the gate electrode by the barrier thickness d, has a concentration n_s and mobility μ_0, while that beneath the length $(L_{SG} - \Delta)$ and $(L_{DG} - \Delta)$ has a sheet resistivity ρ_{s0} associated with the maximum 2-DEG concentration n_{s0}. The V_{GS} modulates n_s, μ_0, and Δ.

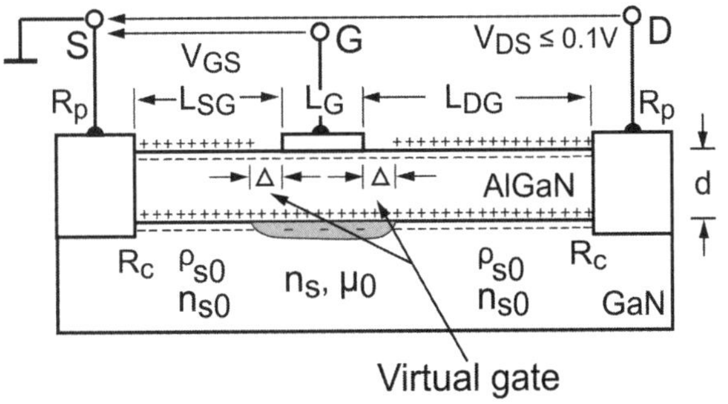

Figure 20.12 Cross-section of a general HEMT showing dimensions, bias conditions and extracted parameters. (After [33].)

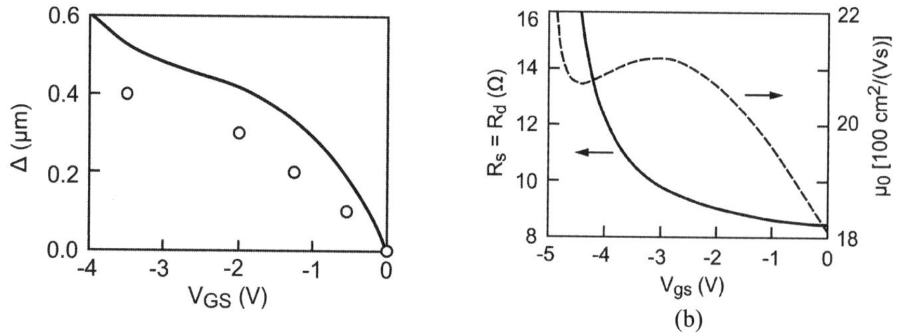

Figure 20.13 (a) Extracted virtual gate lateral extension Δ, and (b) extracted mobility μ_0 and $R_s = R_d$ versus V_{GS}. Circles denote results from [34]. (After [33].)

With $R_{CP} = R_C + R_P$, where R_P denotes the contact resistance between the electrode metallization and the measurement probe, and R_C the contact resistance between the 2-DEG and source/drain metallization, the resistances R_s and R_d can be expressed as

$$R_{s,d} = [\rho_{s0}(L_{SG,DG} - \Delta) + (q n_s \mu_0)^{-1}\Delta] W_G^{-1} + R_{CP}$$

n_s is approximated from gate capacitance estimation (see Section 4.7.2). Other values, μ_0, Δ and R_{CP} are obtained by comparison of two devices located nearby on the same substrate with different L_{SG} and L_{DG} and biased at the same V_{GS}, V_{DS}.

Pradeep et al. point out that in GaAs devices Δ has negligible value and $R_s = R_d$ is constant, underlining the absence of traps and virtual gate effect. On the other hand, in the GaN device, $R_s = R_d$ rises for more negative V_{GS} because the electrons from the gate fill the AlGaN surface traps over longer and longer Δ (see Figure 20.13).

Similar results have been reported by Ghosh et al. Based on TCAD simulation, it was concluded that nonlinear source and drain resistances increase with drain current, which is due to the saturation of electron velocity in this region. Also in this case the geometric distance between source and gate (and gate and drain, respectively) electrodes is defined as the lateral extension of the resistance region [35].

Nasser et al. [36] performed TCAD simulations and compared these with S-parameter measurements. It has been pointed out that a significant bias dependence of R_s and R_d was found due to the electric field dependence of the mobility in agreement with [35].

20.5 FREQUENCY AND TEMPERATURE DEPENDENCE OF SERIES RESISTANCES

Two further aspects should be mentioned briefly that affects the resistance value and moreover may also give rise to some frequency dependence, i.e. surface traps and surface passivation.

In [37], the influence of Si_3N_4 passivation on microwave power performance of SiC MESFETs was investigated. Fabricated 2-μm × 200-μm devices were measured before and after passivation. Small-signal circuit elements were extracted using the method after Dambrine et al. [10]. The source and drain resistances tended to be a function of frequency after passivation. They could be represented by the parallel components of capacitor and resistor. At low frequency of 4 GHz (neglecting the capacitive influence) the source resistance changed from 12.1Ω to 10.4Ω and drain resistance from 20.0Ω to 17.3Ω.

As is widely accepted, surface traps cause dispersion. Mukherjee et al. [38] reported on a physics-based analytical model of SiC MESFET incorporating trapping and thermal effects. Surface trapping effects have been modeled by using a gate and drain-voltage dependent drain resistance. As the surface trapping effect is insignificant in the ungated region between gate and source because of lower electric field, the source resistance R_s is fairly constant. In contrast, the drain resistance R_d shows considerable variation with the control voltages due to the high electric field in the drain-gate region, having a strong influence on the surface trapping situation. R_d enhances with increasing V_{DS} and decreases with increasing V_{GS}. Increasing V_{DS} results in higher electric field strength, so that the concentration of surface-trapped electrons also increases. Thus, a virtual gate [14] is formed that reduces drain current flow, which can be equivalently modeled by an increasing drain resistance.

An analytical model incorporating bulk and interface trapping effects was presented in [39] and its impact on the DC IV characteristic was investigated. The 4H-SiC MESFET device was assumed to be passivated on the ungated regions between source-gate and gate-drain of the device with an isolator, typically silicon dioxide (SiO_2). Concerning bulk resistances, surface trapping effects at the channel-dielectric interface resulted in the reduction in the drain current under static conditions.

Either with an unpassivated or passivated surface, the surface trapping induced depletion layer extends into the channel between the source-gate and gate-drain modulating the source and drain parasitic resistances R_s and R_d.

An analytical model inspired by gradual channel approximation was used, considering a linear variation of the potential at the interface between source and the gate. The model results clearly show that the series resistance (R_s) strongly depends on an applied bias similar to GaAs MESFETs [15] as well as on the surface conditions. A similar behavior is observed in the case of drain series resistance (R_d).

The self-heating in GaN HEMTs needs to be taken into account for high power applications [40]. The low field mobility in GaN HEMTs have been shown to have a large temperature dependence. In [41] the extracted series resistance showed a significant increase with temperature between 74K and 406K, which is assumed to be correlated with an observed corresponding decrease in electron mobility

20.6 SUMMARY

In this chapter, various approaches were briefly described to determine the series resistances of GaAs and GaN field-effect transistors. Theoretical results depend on the definition of the source (or drain) access area and on the device structure. Experimental studies frequently apply DC measurements in gate conduction mode. S-parameter measurements prefer pinch-off or forward gate mode of operation. The major limitation of these approaches is that the determined resistance values are derived from operating conditions that may be less relevant for the final application.

The concept of random optimization with adaptive starting interval sizes in Chapter 19, tested on GaAs FETs, has provided very good extraction results for most of the model parameters. Nevertheless, it was also recognized that consistent (physically meaningful) extraction of less sensitive resistances remained a challenge.

A refined extraction procedure was developed based on measurement data at two discrete frequencies in a given bias point, whereby a lower frequency is fixed and a second one is scanned over the measured frequency range. The extracted values of the resistances thus obtained clearly indicate reliable measurement regimes with less measurement uncertainties indicated by the perfect convergence of the minima of all four suberror functions of individual S-parameter coefficients. Using these particular frequencies successful extraction results have been obtained both in the pinch-off and saturation regions.

Regarding GaAs FETs, it has been shown that there is a wide range of control voltages in which the series resistances change only slightly. Thus, in particular, operating in the saturation region, it is justifiable to assume bias-independent resistances to simplify microwave circuit design and optimization.

Based on S-parameter measurements bias dependency of series resistances in GaN-HEMTs have been found to provide similar behavior as given for GaAs HEMTs. In particular, strong decrease of the resistance value has been observed in the direction of forward gate operation, which recalls parasitic MESFET phenomenon. However, contrasting published results predicting a strong increase of parasitic bulk resistances with growing channel current refer to the entire access regions between gate and source and gate and drain, respectively.

Frequency and temperature dependence of resistances were briefly addressed.

REFERENCES

[1] Hower, P. L., and N. G. Bechtel, "Current Saturation and Small-Signal Characteristics of GaAs Field-Effect Transistors," *IEEE Trans.*, Vol. ED-20(3), 1973, pp. 213–220.

[2] Fukui, H., "Determination of the Basic Device Parameters of a GaAs MESFET," *The Bell System Technical Journal*, Vol. 58(3), 1979, pp. 771–797.

[3] Williams, R. E., *Gallium Arsenide Processing Techniques*, Dedham: Artech House, Inc., 1984.

[4] Lee, K., M. Shur, K. W. Lee, T. Vu, P. Roberts, and M. Helix, "A New Interpretation of 'End' Resistance Measurements," *IEEE Electron Device Lett.*, Vol. EDL-5(1), 1984, pp. 5–7.

[5] Lee, K. W., K. Lee, M. S. Shur, T. T. Vu, P. C. T. Roberts, and M. J. Helix, "Source, Drain, and Gate Series Resistances and Electron Saturation Velocity in Ion-Implanted GaAs FET's," *IEEE Trans.*, Vol. ED-32(5), 1985, pp. 987–992.

[6] Yang, L., and S. Long, "New Method to Measure the Source and Drain Resistance of the GaAs MESFET," *IEEE Electron Device Lett.*, Vol. EDL-7(2), 1986, pp. 75–77.

[7] Liu, S.-M. J., S.-T. Fu, M. Thurairaj, and M. B. Das, "Determination of Source and Drain Series Resistances of Ultra-Short Gate-Length MODFET's," *IEEE Electron Device Lett.*, Vol. EDL-10(2), 1989, pp. 85–87.

[8] Baier, S. M., M. Shur, K. Lee, N. C. Cirillo, and S. A. Hanka, "FET Characterization Using Gated-TLM Structure," *IEEE Trans.*, Vol. ED-32, 1985, pp. 2824–2829.

[9] Ladbrooke, P. H., *MMIC Design: GaAs FETs and HEMTs*, Artech House, Inc., 1989.

[10] Dambrine, G., A. Cappy, F. Heliodore, and E. Playez, "A New Method for Determining the FET Small-Signal Equivalent Circuit," *IEEE Trans.*, Vol. MTT-36(7), 1988, pp. 1151–1159.

[11] Costa, J. C., M. Miller, M. Golio, and G. Norris, "Fast, Accurate, On-Wafer Extraction of Parasitic Resistances and Inductances in GaAs MESFETs and HEMTs," *IEEE MTT-S Int. Microwave Symp. Digest*, 1992, pp. 1011–1014.

[12] Kompa, G, and M. Novotny, "Highly Consistent FET Model Parameter Extraction Based on Broadband S-Parameter Measurements," *IEEE MTT-S Int. Microwave Symp. Dig.*, 1992, pp. 293–296.

[13] Tsai, Y.-T., and T. A. Grotjohn, "Source and Drain Resistance Studies of Short-Channel MESFET's Using Two-Dimensional Device Simulators," *IEEE Trans.*, Vol. ED-37(3), 1990, pp. 775–780.

[14] Kompa, G., *Basic Properties of III-V Devices–Understanding Mysterious Trapping Phenomena*, Kassel, Kassel University Press GmbH, 2014.

[15] Byun, Y. H., M. S. Shur, A. Peczalski, and F. L. Schuermeyer, "Gate-Voltage Dependence of Source and Drain Series Resistances and Effective Gate Length in GaAs MESFET's," *IEEE Trans.*, Vol. ED-35(8), 1988, pp. 1241–1246.

[16] Dutta, S., "Impact of Parasitic Resistances on the Electrical Characteristics of a SiC MESFET," *Superlattices and Microstructure*, Vol. 110, 2017, pp. 10–18.

[17] Menozzi, R., L. Selmi, P. Gandolfi, and B. Riccò, "Extraction of the Series Resistances and Effective Channel Length of GaAs MESFETs by Means of Electrical Methods: A Numerical Study," *IEEE International Electron Devices Meeting* (IEDM), 1991, pp. 341–344.

[18] Menozzi, R., P. Cova, and L. Selmi, "Experimental Application of a Novel Technique to Extract Gate Bias Dependent Source and Drain Parasitic Resistances of GaAs MESFETs," *Solid-State Electronics*, Vol. 36(7), 1993, pp. 1083–1084.

[19] Selmi, L., R.Menozzi, P. Gandolfi, and B. Riccò, "Numerical Analysis of the Gate Voltage Dependence of the Series Resistances and Effective Channel Length in Submicrometer GaAs MESFET's," *IEEE Trans.*, Vol. ED-39(9), 1992, pp. 2015–2020.

[20] Kompa, G., "Reliable Extraction of Small-Signal Elements of a Generalized Distributed FET Model," *IEEE MTT-S Int. Microwave Symp. Dig.*, Baltimore, USA, 1998, Paper TUA-8, pp. 291–294.

[21] Kompa, G., and M. Novotny, "Frequency-Dependent Measurement Error Analysis and Refined FET Model Parameter Extraction Including Bias-Dependent Series Resistors," *IEEE Workshop on Experimentally Based FET Device Modelling & Related Nonlinear Circuit Design*, Kassel, Germany, July 17-18, 1997, pp. 6.1–6.16, https://archive.org/details/DTIC_ADA331468/page/n279.

[22] Novotny, M., and G. Kompa, "Unique and Physically Meaningful Extraction of the Bias-Dependent Series Resistors of a 0.15 μm PHEMT Demands Extremely Broadband and Highly Accurate Measurements," *IEEE MTT-S Int. Microwave Symp. Digest*, 1996, pp. 1715–1718.

[23] Lin, F., and G. Kompa, "FET Model Parameter Extraction Based on Optimization with Multiplane Data-Fitting and Bidirectional Search - a New Concept," *IEEE Trans.*, Vol. MTT-42, 1994, pp. 1114-1121.

[24] Schlechtweg, M., *Broadband characterization and modelling of GaAs-MESFETs and (AlGa)As/GaAs-MODFETs up to 40 GHz* (in German), Doctoral Thesis, Kassel University, 1989.

[25] Lin, F., J. Cao, P. S. Kooi, and M. S. Leong, "Extraction of Bias-Dependent Source and Gate Resistance from Measured S-Parameters under All Bias Conditions," *Journal of Electromagnetic Waves and Applications*, Vol. 11, 1997, pp. 1103–1119.

[26] Sischka, F., *IC-CAP Modeling Handbook*, Agilent Technologies, October 2010.

[27] Lott, U., "Sensitivity Analysis of the Millimeter-Wave HEMT Performance Parameters f_t and f_{max} to Errors in the Equivalent Circuit Elements," *IEEE MTT-S Int. Microwave Symp. Digest*, 1994, pp. 857–860.

[28] Schlechtweg, M., private communications, 1997.

[29] Feuer, M. D., "Two-Layer Model for Source Resistance in Selectively Doped Heterojunction Transistors," *IEEE Trans.*, Vol. ED-32, 1985, pp. 7–11.

[30] Lee, S.J., and C.R. Crowell, "Parasitic Source and Drain Resistance in High-Eelectron-Mobility Transistors," *Solid-State Electronics*, Vol. 28, 1985, pp. 659–668.

[31] Oxley, C. H., "Method for Measuring Source Resistance R_s in Saturation Region of GaN HEMT Device Over Bias Conditions (V_{gs}, V_{ds})," *Electronics Lett.*, Vol. 40(5), 2004, pp. 344–346.

[32] Palacios, T., S. Rajan, A. Chakraborty; S. Heikman, S. Keller, S. P. DenBaars, and U. K. Mishra, "Influence of the Dynamic Access Resistance in the g_m and f_T Linearity of AlGaN/GaN HEMTs," *IEEE Trans.*, Vol. ED-52(10), 2005, pp. 2117–2123.

[33] Pradeep, D., M. Amit,and S. Karmalkar, "DC Extraction of Gate Bias-Dependent Parasitic Resistances and Channel Mobility in an HEMT," *IEEE Electron Device Lett.*, Vol. 37(11), 2016, pp. 1403–1406.

[34] Kasai, S., J. Kotani, T. Hashizume, and H. Hasegawa, "Gate Control, Surface Leakage Currents, and Peripheral Charging in AlGaN/GaN Heterostructure Field Effect Transistors Having Nanometer-Scale Schottky Gates," *Journal of Electronic Materials*, Vol. 35(4), 2006, pp. 568–575.

[35] Ghosh, S., S. A. Ahsan, and Y. S. Chauhan, "Modeling of Source/Drain Access Resistances and Their Temperature Dependence in GaN HEMTs," 2016 *IEEE Int. Conf. on Electron Devices and Solid-State Circuits (EDSSC)*, 2016, pp. 247–250.

[36] Nasser, C., D. Ritter, and M. Rudolph, "Bias Dependence of the Access Resistance in GaN HEMTs," *22nd Int. Microwave and Radar Conf. (MIKON)*, 2018, pp. 271–273.

[37] Cha, H.-Y., Y. C. Choi, R. M. Thompson, V. Kaper, J. R. Shealy, L. F. Eastman, and M. G. Spencer, "Influence of Si_3N_4 Passivation on Surface Trapping in SiC Metal-Semiconductor Field-Effect Transistors," *Journal of Electronic Materials*, Vol. 33(8), 2004, pp. 908–911.

[38] Mukherjee, S. S., S. S. Islam, and R. J. Bowman, "An Analytical Model of SiC MESFET Incorporating Trapping and Thermal Effects," *Solid-State Electronics*,Vol. 48, 2004, pp. 1709–1715.

[39] Rao, M. H. L., and N. V. L. N. Murty, "An Improved Analytical Model of 4H-SiC MESFET Incorporating Bulk and Interface Trapping Effects," *Journal of Semiconductors*, 2015, pp. 014004/1–12.

Chapter 21

Model Parameter Extraction with Measurement-Correlated Parameter Starting Values

21.1 INTRODUCTION

The main features of the model parameter extraction method based on the principle of measurement-correlated parameter starting values have already been discussed in Chapter 19 and applied to a three-shell equivalent network. Both effective capacitive and effective inductive parameters were introduced and identified by S-parameter data measured at rather low and moderate frequencies. Extrinsic gate-to-drain capacitances were temporarily ignored. The distribution of the effective capacitances among the individual model capacitances were estimated. Knowing the distributed capacitive network, the model inductances could be determined from the effective ones. The extracted model elements were used to define starting value intervals for a repetitive random optimization procedure. The resistance search intervals were empirically defined. The results vividly illustrate the evolution of the model parameter extraction process showing, in particular, the correlation between distributed branch capacitances and their mutual interactions.

In the following, the optimizer-based model parameter extraction method using measurement-correlated parameter starting values is extended. Beyond the empirical choice of initial capacitive model elements, automated finding of appropriate capacitance ratios for the optimization process is discussed. Resistances are determined from cold reverse S-parameter measurements instead of cold pinch-off or cold forward measured data.

Measured cold pinch-off S-parameter data is used to generate physics-based starting values. The high quality of the starting values will then be demonstrated in terms of S-parameter fitting. The need for random starting values is therefore eliminated and only a final fine-tuning optimization will be required. A parameter optimization based on the leaping simplex algorithm (see Section 12.7.5) will be carried out to refine the model parameters for demonstration.

A generalized distributed 22-element small-signal FET equivalent model, illustrated in Figure 21.1(a), is investigated. Under cold pinch-off condition, the equivalent circuit can be simplified as shown in Figure 21.1(b), thereby reducing the number of unknown parameters considerably. Optimization-based extraction of the model parameters involves three steps: the generation of high-quality starting values, formulation of the objective function, and the search of optimal model parameter values through optimization. In the following, a detailed description of these steps will be given.

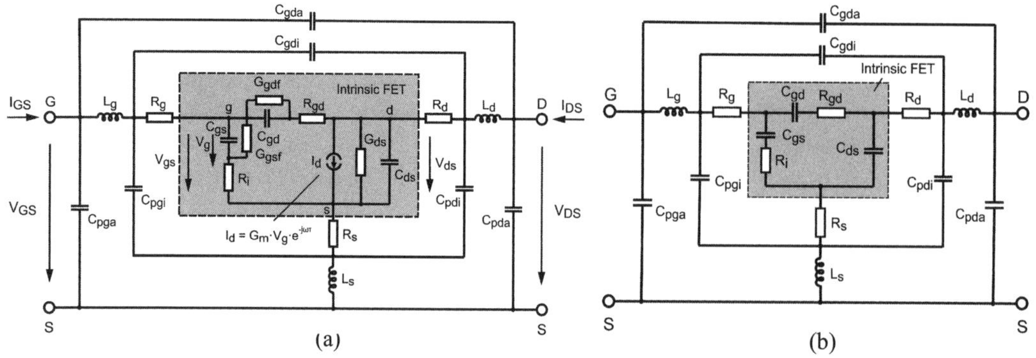

Figure 21.1 (a) Distributed 22-element small-signal FET equivalent model. G_{gsf} and G_{gdf} denote gate-current leakage or gate forward/breakdown effects. (b) Related 17-element pinch-off FET model.

A similar complex circuit topography has been treated by Shealy et al. [1] where the intermediate capacitance C_{pgi} denotes an on-mesa capacitance that has been effectively absorbed into the pad capacitances with little error, as stated by the authors.

21.2 MODEL PARAMETER EXTRACTION CONDITIONS

Since the various model elements are correlated, the starting value estimation should be carried out in a way that takes this correlation into account. As the number of optimization variables is rather large, it is necessary to define capacitance ratio values for the optimization process. As will be discussed, these capacitance ratios can be empirically defined (Section 21.2.1) or directly derived from top-view images of the fabricated device (Section 21.2.2).

21.2.1 Heuristically Defined Capacitance Ratio Values

In the following, empirically defined capacitance ratios for GaAs FET and GaN HEMT devices are reviewed and interpreted. Applying a capacitance ratio scanning method, the optimal solution for each ratio is found, confirming the previously published ratio values. This technique may also be generalized and automated with microwave transistors such as HBT, LDMOS, FinFET, and others.

21.2.1.1 Capacitance Ratios for GaAs FETs

Investigating the capacitance ratio of C_{gd}/C_{gs} for various in-fixture and on-wafer measured GaAs MESFET and HEMT devices with gate lengths between 0.15 and 0.5 μm and gate widths between 120 μm and 350 μm (two-finger devices with typically 20-mW output power), it could be followed that this ratio equals approximately 1 as indicated in Table 21.1.This can be understood so far that, under cold pinch-off operation, the depletion region extensions on both sides of the gate electrode are assumed to be equal. Furthermore, the pad capacitances may be assumed to be equal (i.e. $C_{pga} \approx C_{pda}$) justified by similar gate and drain contacting pads of the investigated transistors. Summarizing, in an earlier work Mwema has

Table 21.1
Gate-Drain/Gate-Source Capacitance Ratios under Pinch-off Condition of GaAs MESFETs and HEMT*

Device	MESFET					HEMT	
L_g (μm)	0.5	0.55	0.55	0.5	0.4	0.3	0.15
w_g (μm)	200	250	300	350	300	200	120
C_{gd}/C_{gs}	1.10	0.99	1.01	1.09	1.05	1.09	1.16

*From: [3].

proposed the following capacitance relations [2, 3]:

$$C_{pga} \approx C_{pda} \tag{21.1}$$

$$C_{gs} \approx C_{gd} \tag{21.2}$$

$$C_{pdi} \approx C_{pda} \tag{21.3}$$

21.2.2.2 Capacitance Ratios for GaN Power HEMTs

Jarndal investigated AlGaN/GaN power HEMTs [4]. He employed the extended 22-element model in Figure 21.1 and modified the capacitance relations according to the more complex gate periphery and altered technology. Taking into account the experiences encountered with GaAs FETs, he proposed the following modified assumptions about the capacitance ratios [4]:

$$C_{pga} \approx C_{pda} \tag{21.4}$$

$$C_{gs} \approx C_{gd} \tag{21.5}$$

$$C_{pdi} = 3 \cdot C_{pda} \tag{21.6}$$

$$C_{gdi} = 2 \cdot C_{gda} \tag{21.7}$$

The distributed interelectrode capacitance C_{pdi} between drain and source fingers was observed to be much larger than the pad capacitance of the drain electrode. Analyzing small-size GaN HEMT devices the assumption $C_{pdi} \approx 3C_{pda}$ in (21.6) had proved very convenient to minimize the error between simulated and measured S-parameters and to reduce the problem of getting trapped in local minima [5, 6]. Relation (21.7) indicates that the gate-drain interelectrode capacitance (C_{gdi}) is larger than the interelectrode capacitance between the open end of drain finger and gate pad/gate bus electrode (C_{gda}) (see e.g., Figure 21.4).

21.2.1.3 Verification of Empirically Defined Capacitance Ratios by Scanning Technique

Recently published works suggested that the proposed capacitance ratio values are not general and that the parameter extraction could not be directly applied to different devices [7, 8]. Other works, like [9], reported diverse values of the same ratios and also used different capacitance ratios. Moreover, in [10] it is stated that the definition of proposed capacitance ratios requires user intervention to assume suitable initial values for the interelectrode capacitances, thus implying that finding suitable capacitance ratio values for different devices requires the intervention of an experienced modeler.

To address these objections, the capacitance ratio values proposed in [5] were thoroughly analyzed [11]. In the extensive analysis, the capacitance ratios C_{gs}/C_{gd}, C_{pga}/C_{pda}, C_{gdi}/C_{gda}, and C_{pdi}/C_{pda} were scanned over a large value interval to study the optimal value for any of the defined capacitance ratios. For the analysis, the same measured data of the same AlGaN/GaN HEMT was employed as in [5], which was a 0.1-mm gate device with two gate fingers and unit gate width of 50 μm. The device was air-bridge- and field-plate-free. The scanning was split into six tasks. In each task, two capacitance ratios were fixed in the proposed values ((21.4) to (21.7)) and the two other ratios were scanned from 0.01 to 100 (25 logarithmically-spaced steps).

The scanning results in [11] conclusively confirm the usefulness of the proposed capacitance ratios for the regarded device. However, it should be noted that different technological aspects such as charge-trapping and passivation [12] could affect the adopted ratio values. For example, different works have documented that, in modern GaN HEMTs, the values of C_{gd} and C_{gs} are influenced by charge-trapping and passivation of surface layers [13-15] or by field-plate structures [16-19]. For instance, in [13] two versions of the same 0.2-mm AlGaN/GaN HEMT were studied, one version with a SiN passivation layer and another without it. Then, under pinch-off voltage condition, the C_{gs} value had obviously become larger after passivation. The influence of field plates on C_{gd} and C_{gs} has been discussed in [19] (see Section 6.2.5), studying two versions of the same wide-bandgap device, one version with a source-connected field plate and another without it. It has been found that the field- plate device has a lower C_{gd} value of 46.6 fF/mm in comparison with 85.9 fF/mm for the device without a field plate.

In conclusion, charge-trapping effects, passivation layers, and field-plate structures change the C_{gs}/C_{gd} values with respect to expectations based purely on the gate-drain and gate-source distances and their ratios. Even with the gate centered between source and drain, assuming $C_{gs}/C_{gd} = 1$ without further technological data of the device may provide only a rough approximation of this capacitance ratio value.

Figure 21.2 shows the error function values for different combinations of fixed/scanned capacitance ratios after optimization using the proposed values in (21.4) - (21.7) denoted by a black dot in the figures, which, in all cases, is located in a broad area with minimum values of the error function [11]. The data suggests that the definition of the capacitance ratios is less crucial.

The capacitance ratio values defined in (21.4) - (21.7) have also proven to be useful in the case of more complex AlGaN/GaN HEMTs up to 3.2-mm gate width including air bridges and gate and drain bus bars [5, 6, 20]. Nevertheless, it would be desirable to generalize the determination of the capacitance ratio values for a device of interest. This will be discussed in the following sections.

21.2.2 Reliable Determination of Capacitance Ratio Values from Top-View Device Images

The surface of a device provides useful information about the structural arrangement of

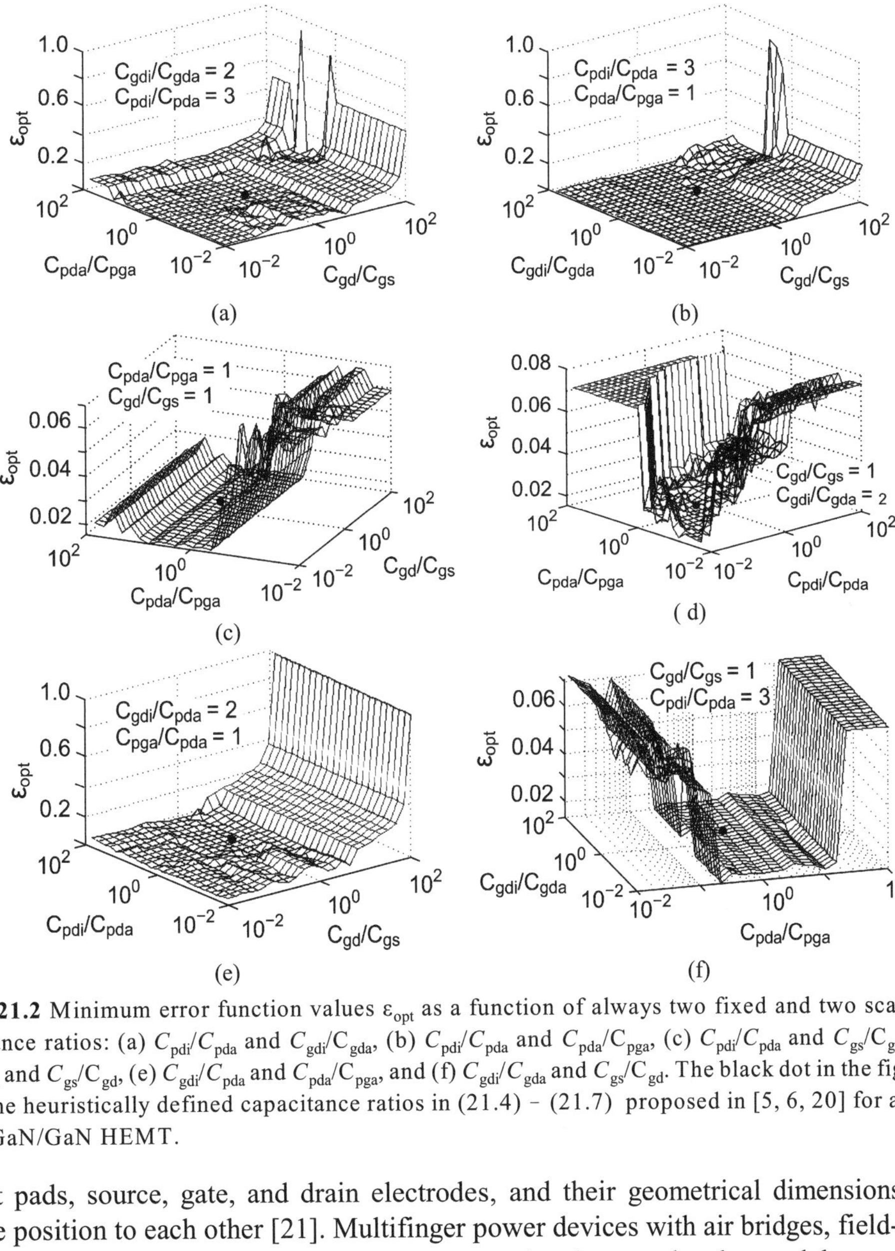

Figure 21.2 Minimum error function values ε_{opt} as a function of always two fixed and two scanned capacitance ratios: (a) C_{pdi}/C_{pda} and C_{gdi}/C_{gda}, (b) C_{pdi}/C_{pda} and C_{pda}/C_{pga}, (c) C_{pdi}/C_{pda} and C_{gs}/C_{gd}, (d) C_{gdi}/C_{gda} and C_{gs}/C_{gd}, (e) C_{gdi}/C_{pda} and C_{pda}/C_{pga}, and (f) C_{gdi}/C_{gda} and C_{gs}/C_{gd}. The black dot in the figures marks the heuristically defined capacitance ratios in (21.4) - (21.7) proposed in [5, 6, 20] for a 0.1-mm AlGaN/GaN HEMT.

contact pads, source, gate, and drain electrodes, and their geometrical dimensions and relative position to each other [21]. Multifinger power devices with air bridges, field-plate structures and distributed gate and source contacting bars make the model parameter extraction more sophisticated. The question arises whether the top-view analysis of a transistor die can provide useful data on parasitic capacitances, thus supporting the determination (estimation) of capacitance value ratios for the distributed equivalent device model.

Zamudio [11] has thoroughly analyzed top-view images of AlGaN/GaN power HEMTs using a commercial digital camera and the built-in microscope of the on-wafer probing system (and in addition a microscope with higher zooming factor) to gauge the horizontal

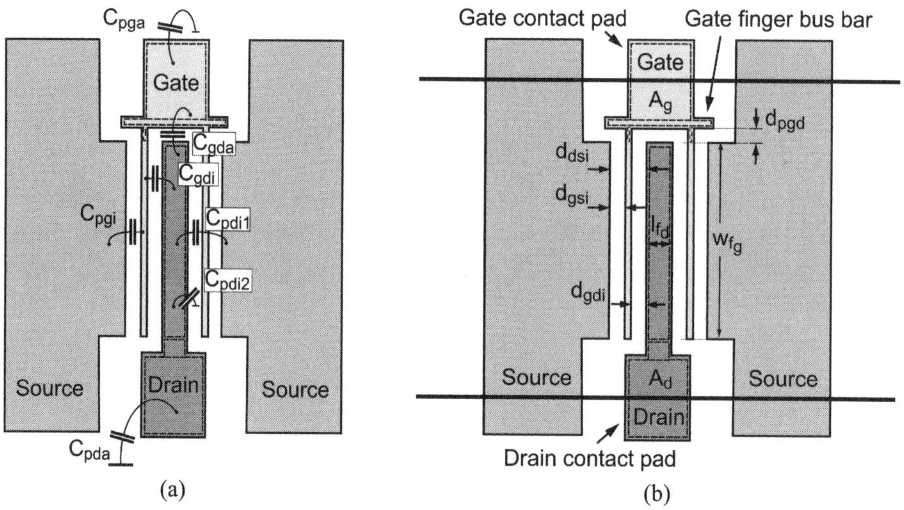

Figure 21.3 Top-view of two-finger FET device (not to scale): (a) parasitic capacitances, (b) geometrical dimensions. The lines to the right denote the network analyzer calibration reference planes.

area of contacting pads, bus bars, and gate fingers contributing to capacitive effects with the underlying metallization. Additionally, due to the finite thickness of the metallization layer, capacitive coupling also occurs between vertical surfaces of contiguous electrodes. Figure 21.3 illustrates the simplified metallization pattern of a two-finger FET device. Figure 21.3(a) indicates the location of the various parasitic capacitances and Figure 21.3(b) denotes relevant geometrical dimensions necessary to estimate their capacitance values.

As illustrated in Figure 21.1, the capacitances C_{pga}, C_{pda}, and C_{gda} form the outer capacitive shell of the small-signal FET equivalent circuit. C_{pga} represents the capacitance of the contacting pad and bus bar of the gate to the backside metallization of the substrate. Similarly, C_{pda} represents the capacitance of the contacting pad and bus bar of the drain to the backside metallization of the substrate. C_{gda} describes the capacitive effect between the contacting pad (or the bus bar) of the gate and the end of the drain electrode.

The intermediate capacitive shell comprises the capacitances C_{pgi}, C_{pdi}, and C_{gdi}. As can be seen from Figure 21.3(a), C_{pgi} indicates the capacitive effect between gate finger and contiguous source electrode (or finger). C_{pdi} is composed of two parts: C_{pdi1} and C_{pdi2}. Similar to C_{pgi}, C_{pdi1} describes the capacitive effect between the drain finger and contiguous source electrode (or finger). C_{pdi2} characterizes the capacitance of drain finger and backside metallization of the substrate.

The value of C_{pga}/C_{pda} is expressed by the ratio of two planar areas, A_g/A_d. A_g and A_d are the top-view areas related to the contacting pad and bus bar of the gate and of the drain electrodes, respectively. This ratio is readily estimated from top-view of the device finger periphery.

The capacitance $C_{pdi} = C_{pdi1} + C_{pdi2}$ includes two effects. C_{pdi1} considers the capacitive effect between electroplated drain and source electrodes above the actual electrodes, and C_{pdi2} considers the capacitance of the total horizontal drain electrode area to the underlying metallization.

The ratio C_{pda}/C_{pdi2} is related to A_d/A_{fd}, where A_{fd} denotes the drain finger area. Then, with the number of drain fingers, n_{fd}, the capacitance ratio C_{pda}/C_{pdi2} can be expressed as

$$\frac{C_{pda}}{C_{pdi2}} \approx \frac{A_d}{n_{f_d} \cdot A_{f_d}} \tag{21.8}$$

Under assumption that the metallization thickness of the drain fingers and the gate bus are equal (i.e., $t_{fd} \approx t_{bus,g}$) then the capacitance ratio C_{pdi}/C_{gda} reads approximately

$$\frac{C_{pdi1}}{C_{gda}} \approx \frac{n_{f_g}}{n_{f_d}} \cdot \frac{w_{f_g}}{l_{f_d}} \cdot \frac{d_{pgd}}{d_{dsi}} \tag{21.9}$$

where $n_{f,g}$ and $n_{f,d}$ are the number of gate and drain electrodes, respectively (typically $n_{f,g}/n_{f,d} = 2$). $w_{f,g}$ and $l_{f,d}$ are the width of the gate and the length of the drain electrode, respectively. d_{dsi} is the spacing between the gate finger and the source electrode (or source finger), and d_{pgd} is the spacing between the drain finger open end and the gate bus electrode.

C_{pgi} relates to the capacitance between gate finger and source electrode (or source finger), and similarly, C_{gdi} originates in the capacitive effect of gate and drain finger. The ratio C_{pgi}/C_{gdi} can be approximated as

$$\frac{C_{pgi}}{C_{gdi}} \approx \frac{\varepsilon_{gsi}}{\varepsilon_{gdi}} \cdot \frac{d_{gdi}}{d_{gsi}} \tag{21.10}$$

where ε_{pgi} and ε_{gdi} are the dielectric constants of the respective regions. d_{gdi} and d_{gsi} are the gate-drain finger and gate source electrode (or source finger) spacing.

Multifinger ($n_{fg} > 2$) devices with GSG (ground-zero-ground) probe pads require air bridges to connect the drain or source contacts in the device. A schematic representation of various interdigitated (or comb) layouts (with six gate fingers) is depicted in Figure 21.4. As can be seen, in Figure 21.4(b) the crossover capacitance C_{pdi1} is strongly increased by the inner source pads connected by air bridges, thus contributing to the source inductance and drain-to-source capacitance.

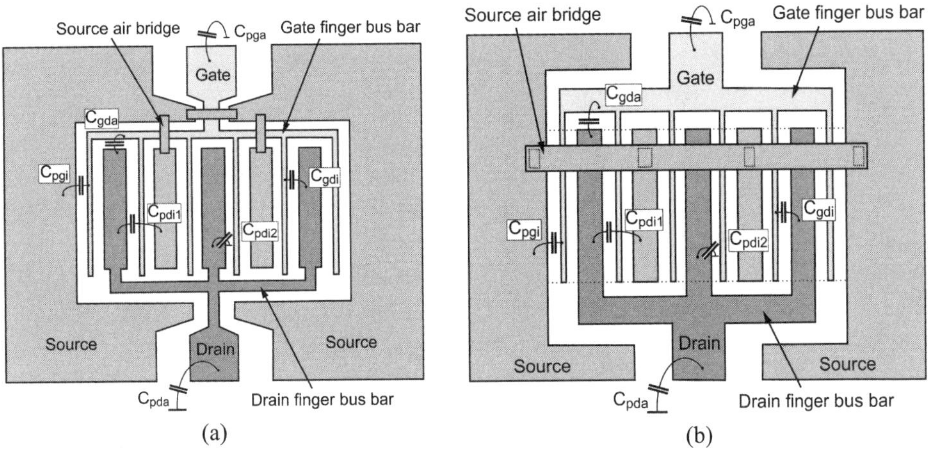

Figure 21.4 Top view of multifinger FET device with different source connections (not to scale): (a) Source air bridges traversing the gate bus, and (b) source air bridge traversing gate and drain finger.

Table 21.2
Reliably Determinable Capacitance Ratio Values of AlGaN/GaN Power Devices with/without Field Plates and Air Bridges

Source air bridges?	No		Yes (Gate Side)		Yes (Drain Side)	
Field-plate structures?	No	Yes	No	Yes	No	Yes
Capacitance ratios reliably evaluated from top-view images	$\frac{C_{pda}}{C_{pga}}$, $\frac{C_{pda}}{C_{pdi2}}$, $\frac{C_{pdi1}}{C_{gda}}$, $\frac{C_{pgi}}{C_{gdi}}$	$\frac{C_{pda}}{C_{pga}}$, $\frac{C_{pda}}{C_{pdi2}}$	$\frac{C_{pdi1}}{C_{gda}}$, $\frac{C_{pda}}{C_{pdi2}}$, $\frac{C_{pgi}}{C_{gdi}}$	$\frac{C_{pda}}{C_{pdi2}}$	$\frac{C_{pdi1}}{C_{gda}}$, $\frac{C_{pgi}}{C_{gdi}}$	None

From: [11].

With respect to the device technology studied, Table 21.2 summarizes the capability to evaluate capacitance ratios reliably. As only capacitance ratios are required, dielectric constant and distance between the metallization layers are not important.

In [11] Zamudio has applied the proposed procedure to 2-mm and 3.2-mm AlGaAs/GaN HEMTs. The results for the 3.2-mm device will be described briefly.

The top-view schematic of the analyzed 3.2-mm device is shown in Figure 21.4(a). Source air bridges are located at the gate side. Field plate structures (at the gate side) are not present. According to Table 21.2, C_{pdi1}/C_{gda}, C_{pda}/C_{pdi2}, and C_{pgi}/C_{gdi} can be reliably evaluated.

According to (21.8), $C_{pda}/C_{pdi2} \approx A_{pda}/A_{pdi2}$, which is obtained as 1.02. Due to (21.10), C_{pgi}/C_{gdi} can be expressed by the respective distance ratio d_{gsi}/d_{gdi}, which is gauged as 4.25.

C_{pdi1} addresses the total capacitance between drain fingers and source electrode, and C_{gda} represents the total capacitive effect between the ends of the drain fingers and the contacting pad (or the bus bar) of the gate. With estimated values of $d_{pgd}/d_{pdi} \approx 1.28$, $w_{f,g}/l_{f,d} \approx 7.04$ and $n_{f,g}/n_{f,d} = 2$, the capacitance ratio is obtained as $C_{pdi1}/C_{gda} \approx 18.03$ applying (21.9).

Further capacitance ratio values required for the preliminary determination of the model parameter values of the distributed model are obtained by the scanning method, as discussed in Section 21.2.1. The scanning process concerns the capacitances C_{pgi}, C_{pdi}, C_{ds} and C_{gs}/C_{gd}. With the measured branch capacitances

$$C_{gs0} = C_{pga} + C_{pgi} + C_{gs} \tag{21.11}$$

$$C_{ds0} = C_{pda} + C_{pdi} + C_{ds} \tag{21.12}$$

$$C_{gd0} = C_{gda} + C_{gdi} + C_{gd} \tag{21.13}$$

the following equations can be used to determine all capacitances of the distributed model reliably:

$$C_{pda} = C_{ds0} - C_{pdi} - C_{ds} \tag{21.14}$$

$$C_{pdi2} = C_{pda}/(C_{pda}/C_{pdi2}) \tag{21.15}$$

$$C_{pdi1} = C_{pdi} - C_{pdi2} \tag{21.16}$$

$$C_{gda} = C_{pdi1}/(C_{pdi1}/C_{gda}) \tag{21.17}$$

$$C_{gdi} = C_{pgi}/(C_{pgi}/C_{gdi}) \tag{21.18}$$

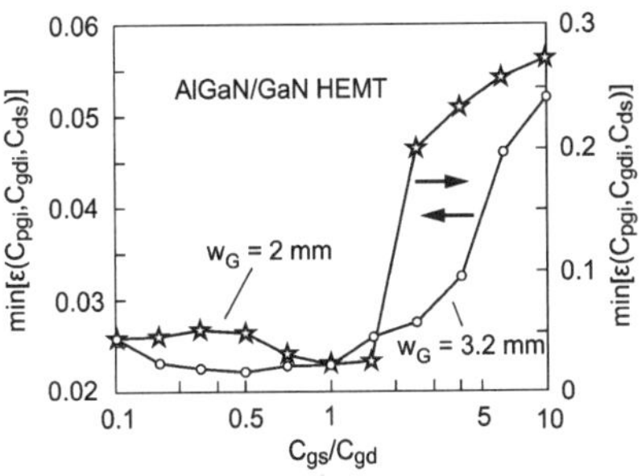

Figure 21.5 Minimum error function value min[ε(C_{pgi}, C_{gdi}, C_{ds})] dependent on C_{gs}/C_{gd} ratio for the analyzed 2-mm and 3.2-mm AlGaN/GaN HEMT. (After [11].)

$$C_{gd} = C_{gd0} - C_{gdi} - C_{gda} \tag{21.19}$$

$$C_{gs} = (C_{gs}/C_{gd})C_{gd} \tag{21.20}$$

$$C_{pga} = C_{gs0} - C_{pgi} - C_{gs} \tag{21.21}$$

The solution of (21.14) – (21.21) leads to the extraction of all model parameters under cold pinch-off conditions.

The next step is to find the minimum error function value by scanning the values of C_{pgi}, C_{pdi}, C_{ds}, and C_{gs}/C_{gd}. As proposed in [11], C_{gs}/C_{gd} ratio is chosen as a controlled fixed parameter whose value varies from 0.1 to 10. For each defined capacitance ratio value, the error function values of scanned capacitive parameters, ε(C_{pgi}, C_{pdi}, C_{ds}), are plotted as stacked 3D graphs, where one variable acts as a curve parameter. For example, for a given C_{gs}/C_{gd} ratio the error function is mapped over the plane formed by C_{pgi} and C_{pdi} with C_{ds} as a parameter. The minimum error function values as a function of the respective capacitance ratio value for the investigated 2-mm and 3.2-mm device are depicted in Figure 21.5. It turns out that the optimum C_{gs}/C_{gd} ratio value for the 2-mm and 3.2-mm AlGaN/GaN HEMT is about 1.0 and 0.36, respectively. Table 21.3 summarizes the model parameter values of the distributed model originating from the optimum C_{gs}/C_{gd} ratio and corresponding C_{pgi}, C_{pdi} and C_{ds} values.

21.2.3 Definition of Pinch-Off Voltage

The model parameter extraction of FET devices relies on accurate cold pinch-off S-parameters at the bias point (V_{DS} = 0V, V_{GS} = V_{th}), where V_{th} is denoted as threshold voltage. V_{th} represents the transition from the nonzero drain current region into the pinch-off region.

In classical textbook definitions, V_{th} is identified in the static-DC IV transfer characteristic as the voltage at which linear extrapolation of I_{DS} intercepts the V_{GS}-axis. The V_{DS}-curve that must be extrapolated corresponds to the one with the largest slope; that is, the V_{DS}-curve related to the peak transconductance G_m. Figure 21.6 shows the static-DC transfer characteristics for a measured 2-mm and 3.2-mm AlGaN/GaN HEMT device. The graphical evaluation of the maximum slope of the transition curves is depicted as a thick dashed line. Then, V_{th} values are received as –1.4V for the 2-mm device and –3.3V for the 3.2-mm device.

Table 21.3
Optimal (Starting) Model Parameter Values of a 3.2-mm AlGaN/GaN HEMT
Generated by a Scanning Process of C_{gs}/C_{gd}, C_{pgi}, C_{pdi}, and C_{ds}

High-Quality Capacitance Values for Final Optimization (fF)										
C_{pga}	C_{pgi}	C_{gs}	C_{pda}	C_{pdi}	C_{ds}	C_{pdi1}	C_{pdi2}	C_{gda}	C_{gdi}	C_{gd}
201.38	523.17	579.14	99.74	259.33	77.89	161.55	97.78	8.96	123.10	1614.98

Intrinsic Inductances (pF) and Resistances (Ω)						Intrinsic Resistances (Ω)	
L_g	L_d	L_s	R_g	R_d	R_s	R_i	R_{gd}
90.10	187.27	0.59	3.79	1.43	0.41	0.00	0.00

From: [11].

Zamudio has pointed out [11] that the voltage V_{th} estimated with static-DC characterization is usually not precise enough for reliable model parameter extraction, and therefore a set of S-parameters is measured with V_{GS} varying around V_{th} (V_{DS} = 0V).

Then, the S-parameters are converted to Y-parameters. The obtained variations of $Im[Y_{11}]$, $Im[Y_{12}]$, and $Im[Y_{22}]$ with respect to V_{GS} around V_{th} are shown in Figure 21.7.

The V_{po} value was selected as the first V_{GS} value in the pinch-off transition region around V_{th} at which the *Y*-parameter curves became approximately constant (i.e., when the relative variation from one V_{GS} value to the next was lower than 10%). According to the results observed in Figure 21.7, the selected V_{po} value for the 3.2-mm device is −4.2V (marked in the graphs of Figure 21.7). Using a similar analysis with measurements of the 2-mm device a V_{po} value of −2.0V was determined.

21.2.4 Definition of Minimum Measurement Range

The optimal frequency range of measured S-parameter data for the extrinsic parameter extraction of the distributed small-signal FET model in Figure 21.9 depends on the device size (gate periphery). Jarndal [4] found empirically that the required measurement frequency range for reliable extraction of the distributed small-signal model parameters has been proven

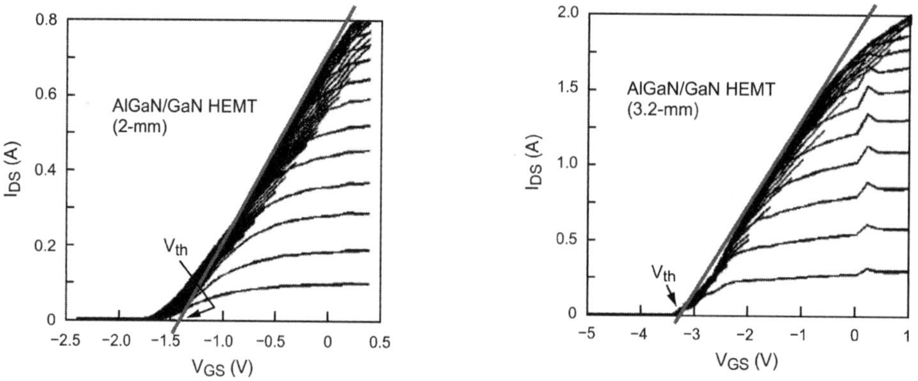

(a) (b)

Figure 21.6 Static-DC transfer characteristics of (a) the 2-mm, and (b) 3.2-mm AlGaN/GaN HEMT. The graphical estimation of V_{th} = −1.4V and V_{th} = −3.3V, respectively, is sketched (thick dashed line).

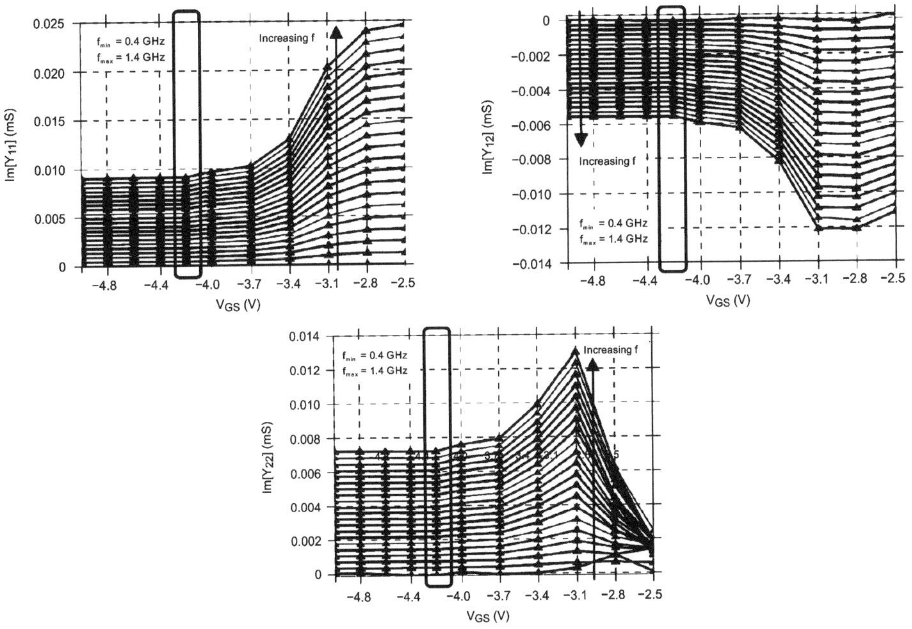

Figure 21.7 Low-frequency Y-parameters of a 3.2-mm AlGaN/GaN HEMT as a function of V_{GS} around V_{th} obtained from measured S-parameters (f = 0.4 GHz - 1.4 GHz). Evaluated pinch-off voltage V_{po} is framed. (After [11].)

to be 60 GHz for a 2×50-μm HEMT device. This frequency, however, reduced to 20 GHz for a 1-mm device and becomes smaller for larger devices.

Regarding the measured S-parameter curves in Figure 21.8 (only S_{12} and S_{22} are shown), the measured data of a 4-mm HEMT device up to about 5 GHz has strong similarity with the respective response of a 1-mm device up to 20 GHz. In this case one can conclude that the optimal extraction frequency range for a 4-mm device may also decrease to around 5 GHz.

The relation between optimal frequency range of extraction and gate periphery of AlGaN/GaN HEMT devices in the range of 0.1 mm and 4 mm is depicted in Figure 21.9 [5]. As can be seen, there is remarkable decrease for the optimal extraction frequency range with increasing device gate width.

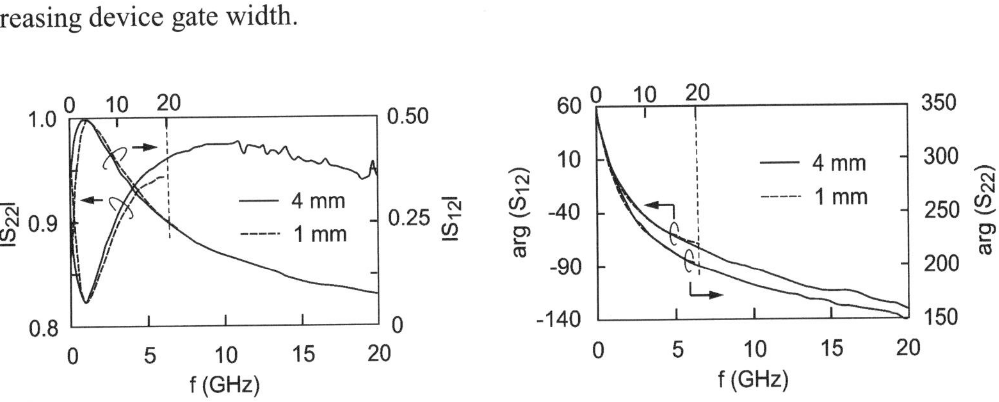

Figure 21.8 Pinch-off S-parameter comparison of a 1-mm and a 4-mm gate-width AlGaN/GaN HEMT (lower and upper frequency axis belong to the 4-mm and 1-mm device, respectively (after [20]).

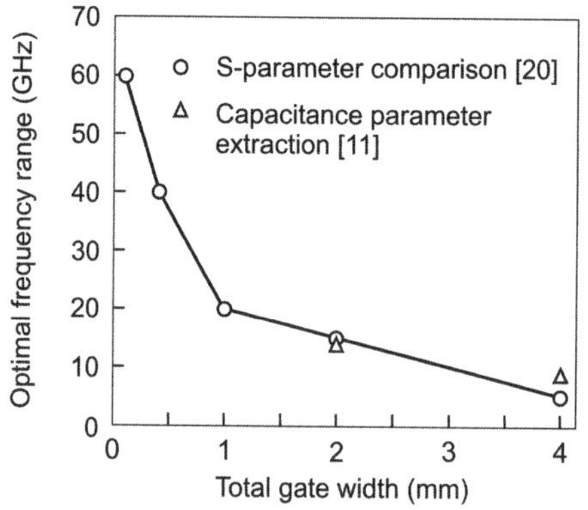

Figure 21.9 Optimal frequency range for reliable model parameter extraction of AlGaN/GaN HEMT devices for investigated gate widths in the range from 0.1 mm to 4 mm.

In [11], the optimal frequency range and gate periphery relation was confirmed. Model parameter extraction was performed using different frequency range, for different sizes. Thus, depending on the reliability of the extracted model parameters, the optimal frequency could be determined.

The resulting extracted values of drain-source and gate-source capacitance parameters with respect to the frequency range of extraction are shown in Figure 21.10 for the 2-mm device and in Figure 21.11 for the 3-mm device. The results indicate that the optimal frequency ranges are between 14 and 17 GHz and between 8 and 10 GHz, respectively, for the studied large-size devices. These frequencies agree well with the values that could be expected from the inspection of Figure 21.9.

21.2.5 Definition of Object Function

Model parameter extraction from S-parameter measurements is a nonlinear mutidimensional optimization problem whose objective function is likely to have multiple local minima. In order to reduce the dimension of the objective function, decomposition of the function into

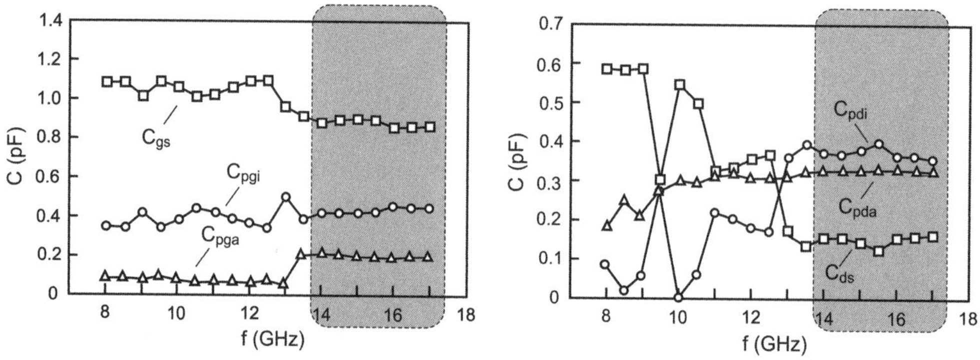

Figure 21.10 Values of gate-source (left) and drain-source (right) capacitance parameters extracted for the model of the 2-mm device with different frequency ranges of the S-parameter data. The optimal frequency range is shaded in gray.

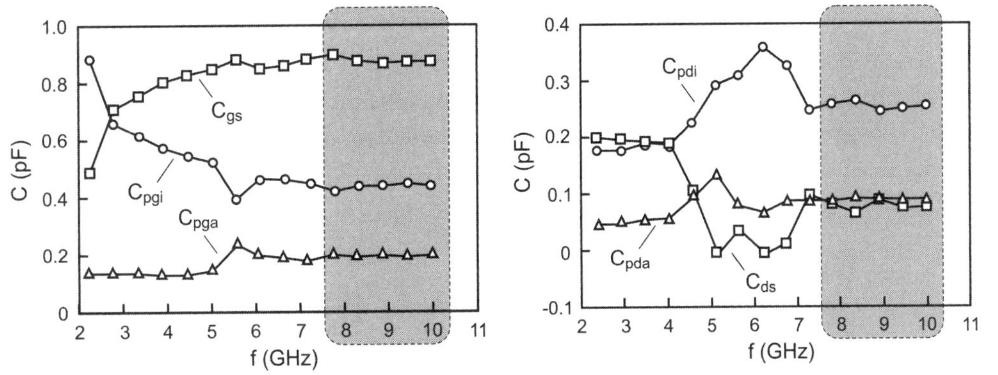

Figure 21.11 Values of gate-source (left) and drain-source (right) capacitance parameters extracted for the model of the 3.2-mm device with different frequency ranges of the S-parameter data. The optimal frequency range is shaded in gray.

smaller subfunctions has been proposed (see Chapter 15). The principle behind the decomposition technique is that the different model parameters can be associated with a particular S-parameter coefficient. However, the model parameters are related to the immittance functions so that parameter conversion from S- to either Y- or Z-parameters is necessary. Since one immittance function is a function of the four S-parameters of a two-port network, error in any of the S-parameters will corrupt the whole immittance parameter set through error propagation. Coupled with the strong correlations among the model parameters, error function decomposition becomes rather questionable [3].

The issue of measurement uncertainty is particularly critical for some bias conditions such as cold pinch-off, under which the S-parameter measurement is bound to have high uncertainties (see Section 11.3.1.1 [22]). This means that parameter extraction for all model parameters should be carried out simultaneously which requires a careful formulation of the objective function to avoid the local minimum problem.

The vector or matrix norm provides a measure for the magnitude of a vector or matrix. The norm is defined as a scalar function that satisfies the property of positivity, homogeneity, and the triangular inequality. As discussed in [3], there are many functions that qualify as a norm; for example, Frobenius or L_F-norm, the maximum or L_M-norm, the column or L_1-norm and the row or L_∞-norm. Testing the various functions for the determination of the extrinsic capacitance C_{pga} of a 15-element model by a scanning procedure, it was found that the L_1- and L_∞-norms exhibited the sharpest minimum of the residual error function value [3].

Regarding model parameter extraction from S-parameter data, the magnitude of the complex error $\varepsilon = \mathrm{Re}(\varepsilon) + \mathrm{Im}(\varepsilon)$ between the measured and simulated S-parameters may be written as

$$|\varepsilon_k| = \sqrt{(\mathrm{Re}[\varepsilon_k])^2 + (\mathrm{Im}[\varepsilon_k])^2}, \qquad k = 1, 2, ..., N \tag{21.22}$$

where N is the number of data points. The real and imaginary part of the error at each point, k, in terms of the measured value $S_{k,\,\mathrm{meas}}$ and its simulated value $S_{k,\,\mathrm{sim}}$ is given by

$$\mathrm{Re}[\varepsilon_k] = \mathrm{Re}[S_{k,meas}] - \mathrm{Re}[S_{k,sim}] \tag{21.23}$$

$$\mathrm{Im}[\varepsilon_k] = \mathrm{Im}[S_{k,meas}] - \mathrm{Im}[S_{k,sim}] \tag{21.24}$$

The particular formulation for the absolute magnitude of the error term assumes that there is no correlation between the real and imaginary parts of the error term, which is, in general, not valid for complex computation [3]. Furthermore, even with negligible correlation, large error terms will dominate the error through the quadratic function, thus compromising the robustness offered by the L_1-norm. This problem may be avoided by using the pseudo-absolute value [23] expressed as

$$|\varepsilon_k| = |\mathrm{Re}[\varepsilon_k]| + |\mathrm{Im}[\varepsilon_k]|, \qquad k = 1, 2, ..., N \tag{21.25}$$

A further enhancement of the objective function may result from weighting, which may be based on the measurement characteristics of the VNA. For example, in the measurement of the reflection coefficient, the uncertainty increases with the magnitude of the reflection coefficient (see Section 11.3.1.1 [22]). Thus, it may be reasonable to weight the residuals in a way that de-emphasizes the data points bound to have high uncertainties. The error term may thus, be expressed as

$$|\varepsilon_{ij}| = \frac{|\mathrm{Re}(\delta S_{ij,n})| + |\mathrm{Im}(\delta S_{ij,n})|}{W_{ij}}, \qquad i, j = 1, 2;\ n = 1, 2, ..., N \tag{21.26}$$

where

$$W_{ij} = \begin{cases} \max[|S_{ij}|] & \text{for } i, j = 1, 2 \text{ with } i \neq j \\ W_{ii} = 1 + |S_{ii}| & \text{for } i = 1, 2 \end{cases} \tag{21.27}$$

and N is the total number of data points. This weighting de-emphasizes data regions with high reflection and/or low transmission coefficients. The scalar error is then expressed as

$$\varepsilon_S = \frac{1}{N}\sum_{n=1}^{N} \|\tilde{\varepsilon}(f_n)\|_1 \tag{21.28}$$

where

$$\tilde{\varepsilon}(f_n) = \begin{bmatrix} \varepsilon_{11}(f_n) & \varepsilon_{12}(f_n) \\ \varepsilon_{21}(f_n) & \varepsilon_{22}(f_n) \end{bmatrix} \tag{21.29}$$

defined at each frequency point n. The norm index 1 in (21.28) denotes l_1-norm.

An objective function that is based on minimizing the fitting error on the S-parameters alone may not necessarily lead to physically relevant parameter values [24]. Often a good fit on the S-parameters may still be achieved so long as the total capacitance in the data is preserved [25]. Other transistor performance measures, depending on the final application, should also be considered to further enhance the objective function. For example, in amplifier and oscillator design, the output and input impedance, the device gain, and stability factor are important for the design of the embedding and matching networks. In the ideal case, these factors should be determined in an independent way; however, they may be expressed as functions of the S-parameters and fitted during the optimization.

The transistor output μ-stability factor at each frequency is expressed in terms of the S-parameters as [26, 27]

$$\mu = \frac{1 - |S_{22}|^2}{|S_{11} - S_{22}^*\Delta_s| + |S_{12}S_{21}|} > 1 \tag{21.30}$$

where S_{22}^* implies the complex conjugate and Δ_s the determinant of the S-parameter matrix at each frequency. The fitting error of the stability factor may then be written as

$$\varepsilon_{\mu} = \frac{1}{N}\sum_{k=1}^{N}\left|\mu_{k,m} - \mu_{k,c}\right| \tag{21.31}$$

where μ_{meas} and μ_{calc} are the output stability factors computed from the measured and model-generated S-parameters, respectively.

With regard to the device gain, several gains may be defined but the maximally efficient gain defined by Kotzebue is the most suitable one, since it remains finite even for an unstable device [28]. This gain may be defined at each frequency as [29]

$$G_{ME} = \frac{G - 1}{\ln G} \tag{21.32}$$

where G is the small-signal common source transducer gain ($G = |S_{21}|^2$ in a 50Ω system). The error in the gain may thus be expressed as

$$\varepsilon_{GME} = \frac{1}{N}\sum_{k=1}^{N}\left|G_{k,m} - G_{k,c}\right| \tag{21.33}$$

where $G_{k,m}$ and $G_{k,c}$ are the gains computed from the measured and simulated S-parameters at kth frequency point. The three error components may then be combined as [3, 5, 6, 11, 20]

$$|\varepsilon| = \sqrt{(\varepsilon_S^2 + \varepsilon_{\mu}^2 + \varepsilon_{GME}^2)/3} \tag{21.34}$$

21.3 CONCEPT OF MEASUREMENT-CORRELATED STARTING VALUE GENERATION

The finding of good starting values of model parameters for the model parameter optimization procedure was the subject of permanent research activities in our microwave group over a long period of more than three decades. The idea of high-quality measurement-correlated starting values directly determined from measured device data was first proposed by the author [24, 30]. In the following time the extraction concept was deepened and refined.

Mwema investigated the 18-element model (without gate drain parasitic capacitances). He introduced the model parameter scanning approach, described in Section 21.5 (see also Section 16.4.2), to find good starting values for final parameter optimization. The distribution of the parasitic capacitances, the capacitance ratios, were empirically defined for GaAs devices [3].

Jarndal [4] extended the model parameter extraction to a 20-element model including gate +rain parasitic capacitances to model the complex topology of high-power GaN devices. From experience he suggested modified capacitance ratio values for GaN devices.

Zamudio [11] thoroughly analyzed a 22-element equivalent circuit and confirmed the capacitance ratios proposed by Jarndal. In extension he gained reliable capacitance ratios from the top view of fabricated devices. Furthermore, he reviewed and refined the extraction of series inductances and resistances. Thus, over time a trustworthy and reliable model parameter extraction strategy could be elaborated. It is a step-by-step extraction procedure that is easily comprehensible and very flexible to use. The key point is to find reasonable starting model parameter values, which from the very beginning of the optimization process already roughly approximate the final solution. This concept stands in complete contradiction to other approaches following global search techniques with arbitrarily defined starting values that may be located far away from the actual solution, and thus, being exposed to high risk of getting trapped in local minima.

21.3.1 Description of the Extraction Algorithm

The flow diagram in Figure 21.12 summarizes the various contributions of the microwave group members to the measurement-correlated starting value generation technique. The analytical parameter extraction is primarily based on cold pinch-off S-parameter measurements. Later on cold-forward and cold-reverse S-parameter data were used to gain more reliable values for the parasitic series resistances (see Section 21.8). Having found high-quality starting values for the extrinsic model parameters, refined extracted values are then obtained on the basis of a mathematical optimization process.

Figure 21.12 presents the flow diagram of the systematic search algorithm for high-quality measurement-correlated starting values of the model parameters. From device inspection and experiment the proposed algorithm makes an assumption of some capacitance ratios in order to reduce the number of unknown model parameters of the distributed 20-element model. Later on, in [11], an improved generation of capacitance ratio values was proposed that permits a parameter extraction that is more generally applicable and delivers an increased reliability of the extracted values of extrinsic model parameters.

Besides the assumption of adequate capacitance ratio values, sweeps of values on capacitance parameters are required to unequivocally solve the equation system set up for the parasitic model parameters.

The technique starts with a conversion of the S-parameters, measured under cold pinch-off, to Y-parameters. The total branch capacitances denoted as C_{gs0}, C_{ds0} and C_{gd0} can be determined from the imaginary part of the low-frequency Y-parameter data. Considering the distributed network in Figure 21.1, the branch capacitances are composed of the extrinsic shell of capacitances C_{pga}, C_{pda}, and C_{gda}, intermediate shell capacitances C_{pgi}, C_{pdi} and C_{gdi}, and the intrinsic shell capacitances C_{gs}, C_{ds} and C_{gd}. Thus, the total branch capacitances can be written as

$$C_{gs0} = C_{gs} + C_{pgi} + C_{pga} \tag{21.35}$$

$$C_{ds0} = C_{ds} + C_{pdi} + C_{pda} \tag{21.36}$$

$$C_{gd0} = C_{gd} + C_{gdi} + C_{gda} \tag{21.37}$$

Equations (21.35) - (21.37) comprise nine unknowns. With fixed four capacitance ratios (see Section 21.2.1, defined for GaN power HEMTs),

$$C_{pda} \approx C_{pga} \tag{21.38}$$

$$C_{gdi} \approx 2C_{gda} \tag{21.39}$$

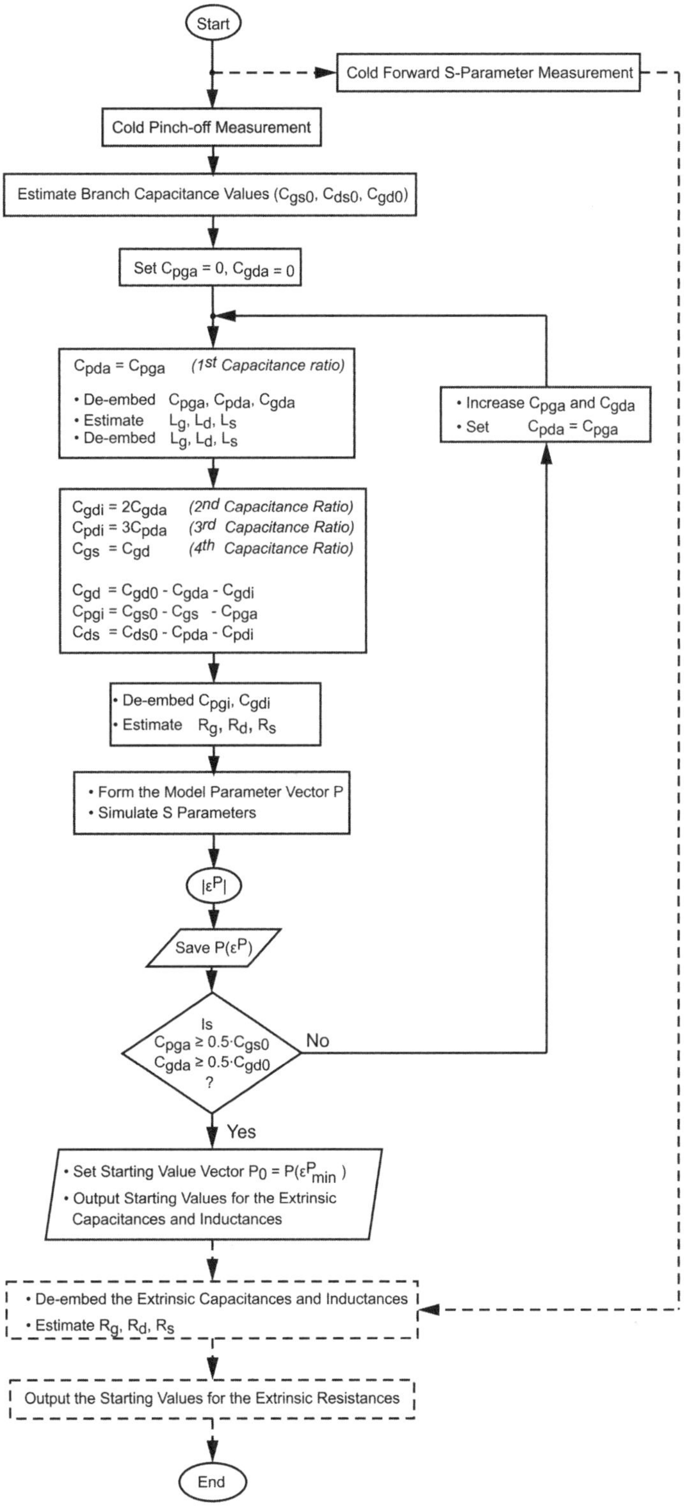

Figure 21.12 Flow diagram of the model parameter starting value generation algorithm [3, 4, 11].

$$C_{pdi} \approx 3C_{pda} \tag{21.40}$$

$$C_{gs} \approx C_{gd} \tag{21.41}$$

and assigning incrementally increasing values for C_{pga} and C_{gda} during the starting value generation process, the following residual three relations complete the set of equations required:

$$C_{pgi} = C_{gs0} - C_{gs} - C_{pga} \tag{21.42}$$

$$C_{ds} = C_{ds0} - C_{pdi} - C_{pda} \tag{21.43}$$

$$C_{gd} = C_{gd0} - C_{gdi} - C_{gda} \tag{21.44}$$

With known extrinsic capacitances, C_{pga}, C_{pda}, and C_{gda} are de-embedded. The stripped Y-parameters are then converted to Z-parameters and the inductances, L_g, L_d, and L_s, are estimated from the imaginary part. The intermediate capacitances C_{pgi}, C_{pdi}, and C_{gdi} are determined from (21.39), (21.40), and (21.42). After de-embedding C_{pgi} and C_{gdi}, the parasitic series resistances R_g, R_d, and R_s can be determined. Finally the S-parameters are simulated with the estimated set of model parameters, denoted as model parameter vector P, and compared with the measured data. The residual fitting error, ε^P, and the corresponding model parameter vector $P(\varepsilon)$ are stored.

The next scanning cycle starts by increasing the value of C_{pga} and C_{gda} incrementally. When $C_{pga} = C_{gs0}/2$ and $C_{gda} = C_{gd0}/2$ the scanning loop is stopped (see Figure 21.12) and the parameter vector $P_0 = P(\varepsilon_{min})$ corresponding to the lowest residual error ε_{min} is taken as the appropriate starting value vector.

The flow diagram of the model parameter starting value generation algorithm offers an alternative estimation of parasitic series resistances from cold-forward and cold-reverse S-parameter measurement, which will be discussed in Section 21.8.

21.3.2 Determination of Series Inductances and Resistances at Pinch-Off

Figure 21.13 represents the pinch-off FET model in which the intrinsic transistor is represented by a T-network. In deriving this network from the 17-element pinch-off model in Figure 21.1(b), the capacitances C_{pgi}, C_{pdi}, and C_{gdi} have been absorbed in C_{gs}, C_{ds}, and C_{gd}, respectively. This is valid if the series resistances are small such that the gate and drain inductances see the combined effect of the parallel combinations of C_{pgi} and C_{gs}, C_{pdi} and C_{ds} and C_{gdi} and C_{gd}.

Thus, assuming that the outer capacitances C_{pga}, C_{pda}, and C_{gda} are known, these can be de-embedded. Then the stripped Z-parameters of the residual network may be written as

$$Z_{11} = R_g + R_s + j\omega(L_g + L_s) + \frac{1}{j\omega}\left(\frac{1}{C_g} + \frac{1}{C_s}\right) + \delta Z_g \tag{21.45}$$

$$Z_{22} = R_d + R_s + j\omega(L_d + L_s) + \frac{1}{j\omega}\left(\frac{1}{C_d} + \frac{1}{C_s}\right) + \delta Z_d \tag{21.46}$$

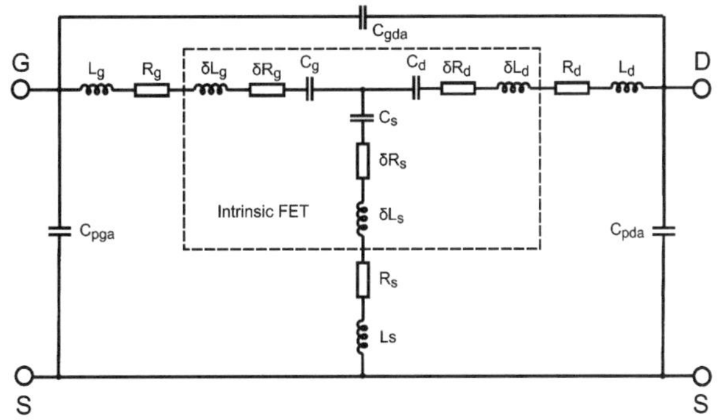

Figure 21.13 The T-network representation of a cold pinch-off FET equivalent circuit.

$$Z_{12} = Z_{21} = R_s + j\omega L_s + \frac{1}{\omega C_s} + \delta Z_s \tag{21.47}$$

where

$$\delta Z_g = \delta R_g + \delta R_s + j\omega(\delta L_g + \delta L_s) \tag{21.48}$$

$$\delta Z_d = \delta R_d + \delta R_s + j\omega(\delta L_d + \delta L_s) \tag{21.49}$$

$$\delta Z_s = \delta R_s + j\omega \delta L_s \tag{21.50}$$

δZ_g, δZ_d and δZ_s represent correction terms related to the intrinsic parameters of the model.

Ignoring the correction terms at the first and multiplying the Z-parameters by ω, the imaginary parts in (21.45) to (21.47) read:

$$\text{Im}[\omega Z_{11}] = (L_g + L_s)\omega^2 - \left(\frac{1}{C_g} + \frac{1}{C_s}\right) \tag{21.51}$$

$$\text{Im}[\omega Z_{22}] = (L_d + L_s)\omega^2 - \left(\frac{1}{C_d} + \frac{1}{C_s}\right) \tag{21.52}$$

$$\text{Im}[\omega Z_{12}] = L_s\omega^2 - \frac{1}{C_s} \tag{21.53}$$

Hence the values of L_g, L_d, and L_s can be extracted from the slope of $\text{Im}(\omega\Delta Z)$ versus ω^2-curve (Figure 21.14), where ΔZ equals $(Z_{11}-Z_{12})$, $(Z_{22}-Z_{12})$, and Z_{12}, respectively.

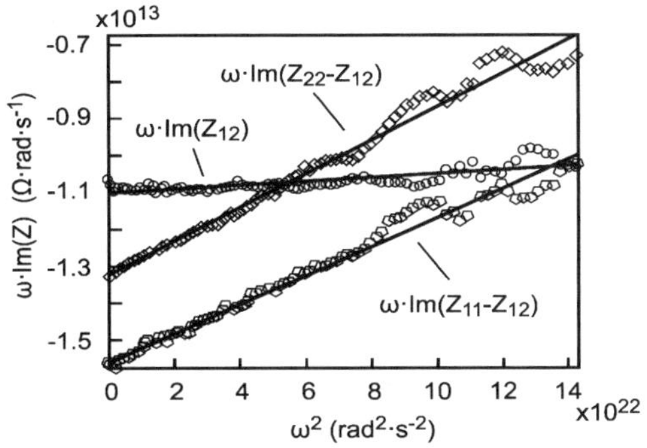

Figure 21.14 Inductance estimation from the measured cold pinched-off stripped Z-parameters for a 0.5-μm AlGaN/GaN HEMT with a 2×50 μm gate width (f_{max} = 60 GHz) [4], [5].

In the next step the estimated values of inductances described above and the inter-electrode capacitances (C_{pgi}, C_{pdi}, C_{gdi}) are de-embedded. It is reported in [3] that the incomplete de-embedding of the outer capacitances and the inductances introduce a nonlinear frequency dependence in the real part of de-embedded Z-parameters. However, by multiplying the de-embedded Z-parameters by ω^2, this effect is reduced. Ignoring the correction terms and multiplying the de-embedded Z-parameters by ω^2 and then taking the real part of these Z-parameters yields

$$\omega^2 \mathrm{Re}[Z_{11}] = \omega^2 (R_g + R_s) \tag{21.54}$$

$$\omega^2 \mathrm{Re}[Z_{22}] = \omega^2 (R_d + R_s) \tag{21.55}$$

$$\omega^2 \mathrm{Re}[Z_{12}] = \omega^2 R_s \tag{21.56}$$

By linear regression, the values of $(R_g + R_s)$, $(R_d + R_s)$, and R_s can be extracted from the slope of $\omega^2 \cdot \mathrm{Re}[Z_{ij}]$ $(i,j = 1,2)$ versus ω^2-curves.

21.3.3 Error Analysis

Under assumption that $\omega C_e (R_i + R_{gd}) \ll 1$, the correction terms for the gate, drain, and source inductance appearing in (21.48) – (21.50) can be approximated in terms of the intrinsic parameters (see Appendix B in [3] for detailed derivation):

$$\delta L_g \approx -\left(1 - \frac{C_{gd} R_{gd}}{C'_{ds} R_i}\right) \frac{C_e^2}{C_{gd}} R_i^2 \tag{21.57}$$

$$\delta L_d \approx -\left(1 + \frac{R_{gd}}{R_i}\right) \frac{C_e^2}{C'_{ds}} R_i R_{gd} \tag{21.58}$$

$$\delta L_s \approx -\left(1 + \frac{R_{gd}}{R_i}\right) \frac{C_e^2}{C'_{ds}} R_{gd}^2 \tag{21.59}$$

where

$$\frac{1}{C_e} = \frac{1}{C'_{gs}} + \frac{1}{C'_{ds}} + \frac{1}{C_{gd}} \tag{21.60}$$

and

$$C'_{gs} = C_{gs} + C_{pgi}, \qquad C'_{ds} = C_{ds} + C_{pdi} \tag{21.61}$$

As can be seen, L_g will be either over- or underestimated, depending on the value of $C_{gd}R_{gd}/(C'_{ds}R_i)$. Both L_d and L_s will always be underestimated if R_i and R_{gd} are taken into account.

The magnitude of the inductance correction terms for simulated S-parameters, with $C'_g = 50$ fF, $C_{gd} = 40$ fF, $C'_{ds} = 30$ fF, and $R_i = R_{gd} = 3\Omega$, are calculated as $\delta L_g \approx 0.01$ pH, $\delta L_d \approx$

-0.1 pH and $\delta L_s \approx -0.1$ pH. These are small quantities confirming that the inductances are relatively insensitive to the resistance values of R_i and R_{gd}, and that inductance estimates are acceptable even when these resistances are ignored. This, however, applies only to a limited extent with regard to L_s which is usually on the order of a few picohenrys in modern FETs for microwave and millimeter-wave applications.

Mwema thoroughly analyzed the quality of estimated values for the parasitic series resistances R_g, R_d, and R_s [3]. They may be approximately estimated from the real parts of (21.45) - (21.47). However, from experience, where C_{pgi}, C_{pdi}, and C_{gdi} are connected in the model relative to these resistances, these are important to take into account, and hence an estimate for those capacitances will be needed. Based on simulated S-parameter data, the real part of the residual Z-parameters have been shown to be related to the series resistances after having de-embedded the intermediate parasitic capacitances.

As can be seen in Table 21.3, based on fictitious S-parameter data ($C_{pdi} = C_{pgi} = 10$ fF), the estimated resistance values, neglecting the resistance correction terms δR_μ (μ = g, d, s), are larger than their corresponding nominal values even after de-embedding the correct capacitance and inductance values from their perfect S-parameter data. This discrepancy can be atttributed to the intrinsic resistances R_i and R_{gd} as will be shown next.

As derived in [3], the resistive correction terms δR_g, δR_d, and δR_s appearing in (21.48)-(21.50), may be written in terms of the intrinsic parameters as follows:

$$\delta R_g \approx \left\{1 + \frac{C_{gd}}{C_{ds}} + \frac{\tau_{gd}}{\tau_{gs}}\left(\frac{C_{gd}}{C_{ds}} + \frac{C_{gd}}{C_{gs}}\right)\right\}\left(\frac{C_e}{C_{gd}}\right)^2 R_i \tag{21.62}$$

$$\delta R_d \approx \frac{C_{gs}}{C_{ds}}\left\{1 + \frac{C_{gd}}{C_{ds}} - \frac{C_{gs}}{C_{gd}} - \frac{\tau_{gd}}{\tau_{gs}}\left(1 - \frac{C_{gd}}{C_{gs}}\right)\right\}\left(\frac{C_e}{C_{gd}}\right)^2 R_i \tag{21.63}$$

$$\delta R_s \approx \frac{C_{gd}}{C_{ds}}\left\{1 + \frac{C_{gd}}{C_{ds}} - \frac{\tau_{gd}}{\tau_{gs}}\right\}\left(\frac{C_e}{C_{gd}}\right)^2 R_i \tag{21.64}$$

with

$$\frac{1}{C_e} = \frac{1}{C_{gs}} + \frac{1}{C_{ds}} + \frac{1}{C_{gd}} \tag{21.65}$$

and

$$\tau_{gs} = C_{gs}R_i, \qquad \tau_{gd} = C_{gd}R_{gd} \tag{21.66}$$

Assuming $C_{gs} = 40$ fF, $C_{gd} = 40$ fF, $C_{ds} = 20$ fF, and $R_i = R_{gd} = 3\Omega$, the correction terms deliver approximately $\delta R_i = \delta R_s = 0.75\Omega$, and $\delta R_g = 1.1\Omega$. These values are in agreement with the estimated values from fictitious S-parameter data in Table 21.4, obtained after

Table 21.4

Overestimation of Estimated Series Resistance Values from Simulated S-Parameters

Series Resistance	Simulated Resistance Value	Estimated Resistance Value
R_g (Ω)	4.0	5.1
R_d (Ω)	3.0	3.75
R_s (Ω)	3.0	3.75

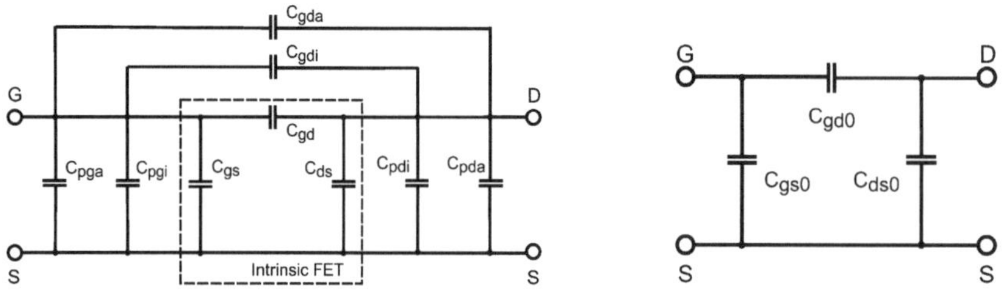

Figure 21.15 The distributed cold pinch-off equivalent circuit of an FET at low frequency (a), and its simplified network with total gate-source, gate-drain and drain-source branch capacitances (b).

de-embedding the correct values of C_{pga}, C_{pda}, C_{pgi}, and C_{pdi}. Note, when $C_{gd} = C_{gs}$, then δR_d in (21.63) will be reduced to $\delta R_d = R_i/(1 + 2C_{ds}/C_{gd})^2$ and thereby, δR_d will be not affected by R_{gd}. Upon closer inspection of (21.62) - (21.64), the resistive error terms are all positive (i.e., the extracted resistances R_g, R_s, and R_d are overestimated). This becomes clearer assuming that C_{ds} is small compared to either C_{gs} or C_{gd}. In this case the limiting values for the correction terms in (5.20) - (5.22) read [3]

$$\lim_{C_{ds}\to 0} \delta R_g \approx \left(1 + \frac{\tau_{gd}}{\tau_{gs}}\right)\frac{C_{ds}}{C_{gd}}R_i \tag{21.67}$$

$$\lim_{C_{ds}\to 0} \delta R_d \approx \frac{C_{gs}}{C_{gd}}R_i \tag{21.68}$$

$$\lim_{C_{ds}\to 0} \delta R_s \approx R_i \tag{21.69}$$

It can be concluded that with $C_{ds} \ll C_{gd}$, the correction term in (21.67) has a minor impact on the determined value of the gate resistance R_g. However, depending on the actual value of R_i, ignoring the correction terms in case of R_s and R_d can lead to uncertain values, which are less close to expectations. Nevertheless, the rough estimates of R_s and R_d values, which denote a lower limit, are entirely reasonable to be used for a final optimization procedure as demonstrated in Table 21.4.

21.4 ESTIMATION OF TOTAL BRANCH CAPACITANCES FROM LOW-FREQUENCY MEASUREMENTS

Regarding the distributed model in Figure 21.1, at low frequencies the inductive and resistive components in series with the capacitances may be neglected since their impedance will be small compared to that of the capacitance. In this case, the lumped model can be approximated by a capacitive network as shown in Figure 21.15. The Y-parameters of this equivalent circuit can be written as

$$Y_{11} = j\omega(C_{gs0} + C_{gd0}) \tag{21.70}$$

$$Y_{22} = j\omega(C_{ds0} + C_{gd0}) \tag{21.71}$$

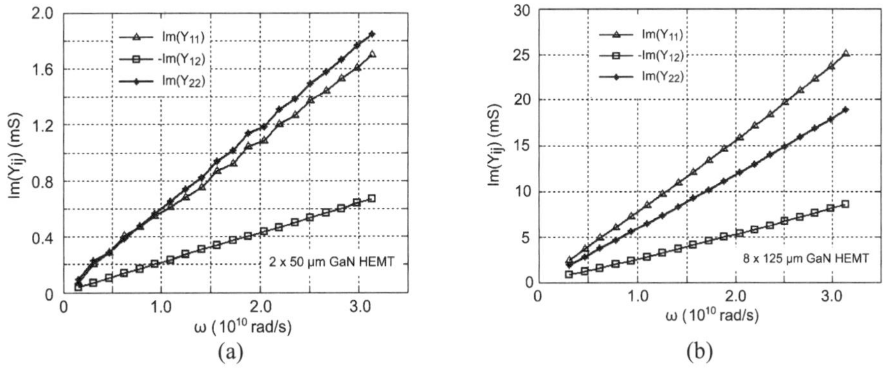

Figure 21.16 Measured low-frequency imaginary parts of admittance parameters $\mathrm{Im}(Y_{ij})$ under cold pinch-off bias conditions in a frequency range of 0.5 GHz up to 5 GHz. (a) 2 × 50 μm AlGaN/GaN HEMT, and (b) 8 × 125 μm AlGaN/GaN HEMT. (After [31].)

$$Y_{12} = Y_{21} = -j\omega C_{gd0} \tag{21.72}$$

where

$$C_{gs0} = C_{pga} + C_{pgi} + C_{gs} \tag{21.73}$$

$$C_{ds0} = C_{pda} + C_{pdi} + C_{ds} \tag{21.74}$$

$$C_{gd0} = C_{gda} + C_{gdi} + C_{gd} \tag{21.75}$$

Equations (21.70) – (21.72) allow the total gate-source, drain-source, and the gate-drain capacitances to be directly estimated from the low-frequency admittance matrix. This can be done by plotting the imaginary parts of Y_{11}, Y_{12}, and Y_{22} against ω and determining the capacitances from the slope of the linear plots.

An example for a small- and large-size AlGaN/GaN HEMT is given in Figure 21.16. From the slope of the $\mathrm{Im}(Y_{11})$, $\mathrm{Im}(Y_{12})$, and $\mathrm{Im}(Y_{22})$ versus ω plots the total gate-source, gate-drain, and drain-source capacitances for the small-size device are $C_{gs0} = 31.7$ fF, $C_{gd0} = 21.3$ fF, $C_{ds0} = 38.0$ fF. The respective large-size GaN HEMT capacitances read $C_{gs0} = 795$ fF, $C_{gd0} = 274$ fF, and $C_{ds0} = 599$ fF.

21.5 ESTIMATION OF DISTRIBUTED CAPACITANCES

We consider the active 18-element device model presented in Figure 21.17(a), where the outer and intermediate feedback capacitances C_{gdi} and C_{gda} as well as the gate-forward conductances G_{gsf} and G_{gdf} are not considered. Figure 21.17(b) shows the corresponding 15-element pinch-off model, which is now the subject under discussion.

For the determination of the inductance and resistance starting values, C_{pga}, C_{pda}, C_{pgi}, and C_{pdi} are assumed to be known. However, from (21.70) – (71.72) in Section 21.4, only the total gate-source, drain-source, and gate-drain capacitances can be determined from the low frequency pinch-off S-parameters converted to Y-parameters. Several methods for determining the contribution of the outer capacitances, C_{pga} and C_{pda}, have been reported [25, 32–34]. But these methods tend to overestimate these capacitances [3]. Since C_{pga} and C_{pda}

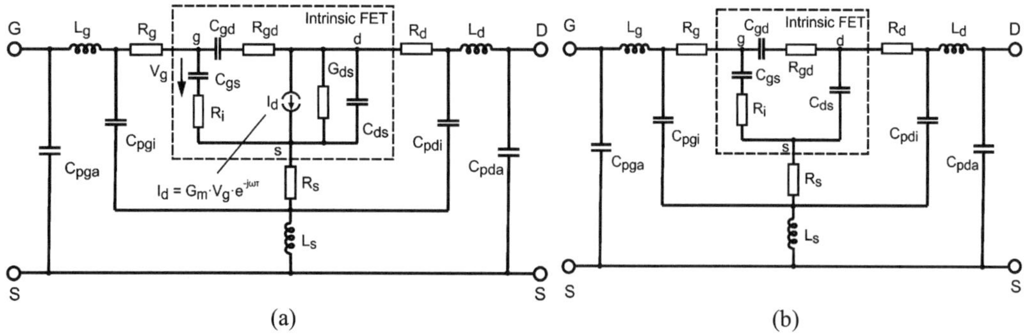

Figure 21.17 An 18-element equivalent FET model: (a) active mode, and (b) cold pinch-off operation mode.

are associated with the measurement environment; that is, whether the transistor is characterized on-wafer or in-fixture, it is reasonable to assume that $C_{pga} \approx C_{pda}$ and therefore only one capacitance needs to be determined. However, minor differences in the two capacitances can be expected due to slight differences in the mounting of the two halves of the test fixture (see [35, 36]). Regarding on-wafer measurement, the positioning of the probe tips may differ from calibration conditions.

Figure 21.18 demonstrates how to find the upper limit of scanning range of extrinsic capacitance C_{pga} (= C_{pda}). Assuming fixed values of the total gate-source capacitance C_{gs0} obtained from low-frequency Y-parameters of the device and varying the C_{pga} value by scanning, extrinsic capacitances can be de-embedded from Y-parameters converted from simulated S-parameter data. Then the inductance values can be computed from the linearized imaginary parts of Z-parameters converted from de-embedded Y-parameters. It can be seen that the inductance values change their sign at a value of approximately two-thirds of C_{gs0} (i.e., $C_{pga} \approx 2/3C_{gs0}$). This value of C_{pga} therefore sets the upper bound for the scanning range of C_{pga} (C_{pda}). In the scanning algorithm a practical value of $C_{pga} = C_{pda} = 0.5C_{gs0}$ was implemented (see Figure 21.12).

The inductances embed the effective capacitances $C_{gs} + C_{pgi}$, C_{gd} and $C_{ds} + C_{pdi}$, respectively, so that both C_{pgi} and C_{pdi} need not be known for inductance determination. However, when the gate is located symmetrically between the source and the drain, the depletion region under the gate will be uniform below pinch-off, and C_{gs} is expected to be equal to C_{gd}. Thus, C_{pgi} can be determined from the total gate capacitance below pinch-off as

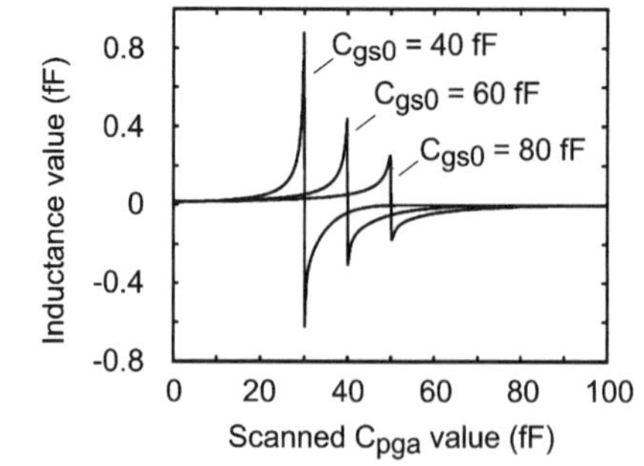

Figure 21.18 Determined gate inductance from simulated S-parameters as function of scanned $C_{pga} = C_{pda}$ value [3].

Table 21.5
Specifications of Transistors*

Product/Foundry	Type	Gate Length (μm)	Width (μm)	On-Wafer/In-Fixture
MGFC 1423	MESFET	0.4	300	In-fixture
JS8902-AS	HEMT	0.3	200	In-fixture
IAF	pHEMT	0.15	120	On-wafer
Plessey	MESFET	0.5	200, 250 300, 350	On-wafer

* Investigated after [3].

$$C_{pgi} = C_{gs0} - C_{gd} - C_{pga} \quad \text{with} \quad C_{gs} = C_{gd} \tag{21.76}$$

For the drain-side of the FET, the drain-source capacitance, C_{ds}, consists of the capacitance formed by the two slabs of bulk semiconductor material on either side of the depletion layer under the gate and the current through the substrate [37, 38]. Since substrate current flow occurs only at high drain voltages, in many cases, C_{ds} is simply set to zero, which is obviously questionable because of the contribution from the separation of the drain and source by the depletion layer under the gate. From (21.68) and (21.69), it can be seen that R_s and R_d will be overestimated by $\delta R_s = \delta R_d \approx R_i$ if C_{ds} (with $C_{gs} \approx C_{gd}$) is in fact set to zero. Furthermore, $C_{ds} = (C_{ds0} - C_{pda})/2$ could be chosen. However, as pointed out by Mwema, setting $C_{pdi} = C_{pda}$ has been found to lead to superior resistance starting values.

As discussed in Section 21.2.1, Mwema has proposed the following capacitance relations concerning GaAs FET devices:

$$C_{pga} \approx C_{pda}, \quad C_{gs} \approx C_{gd}, \quad C_{pdi} \approx C_{pda} \tag{21.77}$$

With the defined capacitance relations (21.77) and the measured branch capacitances,

$$C_{gs0} = C_{pga} + C_{pgi} + C_{gs} \tag{21.78}$$

$$C_{ds0} = C_{pda} + C_{pdi} + C_{ds} \tag{21.79}$$

$$C_{gd0} = C_{gd} \tag{21.80}$$

the following equations can be used to determine all capacitances of the distributed model in Figure 21.18(b) reliably:

$$C_{pgi} = C_{gs0} - C_{pga} - C_{gs} \tag{21.81}$$

$$C_{ds} = C_{ds0} - 2C_{pda} \tag{21.82}$$

The starting values for the model parameters were estimated by scanning $C_{pga} = C_{pda}$ for measured S-parameters for four transistors from different foundries with specifications tabulated in Table 21.4. The maximum frequency for the IAF pHEMT data was 120 GHz while it was 40 GHz for the other listed transistors. The variation of the residual fitting error using the starting value vector corresponding to each value of C_{pga} is presented in Figure 21.19. Also shown on the graph are the automatically detected minimum of the error and the corresponding capacitance value. For the pHEMT data, there are multiple local minima,

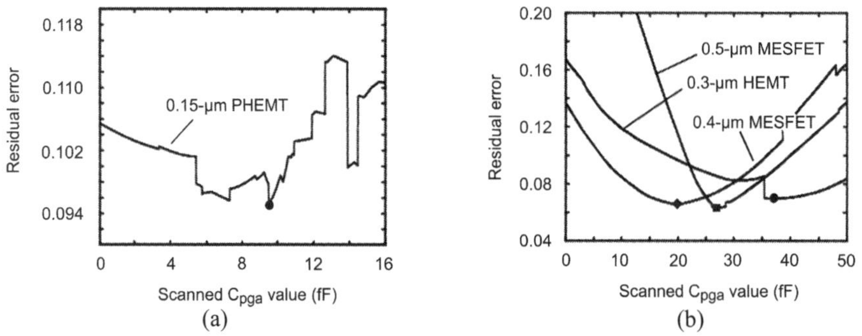

Figure 21.19 Variation of the fitting error with the gate parasitic capacitance $C_{pga} = C_{pda}$. (a) IAF 0.15-μm pHEMT (2×60 μm), and (b) Plessey 0.5-μm MESFET (2×175 μm), JS8902-AS 0.3-μm HEMT (200 μm), MGFC 1423 0.4-μm MESFET (300 μm) [3].

suggesting a need for carefully setting the capacitance scanning range. For the other transistors, a clear minimum in the error C_{pga} is evident with the implication that this capacitance estimate is likely to be close to the correct capacitance value.

Jarndal [4] has extended the scanning procedure and applied it to an expanded 20-element FET model, which reduces to a 17-element model under cold pinch-off operation conditions, as depicted in Figure 21.1(b). In contrast to Figure 21.17(b) this additionally comprises the gate-to-drain feedback capacitances C_{gdi} and C_{gda}. Thus, the number of unknown capacitances has increased by two, representing, nine capacitive model elements in total. Taking into account the experiences of Mwema, he proposed the following assumptions about the capacitance ratios (see Section 13.2.1.1):

$$C_{pga} \approx C_{pda} \tag{21.83}$$

$$C_{gs} \approx C_{gd} \tag{21.84}$$

$$C_{pdi} = 3 \cdot C_{pda} \tag{21.85}$$

$$C_{gdi} = 2 \cdot C_{gda} \tag{21.86}$$

The three measured branch capacitances comprise nine unknown distributed capacitances:

$$C_{gs0} = C_{pga} + C_{pgi} + C_{gs} \tag{21.87}$$

$$C_{ds0} = C_{pda} + C_{pdi} + C_{ds} \tag{21.88}$$

$$C_{gd0} = C_{gda} + C_{gdi} + C_{gd} \tag{21.89}$$

With additional two scanned quantities C_{pga} and C_{gda} and four defined capacitance relations (21.83) - (21.86), the following equations can be used to determine all capacitances of the distributed model in Figure 21.17(b) reliably:

$$C_{ds} = C_{ds0} - 4 \cdot C_{pga} \tag{21.90}$$

$$C_{gd} = C_{gd0} - 3 \cdot C_{gda} \tag{21.91}$$

$$C_{pgi} = C_{gs0} - C_{pga} - C_{gd} \tag{21.92}$$

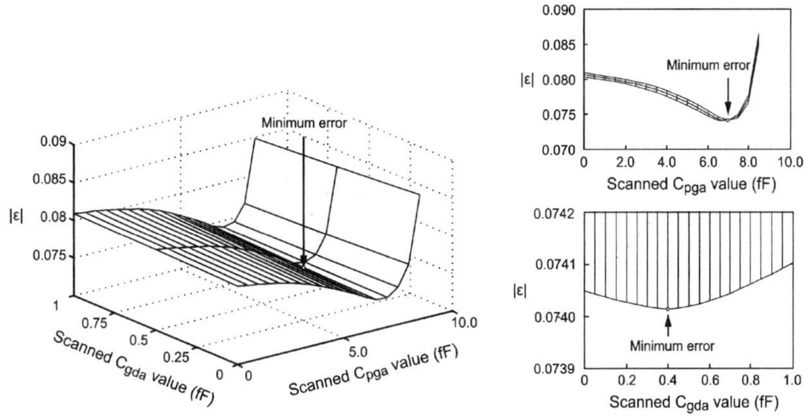

Figure 21.20 Residual error between measured and simulated S-parameters versus C_{pga} and C_{gda} for a 0.5-μm gate AlGaN/GaN HEMT with a 2×50 μm gate width. (After [4].)

The resulting estimated parameters are used to simulate the device S-parameters, which are then compared with the measured ones to calculate the residual fitting error (ε). The outer capacitances ($C_{pga} = C_{pda}$ and C_{gda}) are incremented by a fixed amount, and the procedure is repeated until $C_{pga} = C_{pda}$ is equal to $0.5C_{gs0}$ and C_{gda} equal to $0.5C_{gd0}$. The vector of model parameters $P(\varepsilon_{min})$ corresponding to the lowest error ε_{min} is then taken as the appropriate starting value.

Figure 21.20 shows the scanned C_{pga} and C_{gda} values and the corresponding residual error. The arrows in this figure indicate the minimum error and the related values of C_{pga} and C_{gda}.

21.6 ESTIMATION OF INDUCTANCES BY LINEAR CURVE FITTING

This section briefly recalls the commonly used estimation of inductances using linear curve fitting of Z-parameters de-embedded from extrinsic parasitic capacitances. The extrinsic inductance extraction was first outlined in [3] and used the equivalent circuit in Figure 21.13 after having de-embedded the extrinsic capacitances between the gate, drain, and source electrodes. Then, as described in Section 21.3.2, L_g, L_d and L_s are given by linear fitting the elements of the corresponding impedance matrix with equations of the form $\mathrm{Im}[Z] \approx \omega L - 1/\omega C$ and neglecting any δ-correction term. Figure 21.14 illustrates an example of the estimation of the inductances of a 0.1-mm AlGaN/GaN HEMT in the frequency range up to 60 GHz. Here $\omega \cdot \mathrm{Im}[Z_{11}-Z_{12}]$, $\omega \cdot \mathrm{Im}[Z_{22}-Z_{12}]$ and $\omega \cdot \mathrm{Im}[Z_{12}]$ indicate the gate inductance L_g, the drain inductance L_d and the source inductance L_s, respectively.

21.7 CLOSED-FORM ANAYLTICAL DETERMINATION OF INDUCTANCES

As already indicated, the series inductances L_g, L_d, and L_s are commonly determined by linear curve fitting of measured S-parameters after having de-embedded the extrinsic

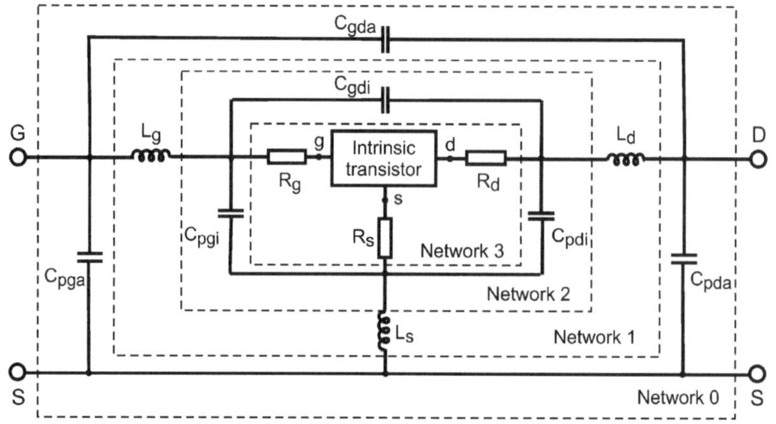

Figure 21.21 Extrinsic element network of the small-signal equivalent circuit under investigation. The dashed lines enclose the four networks resulting from successively adding the four shells of the extrinsic element network to the intrinsic transistor.

parasitic capacitances C_{pga}, C_{pda}, and C_{gda}. As discussed in Section 21.3.3, this practice includes some approximations in so far that the inductance correction terms have been neglected. Nevertheless, the fitted inductance values may serve as high-quality starting values for an extraction procedure based on overall optimization.

Recently, Zamudio and Kompa have developed a fully analytical inductance extraction procedure under the assumption that the distributed capacitance network is known [39]. The formulation of the extraction refers to the extrinsic element network of the equivalent circuit in Figure 21.21. The intrinsic part of transistor is shown in Figure 21.22.

Under cold pinch-off operating conditions (i.e., $V_{DS} = 0\text{V}$ and $V_{GS} < V_p$) capacitive effects dominate so that the circuit of the intrinsic transistor in Figure 21.21 simplifies to a pure capacitive model topology [5, 40, 41], like those in Figure 21.23. The pi- and T-networks in Figure 21.23 are interchangeable, and their elements are related by conversion formulae known as pi-T or Δ-Y transformations.

Considering the extrinsic element network in Figure 21.21 and the intrinsic-transistor equivalent circuit in Figure 21.22, the overall admittance matrix for cold pinch-off is described by the following matrix equation:

$$Y_0 = \{[(jE + R)^{-1} + jF]^{-1} + j\Lambda\}^{-1} + jP \tag{21.93}$$

where the matrices P, R, Λ, F, and E are defined as

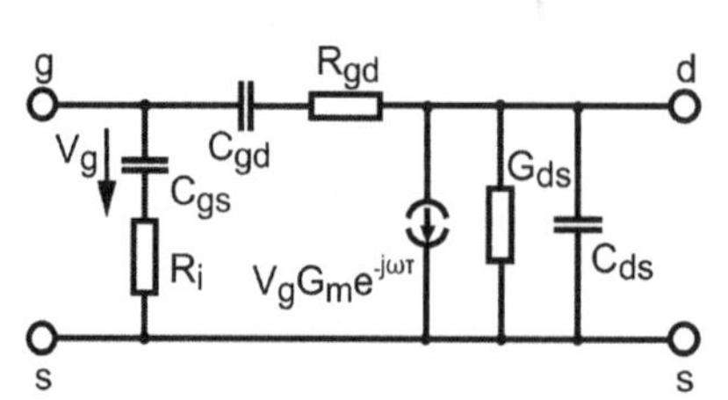

Figure 21.22 Model used for the intrinsic transistor.

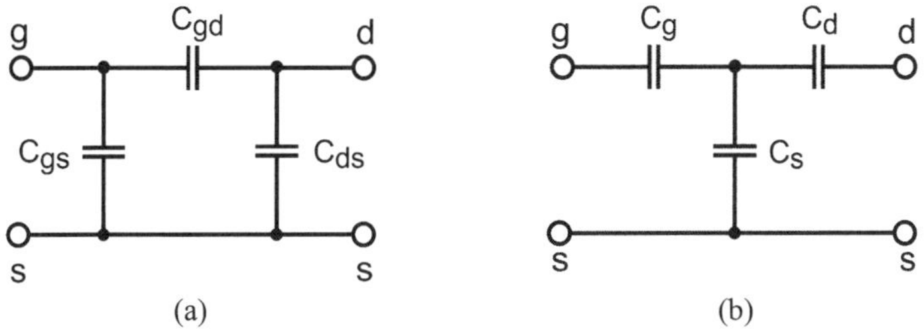

Figure 21.23 Intrinsic capacitive network under cold pinch-off: (a) pi-network, and (b) equivalent T-network.

$$P \equiv \omega \begin{bmatrix} C_{pga} + C_{gda} & -C_{gda} \\ -C_{gda} & C_{pda} + C_{gda} \end{bmatrix} \tag{21.94}$$

$$R \equiv \begin{bmatrix} R_g + R_s & R_s \\ R_s & R_d + R_s \end{bmatrix} \tag{21.95}$$

$$\Lambda \equiv \omega \begin{bmatrix} L_g + L_s & L_s \\ L_s & L_d + L_s \end{bmatrix} \tag{21.96}$$

$$F \equiv \omega \begin{bmatrix} C_{pgi} + C_{gdi} & -C_{gdi} \\ -C_{gdi} & C_{pdi} + C_{gdi} \end{bmatrix} \tag{21.97}$$

$$E \equiv -\frac{1}{\omega} \begin{bmatrix} C_g^{-1} + C_s^{-1} & C_s^{-1} \\ C_s^{-1} & C_d^{-1} + C_s^{-1} \end{bmatrix} \tag{21.98}$$

Rearranging (21.93) to contain all the extrinsic reactive parameters of the equivalent circuit at one side of the equation results in

$$\{[(Y_0 - jP)^{-1} - j\Lambda]^{-1} - jF\}^{-1} = jE + R \tag{21.99}$$

Taking only the imaginary parts of (21.98)

$$Im\langle\{[(Y_0 - jP)^{-1} - j\Lambda]^{-1} - jF\}^{-1}\rangle = E \tag{21.100}$$

Note that the only unknown matrix in (21.100) is the one that contains the inductance parameters. Therefore, the inductance extraction basically consists of solving the complex-valued nonlinear matrix equation (21.99).

As presented in Appendix E, it can be shown that (21.100) is actually interchangeable with the following equation:

$$X + A_1 X^{-1} A_1^T = Q \tag{21.101}$$

where

$$Q \equiv -(E^{-1} - F)^{-1} - F^{-1} \tag{21.102}$$

$$X \equiv \Lambda - B_1 - F^{-1} \tag{21.103}$$

$$A_1 \equiv Re[(Y_0 - jP)^{-1} \tag{21.104}$$

$$A_2 \equiv Im[(Y_0 - jP)^{-1} \tag{21.105}$$

The expression in (21.101) is a case of a general type of nonlinear matrix equation known as the algebraic Riccati equation, in which the matrix X is the only unknown and the matrices Q and A_1 are equation parameters. When X, Q, and A_1 are real-valued square matrices, closed-form solutions for (21.101) have been outlined in [40] based on previous methods to solve other cases of the algebraic Riccati equation. After Adam et al. [42], the solution method for (21.100) requires the calculation of the eigenvectors V_1, V_2, V_3, and V_4 of the following matrix:

$$\Phi \equiv \begin{bmatrix} I & -A_1^{-1}QA_1^{-1} \\ Q & I - QA_1^{-1}QA_1^{-1} \end{bmatrix} \tag{21.106}$$

Then, the calculated eigenvectors are used to constitute the columns of the following matrix

$$W = \begin{bmatrix} W_{11} & W_{12} \\ W_{21} & W_{22} \end{bmatrix} \equiv [V_1\ V_2\ V_3\ V_4] \tag{21.107}$$

Note that W is a 4×4 matrix, since each eigenvector in (21.106) has four values. Therefore, the elements of W described as W_{11}, W_{12}, W_{21} and W_{22} are in fact 2-by-2 matrices.

The eigenvectors V_1, V_2, V_3, and V_4 are ordered in W by descending magnitude of their eigenvalues, from left to right. After [40], the maximal solution of (21.101), called X_1, is expressed with respect to the elements of W as

$$X_1 = W_{21}W_{11}^{-1} \tag{21.108}$$

When the value of X_1 is inserted in (21.103) and it is combined with the knowledge of B_1 and F, the calculation of Λ is simple, which completes the extraction of the inductance values.

Moreover, according to [42], (21.101) has up to six real-valued solutions, formed by combinations W_{11}, W_{12}, W_{21}, and W_{22} different than (21.106), but the solution represented by (21.108) is the unique positive definite solution. Thus, (21.108) represents the solution that is of interest for the inductance extraction, because the positive definiteness of the matrix Λ implies $L_d \geq L_s$, $L_g \geq L_s$ and $L_s \geq 0$, which are generally true conditions of the physical structure in field-effect transistors.

21.7.1 Measurement-Like S-Parameters Based on VNA Uncertainty Specifications

The extraction uncertainty of the analytical estimation of series inductances was first tested on the basis of ideal (fictitious) S-parameter data. Used specified values of model parameters in Figure 21.21 are listed in Table 21.6. For the uncertainty analysis, 201 frequency points

Table 21.6
Element Values Used to Generate Ideal S-Parameters for the Uncertainty Analysis of the Inductance Extraction

$C_{pga} = 80$ fF	$C_{pda} = 70$ fF	$C_{gda} = 10$ fF	$L_g = 100$ pH	$R_g = 5\Omega$
$C_{pgi} = 90$ fF	$C_{pdi} = 60$ fF	$C_{gdi} = 20$ fF	$L_d = 600$ pH	$R_d = 3\Omega$
$C_{gs} = 30$ fF	$C_{ds} = 50$ fF	$C_{gd} = 100$ fF	$L_s = 10$ pH	$R_s = 1\Omega$

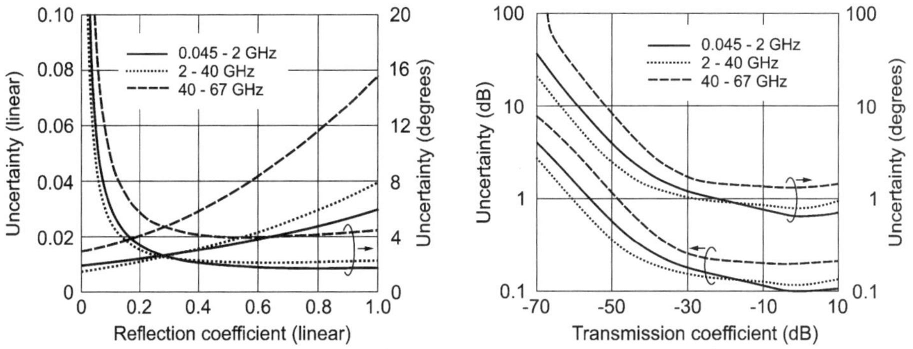

Figure 21.24 Measurement uncertainties of S-parameters from VNA specifications (PNA N5250A with N4694A Cal Kit). (After [43].)

with equal spacings from 45 MHz to 60 GHz were employed.

In S-parameter measurements, the instrument errors are related to the uncertainties of the VNA that is employed. Models for the VNA measurement uncertainty were used in [20, 22, 23] to generate artificial measurements with added instrument noise, based on a single measurement of S-parameters. Figure 21.24 shows the uncertainty models we used. They are specifications of a VNA of type PNA N5250A by Agilent [22].

The uncertainty model in Figure 21.24 are worst-case specifications that correspond to the 3σ confidence interval where σ is the standard deviation. In this case 99.7% of measurement results will be within the uncertainty limits specified by the system manufacturer. Thus, the uncertainty (symbol U) of measured magnitude and phase of scattering coefficients, denoted by $|U_{\mu\nu}|$, $(\mu,\nu) = (1,2)$, and $\angle U_{\mu\nu}$, can be expressed by the standard deviation σ as $U = 3\sigma$. Then for the magnitude and phase of U we can write

$$\sigma_{\mu\nu}^{m} = \frac{|U_{\mu\nu}^{m}|}{3}, \quad \sigma_{\mu\nu}^{p} = \frac{\angle U_{\mu\nu}^{p}}{3} \tag{21.109}$$

A total of 5,000 random values with normal distribution and standard deviation $\sigma_{\mu\nu}^{\mathrm{m}}$ were generated with a computer program to construct the magnitude of system noise, denoted $|N_{\mu\nu}|$, estimated to be added to each $|S_{\mu\nu}|$.

An additional 5,000 random values with normal distribution and standard deviation equal

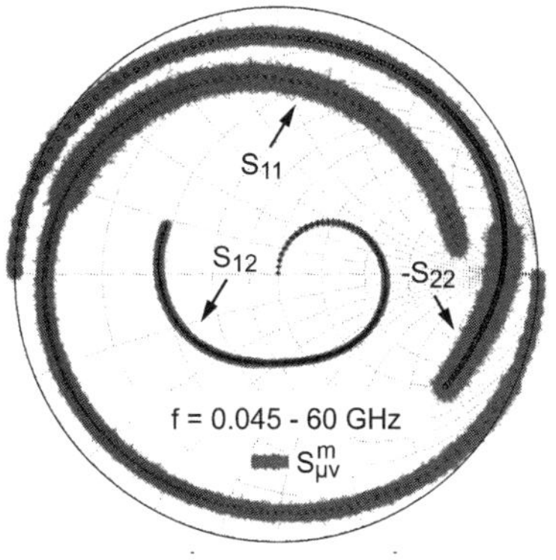

Figure 21.25 Fictitious (symbols x, +,○) and noisy S-parameters (symbol ▬)

to $\sigma^{p}_{\mu\nu}$ were generated to form the phase of system noise, termed $\angle N_{\mu\nu}$, estimated to be added to each $\angle S_{\mu\nu}$. The measurement-simulated S-parameter, $S^{n}_{\mu\nu}$, is expressed in terms of the corresponding noisy pair ($|N_{\mu\nu}|$, $\angle N_{\mu\nu}$) and the noiseless fictitious S-parameter as follows:

$$S^{m}_{\mu\nu} = (|S_{\mu\nu}| + |N_{\mu\nu}|)\exp(j[\angle S_{\mu\nu} + \angle N_{\mu\nu}]) \quad (21.110)$$

21.7.2 Validation of Inductance Determination Based on Noisy S-Parameters

The sets of measurement-simulated and ideal S-parameters that were constructed for the uncertainty analysis of the inductance extraction are shown in Figure 21.25.

The 5,000 triplets of inductance values L_g, L_d and L_s that were calculated with the proposed inductance extraction and the 5,000 sets of S^n are shown in Figure 21.26, together with the values extracted using the set of fictitious noiseless S-parameters. The results are also presented as histograms of probability distribution in Figure 21.27.

The inductance values extracted with the 5000 sets of noisy S-parameters S^n vary around the results that were obtained from the fictitious (noiseless) S-parameter set, which in turn agree well with the original inductance values listed in Table 21.6.

The histograms in Figure 21.27 show normal distributions and are centered on the values extracted from the ideal set of S-parameters. This figure is an illustration of the extraction uncertainties related to specific confidence intervals. For example, the numerical uncertainties observable for 99.7% confidence interval were approximately 0.2 pH for L_s = 10 pH, 0.5 pH for L_d = 60 pH, and 0.4 pH for L_g = 100 pH, which results in extraction uncertainties of 2%, 0.83%, and 0.4% relative to the actual value of the respective inductance effect. Thus, the worst relative uncertainty for the inductance extraction was 2%.

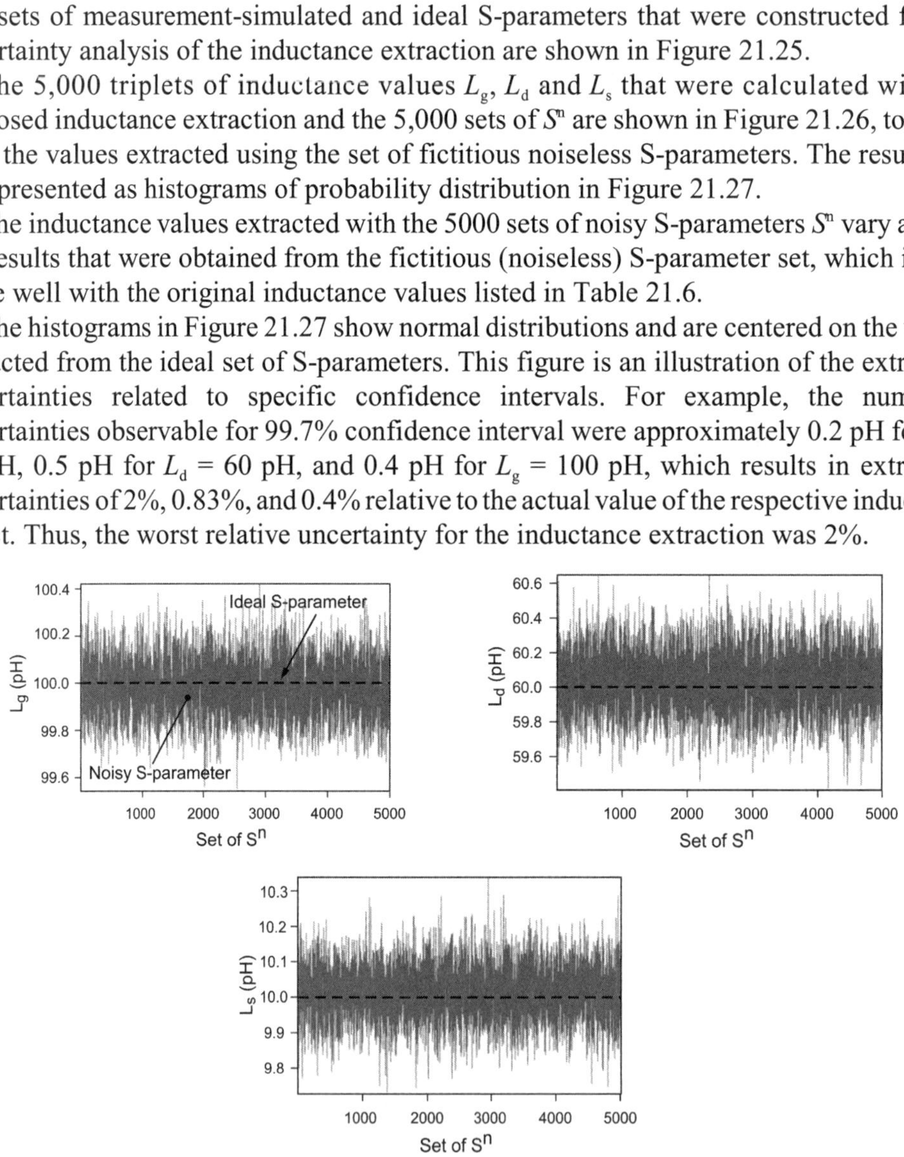

Figure 21.26 Inductance values obtained using the proposed extraction with all the generated 5,000 sets of noisy S-parameters. Dashed line indicates extracted values based on noiseless fictitious S-parameters.

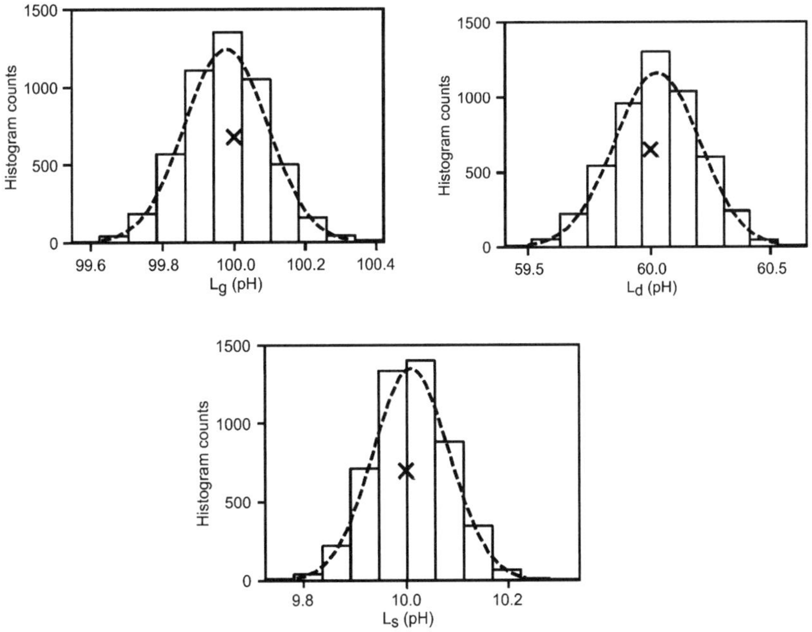

Figure 21.27 Histograms of inductance extraction results from using the sets S^n. Normal distributions fitted to each histogram are shown with dashed lines and values extracted with the ideal set of S-parameter are denoted with × markers.

The inspection of the histograms in Figure 21.28 indicates that the probability distributions of extraction uncertainty were dominated by the normal distribution of the system error (measurement uncertainty). As a result, the contribution to the probability distribution of the uncertainty that is due to the proposed extraction method was relatively insignificant.

Figure 21.28 illustrates the inductance extraction versus frequency using both ideal S-parameters and chosing one of the 5,000 data sets indiscriminately simulating noisy data.

As shown in Figure 21.28, the inductance values are independent of frequency as

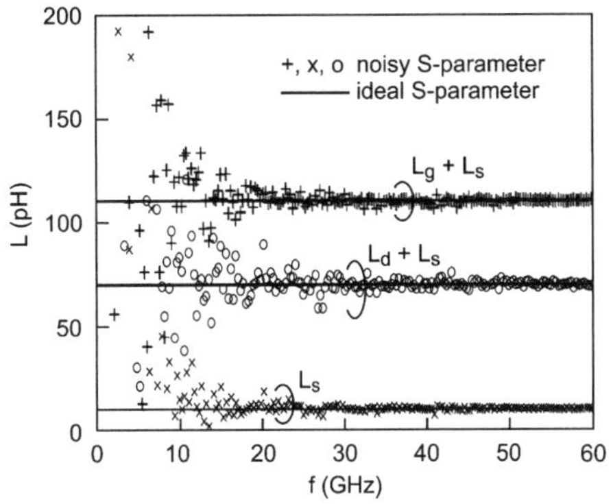

Figure 21.28 Comparison of frequency variation of inductance values extracted on the basis of ideal S-parameters and using one set of synthesized noisy S-parameters.

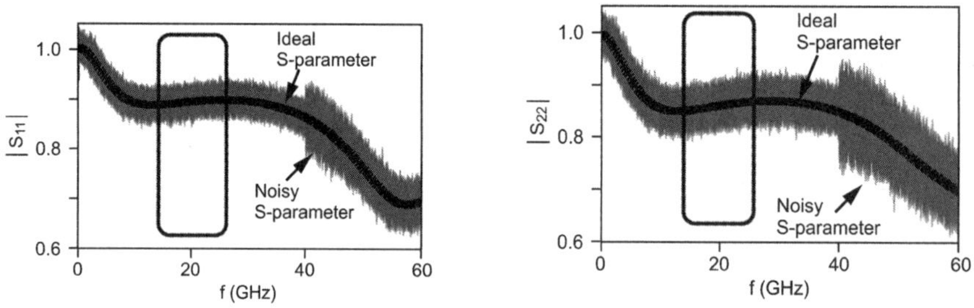

Figure 21.29 Synthesized ideal and noisy $|S_{11}|$ and $|S_{22}|$. The rounded rectangles illustrate the onset of the inductance effects; that is, the minimum frequency range of reliable extraction. The noise step at 40 GHz indicates the increased uncertainty specified for the VNA system used [43].

expected. The extraction results from the set of ideal S-parameters are constant and those from the sets of noisy data exhibit a negligible variation from 20 to 60 GHz, representing mid and high frequencies.

Note that below 20 GHz, the onsets of inductance effects are masked by the strong capacitive effects related to the cold pinch-off bias and the extraction is not recommended in such frequency range. For the synthetic noisy-like S-parameters the onset of the inductance effects is found from the $|S_{11}|$ or $|S_{22}|$ of the data used for extraction. This onset denotes the frequency range to start reliable extraction (in this case around 20 GHz) and is indicated by the frequency range where the $|S_{11}|$ or $|S_{22}|$ traces first exhibit an increase, as illustrated in Figure 21.29.

21.8 ESTIMATION OF RESISTANCES

In this section we discuss different approaches for the estimation of the parasitic gate-resistance R_g, the drain-resistance R_d, and source-resistance R_s. Adequate values can be obtained (1) by cold pinch-off, (2) cold forward, and (3) cold reverse measurements. The approaches are briefly described and commented.

21.8.1 Cold Pinch-Off Measurement

It is reported in [3] that the real parts of the Z-parameters in (21.45) - (21.47) are related to the series resistances after de-embedding the interelectrode capacitances. It is also reported that the incomplete de-embedding of the outer capacitances and the inductances introduce nonlinear frequency dependence in the real part of de-embedded Z-parameters (see Section 21.3.2). By multiplying the de-embedded Z-parameters by ω^2, this effect is reduced. Ignoring the correction terms δR_g, δR_d, and δR_s in (21.48) - (21.50) and multiplying the de-embedded Z-parameters by ω^2 and then taking the real part of these Z-parameters yields

$$\omega^2 \mathrm{Re}[Z_{11}] = \omega^2 (R_g + R_s) \tag{21.110}$$

$$\omega^2 \mathrm{Re}[Z_{22}] = \omega^2 (R_d + R_s) \tag{21.111}$$

$$\omega^2 \mathrm{Re}[Z_{12}] = \omega^2 R_s \tag{21.112}$$

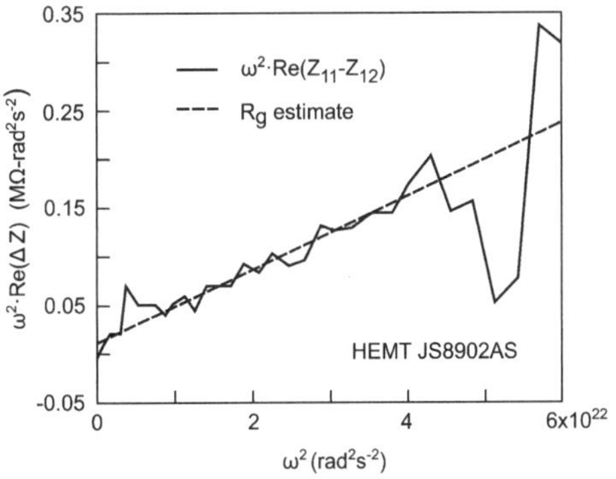

Figure 21.30 Gate resistance estimation from cold pinch-off S-parameters for in-fixture measured HEMT JS8902AS (f_{max} = 40 GHz). (After [3].)

By linear regression, the values of ($R_g + R_s$), ($R_d + R_s$), and R_s can be extracted from the slope of $\omega^2 \cdot \text{Re}[Z_{ij}]$ (i, j = 1,2) versus ω^2-curves. The frequency variation of the real parts of the Z-parameters of the JS8902-AS HEMT for the determination of the gate resistance R_g estimate is shown in Figure 21.30.

21.8.2 Cold Forward Measurement

Cold forward S-parameter measurements were first proposed by Chigaeva et al. [44] and have been used in several works [5, 7, 44–50]. According to [47], cold forward S-parameter measurements are characterized by: (1) V_{DS} = 0V and a positive V_{GS} value, (2) a high value of I_{GS}, and (3) a positive imaginary part of S_{11} in the entire frequency range measured.

Experiments by Zamudio [11] with different GaN HEMTs showed that the positive V_{GS} value required to produce predominant inductive behavior of S_{11} is generally higher than the values applied usually for GaAs devices. Such an observation coincides with the findings published in other works of GaN device modeling [44, 50], where this was explained by the higher differential resistance of the gate that is typical for GaN transistors.

Further experiments with similar GaN devices of different manufacturers demonstrated that the gate differential resistance (and thus the positive V_{GS} value needed to produce dominant inductive behavior of S_{11}) varies greatly among devices of different fabrication processes [11].

In Figure 21.31 measured S_{11} data of a 0.1-mm AlGaN/GaN HEMT fabricated by the Ferdinand-Braun Institut für Höchstfrequenztechnik (FBH), Berlin, are plotted for various gate-source voltages (V_{DS} = 0V). It can be seen that the S_{11} traces exhibit a decreasing capacitive behavior with increasing V_{GS} values until inductive behavior prevails at V_{GS} = 1.74V. Figure 21.32 shows an example of reliable resistance estimation from cold forward measurement of a 0.5-μm AlGaN/GaN HEMT with a 2×50 μm gate width.

However, as pointed out in [45], the positive V_{GS} value required to produce predominantly inductive behavior of S_{11}, and the associated high I_{GS} could damage or destroy the Schottky contact of the gate, causing long-term or irreversible harm to the device response. Instead of cold forward, an alternative bias condition for the extraction of extrinsic resistance parameters is proposed in the following section.

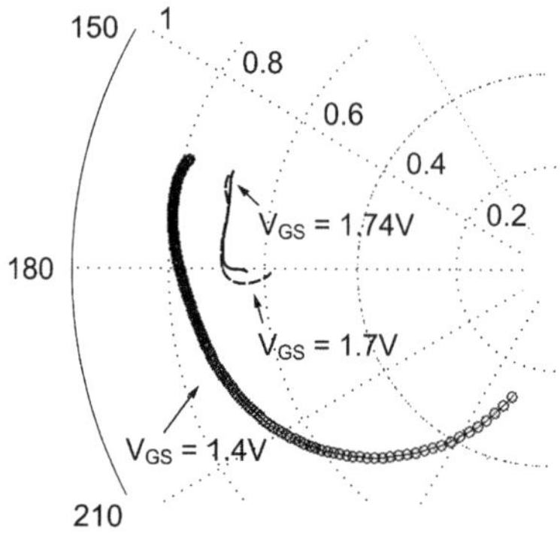

Figure 21.31 Cold forward measurement ($V_{DS} = 0V$) of a 0.1-mm AlGaN/GaN HEMT in the frequency range from 0.4 GHz to 40 GHz with increasing positive V_{GS} value to reach predominantly inductive behavior of S_{11}. (After [11].)

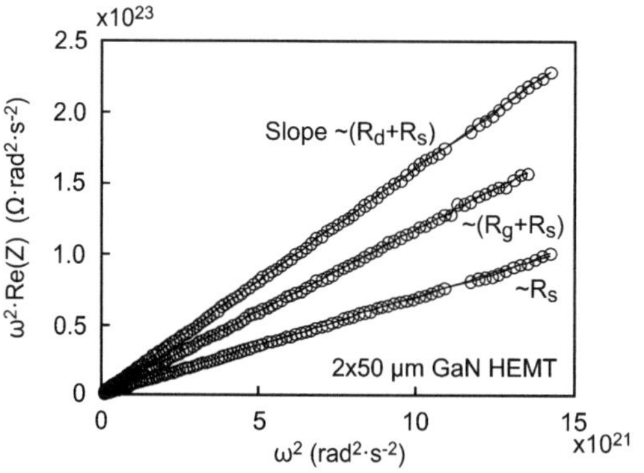

Figure 21.32 Resistance estimation from measured cold forward Z-parameters for a 0.5-μm AlGaN/GaN HEMT with a 2×50 μm gate width (upper frequency $f_{max} = 20$ GHz). (After [4].)

21.8.3 Standard Cold Reverse Measurement

As previously outlined, the cold forward biasing for resistance determination has some drawbacks such as the variability of the required V_{GS} value [8, 45, 47] and the inherent high I_{GS}, which may cause damage to the device or its performance [45].

Resistance extractions from S-parameters measured at $V_{DS} = 0V$ and $V_{pinch\text{-}off} < V_{GS} < V_{bi}$, which are safer than cold forward, have been reported [8, 45, 47, 48]. But these extractions are incompatible with the circuit topology in Figure 21.21because their models lack parameters for the interelectrode capacitive effects.

21.8.4 Modified Cold Reverse Measurement

In the following, a new extraction method for the resistances is proposed, which includes some modification of the standard cold reverse measurement technique. In this case, the measurement is restricted to negative V_{GS} values (i.e., $V_{pinch\text{-}off} < V_{GS} < 0V$ ($V_{DS} = 0V$)). For such a bias point the intrinsic transistor can be modeled by the network depicted in Figure

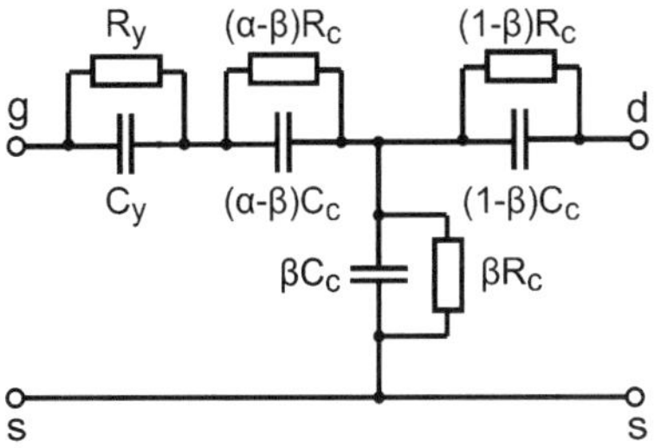

Figure 21.33 Equivalent circuit of the intrinsic transistor at cold reverse biasing ($V_{DS} = 0V$, $V_{pinch\text{-}off} < V_{GS} < 0V$).

21.33. The total channel impedance comprises the channel capacitance C_c and the channel resistance R_c. α and β are factors <1, which indicate the distribution among gate, drain, and source channel regions. R_y and C_y model the Schottky contact.

Thus, the corresponding impedance matrix of network 3 in Figure 21.22, which is called ς here, is given by the following expression:

$$[\varsigma] = \begin{bmatrix} R_s + R_g + Z_k + \alpha Z_c & R_s + \beta Z_c \\ R_s + \beta Z_c & R_s + R_d + Z_c \end{bmatrix} \tag{21.113}$$

[7, 46, 48, 50–52]. Z_c represents the impedance of the conduction channel where the constants α and β describe the channel distribution among gate, drain and, source regions. Z_c reads

$$Z_c = \frac{R_c}{1 + j\omega C_c R_c} = \frac{R_c - j\omega C_c R_c^2}{1 + \omega^2 C_c^2 R_c^2} \tag{21.114}$$

Z_k is the impedance of the Schottky contact and can be written as

$$Z_k = \frac{R_k}{1 + j\omega C_k R_k} = \frac{R_k - j\omega C_k R_k^2}{1 + \omega^2 C_k^2 R_k^2} \tag{21.115}$$

Biasing the transistor with $V_{GS} \geq 0V$, the channel is minimized, causing the frequency-varying term in (21.114) to rapidly vanish in the GHz region, reducing (21.113) to $R_s + R_g + \alpha R_{ch}$, $R_s + \beta R_{ch}$ and $R_s + R_d + R_{ch}$. A drawback of this bias condition for resistance parameter extraction is that the above three terms are easily affected either by the frequency dependency of the channel resistance that occurs in devices prone to dispersive behavior, or by the bias dependency of the channel resistance that appears for small fluctuations of V_{DS} and V_{GS} (in mV) from the nominal values supplied.

In the case with $V_{GS} \leq V_{pinch\text{-}off}$, the channel resistance is maximized, reducing the imaginary parts of (21.114) to capacitive effects and causing large frequency-varying terms in the real parts of (21.113). A drawback of this bias condition, in relation to the resistance parameter extraction is that those large frequency-varying terms typically mask the effects of R_g, R_s and R_d.

In cold reverse bias conditions, with V_{GS} less negative than $V_{pinch\text{-}off}$, the channel resistance is decreased from its highest values and this allows the effects of R_g, R_s, and R_d to be clearly visible in certain frequency ranges. At the same time, the related R_c is now low enough to make the term $\omega^2 C_c^2 R_c^2$ negligible with respect to unity, and therefore cold reverse bias conditions maintain the effect of C_c in (21.114). Keeping the effect of C_c reduces the variations of the real parts of (21.113) due to dependencies of the channel resistance that are not considered in the present small-signal circuit (e.g., frequency or bias dependencies or temperature-related effects).

The problem of the cold reverse condition is formulating the parameter extraction to find the values of α, β, R_c, C_c, R_k, and C_k, in order to separate the frequency-varying terms in (21.113) from the values of R_g, R_s, and R_d. The required formulation is explained next.

First, in the cold reverse bias condition the diode-like effect of the gate Schottky contact is reverse-biased, so its conductance is negligible, which means that R_k is expected to be large, as has been reported in [7, 8] (in the order of several kΩ [7]). Then, Z_k in (21.115) is approximately reduced to $-1/(\omega C_k)$, as it appears in other similar models described in [8, 45, 47, 52]. As a result, the real and imaginary parts of components of ζ (21.113) are

$$\mathrm{Re}[\zeta_{11}] \approx R_g + R_s + \frac{\alpha R_c}{1 + \omega^2 C_c^2 R_c^2} \tag{21.116}$$

$$\mathrm{Re}[\zeta_{12}] \approx R_s + \frac{\beta R_c}{1 + \omega^2 C_c^2 R_c^2} \tag{21.117}$$

$$\mathrm{Re}[\zeta_{22}] \approx R_d + R_s + \frac{R_c}{1 + \omega^2 C_c^2 R_c^2} \tag{21.118}$$

$$\mathrm{Im}[\zeta_{11}] \approx -\frac{1}{\omega C_k} - \frac{\alpha R_c}{1 + \omega^2 C_c^2 R_c^2} \tag{21.119}$$

$$\mathrm{Im}[\zeta_{12}] \approx -\frac{\beta R_c}{1 + \omega^2 C_c^2 R_c^2} \tag{21.120}$$

$$\mathrm{Im}[\zeta_{22}] \approx -\frac{\omega C_c R_c^2}{1 + \omega^2 C_c^2 R_c^2} \tag{21.121}$$

Now, note that (21.118) is written in terms of (21.121) as follows:

$$\mathrm{Re}[\zeta_{22}] = R_d + R_s - \frac{\mathrm{Im}[\zeta_{22}]}{\omega C_c R_c} \tag{21.122}$$

Multiplying by $\omega/\mathrm{Im}[\zeta_{22}]$, (21.122) reads

$$\frac{\omega \mathrm{Re}[\zeta_{22}]}{\mathrm{Im}[\zeta_{22}]} = -\frac{1}{C_c R_c} + \frac{\omega}{\mathrm{Im}[\zeta_{22}]}(R_d + R_s) \tag{21.123}$$

The sum $R_s + R_d$ is obtained as the slope of the linear approximation to the plot of (21.123) with $\omega\mathrm{Re}[\zeta_{22}]/\mathrm{Im}[\zeta_{22}]$ as the y-axis and $\omega/\mathrm{Im}[\zeta_{22}]$ as the x-axis.

Combining (21.117) and (21.121) delivers the relation, which is similar to (21.123)

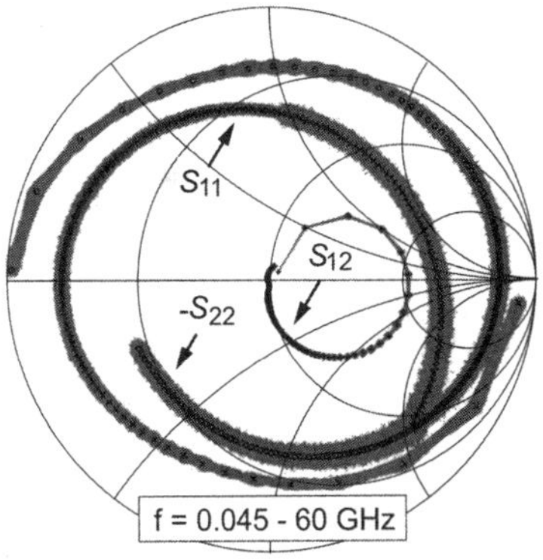

Figure 21.34 Generated ideal (symbols x, +,○) and noisy S-parameters (marked by ▬) for the uncertainty analysis of the resistance extraction: 201 points equally distributed from 45 MHz to 60 GHz [39].

$$\frac{\omega \mathrm{Re}[\zeta_{12}]}{\mathrm{Im}[\zeta_{22}]} = -\frac{1}{\beta C_c R_c} + \frac{\omega}{\mathrm{Im}[\zeta_{22}]} R_s \tag{21.124}$$

Thus, R_s is received as the slope of the linear fit to the plot of (21.124) with the y-axis being $\omega\mathrm{Re}[\zeta_{12}]/\mathrm{Im}[\zeta_{22}]$ and the x-axis $\omega/\mathrm{Im}[\zeta_{22}]$.

Combining (21.116) and (21.121), we get

$$\frac{\omega \mathrm{Re}[\zeta_{11}]}{\mathrm{Im}[\zeta_{22}]} = -\frac{1}{\alpha C_c R_c} + \frac{\omega}{\mathrm{Im}[\zeta_{22}]} (R_g + R_s) \tag{21.125}$$

In this case, the sum $R_g + R_s$ is obtained as the slope of the linear approximation to the plot of (21.125) with the y-axis being $\omega\mathrm{Re}[\zeta_{11}]/\mathrm{Im}[\zeta_{22}]$ and the x-axis $\omega/\mathrm{Im}[\zeta_{22}]$.

Similar to Section 21.7.2 ideal and fictitious noisy S-parameters have been generated to analyze the extraction uncertainty of resistances according to the proposed method. In addition to the model parameter values in Table 21.6, the parameter values chosen for the intrinsic transistor model in Figure 21.33, are listed in Table 21.7. Note that a channel resistance, R_c, with frequency variation (varying from 4 to 11.75 Ω) has been defined to highlight the robustness of the extraction.

The resistance values R_g, R_d, and R_s that were obtained with the proposed resistance extraction as a function of the 5,000 sets of measurement-like S-parameters S^n are shown in Figure 21.35, together with the values extracted using the set of fictitious noiseless S-parameters. The results are also presented as histograms of probability distribution in Figure 21.36.

Table 21.7

Chosen Parameter Values for the Intrinsic Transistor*

$C_k = 3$ pF	$\alpha = 0.3$	$R_k = 1\mathrm{k}\Omega$
$C_c = 1$ pF	$\beta = 0.5$	$R_c = 4\Omega + \sqrt{f(\mathrm{GHz})}\ \Omega$

*See Figure 21.33.

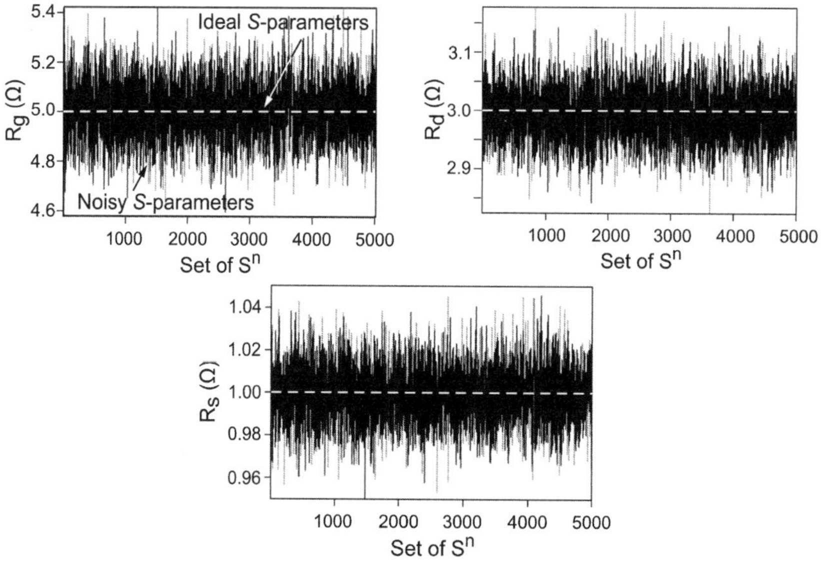

Figure 21.35 Resistance values obtained using the proposed extraction with all the generated 5,000 sets of noisy S-parameters. The dashed line indicates extracted values based on ideal (noiseless) S-parameters.

The resistance values extracted with the 5,000 sets of constructed noisy S-parameters fluctuate around the resistance values that were obtained from the ideal S-parameter set, and the latter are in good agreement with the original resistance values listed in Table 21.6. The histograms of Figure 21.36 present normal distributions and are centered on the values extracted from the ideal S-parameter set.

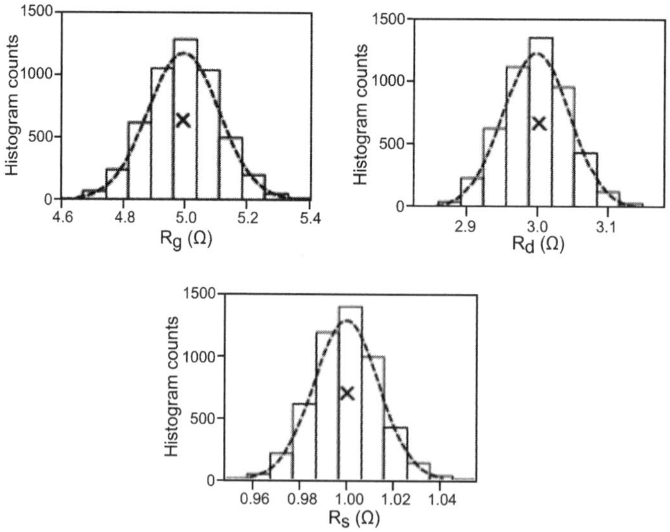

Figure 21.36 Histograms of resistance extraction results from using the sets S^n. Normal distributions fitted to each histogram are shown with dashed lines and values extracted with the ideal set of S-parameters are denoted with × markers.

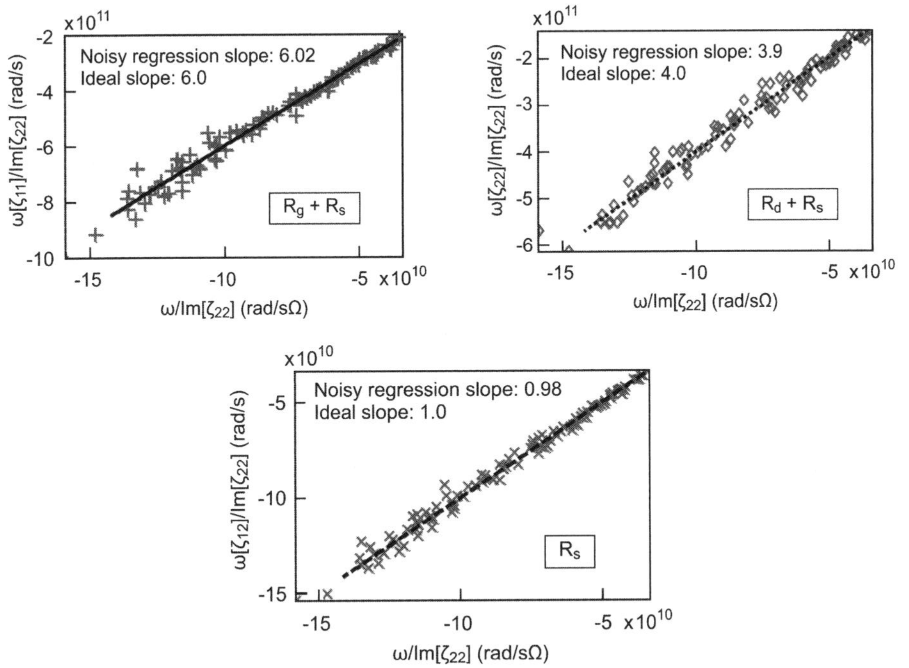

Figure 21.37 Frequency variation of the resistance values extracted with one set of synthetic noisy and ideal S-parameters.

Figure 21.36 shows the extraction uncertainties related to specific confidence intervals. For example, the numerical uncertainties observable for a 99.6% confidence interval were about 0.04Ω for R_s = 1Ω, 0.1Ω for R_d = 3Ω and 0.3Ω for R_g = 5Ω, resulting in extraction uncertainties of 4%, 3.33%, and 6% relative to the actual value of the respective resistance effect. Thus, the worst relative extraction uncertainty was 6%.

Histograms of resistance extraction results from using the sets of S^n. Normal distribution fitted to each histogram is shown with dashed lines and values extracted with the ideal set of S-parameters are denoted with X markers.

21.9 MODEL PARAMETER EXTRACTION WITH MEASURED S-PARAMETERS

This section presents various extraction examples concerning both in-fixture and on-wafer measurements. Extraction method used in Section 21.9.1 is based on the estimation of inductance values applying linear curve fitting. Generally, this approach will deliver very good extraction results. Nevertheless, these results may be considered as high-quality starting values for a subsequent optimization step for model parameter value refinement.

In Section 21.9.2 extraction results, based on the closed-form analytical determination of inductances are discussed. It will be demonstrated that in view of the measurement uncertainty of commercially available VNAs, the extracted results are fully acceptable so that any additional optimization is no longer needed.

21.9.1 Extraction Based on Linear Curve Fitting of Inductances

This section describes extraction results using linear curve fitting of inductances as described in Section 21.6.1. Section 21.6.2 treats a refinement of the extracted model parameters on the basis of an optimization procedure.

21.9.1.1 High-Quality Starting Model Parameter Values

In this section the high quality of extracted model parameter values is demonstrated for GaAs technology based FET devices (MESFET, HEMT) and GaN technology based power HEMTs, measured both in-fixture and on-wafer. Note that these estimated parameter values are not the final extracted results. They are only used as starting (initial) values for a further refinement optimization process. Because of the high-quality parameter values, global optimization algorithms are no longer needed. Thus, the risk of getting trapped in a local minimum has been extremely lowered. Therefore local optimizers can be applied as broadly discussed in Chapter 12. As pointed out in Section 12.7.5, the developed modified simplex algorithm has been thoroughly tested and proven reliable.

The good fitting on measured S-parameters for a 0.15-μm 2×60 μm GaAs HEMT (on-wafer measured with three-band VNA up to 120 GHz) using the estimated parameter values is demonstrated in Figure 21.38. The frequency variation of the S-parameters derived from the estimated values closely follow the S-parameters, thus confirming the high quality of the extracted parameter values. Note that, after subsequent parameter value refinement by optimization nearly the same curves (dotted circles) have been obtained. A slight difference between estimated and optimized calculation can only be observed with the $|S_{11}|$ curve.

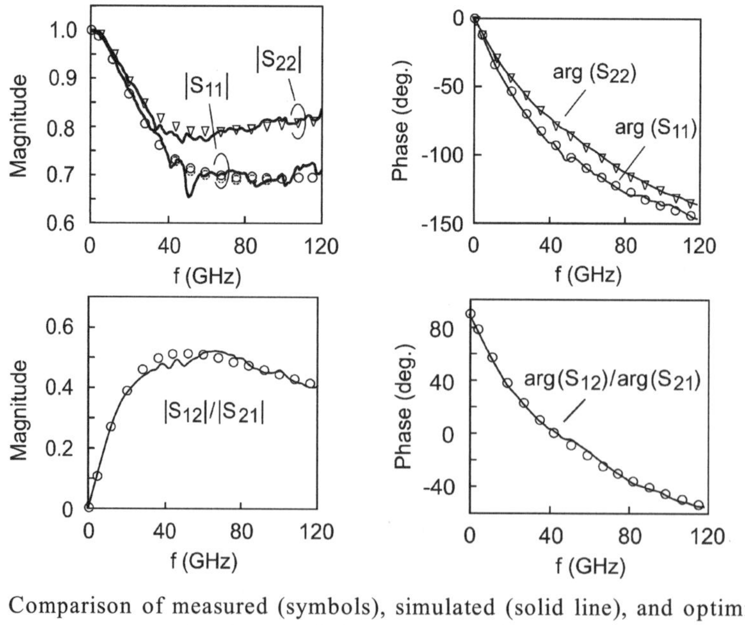

Figure 21.38 Comparison of measured (symbols), simulated (solid line), and optimized (refined) (dotted circles) S-parameters of 0.15-μm pHEMT with 2×60 μm gate width, applying extracted measurement-correlated estimated model parameter values [3].

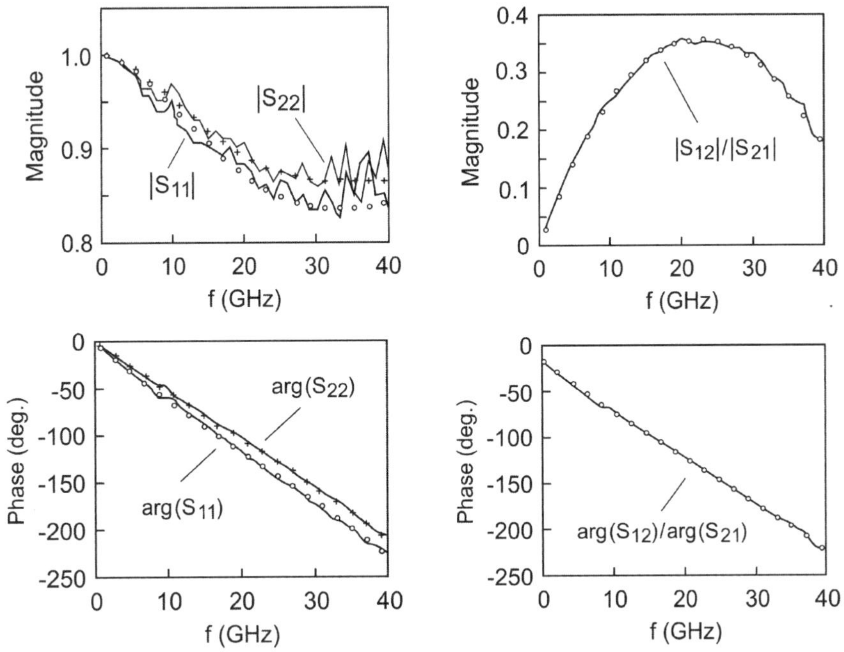

Figure 21.39 Fitting on the JS8902AS HEMT S-parameter data using measurement correlated starting values (18-element model, f_{max} = 40 GHz).

Figure 21.39 shows an example for an in-fixture measured 0.3-µm gate HEMT (200 µm gate width) of type JS8902-AS [53]. The agreement of measurement and simulation based on estimated model parameter values is in this case also very good. Although not explicitly plotted, the fitting for other investigated transistor types in Table 21.4 was similar [3].

In Figure 21.40 measured and simulated cold pinch-off S-parameters of the 2-mm and 3.2-mm AlGaN/GaN HEMTs are shown. The magnitude and phase of S_{11} and S_{12} are shown. Comparison of simulated and measured S_{22} and S_{21} delivers similar results. Simulation was based on estimated model parameter values obtained from measurement-correlated extraction procedure. Note in this case the extraction of the inductances was based on linear curve fitting described in Section 21.6. Although the parameter values are only provisional, the agreement for the 2-mm HEMT is excellent. In the case of the 3.2-mm, the transition from the capacitive behavior at lower frequencies to the higher inductive property really needs further optimization. However, applying analytical determination of the inductances as described in Section 21.9.2, a nearly perfect agreement between simulated and measured S-parameters is obtained as demonstrated in Figure 21.42, so that a final optimization for extraction refinement can be waived.

21.9.1.2 Model Parameter Value Refinement by Optimization

In this section the result of the optimized value for each model parameter is presented. Model parameter optimization has been done based on the principle of the bidirectional search technique described in Chapter 16 [54]. The closeness of the generated starting parameter

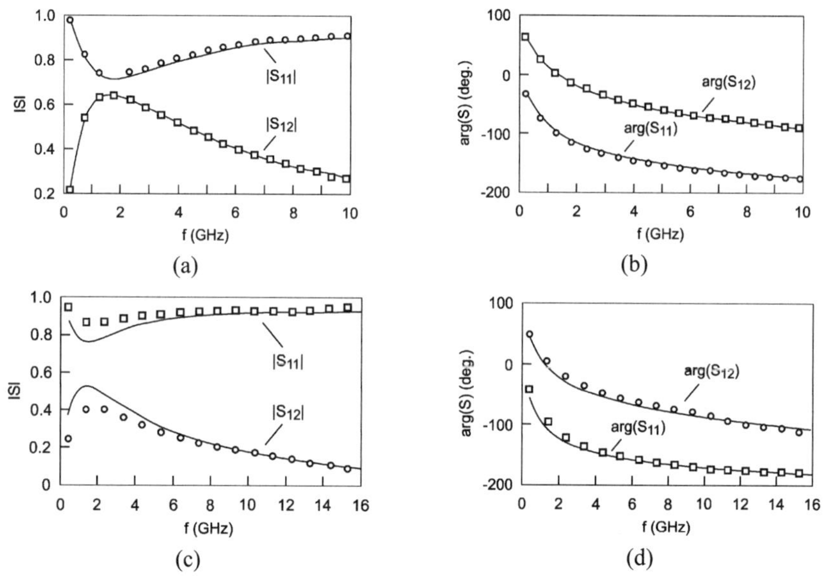

Figure 21.40 S-parameters measured in cold pinch-off. Above 2-mm device: (a) magnitude of S_{11} and S_{12} and (b) phase of of S_{11} and S_{12}; below 3.2-mm device: (c) magnitude of S_{11} and S_{12} and (d) phase of of S_{11} and S_{12} [11].

values to the true values exhibits a strongly shrunken searching space with hopefully no severe local minima in addition to the global one (notice noisy data can indeed cause many local minima within the nearest neighborhood of a strong global minimum). The bidirectional search algorithm starts with the derived estimated parameter values. Then these are improved iteratively. At each iteration through the optimization process, the extrinsic parameters are assigned suitable values and then de-embedded from the measured data to determine the intrinsic Y-parameters. The intrinsic model parameters are then estimated by means of data fitting from the de-embedded measurements. The whole estimated model parameters are then used to fit the measured S-parameters. This process is continued to find the optimal model parameters. The modified simplex optimization algorithm described in section 12.7.5 is used to minimize the objective function (Section 21.2.5). The object function used was according to (21.34).

It was already shown in Figure 21.38 the excellent agreement of the optimized S-parameters with the measured data for an on-wafer measured 0.1-mm AlGaAs/GaAs HEMT over a frequency range of 120 GHz.

In Table 21.8 the estimated (initial) and optimized device parameters of a 0.1-mm AlGaN/GaN HEMT are listed. The similarity of estimated and optimized model parameters is very satisfying. This is, in particular, valid for the series resistances, which, in this case, were extracted under cold forward operation as described in Section 21.8.2.

Zamudio [11] applied the measurement-correlated extraction procedure to 2-mm and 3.2-mm AlGaN/GaN power devices applying top-view device image analysis. The frequency range for extraction was 10 and 16 GHz, respectively, which is in accordance with the optimal ranges specified in section 21.2.4 (Figure 21.9). Table 21.9 indicates the optimized

Table 21.8

Starting and Optimized Model Parameter Values of a 0.5-μm AlGaN/GaN HEMT*

Model	Initial	Optimized	Model	Initial	Optimized
C_{pga} / fF	7.00	9.97	L_g / pH	39.50	46.55
C_{pda} / fF	7.00	7.13	L_d / pH	46.40	47.90
C_{gda} / fF	0.40	0.47	L_s / pH	5.25	6.25
C_{pdi} / fF	21.00	29.42	R_g / Ω	5.20	4.80
C_{pgi} / fF	4.90	7.09	R_d / Ω	9.20	11.80
C_{gdi} / fF	0.80	0.86	R_s / Ω	6.60	5.47
C_{gs0} / fF	27.81	32.44	C_{gs} / fF	20.37	15.38
C_{ds0} / fF	36.47	36.55	C_{ds} / fF	8.47	0.00
C_{gd0} / fF	21.21	21.50	C_{gd} / fF	20.01	20.17

*From: [4].

Table 21.9

Measurement-Correlated Starting Parameter Values and Final Values after Refined Optimization (2-mm/3-mm GaN HEMT Above/Below)*

Model Parameter	Initial Value	Optimized Value	Model Parameter	Initial Value	Optimized Value
C_{pga} / fF	201.38	203.87	L_g / pH	90.10	83.89
C_{pgi} / fF	523.17	567.27	L_d / pH	187.27	168.26
C_{gs} / fF	579.14	603.86	L_s / pH	0.59	8.22
C_{pda} / fF	99.74	115.48	R_g / Ω	1.69	1.60
C_{pdi} / fF	259.33	213.15	R_d / Ω	2.21	2.05
C_{ds} / fF	77.89	119.46	R_s / Ω	0.50	0.14
C_{gda} / fF	8.96	10.21	R_i / Ω	—	1.63
C_{gdi} / fF	123.10	130.87	G_{ds} / mS	—	50.08
C_{gd} / fF	1581.37	1581.37			
C_{gs0} / fF	1328.41	1375.00			
C_{ds0} / fF	478.53	448.09			
C_{gd0} / fF	1713.43	1722.45			

Model Parameter	Initial Value	Optimized Value	Model Parameter	Initial Value	Optimized Value
C_{pga} / fF	190.60	232.04	L_g / pH	42.38	49.81
C_{pgi} / fF	442.74	392.99	L_d / pH	77.19	99.62
C_{gs} / fF	875.42	1037.21	L_s / pH	17.12	17.83
C_{pda} / fF	157.02	238.48	R_g / Ω	2.58	2.68
C_{pdi} / fF	329.75	319.51	R_d / Ω	1.86	2.29
C_{ds} / fF	353.29	150.33	R_s / Ω	0.39	0.31
C_{gda} / fF	0.00	0.00	R_i / Ω	—	—
C_{gdi} / fF	0.00	0.00	G_{ds} / mS	—	—
C_{gd} / fF	875.42	1021.54			
C_{gs0} / fF	1508.76	1662.24			
C_{ds0} / fF	840.06	708.32			
C_{gd0} / fF	875.42	1021.54			

*From: [11].

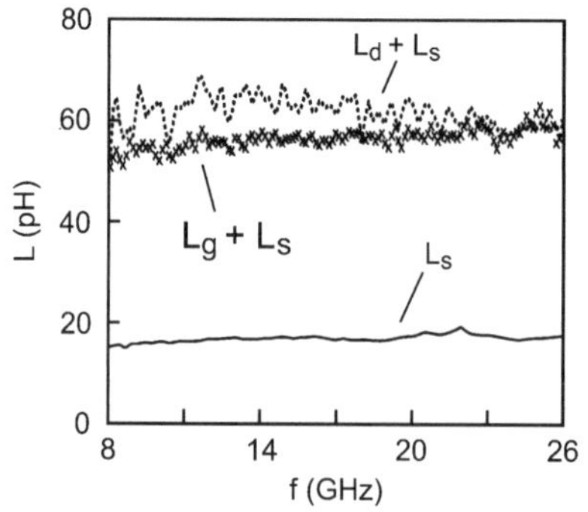

Figure 21.41 Inductance extraction results with measured S-parameters of a 1.2-mm GaN HEMT at V_{DS} = 0V and $V_{GS} < V_{pinch\text{-}off}$ [39].

model parameter values of a 2-mm and 3.2-mm AlGaN/GaN power device in addition to the estimated values using resistance extraction with modified cold reverse measurement technique, as proposed in Section 21.8.4. Again the similarity with the estimated values underlines the high quality of the starting values for the optimization process. Most important, the various extraction results are of consistent quality. This may essentially reduce the common queasy feeling in model parameter extraction tasks, always posing the question about the trustworthiness of extracted results, and this, in particular, in view of the physical relevance.

Studying the extracted model elements in Tables 21.8 and 21.9 more closely it becomes obvious that the capacitances of the intermediate topology shells provide significant contributions to the total branch capacitances and the outer feedback capacitances C_{gda} of the investigate GaN power devices are of less relevance. In other words, the assumption that the interelectrode capacitances are always relatively small, and thus can be neglected, could not be confirmed [8].

21.9.2 Extraction Based on Analytical Determination of Inductances

The proposed extraction strategy for inductances (Section 21.7) and resistances (section 21.8.4) were used to generate the model of a real 1.2-mm GaN HEMT (four gate fingers of 0.3 mm width each, manufactured by Fraunhofer-IAF, Freiburg I. Br., Germany) using cold pinch-off S-parameter measurements at V_{DS} = 0V, $V_{GS} < V_{pinch\text{-}off}$. The low-frequency parts of the measurements (below 1 GHz) were used to calculate the total capacitances at the device ports C_{gs0}, C_{ds0}, and C_{gd0}, defined in Section 21.3.1 ((21.35) - (21.37)). In this case the values for C_{pga}, C_{pda}, and C_{gda} were not obtained by scanning technique as described, for example, in Section 21.5, but were obtained using cold pinch-off S-parameter measurements of devices of the same fabrication technology but different gate peripheries, by a fully analytical approach [21]. The rest of the capacitance values were found with an analytical search method, as in [5, 40, 41]. All the capacitance values that were extracted for the 1.2-mm GaN HEMT are listed in Table 21.8. Inductance values were extracted with cold pinch-off S-parameter data and proposed inductance extraction in Section 21.6.3. Resistance values were extracted with measured S-parameter data (V_{DS} = 0V, and V_{GS} = −1.6V) and proposed

Table 21.10

Extracted Model Element Values from Measured S-Parameter Data of a 1.2-mm GaN HEMT

Capacitance values extracted previously with low-frequency cold pinch-off S-parameter measurements using [3, 9–11].	C_{pga} = 23.67 fF C_{pgi} = 590.23 fF C_{gs} = 406.88 fF	C_{pda} = 177.40 fF C_{pdi} = 325.62 fF C_{ds} = 13.45 fF	C_{gda} = 0.00 fF C_{gdi} = 0.23 fF C_{gd} = 499.18 fF
Analytical extraction of inductance values	L_g = 40.59 pH	L_d = 42.22 pH	L_s = 17.27 pH
Analytical extraction of resistance values	R_g = 0.92Ω	R_d = 0.52Ω	R_s = 2.24Ω

resistance extraction.

The inductance extraction proposed in Section 21.7 was applied with cold pinch-off S-parameter measurements of the 1.2-mm device, and the extraction results versus frequency are presented in Figure 21.41 and listed in Table 21.10.

The inductance extraction results from measured S-parameters show negligible frequency dependence, similar to that observed in the results with constructed noisy S-parameters (see Section 21.7.2). The frequency where reliable extraction starts, 8 GHz in this case, was found identifying the frequency range where $|S_{11}|$ and $|S_{22}|$ traces of measured cold pinch-off S-parameters begin to increase, as illustrated in Figure 21.42. Moreover, the similarity of the inductance values of the gate and drain, as well as the relatively lower value of the source inductance, correspond well to previous modeling experiences of comparable GaN HEMTs [11, 38, 39].

Using the extracted values of Table 21.10, the extrinsic reactive effects were stripped off from the S-parameters of the 1.2-mm device measured at the bias point V_{DS} = 0V and V_{GS} =−1.6V in order to apply the resistance extraction proposed. The results obtained from this extraction are presented in Figure 21.43 and listed in Table 21.10. The resistance extraction results show linear behavior with normalized frequency, in a similar way as observed previously in the results from fictitious noisy S-parameters in Figure 21.37.

As a further test of model verification, the S-parameters of the circuit in Figure 21.21 were built with the values of Table 21.10, finding an excellent agreement with the cold pinch-off S-parameter measurements, as depicted in Figure 21.42.

21.10 DETERMINATION OF INTRINSIC MODEL PARAMETERS

After de-embedding the extrinsic parameters discussed in the previous sections, the bias dependent intrinsic parameters can be extracted. An efficient technique has been developed for the extraction optimal values for the intrinsic elements. In this technique, the intrinsic Y-parameters are formulated in a way where the optimal intrinsic element value can be extracted using simple linear data fitting. The method was originally proposed by Mwema [3] treating the intrinsic eight-element model. Afterwards, Jarndal [4] applied the analysis to an extended 10-element model comprising additionally the gate-forward conductances G_{gsf} and G_{gdf}, as shown in Figure 21.44. In the following the approach given in [4] is briefly described.

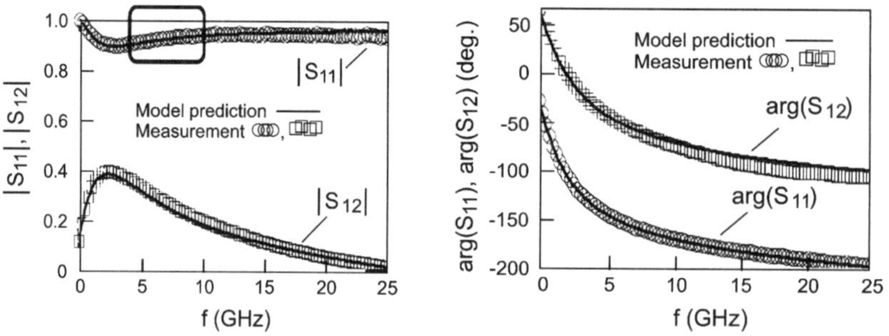

Figure 21.42 Model prediction (solid line) and cold pinch-off S-parameter measurements. The rounded rectangle highlights the onset of inductance effects; that is, the range of the minimum frequency reliable extraction [39].

The admittance of the intrinsic gate-source branch Y_{gs} reads

$$Y_{gs} = Y_{11} + Y_{12} = \frac{G_{gsf} + j\omega C_{gs}}{1 + R_i G_{gsf} + j\omega R_i C_{gs}} \tag{21.126}$$

By defining a variable D_{gs} as

$$D_{gs} = \frac{|Y_{gs}|^2}{\text{Im}[Y_{gs}]} = \frac{G_{gsf}^2}{\omega C_{gs}} + \omega C_{gs} \tag{21.127}$$

C_{gs} is obtained from the slope of the fitted curve to measured data. A similar equation is formulated in terms of Y_{gd}, C_{gd}, and G_{gdf}. As pointed out in [11], (21.127) is approximately valid. However, when $R_i G_{gsf} \ll 1$ and $(\omega C_{gs} R_i)^2 \ll 1$, then (21.127) can be reliably fitted with a straight line (ωD_{gs} versus ω^2). By redefining D_{gs} as

$$D_{gs} = \frac{|Y_{gs}|^2}{\text{Im}[Y_{gs}]} = \frac{G_{gsf}(1 + R_i G_{gsf})}{\omega C_{gs}} + \omega R_i C_{gs} - j \tag{21.128}$$

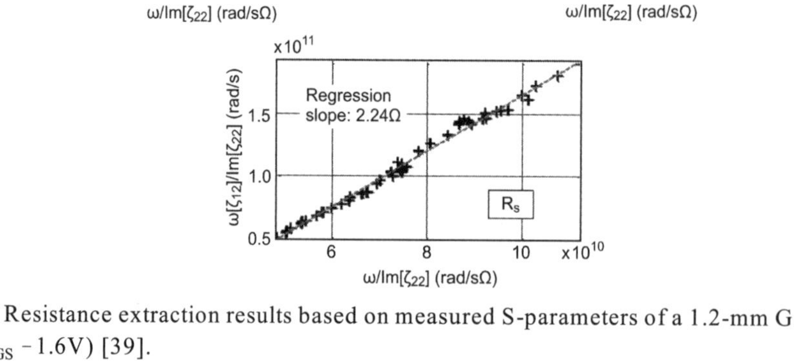

Figure 21.43 Resistance extraction results based on measured S-parameters of a 1.2-mm GaN HEMT (V_{DS} = 0V, V_{GS} - 1.6V) [39].

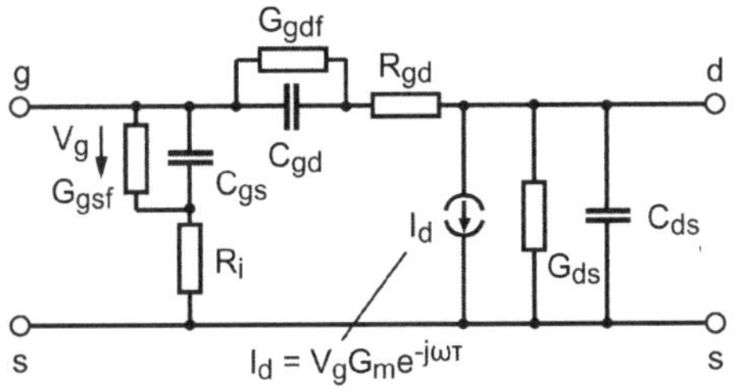

Figure 21.44 Equivalent circuit of the intrinsic transistor at active biasing including the gate-forward condition.

R_i can be determined from the plot of the real part of ωD versus ω^2 by linear fitting. G_{gsf} can be determined from the real part of Y_{gs} at low frequencies (typically in the megahertz range or below). The admittance for the intrinsic gate-drain branch Y_{gd} is given by

$$Y_{gd} = -Y_{12} = \frac{G_{gdf} + j\omega C_{gd}}{1 + R_{gd}G_{gdf} + j\omega R_{gd}C_{gd}} \tag{21.129}$$

Under assumption that $R_{gd}G_{gdf} \ll 1$ and $(\omega C_{gd}R_{gd})^2 \ll 1$, then the same procedure given in (21.127) and (21.128) can be used for extracting C_{gd}, R_{gd}, and G_{gdf}.

The admittance of the intrinsic transconductance branch Y_{gm} can be expressed as

$$Y_{gm} = Y_{21} - Y_{12} = \frac{G_m e^{-j\omega\tau}}{1 + R_i G_{gsf} + j\omega C_{gs}} \tag{21.130}$$

By defining a variable D_{gm} as

$$D_{gm} = \left|\frac{Y_{gs}}{Y_{gm}}\right|^2 = \left(\frac{G_{gsf}}{G_m}\right)^2 + \omega^2\left(\frac{C_{gs}}{G_m}\right)^2 \tag{21.131}$$

G_m can be determined from the slope of the curve for D_{gm} versus ω^2 by linear data-fitting.

By redefining D_{gm}, using (21.126) and (21.130), as

$$Y_{ds} = Y_{i,22} + Y_{i,12} = G_{ds} + j\omega C_{ds} \qquad D_{gm} = (G_{gsf} + j\omega C_{gs})\frac{Y_{gm}}{Y_{gs}} = G_m e^{-j\omega\tau} \tag{21.132}$$

τ can be determined from the plot of the phase of D_{gm} versus ω by linear data-fitting. The admittance of the intrinsic drain-source branch Y_{ds} can be expressed as

$$Y_{ds} = Y_{i,22} + Y_{i,12} = G_{ds} + j\omega C_{ds} \tag{21.133}$$

C_{ds} can be extracted from the plot of the imaginary part of Y_{ds} versus ω by linear data-fitting. Due to the frequency-dependent effect in the output conductance, G_{ds} is determined from the curve of $\omega \cdot \mathrm{Re}[Y_{ds}]$ versus ω by linear fitting, instead of the whole range, for optimal extraction of the intrinsic parameters. Further refinement can be done by using only the more linear part of the curves, instead of the whole range, for optimal extraction of the intrinsic parameters. A searching procedure for determining the more linear part of the curves is developed. This procedure is based on the correlation coefficient as a measure to define the more linearly correlated data range. Intrinsic parameter extraction using the proposed

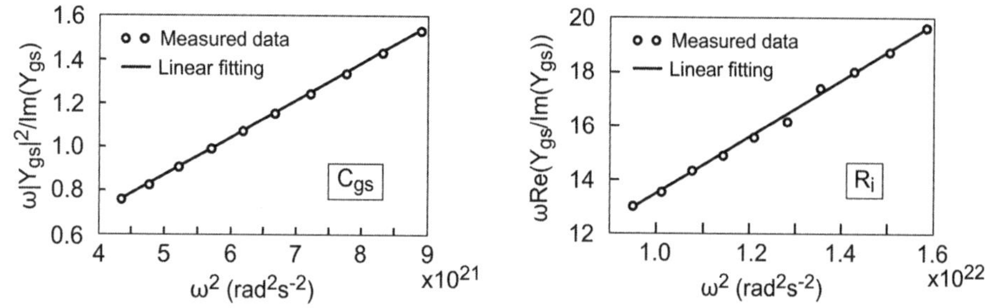

Figure 21.45 Fitting of highly correlated measured data of a 0.5-μm gate AlGaN/GaN HEMT with 2×50 μm gate-width, at V_{GS} = -2V and V_{DS} = 20V, for extraction of C_{gs} and R_i [4].

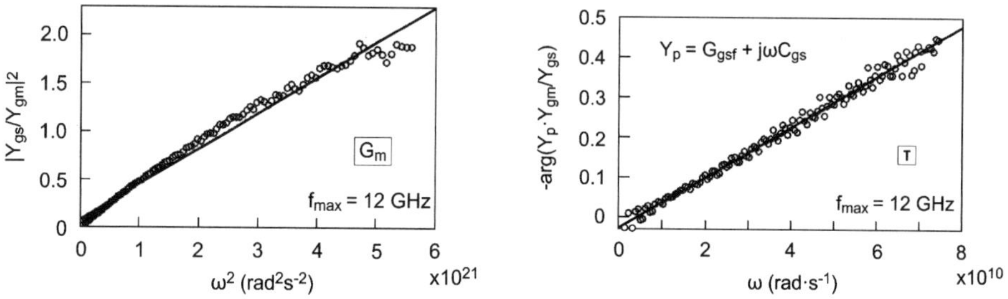

Figure 21.46 Data-fitting of the highly correlated measured data of a 2-mm AlGaN/GaN HEMT (V_{GS} = -1.5V, V_{DS} = 5.0V) for extraction G_m and τ. (After [11].)

procedure gives an accurate determination for the intrinsic element value even for significant frequency dependency of this element [4].

Figure 21.45 shows the results of optimal frequency range determination for extracting C_{gs} and R_i of a 0.5-μm gate AlGaN/GaN HEMT with 2×50 μm gate-width at V_{GS} = -2V and V_{DS} = 20V. Figure 21.46 shows exemplarily the extracted result for G_m and τ of a 2-mm AlGaN/GaN HEMT operated at V_{GS} = -1.5V and V_{DS} = 5.0V.

Figures 21.47–21.50 present extracted optimized intrinsic parameters versus extrinsic bias voltages of AlGaN/GaN HEMTs of different sizes.

From the device physics point of view, C_{gs} appears as a parallel-plate capacitance whose two plates are formed by the gate metal and the 2DEG channel charge [56]. The effective separation of this capacitor is determined by the depletion layer depth under the gate, which extends all the way to heterojunction at high negative gate voltage. Therefore, rapid decrease for C_{gs} can be noted in Figure 21.47 in the pinch-off region ($V_{gs} \leq -4$V). The lateral electric field established by the drain voltage accelerates charge carriers in the channel to scatter into the barrier layers. This results in reduction of the layer depth, and therefore gradual increase for C_{gs} with increasing the drain voltage is observed around the grounded gate voltage (V_{GS} = 0V).

The C_{gd} capacitance is originated in the extension of the depletion region into the gate-drain space. The extension increases with increasing drain voltage. Therefore, smaller values for C_{gd} are obtained with increasing drain voltage, as shown in Figure 21.47. As can also be seen in this figure, C_{gd} increases rapidly with increasing gate voltage under low drain voltage conditions. Under these conditions the depletion region is diminished and uniformly distributed under the gate. Therefore, the gate capacitance can be split into two

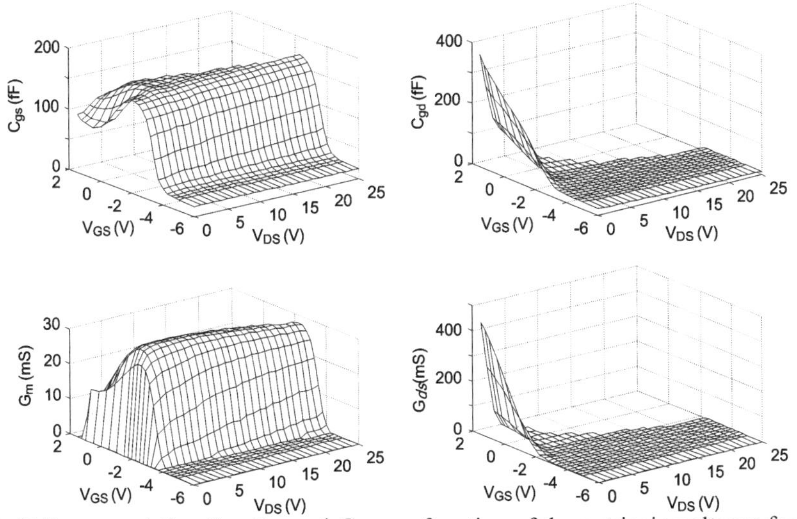

Figure 21.47 Exctracted C_{gs}, C_{gd}, G_m, and G_{ds} as a function of the extrinsic voltages for a 0.5-µm gate AlGaN/GaN HEMT with 2×50 µm gate-width [4].

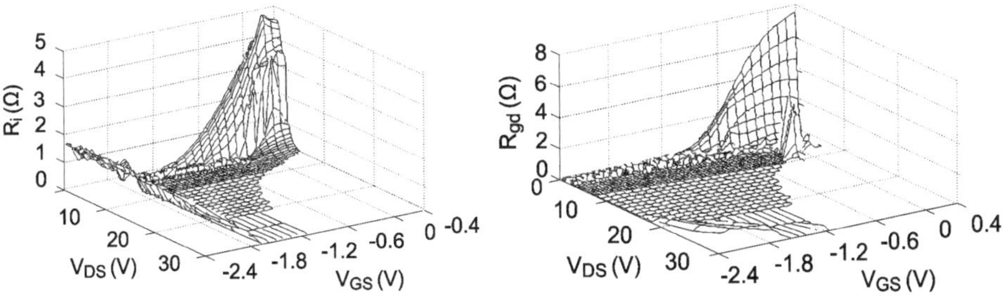

Figure 21.48 Extracted bias dependency of R_i and R_{gd} for a 2-mm AlGaN/GaN HEMT[11].

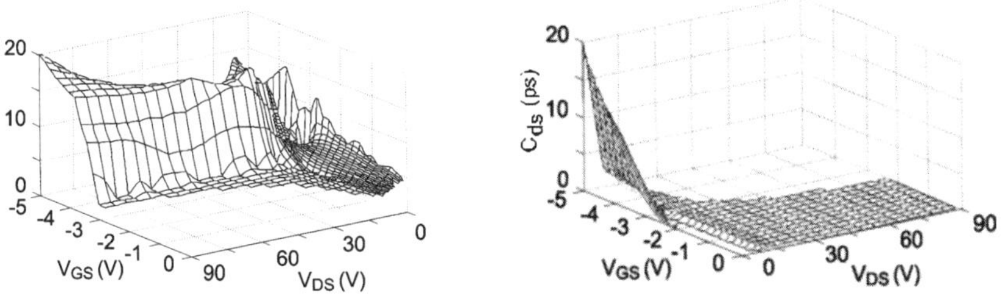

Figure 21.49 Extracted bias dependency of τ and C_{ds} for a 3-mm AlGaN/GaN HEMT [11].

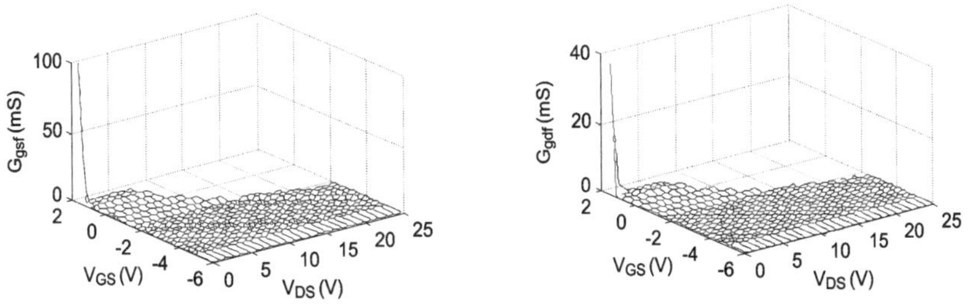

Figure 21.50 Exctracted bias dependency of G_{gsf} and G_{gdf} for a 100-µm AlGaN/GaN HEMT [4].

approximately equal capacitances, C_{gd} and C_{gs}.

G_m is related to the channel charge density and electron velocity. These two factors are mainly controlled by the gate and drain voltage, respectively. Therefore, significant increase for G_m can be seen in Figure 21.47 with increasing gate voltage. Due to the linear variation of the electron velocity under low electric field strengths along the channel [57], G_m increases linearly with increasing drain voltage in the ohmic region ($V_{DS} < 5V$). In the saturation region ($V_{DS} > 5V$), G_m becomes constant as the electron velocity saturates at higher channel electric field.

Physically, G_{ds} models the variation of the drain current with the drain voltage. Thus, small values for G_{ds} in the saturation region can be observed in Figure 21.47. Increasing the gate voltage increases the channel charge density and thus the drain current. This increase becomes remarkable in the ohmic region due to the reduction of the depletion layer. Therefore, a corresponding increase for G_{ds} can be noted in this region with increasing the gate voltage.

R_i models the undepleted part (low-field region) of the channel under the gate, through which the gate-source capacitance is charged. The value of R_i is equal to the ratio of the potential drop in this channel part and the channel current, which is approximately equal to the drain current [57]. Therefore, it is expected that the value of R_i should be high in the low drain current region, as shown in Figure 21.48.

Charging resistance R_{gd} is included in the model mainly to simulate the symmetrical distribution of the depletion region under the gate in the ohmic-forward region ($V_{DS} < 5V$ and $V_{GS} > 0V$). Therefore, it is expected that R_{gd} should have the same behavior as R_i in this region, as shown in Figure 21.48.

From the physics point of view, τ is argued to characterize the transit time of electrons in the channel under the depletion region [57]. Extension of the depletion region in the gate-drain area, due to an increase of the drain-gate voltage, increases the required transit time for the electron. Therefore, τ is expected to be increased with increasing drain voltage or decreasing gate voltage, as can be observed in Figure 21.49.

The physical origin of C_{ds} can be attributed to the high-field part of the depletion layer, which separates the source and drain electrodes in an electrostatic sense [56]. Therefore, higher values for C_{ds} will be obtained in the ohmic region due to the reduction of the high-field part, as can be seen in Figure 21.49.

G_{gsf} and G_{gdf} model the conduction current through the gate-source and gate-drain Schottky diodes. Therefore, they have large values only when the gate voltage is beyond the diodes turn-on voltage ($V_{GS} \approx 1.5V$), as can be seen in Figure 21.50.

21.11 SMALL-SIGNAL MODEL VERIFICATION

The small-signal modeling is verified by comparing measured and simulated S-parameters, where the simulation is based on the optimized model parameters based measurement-correlated starting parameter values. In the following, the results of AlGaN/GaN HEMTs of different sizes (0.1-mm, 1-mm, 3.2-mm total gate width) are presented.

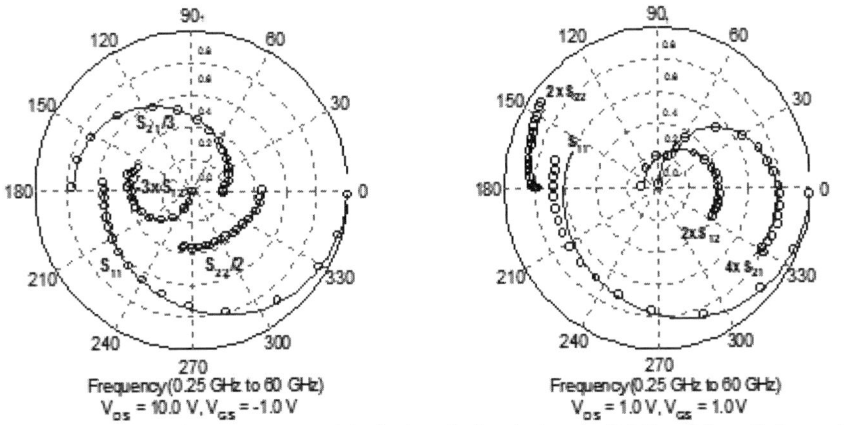

Figure 21.51 Comparison of measurement (circles) and simulation (solid lines) for a 0.5-μm AlGaN/GaN HEMT (2×50 μm gate-width) [4].

Figure 21.51 shows the result of S-parameter simulation at different bias points, in linear and saturation regions, over a wide frequency range. Very good agreement is achieved between measurement and simulation.

Figure 21.52 illustrates the good agreement between measured and simulated S-parameter data for a 1-mm AlGaN/GaN HEMT at different bias points (i.e., in saturated and ohmic

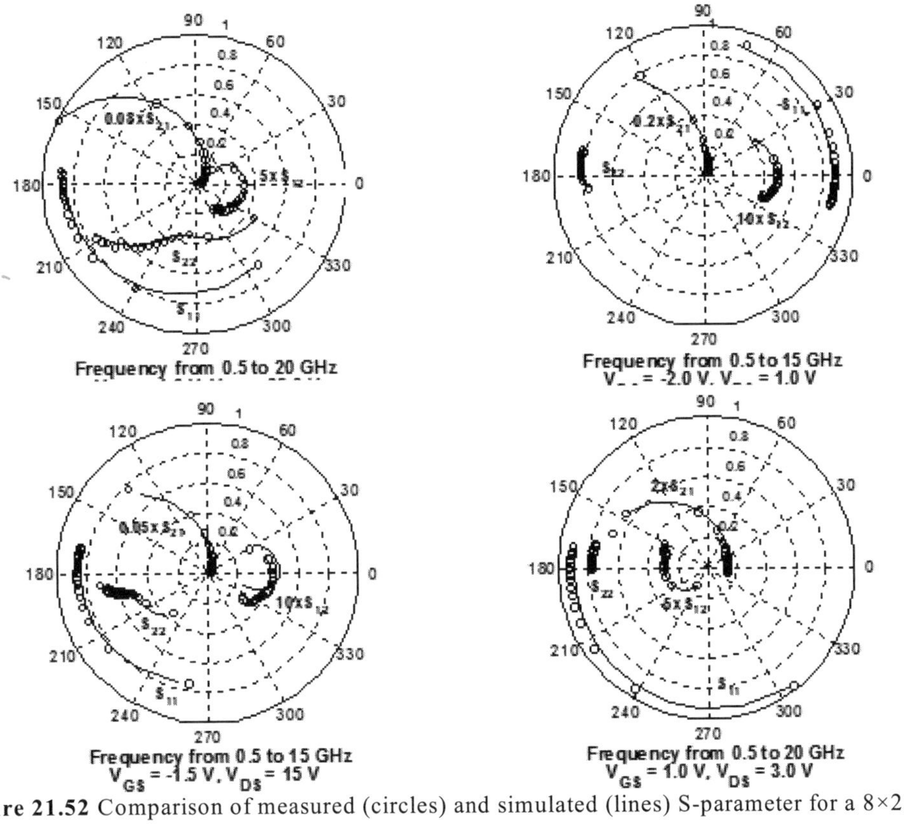

Figure 21.52 Comparison of measured (circles) and simulated (lines) S-parameter for a 8×250-mm AlGaN/GaN HEMT under bias condition in: (a) saturation region, and (b) ohmic region [4].

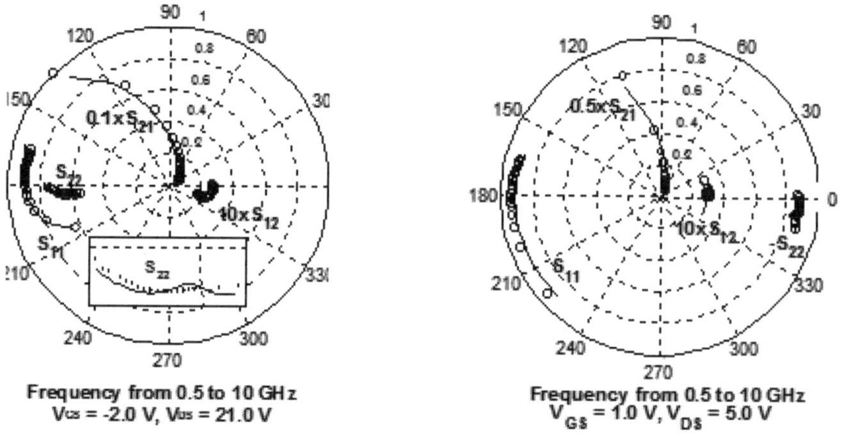

Figure 21.53 Comparison of measured (circles) and simulated (lines) S-parameters for a 16×250-μm AlGaN/GaN HEMT biased in the saturation region (a) and ohmic region (b) [4].

region). As can be seen, the extracted model can accurately simulate the measured S-parameters. Also it can predict the kink effect in the S_{22} trace, which occurs in larger size FETs and is related to the frequency dependence of the device output impedance [58].

Furthermore, the result for a 4-mm AlGaN/GaN HEMT is depicted in Figure 21.53. Again ohmic and saturated bias points have been chosen. As in the previous cases, excellent agreement is observed between simulation and measurement. This is also observed with a 3.2-mm AlGaN/GaN HEMT device with selected three bias points, widely dispersed in the output I–V plane.

21.12 SUMMARY

In this chapter, new model parameter extraction strategies focusing on small-signal field-effect transistor models with distributed capacitive effects were treated. The extraction concept was described, pointing out that the extraction should rely on a single set of measured S-parameter data because of the intercorrelations between all model parameters. Total gate-source, drain-source, and gate-drain branch capacitances can be obtained from low-frequency S-parameter measurements. It was shown that the partitioning of the effective capacitances into contact, interelectrode, and intrinsic capacitances turns out to be rather challenging. Test structures and/or field simulation techniques could be used if accessible. However, in this case, capacitance ratios were defined that can be estimated from surface inspection of fabricated device or determined by a scanning procedure for each capacitance ratio.

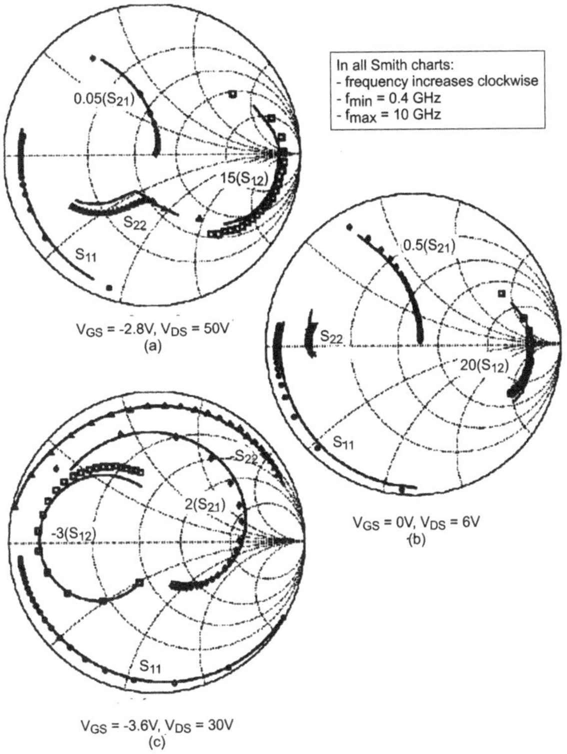

Figure 21.54 Comparison of measured (circles) and simulated (lines) S-parameters for a 3.2-mm AlGaN/GaN HEMT, on active bias points in (a) saturation, (b) high-V_{GS}, low-V_{DS}, and (c) pinch-off regions [11].

With the capacitance distribution known, inductances and resistances can be found in the usual way by de-embedding outer capacitances and by linear curve fitting of de-embedded device response. However, linear regression presupposes that the relationship between the dependent and independent variables is linear. Slope and intercept are the estimates. Due to measurement errors (data spread), limited (frequency) range of data and nonlinear frequency dependence, the estimated slope (smoothing effect) and intercept point of the final regression curve should be verified by visual inspection, which makes human intervention necessary.

The main focus of this chapter was on novel extraction techniques for the extrinsic series inductances and resistances of the small-signal equivalent circuit with distributed capacitive effects. Inductances are extracted in a fully closed form. For the determination of series resistances a new approach was presented taking into account the distributed nature of the capacitive effects in the small-signal model. The parameter extractions are based on S-parameter measurements and follow a matrix formulation. Their implementation in programming languages for numerical computing (e.g., MATLAB, Scilab, or C) is straightforward. Notwithstanding the relatively easy programmability of these extractions, it is always advisable to aid the extractions with a visual user inspection for the frequency dependence of the extracted parameter values. Nevertheless, in general, optimization or iterative algorithms are no longer needed for the extrinsic parameter extractions, since even the linear regression has an established formulation in a closed form. The extracted parameter values are primed to be used as a physically meaningful starting point for further refining optimization runs applied in the development of more complex or demanding modeling tasks such s large-signal modeling.

The uncertainties of the proposed extraction procedures were statistically analyzed. The probability distributions of the extracted values proved that the contributions to uncertainty by the extraction procedures are comparatively insignificant with respect to the measurement uncertainty specifications of commercial network analyzers.

The extracted results for the model parameter values of the distributed equivalent circuit confirm that the interelectrode capacitances of large-size gate-periphery GaN HEMTs contribute significantly to the effective branch capacitances and should not be ignored.

REFERENCES

[1] Shealy, J. R., J. Wang, and R. Brown, "Methodology for Small-Signal Model Extraction of AlGaN HEMTs," *IEEE Trans.*, Vol. ED-55(7), 2008, pp. 1603–1613.

[2] Mwema, W. N., and G. Kompa, "A New Simplified and Reliable HEMT Approach Using Pinched Cold FET S-Paraameters," *IEEE MTT-S Int. Microwave Symp. Digest*, 2000, pp. 1393–1395.

[3] Mwema, W. N., *A Reliable Optimisation-Based Model Parameter Extraction*, Doctoral Thesis, University of Kassel (2002), Kassel University Press GmbH, 2002.

[4] Jarndal, A. H., *Large-Signal Modeling of GaN Device for High Power Amplifier Design*, Doctoral Thesis (2006), University of Kassel, Kassel University Press GmbH, 2007.

[5] Jarndal, A., and G. Kompa, "A New Small-Signal Modeling Approach Applied to GaN Devices," *IEEE Trans.*, Vol. MTT-53, 2005, pp. 3440–3447.

[6] Jarndal, A., and G. Kompa, "A New Small-Signal Model Parameter Extraction Method Applied to GaN Devices," *IEEE MTT-S Int. Microwave Symp. Dig.*, Long Beach, CA, 2005.

[7] Brady, R. G., C. H. Oxley, and T. J. Brazil, "An Improved Small-Signal Parameter-Extraction Algorithm for GaN HEMT Devices," *IEEE Trans.*, Vol. MTT-56(7), 2008, pp. 1535–1544.

[8] Fan, Q., J. H. Leach, and H. Morkoc, "Small Signal Equivalent Circuit Modeling for AlGaN/GaN HFET: Hybrid Extraction Method for Determining Circuit Elements of AlGaN/GaN HFET," *Proceedings of the IEEE*, Vol. 98(7), 2010, pp. 1140–1150.

[9] Teng, M., H. Vue, C. Chi, and M. Xiaohua, "A New Small-Signal Model for Asymmetrical AlGaN/GaN HEMTs,“ *Journal of Semiconductors*, Vol. 31(6), 2010, pp. 064002/1–5.

[10] Jarndal, A., "Genetic Algorithm Based Extraction Method for Distributed Small-Signal Model of GaN HEMTs," *Proceedings of the IEEE International Conference on Semiconductor Electronics (ICSE)*, 2010, pp. 41–44.

[11] Zamudio Flores, J. A., *Device Characterization and Modelling of Large-Size GaN HEMTs*, Doctoral Thesis (2012), University of Kassel, Kassel University Press GmbH, 2012.

[12] Kompa, G., *Basic Properties of III-V Devices–Understanding Mysterious Trapping Phenomena*, Kassel, Kassel University Press GmbH, 2014.

[13] Liu, Z. G., G. I. Ng, S. Arulkumaran, Y. K. T. Maung, K.L. Teo, S. C. Foo, and V. Sahmuganathan, "Comprehensive Study on the Bias-Dependent Equivalent.Circuit Elements Affected by PECVD SiN Passivation in AlGaN/GaN HEMTs," *IEEE-Trans.*, Vol. ED-58(2), 2011, pp. 473–479.

[14] Mizutani, T., Y. Ohno, M. Akita, S. Kishimoto, and K. Maezawa, "A Study on Current Collapse in AlGaN/GaN HEMTs Induced by Bias Stress," *IEEE Trans.*, Vol. ED-50(10), 2003, pp. 2015–2020.

[15] Tajima, Y., "100 W GaN HEMT Modeling," *Microwave Journal*, Vol. 50(5), 2007, pp. 204–214.

[16] Karmalkar, S., and U. K. Mishra, "Enhancement of Breakdown Voltage in AlGaN/GaN High Electron Mobility Transistors Using a Field Plate," *IEEE Trans.*, Vol. ED-48(8), 2001, pp. 1515–1521.

[17] Saito, W., Y. Kakiuchi, T. Nitta, Y. Saito, T. Noda, H. Fujimoto, A. Yoshioka, T. Ohno, and M. Yamaguchi, "Field-Plate Structure Dependence of Current Collapse Phenomena in High-Voltage GaN-HEMTs," *IEEE Trans.*, Vol. MTT-31(7), 2010, pp. 659–661.

[18] Koudymov, A., V. Adivarahan, J. Yang, G. Simin, and M. A. Khan, "Mechanism of Current Collapse Removal in Field-Plated Nitride HFETs," *IEEE Electron Device Lett.*, Vol. 26(10), 2005, pp. 704–706.

[19] Sriram, S., H. Hagleitner, D. Namishia, T. Alcorn, T. Smith, and B. Pulz, "High-Gain SiC MESFETs Using Source-Connected Field-Plates," *IEEE Electron Device Lett.*, Vol. EDL-30(9), 2009, pp. 952–953.

[20] Jarndal, A., and G. Kompa, "An Accurate Small-Signal Model for AlGaN-GaN HEMT Suitable for Scalable Large-Signal Model Construction," *IEEE Microwave and Wireless Components Lett.*, Vol. 16 (6), 2006, pp. 333–335.

[21] Schwantuschke, D., M. Seelmann-Eggebert, P. Brückner, R. Quay, M. Mikulla, O. Ambacher, and I. Kallfass, "A Fully Scalable Compact Small-Signal Modeling Approach for 100 nm AlGaN/GaN HEMTs," *Proc. 8th European Microwave Integrated Circuits Conference (EuMIC)*, 2013, pp. 284-287.

[22] Hewlett-Packard Company, *System Manual HP8510B Network Analyzer, P/N 08510-90074*, Santa Rosa, CA, July 1987.

[23] Hingham, N. J., "Accuracy and Stability of Numerical Algorithms," *Society for Industrial and Applied Mathematics (SIAM)*, Philadelphia, 1996.

[24] Kompa, G., and M. Novotny, "Frequency-Dependent Measurement Error Analysis and Refined FET Model Parameter Extraction Including Bias-Dependent Series Resistors," *IEEE Workshop on Experimentally Based FET Device Modelling & Related Nonlinear Circuit Design, Kassel*, Germany, July 17-18, 1997, pp. 6.1–6.16, https://archive.org/details/DTIC_ADA331468/page/n279.

[25] Anholt, R., and S. Swirhun, "Equivalent-Circuit Parameter Extraction for Cold GaAs MESFET's," *IEEE Trans.*, Vol. MTT-39(7), 1991, pp. 1243–1247.

[26] Edwards, M. L., and J. H. Sinsky, "A New Criterion for Linear Two-Port Stability Using Single Geometrically Derived Parameter," *IEEE Trans.*, Vol. MTT-40(12), 1992, pp. 2303–2311.

[27] Lombardi, G., and B. Neri, "Criteria for the Evaluation of Unconditional Stability of Microwave Linear Two-Ports: A Critical Review and New Proof," *IEEE Trans.*, Vol. MTT-47(6), 1999, pp. 746–751.

[28] Kotzebue, K. L., "Maximally Efficient Gain: A Figure of Merit for Linear Active 2-Ports," *Electronics Lett.*, Vol. 12(19), 1976, pp. 490–491.

[29] Johnson, K. M., "Large Signal GaAs MESFET Oscillator Design," *IEEE Trans.*, Vol. MTT-27(3), 1979, pp. 217–227.

[30] Kompa, G., "Reliable Extraction of Small-Signal Elements of a Generalized Distributed FET Model," *IEEE MTT-S Int. Microwave Symp. Dig.*, Baltimore, USA, 1998, Paper TUA-8, pp. 291–294.

[31] Mengistu, E. S., *Large-Signal Modeling of GaN HEMTs for Linear Power Amplifier Design*, Doctoral Thesis, University of Kassel, 2007.

[32] White, P. M., and R. M. Healy, "Improved Equivalent Circuit for Determination of MESFET and HEMT Parasitic Capacitances from 'Coldfet' Measurements," *IEEE Microwave and Guided Wave Lett.*, Vol. 3, 1993, pp. 453–454.

[33] Kim, C.-H., K.-S. Yoon, J.-W. Yang, J.-H. Lee, C.-S. Park, J.-J. Lee, and K.-E. Pyun, "A New Extraction Method to Determine Bias-Dependent Source Series Resistance in GaAs FET's," *IEEE Trans.*, Vol. MTT-46(9), 1998, pp. 1242–1250.

[34] Yanagava, S., H. Ishihara, and M. Ohtomo, "Analytical Method for Determining Equivalent Circuit Parameters of GaAs FET′s," *IEEE Trans.*, Vol. MTT-44(10), 1996, pp. 1637–1641.

[35] Kompa, G., M. Schlechtweg, and F. van Raay, "Precisely Calibrated Coaxial-to-Microstrip Transitions Yield Improved Performance in GaAs FET Characterization," *IEEE Trans.*, Vol. MTT-38, 1990, pp. 62–68.

[36] Kompa, G., "Small- and Large-Signal Characterization Methods, Modelling and Verification," *Digest IEEE Workshop on Measurement Techniques for Microwave Device Characterization and Modelling*, Stuttgart, Germany, April 23, 1990, pp. 67–97.

[37] Ladbrooke, P. H., and S. R. Blight, "Low-Field Low-Frequency Dispersion of Transconductance in GaAs MESFET's with Implications for Other Rate-Dependent Anomalies," *IEEE Trans.*, Vol. MTT-35, 1988, pp. 257–267.

[38] Rohdin, H., N. Moll, C.-Y. Su, and G. S. Lee, "Interfacial Gate Resistance in Schottky-Barrier-Gate Field-Effect Transistors," *IEEE Trans.*, Vol. ED-45, 1998, pp. 2407–2416.

[39] Zamudio Flores, J. A., and G. Kompa, "Closed-Form Extraction Strategy of Physically-Meaningful Parameters of Small-Signal HEMT Models with Distributed Parasitic Capacitive Effects," in preparation.

[40] Zamudio Flores, J. A., and G. Kompa, "Improved Measurement-Based Extraction Algorithm of a Comprehensive Extrinsic Element Network for Large-Size GaN HEMTs," *Proc. European Microwave Integrated Circuits Conf. (EuMIC)*, 2010, pp. 250– 253.

[41] Zamudio Flores, J. A.,S. Dahmani, and G. Kompa, "Large-Signal Modeling of Large-Size GaN HEMTs with a Comprehensive Extrinsic Elements Extraction Algorithm," *Int. J. Microw. Wirel. Technol.*, Vol. 2(1), 2010, pp. 63– 73.

[42] Adam, M., G. Tziallas, and F. Sanida, "Riccati Equation Solution Method for the Computation of the Solutions of X + ATX - 1 = Q and X - ATX - IA = Q," *Open Appl. Informatics J.*, Vol. 3(22), 2009, pp. 22–33.

[43] Keysight Technologies, Inc., *Downloadable Vector Network Analyzer Uncertainty Calculator*, 2015. [Online]. Available: http:///www.keysight.com/main/software.jspx?id =1000000418:epsg:sud.

[44] Chigaeva, E., W. Walthes, I. Dettman, N. Wieser, W. Vogel, H. Roll, F. Scholz, and H. Schweizer, "Advanced Large-Signal Modeling of GaN-HEMTs," *Proc. IEEE Lester Eastman Conf. on High Performance Devices*, August 2002, pp. 115– 180.

[45] Chen, G., V. Kumar, R. Schwindt, and I. Adesida, "A Low Gate Bias Model Extraction Technique for AlGaN/GaN HEMTs," *IEEE Trans.*, Vol. MTT-54(7), 2006, pp. 2949–2953.

[46] Crupi, G., D. Xiao, D. M. M.-P. Schreurs, E. Limiti, A. Caddemi, W. De Raedt, and M. Germain, "Accurate Multibias Equivalent-Circuit Extraction for GaN HEMTs," *IEEE Trans.*, Vol. MTT-54(10), 2006, pp. 3616–3622.

[47] Zarate-de Landa, A., J. Zuniga-Juarez, J. Loo-Yau, J. Reynoso-Hernandez, M. Maya-Sanchez, and J. del Valle-Padilla, "Advances in Linear Modeling of Microwave Transistors," *IEEE Microwave Magazine*, Vol. 10(2), 2009, pp. 100–111, 146.

[48] Jeon, M.-Y., B.-G. Kim, Y.-J. Jeon, and Y.-H. Jeong, "A Technique for Extracting Small-Signal Equivalent-Circuit Elements of HEMTs," *IEICE Trans. on Electronics* (Special Issue on High-Frequency/Speed Devices in the 21st Century), Vol. 82(11), 1999, pp. 1968–1976.

[49] Lu, J., Y. Wang, L. Ma, and Z. Yu, "A New Small-Signal Modeling and Extraction Method in AlGaN/GaN HEMTs," *Solid-State Electronics*, Vol. 52(1), 2008, pp. 115–120.

[50] Manohar, S., A. Narayanan, A. Keerti, A. Pham, J. Brown, R. Borges, and K. Linthieum., "Characteristics of Microwave Power GaN HEMTs on 4-Inch Si Wafers," *IEEE MTT-S Int. Microwave Symp. Digest*, 2002, pp. 449–452.

[51] Dambrine, G., A. Cappy, F. Heliodore, and E. Playez, "A New Method for Determining the FET Small-Signal Equivalent Circuit," *IEEE Trans.*, Vol. MTT-36, 1988, pp. 1151–1159.

[52] Caddemi, A., G. Crupi, and N. Donato, "Microwave Characterization and Modeling of Packaged HEMTs by a Direct Extraction Procedure Down to 30 °K," *IEEE Trans.*, Vol. IM-55(2), 2006, pp. 465–470.

[53] Werthof, A., *Experimental Modeling of Active Devices for the Simulation of Nonlinear Microwave Circuits* (in German), Doctoral Thesis, University of Kassel, 1994; published by VDI Verlag, Düsseldorf, Fortschritt-Berichte VDI, Reihe 9 (Elektronik), Nr. 200, 1995.

[54] Lin, F., and G. Kompa, "FET Model Parameter Extraction Based on Optimization With Multiplane Data-Fitting and Bidirectional Search—A New Concept," *IEEE Trans.*, Vol. MTT-42, 1994, pp. 1114–1121.

[55] Novotny, M., and G. Kompa, "Unique and Physically Meaningful Extraction of the Bias-Dependent Series Resistors of a 0.15 μm PHEMT Demands Extremely Broadband and Highly Accurate Measurements," *IEEE MTT-S Int. Microwave Symp. Digest*, 1996, pp. 1715–1718.

[56] Rohdin, H., A. Nagy, V. Robbins, C.-Y. Su, A. S. Wakita, J. Seeger, T. Hwang, P. Chye, P. E. Gregory, S. R. Bahl, F. G. Kellert, L. G. Studebaker, D. C. D'Avanzo, and S. Johnsen, "0.1-μm Gate-Length AlInAs/GaInAs/GaAs MODFET MMIC Process for Applications in High-Speed Wireless Communications," *Hewlett-Packard Journal*, Vol. 49, February 1998, pp. 1–37.

[57] Ladbrooke, P. H., *MMIC Design: GaAs FETs and HEMTs*, Artech House, Inc., 1989.

[58] Golio, J. M., *Microwave MESFETs & HEMTs*, Artech House, Inc., 1989.

Chapter 22

Basics of Nonlinear FET Modeling

In this section building blocks of equivalent circuits of transistors are described. Depending on the mode of operation we distinguish between small- and large-signal transistor models. The basic interrelations between small- and large signal model elements are discussed.

22.1 TERMS AND DEFINITIONS

22.1.1 Types of Model Elements in Equivalent Circuits

The electrical equivalent network of a microwave transistor comprises inductance, capacitance, and resistance model elements.

Inductances are commonly assumed to be linear. They equivalently describe the electrical properties of, for example, short high impedance (planar) connection lines (see Section 6.6.1 in [1]) or short length of bondwires (Section 6.6.2). Furthermore, a via hole (Section 6.6.3) is modeled as a linear inductance element.

Capacitances can reveal both linear and nonlinear properties. Contact pads or bond pads are described as parasitic linear model elements. They provide electrical access to the device to apply the DC voltages to operate the device, but, in particular, they are necessary for in-fixture or on-wafer device testing. Nonlinear capacitances are dependent on the applied voltage, such as the reverse-biased metal-semiconductor contact [2], which represents a capacitor configuration. The thickness of the depletion layer (space charge region) between the metal contact and the *n*-doped conductive region varies with the applied electrode voltage.

Resistances are mostly linear elements and frequency-independent. Regarding the metal stripe of the gate electrode, skin effect may become effective and noticeable only at higher frequencies of about 100 GHz (see Section 6.2.3). The ohmic access regions on both sides of the gate (channel) region are basically linear. However, the bias-dependent variation of the depletion region beneath, and in particular, at the edges of the gate electrode indeed affects the adjoined bulk resistance extent so that source and drain resistances may be assumed to be bias-dependent (see Chapter 20). However, as was shown, in the saturation region these resistances are approximately constant, so that regarding circuit design, the parasitic bulk resistances are usually assumed to be linear. In contrast, the channel conductance is strongly dependent on the bias conditions, and thus is nonlinear.

The value of a passive nonlinear model element depends on the voltage applied. Active one-dimensional dependent nonlinear model elements, such as nonlinear current sources, are characterized in that they are under control of a voltage somewhere else in the circuit. When in addition the voltage drop across the current source influences the current output, it is called a two-dimensional controlled (dependent) current source. Under the small-signal operation condition, the current source is characterized by its transconductance, which provides an output current in dependence of an input (control) voltage somewhere else in the circuit. Similarly, transcapacitances are small-signal elements and related to dependent nonlinear charge sources. They change their capacitance value with voltage change between nodes somewhere else in the circuit. In the two-dimensional case, both transconductancs and transcapacitances are additionally dependent on the voltage drop across their own nodes.

22.1.2 Quasi-Static Assumption

The *quasi-static assumption* plays a key role in the derivation of a large-signal transistor model, and thereby is briefly discussed. In a field-effect transistor structure two-dimensional voltage-controlled nonlinear current and charge functions $I(V_1,V_2)$ and $Q(V_1,V_2)$ exist as network state functions. Regarding only the one-dimensional case for simplification, then the relationship between $I(V)$ or $Q(V)$ and the control voltage V can be uniquely described by a nonlinear characteristic as illustrated in Figure 22.1.

In Figure 22.1(a) the functional dependence between a current $I(t)$ and a voltage $V(t)$ is plotted for measured time points t_1, t_2, and t_3. It is assumed that the repetition of the measurement at different time points will not change the shape of the curve (i.e., the function is time-independent). In this case neither the chronological order nor the degree of dynamic signal variation between the indicated voltage values $V(t_1)$, $V(t_2)$, and $V(t_3)$ have any influence on the respective instantaneous value of the voltage-related current value. The nonlinear interrelation between current and voltage is entirely defined by the *static* characteristics; that is, the parameter time does not play any role for the fundamental interrelation of both variables, and can therefore be disregarded in the characterization of the nonlinearity (Figure 22.1(b)).

Such a transfer function derived from static measurement is still valid for signals at sufficiently low frequencies of the control voltage $V(t)$. That means, in the quasi-static operation region, it does not play any role with which velocity the control voltage is moving along the characteristic curve. The static $I(V)$ characteristic in Figure 22.1(b) could principally also be a $Q(V)$ characteristic. Thus, $I(t)$ and $Q(t)$, respectively, are only *indirectly* dependent on time by the instantaneous value of the control voltage. Thus, differentiations of the state functions with respect to time have an effect only on the control voltages, and not on the static (quasi-static) characteristics.

Dispersion effects commonly lead to frequency-dependent variations of the characteristic.

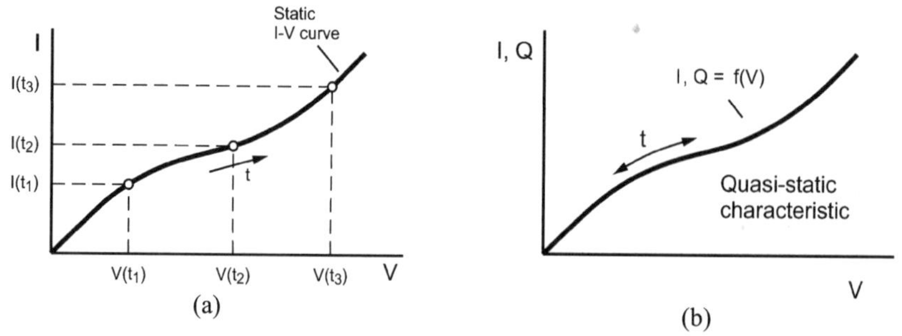

Figure 22.1 Static (quasi-static) I(V) characteristic: (a) unique time-invariant interrelation between I and V, and (b) quasi-static characteristic of one-dimensional network state functions $I(V)$ and $Q(V)$.

In time-domain (in the case of fast digital changes of the state of the FET) dispersive properties can manifest themselves by hysteresis loops in the output characteristic.

Temperature effects can alter the transfer characteristics at very low frequencies in contrast to static operation. In semiconductor devices a low-frequency dispersion can, for example, originate in the charging and discharging of traps. The trapping time constants lie typically in the millisecond range [2]. Trapping centers are intermediate energy levels in a semiconductor due to lattice imperfections, which mainly exist in semiconductor interfaces (including a vacuum-semiconductor surface interface). Electrons are captured and released again after some dwell time. At higher frequencies the traps can no longer follow the signal variations. Then they become *frozen* and ineffective. At very high frequencies (above 100 GHz) electron inertia effects can cause (electron) dispersion.

22.1.3 Definition of Voltages

Regarding the periodic signal waveform in Figure 22.2, we may introduce different voltage definitions. V_0 is the DC bias voltage. The time-varying voltage $V(t)$ denotes the *instantaneous* voltage, and $\overline{V(t)}$ determines its mean value. With reference to the bias point, we may define $V_{\sim}(t)$ as a large-signal voltage; that is, time-dependent voltage around a given bias point. The purely alternating voltage around the mean value $\overline{V(t)}$ is termed $V_{\approx}(t) = \overline{V(t)} - V(t)$. A small-signal voltage is written as $\tilde{V}(t)$.

22.1.4 Modes of Operation

The following modes of operation may be distinguished:

Static (DC) Mode of Operation
In the DC case, the nonlinear properties of a field-effect transistor are described by the nonlinear DC drain-source current function $I_{ds} = I_{ds}(V_{gs}, V_{ds})$. The dependence of the drain current on the gate-source control voltage for constant V_{ds} values characterizes the forward transfer characteristics. When $I_{ds} = I_{ds}(V_{gs}, V_{ds})$ is plotted over V_{ds} for constant V_{gs} values, then the static output characteristics of the transistor occurs (Figure 22.3).

Quasi-Static Mode of Operation
Quasi-static mode of operation denotes a slowly varying voltage stimulus of a device so that all device-related electrical and thermal time constants are small with respect to the time constant of the voltage change. This means that the device is able to follow the voltage

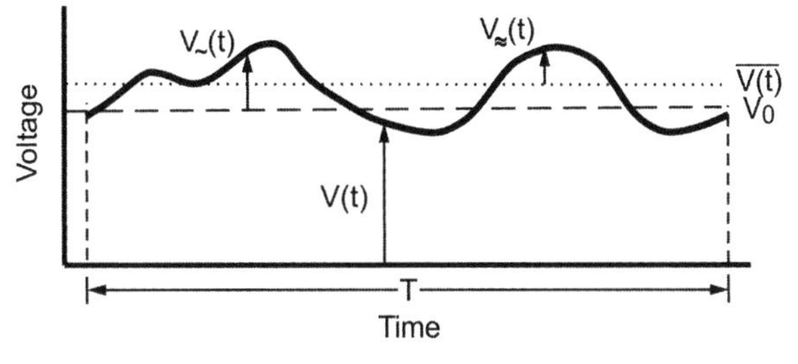

Figure 22.2 Periodic voltage signal waveform and voltage definitions.

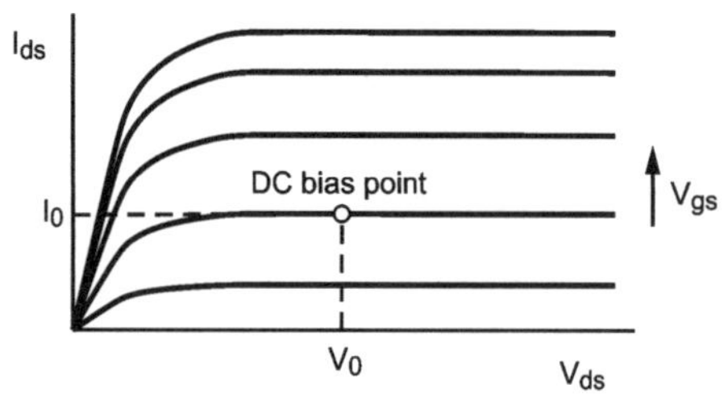

Figure 22.3 Static DC $I(V)$ output characteristics of a FET device.

variation $V(t)$ without any delay always on the basis of a thermodynamic state of equilibrium. For sufficiently low signal frequencies, the static $I(V)$ curve also describes the time dependence $I(t) = f[V(t)]$. The quasi-static approach for nonlinear FET modeling was first proposed by Willing and Rauscher [3]. Figure 22.4 depicts a quasi-static trajectory with harmonic voltage and current signals.

RF Mode of Operation

At RF frequencies the time-dependent variation of the state variables is fast so that the bias-dependent RF parameters become relevant. We may distinguish

(1) RF *small-signal mode of operation*: Under small-signal RF operation conditions a small-signal excitation is given in a bias point. Under assumption of a dispersionless $I(V)$ characteristic, the RF parameters can be derived from the static curves by Taylor series expansion.

(2) RF *large-signal mode of operation*: Under large-signal excitation in a given bias point (note that the dynamic effective bias point commonly differs from the DC bias point). The large-signal RF parameters are bias-dependent and dependent on the instantaneous value of the control voltages. In the dispersionless case, the bias-dependent transconductance and channel conductance can be directly derived from the nonlinear $I(V)$ characteristics.

Pulsed DC Mode of Operation

The drain current function is dependent on the duty cycle; that is, the ratio of the pulse to cycle duration (caused by the charging and discharging time constant of traps). Pulsed DC measurements are used to characterize the temperature and electrical dispersion of drain current source of FET devices (see Section 9.2).

Pulsed RF Mode of Operation

There are commercial and aerospace/defense applications that require testing using a pulsed-

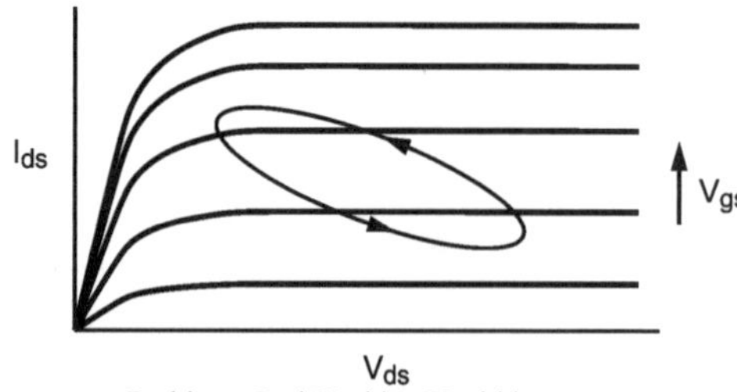

Figure 22.4 Quasi-static trajectory $I_{ds}(t) = I_{ds}(V_{gs}(t), V_{ds}(t))$ in the I-V plane.

RF stimulus [4]. This applies, for example, to some components of radar systems to ensure their functionality in the final environment. Furthermore, high-power (packaged) devices may suffer from too strong heating in case of DC characterization, and thereby being exposed to the risk of thermal damage. Pulsed-RF device operation and characterization will not be further dealt with in this book.

22.2 NONLINEAR EQUIVALENT CIRCUIT ELEMENTS

In this section we examine the relationship between the large-signal and small-signal characterizations, and in particular show how the small-signal characterization can be derived from the large-signal one.

An important question involves the precise definitions of the small-signal conductance (or resistance) and capacitance of linear elements. For example, the large-signal *I-V* characteristic of a linear conductance is given by Ohm's law, $V = R \cdot I$. But suppose a current-controlled nonlinear resistor were used in an application where a small-signal AC current is applied and a DC control current I_0 exists. The AC component of its voltage should be given by $\tilde{V}(t) = \tilde{R}(t) \cdot \tilde{I}(t)$, where $\tilde{V}(t)$ and $\tilde{I}(t)$ are the small-signal voltage and current, respectively. Figure 2.1 illustrates this case, and it is clear that

$$\tilde{V}(t) = \left.\frac{dV}{dI}\right|_{I=I_0} \tilde{I}(t) \tag{22.1}$$

so

$$\tilde{R}(I_0) = \left.\frac{dV}{dI}\right|_{I=I_0} \tag{22.2}$$

Of course, (22.1) is exact only as the magnitude of $\tilde{I}(t)$ approaches zero. This definition of resistance is called the *incremental resistance* and is valid in small-signal quasi-linear analysis.

The Substitution Theorem [5]

Figure 22.6(a) shows a linear voltage-controlled current source connected to a network. Its current is $G \cdot V$, where V, the control voltage, is the voltage at its terminals. The current is clearly unchanged if a conductance of value G is substituted for the controlled source. The

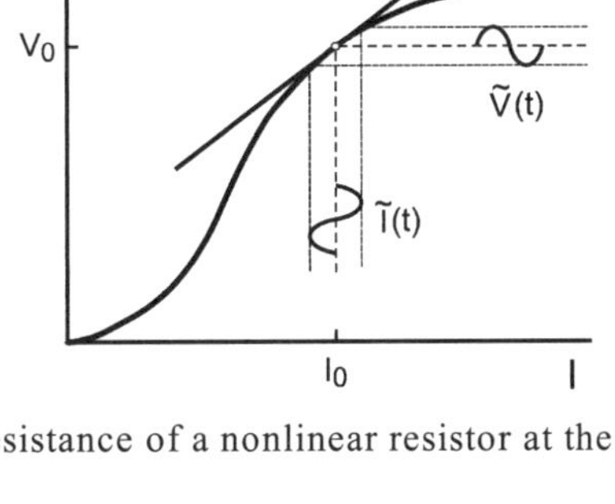

Figure 22.5 Incremental resistance of a nonlinear resistor at the DC bias point I_0. (After [5].)

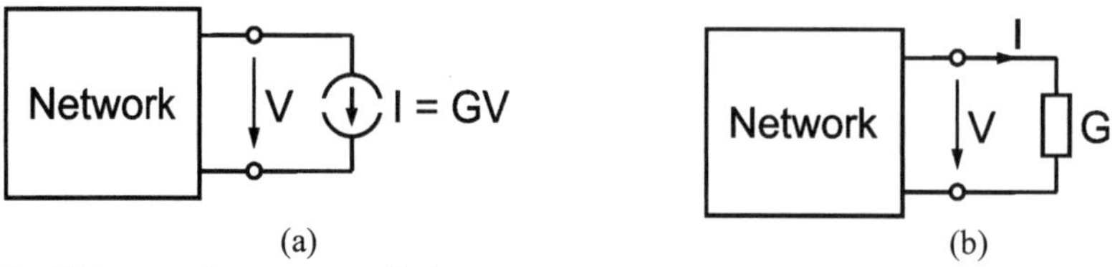

Figure 22.6 Linear voltage-controlled current source connected to a network (a) and its conductance equivalence (b).

same is true if the IV relationship of the source is a more complicated function of voltage: a conductance having the same IV characteristic can be substituted, and the representation are equivalent.

We can now state the substitution theorem precisely: A linear or nonlinear resistive circuit element having the characteristic $I=f(V)$ is equivalent to a controlled current source having the same characteristic, wherein V is the terminal voltage. Although this definition refers to large-signal V and I, the substitution theorem is equally applicable to a small-signal incremental characteristic. It is also applicable by analogy to capacitive elements of current-controlled elements.

One important application of the substitution theorem is shown in Figure 22.7, where the IV characteristic of a nonlinear conductance is described by the power series $I=I(V)=G_1V + G_2V^2 + G_3V^3 + \dots$. The nonlinear element can be described by an equivalent circuit that includes a linear conductance G_1 and controlled current sources representing the higher-degree terms in the series. Of course, the linear component G_1V could also be represented by a current source if it were more convenient to do so.

22.3 NONLINEAR MODEL CONDUCTANCE

A nonlinear conductive element, whether a controlled source or two-terminal, can be described by an IV function having the form $I = I(V_1, V_2, \dots)$. The notation of a current-controlled element, $V= V(I_1, I_2, \dots)$ is rarely used [5].

22.3.1 Single-Controlled Model Conductance

Figure 22.8(a) illustrates a nonlinear voltage-controlled conductance and its IV characteristic. In a two-terminal voltage-controlled conductance, the current $I(t)$ is a function of the instantaneous voltage $V(t)$. Its voltage-current characteristics can then be expressed as

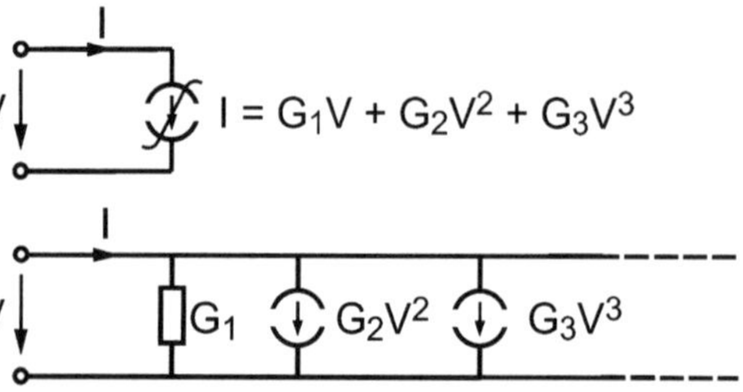

Figure 22.7 shows a nonlinear voltage-controlled current source and its nonlinear equivalence.

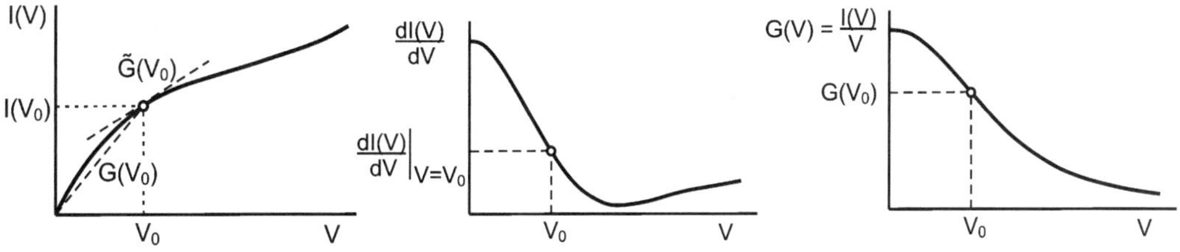

(a) (b) (c)

Figure 22.8 Various presentations of a nonlinear conductive element: (a) $I(V)$ characteristic (nonlinear (DC) conductance $G(V)$ and small-signal conductance $\tilde{G}(V)$ at a bias point V_0 included), (b) incremental conductance $\mathrm{d}I(V)/\mathrm{d}V$ corresponding to small-signal conductance $\tilde{G}(V)$, and (c) nonlinear conductance $G(V)$.

$$I(t) = I(V(t)) \qquad \text{for all } t \tag{22.3}$$

We assume that $V(t)$ is composed by a DC control voltage V_0 (bias voltage) and an incremental voltage $\Delta V(t)$. We can expand the current in a Taylor series around V_0 to determine its small AC part [5]:

$$I(V_0 + \Delta V) = I(V_0) + \left.\frac{dI(V)}{dV}\right|_{V=V_0} \Delta V + \left.\frac{d^2I(V)}{dV^2}\right|_{V=V_0} \frac{\Delta V^2}{2!} + \left.\frac{d^3I(V)}{dV^3}\right|_{V=V_0} \frac{\Delta V^3}{3!} + \ldots \tag{22.4}$$

where the indication of time dependence (t) has been deleted from $\Delta V(t)$ for simplicity. We can assume that $\Delta V \ll V_0$ and that the nonlinearity is weak enough so that the series converges. Then the small-signal current ΔI is

$$\Delta I = I(V_0 + \Delta V) - I(V_0) = \left.\frac{dI(V)}{dV}\right|_{V=V_0} \Delta V + \left.\frac{d^2I(V)}{dV^2}\right|_{V=V_0} \frac{\Delta V^2}{2!} + \left.\frac{d^3I(V)}{dV^3}\right|_{V=V_0} \frac{\Delta V^3}{3!} + \ldots \tag{22.5}$$

As pointed out in [5] in (22.5) the incremental current ΔI is the total small-signal current, including DC as well as small-signal AC components because the even-degree terms in (22.5) introduce them. Under the stated assumptions, the change in the DC operating point is small compared to $I(V_0)$, so the DC components resulting from ΔV^2, ΔV^4, ... are usually negligible. In the quasi-linear case, the terms of degree greater than one are assumed to be negligible, so

$$\frac{\Delta I}{\Delta V} \cong \left.\frac{dI(V)}{dV}\right|_{V=V_0} = \left.\frac{\tilde{I}}{\tilde{V}}\right|_{V=V_0} = \tilde{G}(V_0) \tag{22.6}$$

where $\mathrm{d}I(V)/\mathrm{d}V$ is the incremental conductance at the bias voltage V_0. The small-signal $\tilde{G}(V_0)$ describes a corresponding linear conductance at V_0 that has the same value. Clearly, $\tilde{G}(V_0)$ depends only on the voltage at which the nonlinear resistor is biased. Generally, in the following, quantities with a tilde reveal quasilinear operation with, typically, sinusoidal time-varying signals. The graphical interpretation of (22.6) is included in Figure 22.8(a). Figure 22.9 indicates the symbolic representation.

However, the nonlinear terms in (22.4) indicate that harmonics and intermodulation

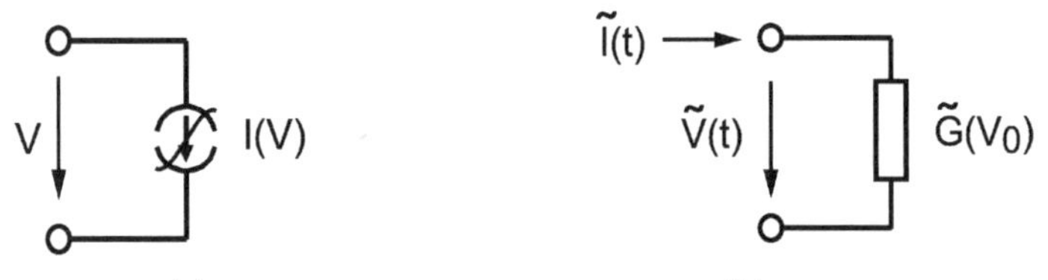

Figure 22.9 Symbolic representation of a nonlinear voltage controlled current source (a) and its linearized small-signal correspondence (b).

products are generated in a nonlinear resistor even under a small-signal operation condition [6]. In the nonlinear case, the total small-signal current ΔI can be expressed as [5]

$$\Delta I = \tilde{G}_1 \Delta V + \tilde{G}_2 \Delta V^2 + \tilde{G}_3 \Delta V^3 + \ldots \tag{22.7}$$

Figure 22.10 shows the model for the nonlinear conductance under small-signal operation condition. The linear term, $\tilde{G}_1$, in (22.7) is the small-signal conductance. The small-signal element $\tilde{G}(V)$ is obtained experimentally as function of external bias voltages by using standard curve-fitting techniques to model sets of small-signal device S-parameters measured over ranges of microwave frequencies and different bias conditions. With (22.6),

$$dI(V) = \tilde{G}(V)dV \tag{22.8}$$

the current $I(V)$ is obtained as the integral over the small-signal conductance $\tilde{G}(V)$:

$$I(V) = \int_V \tilde{G}(V)dV \tag{22.9}$$

The integration reference point frequently agrees with the bias point. It can, however, also be chosen in any other point of the IV plane to reduce the computing time for simulation, using, for example table-based FET models. In this case the nonlinear response comprises line integrals over the bias dependent small-signal elements from the defined integration reference point and the instantaneous control voltage values. First we regard the point of origin as the integration reference point. Equation (22.9) then reads explicitly

$$I(V) = \int_0^V \tilde{G}(V)dV' \tag{22.10}$$

When an integration point $V_0 > 0$ is chosen, then the line integral can be split into two partial integrals:

$$I(V) = \int_0^{V_0} \tilde{G}(V)dV' + \int_{V_0}^{V} \tilde{G}(V)dV' \tag{22.11}$$

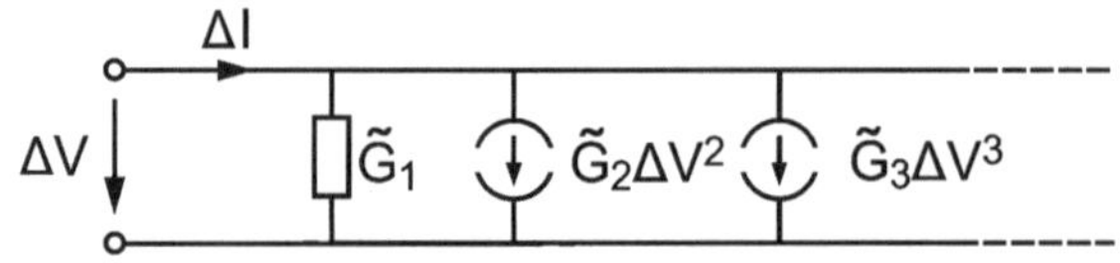

Figure 22.10 Small-signal nonlinear equivalent circuit of the nonlinear conductance.

The first integral term describes a constant reference current I_0,

$$I(V_0) = \int_0^{V_0} \tilde{G}(V) dV' = I_0 \tag{22.12}$$

Rauscher [7, 8] expressed the instantaneous current through a nonlinear conductance as the product of the instantaneous element value and the instantaneous value of the voltage directly associated with the element:

$$I(t) = G\big(V(t)\big) \cdot V(t) \tag{22.13}$$

where G stands symbolically for a nonlinear (quasi-static DC) conductance as depicted in Figure 22.8(c), and $V(t)$ represents the associated voltage for the nonlinear element. In the expression for $G(V)$ the time variable does not appear explicitly. Any time dependence of the instantaneous element value is, therefore, solely due to the time dependence of V. A nonlinear circuit element of this kind is termed *time invariant*. The postulate that the model element be time invariant is fundamental to the quasi-static approach.

Thus, a nonlinear conductance $G(V)$ may be defined as

$$G(V) = \frac{I(V)}{V} \tag{22.14}$$

which is the ratio of current and voltage at a given voltage value, as illustrated in Figure 22.8(c). A relationship between the nonlinear conductance $G(V)$ and small-signal linear conductance $\tilde{G}(V)$ is simply derived from (22.14) by differentiation,

$$\frac{dI(V)}{dV} = G(V) + \frac{dG(V)}{dV} V \tag{22.15}$$

With $dI(V)/dV = \tilde{G}(V)$, (22.15) reads

$$\tilde{G}(V) = G(V) + V \frac{dG(V)}{dV} \tag{22.16}$$

Thus, the small-signal conductance $\tilde{G}(V)$ and the nonlinear quasi-DC conductance $G(V)$ are related through an ordinary differential equation (or, respectively, a partial differential equation in the case of two-dimensional conductance).

Figure 22.11 illustrates a nonlinear quasi-DC conductance $G(V)$ and its linearized small-signal conductance $\tilde{G}(V)$ at a bis point V_0.

As discussed by Rauscher, the static function $G(V)$ can be derived from the experimentally determined bias dependence of the small-signal conductance. To solve the equation, $\tilde{G}(V)$ is approximated by a polynomial expression of the form

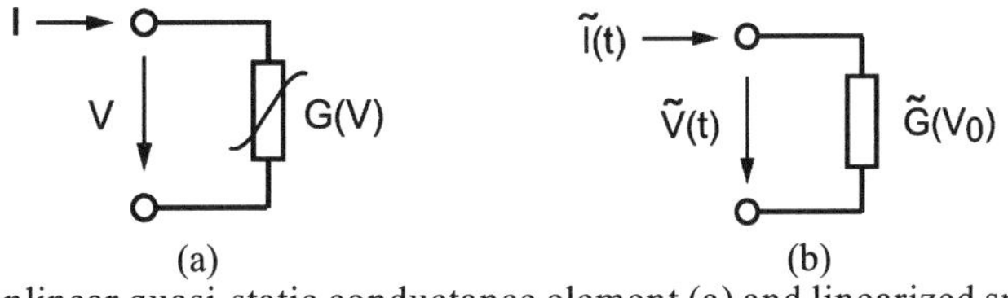

Figure 22.11 Nonlinear quasi-static conductance element (a) and linearized small-signal conductance at bias point V_0 (b).

$$\tilde{G}(V) = \sum_{k=1} \gamma_k V^{k-1} \tag{22.17}$$

A similar polynomial expression is defined for $G(V)$,

$$G(V) = \sum_{k=1}^{n} c_k V^{k-1} \tag{22.18}$$

Applying (22.16) delivers

$$\tilde{G}(V) = \sum_{k=1}^{n} c_k k V^{k-1} \tag{22.19}$$

By coefficient comparison with (22.17) we get $c_k = \gamma_k/k$. Substituted in (22.19) the nonlinear static conductance is readily obtained to be

$$G(V) = \sum_{k=1} \frac{\gamma_k}{k} V^{k-1} \tag{22.20}$$

22.3.2 Multicontrolled Model Conductance

Often a nonlinear conductive model element is controlled by more than one voltage. This is, for example, valid for the drain current in a FET equivalent circuit, which depends not only on the drain terminal voltage (direct control) but also on the gate voltage (indirect control). Thus, assuming $I = I(V_1, V_2)$, the current can be expanded in a two-dimensional Taylor series. After subtracting the DC current component, the following expression is obtained [5]:

$$\begin{aligned} \Delta I = {} & \frac{\partial I(V_1,V_2)}{\partial V_1}\Delta V_1 + \frac{\partial I(V_1,V_2)}{\partial V_2}\Delta V_2 \\ & + \frac{1}{2}\left(\frac{\partial^2 I(V_1,V_2)}{\partial V_1^2}\Delta V_1^2 + 2\frac{\partial^2 I(V_1,V_2)}{\partial V_1 \partial V_2}\Delta V_1 \Delta V_2 + \frac{\partial^2 I(V_1,V_2)}{\partial V_2^2}\Delta V_2^2\right) \\ & + \frac{1}{6}\left(\frac{\partial^3 I(V_1,V_2)}{\partial V_1^3}\Delta V_1^3 + 3\frac{\partial^3 I(V_1,V_2)}{\partial V_1 \partial V_2^2}\Delta V_1 \Delta V_2^2 + 3\frac{\partial^3 I(V_1,V_2)}{\partial V_1^2 \partial V_2}\Delta V_1^2 \Delta V_2 + \frac{\partial^3 I(V_1,V_2)}{\partial V_2^3}\Delta V_2^3\right) + \ldots \end{aligned} \tag{22.21}$$

where the derivatives are evaluated at the bias point of V_1 and V_2, V_{10} and V_{20}, respectively. In the small-signal, quasi-linear case, the high-degree terms in (2.10) are neglected, which results in

$$\begin{aligned} \tilde{I} &= \left.\frac{\partial I(V_1,V_2)}{\partial V_1}\right|_{V_{10},V_{20}} \tilde{V}_1 + \left.\frac{\partial I(V_1,V_2)}{\partial V_2}\right|_{V_{10},V_{20}} \tilde{V}_2 \\ &= \tilde{G}_1(V_{10},V_{20})\tilde{V}_1 + \tilde{G}_2(V_{10},V_{20})\tilde{V}_2 \end{aligned} \tag{22.22}$$

where $\tilde{G}_1(V_1, V_2)$ and $\tilde{G}_2(V_1, V_2)$ denote, respectively, the direct and indirect controlled small-signal conductances at port 1, as illustrated in Figure 22.12.

As already mentioned, a practical example is the drain current source in a FET equivalent circuit shown in Figure 22.13. In this case, the drain current I is a function of both the gate voltage V_{gs} (indirect control voltage) and the (direct controlled or terminal) drain voltage V_{ds};

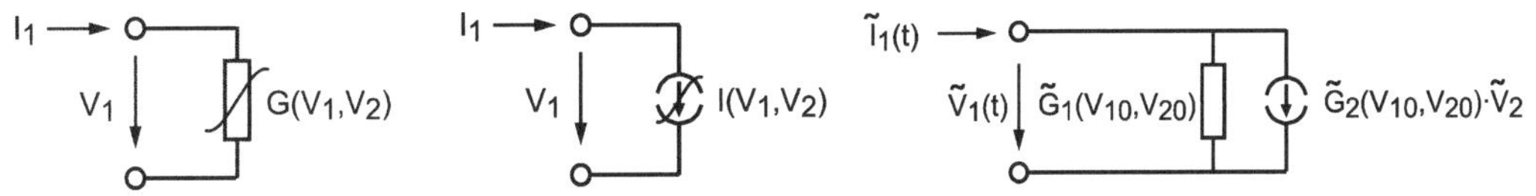

Figure 22.12 Twofold controlled nonlinear current source $I(V_1, V_2)$ and its small-signal equivalence.

thus, $I = I(V_{gs}, V_{ds})$. In this case, V_{ds} is applied to the current source and V_{gs} is a node voltage elsewhere in the circuit.

Referring to (22.22), the small-signal drain current can be written as

$$\tilde{I}_{ds} = \tilde{G}_m(V_{gs0}, V_{ds0})\tilde{V}_{gs} + \tilde{G}_{ds}(V_{gs0}, V_{ds0})\tilde{V}_{ds} \tag{22.23}$$

$\tilde{G}_{ds}(V_{gs0}, V_{ds0})$ denotes the bias-dependent small-signal output conductance of the FET device. The indirect controlled small-signal element $\tilde{G}_m(V_{gs0}, V_{ds0})$ is termed transconductance of the nonlinear current source.

In an extension to (22.13) a nonlinear two-dimensional conductance $G(V_1, V_2)$ could be defined [7] so that the instantaneous current through this nonlinear conductance (or transconductance) element is expressed in the form:

$$I(t) = G\left[V_1(t), V_2(t)\right]\cdot V(t) \tag{22.24}$$

Discarding the time dependence, (11.1x) may be expanded in a Taylor series in the bias point (V_{10}, V_{20}):

$$\begin{aligned} I &= I(V_1,V_2) = G(V_1,V_2)V_1 \\ &= G(V_{10},V_{20})V_{10} + \left.\frac{\partial\left[G(V_1,V_2)V_1\right]}{\partial V_1}\right|_{V_{10},V_{20}}\tilde{V}_1 + \left.\frac{\partial\left[G(V_1,V_2)V_1\right]}{\partial V_2}\right|_{V_{10},V_{20}}\tilde{V}_2 \\ &= G(V_{10},V_{20})V_{10} + \left[G(V_{10},V_{20}) + V_1\left.\frac{\partial G(V_1,V_2)}{\partial V_1}\right|_{V_{10},V_{20}}\right]\tilde{V}_1 \\ &\quad + V_1\left.\frac{\partial G(V_1,V_2)}{\partial V_2}\right|_{V_{10},V_{20}}\tilde{V}_2 \end{aligned} \tag{22.25}$$

Coefficient comparison with (22.23) yields

$$\tilde{G}(V_{10},V_{20}) = G(V_{10},V_{20}) + V_{10}\left.\frac{\partial G(V_1,V_2)}{\partial V_1}\right|_{V_{10},V_{20}} \tag{22.26}$$

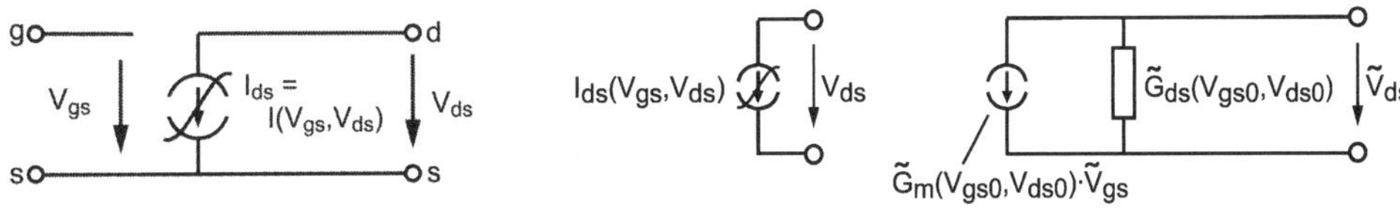

Figure 22.13 Nonlinear FET drain current source $I_{ds} = I(V_{gs}, V_{ds})$ as an example for twofold controlled nonlinear element.

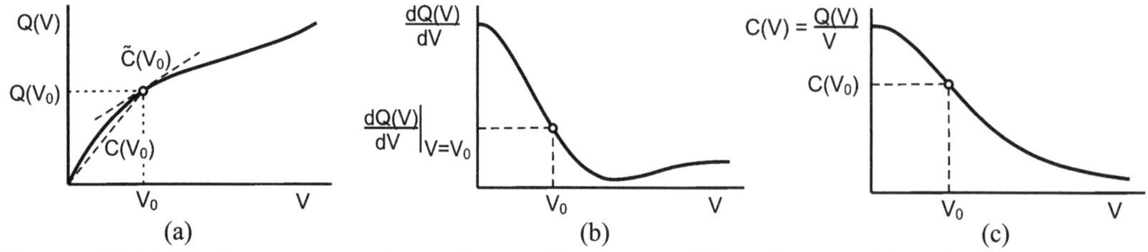

Figure 22.14 Various presentations of a nonlinear capacitive element: (a) $Q(V)$ characteristic (nonlinear (DC) capacitance $C(V)$ and small-signal conductance $\tilde{C}(V)$ at a bias point V_0 included), (b) incremental capacitance $dQ(V)/dV$ corresponding to small-signal capacitance $\tilde{C}(V)$, and (c) nonlinear capacitance $C(V)$.

$$\tilde{G}_m(V_{10},V_{20}) = V_{10}\frac{\partial G(V_1,V_2)}{\partial V_2}\bigg|_{V_{10},V_{20}} \tag{22.27}$$

where $\tilde{G}(V_{10}, V_{20})$ is the bias-dependent (directly controlled) small-signal conductance, and $\tilde{G}_m(V_{10}, V_{20})$ is the bias-dependent (indirectly controlled) small-signal transconductance. Figure 22.12(a) illustrates the circuit symbol for a two-dimensional nonlinear quasi-static conductance and Figure 22.12(c) its small-signal approximation in a bias point (V_{10}, V_{20}).

22.4 NONLINEAR MODEL CAPACITANCE

A nonlinear capacitive element is described by a Q-V function having the form $Q = Q(V_1,V_2,...)$.

22.4.1 Single-Controlled Model Capacitance

Figure 22.14(a) illustrates a nonlinear voltage-controlled capacitor and its Q-I characteristic. The charge $Q(t)$ is a function of the instantaneous voltage $V(t)$ across its terminals. The charge-voltage characteristic can therefore be expressed as

$$Q(t) = Q(V(t)) \quad \text{for all } t \tag{22.28}$$

Similar to (22.4), equation (22.28) can be expanded in a Taylor series near V_0 as

$$Q(V_0 + \Delta V) = Q(V_0) + \frac{dQ}{dV}\Delta V + \frac{1}{2}\frac{d^2Q}{dV^2}\Delta V^2 + \frac{1}{6}\frac{d^3Q}{dV^3}\Delta V^3 + ... \tag{22.29}$$

or

$$\Delta Q = Q(V_0 + \Delta V) - Q(V_0) = \frac{dQ}{dV}\Delta V + \frac{1}{2}\frac{d^2Q}{dV^2}\Delta V^2 + \frac{1}{6}\frac{d^3Q}{dV^3}\Delta V^3 + ... \tag{22.30}$$

where all derivatives are computed at V_0, and therefore are not functions of ΔV. The incremental current ΔI through the capacitor is given by

$$\Delta I = \frac{d(\Delta Q)}{dt} = \left(\frac{dQ}{dV} + \frac{d^2Q}{dV^2}\Delta V + \frac{1}{2}\frac{d^3Q}{dV^3}\Delta V^2 + ...\right)\frac{d(\Delta V)}{dt} \tag{22.31}$$

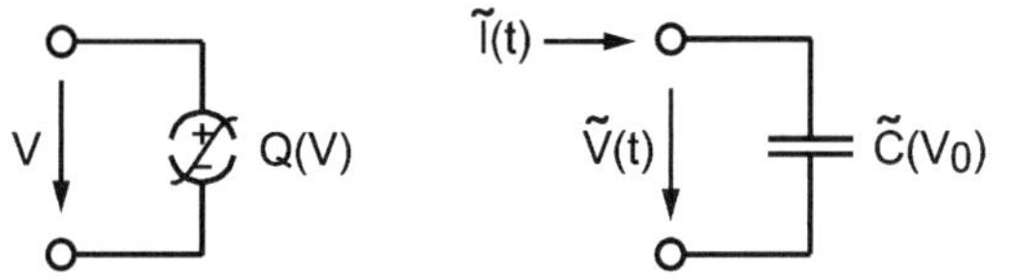

Figure 22.15 Nonlinear charge source $Q(V)$ and its linearized small-signal capacitance $\tilde{C}(V_0)$ at bias point V_0.

Assuming a quasilinear operation condition, the terms of degree greater than one are assumed to be negligible, so

$$\frac{\Delta Q}{\Delta V} \cong \left.\frac{dQ(V)}{dV}\right|_{V=V_0} = \left.\frac{\tilde{Q}}{\tilde{V}}\right|_{V=V_0} = \tilde{C}(V_0) \tag{22.32}$$

where $dQ(V)/dV$ is the incremental capacitance at the bias voltage V_0. The small-signal $\tilde{C}(V_0)$ describes a corresponding linear capacitance at V_0 that has the same value. Clearly, $\tilde{C}(V_0)$ depends only on the voltage at which the nonlinear capacitor is biased. A graphical interpretation of (22.32) is included in Figure 22.16(a). Figure 22.17 indicates the symbolic representation of a nonlinear voltage controlled charge source and its linearized small-signal capacitance $\tilde{C}(V_0)$ at bias point V_0.

The current through the capacitor is found by taking the time-derivative of the charge upon the capacitor:

$$I(t) = \frac{dQ(t)}{dt} = \frac{dQ(V(t))}{dV}\cdot\frac{dV(t)}{dt} = \tilde{C}(V(t))\frac{dV(t)}{dt} \tag{22.33}$$

We assume that $V(t)$ is composed by a DC control voltage V_0 (bias voltage) and an incremental voltage $\Delta V(t)$. We can expand the current in a Taylor series around V_0 to determine its small AC part [5]:

$$I(V_0 + \Delta V) = I(V_0) + \frac{d}{dt}\left(\left.\frac{dQ(V)}{dV}\right|_{V=V_0}\Delta V + \left.\frac{d^2Q(V)}{dV^2}\right|_{V=V_0}\frac{\Delta V^2}{2!} + \left.\frac{d^3Q(V)}{dV^3}\right|_{V=V_0}\frac{\Delta V^3}{3!} + \ldots\right) \tag{22.34}$$

where the indication of time dependence (t) has been deleted from $\Delta V(t)$ for simplicity. We can assume that $\Delta V \ll V_0$ and that the nonlinearity is weak enough so that the series converges. Then the small-signal current ΔI is

$$\Delta I = I(V_0 + \Delta V) - I(V_0) = \frac{d}{dt}\left(\left.\frac{dQ(V)}{dV}\right|_{V=V_0}\Delta V + \left.\frac{d^2Q(V)}{dV^2}\right|_{V=V_0}\frac{\Delta V^2}{2!} + \left.\frac{d^3Q(V)}{dV^3}\right|_{V=V_0}\frac{\Delta V^3}{3!} + \ldots\right) \tag{22.35}$$

In the quasilinear case, the terms of degree greater than one are assumed to be negligible, so

$$\frac{\Delta I}{d/dt(\Delta V)} \cong \left.\frac{dQ(V)}{dV}\right|_{V=V_0} = \left.\frac{\tilde{I}}{d\tilde{V}/dt}\right|_{V=V_0} = \tilde{C}(V_0) \tag{22.36}$$

where $dQ(V)/dV$ is the incremental capacitance at the bias voltage V_0. The small-signal capacitance $\tilde{C}(V_0)$ describes a corresponding linear capacitance at V_0 which has the same value. Clearly, $\tilde{C}(V_0)$ depends only on the voltage at which the nonlinear capacitor is biased.

Thus, under small-signal operation condition the capacitive current reads

$$\tilde{I} = \tilde{C}(V_0)\cdot d\tilde{V}/dt \tag{22.37}$$

As the small-signal capacitance is exactly equal to the incremental capacitance at the bias voltage, this fact can be used in parameter determination since the small-signal capacitor is simpler to measure. With (22.36)

$$dQ(V) = \tilde{C}(V)dV \tag{23.38}$$

the charge $Q(V)$ is obtained as the integral over the small-signal capacitance $\tilde{C}(V)$:

$$Q(V) = \int_V \tilde{C}(V)dV \tag{23.39}$$

The integration reference point frequently agrees with the bias point. It can, however, also be chosen in any other point of the Q-V plane to reduce the computing time for simulation, using e.g. table-based FET models. In this case the the nonlinear response comprises line integrals over the bias dependent small-signal elements from the defined integration reference point and the instantaneous control voltage values. First we regard the point of origin as the integration reference point. Equation (22.39) then reads explicitly

$$Q(V) = \int_0^V \tilde{C}(V)dV' \tag{22.40}$$

When an integration point $V_0 > 0$ is chosen, then the line integral can be split into two partial integrals:

$$Q(V) = \int_0^{V_0} \tilde{C}(V)dV' + \int_{V_0}^{V} \tilde{C}(V)dV' \tag{22.41}$$

The first integral term describes a constant reference charge Q_0,

$$Q(V_0) = \int_0^{V_0} \tilde{C}(V)dV' = Q_0 \tag{22.42}$$

A nonlinear (DC) capacitance may also be defined as $C(V) = Q(V)/V$, which is the ratio of charge and voltage at a given voltage value, as illustrated in Figure 22.14(c). Figure 22.16 illustrates the model of a nonlinear capacitance and its linearized small-signal element. The current through the capacitor is given as

$$\begin{aligned} I(V(t)) &= \frac{dQ(V(t))}{dt} = \frac{d[C(V(t))\cdot V(t)]}{dt} = C(V(t))\frac{dV(t)}{dt} + V(t)\frac{dC(V)}{dV}\cdot\frac{d(V(t)}{dt} \\ &= \left[C(V(t)) + V(t)\frac{dC(V(t))}{dV(t)}\right]\frac{dV(t)}{dt} \end{aligned} \tag{22.43}$$

Comparison with (22.33) provides a relationship between the nonlinear (DC) capacitance

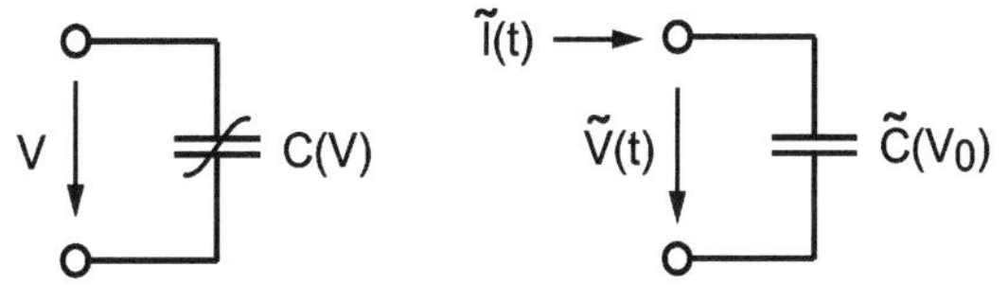

Figure 22.16 Nonlinear quasi-static capacitor (Rauscher) and its small-signal approximation.

$C(V)$ and the small-signal capacitance $\tilde{C}(V)$,

$$C(V) + V\frac{dC(V)}{dV} = \tilde{C}(V) \tag{22.44}$$

The calculation of the large-signal current through a capacitor is sometimes a source of confusion whether taking the nonlinear or small-signal capacitance value. However, as has become evident by (22.44), both definitions [equations (22.33) and (22.43)] lead to the same result. In any case, computing the current directly as the derivative of the charge and bypassing the definition of capacitance avoids any confusion.

22.4.2 Multicontrolled Model Capacitance

In addition to the voltage V_1 across its terminals, the charge Q is now assumed to be also dependent on a voltage V_2 elsewhere in the circuit. Then we obtain from Taylor series expansion (similar to (22.21)):

$$\begin{aligned}\Delta Q = {} & \frac{\partial Q(V_1,V_2)}{\partial V_1}\Delta V_1 + \frac{\partial Q(V_1,V_2)}{\partial V_2}\Delta V_2 \\ & + \frac{1}{2}\left(\frac{\partial^2 Q(V_1,V_2)}{\partial V_1^2}\Delta V_1^2 + 2\frac{\partial^2 Q(V_1,V_2)}{\partial V_1 \partial V_2}\Delta V_1 \Delta V_2 + \frac{\partial^2 Q(V_1,V_2)}{\partial V_2^2}\Delta V_2^2\right) \\ & + \frac{1}{6}\left(\frac{\partial^3 Q(V_1,V_2)}{\partial V_1^3}\Delta V_1^3 + 3\frac{\partial^3 Q(V_1,V_2)}{\partial V_1 \partial V_2^2}\Delta V_1 \Delta V_2^2 + 3\frac{\partial^3 Q(V_1,V_2)}{\partial V_1^2 \partial V_2}\Delta V_1^2 \Delta V_2 + \frac{\partial^3 Q(V_1,V_2)}{\partial V_2^3}\Delta V_2^3\right) + \ldots\end{aligned} \tag{22.45}$$

where the derivatives are evaluated at the bias point of V_1 and V_2, V_{10} and V_{20}, respectively. In the small-signal, quasilinear case, the high-degree terms in (22.45) are neglected, which results in

$$\Delta Q = \left.\frac{\partial Q(V_1,V_2)}{\partial V_1}\right|_{V_{10},V_{20}}\Delta V_1 + \left.\frac{\partial Q(V_1,V_2)}{\partial V_2}\right|_{V_{10},V_{20}}\Delta V_2 \tag{22.46}$$

With

$$\tilde{C}_1(V_{10},V_{20}) = \left.\frac{\partial Q(V_1,V_2)}{\partial V_1}\right|_{V_{10},V_{20}} \tag{22.47}$$

and

$$\tilde{C}_2(V_{10},V_{20}) = \left.\frac{\partial Q(V_1,V_2)}{\partial V_2}\right|_{V_{10},V_{20}} \tag{22.48}$$

(22.46) reads

$$\Delta Q = \tilde{C}_1(V_{10},V_{20})\Delta V_1 + \tilde{C}_2(V_{10},V_{20})\Delta V_2 \tag{22.49}$$

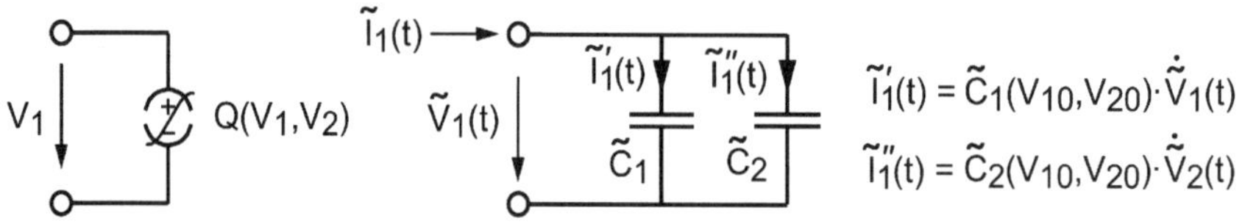

Figure 22.17 (a) Nonlinear capacitor with double controlled charge $Q(V_1,V_2)$, and (b) its linearized small-signal model at a bias point (V_{10}, V_{20}) with terminal capacitance $\tilde{C}_1(V_{10}, V_{20})$ and transcapacitance $\tilde{C}_2(V_{10}, V_{20})$.

$\tilde{C}_1(V_{10}, V_{20})$ and $\tilde{C}_2(V_{10}, V_{20})$ denote the incremental expansion terms in the considered bias point (V_{10}, V_{20}). Assuming very small incremental voltages ΔV_1 and ΔV_2, the $Q(V_1, V_2)$ function around the operating point can be approximated to be linear with respect to the voltages so that (22.49) can be written under small-signal operation condition as (see Figure 22.17)

$$\tilde{Q} = \tilde{C}_1(V_{10},V_{20})\tilde{V}_1 + \tilde{C}_2(V_{10},V_{20})\tilde{V}_2 \tag{22.50}$$

Thus, the nonlinear charge source is split into two capacitive parts. $\tilde{C}_1(V_{10}, V_{20})$ represents a direct controlled capacitance (its value depends only on the terminal voltage), whereas $\tilde{C}_2(V_{10}, V_{20})$ is under control of a voltage elsewhere in the circuit. Terms involving voltages other than the capacitor's terminal voltages are usually called transcapacitances and are frequently denoted by C_m.

The current through a nonlinear capacitor characterized by $Q = Q(V_1, V_2, \ldots)$ is calculated by taking the time-derivative of the charge upon the capacitor:

$$I(t) = \frac{dQ\big(V_1(t),V_2(t)\big)}{dt} = \frac{\partial Q(V_1,V_2)}{\partial V_1}\frac{\partial V_1(t)}{\partial t} + \frac{\partial Q(V_1,V_2)}{\partial V_2}\frac{\partial V_2(t)}{\partial t} + \ldots \tag{22.51}$$

Restriction to double controlled nonlinear capacitor results in

$$I(t) = C_1(V_1,V_2)\frac{\partial V_1(t)}{\partial t} + C_2(V_1,V_2)\frac{\partial V_2(t)}{\partial t} \tag{22.52}$$

Under the small-signal operation condition we get with (22.50):

$$\tilde{I}(t) = \frac{d\tilde{Q}(t)}{dt} = \tilde{C}_1(V_{10},V_{20})\frac{d\tilde{V}_1}{dt} + \tilde{C}_2(V_{10},V_{20})\frac{d\tilde{V}_2}{dt} = \tilde{I}_1'(t) + \tilde{I}_1''(t) \tag{22.53}$$

As for the single controlled capacitor, the double controlled nonlinear (DC) capacitance may also be defined as $Q(V_1, V_2) = C(V_1, V_2) \cdot V_1$, where V_1 may be the voltage drop upon the capacitor and V_2 the remote control voltage.

Taylor series expansion around the bias point (V_{10}, V_{20}) delivers for the incremental charge $\Delta Q(V_{10}, V_{20})$

$$\Delta Q(V_{10},V_{20}) = \left(C(V_{10},V_{20}) + V_1 \left.\frac{\partial C(V_1,V_2)}{\partial V_1}\right|_{V_{10},V_{20}} \right)\Delta V_1 + V_1 \left.\frac{\partial C(V_1,V_2)}{\partial V_2}\right|_{V_{10},V_{20}} \Delta V_2 \tag{22.54}$$

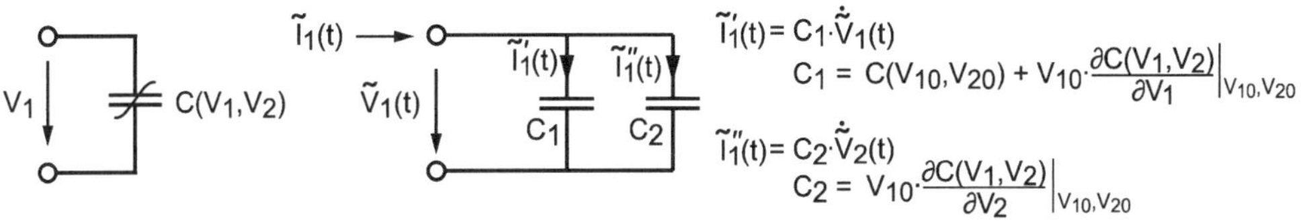

Figure 22.18 (a) Nonlinear double controlled capacitor $C(V_1,V_2)$, and (b) its linearized small-signal equivalent circuit.

Comparison with (22.49) delivers a relationship between the nonlinear capacitance $C(V_1, V_2)$ in a bias point and corresponding small-signal capacitances:

$$\tilde{C}_1(V_{10},V_{20}) = C(V_{10},V_{20}) + V_{10}\left.\frac{\partial C(V_1,V_2)}{\partial V_1}\right|_{V_{10},V_{20}} \tag{22.55}$$

$$\tilde{C}_2(V_{10},V_{20}) = V_{10}\left.\frac{\partial C(V_1,V_2)}{\partial V_2}\right|_{V_{10},V_{20}} \tag{22.56}$$

Based on the definition that the current is the time derivative of the product of nonlinear capacitance and terminal voltage, then the small-signal current $\tilde{I}(t)$ turns out to be

$$\tilde{I}(t) = \left(C(V_{10},V_{20}) + V_{10}\left.\frac{\partial C(V_1,V_2)}{\partial V_1}\right|_{\substack{V_1=V_{10}\\ V_2=V_{20}}} \right)\tilde{V}_1 + V_{10}\left.\frac{\partial C(V_1,V_2)}{\partial V_2}\right|_{\substack{V_1=V_{10}\\ V_2=V_{20}}}\tilde{V}_2 \tag{22.57}$$

22.5 SUMMARY

In this chapter basic terms frequently used in connection with transistor modeling were discussed. This concerns the types of model elements in equivalent circuits of field-effect transistor devices, the modes of operation, and the definition of voltages of time-varying signal waveforms. One- and two-dimensional nonlinear current and charge sources were described and their interrelation to small-signal model elements provided.

REFERENCES

[1] Kompa, G., *Practical Microstrip Design and Applications*, Norwood: Artech House, Inc., 2005.

[2] Kompa, G., *Basic Properties of III-V Devices–Understanding Mysterious Trapping Phenomena*, Kassel, Kassel University Press GmbH, 2014.

[3] Rauscher, C., and H. A. Willing, "Simulation of Nonlinear Microwave FET Performance Using a Quasi-Static Model," *IEEE Trans.*, Vol. MTT-27(10), 1979, pp. 834–840.

[4] Keysight Technologies, *Pulsed-RF S-Parameter Measurements with the PNA Microwave Network Analyzers Using Wideband and Narrowband Detection*, Application Note, July 31, 2014.

[5] Maas, S. A., *Nonlinear Microwave and RF Circuits*, 2nd Ed., Artech House, Inc. 2003.

[6] Rodrigues, P. J. C., *Computer-Aided Analysis of Nonlinear Microwave Circuits*, Norwood: Artech House, Inc., 1998.

[7] Willing, H. A., C. Rauscher, and P. de Santis, "A Technique for Predicting Large-Signal Performance of a GaAs MESFET," *IEEE Trans.,* Vol. MTT-26, 1978, pp. 1017–1023.

Chapter 23

Non-Quasi-Static Transistor Model

23.1 INTRODUCTION

In the following a nonlinear non-quasi-static field-effect transistor model is developed successively following the flowchart in Figure 23.1. The *static* or DC model refers ideally to zero frequency operation. Extrinsic model parameters, the extrinsic transconductance G_M, and extrinsic drain-source (channel) conductance G_{DS} can be obtained by DC measurement of the IV characteristics. Knowing the parasitic series resistances R_g, R_s, and R_d (see Chapter 6) the corresponding intrinsic parameters G_m and G_{ds} can be determined. Although the DC model parameters are strictly valid only for the DC case, traditional (popular) nonlinear DC models discussed in Chapter 25 have also been applied with more or less success beyond this theoretical limit. Note that several classical models have been revised over time, implementing dispersion effects, for example, by adding a simple R–C series branch in parallel to the output port of the device, which makes the output conductance frequency dependent [1–3].

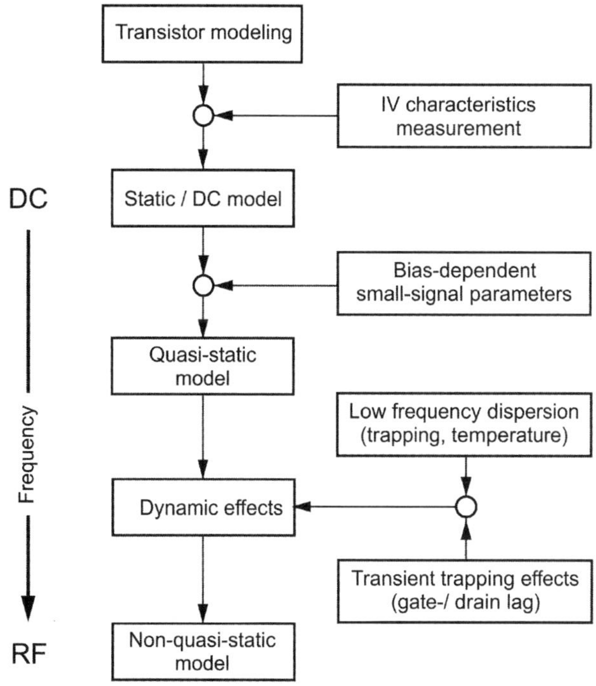

Figure 23.1 Schematic illustration of conceptual non-quasi-static transistor model design.

The *quasi-static* mode of operation was already discussed in the previous chapter. The main feature is characterized by the assumption that the control voltages vary sufficiently slowly so that at any time the device remains in electrical and thermal equilibrium. Dynamic trapping or dynamic thermal phenomena (memory effects) do not yet have any significant impact on the device performance. Nonlinear models are derived from the measurement of S-parameters in a large number of bias points distributed over the entire range of the I–V output plane. The nonlinear model elements are obtained by integration over the bias-dependent small-signal model elements. With increasing operation frequency electrical and thermal dynamic effects occur, which have an adverse impact on the device performance such as current collapse [4]. Models that include electrical and thermal trapping effects are also termed nonlinear *electrothermal* transistor models. Additionally, DC-pulsed characterization of transistors have become very important for the drain current source characterization. It has been found that gate- and drain-lag dispersion phenomena [5–7] must be taken into account to complete the electro-thermal transistor model [8]. Thus, recent transistor modeling concepts include gate- and drain-lag submodels with separate trapping and detrapping functions. Indeed, the charging and decharging time constants of trap centers are very different; trapping has proven much faster than the detrapping process [9]. Nevertheless, the implementation of such differentiated models entails greater effort so that frequently a simplified gate-lag (and respective drain-lag) model with a single trapping time constant is still preferred.

23.2 TRANSISTOR BEHAVIORAL MODEL

Compact transistor models are represented by equivalent circuit topologies with diverse model elements. Based on measurements the model element values are most likely to be determined by optimization. In contrast, behavioral models are characterized by mathematical equations with respect to their input/output ports. The detailed physical construction of the device and respective intrinsic electrical behavior is of minor importance.

In the following the constitutive relations of a generalized nonlinear electrical N-port network are derived and applied to the two-port representation of a common source field-effect transistor [10].

23.2.1 Nonlinear N-Port Network

Figure 23.2 illustrates a nonlinear N-port network, which is assumed to describe the nonlinear electrical behavior of a physical device with $2N$ external terminals grouped into N ports. For the characterization of the N-port we choose the voltage-current description. This approach is preferred, when the network consists of lumped elements. This is in particular given for a microwave field-effect transistor under discussion as long as the physical dimensions of the device are much smaller than the operating wavelength.

The ith port ($i = 1, 2, ..., N$) is associated with the port voltage $V_i(t)$ between the terminals $(i - i')$ and the port current $I_i(t)$ entering one terminal of port (i). The voltage and currents are

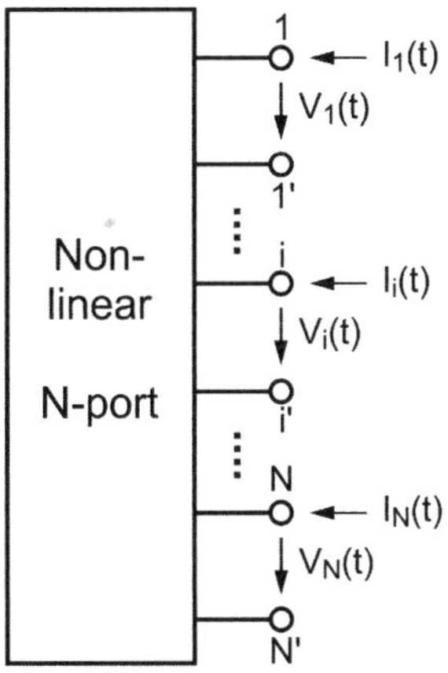

Figure 23.2 Transistor described as nonlinear N-port network.

real-valued functions of time. The nonlinear N-port is uniquely described by $2N$ relations. N equations directly follow from the definition that the sum of currents entering through the two terminals of each port becomes zero. Further N relations, termed the *constitutive relations* of the N-port [11], involve only currents and voltages of the ports.

It is assumed that the N-port current vector $\boldsymbol{I}(t)$ is not only dependent on the N-port voltage vector $\boldsymbol{V}(t)$ but also on its time-derivatives [12, 13]. Thus, we can write

$$\boldsymbol{I}(t) = \boldsymbol{\Phi}\left[\boldsymbol{V}(t), \dot{\boldsymbol{V}}(t), \ddot{\boldsymbol{V}}(t), ...\right] = \boldsymbol{\Phi}\left[h^{(0)}\boldsymbol{V}(t),\ h^{(1)}\boldsymbol{V}(t),\ ...,\ h^{(k)}\boldsymbol{V}(t),\ ...\right] \tag{23.1}$$

with $\boldsymbol{I} = (I_1, I_2, ..., I_N)^T$ and $\boldsymbol{V} = (V_1, V_2, ..., V_N)^T$. The superscript T denotes the transpose of the matrix. $h^{(k)} = d^{(k)}/dt^{(k}$ denotes the time-differential operator of order k. $\boldsymbol{\Phi}$ signifies the N-port constitutive function, which we assume to be time-independent. This fundamental restriction is known as the quasi-static condition (see Section 22.1.2). In this case (23.1) can be written as

$$\boldsymbol{I}(t) = \sum_{k=0}^{K} h^{(k)}\boldsymbol{\Phi}^{(k)}[\boldsymbol{V}(t)] \tag{23.2}$$

taking into account a number of K voltage derivatives of lowest order. $\boldsymbol{\Phi}^{(k)}$ ($k = 0, 1, ..., K$) is the quasi-static N-port constitutive function of kth order. Keeping in mind that the time-differential operator is onlyacting on thevoltage $\boldsymbol{V}(t)$, the total differential of $\boldsymbol{I}(t)$in (23.2) in a fixed reference point $\boldsymbol{V}_0 = (V_{10}, V_{20}, ..., V_{N0})^T$ reads with $\boldsymbol{V}_h(t)$ defined in (F.6), Appendix F:

$$\begin{aligned} d\boldsymbol{I}(\boldsymbol{V}(t)) &= \sum_{k=0}^{K} d\left(h^{(k)}\boldsymbol{\Phi}^{(k)}(\boldsymbol{V}(t))\right) \\ &= \sum_{k=0}^{K} \left.\frac{\partial\boldsymbol{\Phi}^{(k)}}{\partial\boldsymbol{V}}\right|_{\boldsymbol{V}_0} d\boldsymbol{V}_h(t) = \sum_{k=0}^{K} \boldsymbol{\varphi}^{(k)}(\boldsymbol{V}_0)\, d\boldsymbol{V}_h(t) \end{aligned} \tag{23.3}$$

with $\boldsymbol{\varphi}^{(k)}(\boldsymbol{V}_0)$ being the voltage derivative of $\boldsymbol{\Phi}^{(k)}(\boldsymbol{V}_0)$ in the considered quiescent point. In the frequency domain, (23.3) has the form:

$$d\boldsymbol{I}(\omega,\boldsymbol{V}_0) = \sum_{k=0}^{K} (j\omega)^k \boldsymbol{\varphi}^{(k)}(\boldsymbol{V}_0)\, d\boldsymbol{V}(\omega,\boldsymbol{V}_0) \tag{23.4}$$

This relation corresponds to the small-signal current-voltage representation of a nonlinear N-port in the referred bias point, which is characterized by the well-known frequency-dependent small-signal conductance matrix $\boldsymbol{Y}(\omega, \boldsymbol{V}_0)$:

$$d\boldsymbol{I}(\omega,\boldsymbol{V}_0) = \boldsymbol{Y}(\omega,\boldsymbol{V}_0)\,d\boldsymbol{V}(\omega,\boldsymbol{V}_0) \tag{23.5}$$

and compared with (23.4), it follows

$$\boldsymbol{Y}(\omega,\boldsymbol{V}_0) = \sum_{k=0}^{K} (j\omega)^k \boldsymbol{\varphi}^{(k)}(\boldsymbol{V}_0) \tag{23.6}$$

$\boldsymbol{Y}(\omega, \boldsymbol{V}_0)$ is usually known from broadband measurement of the scattering coefficient. Then with (23.6) the series expansion terms $\boldsymbol{\varphi}^{(k)}(\boldsymbol{V}_0)$ ($k = 0, 1, ..., K$), can be determined by optimization. This is done for further discrete values of $\boldsymbol{V}$ covering the whole expected dynamic range $\boldsymbol{V}(t)$. Finally, the N-port constitutive function of kth order $\boldsymbol{\Phi}^{(k)}(\boldsymbol{V})$ ($k = 0, 1, ..., K$) can simply be obtained for any value of $\boldsymbol{V}$ by integration over $\boldsymbol{\varphi}^{(k)}(\boldsymbol{V})$.

23.2.2 Nonlinear Two-Port

Active devices used in a circuit design can often be treated as two-ports, like the discussed microwave field-effect transistor in common source configuration. Therefore, in the following, the relations for a two-port are given in detail. From (23.3) and with the notation as defined in Appendix F the total differential of the two-port constitutive function of kth order $\Phi_i^{(k)}(V_1,V_2)$ ($i = 1, 2$; $k = 0, 1, ..., K$) reads:

$$\begin{aligned} d\Phi_i^{(k)}(V_1,V_2) &= \frac{\partial \Phi_i^{(k)}(V_1,V_2)}{\partial V_1} dV_1 + \frac{\partial \Phi_i^{(k)}(V_1,V_2)}{\partial V_2} dV_2 \\ &= \varphi_{i,1}^{(k)}(V_1,V_2)dV_1 + \varphi_{i,2}^{(k)}(V_1,V_2)dV_2 \qquad (i = 1,2;\ k = 0,1,...,K) \end{aligned} \tag{23.7}$$

The solution of (23.7) is given by the contour integral

$$\Phi_i^{(k)}(V_1,V_2) = \Phi_i^{(k)}(V_{10},V_{20}) + \int_{(V_{10},V_{20})}^{(V_1,V_2)} \varphi_{i,1}^{(k)}(V_1',V_2)dV_1' + \varphi_{i,2}^{(k)}(V_1,V_2')dV_2' \qquad (i = 1,2;\ k = 0,1,...,K) \tag{23.8}$$

which is path-independent, if the integrability condition

$$\frac{\partial}{\partial V_2}\left(\frac{\partial \Phi_i^{(k)}}{\partial V_1}\right) = \frac{\partial}{\partial V_1}\left(\frac{\partial \Phi_i^{(k)}}{\partial V_2}\right) \qquad (i = 1,2;\ k = 0,1,...,K) \tag{23.9}$$

is fulfilled. Therefore, the integral path can simply be chosen along the coordinate axes. This leads to the integral equations

$$\Phi_i^{(k)}(V_1,V_2) = \Phi_i^{(k)}(V_{10},V_{20}) + \int_{V_{10}}^{V_1} \varphi_{i,1}^{(k)}(V_1',V_{20})dV_1' + \int_{V_{20}}^{V_2} \varphi_{i,2}^{(k)}(V_1,V_2')dV_2' \qquad (i = 1,2;\ k = 0,1,...,K) \tag{23.10}$$

$$\Phi_i^{(k)}(V_1,V_2) = \Phi_i^{(k)}(V_{10},V_{20}) + \int_{V_{10}}^{V_1} \varphi_{i,1}^{(k)}(V_1{}',V_2)dV_1' + \int_{V_{20}}^{V_2(t)} \varphi_{i,2}^{(k)}(V_{10},V_2')dV_2' \qquad (23.11)$$
$$(i = 1,2;\ k = 0,1,\ldots,K)$$

which can be solved under the assumption of continuity of the partial derivatives. In the following, integral (23.10) will be used. The incremental terms $\varphi_{i,j}^{(k)}(V_1,V_2)$ ($i = 1, 2$; $k = 0, 1, \ldots, K$) are determined from (23.6), which reads for the two-port [see (F.11)]:

$$Y_{ij}(\omega,V_{10},V_{20}) = \sum_{k=0}^{K} (j\omega)^{(k)} \varphi_{i,j}^{(k)}(\omega,V_{10},V_{20}) \qquad (i,j = 1,2) \qquad (23.12)$$

It is obvious from (23.12) that the frequency characteristics of the small-signal Y-parameters strongly influence the number K of required expansion terms comprising the incremental port state functions $\varphi_{i,j}^{(k)}$. However, the prior knowledge of the extrinsic parasitic linear model network permits the reduction of the nonlinear analysis to the inner part of the FET device. As a consequence the degree of the power series in (23.12) is considerably reduced. The intrinsic bias-dependent admittance matrix $\boldsymbol{Y}_{int}$ can be calculated from the measured device data by applying the deembedding procedure. The extrinsic outer part of the transistor is described by a parasitic bias-independent admittance matrix $\boldsymbol{Y}_{p1}$ covering the capacitances C_{pg} and C_{pd}, and a bias-independent impedance matrix $\boldsymbol{Z}_{p2}$ taking into account the extrinsic R–L series circuits. With $\boldsymbol{Y}_{p1}$ and $\boldsymbol{Z}_{p2}$ known, the internal admittance matrx $\boldsymbol{Y}_{int}$ can then be calculated from the measured scattering matrix $\boldsymbol{S}$ in the transistor inner reference planes (g–s) and (d–s):

$$\boldsymbol{Y}_{int} = \left(\left[\boldsymbol{Y} - \boldsymbol{Y}_{pl} - (\boldsymbol{E} + \boldsymbol{S})^{-1} (\boldsymbol{E} - \boldsymbol{S})\right]^{-1} - \boldsymbol{Z}_{p2}\right)^{-1} \qquad (23.13)$$

23.3 QUASI-STATIC TRANSISTOR MODEL

In a first modeling step, it is assumed that the operating frequency is rather low (i.e., $\omega \ll 1/\tau_{tr}$, where τ_{tr} is the trapping time constant with typical values in the millisecond range [14]). Another similar low-frequency inertia phenomenon originates from slow temperature transients. Under these assumptions the frequency dispersive FET transconductance $G_m(\omega)$, and channel conductance $G_{ds}(\omega)$ as well, equal the DC value:

$$G_m(\omega) \approx G_{m,DC},\ G_{ds}(\omega) \approx G_{ds,DC} \qquad (\omega \ll 1/\tau_{tr}) \qquad (23.14)$$

With $\omega\tau \ll 1$, where τ is the carrier propagation time under the gate, it follows that the gate-source voltage delay for the drain current control is negligible. Also, relaxation phenomena originating from series connections of capacitors and resistors in the branches of the inner equivalent pi network can be neglected at low frequencies [$\omega\tau_\mu = \omega R_\mu C_\mu \ll 1$ (μ = gs, gd, ds)]. Based on these low-frequency assumptions a simplified equivalent network for the inner FET can be defined as shown in Figure 23.3. G_{gs} models the small-signal conductance of the gate-source diode in gate forward operation (or gate leakage current), and G_{gd} describes the small-signal conductance of the gate-drain diode in the breakdown operation region. The Y-

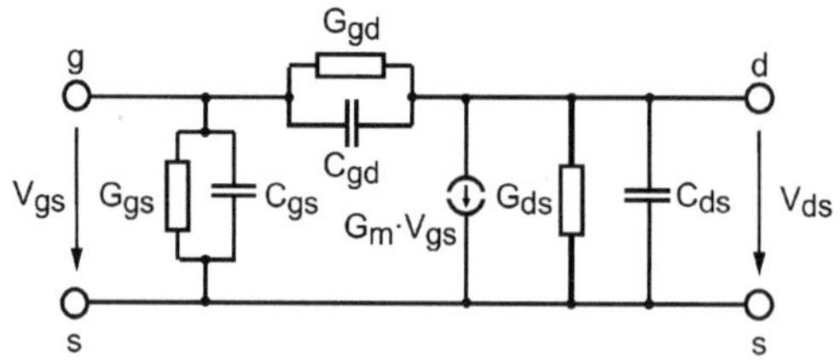

Figure 23.3 Intrinsic quasi-static linear FET equivalent circuit.

parameters are given as

$$\boldsymbol{Y} = \begin{bmatrix} G_{gs} + G_{gd} + j\omega(C_{gs} + C_{gd}) & -G_{gd} + j\omega(-C_{gd}) \\ G_m - G_{gd} + j\omega(-C_{gd}) & G_{ds} + G_{gd} + j\omega(C_{ds} + C_{gd}) \end{bmatrix} \tag{23.15}$$

Comparing with (23.12), the linear frequency dependent admittance parameters (23.15) relate to the incremental state functions as follows:

$$(Y) = \begin{pmatrix} \varphi_{1,1}^{(0)} + j\omega\varphi_{1,1}^{(1)} & \varphi_{1,2}^{(0)} + j\omega\varphi_{1,2}^{(1)} \\ \varphi_{2,1}^{(0)} + j\omega\varphi_{2,1}^{(1)} & \varphi_{2,2}^{(0)} + j\omega\varphi_{2,2}^{(1)} \end{pmatrix} \tag{23.16}$$

According to (23.9) the constitutive functions $\Phi_i^{(k)}$ ($i = 1, 2$; $k = 0, 1$) yield:

$$\Phi_1^{(0)}(V_1,V_2) = \Phi_1^{(0)}(V_{10},V_{20}) + \int_{V_{10}}^{V_1} \varphi_{1,1}^{(0)}(V_1',V_{20})\,dV_1' + \int_{V_{20}}^{V_2} \varphi_{1,2}^{(0)}(V_1,V_2')\,dV_2' \tag{23.17}$$

$$\Phi_1^{(1)}(V_1,V_2) = \Phi_1^{(1)}(V_{10},V_{20}) + \int_{V_{10}}^{V_1} \varphi_{1,1}^{(1)}(V_1',V_{20})\,dV_1' + \int_{V_{20}}^{V_2} \varphi_{1,2}^{(1)}(V_1,V_2')\,dV_2' \tag{23.18}$$

$$\Phi_2^{(0)}(V_1,V_2) = \Phi_2^{(0)}(V_{10},V_{20}) + \int_{V_{10}}^{V_1} \varphi_{2,1}^{(0)}(V_1',V_{20})\,dV_1' + \int_{V_{20}}^{V_2} \varphi_{2,2}^{(0)}(V_1,V_2')\,dV_2' \tag{23.19}$$

$$\Phi_2^{(1)}(V_1,V_2) = \int_{V_{10}}^{V_1} \varphi_{2,1}^{(1)}(V_1',V_{20})\,dV_1' + \int_{V_{20}}^{V_2} \varphi_{2,2}^{(1)}(V_1,V_2')\,dV_2' \tag{23.20}$$

With

$V_1 = V_{gs}$ and $V_2 = V_{ds}$ and

$\Phi_1^{(0)}(V_1,V_2) = I_1(V_1,V_2) = I_g(V_{gs},V_{ds})$,

$\Phi_1^{(1)}(V_1,V_2) = Q_1(V_1,V_2) = Q_g(V_{gs},V_{ds})$,

$\Phi_2^{(0)}(V_1,V_2) = I_2(V_1,V_2) = I_d(V_{gs},V_{ds})$,

$\Phi_2^{(1)}(V_1,V_2) = Q_2(V_1,V_2) = Q_d(V_{gs},V_{ds})$,

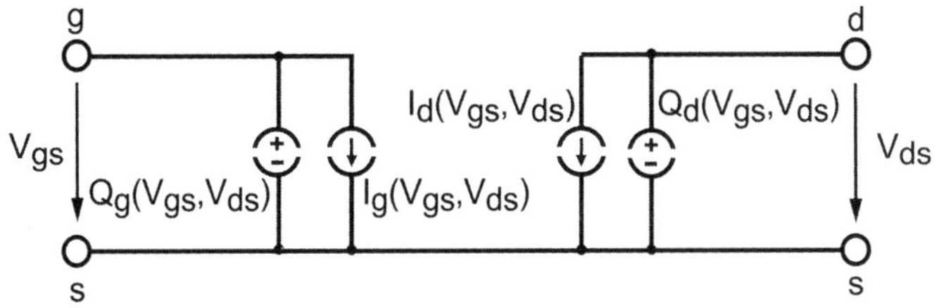

Figure 23.4 Intrinsic quasi-static nonlinear FET equivalent network.

the gate-drain current and charge sources read

$$I_g(V_{gs},V_{ds}) = I_g(V_{V_{gs0}},V_{ds0}) + \int_{V_{gs0}}^{V_{gs}} [G_{gs}(V'_{gs},V_{ds0}) + G_{gd}(V'_{gs},V_{ds0})]\, dV'_{gs} + \int_{V_{ds0}}^{V_{ds}} [-G_{gd}(V_{gs},V'_{ds})]\, dV'_{ds} \tag{23.21}$$

$$Q_g(V_{gs},V_{ds}) = Q_g(V_{gs0},V_{ds0}) + \int_{V_{gs0}}^{V_{gs}} [C_{gs}(V'_{gs},V_{ds0}) + C_{gd}(V'_{gs},V_{ds0})]\, dV'_{gs} + \int_{V_{ds0}}^{V_{ds}} [-C_{gd}(V_{gs},V'_{ds})]\, dV'_{ds} \tag{23.22}$$

$$I_d(V_{gs},V_{ds}) = I_{ds}(V_{gs0},V_{ds0}) + \int_{V_{gs0}}^{V_{gs}} [G_m(V'_{gs},V_{ds0}) - G_{gd}(V'_{gs},V_{ds0})]\, dV'_{gs} + \int_{V_{ds0}}^{V_{ds}} [G_{ds}(V_{gs},V'_{ds}) + G_{gd}(V_{gs},V'_{ds})]\, dV'_{ds} \tag{23.23}$$

$$Q_d(V_{gs},V_{ds}) = \int_{V_{gs0}}^{V_{gs}} [-C_{gd}(V'_{gs},V_{ds0})]\, dV'_{gs} + \int_{V_{ds0}}^{V_{ds}} [C_{ds}(V_{gs},V'_{ds}) + C_{gd}(V_{gs},V'_{ds})]\, dV'_{ds} \tag{23.24}$$

Figure 23.4 shows the intrinsic quasi-static large-signal FET model.

23.3.1 Symmetric Equivalent Pi Network

Defining a symmetric nonlinear equivalent pi-network with shunted current and charge sources in each branch, as illustrated in Figure 23.5, a small-signal model topology (Figure 23.3) can be consistently derived. The restrictions are as follows. C_{gs} and G_{gs} are predominantly dependent on the gate-source voltage and C_{ds} on the drain-source voltage. This is usually a valid approximation for FETs in the saturation region. The integral equations for

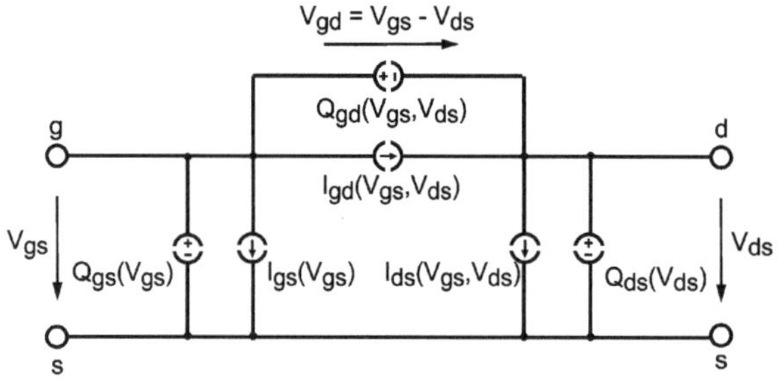

Figure 23.5 Symmetric equivalent pinetwork.

the current and charge sources are then as follows

$$I_{gs}(V_{gs}) = I_{gs}(V_{gs0}) + \int_{V_{gs0}}^{V_{gs}} G_{gs}(V'_{gs})\, dV'_{gs} \tag{23.25}$$

$$Q_{gs}(V_{gs}) = Q_{gs}(V_{gs0}) + \int_{V_{gs0}}^{V_{gs}} C_{gs}(V'_{gs})\, dV'_{gs} \tag{23.26}$$

$$\begin{aligned} I_{gd}(V_{gs},V_{ds}) &= I_{gd}(V_{gs0},V_{ds0}) \\ &+ \int_{V_{gs0}}^{V_{gs}} G_{gd}(V'_{gs},V_{ds0})\, dV'_{gs} + \int_{V_{ds0}}^{V_{ds}} [-G_{gd}(V_{gs},V'_{ds})]\, dV'_{ds} \end{aligned} \tag{23.27}$$

$$\begin{aligned} Q_{gd}(V_{gs},V_{ds}) &= Q_{gd}(V_{gs0},V_{ds0}) \\ &+ \int_{V_{gs0}}^{V_{gs}} C_{gd}(V'_{gs},V_{ds0})\, dV'_{gs} + \int_{V_{ds0}}^{V_{ds}} [-C_{gd}(V_{gs},V'_{ds})]\, dV'_{ds} \end{aligned} \tag{23.28}$$

$$\begin{aligned} I_{ds}(V_{gs},V_{ds}) &= I_{ds}(V_{gs0},V_{ds0}) \\ &+ \int_{V_{gs0}}^{V_{gs}} G_{m}(V'_{gs},V_{ds0})\, dV'_{gs} + \int_{V_{ds0}}^{V_{ds}} G_{ds}(V_{gs},V'_{ds})\, dV'_{ds} \end{aligned} \tag{23.29}$$

$$Q_{ds}(V_{ds}) = Q_{ds}(V_{ds0}) + \int_{V_{ds0}}^{V_{ds}} C_{ds}(V'_{ds})\, dV'_{ds} \tag{23.30}$$

23.3.2 Asymmetric Equivalent Pi Network

Figure 23.6 shows an asymmetric equivalent pi network. In this case it is assumed that the nonlinear gate-source current I_{gs} is only a function of the gate-source voltage V_{gs}. This large-signal model topology is well adapted to the small-signal model structure. The current source in the feedback arm of the nonlinear model can be directly identified with the experimentally observed breakdown current between gate and drain. $I_{ds}(V_{gs},V_{ds})$ is directly correlated to the voltage controlled drain current source, and $I_{gs}(V_{gs})$ to the gate-source diode characteristics. As our experience shows, this model topology is well suited for consistency analysis and

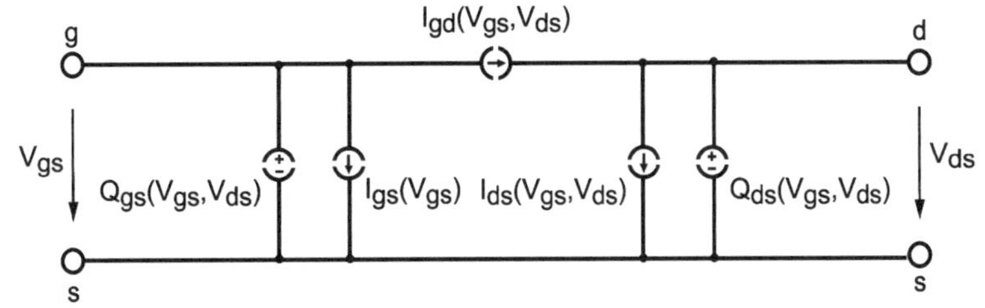

Figure 23.6 Asymmetric equivalent pi network.

direct large-signal model parameter extraction as well [15, 16]. Therefore, current and charge source functions are given in detail:

$$I_{gs}(V_{gs}) = I_{gs}(V_{gs0}) + \int_{V_{gs0}}^{V_{gs}} G_{gs}(V'_{gs})\, dV'_{gs} \tag{23.31}$$

$$I_{gd}(V_{gs},V_{ds}) = I_{gd}(V_{gs0},V_{ds0}) + \int_{V_{gs0}}^{V_{gs}} G_{gd}(V'_{gs},V_{ds0})\, dV'_{gs} + \int_{V_{ds0}}^{V_{ds}} [-G_{gd}(V_{gs},V_{ds})]\, dV'_{ds} \tag{23.32}$$

$$I_{ds}(V_{gs},V_{ds}) = I_{ds}(V_{gs0},V_{ds0}) + \int_{V_{gs0}}^{V_{gs}} G_{m}(V'_{gs},V_{ds0})\, dV_{gs} + \int_{V_{ds0}}^{V_{ds}} G_{ds}(V_{gs},V_{ds})\, dV'_{ds} \tag{23.33}$$

$$\begin{aligned} Q_{gs}(V_{gs},V_{ds}) = Q_{gs}(V_{gs0},V_{ds0}) &+ \int_{V_{gs0}}^{V_{gs}} [C_{gs}(V'_{gs},V_{ds0}) + C_{gd}(V'_{gs},V_{ds0})]\, dV'_{gs} \\ &+ \int_{V_{ds0}}^{V_{ds}} [-C_{gd}(V_{gs},V'_{ds})]\, dV'_{ds} \end{aligned} \tag{23.34}$$

$$\begin{aligned} Q_{ds}(V_{gs},V_{ds}) = Q_{ds}(V_{gs0},V_{ds0}) &+ \int_{V_{gs0}}^{V_{gs}} [-C_{gd}(V'_{gs},V_{ds0})]\, dV'_{gs} \\ &+ \int_{V_{ds0}}^{V_{ds}} [C_{ds}(V_{gs},V'_{ds}) + C_{gd}(V_{gs},V'_{ds})]\, dV'_{ds} \end{aligned} \tag{23.35}$$

Integrability condition (23.9) read as follows: The incremental transconductance and channel conductance are defined as

$$\tilde{G}_m(V_{gs0},V_{ds0}) = \left.\frac{\tilde{I}(V_{gs},V_{ds})}{\tilde{V}_{gs}}\right|_{V_{gs0},V_{ds0}} = \left.\frac{\partial I(V_{gs},V_{ds})}{\partial V_{gs}}\right|_{V_{gs0},V_{ds0}} \tag{23.36}$$

$$\tilde{G}_{ds}(V_{gs0},V_{ds0}) = \left.\frac{\tilde{I}(V_{gs},V_{ds})}{\tilde{V}_{ds}}\right|_{V_{gs0},V_{ds0}} = \left.\frac{\partial I(V_{gs},V_{ds})}{\partial V_{ds}}\right|_{V_{gs0},V_{ds0}} \tag{23.37}$$

Partial differentiation of $\tilde{G}_m$ with respect to V_{ds} results in

$$\frac{\partial \tilde{G}_m}{\partial V_{ds}} = \frac{\partial}{\partial V_{ds}}\left[\frac{\partial I(V_{gs},V_{ds})}{\partial V_{gs}}\right] \tag{23.38}$$

and $\tilde{G}_{ds}$ with respect to V_{gs} gives

$$\frac{\partial \tilde{G}_{ds}}{\partial V_{gs}} = \frac{\partial}{\partial V_{gs}}\left[\frac{\partial I(V_{gs},V_{ds})}{\partial V_{ds}}\right] \tag{23.39}$$

The mixed derivatives in (23.38) and (23.39) are equal when the derivatives in the regarded point and region, respectively, are continuous. In this case, the sequence of the partial differentiation can be interchanged. Then it follows for the relationship between the model conductance and the model transconductance

$$\frac{\partial \tilde{G}_m}{\partial V_{ds}} = \frac{\partial \tilde{G}}{\partial V_{gs}} \tag{23.40}$$

Thus, it has been shown that $\tilde{G}_m$ and $\tilde{G}$ are interdependent functions being under control of the voltages V_{gs} and V_{ds}.

23.4 DISPERSIVE DRAIN CURRENT MODEL

For further extension of the validity of frequency range of the nonlinear model, low-frequency dispersion of the transconductance, and channel conductance as well, must be taken into account (Figure 23.7). Such dispersion effects originate from trapping centers near the semiconductor surface and buffer layer [10]. As is known, the traps are frozen at frequencies above kilohertz (respectively, megahertz), meaning that the frequency-dependent transconductance and channel conductance have reached the RF limit values,

$$\begin{aligned} G_m(\omega) &\approx G_{m,RF} = \text{const.} \\ G_{ds}(\omega) &\approx G_{ds,RF} = \text{const.} \qquad (\omega > 1/\tau_{tr}) \end{aligned} \tag{23.41}$$

Equation (23.41) is valid for device operation in class A with limited signal excitation level. In this case the distortion is low and the trapping condition is stable with a constant (quiescent) bias point. This will also not dramatically change with higher signal amplitudes as the distorted signal waveform is symmetric and the dynamic bias point (time-averaged control voltages) and the DC bias point are nearly equal. The situation differs dramatically when other modes of operation are selected such as, for example, class B, AB, or C. Now the signal waveforms are no longer symmetric. The distortion grows with increasing input power level providing a stronger shift of the dynamic (effective) bias point. However, this in turn involves slow changes of the trap situation in the device, independent of the question of self-heating phenomena.

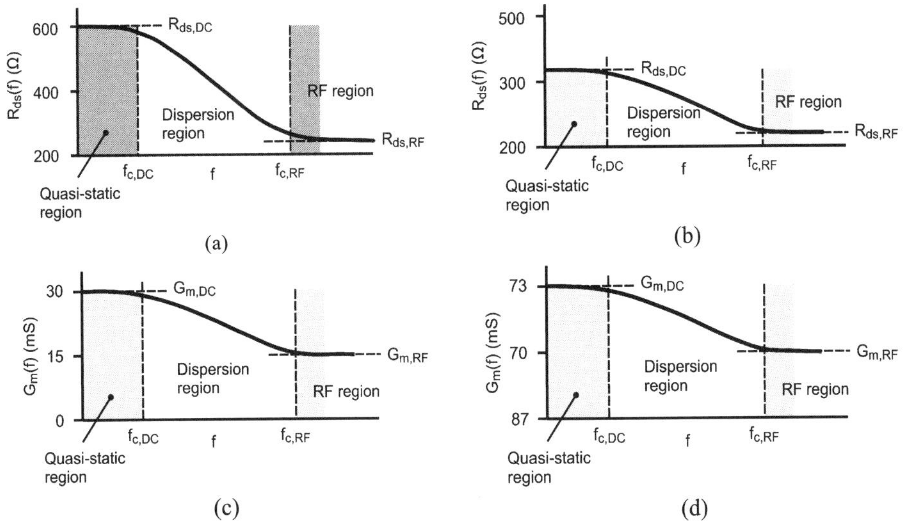

Figure 23.7 Dispersion of channel conductance $G_{ds} = 1/R_{ds}$ and transconductance G_m of a 300-μm MESFET of type MGFC 1423. RF and DC limit values were measured. (a), (c): $V_{gs0} = -0.75V$, $V_{ds0} = 2V$; (b), (d): $V_{gs0} = -0.25V$, $V_{ds0} = 2V$.

23.4.1 Drain Current Source with Bias-Dependent Conductances

Originally, the formulation of a dispersive drain current source was based on an extension of the quasi-static approach. It was common to characterize the device by DC and S-parameter measurements. Over time pulsed-DC measurements were applied, because it was possible to clearly distinguish between temperature, surface/ buffer trapping, and gate/drain lag phenomena. This section provides a short review about the initial approaches for drain current source extensions taking into account trapping-induced dispersion effects is given [10, 17-19].

Similar to the quasi-static expression of the low-frequency drain current source (23.29) and choosing the averaged dynamic operating voltages $[\overline{V_{gs}(t)}, \overline{V_{ds}(t)}]$ as the reference point for integration [10], the nonlinear RF drain current source can be written as follows:

$$I_{ds,RF}[V_{gs}(t),V_{ds}(t)] = I_{ds,RF}[\overline{V_{gs}(t)},\overline{V_{ds}(t)}] + \int_{\overline{V_{gs}(t)}}^{V_{gs}(t)} G_{m,RF}[V'_{gs}, \overline{V_{ds}(t)}]\, dV'_{gs} + \int_{\overline{V_{ds}(t)}}^{V_{ds}(t)} G_{ds,RF}[V_{gs}(t), V'_{ds}]\, dV'_{ds} \tag{23.42}$$

where $I_{ds,RF}\,[\overline{V_{gs}(t)}, \overline{V_{ds}(t)}]$ represents a DC current value that results from the time-averaged control voltages $\overline{V_{gs}(t)}$ and $\overline{V_{ds}(t)}$. The bias-dependent incremental DC drain current elements have been substituted by the corresponding RF quantities. As is shown in [10] (23.42) can be reformulated as follows:

$$\begin{aligned} I_{ds,RF}[V_{gs}(t),V_{ds}(t)] = {} & I_{ds,DC}[V_{gs}(t),V_{ds}(t)] \\ & + \int_{V_{gs0}}^{V_{gs}(t)} \left[G_{m,RF}(V'_{gs},V_{ds0}) - G_{m,DC}(V'_{gs},V_{ds0})\right] dV'_{gs} \\ & + \int_{V_{ds0}}^{V_{ds}(t)} \left\{G_{ds,RF}[V_{gs}(t),V'_{ds}] - G_{ds,DC}[V_{gs}(t),V'_{ds}]\right\} dV'_{ds} \\ & - \left\{ \int_{V_{gs0}}^{\overline{V_{gs}(t)}} \left[G_{m,RF}(V'_{gs},V_{ds0}) - G_{m,DC}(V'_{gs},V_{ds0})\right] dV'_{gs} \right. \\ & \left. + \int_{V_{ds0}}^{\overline{V_{ds}(t)}} \left[G_{ds,RF}[\overline{V_{gs}(t)},V'_{ds}] - G_{ds,DC}[\overline{V_{gs}(t)},V'_{ds}]\right] dV'_{ds} \right\} \end{aligned} \tag{23.43}$$

We introduce a bias-dependent dispersive difference transconductance $\Delta G_{\mathrm{m}}(V_{\mathrm{gs}},V_{\mathrm{ds}})$ and dispersive difference channel conductance $\Delta G_{\mathrm{ds}}(V_{\mathrm{gs}},V_{\mathrm{ds}})$ as well, defined as

$$\Delta G_m(V_{gs},V_{ds}) = G_{m,RF}(V_{gs},V_{ds}) - G_{m,DC}(V_{gs},V_{ds}) \tag{23.44}$$

$$\Delta G_{ds}(V_{gs},V_{ds}) = G_{ds,RF}(V_{gs},V_{ds}) - G_{ds,DC}(V_{gs},V_{ds}) \tag{23.45}$$

Substituting ΔI_{ds} for the first two integral terms in (23.43) and $\Delta\overline{I_{\mathrm{ds}}}$ for the second ones yields

$$I_{ds,RF}\left(V_{gs}(t),V_{ds}(t)\right) = I_{ds,DC}\left(V_{gs}(t),V_{ds}(t)\right) + \Delta I_{ds}\left(V_{gs}(t),V_{ds}(t)\right) + \Delta\overline{I_{ds}}\left(V_{gs}(t),V_{ds}(t)\right) \tag{23.46}$$

This kind of formulation for the drain current is very instructive and informative for the rating of a given FET device with respect to its dispersive properties. The first term ($I_{\mathrm{ds,DC}}$) implies the quasi-static $I(V)$ characteristics of an ideal dispersionless device. The second one (ΔI_{ds}) describes an additional current contribution originating from the *instantaneous* transconductance and channel conductance dispersion. The term $\Delta\overline{I_{\mathrm{ds}}}$ describes an additional dynamic shifting of the quiescent bias point with respect to dispersion phenomena. Practically, the calculation of this memory term requires the knowledge of the steady-state signal waveform of the large-signal control voltages. Physically seen, a change of the average value of the signal waveform results in an altered trap situation in the semiconductor device, and thus provides shifting of the quiescent bias point.

As can be seen from (23.42), the difference transconductance and channel conductance are the fundamental dispersion parameters, which characterize a dispersive drain current source. Therefore, we have analyzed the dispersion properties of various FETs. Figure 23.8(a) shows the measured difference transconductance ΔG_{m} of a GaAs MESFET FSX03X. Although some ripples can be observed in the 3D-graph, ΔG_{m} may be roughly described by a plane surface, which is slightly inclined with respect to the V_{gs}-axis. Moreover, only negative ΔG_{m} values exist.

Completely different ΔG_m characteristics have been measured for a HEMT device of type

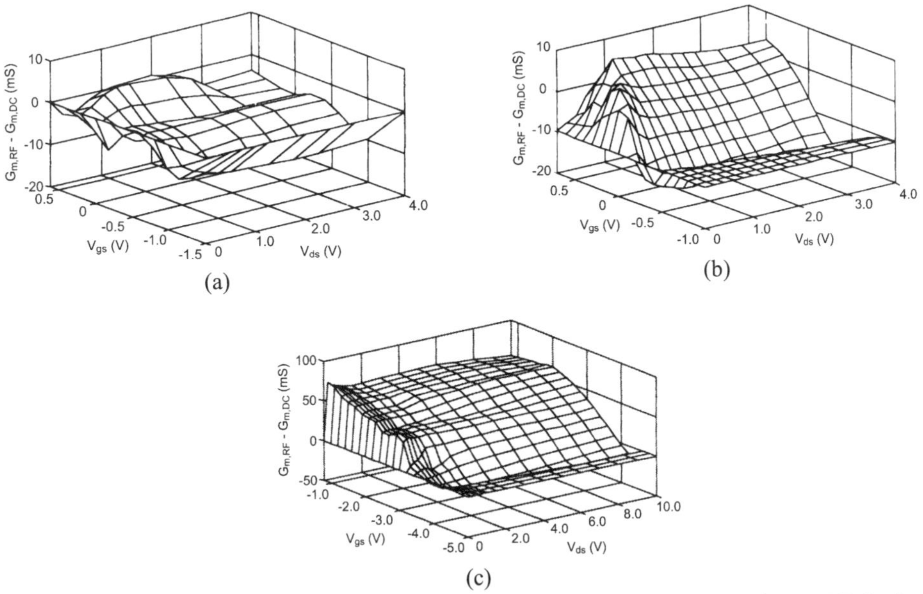

Figure 23.8 Measured bias-dependent difference transconductance ΔG_m for various FET devices: (a) GaAs-MESFET FSX03X, (b) HEMT JS8902-AS, and (c) Power-GaAs-MESFET JS8836A-AS.

JS8902-AS (Figure 23.8(b)). ΔG_m is again approximately independent on the drain-source voltage. Outside of the pinch-off region the point lattice of the extracted data forms a plane surface, which is much more inclined than in the previous case of a MESFET. Measurements have also shown that the difference channel conductance (not shown) is much lower than the ΔG_m values for low-noise MESFETs and HEMTs in the saturation region. This means that the corresponding integral terms in (23.43) are of minor influence on the RF nonlinear drain current source.

Figure 23.8(c) shows an example of a power MESFET (JS8836A-AS). As can be seen, ΔG_m is approximately independent on V_{ds}, but the decrease with V_{gs} toward pinch-off is much stronger. It has turned out from measurements that the difference channel conductance in the saturation region is of the same order as ΔG_m. Thus, regarding power FETs, ΔG_{ds} contributes significantly to the RF *I-V* characteristics.

The discussed dispersion-induced shifting of the dynamic operating point may be remarkable with large modulation signals so that nonlinear FET models should account care for this dynamic effect.

With the approximation $\overline{V_{gs}(t)} \approx V_{gs0}$ and , $\overline{V_{ds}(t)} \approx V_{ds0}$, (23.43) simplifies to

$$\begin{aligned} I_{ds,RF}\left(V_{gs}(t),V_{ds}(t)\right) \approx\ & I_{ds,DC}\left(V_{gs}(t),V_{ds}(t)\right) \\ & + \int_{V_{gs0}}^{V_{gs}(t)} \left(G_{m,RF}(V_{gs}',V_{ds0}) - G_{m,DC}(V_{gs}',V_{ds0})\right) dV_{gs}' \\ & + \int_{V_{ds0}}^{V_{ds}(t)} \left(G_{ds,RF}(V_{gs}(t),V_{ds}') - G_{ds,DC}(V_{gs}(t),V_{ds}')\right) dV_{ds}' \end{aligned} \tag{23.47}$$

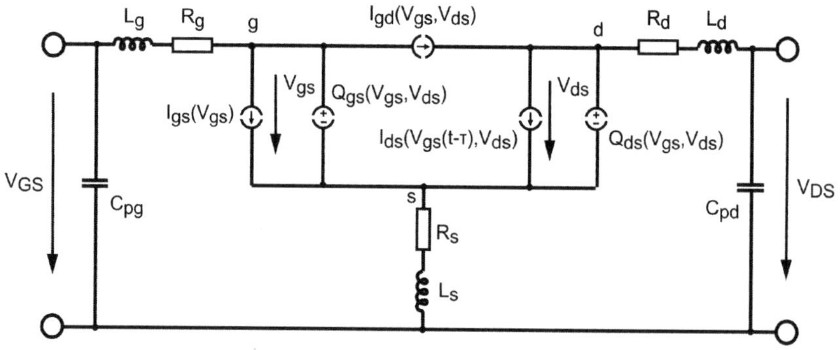

Figure 23.9 Dispersive FET model.

and with (23.30)

$$I_{ds,RF}\left(V_{gs}(t),V_{ds}(t)\right) \approx I_{ds,DC}(V_{gs0},V_{ds0}) + \int_{V_{gs0}}^{V_{gs}(t)} G_{m,RF}(V'_{gs},V_{ds0})\,dV'_{gs} + \int_{V_{ds0}}^{V_{ds}(t)} G_{ds,RF}(V_{gs}(t),V'_{ds})\,dV'_{ds} \tag{23.48}$$

Equation (23.48) was originally proposed by Willing and Rauscher in 1979 [17]. Analyzing measured DC and RF device response data, no strong difference in the element values of the DC and RF transconductance had been observed in the drain current saturation region (Figure 23.7(d)). Thus, low-frequency dispersion phenomena had not been discussed in depth by the authors. As electron inertia becomes relevant only at highest frequencies (e.g., above 100 GHz for GaAs material) the modeling of the drain current source based on (23.48), was mistakenly termed “quasi-static.”

Figure 23.9 illustrates the dispersive FET model. The inner nonlinear part equals the topology in Figure 23.10. The drain current source is under control of a time-delayed gate-source voltage, originating from the bias-dependent transit time of the charge carriers $\tau(V_{gs},V_{ds})$ under the gate.

A few comments supplementary to (23.42) are appropriate. Root [18] has used a first-order low-pass function to interpolate between the DC and RF drain current models. The dispersive drain current in frequency domain is defined as

$$I_{ds}(\omega) = H(\omega)\cdot I_{ds,DC}(V_{gs},V_{ds}) + (1 - H(\omega))\cdot I_{ds,RF}(V_{gs},V_{ds}) \tag{23.49}$$

with H(ω) being a (purely mathematical) low pass with a cutoff frequency in the dispersion transition region. In time domain (23.49) reads [20]

$$I_{ds}(t) = \overline{I_{ds,DC}(V_{gs},V_{ds})} + I_{ds,RF}(V_{gs},V_{ds}) - \overline{I_{ds,RF}(V_{gs},V_{ds})} \tag{23.50}$$

which is identical to

$$I_{ds}(t) = I_{ds,DC} + (I_{ds,RF} - I_{ds,DC}) - \overline{(I_{ds,RF} - I_{ds,DC})} \tag{23.51}$$

Equation (23.51) is equivalent to the time-domain formulation after Werthof [19]. Very close to (23.50) is the expression [10, 21, 22]

Table 23.1

Charging Time Constant τ_i and Cutoff Frequency $f_{\tau i}$ of Investigated FET Devices

FET Type	$f_{\tau i} = 1/(2\pi R_i C_{gs})$ [GHz]	$\omega\tau_i$ [10^{-3}]	τ_i [ps]
FSX03X	135.5	7.4	1.2
JS8902-AS	64.4	15.5	2.5
JS8836A-AS	79.6	1.3	2.0
MGFC 1423	234.6	4.3	0.7
MGFC 4401	188.9	5.3	0.8

$$I_{ds}(t) = I_{ds,DC}(\overline{V_{gs}},\overline{V_{ds}}) + I_{ds,RF}(V_{gs},V_{ds}) - I_{ds,RF}(\overline{V_{gs}},\overline{V_{ds}}) \tag{23.52}$$

where the control voltages are averaged instead of currents. As pointed out in [20], the difference between (23.50) and (23.52) concerns only the DC component of the current, which is small assuming weak nonlinearities or moderate stimuli from the dynamic (averaged) bias point.

23.4.2 Gate Charging Resistance

Until now the small-signal gate charging resistance R_i (see Section 6.2.4) has not been taken into account. As can be seen from Table 23.1, the extracted values for the charging time constant $\tau_i = R_i C_{gs}$ are very low and lie within a range of 0.7 and 2.5 ps for all investigated FET devices. Therefore, R_i only becomes dominant at higher frequencies or pulsed stimulus, leading to intrinsic delays of the instantaneous gate-source charge $Q_{gs}(t)$ shown in Figure 23.5 and Figure 23.6, respectively. With $R_i \neq 0$ and $\omega\tau \ll 1$ the admittance coefficient Y_{11} in (23.15) and (23.16) would extend to

$$Y_{11} \approx \varphi_{1,1}^{(0)} + j\omega\varphi_{1,1}^{(1)} + (j\omega)^2\varphi_{1,1}^{(2)} = G_{gs} + G_{gd} + j\omega(C_{gs} + C_{gd}) + (j\omega)^2(-C_{gs}\tau_i) \tag{23.53}$$

With $\varphi_{1,1}{}^{(2)} = -C_{gs}\tau_i$ and $\varphi_{1,2}{}^{(2)} = 0$ a constitutive function of second-order, $\Phi_1{}^{(2)}$, which reflects the relaxation behavior of the gate charge source, could be derived using (23.10).

As already mentioned, the R_i induced relaxation of the Q_{gs} charge source has approximately been taken into account as follows. Regarding the C_{gs}–R_i branch, the instantaneous control voltage of the charge equals the instantaneous branch voltage minus the instantaneous voltage drop over the charging resistor. Assuming that R_i is only weakly nonlinear, the voltage drop is approximately $R_i(V_{gs}(t), V_{ds}(t))\cdot I(t)$. In this case intrinsic delays have been taken into account by cascading the bias-dependent small-signal resistor with the nonlinear sources.

As can be seen in Figure 23.10, the gate-source capacitance is series-connected with the charging resistance R_i so that the control voltage V_g follows a step function of V_{gs} at the electrodes g–s with some delay. In the frequency domain, the admittance of the R–C branch reads with $\omega\tau_i = \omega R_i C_{gs} \ll 1$ approximately

$$\begin{aligned} Y_i &= \frac{1}{R_i + 1/j\omega C_{gs}} = j\omega C_{gs}/(1 + j\omega\tau_i) \\ &\approx j\omega C_{gs} - (j\omega)^2 C_{gs}\tau_i \quad \text{with} \quad \omega R_i C_{gs} = \omega\tau_i \ll 1 \end{aligned} \tag{23.54}$$

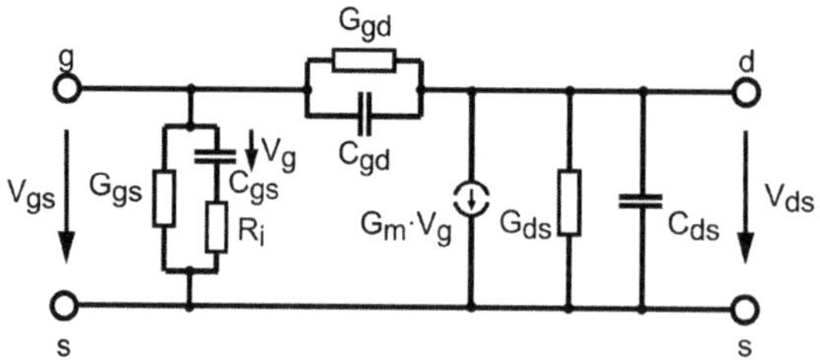

Figure 23.10 Inner quasi-static linear FET model under consideration of time relaxation in the gate branch.

Equation (23.54) indicates a first-order non-quasi-static model. Then with (23.53), the Y-matrix reads

$$\boldsymbol{Y} = \begin{pmatrix} G_{gs} + G_{gd} + j\omega(C_{gs} + C_{gd}) + (j\omega)^2(-C_{gs}\tau_i) & -G_{gd} + j\omega(-C_{gd}) \\ G_m - G_{gd} + j\omega(-C_{gd}) & G_{ds} + G_{gd} + j\omega(C_{ds} + C_{gd}) \end{pmatrix} \tag{23.55}$$

A respective extension of (23.16) gives

$$(Y) = \begin{pmatrix} \varphi_{1,1}^{(0)} + j\omega\varphi_{1,1}^{(1)} + (j\omega)^2\varphi_{1,1}^{(2)} & \varphi_{1,2}^{(0)} + j\omega\varphi_{1,2}^{(1)} \\ \varphi_{2,1}^{(0)} + j\omega\varphi_{2,1}^{(1)} & \varphi_{2,2}^{(0)} + j\omega\varphi_{2,2}^{(1)} \end{pmatrix} \tag{23.56}$$

Recalling (23.10) the state function $\Phi_1^{(2)}$ reads

$$\begin{aligned} \Phi_1^{(2)}(V_1,V_2) &= \Phi_1^{(2)}(V_{10},V_{20}) + \int_{V_{10}}^{V_1} \varphi_{1,1}^{(2)}(V_1',V_{20})dV_1' + \int_{V_{20}}^{V_2} \varphi_{1,2}^{(2)}(V_1,V_2')dV_2' \\ &= \Phi_1^{(2)}(V_{10},V_{20}) - \int_{V_{10}}^{V_1} C_{gs}(V_1',V_{20})\tau_i(V_1',V_{20})dV_1' \end{aligned} \tag{23.57}$$

$\Phi_1^{(2)}(V_1,V_2)$ can be interpreted as a relaxing capacitive gate current. In extension of Figure 23.4, Figure 23.11 shows the quasi-static time relaxation model.

The relaxation time effects of the charge source caused by the charging resistance R_i is often approximately taken into account by connecting the bias-dependent small-signal charging resistance $R_i(V_{gs},V_{ds})$ in series with the gate-source branch. It is

$$\begin{aligned} V_{gs}(t) &= \int^{I_q[V_{gs}(t),V_{ds}(t)]} R_i(V_{gs},V_{ds})dI_q(V_{gs},V_{ds}) + V_g(t) \\ &\approx R_i\big[V_{gs}(t),V_{ds}(t)\big]\cdot I_q\big[V_{gs}(t),V_{ds}(t)\big] + V_g(t) \end{aligned} \tag{23.58}$$

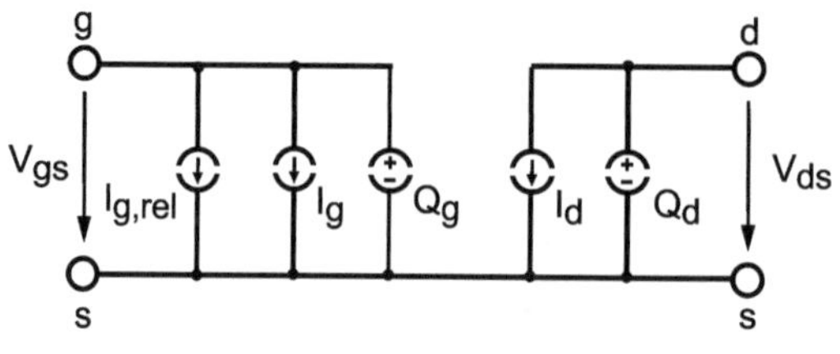

Figure 23.11 Inner quasi-static nonlinear FET time relaxation model.

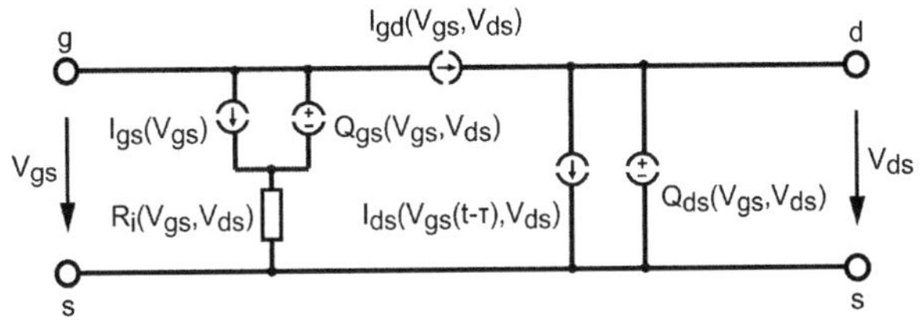

Figure 23.12 Inner quasi-static nonlinear FET time relaxation equivalent circuit of a FET with pi-structure topology.

if $R_i(V_{gs}, V_{ds})$ is only weakly nonlinear, which is commonly fulfilled in the saturation region. $V_g(t)$ is the voltage drop over the gate capacitance. It corresponds to the gate-source voltage $V_{gs}(t)$ reduced by the voltage drop over R_i. $V_g(t)$ controls the gate charge source and the gate and drain current sources.

It can be seen that the source that describes the time relaxation of the charge source is only dependent on the gate-source voltage V_{gs}. Thus, the pi-equivalent circuits in Figure 23.5 and Figure 23.6 can be directly applied if the source $\Phi_1^{(2)}$ additionally is taken into account at the gate-source input port.

Figure 23.12 shows the intrinsic quasi-static nonlinear time relaxation equivalent circuit of a field-effect transistor with pi-structure topology. The intrinsic nonlinear part is similar to the topology given in Figure 23.6. The drain current source is controlled by a time-delayed gate source voltage originating from the bias-dependent propagation time $\tau(V_{gs}, V_{ds})$ of the carriers under the gate. Figure 23.13 depicts the complete dispersive FET model with the parasitic extrinsic part of the network.

23.5 ELECTROTHERMAL DRAIN CURRENT MODELING

The dispersion of a FET transistor is mostly determined by the nonlinear drain current source. We have already discussed a drain current model in Section 23.3 describing electrical dispersion; that is, considering the influence of trapping effects on the IV characteristics, which causes transconductance and channel conductance variation due to signal level dynamic. In addition, the average value of the stimulated signal waveform can differ from the quiescent bias point, which influences the quasi-static trap conditions in the device.

The large-signal FET model described in Section 23.3 was successfully implemented in

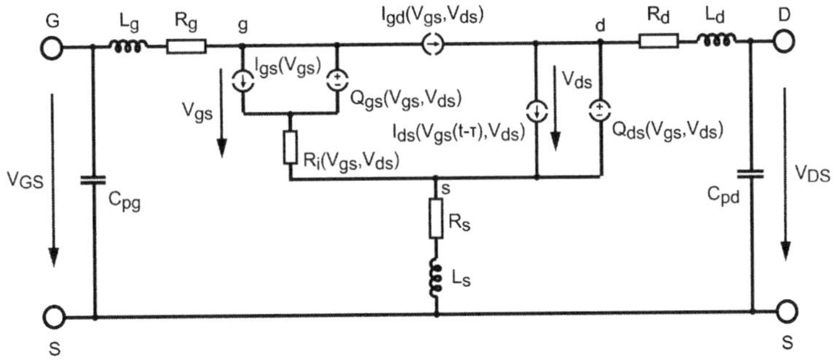

Figure 23.13 Large-signal time relaxation model for a FET.

a home-made harmonic balance simulator and could cover reliably strong dispersion effects of HEMT devices in amplifiers and oscillators [19]. However, as pointed out in [23], it is not straightforward to implement the memory term $\Delta\overline{I}_{ds}$ in (23.46), containing integrals of the dispersion currents over one period, in a commercially available nonlinear circuit simulator such as HP MDS (Microwave Design System) (now superseded by ADS (Advanced Design System)). In 1995 Filicori et al. [21] proposed a comprehensive electro-thermal I_{ds} model that had been proven for easy implementation in commercial simulation tools. Our experience has shown that this model with some minor modifications has met all modeling expectations regarding model consistency and reliability of high-power HEMT devices. In the following this model is presented in detail.

23.5.1 Pulsed-DC Related Drain Current Model

The I_{ds} model outlined briefly in this section is based on the research work reported by Filicori et al. in 1995 [21]. The following assumptions are made. The signal frequencies of gate-source and drain-source voltages, v_{gs} and v_{ds}, respectively, are higher than the upper cut-off frequency of the low-frequency dispersive phenomena (self-heating and charge-trapping effects), but are low enough to neglect microwave reactive effects of the gate charge or propagating times of the electrons in the channel. Therefore, charge-trapping and thermal states do not change as swiftly as the voltage signals and are assumed fixed to their average condition.

The average values of charge-trapping and thermal states are assumed to depend only on the average values of v_{gs}, v_{ds} and p_{ds} (i.e., V_{gs}, V_{ds} and P_{ds}). This assumption implies that there is no significant AC-to-DC conversion in the relation between the signals v_{gs} and v_{ds} and the charge-trapping states. In consequence, as stated in [21], the average value of the charge-trapping states "is not significantly affected by the amplitude or shape of the alternate components ($v_{gs} - V_{gs}$) and ($v_{ds} - V_{ds}$)." This model assumption appears in other I_{ds} models [24–27], and has been confirmed experimentally, as in the results reported in [28–30].

The expression of the I_{ds} model is a first-order Taylor series expansion (i.e., a functional *linearization*) with respect to V_{gs}, V_{ds} and T_{ch}. This assumption implies a linear or mildly nonlinear dependence of the average values of the thermal and charge-trapping states on V_{gs} and V_{ds}. The series expansion is made around the operation condition characterized by the static-DC voltages V_{gs}*, V_{ds}*, and the steady-state channel temperature T_{ch}*.

In [21] Filicori et al. proposed an empirical I_{ds} model for GaAs MESFETs based on pulsed-DC IV measurements. The equation for the dispersive drain-source current reads:

$$\begin{aligned} I_{ds}(v_{gs},v_{ds}) &= I_{ds}^{Iso}(v_{gs},v_{ds}) \\ &+ f_G(v_{gs},v_{ds})\cdot(v_{gs} - V_{gs}) + f_D(v_{gs},v_{ds})\cdot(v_{ds} - V_{ds}) \\ &+ f_{th}(v_{gs},v_{ds})\cdot\Delta T_{ch} \end{aligned} \tag{23.59}$$

where

$$\Delta T_{ch} = T_{ch} - T_{ch}^{*} = R_{th}\cdot P_{ds} + (T_C - T_C^{*}) \tag{23.60}$$

Instantaneous voltages $v_{gs}(t)$ and $v_{ds}(t)$ are referred to the intrinsic ports of the device. Their mean values are denoted by V_{gs} and V_{ds}. Hence, $(v_{gs} - V_{gs})$ and $(v_{ds} - V_{ds})$ describe their alternate (i.e., time-varying) components. P_{ds} is the average dissipated power derived from the intrinsic drain-source port (contribution of the gate port is assumed to be negligible). T_{ch} denotes the average channel temperature and T_C the case temperature referring to the bottom metallic surface of the device (e.g., controlled by the thermo-chuck of the probing station). R_{th} denotes the device thermal resistance, characterizing the difference between the channel temperature T_{ch} and the case temperature T_C with respect to dissipated power P_{ds}.

$I_{ds}^{Iso}(v_{gs}, v_{ds})$ represents a theoretical isothermal DC IV characteristic of an operation with constant channel temperature. The functions $f_D(v_{gs}, v_{ds})$ and $f_G(v_{gs}, v_{ds})$ represent the differences between dynamic and static I_{ds} due to charge trapping for each pair of instantaneous voltages (v_{gs}, v_{ds}) applied to the device. The function $f_{th}(v_{gs}, v_{ds})$ represents the differences of the dynamic I_{ds} with respect to an isothermal DC characteristics due to changes of heat in the device. Variables with the superscript "*" denote initial values.

The I_{ds} model for *constant case temperature* ($T_C = T_C$*) is given by

$$\begin{aligned} I_{ds}(v_{gs},v_{ds}) = {} & I_{ds}^{Iso}(v_{gs},v_{ds}) \\ & + f_G(v_{gs},v_{ds})\cdot(v_{gs} - V_{gs}) + f_D(v_{gs},v_{ds})\cdot(v_{ds} - V_{ds}) \\ & + f_{th}(v_{gs},v_{ds})\cdot R_{th}\cdot P_{ds} \end{aligned} \tag{23.61}$$

For constant case temperature, $f_{th}\cdot R_{th}$ is condensed as f_P and the model does not strictly identify the thermal resistance separately. Then (23.61) becomes

$$\begin{aligned} I_{ds}(v_{gs},v_{ds}) = {} & I_{ds}^{Iso}(v_{gs},v_{ds}) \\ & + f_G(v_{gs},v_{ds})\cdot(v_{gs} - V_{gs}) + f_D(v_{gs},v_{ds})\cdot(v_{ds} - V_{ds}) \\ & + f_P(v_{gs},v_{ds})\cdot P_{ds} \end{aligned} \tag{23.62}$$

Under typical static-DC operation that is nonisothermal, $I_{ds} = I_{DC}$, $v_{gs} = V_{gs}$, $v_{ds} = V_{ds}$, and $P_{ds} = I_{DC}V_{ds}$. Applying such equivalences to (23.61) for the model with constant case temperature, the following expression is received for the static-DC IV characteristics:

$$I_{DC} = \frac{I_{ds}^{Iso}(v_{gs},v_{ds})}{1 - f_P(v_{gs},v_{ds})\cdot V_{ds}} \tag{23.63}$$

23.5.2 Transistor Thermal Model

Equation (23.60) describes the relationship between the change in the channel temperature with respect to the average dissipated power and the change in the case temperature. In the sense of classical heat transfer analysis, the thermal resistance (R_{th} [°C/W]) is defined as the ratio of heat transfer rate ($\dot{Q}$ [W]) and the temperature difference (ΔT [°C]) across a device.

In the adopted I_{ds} model, the thermal resistance is considered constant with respect to P_{ds}, because the dependency of the thermal state (self-heating phenomena) on the static-DC bias point is already represented in the parameter f_{th}. So, R_{th} agrees well with the classical definition.

However, if the electrical power in the device is a function of frequency so that the dissipated power is denoted by $P_{ds}(\omega)$, then this frequency-dependency is passed to the change in the channel temperature, and hence, it is correct to discuss a *thermal impedance* Z_{th}, instead of a thermal resistance R_{th}. In [21], there is no consideration of a thermal impedance or of the frequency-dependency of the dissipated power. The model formulation found in [21] suggests a first-order low-pass filter (a single RC circuit) with the only purpose of calculating P_{ds} (the average value of the dissipated power) from the time-varying power signal p_{ds} [31].

In the general case of the adopted I_{ds} model, the frequency-dependence of the instantaneous dissipated power, $p_{ds}(\omega)$, must be considered for signals v_{gs} and v_{ds} with any given frequency by taking into account the device thermal impedance Z_{th}. When the thermal impedance Z_{th} is considered, the product $R_{th} \cdot P_{ds}$ (expressed explicitly in (23.60) and (23.61), and implicitly in (23.62) and (23.63)) must be reformulated. The electrothermal subcircuit depicted in Figure 23.14, adopted from [32], shows the relationship between the frequency-dependent dissipated power $p_{ds}(\omega)$, the *thermal impedance,* and the produced temperature change ΔT.

In the electrothermal subcircuit of Figure 23.14(a), R_{th} and C_{th} form the thermal impedance Z_{th} and the implied time constant of the dispersive thermal phenomena, τ_{th}, is equal to $R_{th} \cdot C_{th}$.

Dahmani et al. presented a method to calculate the electrothermal parameters using heat-transfer simulations and numerical techniques [33, 34]. The thermal profile of the transistor was obtained using a layer structure physically equivalent to that of the studied devices and implementing the heat equation numerically. The procedure takes into account the different types of heat-transfer mechanisms, the boundary conditions, and the temperature-dependent thermal conductivities of the material layers involved. The thermal resistance is obtained through the calculation of the ratio of the temperature rise in each layer to the dissipated power (for individual fingers or total HEMT structure). The most relevant layers for thermal impedance determination are the SiC substrate layer and the heat-sink related material [35].

Exemplarily Table 23.2 summarizes simulated thermal resistances and Table 23.3

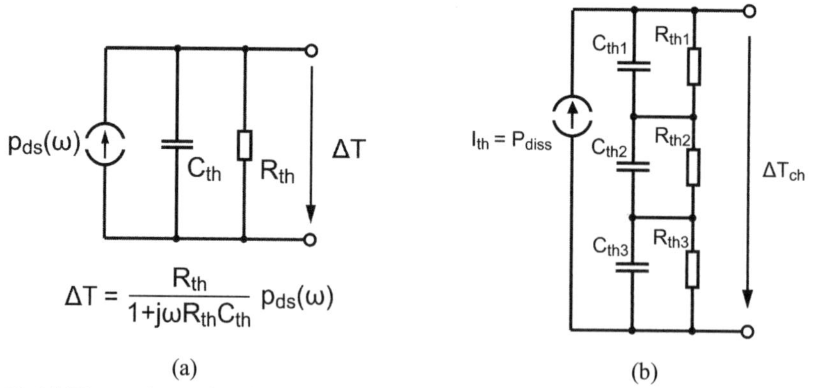

Figure 23.14 Electrothermal subcircuit to determine channel temperature. (a) First-order low-pass filter representation [32], and (b) third-order low-pass filter representation for increased accuracy.

Table 23.2

Thermal Resistances for a 3.2-mm AlGaN/GaN HEMT on SiC†

RC Order	R_{th1}*	R_{th2}*	R_{th3}*	R_{th4}*	R_{th5}*
Order 1	14				
Order 2	4.53	9.47			
Order 3	2.7	4.2	7.1		
Order 5	1.6	1.3	2.8	0.5	7.8

* In °C/W.
† From: [36].

Table 23.3

Thermal Time Constants for a 3.2-mm AlGaN/GaN HEMT on SiC†

RC Order	τ_{th1}*	τ_{th2}*	τ_{th3}*	τ_{th4}*	τ_{th5}*
Order 1	106.6				
Order 2	3.9	222.5			
Order 3	2.0	74.1	1335		
Order 5	0.82	5.8	83.4	41.1	1905

* In μs.
† From: [36].

experimentally determined thermal time constants for a 3.2-mm AlGaN/GaN HEMT on SiC. The thermal capacitance is then obtained by the ratio of τ_{th}/R_{th} represented in Table 23.4.

Figure 23.15 illustrates the enhancement in the large-signal modeling procedure in particular on intermodulation prediction and accuracy with a higher order of electrothermal subcircuit. Simulation results of a 3.2-mm AlGaN/GaN HEMT ($V_{GS0} = -3.4$V, $V_{DS0} = 40$V, $I_{DS0} = 25$ mA $\approx$ 10% of I_{DSS}) are compared between a first-order[37] and a third-order thermal model [36].

Moreover, in their published work, Dahmani et al. have demonstrated that the parameter values of the electrothermal subcircuit calculated using their heat-transfer simulations agree well with values extracted based on the measurements. Table 23.5, redrawn from [34], lists values for R_{th} and τ_{th} calculated by Dahmani et al. using their heat-transfer simulations and also based on measured data.

Table 23.4

Thermal Capacitances for a 3.2-mm AlGaN/GaN HEMT on SiC†

RC Order	C_{th1}*	C_{th2}*	C_{th3}*	C_{th4}*	C_{th5}*
Order 1	7.6				
Order 2	0.86	23.5			
Order 3	0.74	17.6	188		
Order 5	0.51	4.5	29.8	82.2	244.3

* In μF.
† From: [36].

Table 23.5
Simulated and Measured Thermal Resistance R_{th} and Thermal Time Constant τ_{th} of GaN HEMTs with Various Gate Peripheries

GaN HEMT	R_{th} (°C/W)		τ_{th} (μs)	
Gate Periphery	Simulated	Measured	Simulated	Measured
0.5 mm				6.40
0.8 mm	19.6	8.35	29	20
3.2 mm		6.4	85	65

From: [34].

In [31], the τ_{th} and R_{th} values obtained by Dahmani et al. for the studied 3.2-mm device (i.e., τ_{th} = 85 μs and R_{th} = 8.35 °C/W) have been used to complete the electrothermal subcircuit and to implement the I_{ds} model.

For the 2-mm device the layer structure data was unavailable and the heat-transfer simulations could not be reliably implemented to calculate R_{th} and τ_{th}. In consequence, the I_{ds} model for constant case temperature expressed by (23.61) was utilized, where $f_{th}{\cdot}R_{th}$ is denoted as f_P, and the value of the thermal resistance R_{th} is not strictly needed. Then, the effect of R_{th} is properly taken away from the electrothermal subcircuit making R_{th} equal to 1°C/W, which implies that the C_{th} values expressed in W·s/°C must be equal to the thermal time constant τ_{th} value in seconds obtained from transient measurements.

23.6 EXTRACTION OF ELECTROTHERMAL DRAIN CURRENT MODEL

In the following, the extraction of the model parameters of the dispersive I_{ds} current source is described. According to [21], the adopted I_{ds} model assumes that the signal frequencies of v_{gs} and v_{ds} are higher than the upper cutoff frequency of the low-frequency dispersive

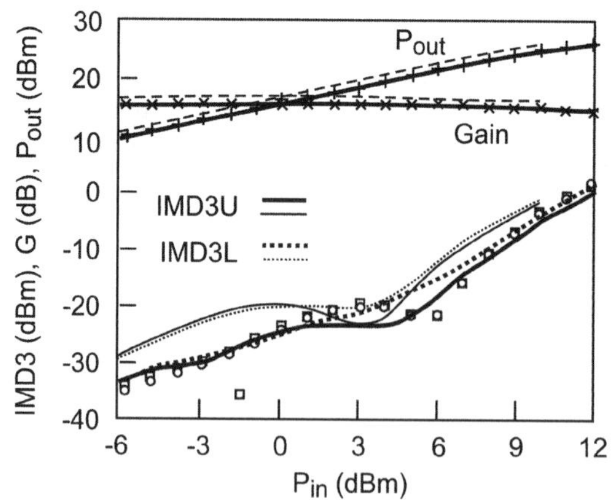

Figure 23.15 Simulation and measurement of a 3.2-mm AlGaN/GaN HEMT using a first-order [37] and third-order [36] electrothermal subcircuit. Two-tone signal measurement was performed in a 50Ω environment (f_0 = 2.14 GHz, Δf = 200 kHz). Symbols denote measurements. Thin lines present simulations after [37] (first-order RC circuit), and thick lines indicate results with third-order model

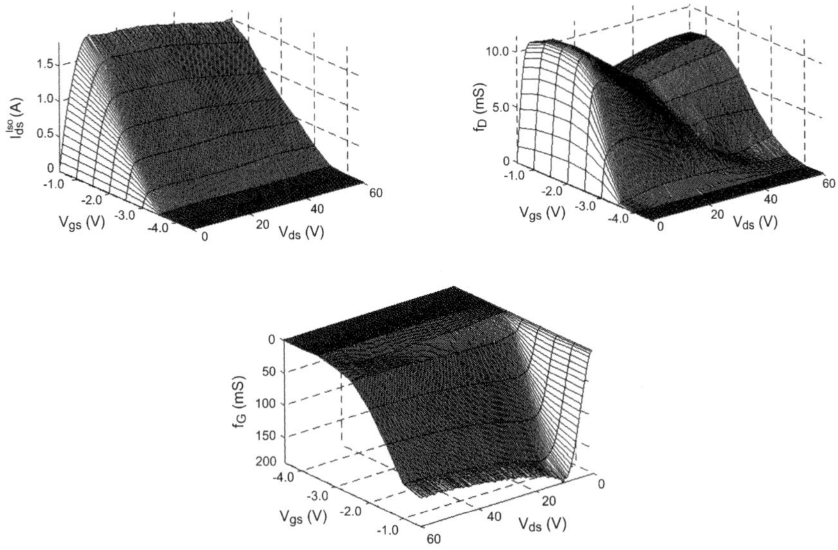

Figure 23.16 Calculated values of isothermal drain current $I_{ds}^{Iso}(v_{gs}, v_{ds})$ and trapping correction functions $f_G(v_{gs}, v_{ds})$ and $f_D(v_{gs}, v_{ds})$ for the 2-mm device [31].

phenomena due to self-heating and charge-trapping effects, but lower than microwave-related effects of junction charge-storage and propagation times. Pulsed-DC IV measurements are performed using voltage pulses for v_{gs} and v_{ds} with related signal frequencies that satisfy this model assumption. Voltage pulse width of 0.1 μs was employed, equivalent to a signal frequency of 10 MHz. The time period between two successive pulses were chosen as 1 ms, equivalent to a pulse repetition frequency of 1 kHz. This results in a duty cycle of 0.01%. The narrow pulse width and low duty cycle ensure that the voltage pulses change faster than the low-frequency dispersive phenomena and minimize heat accumulation from one pulse to the next. As a result, charge trapping and self-heating effects are effectively controlled by the quiescent bias point of the pulsed-DC IV measurement [31].

23.6.1 Extraction of Isothermal Drain Current and Trapping Correction Functions

As in [21, 38–40], voltage-dependent values of $I_{ds}^{Iso}(v_{gs}, v_{ds})$, $f_G(v_{gs}, v_{ds})$, and $f_D(v_{gs}, v_{ds})$ were extracted from pulsed-DC IV measurements on passive quiescent bias points (i.e., with negligible P_{DS}). Under such conditions the thermal-related parameter (f_{th} or f_P) vanishes. The three passive quiescent points employed were: (1) $V_{GS} = 0V$ with $V_{DS} = 0V$, (2) $V_{GS} < V_{th}$ with $V_{DS} = 0V$, and (3) $V_{GS} < V_{th}$ with $V_{DS} \gg 0V$.

The voltage-dependent values of $I_{ds}^{Iso}(v_{gs}, v_{ds})$, $f_G(v_{gs}, v_{ds})$, and $f_D(v_{gs}, v_{ds})$ that were calculated are shown in Figure 23.16 and Figure 23.17 [31]. Voltage dependencies of I_{ds}^{Iso} that are shown in those figures present the expected shape of IV characteristics, but with nearly constant values in the saturation region. This corresponds well with an ideal isothermal IV characteristic, which is the effect represented by I_{ds}^{Iso}.

The parameters $f_G(v_{gs}, v_{ds})$ and $f_D(v_{gs}, v_{ds})$ shown in Figure 23.16 for the 2-mm GaN device

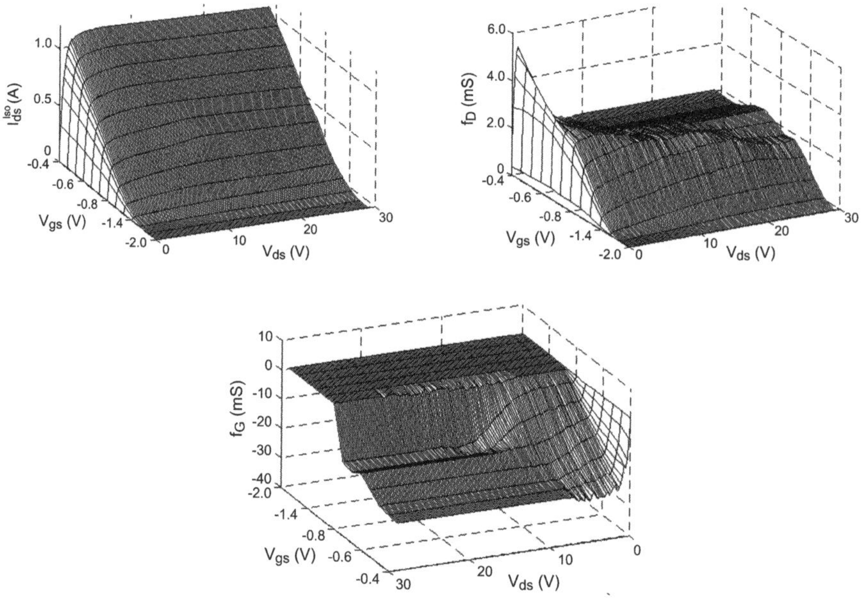

Figure 23.17 Calculated values of $I_{ds}^{Iso}(v_{gs}, v_{ds})$, $f_G(v_{gs}, v_{ds})$ and $f_D(v_{gs}, v_{ds})$ for the 3.2-mm GaN device [31].

have larger values than those shown in Figure 23.17 for the 3.2-mm device. This agrees with observations made during pulsed-DC IV characterization, where charge-trapping effects appeared more pronounced for the 3.2-mm device than for the 2-mm device. Note the device fabrication technology engineered by Nitronex (see Section 4.2.4 in [14]) seems likely to be the main reason for the lower charge-trapping effects observed for the 2-mm device.

23.6.2 Extraction of Thermal Parameters f_{th} and f_p

According to Filicori et al. [21], the thermal-related parameters of the I_{ds} model $f_{th}(v_{gs}, v_{ds})$ or $f_p(v_{gs}, v_{ds})$ are calculated using a pulsed-DC IV measurement without self-heating effects as a reference, typically $V_{GS} = 0$V with $V_{DS} = 0$V, together with pulsed-DC IV measurements on active quiescent bias points (quiescent bias points with nonnegligible P_{DS}).

Actually, in previous I_{ds} models based on the ideas of Filicori et al., such as [15, 27, 31], f_p is calculated using pulsed-DC IV measurements on only one active quiescent bias point. Experience has shown that when modeling large-size devices, the calculation of f_p with pulsed-DC IV measurements on a single active quiescent bias point results in models with strictly localized accuracy [31]. This means that during model verification tests, the accuracy of those models rapidly degraded as the tested bias points moved away of the active quiescent bias point used to calculate f_p.

IV measurements on a single active quiescent bias point results in models with strictly localized accuracy. This means that during model verification tests, the accuracy of those models rapidly degraded as the tested bias points moved away of the active quiescent bias point used to calculate f_p.

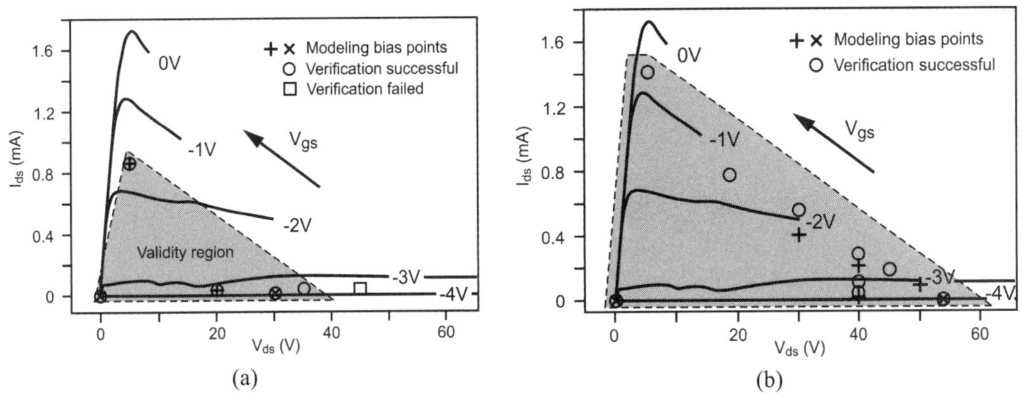

Figure 23.18 Validity regions for pulsed-DC IV verification tests of models developed for the 3.2-mm GaN device: (a) Model validity region considering only two active quiescent bias points for the calculation of *fP*, and (b) expanded model validity region considering additional active quiescent bias points for the calculation of *fP*. (After [31]).

According to [21], the expression of the adopted I_{ds} model is a linearization with respect to V_{gs}, V_{ds} and T_{ch}, around a given operation condition. This model assumption implies a linear or mildly nonlinear dependence of the average values of the thermal states on V_{gs} and V_{ds}.

In reality, if the actual dependency of the average values of the thermal states on the bias condition were as linear as assumed, the use of additional pulsed-DC IV measurements on different active quiescent bias points would not add linearly independent data to the calculation of f_p. Then, similar results could be expected independently of the number of different active quiescent bias points considered to calculate f_p.

However, as reported by Zamudio [31] considering additional pulsed-DC IV measurements on different active quiescent bias points improved the representation of self-heating effects by f_p and the extension of the model validity region. The region containing the bias points where the model is considered to produce accurate simulations is called *model validity region.*

Figure 23.18 depicts validity regions obtained for two modeling trials carried out for the 3.2-mm GaN power device that were tested comparing pulsed-DC IV measurements with model simulations. The only difference between the two models developed was the calculation of f_p. As shown in Figure 23.18, the second modeling trial (Figure 23.18(b)) employed more pulsed-DC IV measurements and with a broader distribution on the IV plane than the first modeling trial part (Figure 23.18(a)). As a result, an expansion of the model validity region was obtained for the second modeling trial with respect to the first one. Meanwhile, the accuracy of the related pulsed-DC IV verification tests was improved or preserved. The expansion of model validity region was directly related to the increasing number of quiescent bias points that were considered to calculate f_p [31].

The high temperatures of operation (high levels of dissipated power) of the studied devices together with device properties related to the fabrication technologies are the reasons considered to explain why the thermal states of the active device do not depend on V_{gs} and V_{ds} as linearly as assumed in the I_{ds} model formulation.

Table 23.6

Quiescent Bias Points with Non-Negligible P_{DS} Employed in the Calculation of $f_{th}(v_{gs}, v_{ds})$ for the 3.2-mm Device and and $f_P(v_{gs}, v_{ds})$ for the 2-mm Device *

GaN HEMT size	V_{GS} (V)	V_{DS} (V)	P_{DS} (W)
3.2-mm	-3.25	50	7.4
	-3.0	40	11.9
	-2.5	30	4.4
2-mm	-1.25	15	1.98
	-1.0	22	4.4
	-0.8	10	3.0
	-1.6	12	0.25
	-1.5	5	0.13

* From: [31].

In consequence, in [31], the voltage-dependent values of $f_{th}(v_{gs}, v_{ds})$ and $f_P(v_{gs}, v_{ds})$ were calculated using pulsed-DC IV measurements on multiple active *quiescent* bias points as listed in Table 23.6. The related pulsed-DC IV characterization results were reported in section 9.2.3.

The I_{ds}-model parameters related to self-heating effects that must be calculated are: (i) f_{th} for the model of the 3.2-mm device with R_{th} known and (ii) f_P for the model of the 2-mm device with R_{th} unknown. The calculated values of those parameters are presented in Figure 23.19.

Note that the values of f_{th} in Figure 23.19(a) and f_P in Figure 23.19(b) were expressed in terms of v_{gs} and v_{ds}, corresponding to the intrinsic voltages. But, pulsed-DC IV characteristics were measured in the extrinsic device ports and are terms of v_{GS} and v_{DS}. Therefore, the pulsed-DC IV measurements were remapped first to the intrinsic device ports using matrix equation (5.3), before applying the parameter extraction, including the V_{GS}, V_{DS} and P_{DS} values that characterize the *quiescent* bias point. Then, each remapped pulsed-DC IV measurement was regridded to an orthogonal $v_{gs} - v_{ds}$ grid. Afterward, interpolation and linear extrapolation were performed to receive an equally spaced full-covered $vgs - v_{ds}$ grid.

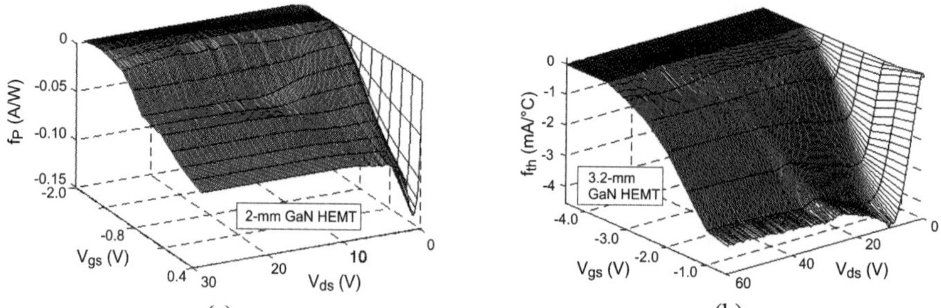

Figure 23.19 Calculated values of $f_{th}(v_{gs}, v_{ds})$ for a 2-mm GaN HEMT and calculated values of $f_P(v_{gs}, v_{ds})$ for a 3.2-mm GaN HEMT device [31].

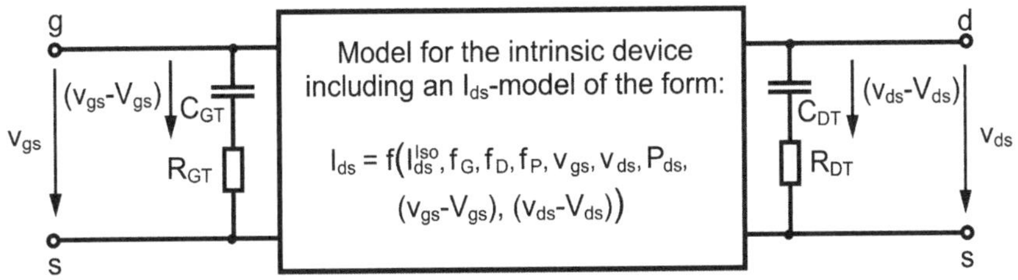

Figure 23.20 Illustration of the RC circuits used in [21] as high-pass filters for the computation of (v_{gs} − V_{gs}) and (v_{ds} − V_{ds}) during simulations, and employed in [38] to denote trapping time constants extracted from measured drain current transients [31].

The critical numerical processes of voltage remapping and regridding were carefully performed as delineated in Section 5.4 for C_{gs}, C_{gd},and C_{ds}.

23.7 EXTRACTION OF TRAPPING AND THERMAL TIME CONSTANTS

Previous I_{ds} models based on the ideas of Filicori et al. used the drain-source and gate-source RC circuits illustrated in Figure 23.20 for the model implementation in CAD [21, 38, 40]. In [21], those RC circuits mainly act as high-pass filters to filter out DC components of v_{gs} and v_{ds}, and to compute in the simulations the I_{ds}-model factors (v_{gs} − V_{gs}) and (v_{ds} − V_{ds}).

The RC circuits shown in Figure 23.20 are suitable to represent time constants of trapping effects, $R_{GT}\cdot C_{GT}$ and $R_{DT}\cdot C_{DT}$. However, values of those parameters were not related to device characterization by Filicori et al. in [15] or in similar I_{ds} models developed later [38, 40]. In [38], a measurement-based extraction of the time constants $R_{GT}\cdot C_{GT}$ and $R_{DT}\cdot C_{DT}$ was presented, using measured drain current transients.

In [31], a similar extraction of $R_{GT}\cdot C_{GT}$ and $R_{DT}\cdot C_{DT}$ was used as in [41], based on i_{DS} transients measured as response to pulsing the quiescent bias, from an initial to pulsed-to point. The initial and pulsed-to quiescent bias points used for extraction of $R_{GT}\cdot C_{GT}$ and $R_{DT}\cdot C_{DT}$ of the 3.2-mm GaN device are illustrated in Figure 23.21. The following initial (cross) and pulsed-to (X) quiescent bias points were chosen for the transient measurements

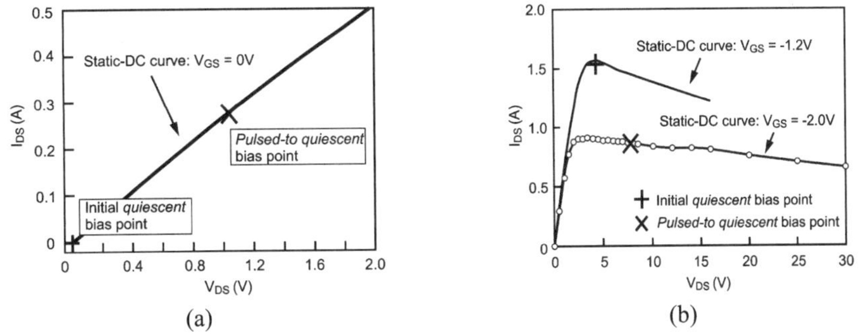

Figure 23.21 Illustration of initial (cross) and pulsed-to (X) quiescent bias points chosen for the transient measurements of the 3.2-mm device: (a) extraction of $R_{GT}\cdot C_{GT}$, and (b) extraction of $R_{DT}\cdot C_{DT}$ [31].

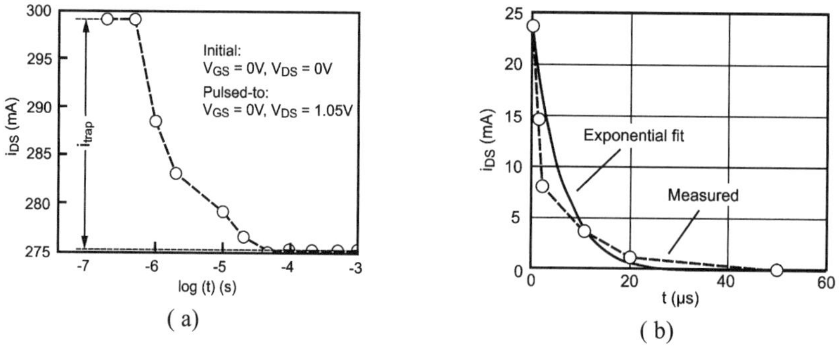

Figure 23.22 (a) Measured i_{DS} transient used to extract $R_{DT}{\cdot}C_{DT}$ of the 3.2-mm device, and (b) separated i_{trap} (dot-dash line) and exponential fit [31].

of the 3.2-mm device used to extract (a) $R_{DT} \cdot C_{DT}$ (V_{GS} = 0V, V_{DS} = 0V, P_{DS} = 0V and V_{GS} = 0V, V_{DS} = 1.05V, $P_{DS} \approx$ 0.31W) and (b) $R_{GT} \cdot C_{GT}$ (V_{GS} = -1.2V, V_{DS} = 4.5V, $P_{DS} \approx$ 5W and V_{GS} = -2V, V_{DS} = 7.5V, $P_{DS} \approx$ 5W).

Figure 23.22 depicts the measured i_{DS} transient employed to extract the time constant $R_{DT}{\cdot}C_{DT}$ of the 3.2-mm GaN device. The selected initial and *pulsed-to quiescent* bias points were (V_{GS} = 0V, V_{DS} = 0V) and (V_{GS} = 0V, V_{DS} = 1.05V), respectively. The duration of the pulse was 1 ms and the pulse was not repeated.Those *quiescent* bias points were selected to isolate the time constant of V_{DS}-related trapping effects, in accordance with the following considerations, as illustrated in Figure 23.21(a) [31]:

- V_{GS}-related trapping effects were not activated because V_{GS} remained constant in 0V for the initial and pulsed-to quiescent bias points.

- Self-heating was considered negligible because the initial and pulsed-to quiescent P_{DS} were 0W and 0.31W (i.e., 0.1 mW/mm), respectively.

- The time constant was related to charge-trapping effects (i.e., electron capture) because decreasing i_{DS} transient was measured for an increased V_{DS} magnitude from the initial to pulsed-to quiescent bias point.

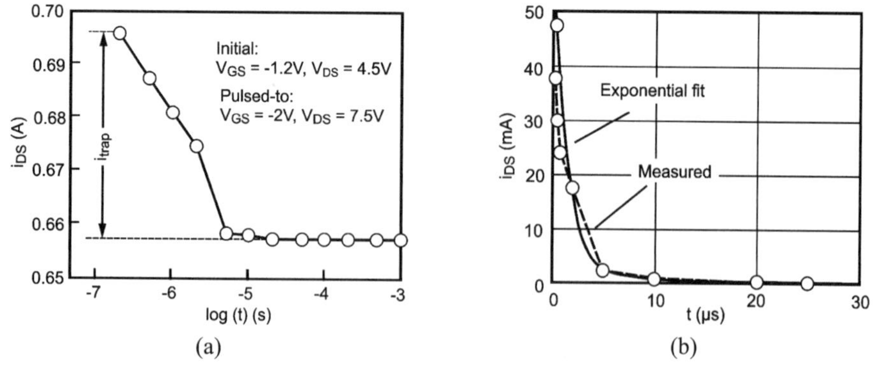

Figure 23.23 (a) Measured i_{DS} transient used to extract $R_{GT}{\cdot}C_{GT}$ of the 3.2-mm device, and (b) separated i_{trap} (dot-dash line) and exponential fit [31].

The time constant $R_{DT}\cdot C_{DT}$ was extracted from the measured transient shown in Figure 23.22 with an exponential fit, assuming that the i_{DS} component attributed to V_{DS}-related trapping effects is described by

$$i_{trap}(t) = I(0)\cdot e^{-t/\tau_{trap}} \tag{23.64}$$

Figure 23.23 depicts the measured i_{DS} transient used to extract $R_{GT}\cdot C_{GT}$ of the 3.2-mm device. The selected initial to pulsed-to quiescent bias points were $V_{GS} = -1.2$V with $V_{DS} = 4.5$V and $V_{GS} = -2$V with $V_{DS} = 7.5$V, respectively. The duration of the V_{GS} and V_{DS} pulses was 1 ms and the pulses were not repeated.Those quiescent bias points were selected to isolate the time constant of V_{GS}-related trapping effects in accordance with the following considerations, as illustrated in Figure 23.21(b):

- V_{DS}-related trapping effects were not considered dominant, because the initial and pulsed-to V_{DS} were in the saturation region ($V_{DS} > V_k$), where current depends mostly on V_{GS}. Moreover, those trapping effects were considered weakly activated, since the 3V-change from initial and pulsed-to V_{DS} was small compared to the 54V-change used to extract the related trapping-effects parameter f_D. Thus, a 3V-change is small compared to the V_{DS} change of the maximum V_{DS}-related trapping effects observed.
- Self-heating effects were not activated P_{DS} for the initial and pulsed-to quiescent bias points.
- The time constant was related to charge-trapping effects (i.e., electron capture) because decreasing i_{DS} transient was measured for an increased V_{GS} magnitude from the initial to pulsed-to quiescent bias point.

The time constant $R_{GT}\cdot C_{GT}$ was extracted from the measured transient shown in Figure 23.22 with an exponential fit, assuming that the i_{DS} component attributed to V_{GS}-related trapping effects is described by (23.64).

23.8 SUMMARY

In this chapter the development of an electrothermal nonlinear model for field-effect transistors was described including gate- and drain-lag characteristics. Starting with a measurement-based bias-dependent small-signal equivalent network a nonlinear large-signal quasi-static model was derived by integration over bias-dependent model parameters derived from S-parameter measurements. This model does not consider any dispersion effect. To extend the validity of the model to higher frequencies, memory effects must be included. These concern thermal and trapping relaxation effects. Special attention was devoted to the modeling of the dispersive drain current source. Simple dispersion models consist of an RC series branch in parallel to the drain-source output terminals, which becomes effective at higher frequencies increasing the output conductance. As the RF conductance is bias dependent, measurement-based extracted models are only valid in a considered bias point. Other dispersion models are composed of three current components: isothermal DC characteristics, a DC-RF dispersion term and a term, which describes the dynamic shifting of the quiescent bias point. The last term was defined on the difference of the averaged RF and DC current (Root model) or on the difference of the averaged RF and DC control voltages as described in this chapter (Filicori model). At low signal levels both models deliver similar results. The usage of pulsed-DC measurements for the characterization of the dispersive drain current source permits a detailed and differentiated analysis of various thermal and electrical dispersion effects.

In addition to trapping dispersion the thermal memory effects were highlighted emphasizing their importance for AlGaN/GaN power devices. It was shown that higher-order electrothermal subcircuits may increase simulation accuracy. The electrothermal current model proposed by Filicori et al. was described in detail. Over time this model was proven to be highly reliable and very robust. It can be easily implemented in commercial software.

REFERENCES

[1] Camacho-Penelosa, C., and C. S. Aitchison, "Modelling Frequency Dependence of Output Impedance of a Microwave MESFET at Low Frequencies," *Electron. Lett.*, Vol. 21, 1985, pp. 528–529.

[2] Golio, J. M., M. G. Miller, G. N. Maracas, and D. A. Johnson, "Frequency-Dependent Electrical Characteristics of GaAs MESFET′s," *IEEE Trans.*, Vol. ED-37, 1990, pp. 1217–1226.

[3] Nsele, S. D., L. Escotte, J.-G. Tartarin, S. Piotrowicz, and S. L. Delage, "Broadband Frequency Dispersion Small-Signal Modeling of the Output Conductance and Transconductance in AlInN/GaN HEMTs," *IEEE Trans.*, Vol. ED-60(4), 2013, pp. 1372–1378.

[4] Binari, S. C., P. B. Klein, and T. E. Kazior, "Trapping Effects in GaN and SiC Microwave FETs," *Proc. of the IEEE*, Vol. 90, 2002, pp. 1048–1058.

[5] Tirado, J. M., J. L. Sánchez-Rojas, and J. I. Izpura," Trapping Effects in the Transient Response of AlGaN/GaN HEMT Devices," *IEEE Trans.*, Vol. ED-54(3), 2007, pp. 410–417.

[6] Jardel, O., F. De Groote, C. Charbonniaud, T. Reveyrand, J. P. Teysier, R, Quere, and D. Floriot, "A Drain-Lag Model for AlGaN/GaN Power HEMTs," *2007 IEEE/MTT-S International Microwave Conference*, 2007, pp. 601– 604.

[7] Tartarin, J. G. G. Astre, S. Karboyan, T. Noutsa, and B. Lambert, "Generation-Recombination Traps in AlGaN/GaN HEMT Analyzed by Time-Domain and Frequency-Domain Measurements: Impact of HTRB Stress on Short Term and Long Term Memory," *2013 IEEE International Wireless Symposium (IWS)*, 2013 (4 pages).

[8] Hefner; A. R., "A Dynamic Electro-Thermal Model for IGBT," *2002 IEEE Industry Applications Society Annual Meeting*, 4.- 9. Oct. 2002, pp. 1094 - 1104.

[9] Grasser, T. (Ed.), *Bias Temperature Instability for Devices and Circuits*, Springer, 2014.

[10] Kompa, G., "Modeling of Dispersive Microwave FET Devices Using a Quasi-Static Approach," *Int. J. of Microwave and Millimeter-Wave Computer-Aided Engineering* (MMWCAE), Vol. 5, No. 3, 1995, pp. 173–194.

[11] Hasler, M., and J. Neirynck, *Nonlinear Circuits*, Norwood: Artech House, 1986.

[12] Madjar, A., and F. J. Rosenbaum, "A Practical AC Large-Signal Model for GaAs Microwave MESFETs,"

IEEE MTT-S Int. Microwave Symp. Digest, 1979, pp. 399–401.

[13] Närhi, T., "Analysis of Strongly Nonlinear Circuits with a Frequency-Domain Method Coupled with a Consistent Large-Signal Model," *IEEE MTT-S Int. Microwave Symp. Digest*, 1993, pp. 633–636.

[14] Kompa, G., *Basic Properties of III-V Devices–Understanding Mysterious Trapping Phenomena*, Kassel, Kassel University Press GmbH, 2014.

[15] Werthof, A., F. van Raay, and G. Kompa, "Direct Nonlinear Power MESFET Parameter Extraction and Consistent Modeling," IEEE MTT-S Int. Microwave Symp. Dig., Atlanta, USA, 1993, pp. 645– 648.

[16] Werthof, A., F. van Raay, and G. Kompa, "Neuartiges Verfahren zur Parameterextraktion nichtlinearer FET Modelle auf der Basis fehlerkorrigierter komplexer Signalspektren," Konferenz für Mikrowelle und Optronik (MIOP), Sindelfingen, Germany, 25.-27. Mai, 1993, S. 106–110.

[17] Rauscher, C., and H. A. Willing, "Simulation of Nonlinear Microwave FET Performance Using a Quasi-Static Model," *IEEE Trans.*, Vol. MTT-27(10), 1979, pp. 834-840.

[18] Root, D. E., S. Fan, and J. Meyer, "Technology-Independent Large-Signal FET Models: A Measurement-Based Approach to Active Device Modeling," *Proc. 15th ARMMS Conference*, September 1991, Paper 5.

[19] Werthof, A., and G. Kompa, "A Unified Consistent DC to RF Large Signal FET Model Covering the Strong Dispersion Effects of HEMT Devices," *22nd European Microwave Conference (EuMC)*, 1992, pp. 1091–1096.

[20] Schmale, I., *Development of a Consistent Electrothermal Large Signal Model for Field-Effect Transistors Used for the Reliable Design of Monolithic Frequency Doublers* (in German), Doctoral Thesis, Kassel University, 1999.

[21] Filicori, F., G. Vannini, A. Santarelli, A. Mediavilla Sánchez, and Y. Newport, "Empirical Modeling of Low-Frequency Dispersive Effects Due to Traps and Thermal Phenomena in III-V FET′ s," *IEEE Trans.*, Vol. MTT-43, 1995, pp. 2972–2981.

[22] Filicori, F, G. Vannini, A. Mediavilla, and A. Tazon, "Modelling of Deviations Between Static and Dynamic Drain Characteristics in GaAs FETs," *23nd European Microwave Conference (EuMC)*, 1993, pp. 454–457.

[23] Schreurs, D., *Measurement Based Modelling of Heterojunction Field-Effect Devices for Non-Linear Microwave Circuit Design*, Doctoral Thesis, Katholieke Universiteit Leuven, 1997.

[24] Chaibi, M., T. Fernandez, J. Rodriguez-Tellez, J. L. Cano, and M. Aghoutane, "Accurate Large-Signal Single Current Source Thermal Model for GaAs MESFET/HEMT," *Electronics Lett.*, Vol. 43(14), 2007, pp. 775–777.

[25] Jeon, K., Y. Kwon, and S. Hong, "A Frequency Dispersion Model of GaAs MESFET for Large-Signal Applications," *IEEE Microwave and Guided Lett.*, Vol. 7(3), 1997, pp. 78–80.

[26] Raffo, A., V. Vadala, G. Vannini, and A. Santarelli, "A New Empirical Model for the Characterization of Low-Frequency Dispersive Effects in FET Electron Devices Accounting for Thermal Influence on the Trapping State," *IEEE MTT-S International Microwave Symposium Digest*, 2008, pp. 1421–1424.

[27] Roh, T., Y. Kim, Y. Suh, W. Park, and B. Kim, "A Simple and Accurate MESFET Channel-Current Model Including Bias-Dependent Dispersion and Thermal Phenomena," *IEEE Trans.*, Vol. MTT-45(8), 1997, pp. 1252–1255.

[28] McGovern, P., J. Benedikt, P. J. Tasker, J. Powell, K. P. Hilton, J. L. Glasper, R. S. Balmer, T. Martin, and M. J. Uren, "Analysis of DC-RF Dispersion in AlGaN/GaN HFET′s Using Pulsed I-V and Time-Domain

Waveform Measurements," *IEEE MTT-S International Microwave Symposium Digest*, 2005, Vol. 2, pp. 4–8.

[29] McGovern, P., P. J. Tasker, J. Powell, K. P. Hilton, J. L. Glasper, R. S. Balmer, T. Martin, and M. J. Uren, "Experimental Investigation of DC-RF Dispersion in AlGaN/GaN HFET′s Using Pulsed I-V and Time-Domain Waveform Measurements," *Int. Conf. on Compound Semiconductor Manufacturing Technology* (CSMANTECH), 2005, Digest Paper 14.14.

[30] Rodriguez-Tellez, J., T. Fernandez, A. Mediavilla, and A. Tazon, "Characterization of Thermal and Frequency-Dispersion Effects in GaAs MESFET Devices," *IEEE Trans.*, Vol. MTT-49(7), 2001, pp. 1352–1355.

[31] Zamudio Flores, J. A., *Device Characterization and Modelling of Large-Size GaN HEMTs*, Doctoral Thesis, Kassel: Kassel University Press GmbH, 2012.

[32] Berroth, M., E. Chigaeva, I. Dettman, N. Wieser, W. Vogel, H. Roll, F. Scholz, and H. Schweizer, "Advanced Large-Signal Modeling of GaN-HEMTs," *Proceedings of the IEEE Lester Eastman Conference on High Performance Devices*, 2002, pp. 172–180.

[33] Dahmani, S., E. S. Mengistu, and G. Kompa, "Electro-Thermal Modeling of Large-Size GaN HEMTs," *Proceedings of the German Microwave Conference (GeMIC)*, 2008, pp. 1–4.

[34] Dahmani, S., E. S. Mengistu, and G. Kompa, "Thermal Model Extraction of GaN HEMTs for Large-Signal Modeling," *Proceedings of the 3rd European Microwave Integrated Circuits Conference (EuMIC)*, 2008, pp. 226–229.

[35] Gaska, R., A. Osinsky, J. W. Yang, and M. S. Shur, "Self-Heating in High-Power AlGaN-GaN HFET′s," *IEEE Electron Device Lett.*, Vol. 19, 1998, pp. 89–91.

[36] Dahmani, S., *Large-Size AlGaN/GaN HEMT Large-Signal Electrothermal Characterization and Modeling for Wireless Digital Communications*, Doctoral Thesis, University of Kassel, 2011.

[37] Jarndal, A. H., *Large-Signal Modeling of GaN Device for High Power Amplifier Design*, Doctoral Thesis (2006), University of Kassel, published by the Kassel University Press GmbH, 2007.

[38] Jarndal, A., and G. Kompa, "Large-Signal Model for AlGaN/GaN HEMT Accurately Predicts Trapping- and Self-Heating Induced Dispersion and Intermodulation Distortion," *IEEE Trans.*, Vol. ED-54, 2007, pp. 2830–2836.

[39] Mengistu, E. S., and G. Kompa, "A Large-Signal Model of GaN HEMTs for Linear High Power Amplifier Design," *Proceedings of the European Microwave Integrated Circuits Conference (EuMIC)*, 2006, pp. 292–295.

[40] Di Giacomo, V., A. Santarelli, F. Filicori, A. Raffo, G. Vannini, R. Aubry, and C. Gaquière, "Low-Frequency Dynamic Drain Current Modeling in AlGaN-GaN HEMTs," *Proceedings of the European Microwave Integrated Circuits Conference (EuMC)*, 2007, pp. 68–71.

[41] Mengistu, E. S., *Large-Signal Modeling of GaN HEMTs for Linear Power Amplifier Design*, Doctoral Thesis, University of Kassel, 2007.

Chapter 24

Large-Signal Measurement Techniques for Device Characterization and Model Verification

Large-signal measurement techniques are needed to analyze the nonlinear electrical performance of microwave transistors to verify derived large-signal models and to analyze and optimize complete nonlinear circuits, such as power amplifiers, oscillators, and multipliers, by purely experimental means.

This chapter describes measurement setups in frequency and time domain focusing on high-power applications with harmonic source- and load-pull facilities.

24.1 OVERVIEW OF LARGE-SIGNAL MEASUREMENT METHODS

As shown in Figure 24.1, large-signal measurement techniques may be roughly divided into three main groups: *large-signal S-parameter*, *pulsed*, and *signal waveform* measurements. Load-pull methods characterize a measurement concept that controls the load impedance to optimize the performance of a device with respect to output power, power efficiency, intermodulation distortion, and so forth. The recent emergence of GaN technology and the provision of high-power GaN field-effect transistors for broadband digital communications has posed new challenges in large-signal device characterization and model verification. This, in particular, refers to high-power capability of measurement setups, multitone and

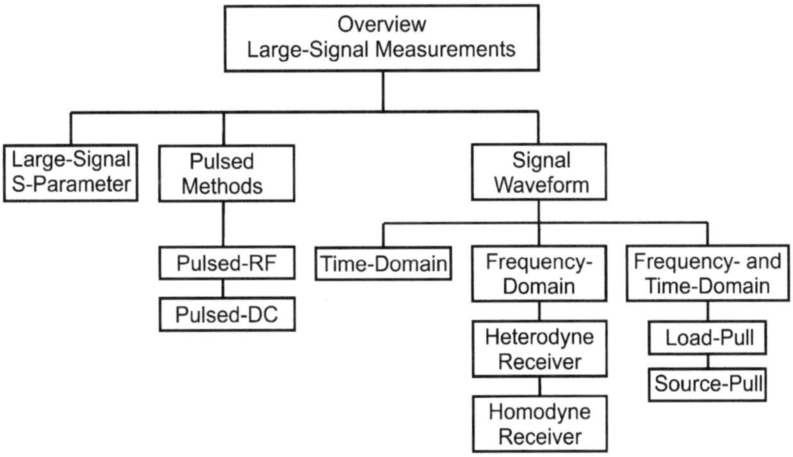

Figure 24.1 Overview of large-signal measurement techniques.

stochastic (e.g., WCDMA) stimulation signals, multiharmonic and multizone source- and load-pulling, and reliable intermodulation distortion characterization including sweet spot phenomena. In [1] a historical review of the evolution of various large-signal measurement techniques is given. Therefore, in the following, the main focus is on recently developed high-power large-signal/nonlinear measurement methods based on frequency and combined frequency/time-domain signal-waveform methods.

24.2 EVOLUTION OF COMBINED FREQUENCY/TIME-DOMAIN SIGNAL WAVEFORM MEASUREMENT

From the author's own industrial experience in the eighties, the oscillator circuit has played a key role in communication and sensor applications. Applying the S-parameter concept at that time, oscillation frequency could be estimated pretty well. However, prediction of oscillator output power commonly failed [2]. This lack of reliable nonlinear circuit design was the strong motivation for our group to develop new large-signal measurement methods for the nonlinear characterization of microwave devices. They were used to verify various nonlinear modeling approaches, but were also utilized to derive nonlinear model elements directly from large-signal measurements.

24.2.1 Pure Signal Waveform Measurement

An early version of an automatic broadband large-signal waveform measurement system combining frequency- and time-domain methods was published in 1989 by van Raay and Kompa [3-6]. The fundamental and harmonic frequency range was 8 and 20 GHz,

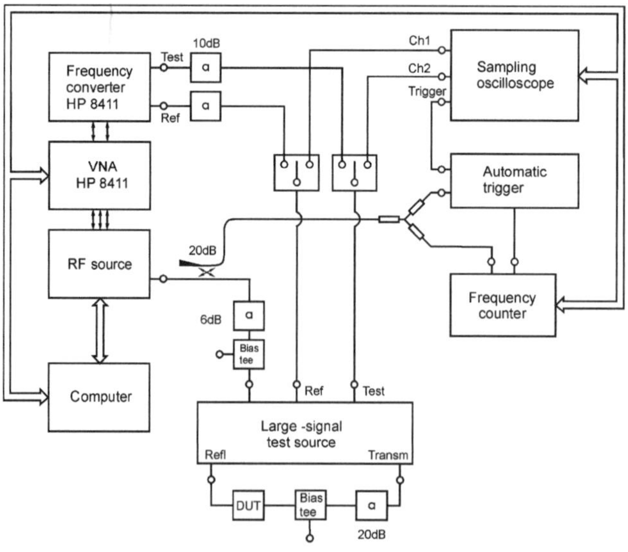

Figure 24.2 First automatic signal waveform measurement setup using vector network analyzer for accurate calibration and wave ratio measurements, and sampling oscilloscope for signal waveform measurement [3].

respectively. As depicted in Figure 24.2, the large-signal measurement system comprises the following key components: VNA, dual-channel sampling oscilloscope, and large signal test set. The vector network analyzer is used for system calibration and for complex wave ratio measurements of the reflected and transmitted fundamental waves. The signal waveforms are recorded by a sampling oscilloscope. The amplitude of the incident wave is determined from the signal waveform by fast Fourier transform (FFT) application. The phase of the incident wave is normalized to zero. All higher harmonic amplitudes and phases are directly obtained from time-domain data. The small-signal hp8743A testset of the HP 8410B network analyzer [7] was modified as a large-signal testset, which transmits full power of the source to the device. At that time the power was limited to 2W.

A large-signal error model has been established describing the systematic measurement errors of the system, and a calibration procedure has been adopted to determine the error model parameters. The calibration has been verified by measuring an ultrafast Schottky diode with different offsets with respect to the defined system reference planes.

Measurement is not only restricted to devices with coaxial input and output ports. With a planar test fixture and using the THLR calibration method reported in [8], it is possible to extend the signal waveform measurements to devices provided for planar circuit installation, such as packaged devices with microstrip leads or chip devices wire-bonded to microstrip connection lines.

Accurate and reliable automatic triggering of sampling oscilloscopes with highly repetitive signals is, as experience has shown, a real design challenge. Generally, the trigger operation is very sensitive, and changing the measuring frequency commonly needs new trigger adjustment. Therefore, for automating the signal waveform measurement set up, a special *automatic trigger circuit* had been developed to synchronize the sampling oscilloscope and microwave source [3]. It comprises an injection-locked FET oscillator with a fundamental frequency of approximately 250 MHz. Locking to the microwave signal is arranged via higher harmonics. Once tuned to a fixed fundamental frequency, harmonic measurements can simply be performed in a stepped-CW mode (i.e., at frequencies which are a multiple of the chosen fundamental frequency).

In 1992 we reported about an *on-wafer one-path two-port* measurement system with an extended 40-GHz harmonic frequency bandwidth [9]. The measurement setup is depicted in Figure 24.3.

The hp71500A microwave transition analyzer (MTA), available since 1991 [10–12], has been used as fundamental and harmonic receiver. Thereby, the first large-signal waveform measurement setup reported in [5] could be simplified considerably with the MTA substituting the vector network analyzer and sampling oscilloscope. Further improvements could be obtained with respect to measurement speed and simplification of the setup firmware. In particular, a trigger circuit for the stand-alone sampling oscilloscope is no longer required. A dynamic accuracy of about 50 dBc up to 8 GHz and 45 dBc up to 12 GHz has been verified without the application of averaging or noise filtering. The on-wafer setup was realized as a fully automatic system under remote control of an external computer. The error-correction algorithm of the earlier setup could be completely overtaken. An hp83650A 50-GHz synthesizer was used as the microwave source.

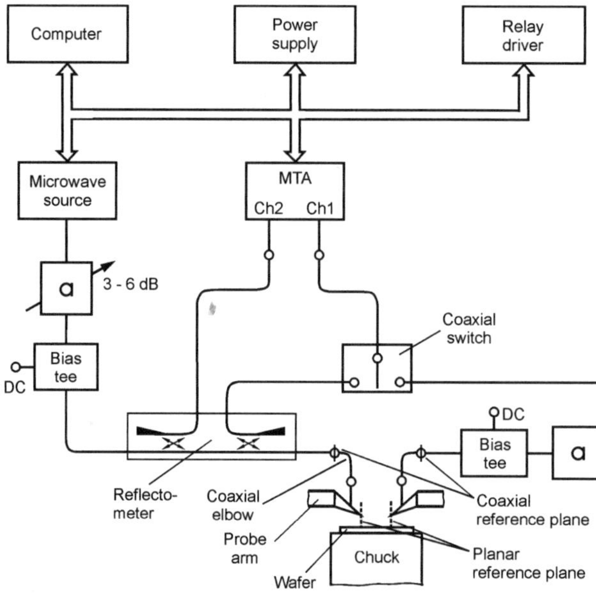

Figure 24.3 On-wafer large-signal waveform measurement system with 40GHz harmonic bandwidth [9].

The key feature of the realized hardware is that the coaxial testset reflectometer and the transmission return port are connected directly to the wafer probes of port 1 and 2 via coax elbows. The reflectometer is mounted on top of the probe arm. This configuration avoids the use of cables between the coaxial testset and the wafer probes, yielding the full hardware performance of the coaxial testset. Optimum hardware directivity and match are of special importance with large-signal measurements because the zigzag reflections between the DUT and source and load cannot be eliminated by the correction algorithm for reason of the nonlinearity of the DUT.

The 40-GHz frequency range of the configuration in Figure 24.3 is mainly limited by the sampling aperture of the MTA and the testset reflectometer hardware. The maximum available power level at the DUT on wafer reference plane is the source power of the synthesizer minus 3- or 6-dB attenuation of the RF input attenuator pad, and additional cable and reflectometer losses that sum up to about 2.0 to 2.5 dB at X-band frequencies so that the maximum RF power at the DUT is about 8 to 9 dBm at 10 GHz.

A two-step calibration procedure is applied (see Section 11.3.1.2). For this purpose, the probe heads and the components of the coaxial testset are separated from each other. In the first step, the system calibration with respect to the coaxial reference planes of the reflectometer and the transmission return path is performed. This calibration is based on common SOLT procedure (Section 10.4.2). Additional measurements are required to evaluate the amplitude and phase tracking of the directional couplers and the receiver tracking errors for fundamental ratio and harmonic measurement [6]. In the second step, the wafer probe heads (including elbows and adapters) are treated as the halves of a noncoaxial test fixture in a formal manner. The probes are calibrated on the hp 8510B 40-GHz VNA system using

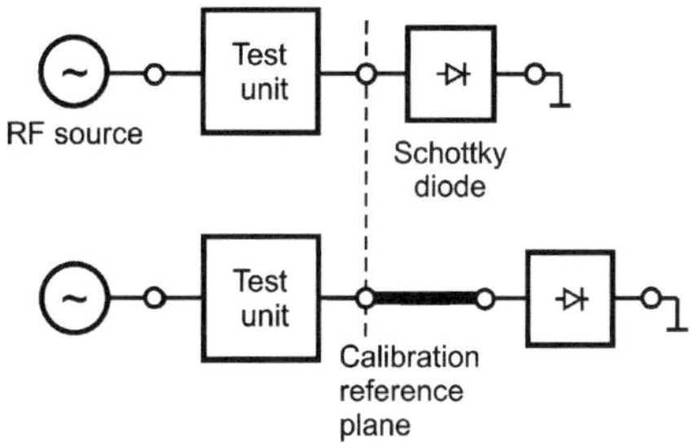

Figure 24.4 Calibration verification with an ultrafast Schottky diode. Tope: The coaxial adapter (coaxial-to-microstrip transition) is directly connected to the test port. Bottom: A short length of transmission line (realized by a coaxial adapter) is inserted between the measurement system and DUT reference plane.

the THLR fixture calibration method [8]. Instead of a sliding load, which is used at higher frequencies in this method, a fixed load is applied over the entire frequency range. The THLR calibration method evaluates the S-parameters of the wafer probes. With the error parameters of the receivers and the coaxial test system and the S-parameters of the probe heads, the raw measurement data obtained in the receiver reference planes can be transformed to the wafer probe reference planes of the DUT. This error correction algorithm yields a precise determination of the four wave spectra being present at the DUT, which is a complete description of its actual operating conditions.

Because of the lack of a nonlinear standard, an ultrafast beam-lead Schottky diode mounted in a microstrip test fixture half with a coaxial adapter is measured with and without a coaxial offset line between reflection test port of the large-signal test set and the reference plane of the test fixture. The two measurement arrangements are shown in Figure 24.4. The earlier 20-GHz signal waveform measurement system used beam lead silicon Schottky diode of type HSCH-5312 [13] with a low junction capacitance of about 0.1 pF (application recommended up to 26 GHz), whereas the calibration of the 40-GHz system was verified with a millimeter-wave zero-bias beam lead GaAs Schottky diode with an extremely low junction capacitance of about 35 fF (recommended application up to 110 GHz) [14].

The reflection response of a Schottky diode as a nonlinear DUT was measured with and without additional offset between the coaxial reference planes of the DUT and the reflectometer test ports. The deviations in the amplitude and phase spectra versus fundamental frequency (Figure 24.5) describe the influence of zigzag reflections between DUT and source. The harmonic phase spectra are normalized to the fundamental phase value. Harmonic phase values in the vicinity of 0 or 180 degrees describe the resistive nonlinear behavior of the diode that is due to its small junction capacitance and indicate that the omission of the sampler phase tracking error is justified up to 40 GHz with the MTA as harmonic receiver.

Figure 24.6 illustrates the effectiveness of the calibration procedure. It depicts both the raw and error-corrected time domain representation of an output signal at 10 GHz fundamental frequency for different input power levels, with four harmonics. With the highest power level, a 10% to 90% risetime and falltime of 15.5 ps is observed.

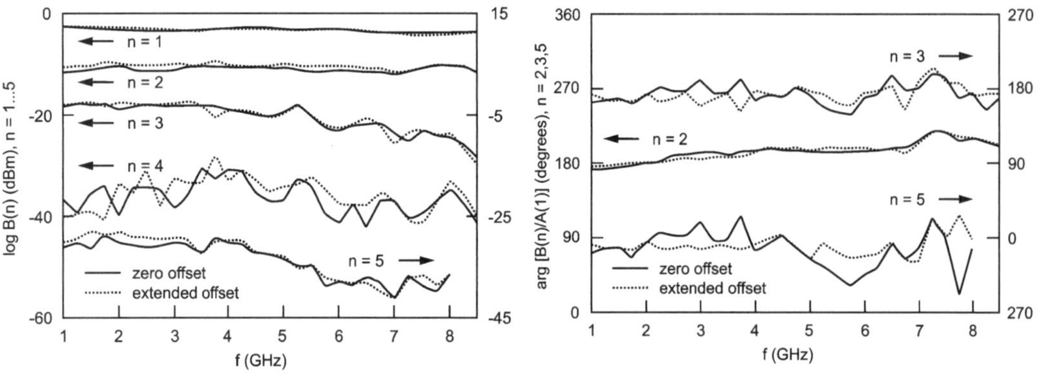

Figure 24.5 Error-corrected reflected harmonic power waves of a Schottky diode as a function of frequency measured with the 40-GHz waveform measurement system (solid lines: zero offset, dashed lines: extended offset). Source mismatching causes zigzag reflections between the source and DUT.

24.2.2 Fundamental Active Load-Pull Measurement

Classical load-pull measurement setups can be divided into two groups depending on whether passive or active tuning elements are used [15]. A general discussion of advantages and drawbacks of these main concepts can be found in [16]. In early load-pull measurement setups passive tuners were used. More recently, active tuning systems were developed: (1) *split-signal type* (e.g., [17, 18]) and (2) the *reflection type* configuration (e.g., [19, 20]). If extended to harmonic load-pull, these concepts lead to rather high tuner hardware complexity (e.g., [20, 21]). Commonly, variable attenuators and phase shifters are used for amplitude and phase setting in active load-pull configurations. In older solutions, a mechanical setting of stepper motors or switches was required for tuning. More modern solutions (e.g., PIN diode based attenuators) are still rather limited in bandwidth and lead to a rather expensive solution if, for example, coverage of several fundamental frequency bands for different applications is required. In 1995, an *electronic load module* (ELM) for the 1.9-GHz frequency band became commercially available [22]. One year later, a harmonic tuning capability was added [23]. For a given reflection coefficient for the fundamental, a large number of tuner states

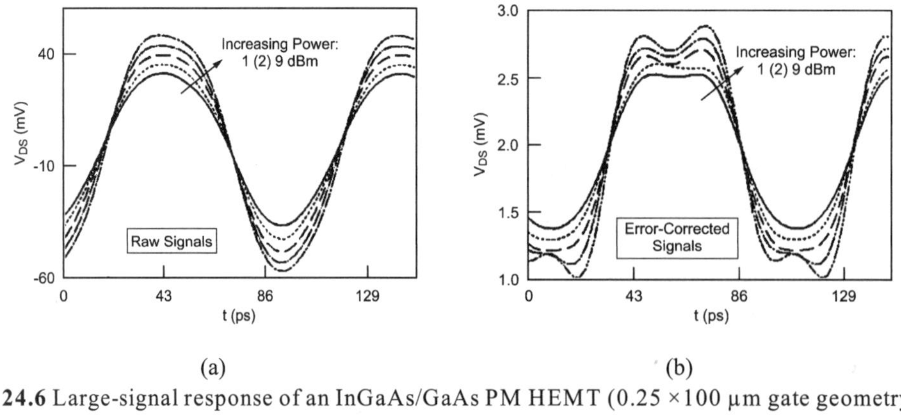

Figure 24.6 Large-signal response of an InGaAs/GaAs PM HEMT (0.25 ×100 μm gate geometry) for 10-GHz excitation: (a) recorded raw data, and (b) error-corrected signal at various power levels.

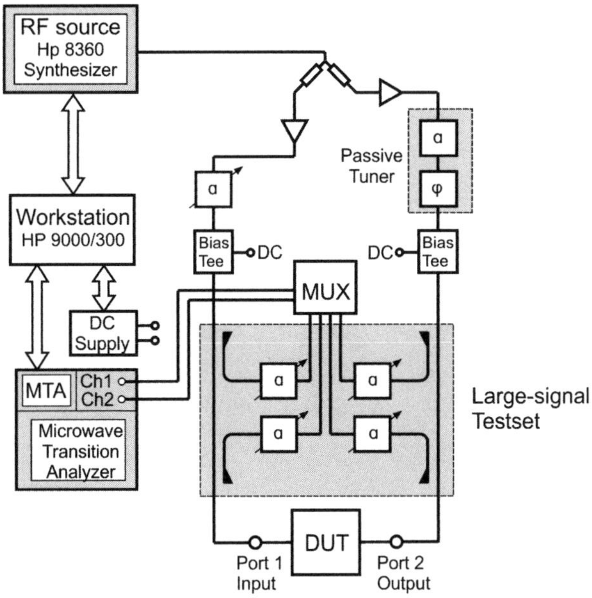

Figure 24.7 Active load-pull signal waveform measurement setup in two-path two-port configuration [24].

exists which, in total, covers the reflection plane for the harmonics within a radius of approximately 0.7. A straightforward tuner setting to given fundamental and harmonic reflection coefficients is not supported.

As will be described in the following, a new concept for an electronic load module was developed that is inherently broadband and allows a direct and independent setting of the reflection coefficients for different harmonics. Besides this, the advantages of the combination of waveform measurements and the harmonic load-pull approach are demonstrated. The key feature is the complete characterization of the operating state of a nonlinear two-port DUT for arbitrary periodic excitation or, respectively, the corresponding driving and loading conditions at both ports of the DUT [15].

In 1996 van Raay et al. [24] presented a new 40-GHz large-signal double reflectometer measurement system combining signal waveform and load-pull characterization. The setup comprises a double-reflectometer testset and a microwave transition analyzer (HP71500A) as the fundamental and harmonic receiver. For the first time, this setup permitted a complete description of the harmonic amplitude and phase spectra under arbitrary loading conditions of the DUT. Figure 24.7 illustrates the block diagram of the *two-path two-port* measurement system with 40-GHz harmonic frequency bandwidth combined with load-pull. The system comprises a 20-GHz synthesizer (RF source), a double-reflectometer testset hardware, and an MTA as the receiver. The four reflectometer monitor outputs are multiplexed to the MTA input channels via a simple 4:2 switch matrix. For optimum hardware performance, both reflectometers are mounted directly on top of the probe arms for on-wafer measurements. In-fixture or coaxial measurements are also possible as well. For fundamental active load-pull testing, the synthesizer output is split and fed to both reflectometers. The load signal is

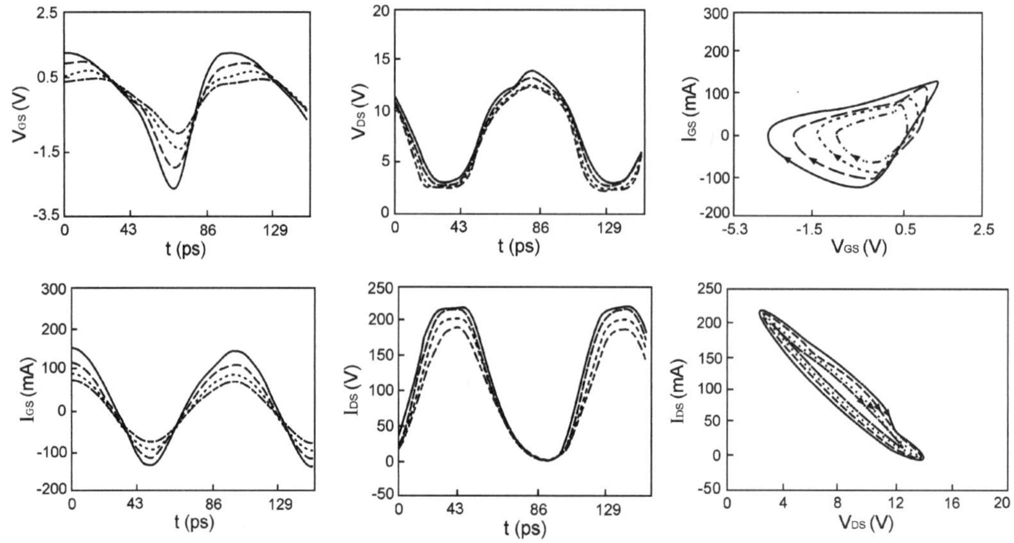

Figure 24.8 Input and load trajectories and voltage current waveforms of 0.3 μm PHEMT with 50-ohm loading at 10 GHz and 16(2)22 dBm input power (DC bias point: V_{GS} = 0V, V_{DS} = 8.6V).

supplied to the DUT via a power amplifier, step attenuator, and a variable delay line for phase shifting. The amplifier assures good isolation between DUT output and input port and maintains adjustment of up to total reflection (Γ = 1) load conditions even for amplifying DUTs. With respect to the MTA and reflectometer specifications, the harmonic frequency limit of the system is 40 GHz. The maximum fundamental input power to the DUT is about 6 dBm without using amplifiers in the source configuration. With respect to the power capability of the testset components, measurements with RF power levels up to 1W CW are possible. The attenuators in front of the switch MUX protect the receiver channels of the MTA from input overload and ensure their operation in the linear region below −10 dBm. These attenuators can be replaced by step attenuators that are switched simultaneously to make for example power sweep measurements over large power ranges (−10 dBm to +30 dBm) feasible.

The realized system is an entirely automatic, computer-controlled setup. As a demonstration of fundamental load-pull, power sweep measurements on a 0.3-μm PHEMT power device with double-recess asymmetric gate (which was manufactured by the Fraunhofer-Institute for Applied Solid-State Physics, Freiburg) were performed at 10 GHz with a passive load tuner adjusted to approximately 50 ohm (Figure 24.8) and toward optimum effective fundamental output power and load (Figure 24.9). The input drive level (up to 25 dBm at 10 GHz) was maintained with a commercial power amplifier between the RF source and the reflectometer input port. The RF power at the DUT's on-wafer input reference plane is precisely adjusted by means of a software leveling loop. In Figure 24.8 and Figure 24.9, the time-domain waveforms and trajectories at input and output are displayed for several input power levels and both aforementioned operating conditions. These error-corrected data refer to the on-wafer device reference planes.

In the tuned condition, the effective fundamental output power of the device is going

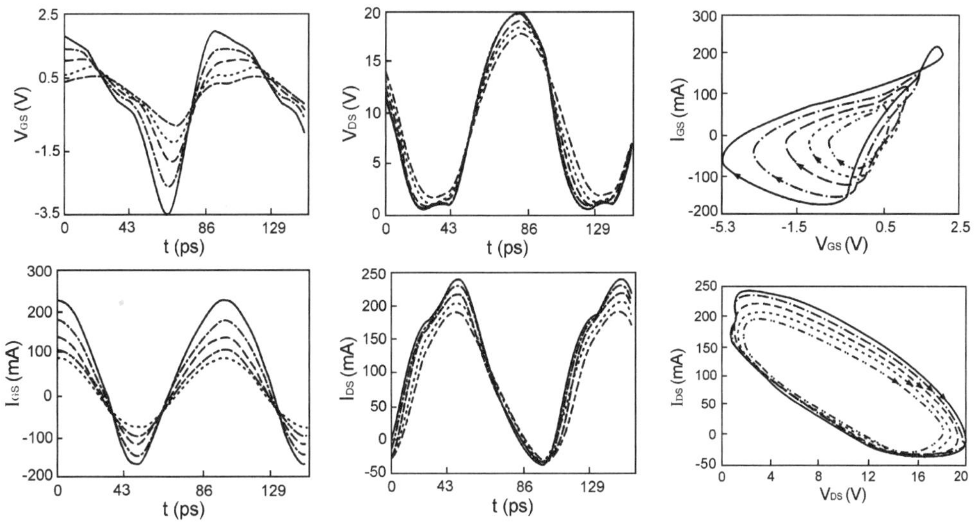

Figure 24.9 Input and load trajectories and voltage current waveforms of 0.3-μm PHEMT tuned to maximum output power at 10 GHz and 17(2)23 dBm input power (DC bias point: V_{GS} = 0.09V, V_{DS} = 10V).

toward 1 W/mm at 10-GHz input frequency. It can be seen from the $I_d(V_{ds})$ trajectory that the 50-ohm load line does not result in the optimum drain voltage swing. The load trajectory is not an ideal line because of the reflections within the port 2 reflectometer, especially for the higher harmonics. Of course, harmonic tuning would be required to remove these reflections. While the 50-ohm load line reaches up to V_{ds} = 15V, a maximum voltage of 20V is observed for the tuned operation. The blowup of the trajectory describes the reactance that compensates the output parasitics of the device. In addition, this reactance explains the negative instantaneous drain currents that occur when the device is pinched off by the gate voltage swing. For the 50-ohm load condition, the current swing stops at the zero current line. The input $I_g(V_{gs})$ trajectories illustrate the gate charging current (or the nonlinear gate-source capacitance) and the gate-source diode characteristic. The gate voltage range of the *tuned* measurement is larger than in the 50-ohm load case because a higher RF input power was applied (25 dBm instead of 22 dBm). In both cases, no breakdown effects are visible. If breakdown would occur, it is expected that, for example the minimum momentary value of the drain current at device pinch-off increases for the highest power levels and/or the instantaneous gate current value at the negative V_{gs} peak becomes negative. It seems that breakdown is more critical at lower operating frequencies because in that case the maximum of V_{ds} coincides with the negative peak of V_{gs}, resulting in a high peak in the gate-drain voltage, which may cause breakdown. For comparison, such a drain-gate breakdown operating condition is illustrated in Figure 24.10, showing the input (I_g vs. V_{gs}) and transfer (I_d vs. V_{gs}) trajectories of an in-fixture measurement of a medium-power MESFET chip (Toshiba JS 8836) within a 50-ohm environment at 1.5 GHz for different input power levels. It can be seen that the I_{ds} - V_{gs} trajectories are only expanding with increasing stimulus power. Conversely, an additional effect is observed with HEMTs. With increasing stimulus power the trajectories are shifting to higher gate-source voltages. It is assumed that the observed phenomenon is caused by trapping effects in the AlGaAs material with a typical aluminum

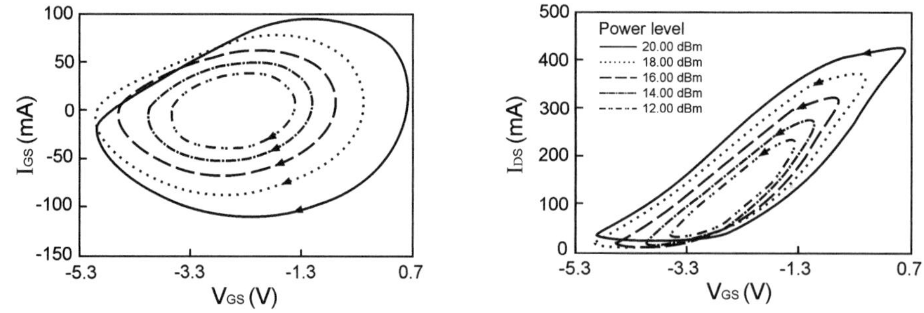

Figure 24.10 Input and transfer trajectories of a nondispersive Toshiba JS8836 medium-power MESFET chip at 1.5 GHz and 12(2)20 dBm input power illustrating onset of drain-gate breakdown (DC bias point: $V_{GS} = 0V$, $V_{DS} = 8.6V$).

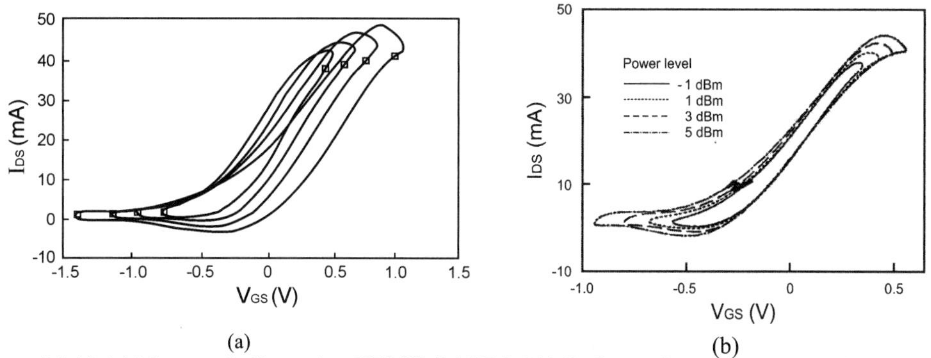

Figure 24.11 (a) Measured dispersive HEMT (MGFC 4401). Operating point: $V_{GS} = -0.1V$, $V_{DS} = 3V$, $I_{DS,DC} = 0.5\ I_{DSS}$, $f_0 = 3.5$ GHz, $P_{in} = 1(2)7$ dBm. (b) Measured nondispersive InGaAs PHEMT (IAF), $P_{in} = -1(2)5$ dBm.

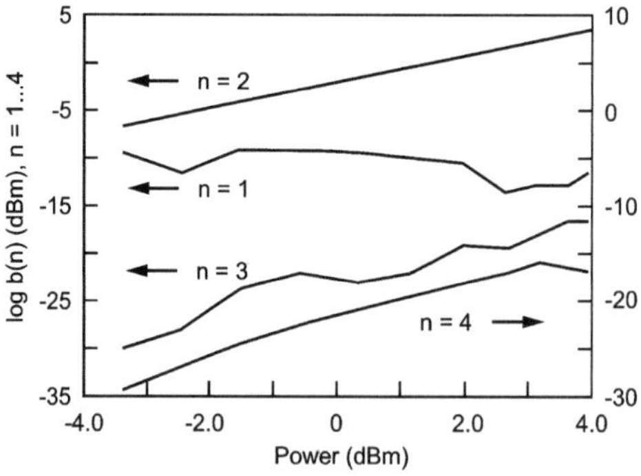

Figure 24.12 Output amplitude spectra at f0 and 2f0 vs. input power for a low-noise HEMT device after tuning to frequency doubler operation [15].

mole concentration of 0.3. Lowering the aluminum content, the power level dependent effective pinch-off shifting has vanished (Figure 24.11(b)) [64].

The fundamental load-pull measurement method, although widely used, does not use any higher harmonic. Harmonics are arbitrarily terminated, which can cause errors (e.g., in the determination of the PAE). The growing level of automation makes the method more user-disposed.

24.2.3 Harmonic Load-Pull Measurement with Electronic Tuner

In 2000 van Raay et al. reported about a waveform-load-pull measurement setup based on an IQ modulator concept [15]. This new concept for an electronic load module was inherently broadband and allowed a direct and independent setting of the reflection coefficients for different harmonics.

24.2.3.1 Electronic Tunable Load

Phase and amplitude control of harmonics in an active load-pull measurement system is traditionally done using electromechanical line stretchers and variable attenuators. These tuning elements are slow to automate and are subject to mechanical wear.

An alternative solution to electromechanical tuners is electronic tunable loads, which have been developed since the nineties. In 1995, ATN [22] reported about a load-pull system for digital mobile radio power amplifiers comprising solid-state impedance tuners specified as ELM directly connected to the input and output of the DUT. The ELM devices are very compact. Details on the functioning of ELM electronics have not been available. Nevertheless, it is interesting to mention that beyond fundamental load-pulling, intermodulation distortion (IM3, IM5, IM7) with two-tone signals can also be treated with the proposed electronic tuners. Worth mentioning as well is the support of modulation formats of digitally modulated signals such as $\pi/4$ differential quadrature phase-shift keying (DQPSK), Gaussian-filtered minimum-shift keying (GMSK) and Gaussian frequency-shift keying (GFSK). ACP measurements are limited to a modulation bandwidth of 60 kHz, which is within the specified GSM modulation bandwidth of 200 kHz.

In digital communication systems the IQ modulator is a key component that provides the modulation of an RF carrier with a digital baseband signal. Various forms of IQ modulation are being employed, moving from QPSK to more multilevel modulation schemes to maximize the amount of information transmitted through a communication channel [25]. Regarding the different modulator structures, the direct conversion IQ modulator has recently attracted much attention because of low hardware complexity and single-chip integration capability [26]. In this case, the RF carrier is directly modulated by baseband I and Q signals [27]. As illustrated in Figure 24.13, it mainly comprises a quadrature hybrid (0°/90° phase splitter), two separate in-phase and quadrature mixers for the divided RF signals, and a combiner for summing the I and Q modulated signals of the in-phase and quadrature channels. At the output a buffer amplifier may follow to compensate for the conversion gain of the mixers and further circuit loss. IQ modulators are classified according to the type of input signals f_1 and f_2 whether these are unbalanced (single-ended) or balanced (differential). If both input signals are single-ended, then it concerns a single-ended IQ modulator. Conversely, balanced input signals denote a double-balanced IQ modulator. The input signals may also be mixed. If, for example, signal with frequency f_1 is balanced and that of f_2 single-ended, the IQ modulator is called f_1-balanced. Conversely, f_2-balanced IQ modulator exhibits a balanced f_2-signal and unbalanced f_1 signal.

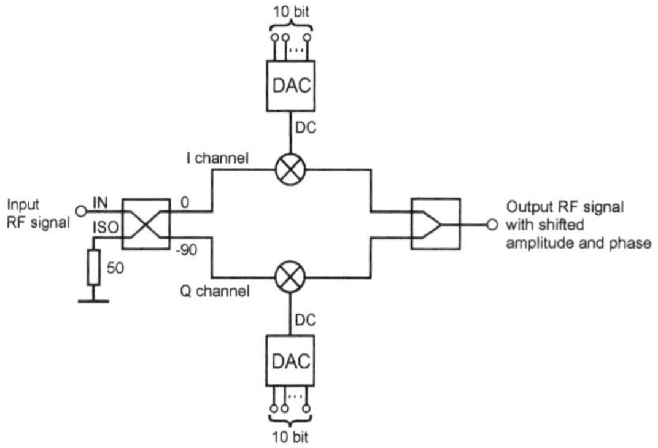

Figure 24.13 Electronically controlled mixer based continuously variable 90-degree vector modulator [29].

24.2.3.2 Single-Ended IQ Modulator

The IQ modulator concept for active tuning in a load-pull measurement system of split-signal type was proposed by Poulin et al. in 1992 [28, 29]. The circuit in Figure 24.13 illustrates a simple single-ended IQ modulator, which can tune phase and amplitude of an RF signal only within one quadrant. Three more modulators are needed to cover the whole phase shift of 360°. Assuming that the RF input signal is a cosine signal with unity amplitude,

$$s(t) = \cos(\omega_0 t) \tag{24.1}$$

then, under assumption of an ideal 0°/90° phase splitter, the output signals in the in-phase channel (I-channel) and quadrature (out-of-phase) channel (Q-channel) become

$$s_I(t) = \frac{1}{2}\cos(\omega_0 t) \tag{24.2}$$

$$s_Q(t) = \frac{1}{2}\sin(\omega_0 t) \tag{24.3}$$

In-phase and quadrature-phase signals are multiplied with the $I(t)$ and $Q(t)$ modulation signals,

$$s_I^M(t) = \frac{1}{2}G\cdot I(t)\cdot \cos(\omega_0 t) \tag{24.4}$$

$$s_Q^M(t) = \frac{1}{2}G\cdot Q(t)\cdot \sin(\omega_0 t) \tag{24.5}$$

with the conversion gain G of the mixers, which is assumed to be equal for simplicity. After summing the in-phase and quadrature signals in (24.4) and (24.5), the composite signal $s_o(t)$ becomes

$$s_o(t) = s_I^M(t) + s_Q^M(t) = \frac{1}{2}G\cdot I(t)\cdot\cos(\omega_0 t) + \frac{1}{2}G\cdot Q(t)\cdot\sin(\omega_0 t) \tag{24.6}$$

In our case, for load pull application, the modulation signals $I(t)$ and $Q(t)$ are simply computer-controlled DC values with 10-bit accuracy. I controls the real part and Q the imaginary part of the output phasor. Therefore, the IQ modulator enables the adjustment of the RF output voltage with arbitrary amplitude and phase signal. Using the trigonometric

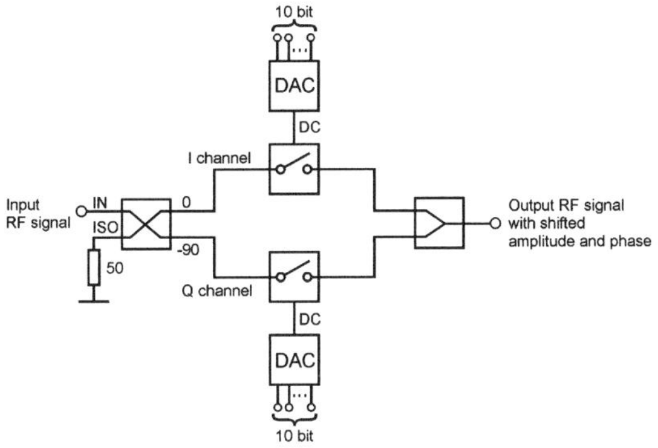

Figure 24.14 Electronically controlled switched 0 or 90 degree vector modulator (after [29]).

identity $\sin(\alpha + \beta) = \sin(\alpha)\cdot\cos(\beta) + \cos(\alpha)\cdot\sin(\beta)$, (24.6) can be rewritten as

$$s_o(t) = \frac{1}{2}G\cdot A\cdot\sin(\omega_0 t + \varphi) \tag{24.7}$$

where

$$A = \sqrt{I^2 + Q^2} \tag{24.8}$$

$$\tan(\varphi) = Q/I, \qquad \varphi = \mathrm{atan}(Q/I) \tag{24.9}$$

To achieve 0 to 360 degrees in phase, switched IQ modulators are needed to obtain either 0 or 90 degrees of phase shift. A switched version is depicted in Figure 24.14 showing similar construction as the continuous one with merely the mixers in Figure 24.13 substituted by switches.

Figure 24.15 shows the block diagram of an IQ modulator covering all four quadrants. It consists of a series connection of a variable vector modulator with three cascaded switched vector modulators and an amplifier to compensate for the insertion loss of the modulator.

Figure 24.16 shows an alternative solution for a continuously variable 90-degree vector modulator using electronically controlled voltage variable attenuators. PIN diodes and FETs could be used as variable resistors. Because of lower cost and being less prone to damage and distortion, Poulin et al. [29] used PIN diodes for the IQ modulator realization. Thus, with the complete modulator circuit in Figure 24.15, they achieved 0 to 360 degrees in phase and a change in amplitude between 0 and 40 dB.

24.2.3.3 Double Balanced IQ Modulator

We have seen that voltage-controlled attenuator based IQ modulators as used by Poulin et al. can cover only one quadrant of the complex plane. Additional three 0°/90° phase switches

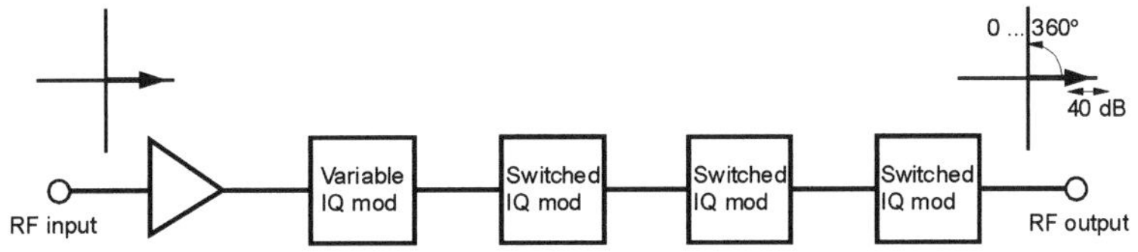

Figure 24.15 Electronically controlled vector modulator achieving 0 to 360 degrees in phase and a change in amplitude beetween 0 and 40 dB. (After [29].)

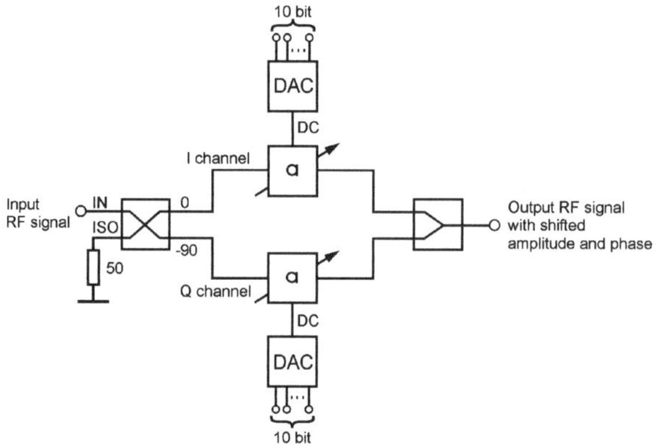

Figure 24.16 Continuously variable 90-degree vector modulator using electronically controlled voltage variable attenuator [29].

are necessary to cover all quadrants. Similar properties would be obtained with proposed single-ended mixers.

In 2000 van Raay et al. reported about a load-pull system using active IQ modulators for amplitude and phase tuning of harmonics [15]. The IQ modulator is given in a balanced configuration resulting in a significant simplification. Figure 24.17 shows a block diagram of the IQ modulator for tunability of the fundamental and first higher harmonic. The circuit can be simply expanded to include further higher harmonic tuning.

The key component of the IQ modulator is an integrated circuit of type PMB 2202 fabricated in silicon BJT technology [30]. The RF frequency range application is from 1.4 GHz to 2.5 GHz, which is convenient for the UMTS downlink frequency band (2.11 GHz - 2.17 GHz). The device includes a direct quadrature modulator and a double-balanced Gilbert cell mixer. The modulator generates two orthogonal carriers that are mixed with the baseband modulation signals I and Q in Gilbert multipliers. The outputs of the Gilbert cells are added and amplified by a linear output stage. For second harmonic generation an external push-pull frequency doubler is used (upper branch in Figure 24.17).

Figure 24.18 shows the top view of the printed circuit of the assembled RF part of the IQ modulator fabricated on PTFE substrate. The circuit size is approximately 65 x 55 mm. The RF output power is 6 dBm at the fundamental frequency (f_0) and -2 dBm at the second harmonic frequency ($2f_0$).

A schematic of the electronic (active) load module is given in Figure 24.17. The IQ modulator is used for magnitude and phase adjustment. The second harmonic is generated in a push-pull doubler circuit. For an equally spaced fundamental frequency schedule, a line with an electrical length of an odd multiple of a quarter-wavelength can be used to generate the Q signal for all frequencies of interest. For measurements within any octave band, a 0°/90° 3-dB hybrid can be used instead. The fundamental and second harmonic signals can be combined with an active circuit or a passive diplexer.

Figure 24.19(a) shows the static IQ modulator transfer function (specified maximum modulation frequency is 400 MHz). It can be seen that the vector magnitude is linear and saturates above 2/3 of the entire IQ control voltage range. Figure 24.19(b) illustrates the orthogonality of a measured IQ modulator. It can be seen that the positive Q channel exhibits

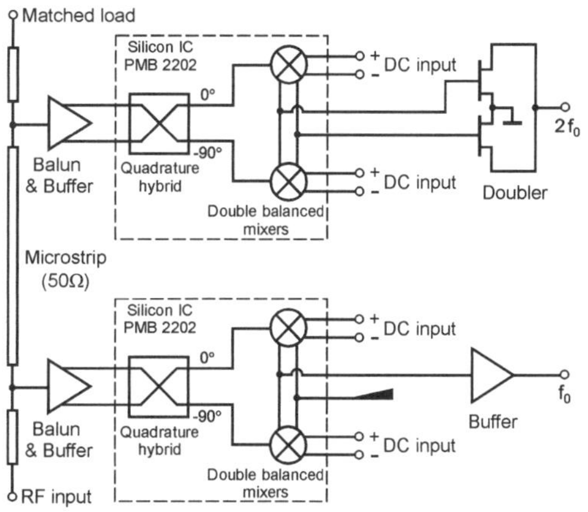

Figure 24.17 Double-balanced IQ modulator for amplitude and phase tuning of first and second harmonic used in a load-pull system of split-signal type [65].

Figure 24.18 Photograph of the assembled RF part of the vector generator [65].

an angular error of approximately 12 degrees. This would not be acceptable in communication application, and commonly integrated IQ modulator chips include some features for correction. However, in our case, such nonidealities are rather noncritical [15].

24.2.3.4 Electronic Harmonic Load-Pull

The waveform measurement system previously described in Section 24.2.2 has been upgraded to a load-pull system by including an *active load module* (ALM) for electronic tuning of the load impedance either manually or automatically (note the originally defined term ELM [15] was reclaimed by ATN demanding copyright).

ALM devices enable fast, electronic tuning without any mechanical part, which is advantageous for example in automatic load-pull contour measurements. The electronic load

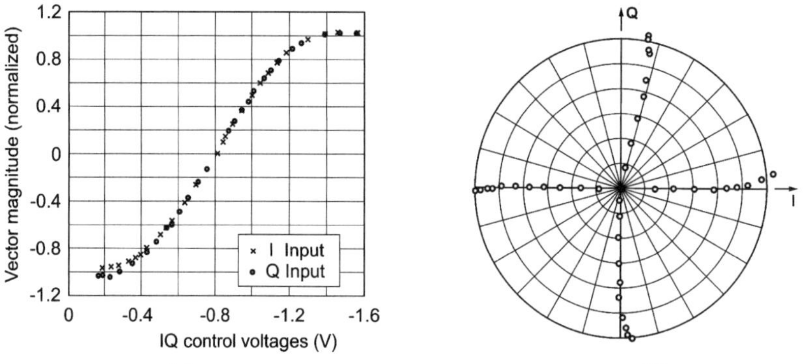

Figure 24.19 Static IQ modulator transfer function (a) and demonstrated orthogonality of the measured IQ modulator (b).

module operates at rather low power levels. The necessary output power can be achieved for example with a commercial broadband amplifier.

It should be emphasized that the source and load configurations are independent from the error model. The error model and the system calibration describe only the testset components (reflectometer and MUX) and the receiver channels. Therefore, the source and load configuration can be built up in accordance to actual measurement needs without recalibration of the system.

For the signal waveform measurement setup in Figure 24.20, the same additional measurements for coupler tracking, receiver ratio tracking and reference channel sampling aperture are applied as in the one-path two-port system described earlier (e.g., [9]).

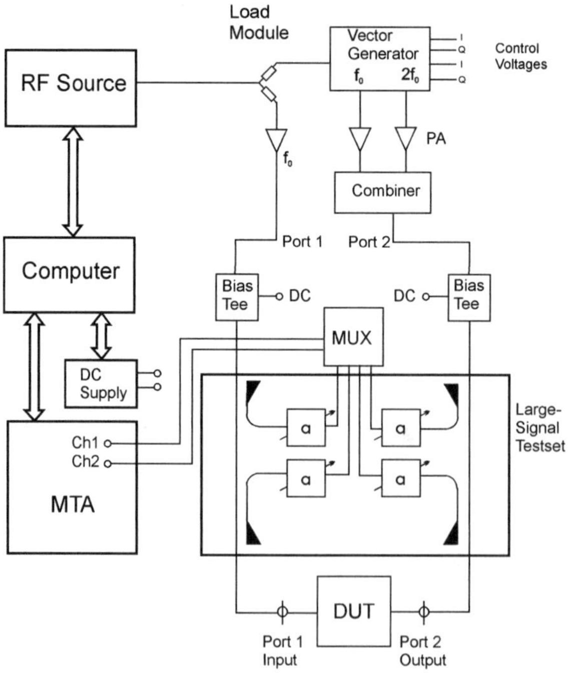

Figure 24.20 Signal waveform measurement setup with harmonic load pulling capability [15].

24.3 OVERVIEW OF HIGH-POWER WIDEBAND SOURCE-/LOAD PULL MEASUREMENT TECHNIQUES

The wireless telecommunication industry has grown at a tremendous pace over the past decade. Today's mobile communication systems require high power for multicarrier capability and broadband operation for high data rate, capacity, and service quality. They rely on high linearity to meet adjacent channel power ratio (ACPR) requirements, also referred to as adjacent channel leakage power ratio (ACLR), and deploy multiband and multistandard so that mobile phones can be used everywhere. A consequence of high spectral efficiency specified by the mobile communication standards is the requirement of highly linear power amplifiers, which are dominantly applied in the base station market.

As compared with the conventional semiconductors such as Si and GaAs, GaN) is less prone to electric field breakdown, which allows it to operate at high voltage [31]. Silicon devices such as Si-LDMOS transistors are used in the existing base station systems and the operating voltage of high-power amplifiers is 28V, which is lower than the base station system power voltage of 48V. This means that the base station systems use circuits to convert the power voltage. If the base stations can reduce the amount of power voltage conversion performed, it will be possible for them to reduce their size and power consumption.

Note that to meet the demands imposed by the new communication standards, the performance of Si-LDMOS is subject to continuous improvements. Currently, silicon LDMOS power transistors are the output devices of choice for base station power amplifiers, but as the limits of operability of these devices are reached, there will be a need for a semiconductor material that can fulfill the high-frequency and high-power requirements of the next generation of wireless technology.

The base station power amplifiers are usually subjected to signals having large peaks, forcing them to operate at backoffs for improved linearity. However, linearity is achieved at the cost of operating the amplifier well below its peak efficiency point. Experimental characterization and modeling of GaN devices with high output powers of 100W and above was a great challenge for any modeler and experimenter at that time. Both national and international funding support was provided to establish an innovative high-power measurement infrastructure for microwave devices. Thus, the author with his research group participated both in the European Network of Excellence Top Amplifier Research Groups in a European Team - TARGET from 2004 - 2007 [32] and in the joint BMBF (German Federal Ministry for Education and Research) project mobileGaN ran from 2006 to 2008 [33].

24.3.1 Setup Costs for High-Power Large-Signal Measurements

It is known that the cost of microwave components increase with rated power. A high-power ultrabroadband passive source- and load-pull system was targeted with the following characteristics. It should cover at least the first, second, and third harmonic frequency range of the UMTS band. In an extended version, it should cover the WiMAX-band also including frequencies up to the third harmonic (zone 1, 2, and 3). To meet the output power of large-

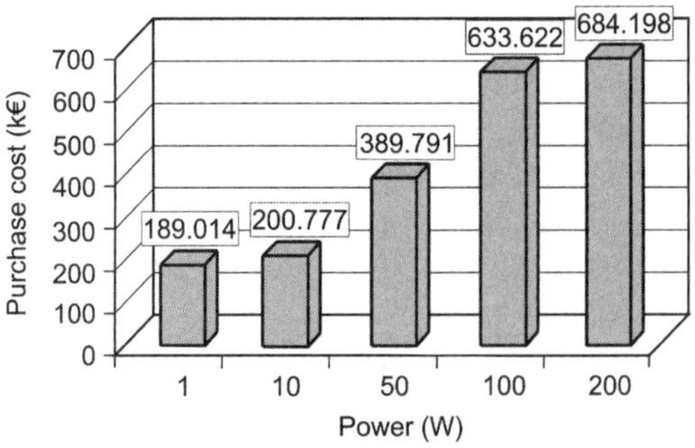

Figure 24.21 Additional purchase cost of an existing 0.5W signal waveform measurement setup (cost estimate 2003) to cover higher-power values.

size GaN power HEMTs, the system was designed for power values up to 100W (CW). The tuning range of the new measurement system should cover UMTS frequency ranges of zones 0 (envelope frequency band), 1, 2, and 3.

Figure 24.21 illustrates the additional cost of an existing 0.5W signal waveform measurement system (see Section 24.2.2) versus RF power to reach the goal. The cost estimation is referred to the year for a limited frequency range of 1.0 GHz to 2.7 GHz (extended signal waveform measurement setup with envelope frequency range included).

24.3.2 Advanced Signal Waveform Measurement Concepts for Wideband Applications

Figure 24.22 shows an overview of the diverse measurement concepts sorted with respect to the signal stimulus. Widely used is single-tone operation giving information about output power and power efficiency. Additional tuning of RF harmonics commonly improves the device performance.

Multitone measurements employ several (at least two) single tones for device stimulation. Very important for device intermodulation characterization is the two-tone measurement

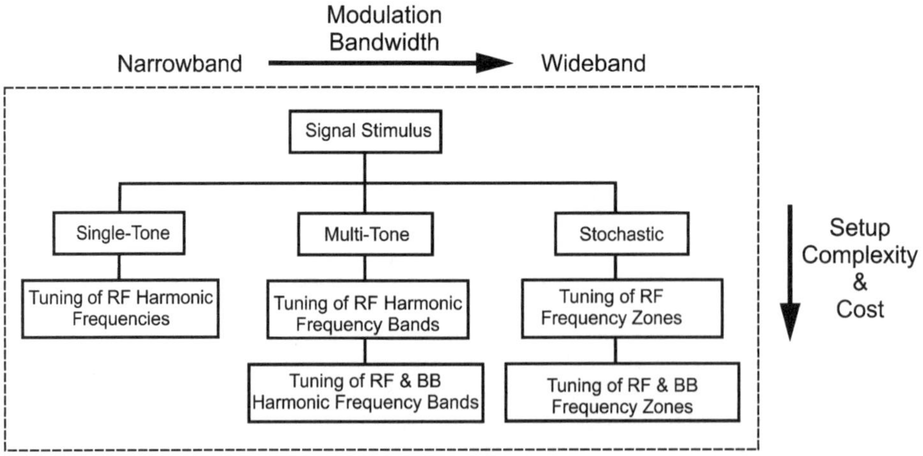

Figure 24.22 Overview of recently developed large-signal measurement techniques with a focus on increased modulation bandwidth of signal stimulus.

method with a standard tone spacing of 278 kHz (GSM standard). However, advanced digital communication modulation schemes require much larger modulation bandwidth. For example, the 3G standard requires 5-MHz modulation bandwidth for WCDMA signals. With a maximum of four channels, a total bandwidth of 20 MHz must be treated. The 4G standard specifies 20 GHz, and future 5G, albeit not yet fixed, may even cover a single-channel modulation bandwidth of 100 MHz. Regarding multicarrier operation, the bandwidth will increase extremely. Thus, the distortion requirements for microwave devices will increase dramatically, which present an enormous challenge for device characterization and measurement in the future.

24.4 HIGH-POWER TIME-DOMAIN SOURCE-/LOAD-PULL MEASUREMENT SETUPS

Substantial improvements have been made to the measurement setup through time. The major improvement was to extend the measurement system to a multiharmonic load- and source-pull system [34, 35].

24.4.1 High-Power Harmonic Source-/Load-Pull System with Passive Envelope Tuning

The multiharmonic source- and load-pull measurement system is depicted in Figure 24.23. The system comprises both RF and envelope source- and load-pulling capabilities. Similar tuning concepts have been used for both the source- and load-pull systems. The source-pull branch presents an open-loop configuration. Source impedance up to third harmonic frequency can be tuned. On the load side the fourth harmonic is additionally included. In the case of simultaneous source- and load-pull operation, a 3-dB power splitter has been used in the system. This is to divide the generator signal into two parts. One part of the spilt signal is fed to the device input through source-pull section and the other one is fed to the device output port through load-pull section. IQ modulators (see Section 24.2.3.3) control the phase and amplitude. Frequency doublers and triplers have been used to extend the capability of the electronic load module up to fourth harmonics (in this case two cascaded doublers are used).

The measurement setup is fully automated. The load impedances for the harmonics can be adjusted precisely and independently to each other (remotely or manually). The fundamental frequency range of the load-pull setup is limited by the used IQ modulator (vector generator), which is 1.4 to 2.5 GHz. The 10W RF power level of the system can be increased with the use of an additional driver amplifier. Measurements have been reported in [34, 36–38].

24.4.2 Issues Arising with Wideband Load-Pull Terminations

Two-tone measurements are used to characterize the intermodulation distortion (IMD) of a power transistor. Typically a tone spacing of 10 kHz is used for device IMD specification on the manufacturer's datasheet. Regarding narrowband mobile communication services such as the GSM standard with a modulation bandwidth of 200 kHz, the specified IMD data may offer relevant information about the linearity of the device under investigation. However, regarding more advanced network services such as UMTS, UMTS-LTE, and 4G (and upcoming 5G), much larger bandwidths up to 100 MHz occur. Then the question arises to what extent the narrowband IMD data is useful for the design of a linear power amplifier.

Typically, the intermodulation distortion of a device is measured in a 50-ohm environment. A load impedance of 50 ohm can be regarded as frequencyindependent. This means that both the two fundamental tones and the lower and upper intermodulation products, mainly IMD3 and IMD5, experience the same termination value. It may be expected that the 50-ohm termination will not be the optimum load for lowest IMD. Therefore, it is very useful to extend a single-tone load-pull measurement setup for two-tone device characterization. A fundamental problem arises with arbitrary load impedances with an additional reactive part exhibiting load frequency dependence. This means that the termination will be different not only for the fundamental tones but also for any intermodulation distortion product. Regarding narrowband IMD measurement with 10-kHz tone spacing, the assumption of constant termination for all spectral signal lines may be an acceptable approximation so that IMD load pulling will deliver relevant device data. However, regarding wideband operation for UMTS application larger tone spacing is needed. In [39] two-tone load-pull was performed with a frequency spacing of 5 MHz. But the impedance termination was assumed to be the same as for the center frequency. In [40] multitone load-pull was performed with a maximum modulation bandwidth of 3.2 MHz. Also in this case the load was assumed to be the same for all signal components.

In the following, passive and active two-tone termination issues will be discussed with regard to different termination configurations [36]. Software power leveling for the input tones was used to ensure same power values. This is in particular necessary with large-tone spacing due to gain variation of the preamplifier used for device stimulation.

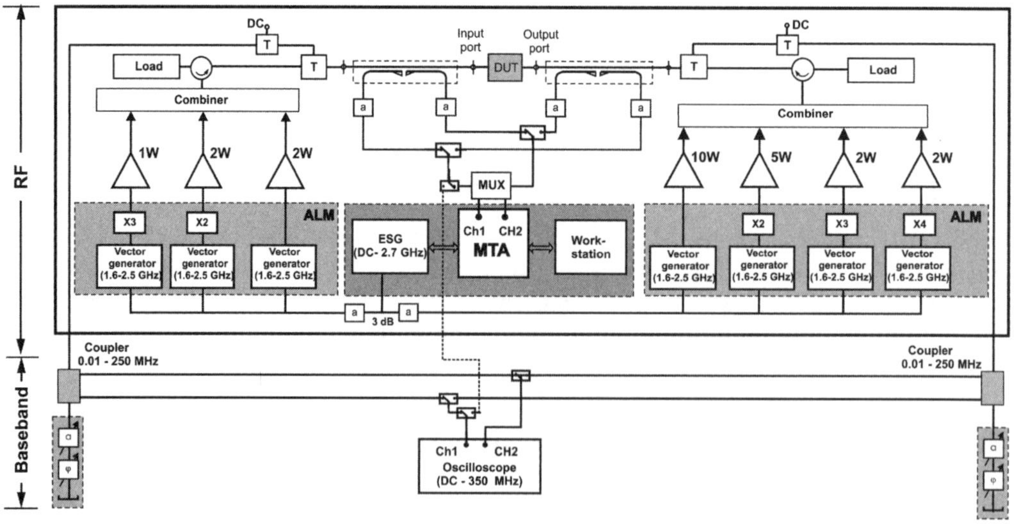

Figure 24.23 High-power active multiharmonic source-/load-pull measurement setup with passive envelope tuning.

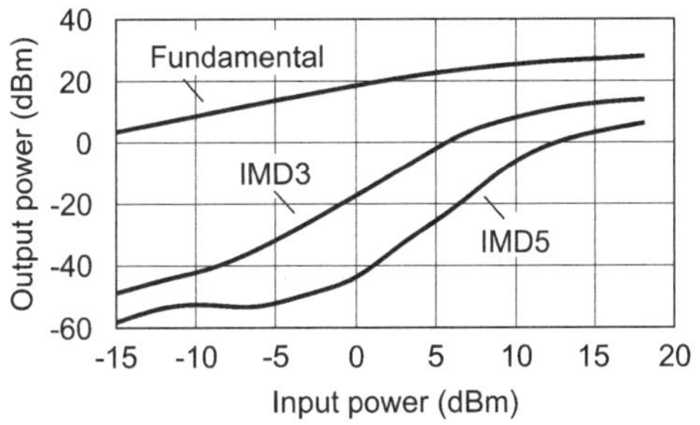

Figure 24.24 Output power of the main tones and the third-, and fifth-order IMD of 1-mmGaN HEMT in 50-ohm environment.

The device under test was an 8 × 125 μm GaN HEMT provided by the Ferdinand-Braun Institute (FBH), Berlin. A frequency spacing of 10 MHz ($f_1 = 2.14$ GHz, $f_2 = 2.15$ GHz) was assumed for the two tones. Figure 24.24 shows the result of a power sweep on the GaN HEMT used in the following experiments. The device is terminated with a 50-ohm load, and operated in class B with supply voltage of 20V and bias current of 150 mA. The input power was 20 dBm so that the device was fully saturated.

A: Two-Tone Load -Pull with Passive Tuner

In Figure 24.25 a classical load-pull termination with a passive tuner fabricated by Maury Microwave was directly connected to the reflectometer to avoid any additional phase shift. Between port 2 of the device and the tuner a reflectometer is inserted to measure the power waves a_2 and b_2, and thus the load reflection coefficient Γ_2. The used setup components have finite physical extension. Assuming a total effective electrical length l_{eff}, and regarding two tones having a frequency spacing Δf, there is a phase difference of the load reflection coefficients for the two tones $\Gamma(f_1)$ and $\Gamma(f_2)$ according to

$$\Delta\varphi = 2\Delta\beta l_{eff} = 2\pi\Delta f l_{eff}/v_p \qquad (24.10)$$

where v_p is the phase velocity of the power waves in the lines.

Due to (24.10) the reflection coefficient of the upper tone exhibits a clockwise phase shift. The upper IMD products show increasing phase shift with increasing order. Correspondingly, the lower terms are shifted counterclockwise (see Figure 24.26). The total variation of the phase of the reflection coefficient across the fundamental tones and the regarded intermodulation products covers an angular range of 60 degrees. The reflection coefficients

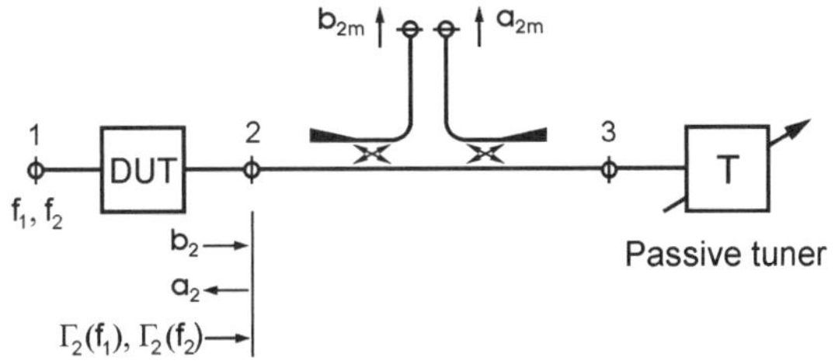

Figure 24.25 Two-tone load-pull with passive tuner.

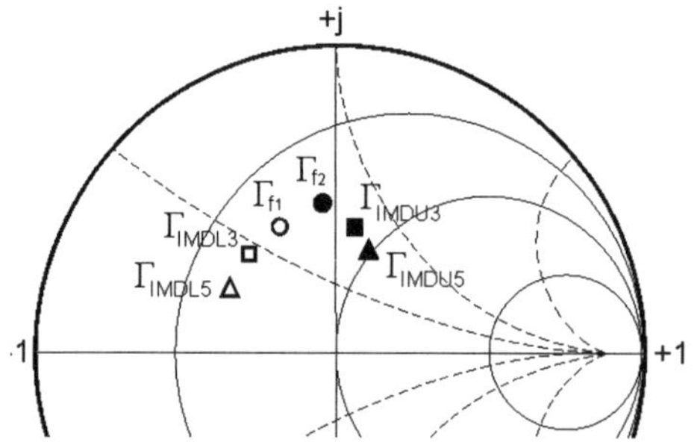

Figure 24.26 Two-tone load-pull with passive tuner provides different load reflection coefficients.

for the fundamental tones read $\Gamma(f_1) = 0.413\angle 105.9°$ and $\Gamma(f_2) = 0.42\angle 93.67°$ (1.2°/MHz). The magnitudes are nearly of equal value. The output power of the tones differ by 1 dB. The intermodulation distortion products of third and fifth order show an asymmetry of 1.95 dB and 1.73 dB, respectively [36, 37].

B: Active Closed Loop

Figure 24.27 shows the investigated active closed-loop load version. With this solution the whole range of passive loads can be covered. As is illustrated, part of the output power of the device is coupled out, amplified, phase shifted, and then again fed to the output port of the DUT. Assuming that all loop components are matched, the reflection coefficient $\Gamma(f)$ reads

$$\Gamma(f) = 10^{(G_T/20dB \;-\; C/20dB \;-\; a/20dB)} e^{-j\left(2\pi l_{loop}\sqrt{\varepsilon_r}f/c_0 \;+\; \varphi_K\right)} \tag{24.11}$$

where G_T is the amplifier transducer gain, C the coupling coefficient, and φ_K the phase of the directional coupler. a is the attenuation of the attenuator, and l_{loop} the effective electrical length of the loop. As can be seen in Figure 24.28, the respective load reflection coefficients, lying on a circle, show strong phase variation of 305 degrees. The center of the circle indicates the eigen-reflection coefficient of the tunable load. The target reflection coefficient for the lower tone was $\Gamma(f_1) = 0.42\angle 105.5°$(4.5°/MHz). The upper tone reflection coefficient turned out to be $\Gamma(f_2) = 0.36\angle 60.15°$. The output power of the two tones differ by 1.64 dB.

C: Electronic Load Module

Figure 24.29 shows a two-signal load-pull system with a balanced IQ modulator (1.6 - 2.5

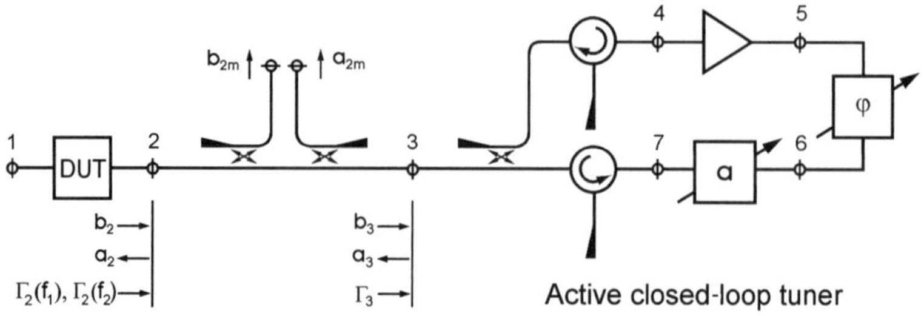

Figure 24.27 Active closed-loop load configuration.

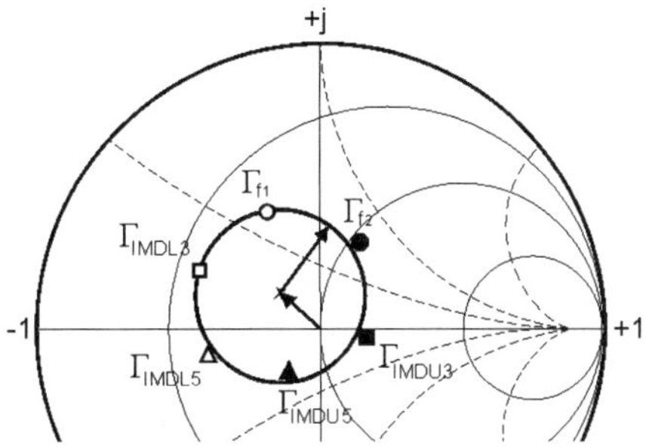

Figure 24.28 Two-tone reflection coefficients of an active closed loop for Γ magnitude of approximately 0.4.

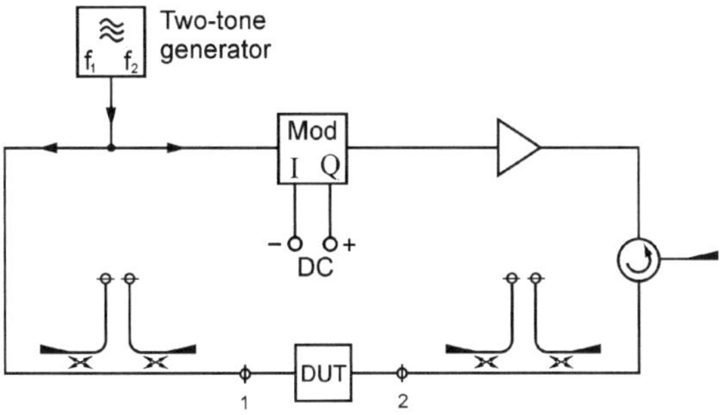

Figure 24.29 Two-tone load-pull with a single balanced IQ modulator.

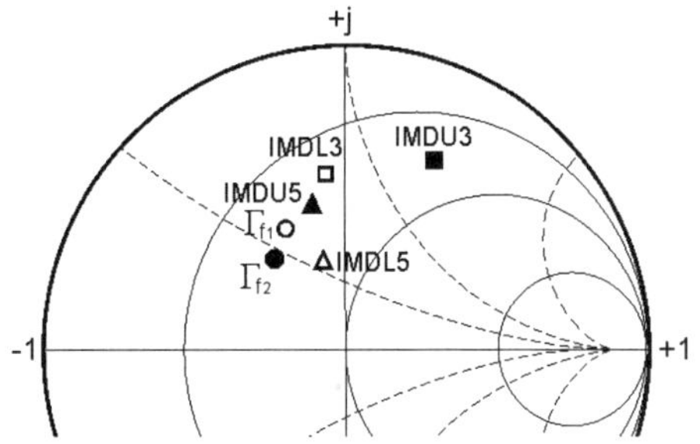

Figure 24.30 Two-tone reflection coefficients of an active tunable load using a balanced IQ modulator.

GHz) in the load branch, as described in Section 24.2.3.3. The RF signal is split into *I* (in phase) and *Q* (quadrature phase) components. By varying the amplitude of the inphase and quadrature phase components by a DC control voltage, the phase and amplitude of the RF signal can be changed independently. As can be seen from Figure 24.30, the target reflection coefficient for the lower tone was reached to be $\Gamma(f_1) = 0.414\angle 104.9°$. Then the upper tone exhibited a reflection coefficient of $\Gamma(f_2) = 0.34\angle 109.15°$ (0.42°/MHz). The output power of the tones differ by 0.62 dB, and the third and fifth intermodulation products differ by 0.43 dB and 2.07 dB, respectively.

Figure 24.31 shows an extended measurement setup using a balanced IQ modulator for each tone. In this case phase and amplitude of each tone can be controlled independently from each other. Figure 24.32 shows the result. We can see that the reflection coefficient of the two tones could be tuned equal to be $\Gamma(f_1) = \Gamma(f_2) = 0.41\angle 105°$. The output power of the

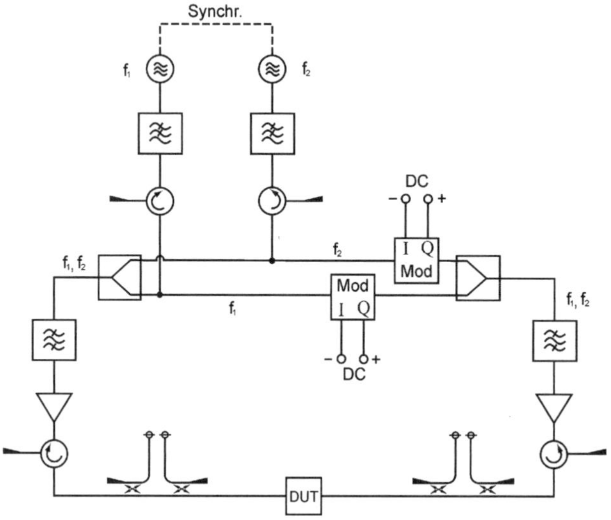

Figure 24.31 Two-tone load-pull with a balanced IQ modulator for each tone.

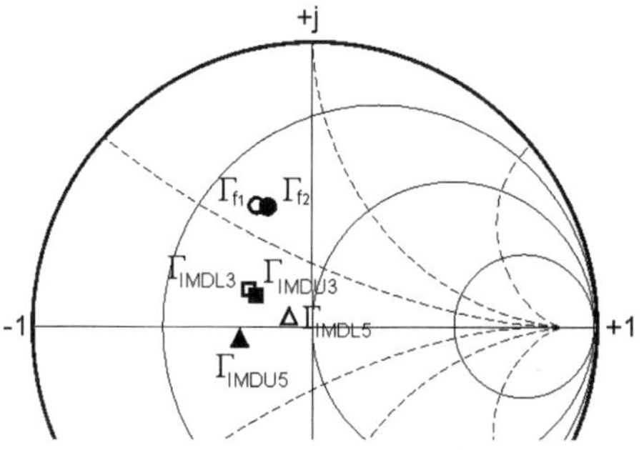

Figure 24.32 Two-tone reflection coefficients of an active tunable load using paired balanced IQ modulators.

two tones were equal. The reflection coefficients for the IMD products indicate a nearly matched condition.

For device verification, an ideal termination for both tones (i.e., equal complex load impedance) would be preferable because the analysis of device intermodulation distortion is less burdened by additional operation conditions.

Regarding the verification of device models, the conditions for simulation and measurement should be the same. Studying commercially available software, such as two-tone simulation with ADS, a constant termination for any zone must be specified. As will be discussed later, this assumption is approximately valid for very small tone spacing in the lower kilohertz frequency range.

In wideband load pull the electrical length of each tone is different, resulting in different termination situation for each tone. This situation is inevitable when designing matching works for amplifiers. But with modeling it is desirable to have an identical loading condition for each tone. Table 24.1 summarizes the results. It is clearly seen that the active closed loop provides the largest difference between the reflection coefficients between the tones, whereas, any deviation can be canceled out using an IQ modulator for each tone.

Table 24.1
Two-Tone Load-Pull Terminations and Their Influence on the Reflection Coefficients of the Two Tones

Termination	$\|[\Gamma(f_1) - \Gamma(f_2)]/\Gamma(f_1)\|$ (°/MHz)
Active closed loop	4.5
Passive tuner	1.2
IQ modulator	0.42
IQ modulator per tone	0

24.4.3 100W Broadband Passive Harmonic and Envelope Source-/Load-Pull Ssytem

Power amplifiers (PAs) used in mobile communication base stations for UMTS applications have to deal with QPSK modulated signals such as WCDMA. These stochastic signals have dynamic ranges up to 12 dB and a bandwidth of 20 MHz in the case of UMTS LTE [41]. Also, such power amplifiers have to operate at high power levels, maintaining both high linearity and efficiency. A successful design of these microwave devices depends on the accuracy of the device model used. As a result, it is imperative to thoroughly verify the CAD model under different source and load terminations. The source-/load-pull technique not only favors model verification but also provides the means to experimentally optimize the input and output matching conditions to either achieve high output power or high efficiency. With reference to more practical applications, it becomes more relevant to employ broadband stochastic stimuli (e.g., WCDMA) accurately reflecting real world telecommunication scenarios. Consequently a source-/load-pull system used for these applications has to fulfill high linearity over a wide frequency range and must be able to deal with high powers.

24.4.3.1 System Description

Figure 24.33 shows the realized passive source-/load-pull system capable of tuning zones 0, 1, 2, and 3 of the UMTS band. A test signal is produced by the generator and injected into the RF path of the measurement system using coupler C3. With the help of a switching matrix and the reflectometers C1 and C2, incident and reflected waves of the input and output port of the DUT can be measured. This allows error-corrected determination of relevant qualities, such as P_{in}, P_{out}, gain and ACPR. During measurement the reflection coefficients of the source tuners can be determined, feeding a signal in reverse direction towards the source tuners through coupler C4. Therefore, a precalibration of the tuners is not needed, decreasing measurement time. The setup is able to handle powers up to 100W CW.

A compact multiharmonic tuner system with a triplexer mounted in front has been developed. The designed slide load tuner consists of an air-filled slabline terminated by a 50Ω load. To enhance the frequency range capable of tuning, the fundamental tuner has been built as a double stub configuration. The maximum achievable tuning range for the first, second, and third harmonic frequency is $|\Gamma_{1st}| \leq 0.938$, $|\Gamma_{2nd}| \leq 0.955$, and $|\Gamma_{3rd}| \leq 0.918$, respectively.

The tuners provide a maximum of 2.3% change of amplitude over the considered

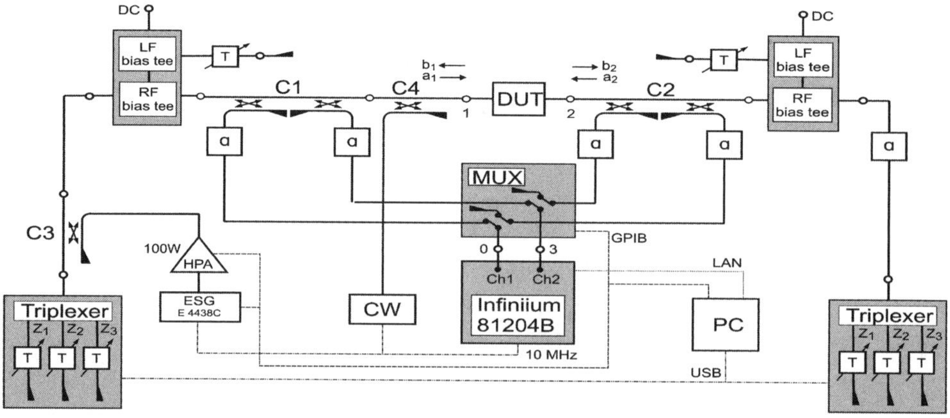

Figure 24.33 Time-domain source-/load-pull system with multiharmonic and envelope tuning [42].

frequency range of 200 MHz. Although the tuner has been designed in a compact way, the transmission lines of the tuner will introduce a phase shift over the desired bandwidth. Future solutions using active electronic load-pull modules can overcome this drawback. For large-signal device model verification with single-tone and two-tone stimuli, all tuners were terminated with 50Ω during the measurement.

The usage of the Agilent DSO81204B Infiniium Ultra Broadband Realtime Oscilloscope [43] as receiver had been decided after thorough analysis of system requirements and concept alternatives. The receiver needed for the measurement setup has to fulfil the following requirements.

Regarding UMTS LTE a modulation bandwidth of 20 MHz must be covered. For future applications such as 4G standard, an extension to 100 MHz should be taken into account. To monitor incident and reflected waves simultaneously a two-channel system is needed. To cover at least zones 1, 2, and 3 of the UMTS band with a center frequency of 2.14 GHz, a receiver frequency range > 6 GHz is required. Regarding a WiMAX case, with a center frequency of 3.8 GHz the receiver has to cover approximately a bandwidth of 12 GHz. ACPR measurements of W-CDMA signals require a dynamic range of at least 45 dBc for the receiver.

For the hardware realization of the measurement concept, no standard solution is known. This is especially valid for the selection of the broadband receiver. Various receivers were thoroughly analyzed and tested for their suitability. The following three receivers were shortlisted: (1) vector signal analyzer, (2) two-channel downconverter, and (3) real time sampling oscilloscope.

(1) The 6-GHz version of the Agilent 89641S Vector Signal Analyzer (VSA) [44] corresponds to the 2.7-GHz vector signal analyzer of type Agilent 89640S [45] but exhibits an extended RF bandwidth of 6 GHz. The system would fulfill the dynamic range needs. However, its modulation bandwidth is not increased and remains 36 MHz. As the center frequency of the UMTS band is 2.14 GHz, the third harmonic frequency is 6.42 GHz. Therefore only source-/load-pull measurements up to the second harmonic could be covered. Regarding WiMAX signals, the fundamental frequency is even higher (center frequency 3.8 GHz). Thus, this receiver solution was no longer pursued.

(2) Agilent N8201A downconverters with 26.5-GHz bandwidth and 200- MHz modulation frequency range [46] could be paired and should be placed in front of the input channels of a 2.7-GHz VSA already in use. One downconverter per channel would be needed. The great advantage is that the whole frequency range under consideration could be covered. As well, the dynamic range needs could be fulfilled. However, there are two strong limitations: First, even though the modulation bandwidth of the downconverter is 200 MHz, the modulation bandwidth of the VSA remains 36 MHz. The second issue is that synchronization of the two separate downconverters must be handled. This could not be definitely assured by the manufacturer. Thus, because of high risk of synchronization this receiver concept was also abandoned.

(3) The Agilent DSO81204B Infiniium Ultra Broadband Realtime Oscilloscope [43] includes the VSA software performing the signal processing. The oscilloscope can cover a frequency range up to 12 GHz and has no modulation bandwidth limitation. It also has four channel inputs of which two channels can be used for memory-intensive measurements, which are given, for example, with stochastic signals such as WCDMA modulated signals. A strong limitation is given by the incorporated 8 bit A/D converter. The signal-to-noise ratio is related to the number of bits according to the relation SNR $= 20 \log_{10}(2^n \times 1.225)$ dB [47]. Thus, in ideal case, for an 8-bit A/D converter the signal to noise ratio is estimated as 50 dB. However, this value decreases due to system noise and with increasing bandwidth. Therefore, the 8-bit A/D-converter decreases effectively to 6 bits, resulting in a signal to noise ratio of only 38 dB. This value would not fulfill the dynamic range requirements of at least 45 dBc. More precise specifications of the system for our special application could not be obtained from the manufacturer. To recall, the 2.7-GHz 89640 VSA system in use has a 12-bit A/D converter corresponding to an ideal value of 74 dB. Using averaging values above 95 dB can be obtained.

In spite of initial skepticism, but having no real alternative receiver solution in sight, the broadband multichannel sampling oscilloscope was thoroughly tested. Contrary to expectations, measurements with single tone, multitone, and WCDMA stimuli have finally shown that the real-time oscilloscope indeed fulfills the required specifications of the measurement setup under investigation. When using averaging, high signal dynamics similar to the 2.7-GHz 89640 VSA system could be obtained.

24.4.3.2 Measurement Results

Measurements on a 3.2-mm high-power GaN HEMT were performed based on a manually tuned fundamental source-/load-pull system configuration. To reduce measurement time and to get an overview of the impedance region, where to search for optimum impedance values, load-pull simulations were done in ADS using a nonlinear model of the DUT as described in Chapter 23.

Figure 24.34 shows P_{out} contours for two-tone device stimulation based on fundamental load-pull measurement with 5-MHz tone spacing corresponding to the bandwidth of a WCDMA signal. The contours are very similar indicating that simple two-tone excitation can

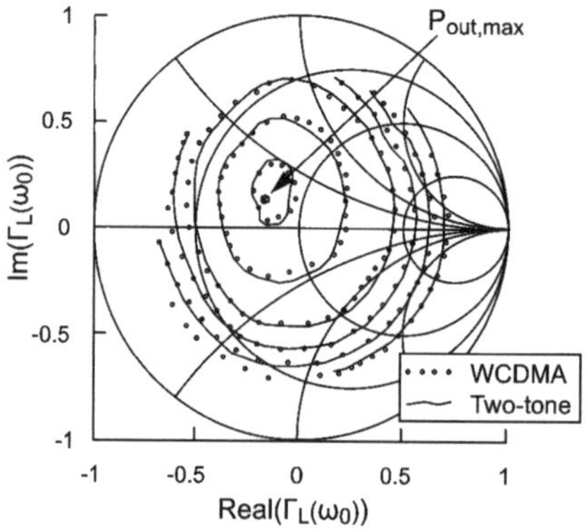

Figure 24.34 P_{out} contours for two-tone (5 MHz-tone spacing) and WCDMA device stimulation based on fundamental load-pull measurement. Reconstructed from [42].

be used for optimum load determination instead of complex and wideband modulation signal.

The measurements were performed using single-tone stimulus with P_{av} = 19 dBm and a center frequency of 2.14 GHz at a bias point with the drain voltage V_{DS} = 40 V, and the gate voltage V_{GS} = −2.855 V, in order to reach high output power levels. As can be seen from Figure 24.34, good agreement between simulated and measured contours is given. Simulation yields an optimum output power of P_{out} = 30.9 dBm with a load impedance of Z_{load} = (28.07 + j34.95) Ω. Load-pull measurement delivered an optimum output power of P_{out} = 34.7 dBm at Z_{load} = (32.25 + j24.0) Ω.

Figure 24.35 shows a comparison of IMD3 and ACP measurements. Also in this case, similarities can be recognized between two-tone and WCDMA stimulation. However, load pull under WCDMA stimulus indicates randomly distributed ACP power with local minima and maxima. One explanation for these differences may be as follows. In the case of the two-tone test, each IMD product exhibits only a single frequency. In contrast, a continuous wideband signal comprises manifold different intermodulation products, which sum up at a considered frequency point [47].

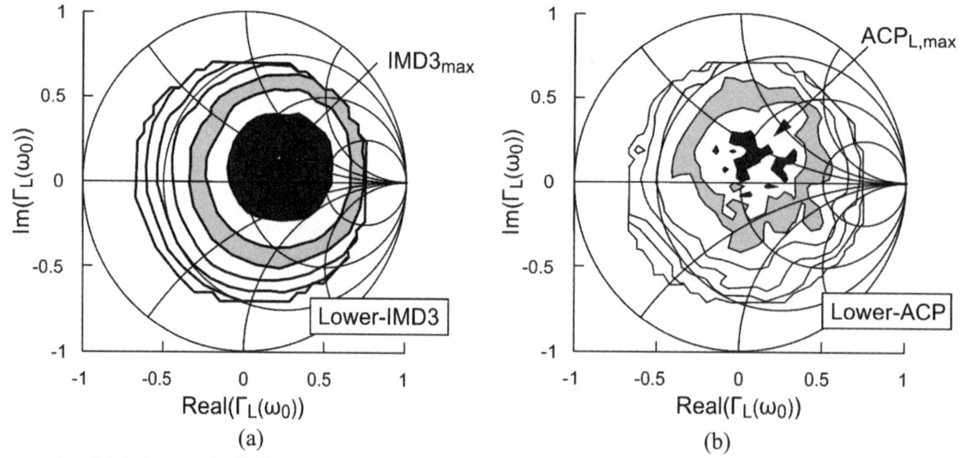

Figure 24.35 (a) Lower-IMD3 and (b) lower-ACP as a function of load impedance in zone 1. From [42].

24.5 HIGH-POWER FREQUENCY-DOMAIN IMD MEASUREMENT SYSTEM INCLUDING ENVELOPE LOAD-PULL

In this section a frequency-domain measurement system is described that was established for intermodulation distortion characterization of power devices including passive envelope load-pull testing.

24.5.1 System Description

The frequency-domain measurement setup established in this work is shown in Figure 24.36. It incorporates a dual-channel 89600 series VSA as receiver. This system is intended for measuring and analyzing active device and/or amplifier nonlinearity within a frequency range from DC to 2.7 GHz with an instantaneous bandwidth of 36 MHz. The dual directional coupler (Amplifier Research DC420 with operational bandwidth of 0.8 to 18 GHz) imposes 50W (CW) limitation in the measured output power.

As depicted in Figure 24.36 the transmitted and reflected complex travelling waves a_1 and b_1 at the input port 1 and b_2 and a_2 and at the output port 2 are monitored by the VSA using a dual-directional coupler. A 100W driver amplifier (Amplifier Research 100S1G4M8 [48] with operational bandwidth of 0.8 to 4.2 GHz) is inserted to generate high linear input power to drive high power DUTs and 1-dB compression point. Envelope source-/load-pull measurements are established by using a combination of RF and LF bias tees cascaded along the low-frequency DC path. It is hence possible to present an arbitrary impedance at the envelope frequency, while being simultaneously decoupled from the impedance at the fundamental and harmonics.

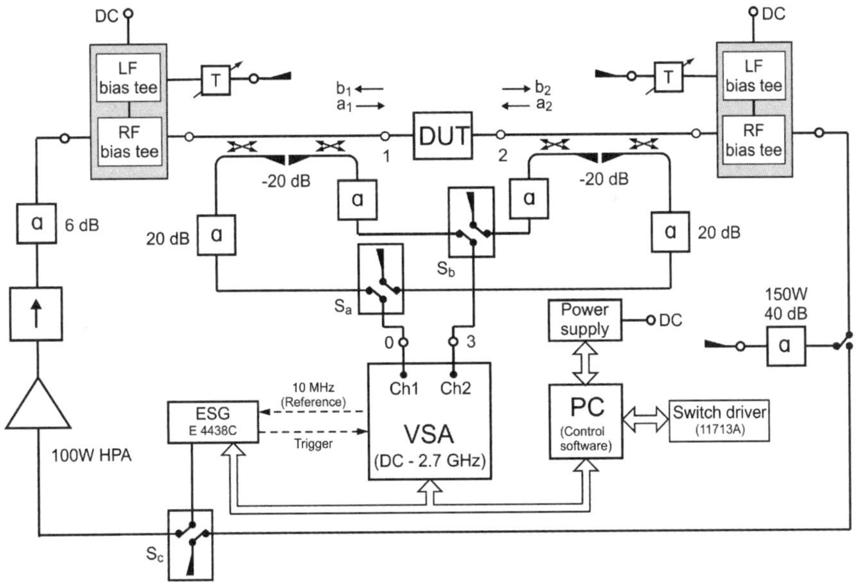

Figure 24.36 Frequency-domain measurement setup for intermodulation distortion characterization of GaN power devices including passive envelope load-pull testing [55].

24.5.2 Envelope and RF Bias Networks

With the upcoming GaN device technology, under wideband dynamic excitation, one of the critical demands imposed on device characterization and power amplifier (PA) design is to restrict the baseband drain impedance to a very low value (preferably $< 1\Omega$) and almost constant over the operational bandwidth. This, otherwise, would lead to drain-bias modulation contributing to the overall spectral distortion. However, in reality, it is difficult to achieve a broadband low-impedance termination [49].

Reported experimental results in [50] confirmed that nonideal DC bias feed, constituting the biasing cable and the DC source, adds series impedance to the biasing circuit, which is generally overlooked during broadband measurements. Such ambiguous IMD3 characterization could mislead in attaining the true device linearity. As a result, improved IMD results based on in-house bias tees, with an excellent broadband IF impedance response delivered by $\lambda/4$ transmission line based bias tee, has been reported in [49].

Reported experimental results confirmed that nonideal DC-bias feed, constituting the biasing cable and the DC source, adds series impedance to the biasing circuit, which is generally overlooked during broadband measurements [51]. Such ambiguous third-order intermodulation distortion (IMD3) characterization could mislead in attaining the true device linearity. Hence, it is paramount to characterize device linearity accurately in order to minimize linearizer design complexity.

Bias tees are among the key components used in the measurement setup, which determine the out-of-band termination. Until recently, the bias tees were used mainly to provide DC bias to the active device during characterization. However, with the introduction of broadband measurement setups, an additional function of the bias tee is to provide a baseband short for the entire operational frequency range.

Drain-bias sensing [52] is one of the few possibilities to experimentally illustrate the implications of baseband termination. Practically, when a nonlinear active device is characterized using multicarrier excitation, due to the nonzero bias supply impedance, an undesired voltage modulation is sensed at the output biasing terminal. This drain-bias modulation would, in turn, add to the device's inband distortion. Through bias sensing, it is possible to detect the envelope component ($\omega_2 - \omega_1$); for example, under two-tone stimulus, where the device under test is operated at class-AB condition for wide carrier spacing.

Gate Bias Network

Apart from providing a DC bias to the device's gate node, the gate bias network has to limit the gate current under breakdown conditions (drain-to gate or gate-to-source), when the device is overdriven.

Further, devices often encounter problems of instability when operating at higher quiescent point, giving rise to oscillations. In other words, the input impedance of the device tends to exhibit negative resistance, leading to signal leakage into the DC path [52, 53]. In order to eliminate the negative resistance, a resistor is placed in the DC path, which is typically estimated as $R = 400/P_{sat}$, where P_{sat} is the saturated output power of the device.

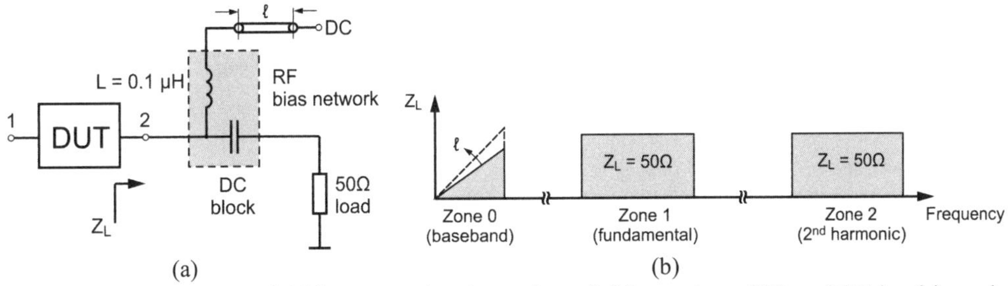

Figure 24.37 (a) Commercial bias network schematic and (b) envelope (IF) and RF load impedance response in a 50Ω environment.

Due to very low gate current flowing through the gate bias supply, there is negligible voltage drop across the resistor, thereby not affecting the gate voltage applied to the FET device. Resistive gate biasing also avoids memory effects at the input, but at high modulation frequencies, gate impedance starts to decrease due to the C_{gs} nonlinearity [54].

Drain Bias Network

The bias network has to satisfy two key requirements. The impedance offered by the DC path of the bias network must be high enough to exhibit an open circuit at the RF and low enough (ideally short) to suppress any low-frequency components, resulting in DC supply modulation, entering the DC path. In the following we will mainly focus on drain biasing issues and its relevance toward IMD/ACPR characterization under broadband multicarrier stimuli for UMTS applications.

Figure 24.37 illustrates the schematic of a commercial drain bias network across relevant frequency zones comprising a series capacitance (DC block) and a parallel inductance, which is RF-grounded via a cable of length l. As indicated the baseband impedance increases monotonically due to the typical inductor component. Additionally, the DC feed of length $l < \lambda/4$, which represents an inductor for baseband frequencies, imposes an electrical offset short. The RF band, however, remains constant, providing 50Ω across both zone 1 and zone 2. Nevertheless, despite solving the DC feed problem through capacitive compensation mechanism, the inherent inductor apparently posed restriction in achieving lower impedance value at envelope frequencies, encouraging to design in-house bias tees relevant for characterizing FET devices for UMTS applications.

For correct intermodulation measurement of investigated GaN HEMT devices, an RF bias tee, showing broadband short for zone 0 frequencies and a 50Ω for RF, should be available. However, there are practical difficulties, as indicated in Figure 24.38, in achieving the

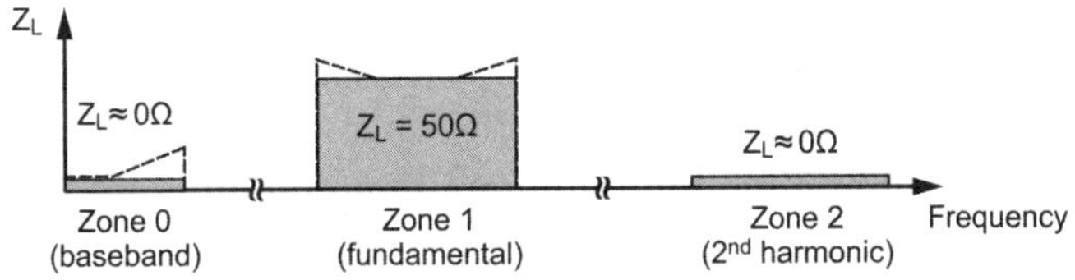

Figure 24.38 Design goal of an in-house bias tee depicting optimized envelope (IF) and RF impedance response in a 50Ω environment.

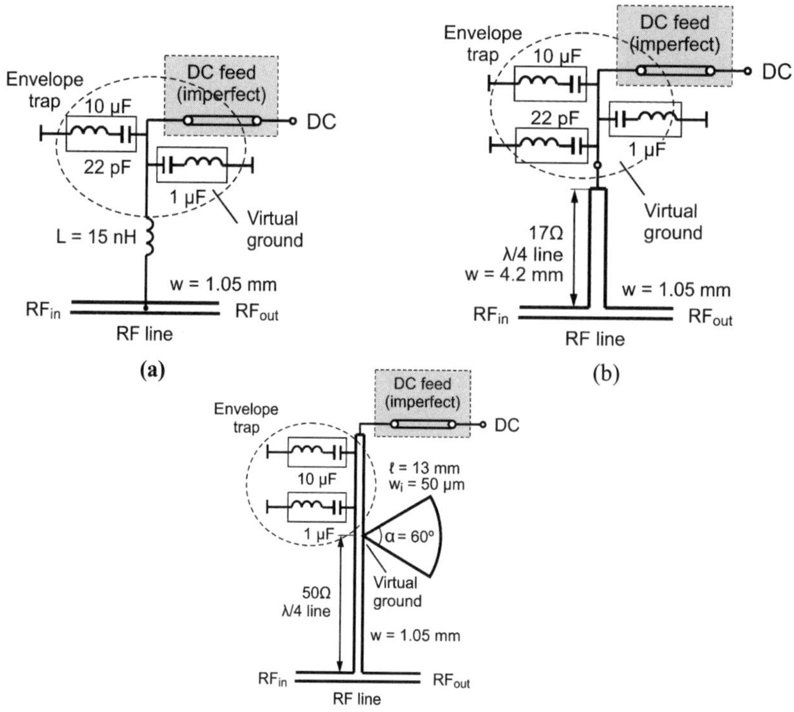

Figure 24.39 Investigated drain bias networks: (a) Bias-T with lumped inductor, (b) bias-T with low-impedance λ/4 short-circuited shunt stub, and (c) bias-T with λ/4 open-circuited radial stub. (After [55].)

desired IF and RF termination properties. Regarding the baseband, the bandwidth of the short termination is limited by the series resonance frequency of the decoupling capacitors. Consequently, for higher modulation bandwidths, the inductance of the (randomly chosen length of) DC supply cable becomes critical. Under the worst condition the envelope impedance can exceed 50Ω. Further, the bandwidth of zone 1 termination is limited by impedance along the bias path, which should provide a high impedance for RF frequencies. Furthermore, the series inductance may provide an impedance of 50Ω for zone 2. Although it is practically difficult to obtain a stopband response across zone 2, it has been proved that the impact of zone 0 termination on device IMD behavior is more critical than zone 2 termination [55, 56]. Therefore, it is likely that the measurement result may not only reflect the intermodulation distortion of the device itself but may be significantly controlled by the imperfect device embedding condition. With these design goals, bias tees of different topologies, shown in Figure 24.39, were investigated at the drain terminal preserving primarily wideband 50Ω response around the operating carrier frequency.

Figure 24.40 illustrates the impedance of the commercial bias network of type HP11612A and in-house RF bias tees under inspection measured using VNA measured across the baseband, fundamental and second harmonic frequencies. The baseband impedances of the respective bias tee variants are measured taking into account even the electrical effects of the DC feed (biasing cable, DC source having negligible influence). Since the focus is toward

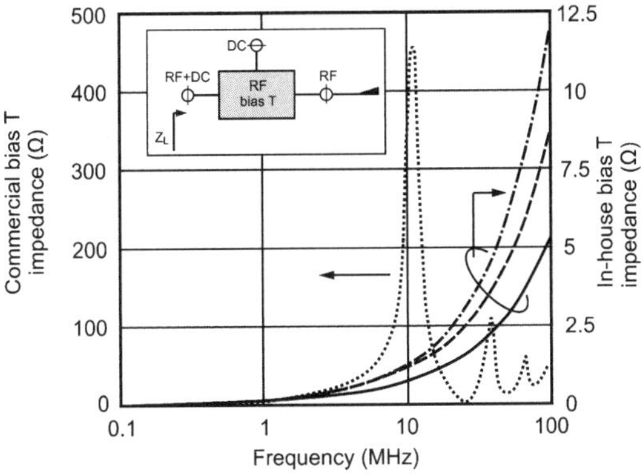

Figure 24.40 Measured baseband impedance for various bias tee configurations. (After [55].)
····· Commercial bias-T — λ/4 Short-circuited line bias-T
- - - λ/4 Open-circuited radial stub bias-T -·-· Lumped inductor bias-T

UMTS base station power amplifier applications, RF bandwidth was mainly considered around an operating frequency of 2.15 GHz.

From Figure 24.40, it can be observed that at 10 MHz, for example, an impedance as high as 500Ω is measured for the commercial bias tee. The reason is evident because of the DC feed length of the cable (l = 2.5m), which was taken into account while measuring bias tee characteristics. For the lumped inductor based bias tee, the decoupling capacitors (1 μF and 10 μF) have successfully held the impedance to a rather constant value with a maximum value of 12Ω measured at 100 MHz within the considered frequency range from DC to 100 MHz. In contrast, the λ/4 short-circuited shunt stub demonstrates an excellent baseband impedance of 5Ω at 100 MHz. In the case of λ/4 open-circuited radial stub based bias tee, a maximum envelope impedance of 9Ω is measured at 100 MHz, which is comparable to the low impedance value of the λ/4 line stub.

In a two-tone signal environment, besides the fundamental frequencies, the RF impedance across the lower and upper IMD3 components should also be principally taken into account during the bias tee design. In this regard, the 22-pF bypass capacitor provides a narrowband short at f_c = 2.16 GHz for λ/4 line.

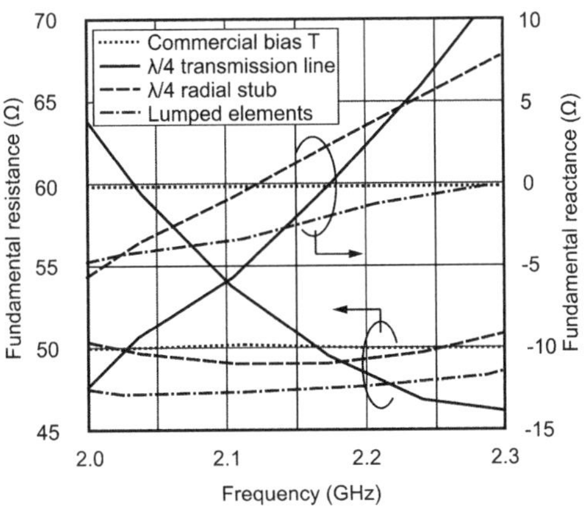

Figure 24.41 Measured fundamental impedance for various bias tee configurations. (After [55].)

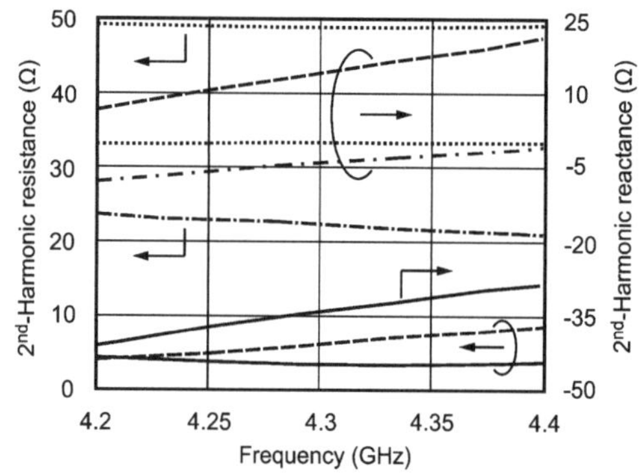

Figure 24.42 Measured second harmonic impedance for various bias tee configurations [55].
····· Commercial bias-T — λ/4 Short-circuited line bias-T
- - - λ/4 Open-circuited radial stub bias-T -·-· Lumped inductor bias-T

Figure 24.41 shows the impedance response for the fundamental frequencies. Indeed, the commercial bias tee demonstrates excellent RF response of constant 50Ω with negligible reactance. Such performance of the RF bias tee is preferred mainly when the DUT is characterized under CW excitation. However, with the increase in test signal complexity, constituting multiple carriers, the envelope impedance response of the bias tee becomes critical. Subsequently, λ/4 open-circuited radial stub based bias tee can be accepted as an optimum solution, which satisfies low out-of-band termination and constant broadband 50Ω at the fundamental frequency with an inevitably minimum reactive term.

From Figure 24.42, it is rather interesting to note the second harmonic termination of a radial stub based bias tee response, which is typically expected to exhibit a short. Its practical implications on active device characterization is not only on sideband distortion suppression but also potentially enhances overall efficiency. This can be principally achieved only when the short is presented at the device reference plane, which equally holds good for the baseband termination [55].

24.5.3 Issues in IMD Measurement Arising with Setup Configuration

It has been shown that not only the construction of the bias-T but also the placement of dual-directional coupler and bias-T relative to each other is of serious impact on measured device intermodulation distortion behavior.

Figure 24.43 shows a section of a load-pull measurement setup. A double-reflectometer is first connected to the device under test recording the outgoing and incident power waves.

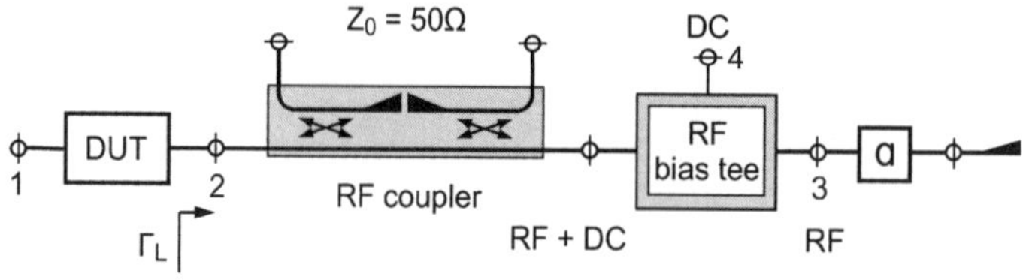

Figure 24.43 Conventional arrangement of directional coupler and bias tee.

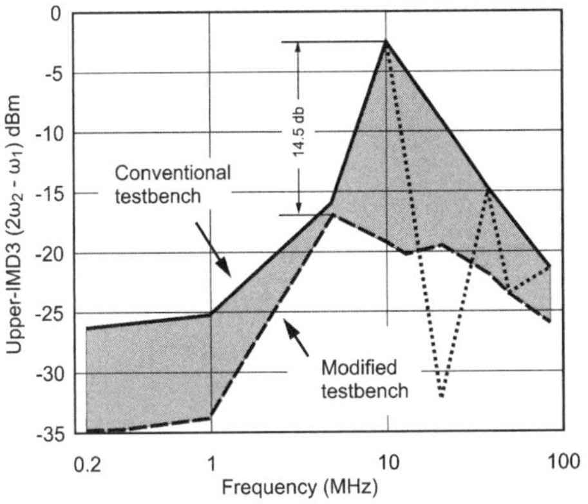

Figure 24.44 Upper IMD3 measurement as a function of carrier spacing variation using in comparison with the conventional and modified testbench arrangement [55]

It follows then an RF bias-T as voltage supply for the drain electrode of the transistor (DUT).

Figure 24.44 shows the measured upper IMD3 of a 0.5-mm GaN HEMT by sweeping the tone spacing from 200 kHz to 80 MHz biased for class-AB operation. Interchanging of coupler and bias-T as shown in Figure 24.45 results in the dashed curve showing dramatic improvement of the measured IMD result of about 14.5 dB, and that occurred only by changing the sequence of setup components. Indeed, the additional coupler length in Figure 24.43 causes a disruptive effect on intermodulation distortion.

Figure 24.46 illustrates the improvement in upper IMD3 measurement based on the proper bias-T design. It is seen that the solution with λ/4 radial stub delivers an improvement of 19 dB with respect to the lower IMD curve in Figure 24.44. That is really impressive.

Figure 24.47 shows the measurement of a 2-mm GaN HEMT under class-C bias condition in a 50Ω environment [55]. The upper IMD3 curve as function of input power shows a clear sweet-spot behavior. It can be seen that the optimized RF testbench (sequence of setup components and optimized bias-T design) reveals the sweet-spot property of the device, whereas the conventional bias-T suppresses this specific device feature independent of the selected order of setup components.

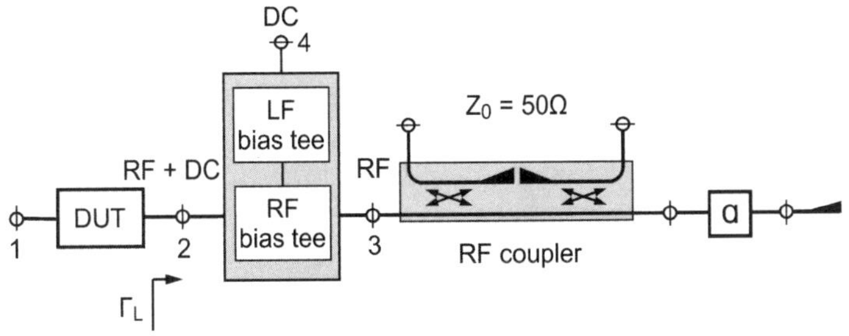

Figure 24.45 Optimal measurement setup arrangement with optimized LF/RF bias tee for intermodulation distortion device characterization.

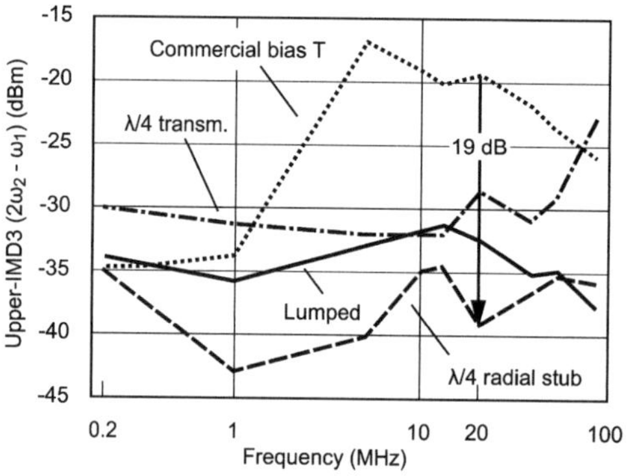

Figure 24.46 Memory-effect characterization of a 0.5-mm GaN HEMT under the modified testbench conditions in a 50Ω environment [55, 56].

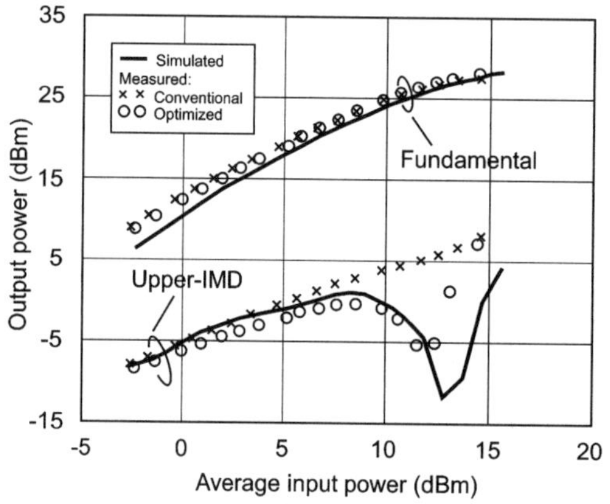

Figure 24.47 Measurement and simulation of output power and upper IMD3 of a 2-mm GaN HEMT under class-C bias condition in a 50Ω environment [55].

24.5.4 System Calibration

Calibration is essential to primarily remove systematic errors of the testbench at the intended measurement frequencies. The systematic errors are often represented using an error model of the measurement system. As shown in Figure 24.48, the raw data measured by the receiver must be transformed to the DUT reference plane by de-embedding the linear test-set parameters.

The calibration of the measurement system is achieved in two steps [55]. First, a two-port SOLT calibration is employed as described in Section 10.4.2. Then, power calibration is performed following the procedure after Benedikt et al.[58] determining mainly the coupling tracking error. Srinidhi has reported that this calibration procedure holds good not only for CW but also when characterizing the DUT under multi-tone excitation [55].

The inherent limitation in implementing merely full 2-port SOLT calibration is that only complex traveling wave ratios can be determined accurately but not the absolute power and phase of the traveling waves. This restriction is due to the unknown transfer function present

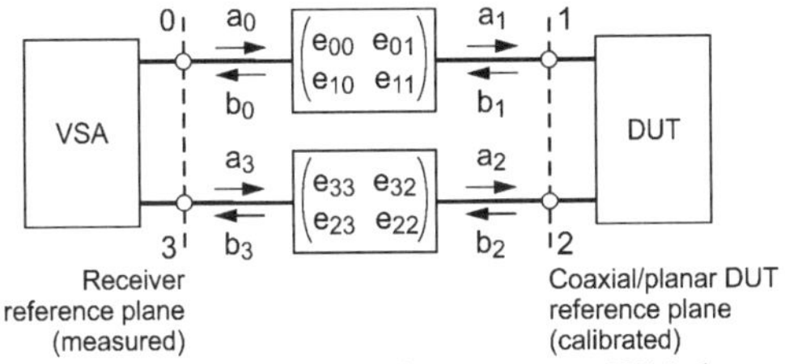

Figure 24.48 Simplified 8-term error model of a two-port VSA-based testbench.

between dependent node a_1 and the independent node a_0, called as coupler tracking e_{10}, shown in Figure 24.49, which is conventionally scaled down to 1 during the error model parameter extraction process.

Power Calibration

The second step, referred to as power calibration, is mainly intended to determine the scaling parameter e_{10} or the coupler tracking [58]. In the case of on-wafer measurements, for instance, the linear testset includes probe heads, a directional coupler, and cables and switches along the path from the DUT reference plane to the receiver's measurement reference plane.

In order to determine e_{10}, a simplified 8-term error model has been considered as shown in Figure 24.49 for obtaining error-corrected large-signal complex traveling waves, neglecting the forward and reverse leakage errors (generally assumed in classical 12-term systematic error model). The error model coefficient e_{10} can be extracted separately by measuring the wave ratio a_1/a_0 with $b_1 = 0$ [58].

One possibility to measure a_1 in the case of on-wafer measurements, could be to directly connect port 1 (coplanar reference plane) in Figure 24.49 to receiver's CH2 (coaxial reference plane), which is impossible [59]. Alternatively, with a coplanar with a coplanar through between port 1 and port 2 and considering the coaxial reference plane by inserting a flexible cable between port 2 and receiver's CH2. The signal flow diagram for the corresponding setup is shown in Figure 24.50, wherein the complex S-parameters of the coaxial cable is predetermined at the required frequency points.

Further, with the measured complex wave ratio measurements of b_{coax}/a_0 ($a_{coax} = 0$), e_{10} can be computed as

$$e_{10} = \frac{S_{30}(1 - e_{11}S_{11})}{S_{21}} \tag{24.12}$$

where

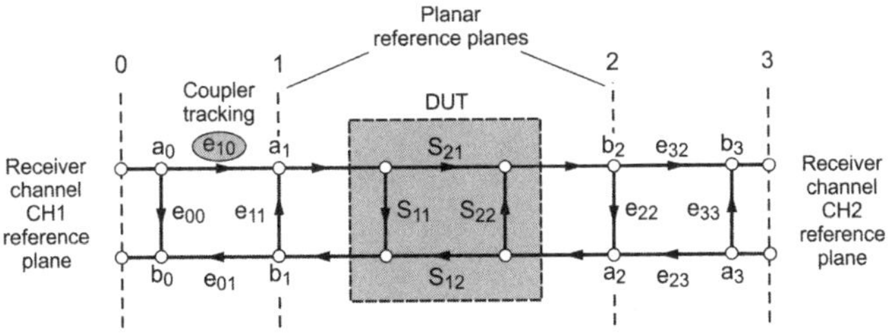

Figure 24.49 Simplified 8-term systematic-error model for vector error-corrected large-signal measurements [55].

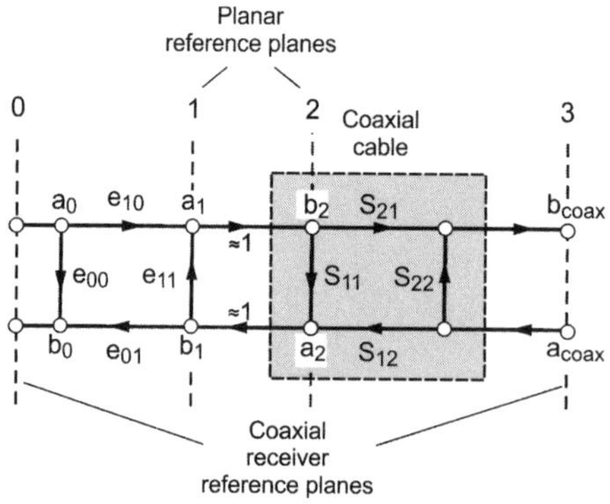

Figure 24.50 Modified error model after coaxial cable insertion.(Planar reference planes 1–2 are connected by a co-planar through line with 200 μm length.)

$$S_{30} = \left. \frac{b_{oax}}{a_0} \right|_{a_{coax}=0} \tag{24.13}$$

Subsequently, the complex power waves at port 1 and port 2, shown in Figure 24.49, can be calculated by using the following expressions:

$$a_1 = e_{10}a_0 + e_{11}b_1 \tag{24.14}$$

$$a_2 = e_{23}a_3 + e_{22}b_2 \tag{24.15}$$

$$b_1 = \frac{b_0 - e_{00}a_0}{e_{01}} \tag{24.16}$$

$$b_2 = \frac{b_3 - e_{33}a_3}{e_{32}} \tag{24.17}$$

Thus, the raw data measured at the receiver's reference plane is transformed to the DUT's reference planes. This calibration procedure indeed holds good not only for CW but also when characterizing the DUT under multitone excitation.

Under WCDMA excitation, the complex aperiodic signal captured by the receiver

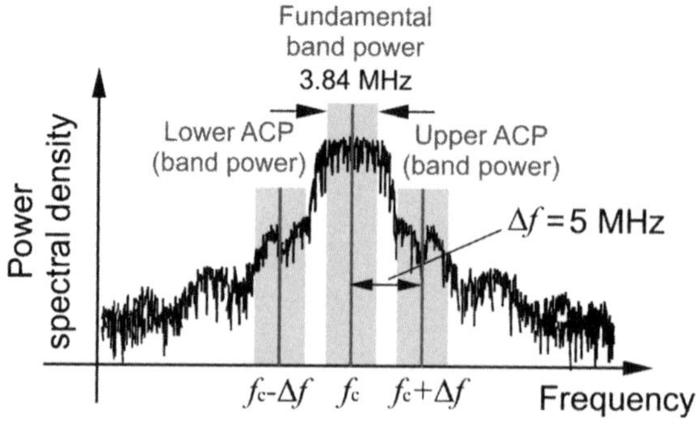

Figure 24.51 3GPP-based UMTS spectral mask of a single-carrier WCDMA spectrum depicting frequencies ($f_c - \Delta f, f_c, f_c + \Delta f$) at which vector error correction is performed.

typically exhibits a continuous spectrum. Subsequently, error-correcting numerous frequency points (depending on the resolution bandwidth) in the relevant channels; namely, fundamental, lower, and upper sideband channels, become impractical, consuming much time during calibration and large-signal measurements. Besides, due to the continuously varying phase at the frequency points, real-time error correction poses another constraint. Hence, a simplified approach has been developed in [55].

With the underlying assumption of approximately constant frequency response of complex test-set parameters over analyzed 3.84 MHz bandwidth, three frequency points (f_1, f_c, f_3 being the center frequencies of the lower ACP band, fundamental band, and higher ACP band), as shown in Figure 24.51, are calibrated, which follows the same procedure described for a multitone case [60]. For vector error correction, the complex raw data at these three frequencies are defined as follows:

- The magnitude of the measured vectors constitute the band powers (summation of powers of all frequency points within the specified band) of lower ACP band, the fundamental band and upper ACP band.
- The corresponding phase is measured at the respective center frequency of each band ($\Phi(f_1)$, $\Phi(f_2)$,$\Phi(f_3)$).

A similar calibration procedure can be adopted for characterizing IMD and ACPR for multicarrier WCDMA stimuli. In this case the center frequency of all relevant bands are taken into account for calibration.

24.5.5 P_{out}-PAE-IMD Load-Pull Optimization

Experimental optimization of bias and matching networks for high linear PA design is one key advantage offered by source-/load-pull characterization apart from CAD model verification. It is hence important to investigate, for instance, to what extent do the bias tees influence IMD load-pull measurements [57].

Conventionally, passive tuners are used in a load-pull system to control impedances. Using an automated in-house multiharmonic mechanical slide tuner, two-tone load-pull characterization of a 2-mm GaN HEMT (Nitronix) was carried out for $\Delta f = 5$ MHz.

Two-tone fundamental load-pull characterization was carried out on a 2-mm GaN HEMT (manufactured by Nitronex), operating in class-AB condition (−1.3V, 15V, $I_{DS0} = 8\%\ I_{DSS}$) with 5-MHz carrier spacing.

Figure 24.52 depicts the P_{out} and PAE contours together with lower and upper IMD products over a wide range of Γ_L (with the intension to obtain an overview on $\Gamma_{L,opt}$ selection strategy) using both conventional and the modified testbench configuration.

In amplifier design, efficiency and linearity parameters often produce a conflicting trade-off. However, practically $\Gamma_{L,opt}$ is chosen near to $P_{out,\,max}$/PAE$_{max}$, almost independent from the linearity performance.

The conventional testbench implemented the commercial RF bias tee of type AU616 while in the modified testbench the radial stub-based in-house RF bias tee was used. With approximately the same optimum load reflection coefficient $\Gamma_{L,opt}(\omega_0)$ at the fundamental frequency, the maximum P_{out} and PAE in Figure 24.52(a) and Figure 24.52(b) are (28 dBm,

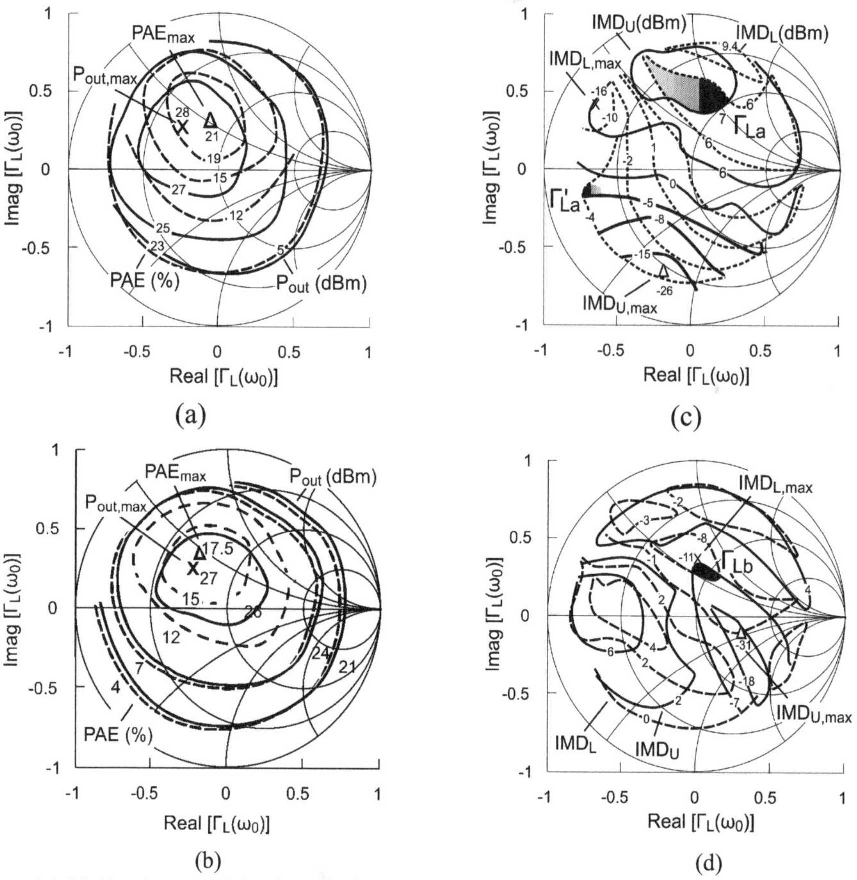

Figure 24.52 Fundamental load-pull characterization of a 2-mm HEMT (Nitronix) under two-tone stimuli using a commercial RF bias tee (AU0616) in a conventional testbench [52] above and using a radial stub-based in-house RF bias tee in the modified testbench [55]. (a) and (b) output power and PAE contours, (c) and (d) upper and lower IMD contours. $f_c = 2.15$ GHz, $\Delta f = 5$ MHz, $V_{GS0} = -1.2$V, $V_{DS0} = 15$V. Conventional testbench: IMD(Γ_{La}) = 6 dBm, optimized testbench: IMD(Γ_{Lb}) = -8 dBm.

21%) and (27 dBm, 17.5%), respectively, measured corresponding to the lower fundamental tone ω_l.

The power amplifiers that are used in telecommunication systems are often designed from the standpoint of high efficiency as being the main design goal. Hence, in practice, it is common to choose $\Gamma_{L,opt}$ for attaining maximum P_{out} and PAE, independent from the IMD response. For instance, by visual inspection of contours in Figure 24.52(c) measured using a conventional testbench, one possibility is to choose Γ'_{La} (shaded overlapping contours of upper and lower IMD), which delivers relatively good linearity with -4 dBm IMD at the cost of a poor PAE of 5% indicated in Figure 24.52(a).

On the other hand, relying on the external linearization techniques, where up to 20-dB IMD cancellation is expected for base station power amplifier applications [61], it is preferable to trade off linearity (IMD ~ 6 dBm) at Γ_{La} shown in Figure 24.52(c), gaining maximum P_{out} and PAE. However, considering contours in Figure 24.52(b) and (d) measured using the modified testbench, at Γ_{lb} = j0.24, where P_{out} and PAE in Figure 24.52(b) are approximately 27 dBm and 17%, respectively, 14-dB IMD suppression is obtained for approximately the same fundamental load termination.

A straightforward application rule is hence demonstrated through which an optimal load can be determined yielding high power output power, PAE, and linearity simultaneously.

24.5.6 Sweet-Spot Measurement

Analysis of a device-sensitive sweet-spot mechanism needs reliable large-signal IMD measurements. With the background of establishing a reliable RF characterization testbench, which was presented in Section 24.5.3, the true IMD sweet-spot behavior in GaN HEMTs can be investigated for its consistency in high linear power amplifier design applications. On the other hand, GaAs HEMTs, which are fundamentally accepted matured semiconductor technology, are additionally considered as a reference to evaluate the sweet-spot reproducibility in GaN HEMTs.

As is known, the smooth turn-on property of the transfer characteristics has direct impact on the overall intrinsic distortion behavior [62]. Underlining this fact, a statistical evaluation of pinch-off voltage V_{pi}, which defines the turn-on behavior, was carried out on 14 samples of 1.2-mm GaAs HEMTs (manufactured by TriQuint Semiconductors) and 6 samples of 2-mm GaN HEMT (manufacured by Nitronex) for examining their respective process consistency.

In order to obtain an unambiguous and consistent definition of V_{pi}, a zero crossing of the third-derivative of drain-source current (i.e., G_{m3} null) was determined for all the devices under investigation, which is close to the position of the ideal threshold. As pointed out in [55], GaAs HEMTs demonstrated a mean pinch-off voltage of $V_{pi} = -2.225$V with a standard deviation of 0.024V, while GaN HEMT exhibited a mean value of $V_{pi} = -1.44$V with a standard deviation of 0.14V. This significant deviation of V_{pi} in the less matured GaN HEMTs indicate a poor repeatability performance,which should have a clear impact on the large-signal sweet-spot characteristics. In the following a few experimental evaluations are summarized.

IMD characteristics of three GaAs HEMTs (V_{pi} = −2.23V, −2.22V, −2.19V) were measured using both a conventional and an optimized testbench. The devices were biased for class-C operation (−2.6V, 9V). As seen in Figure 24.53, primarily good sweet-spot reproducibility in both conventional and modified testbench conditions is obtained. However, a strong IMD asymmetry of 7 dB is observed with the conventional testbench using commercial RF bias tee of type HP11612A manufactured by Hewlett-Packard (Figure 24.53(a)). In contrast, with the improved testbench, the asymmetry between lower and upper IMD is eliminated as shown in Figure 24.53(b) together with remarkable IMD suppression.

Figure 24.54 shows similar results in case of GaN HEMTs, which were biased for class-C operation (−1.6V, 15V). Referring to the upper and lower IMD characteristics measured using a conventional testbench (Figure 24.54(a)), no sweet spots are observed due to the nonzero baseband termination as reported in [62]. Additionally, the sweet-spot reproducibility becomes more dramatic as shown in Figure 24.54(b). This is in accordance with the larger pinch-off voltage distribution of the tested devices (i.e., V_{pi} = −1.37V, −1.2V,

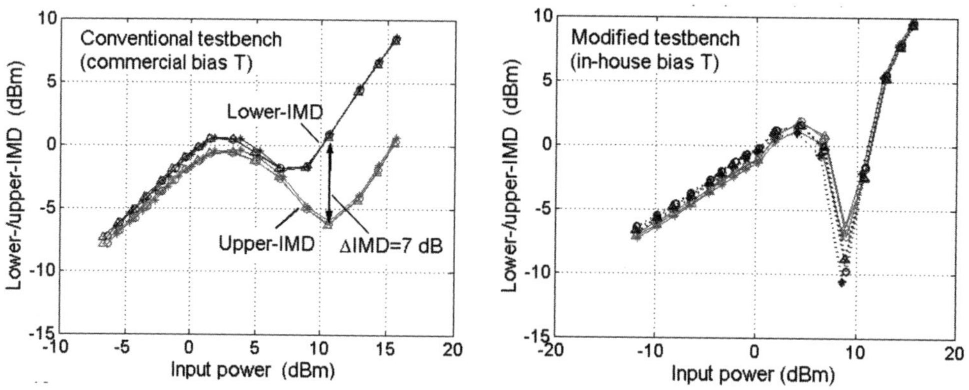

(a) (b)

Figure 24.53 Sweet-spot reproducibility in three GaAs HEMT devices biased for class-C operation (-2.6V, 9V): (a) conventional testbench, (b) modified testbench [55].

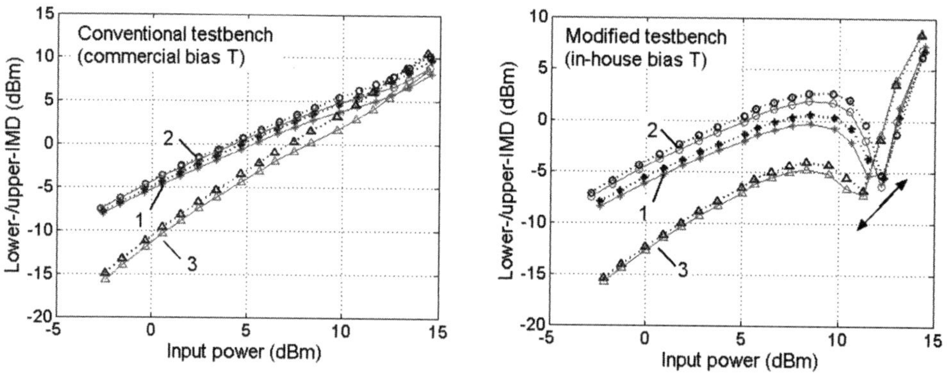

(a) (b)

Figure 24.54 Sweet-spot reproducibility in three GaN HEMT devices biased for class-C operation (-1.6V, 15V).

-1.54V). However, with the optimized setup, the low distortion sweet-spot behavior is revealed and its variation is in agreement with the experimentally determined pinch-off voltages [56].

A 0.5-mm GaN HEMT in a 50Ω environment was excited by a two-carrier WCDMA signal (Test Model 1, 32DPCH) with $\Delta f = 15$ MHz around $f_c = 2.15$ GHz ($V_{GS0} = -4.2$V, 2% I_{DSS}, V_{DS0} =25V). The measured IMR3 characteristics using the conventional and optimized measurement setup are shown in Figure 24.55. The sweet-spot occurrence is evident when in- house RF bias tee is implemented in the modified testbench. In this case, the third-order intermodulation distortion ratio (IMR3) requirement of -45 dBc or lower becomes more vital because of its widespread interference with the adjacent channels. The dip in IMR3 is thus a fundamental device property, which is now revealed since the tee bias has provided a wide baseband and second-harmonic short termination. Furthermore, it is seen that the device exhibits only a weak memory-effect with measured IMD3 asymmetry less than 0.5 dB.

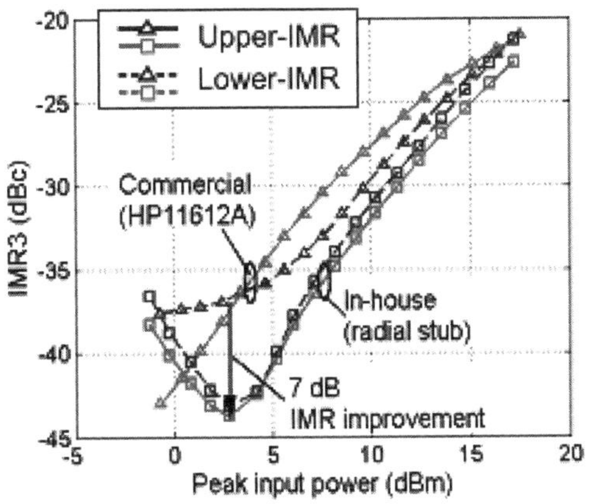

Figure 24.55 IMD3 characterization of a 0.5-mm GaN HEMT in a 50Ω environment stimulated by a two-carrier WCDMA signal.

24.6 SUMMARY

The performance requirements of broadband high-power amplifiers for present wireless communications are permanently increasing driven by the ever more-restrictive specifications of international wireless radio standards (see chapter introduction in [31]). This, in particular, concerns the modulation bandwidth, the linearity (intermodulation distortion (IMD)), and the signal dynamics (peak-to average ratio (PAR)). Because of nonlinear amplification, high-power amplifiers are commonly operated with some power backoff, which drastically lowers the efficiency of the device. The PAE in turn causes additional thermal stress to the amplifier device.

Facing these new challenges, AlGaN/GaN HEMT technology has moved into focus, exhibiting high power density, high temperature operation, and improved linearity. High breakdown voltage allows large drain voltages to be used, leading to high output impedance, resulting in easier matching and lower loss matching circuits [31].

Chapter 23 described the development of power transistor models, which can accurately predict the device properties such as IMD, PAE, output power of the fundamental, and higher harmonics in any bias point in the I-V output plane. Large-signal modeling has been performed on the basis of model parameter extraction from (static) DC, pulsed-DC, and S-parameter measurements.

This chapter has discussed recently developed large-signal measurement methods and established measurement setups that can be used for direct large-signal model parameter extraction and experimental device optimization by waveform engineering.

However, the primary objective of designing discussed measurement systems is the verification of developed large-signal models in the previous chapter under various realistic operating conditions, testing and simulating power devices, for example, with complex WCDMA signals.

The motivation and introduction of signal-waveform measurement techniques was briefly described. Initially, focusing on the characterization of the nonlinear behavior of power devices in a 50-ohm environment, the setup was extended and configured first with passive load-pull capability. Further improvements included increased maximum power (e.g., up to 100W) load-pull extension to higher harmonics and, in particular, also covering the envelope frequency region. Similarly, source pulling was developed.

Commonly, tuning of the input and output impedances is traditionally established with passive devices, such as variable attenuators and phase shifters. The introduction of the IQ (in-phase/quadrature) modulator concept has considerably simplified the setup construction. Amplitude and phase setting are under control of an IQ modulator (vector generator) assigned to each harmonic. The power level of each harmonic can be adjusted by the auxiliary amplifier behind the modulator. In contrast to the measurement concept in [29] frequency upconversion is provided after the amplitude and phase adjustment. The main advantage of this concept is that the IQ modulator, which is the most critical component, needs only to be realized for the fundamental frequency range for any harmonic.

A specific feature of the presented measurement setups was the split-signal type configuration. The forward and backward injected signals to the device-under-test is under control of a single microwave source, feeding all active load modules (IQ modulator plus frequency multiplier). Thus, the equipment costs are rather low. This is in contrast to the active source and load-pull system reported in [66] using a number of Agilent ESG series signal generators for signal magnitude and phase adjustment. Apart from high cost, the synchronization may be challenging (jitter) and may contribute to additional measurement errors.

Measurement setups for characterization of high power devices were realized both in time- and frequency-domain. The frequency-domain measurement setup was established for IMD characterization. It has been shown that the design of bias networks has a large influence on the IMD measurement results. Similarly, the arrangement of common setup components (DUT, input and output bias tees and couplers) has a tremendous effect on the measurement. This has been impressively demonstrated with the measurement of sweet spots.

The problems occurring with wideband multitone load-pulling were discussed and the effects of various load termination configurations quantisized. This has also been pointed out in [67, 68). It has been shown that the IQ modulator concept may be used to avoid any phase delay between the different tones.

The presented measurement setups are modular constructed and may be adapted to match further upcoming wireless radio standards requirements, such as a higher-operation frequency band and much larger modulation bandwidth.

REFERENCES

[1] Kompa, G., F. Lin, M. Novotny, F. van Raay, and A. Werthof, "Recent Advances in Large-Signal Measurements and Modelling of Active Microwave Devices," *Digest of 2nd Int. Workshop on Integrated Nonlinear Microwave and Millimeterwave Circuits* (INMMIC), October 7-9, 1992, University of Duisburg (Duisburg, Germany), pp. 231–245, https://www.researchgate.net/publication/271211407_Recent_ advances_ in_ large-signal_ measurements_and_ modelling_ of_active_microwave_devices.

[2] Lewandowski, A., G. Kompa, W. Mwema, and W. Wiatr, "A 24 GHz PHEMT-Based Oscillator," *Journal of Telecommunications and Information Technology*, Vol. 1, 2003, pp. 15–19.

[3] van Raay, F., *Error-Corrected 20 GHz Signal Waveform Measurement Setup for the Direct Large-Signal Analysis of Microwave Field-Effect Transistors* (in German), Doctoral Thesis, University of Kassel, 1990.

[4] Kompa, G., and van Raay, F., "New Approach in Error-Corrected Large-Signal Measurement of Nonlinear Devices," *Proc. of 4th Exhibition and Conference for Ultra High Frequency Technology, Microwave and Optronics (MIOP '89)*, Sindelfingen, Paper 2A.2, 1989.

[5] Kompa, G., and van Raay, F., "Automatic Large-Signal Measurement System for Nonlinear Device Modelling and Model Verification," *Proc. of 19th Europ. Microwave Conf.*, London, 1989, pp. 587–594.

[6] Kompa, G., and van Raay, F., "Error-Corrected Large-Signal Waveform Measurement System Combining Network Analyzer and Sampling Oscilloscope Capabilities," *IEEE Trans.*, Vol. MTT-38, 1990, pp. 358–365.

[7] Hewlett Packard, *Microwave Network Analyzer Applications*, Application Note 117-1, June 1970.

[8] Kompa, G., M. Schlechtweg, and F. van Raay, "Precisely Calibrated Coaxial-to-Microstrip Transitions Yield Improved Performance in GaAs FET Characterization," *IEEE Trans.*, Vol. MTT-38, 1990, pp. 62–68.

[9] van Raay, F., and G. Kompa, "A New On-Wafer Large-Signal Waveform Measurement System with 40 GHz Harmonic Bandwidth," *IEEE MTT-S Int. Microwave Symp. Digest*, 1992, pp. 1435–1438.

[10] Hewlett Packard, *HP 71500A Microwave Transition Analyzer System, HP 70820A Microwave Transition Analyzer Module*, Technical Data 5091-0792E, USA, 3/91.

[11] Browne, J., "Transition Analyzer Scans Amplitude and Phase of 40-GHz Pulses," *Microwaves RF*, 1991, pp. 150–157.

[12] Coffron, J., M. Oethlefsen, and M. Levernier; "The HP 70820A Microwave Transition Analyzer Module: An Innovative Instruments Architecture," Hewlett Packerd Co., USA, 1991.

[13] Agilent Technologies, *Beam Lead Schottky Diodes for Mixers and Detectors (1–26 GHz)*, Technical Data, HSCH-5300 Series, November 1999.

[14] Keysight Technologies, *HSCH 9161 HSCH 9162, GaAs Detector Diode*, Data Sheet, December 1, 2017.

[15] van Raay, F., and G. Kompa, "Waveform Measurements—The Load-Pull Aspect," *55th ARFTG Conference Digest*, Vol. 37, June 2000, pp. 14–21.

[16] Müller, J.-E., and B. Gyselinclex, "Comparison of Active Versus Passive On-Wafer Load-Pull Characterisation of Microwave and mm-Wave Power devices," *1994 IEEE MTT-S Digest*, 1994, pp. 1077–1080.

[17] Kotzebue, K., T. S. Tan, and D. McQuate, "An 18 to 26.5 GHz Waveguide Load-Pull System Using Active-Load Tuning," *IEEE MTT-S Symposium Digest*, 1987, pp. 453–456.

[18] Takayama, Y., "A New Load-Pull Characterization Method for Microwave Power Transistors," *IEEE MTT-S Int. Microwave Symp. Digest*, 1976, pp. 218–220.

[19] Stancliff, R. B., and D. D. Poulin, "Harmonic Load-Pull," *IEEE MTT-S Int. Microwave Symposium Digest*, 1979, pp. 185–187.

[20] Quéré, R., J. M. Nebus, J. P. Villotte, and J. Obregon, "A Complete, Measurement Based, Methodology for the Extraction and Verification of Nonlinear Models of Active Devices," *IEEE Workshop on Experimentally Based FET Device Modelling & Related Nonlinear Circuit Design*, Kassel, Germany, July 17-18, 1997, pp. 11.1–11.16, https://archive.org/details/DTIC_ADA331468/page/n279.

[21] Larose, R., F. M. Ghannouchi, and R. G. Bosisio, "A New Multi-Harmonic Load-Pull Method for Non Linear Device Characterization and Modeling," *IEEE MTT-S Int. Microwave Symp. Digest*, 1990, pp. 443–446.

[22] ATN Microwave Inc., "Product Feature: A Load-Pull System for Digitial Mobile Radio Power Amplifiers," *Microwave Journal*, Vol. 38(3), March 1995, pp. 116–118.

[23] ATN Microwave Inc., "Product Feature: A Load-Pull System with Harmonic Tuning," *Microwave Journal*, Vol. 39(3), March 1996, pp. 128–132.

[24] van Raay, F., and G. Kompa, "A 40GHz Large-Signal Double-Reflectometer Waveform Measurement System Designed for Load-Pull Applications," *26th EuMC*, Prague, 9-12 September 1996, pp. 657–661.

[25] Sam, B., and P. Halford, "High-Performance Quadrature Modulators for Broadband Wireless Communication," *2001 IEEE Rdio Frequency Integrated Circuits Symposium*, MON2A-3, pp. 17–20.

[26] Tiiliharju, E., *Integration of Boradband Direct-Conversion Quadrature Modulators*, Doctoral Thesis, Helsinki University of Technology, 2006.

[27] Karthaus, U., N. Alomari, G. Bergmann, and H. Schuhmacher, "High Dynamic Range, High Output Power I/Q Modulator in 50 GHz f_t SiGe Technology," *2004 IEEE Radio Frequency Integrated Circuits Symposium*, 2004, pp. 539–542.

[28] Poulin, D. D., J. R. Mahon, and J.-P. Lanteri, "High Power On-Wafer Pulsed Active Load Pull System," *IEEE MTT-S Int. Microwave Symp. Digest*, 1992, pp. 1431–1434.

[29] Poulin, D. D., J. R. Mahon, and J.-P. Lanteri, "A High Power On-Wafer Pulsed Active Load Pull System," *IEEE Trans.*, Vol. MTT-40, 1992, pp. 2412–2417.

[30] Siemens, *ICs for Communications, Mixer DC - 2.5 GHz and Vector Modulator 1.4 - 2.5 GHz, PMB 2202 Version 1.2*, Preliminary Data Sheet, 08.97.

[31] Kompa, G., *Basic Properties of III-V Devices–Understanding Mysterious Trapping Phenomena*, Kassel, Kassel University Press GmbH, 2014.

[32] Devlin, C., T. Brazil, and G. Magerl, "TARGET Network: Microwave Power Amplifiers," *IEEE Microwave Magazine*, April 2006.

[33] BMBF-Joint Project "mobileGaN", Subproject: *Characterization and Modeling of GaN FETs for an Optimal Design of High-Efficient Linear Power Amplifiers for UMTS Applications* (in German), August 2006 - April 2008.

[34] Bunz, B., and G. Kompa, "Active Load Pull with Fourth Harmonic Tuning Based on an IQ Modulator Concept," *33rd European Microwave Conference Proc.*, 2003, pp. 359-361.

[35] Bunz, B., *Design Method of Broadband Frequency Multipliers for a Multiharmonic Source-Pull-Measurement System* (in German), Doctoral Thesis, University of Kassel, 2004; published by Kassel University Press, 2004.

[36] El Din, M. G., B. Bunz, and G. Kompa, "10 W Broadband Load-Pull for GaN/AlGaN Characterization," *Conf. Proc. GeMIC 2005*, 2005, 185–188.

[37] El Din, M. G., *Extension of a Waveform Signal Measurement System towards Two-Tone Signals with Investigation on Linearity of High Power GaN/AlGaN Transistors*, Master Thesis, University of Kassel, Germany, January 2005.

[38] Bunz, B., A. Ahmed, and G. Kompa, "Influence of Envelope Impedance Termination on RF Behaviour of GaN HEMT Power Devices," *13th GAAS Symposium*, Paris, 2005.

[39] Ghannouchi, F. M., G. Zhao, and F. Beauregard, "Simultaneous Load-Pull of Intermodulation and Output Power Under Two-Tone Excitation for Accurate SSPA′ s Design," *IEEE Trans.*, Vol. MTT-42(6), 1994, pp. 929–934.

[40] Hajji, R., F. Beauregard, and F. M. Ghannouchi, "Multitone Power and Intermodulation Load–Pull Characterization of Microwave Transistors Suitable for Linear SSPA's Design," *IEEE Trans.*, Vol. MTT-45(7), 1997, pp. 1093–1099.

[41] UMTS LTE [1]: www.3gpp.org, "Long Term Evolution of the 3GPP Radio Technology," *http://www.3gpp.org/Highlights/LTE/LTE.htm*, Oct. 2007.

[42] Wittwer, B., *Multiharmonic Source- and Load-Pull Measurement System for the Characterization of Active Devices Using WCDMA Signals* (in German), Doctoral Thesis, University of Kassel, 2013; published by Dr. Hut Verlag, München, 2014.

[43] Agilent Technologies, *Agilent Infiniium Oscilloscopes Performance Guide Using 89600 Vector Signal Analyzer Software*, Application Note, March 31, 2011.

[44] Agilent Technologies, *89641S 6 GHz RF Vector Signal Analyzer*, Data Sheet.

[45] Agilent Technologies, *89640S 2.7 GHz RF Vector Signal Analyzer*, Data Sheet.

[46] Keysight Technologies, *N8201A Performance Downconverter LXI Module, 3 Hz to 26.5 GHz*, Data Sheet.

[47] Wittwer, B., E. R. Srinidhi, G. Kompa, and A. Bangert, "ACP and IMD3 Measurement and Comparison Using a Novel Load-Pull System," *Proc. 40th European Microwave Conference*, 2010, pp. 1568–1571.

[48] Amplifier Research Inc., *Amplifier Research 100S1G4M8 Microwave Amplifier, 0.8 - 4.2 GHz, 100W.*

[49] Srinidhi, E. R., A. Jarndal, and G. Kompa, "A New Method for Identification and Minimization of Distortion Sources in GaN HEMT Devices Based on Volterra Series Analysis," *IEEE Electron Device Lett.*, Vol. 28(5), 2007, pp. 343–345.

[50] Srinidhi, E. R., R. Ma, and G. Kompa, "Optimization of Broadband Drain Modulation in GaN HEMT Devices," *IEEE Proc. of Radio and Wireless Symposium (RWS)*, Orlando, 2008, pp. 81–84.

[51] Le Gallou, N., J. M. Nebus, E. Ngoya, and H. Buret, "Analysis of Low Frequency Memory and Influence on Solid State HPA Intermodulation Characteristics," *Proc. IEEE MTT-S Int. Microw. Symp. Dig.*, 2001, pp. 979–982.

[52] Cripps, S. C., *RF Power Amplifiers for Wireless Communications*, Norwood, MA: Artech House, 2006.

[53] Fujitsu, *150-W, 2.11 –2.17 GHz Push-Pull Compact Amplifier for IMT-2000 Base-Station Application*, Application Note 008.

[54] Vuolevi, J. H. K., T. Rahkonen, and J. P. A. Manninen, "Measurement Technique for Characterizing Memory Effects in RF Power Amplifiers," *IEEE Trans.*, Vol. MTT-49, 2001, pp. 1383–1389.

[55] Srinidhi, E. R., *Intermodulation Distortion Modelling and Measurement Techniques for GaN HEMT Characterization*, Doctoral Thesis, Department of High Frequency Engineering, University of Kassel, Kassel, Germany, December 2008.

[56] Pedro, J. C., and N. B. Carvalho, *Intermodulation Distortion in Microwave and Wireless Circuits*, Norwood: Artech House, Inc., 2003.

[57] Srinidhi, E. R., B. Wittwer, R. Ma, and G. Kompa, "Key Aspects for Characterizing Device Inherent IMD in GaN HEMTs," *38th European Microwave Conference (EuMC)*, 2008, pp. 1117–1120.

[58] Benedikt, J., R. Gaddi, P. J. Tasker, and M. Goss, "High Power Time-Domain Measurement System with Active Harmonic Load–Pull for High Efficiency Base Station Amplifier Design," *IEEE Trans.*, Vol. MTT-48(12), 2000, pp. 2617–2624.

[59] van Raay, F., and G. Kompa, "A 40GHz Large-Signal Double-Reflectometer Waveform Measurement System Designed for Load-Pull Applications," *26th EuMC*, Prague, 9-12 September 1996, pp. 657–661.

[60] Srinidhi, E. R., R. Ma, and G. Kompa, "Application Rules for Accurate IMD Characterization in GaN HEMTs," *Proc. of German Microwave Conference (GeMIC)*, 2008, pp. 171–174.

[61] Cripps, S. C., "Ignoring the Obvious: Possibilities for On-Chip Linearization of RFIC Power Amplifiers," *Proc. Gallium Arsenide Applications Symposium (GAAS)*, Milano, Italy, September 2002, pp. 23–27.

[62] Cripps, S. C., *Advanced Techniques in RF Power Amplifier Design*, Norwood, MA: Artech House, 2002.

[63] Carvalho, N. B., and J. C. Pedro, "Large- and Small-signal IMD Behaviour of Microwave Power Amplifiers," *IEEE Trans.*, Vol. MTT-47(12), 1999, pp. 2364–2374.

[64] Kompa, G., "Fundamentals in Successful Experimental Active Device Modelling," *Proc. Workshop on Advanced Microwave Devices, Characterisation and Modelling*, European Microwave Conf., 1993, pp. 22–25.

[65] van Raay, F., and G. Kompa, "Waveform Load-Pull Measurements with Harmonic Tuning," *Proc. Workshop on New Large Signal Characterization and Modeling Techniques for RF and Microwave Circuits and Systems*, 30th European Microwave Conference 2000, Paris, 2000.

[66] Poulin, D. D., J. R. Mahon, and J. P. Lanteri, "A High Power On-Wafer Pulsed Aactive Load Pull System," *IEEE Trans.*, Vol. MTT-40(12), 1992, pp. 2412–2417.

[67] Spirito, M., M. J. Pelk, F. van Rijs, S. J. C. H. Theeuwen, D. Hartskeerl, and L. C. N. de Vreede, "Active Harmonic Load-Pull for On-Wafer Out-of-Band Device Linearity Optimization," *IEEE Trans.*, Vol. MTT-54(12), 2006, pp. 4225–4236.

[68] Marchetti, M., M. J. Pelk, K. Buisman, W. C. E. Neo, M. Spirito, and L. C. N. de Vreede, "Active Harmonic Load-Pull With Reralistic Wideband Communications Signals," *IEEE Trans.*, Vol. MTT-56(12), 2008, pp. 2979–2988.

Chapter 25

Popular Nonlinear FET Models: Capabilities and Limitations

Device modelers and circuit designers in industrial environment commonly trust proven device models and may be less open for new approaches in device modeling. The author remembers a panel session titled "FET Measurement & Modeling—Critical Review of State-of-the-Art, Perspectives for First Pass Success, Current Issues" at Kassel Workshop in 1997 ("Experimentally Based FETModelling & Related Nonlinear Circuit Design") [1] with speakers from industry and RF software companies. New device modeling approaches were discussed simulating, in particular, device dispersion effects and high-power dynamic bias shifting. A representative of the industry commented that a simple Curtice model would work well, although there might be a difference between measurement and simulation. This problem could be simply solved by changing relevant model parameters stepwise until a match is observed with measured data.

Note, at that time power amplifier circuit design focused mainly on the maximization of output power and efficiency. However, since then, the requirements on the performance of today's broadband amplifiers have become more challenging, trading off among contrasting design goals of high output power, high power efficiency, and low intermodulation distortion.

As reviewed in Chapter 2, transistor concepts and transistor technology have been evolved dramatically the last decades, starting with simply constructed GaAs-based MESFETs in the seventies, followed by HEMTs with various III-V semiconductor compound composition in the eighties and GaN-based devices in the nineties. This continuous active device progressing has been a permanent challenge for device modelers to keep pace with technology development and to satisfy circuit designers with device models that are trustworthy and reliable to reach the design target.

Diversification of device construction and semiconductor material have been further challenges. As discussed in Section 2.3.2.2, the formation of a parasitic MESFET channel in HEMTs with increasing gate voltage leads to a typical bell-shaped behavior of the transconductance, which has to be taken into account in nonlinear modeling. Since GaN technology became available around 2000, further issues have arisen originating from its advantageous material properties. The wide bandgap property provides high breakdown voltage, which enables high output power leading to higher temperature operation conditions. In addition, trapping effects provide a serious dispersion problem.

In this chapter, popular industry standard FET models such as the Curtice quadratic model (Curtice-2), Tajima model, Materka-Kapsprak model (Materka), Statz model, TOM model,

Angelov model, Root model, and TOPAS model are tested on the basis of measured data of an AlGaN/GaN power HEMT with 8×125 μm gate width produced by FBH, Berlin. Model verification includes static- and pulsed-DC measurements, and large-signal measurements, such as single and two-tone excitation. Simulations were performed on the basis of ADS model library [2]. The models are available in ADS. The TOPAS model has been implemented in ADS as a SDD (symbolically defined device) component.

A recently cited statement from industry should be noted: "Unfortunately, the industry is in state of confusion regarding nonlinear FET modeling. There are many approaches. All the approaches have strengths and weaknesses. Some of the most popular and well-known models have deep flaws, are seriously outdated, and yet remain popular to this day. There are very few people using analytic models that in some cases are truly invalid. There is a lot resistance to change."

In the following the models will be briefly described, their implementation in commercial software discussed, and simulation results compared with measurements. The chapter will close with some comments on the usefulness of tested models with a focus on GaN power FET simulation.

25.1 WIDELY USED 15-ELEMENT SMALL-SIGNAL EQUIVALENT CIRCUIT

The active part of the transistor is embedded in a passive (linear) network that equivalently describes the contacting situation of the device to the environment, including contact pads, access transmission lines, bulk and contact resistances, and interelectrode capacitances. Figure 25.1 illustrates a widely used 15-element small-signal equivalent network. It comprises the intrinsic part with seven bias-dependent model elements and the extrinsic parasitic network with eight bias-independent parameters (i.e., shunted gate and drain pad capacitances C_{pg} and C_{pd}, respectively) and series gate, drain, and source inductances and resistances L_g, L_d, L_s, R_g, R_d, and R_s, respectively.

Capacitances C_{pg} and C_{pd} are usually extracted from S-parameter measurement under cold pinch-off ($V_{ds} = 0$, $V_{gs} < V_p$) condition at low frequencies. After de-embedding the extrinsic capacitances, inductances and resistances (R_g, R_d, R_s, R_i, L_g, L_d, L_s) can be determined from small-signal S-parameter measurements under cold pinch-off or gate-forward bias conditions at medium frequencies.

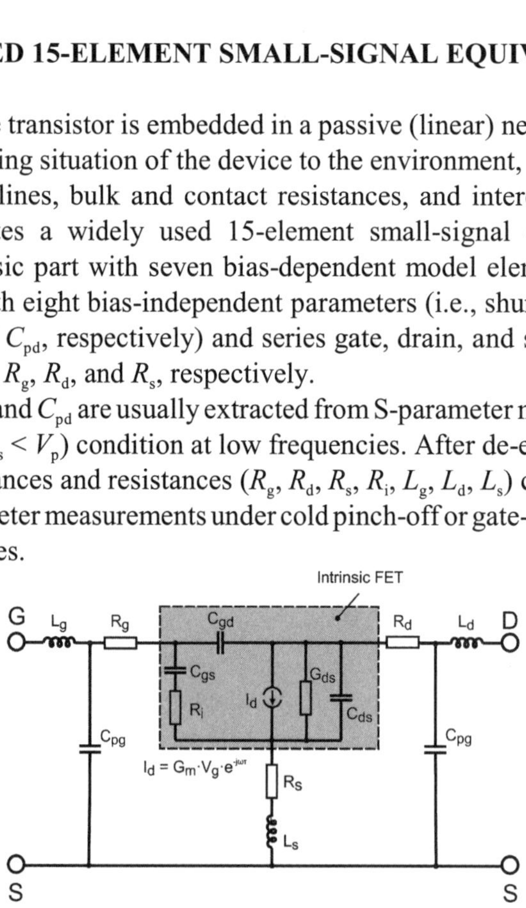

Figure 25.1 Widely used 15-element equivalent circuit of a FET device.

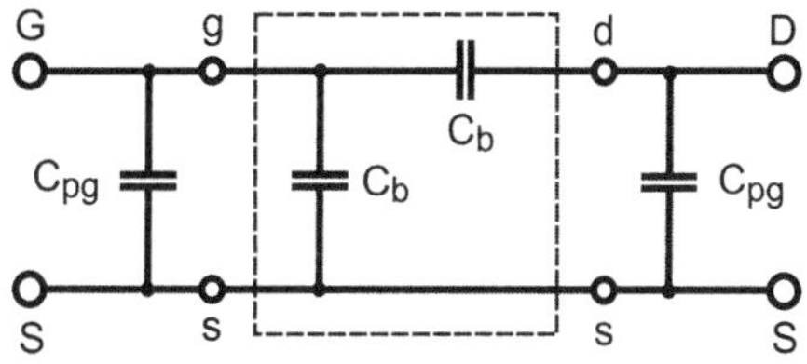

Figure 25.2 Simplified equivalent circuit of a FET device under cold pinch-off biasing for the extraction of C_{pg} and C_{pd} capacitances.

25.1.1 Extraction of Pad Capacitances C_{pg} and C_{pd}

As pointed out by Dambrine et al. [3], under cold bias condition ($V_{DS} = 0$) where the gate voltage V_{GS} is lower than the pinch-off voltage V_p ($V_{GS} < V_p$), the channel region under the gate is totally depleted, so that only fringing capacitances at each side of the gate appear. At low frequencies the resistances and inductances have a negligible influence on the measurement. In this case, the simplified equivalent circuit in Figure 25.2 is adequate. The imaginary parts of the Y-parameters read

$$\mathrm{Im}[Y_{11}] = j\omega(C_{pg} + 2C_b) \tag{25.1}$$

$$\mathrm{Im}[Y_{12}] = \mathrm{Im}[Y_{21}] = -j\omega C_b \tag{25.2}$$

$$\mathrm{Im}[Y_{22}] = j\omega(C_b + C_{pd}) \tag{25.3}$$

Figure 25.3 shows the measured frequency normalized Y parameters, yielding the C_{pg} and C_{pd} pad capacitances.

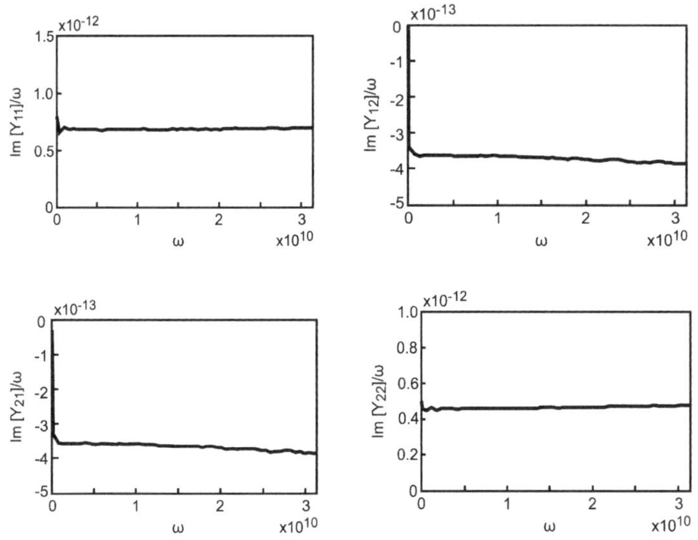

Figure 25.3 Measured normalized imaginary part of Y-parameters of investigated GaN power HEMT at low frequencies under cold pinch-off biasing for the extraction of C_{pg} and C_{pd} capacitances.

25.1.2 Extraction of Parasitic Resistances and Inductances

Dambrine suggested to determine resistances and inductances from cold gate-forward measurements. However, this implies an additional resistance, the channel resistance R_c, to be determined. As the measurement provides only three relations between the four unknowns R_g, R_s, R_d, and R_c, an additional relation is needed to separate the four unknowns. Various approaches were recommended [3]: Resistance measurement after Hower [4] or Fukui [5], gate resistance measurement by test structure, DC measurement of R_s and R_d after Yang et al. [6], or calculation of R_c if the channel technological parameters are known.

Chosing the cold pinch-off bias point ($V_{GS} = V_p$, $V_{DS} = 0$) can eliminate the resistance extraction issue as discussed in Section 21.3.2. In this case, the equivalent network results in a T-network, which can be simply characterized by its Z-matrix. Real and imaginary parts provide the resistive and inductive series elements:

$$\mathrm{Im}[\omega Z_{11}] = (L_g + L_s)\omega^2 - \left(\frac{1}{C_g} + \frac{1}{C_s}\right) \tag{25.4}$$

$$\mathrm{Im}[\omega Z_{22}] = (L_d + L_s)\omega^2 - \left(\frac{1}{C_d} + \frac{1}{C_s}\right) \tag{25.5}$$

$$\mathrm{Im}[\omega Z_{12}] = L_s\omega^2 - \frac{1}{C_s} \tag{25.6}$$

Hence the values of L_g, L_d and L_s can be extracted from the slope of $\mathrm{Im}(\omega\Delta Z)$ versus ω^2-curve (Figure 25.4), where ΔZ equals $(Z_{11}-Z_{12})$, $(Z_{22}-Z_{12})$, and Z_{12}, respectively.

$$\omega^2\mathrm{Re}[Z_{11}] = \omega^2(R_g + R_s) \tag{25.7}$$

$$\omega^2\mathrm{Re}[Z_{22}] = \omega^2(R_d + R_s) \tag{25.8}$$

$$\omega^2\mathrm{Re}[Z_{12}] = \omega^2 R_s \tag{25.9}$$

By linear regression, the values of $(R_g + R_s)$, $(R_d + R_s)$, and R_s can be extracted from the slope of $\omega^2\cdot\mathrm{Re}[Z_{ij}]$ $(i,j = 1,2)$ versus ω^2-curves. In Table 25.1 the determined extrinsic resistive and inductive parameters are listed.

25.1.3 Intrinsic Bias-Dependent Model Elements

Applying the de-embedding procedure in [3], the model parameters of the intrinsic network can be calculated. At low frequencies the Y-parameters can be estimated as [3]

$$y_{11} \approx R_i C_{gs}^2\omega^2 + j\omega(C_{gs} + C_{gd}) \tag{25.10}$$

Table 25.1
Extracted Extrinsic Model Elements

R_g/Ω	R_d/Ω	R_s/Ω	L_g/pH	L_d/pH	L_s/pH	C_{pg}/fF	C_{pd}/fF	C_b/fF
4.63	4.25	2.67	49	45	3.8	62	135	320

$$y_{12} = -j\omega C_{gd} \tag{25.11}$$

$$y_{21} \approx G_m - j\omega\left[C_{gd} + G_m(R_i C_{gs} + \tau)\right] \tag{25.12}$$

$$y_{22} = G_{ds} + j\omega(C_{ds} + C_{gd}) \tag{25.13}$$

From (25.10) to (25.13) the model parameters can be determined as follows:

$$C_{gs} = \frac{Im[Y_{11}] + Im[Y_{12}]}{\omega} \tag{25.14}$$

$$C_{gd} = -\frac{Im[Y_{12}]}{\omega} \tag{25.15}$$

$$C_{ds} = \frac{Im[Y_{22}] + Im[Y_{12}]}{\omega} \tag{25.16}$$

$$G_{ds} = Re[Y_{12}] \tag{25.17}$$

$$R_i = \frac{Re[Y_{11}]}{(Im[Y_{11}] + Im[Y_{12}])^2} \tag{25.18}$$

Figure 25.5 illustrates the extracted bias-dependent intrinsic model parameters for the investigated GaN power HEMT. For the analytical models to be implemented, R_i, G_{ds}, C_{gd} (V_{gd}) and C_{gs} ($V_{gs} = 0$, V_{ds}) must be calculated [2]. For R_i, G_{ds} and C_{ds} the mean values in Figure 25.5 were taken. For C_{gs} ($V_{gs} = 0$, V_{ds}) and C_{gd} (V_{gd}) the one-dimensional characteristics in Figure 25.6 were considered.

The active nonlinear part of the transistor is embedded in the linear parasitic network shown in Figure 25.7, which equivalently describes the contacting situation of the device to the environment, including contact pads, access transmission lines, bulk and contact resistances, and interelectrode capacitances. In addition to Figure 25.1 dispersion is also considered connecting a series R_{lf}–C_{lf} branch in parallel to the drain and source electrodes.

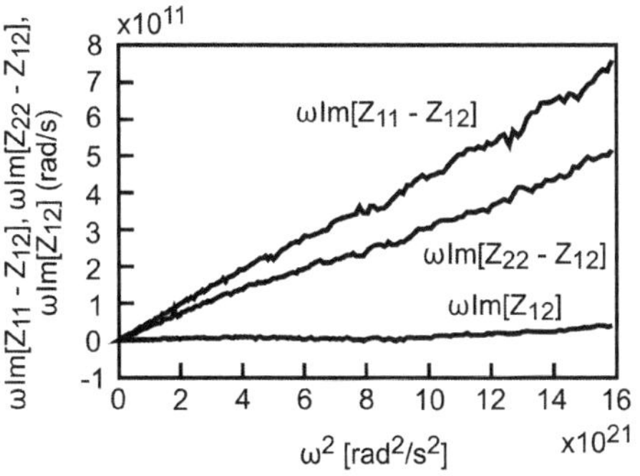

Figure 25.4 Inductance evaluation from measured cold pinch-off Z-parameters for investigated 2×250 μm gate GaN power HEMT (additive capacitive parts not considered).

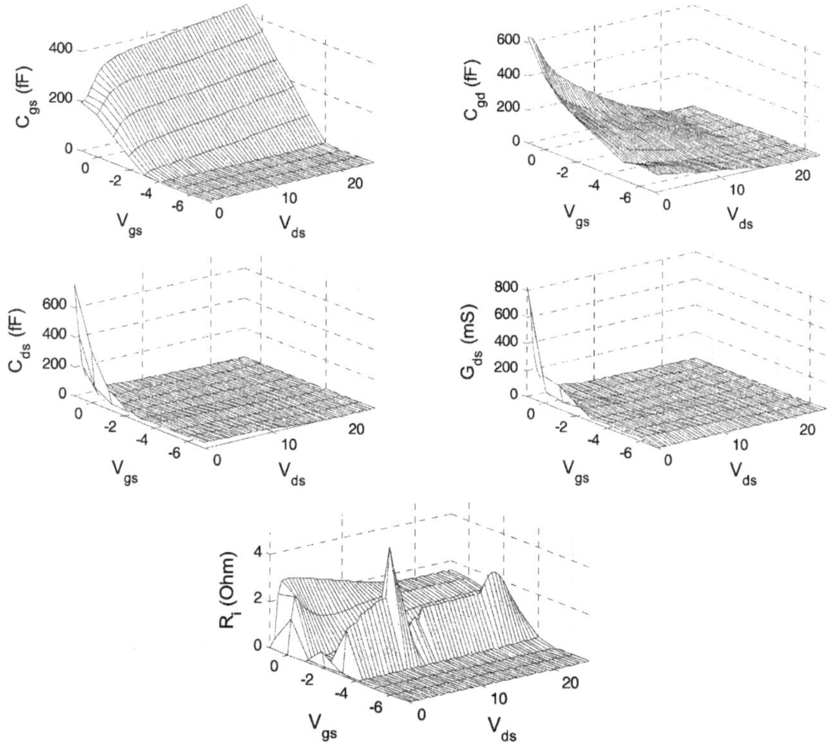

Figure 25.5 Extracted bias-dependent intrinsic model elements of investigated 2×250 μm gate GaN HEMT.

Constant dispersion elements are only valid around a given bias point. Multibias extension could be performed by defining a bias-dependent dispersion resistance $R_{\mathrm{lf}}(V_{\mathrm{gs,DC}}, V_{\mathrm{ds,DC}})$.

Taking into account dispersion effects, an additional dispersion branch comprising a R_{lf}-C_{lf} series circuit is included. At low frequencies this circuit is an open circuit, at high frequencies R_{dsd} is parallel to the channel resistance R_{ds}, increasing the composed output conductance according to dispersion.

As the earlier published nonlinear models were pure DC models, parasitic inductances are missing. Only parasitic resistances R_{g}, R_{d}, and R_{s} have been taken into account to determine the intrinsic control voltages from the applied terminal voltages. As the resistances are assumed to be bias-independent (linear), they are not regarded in the following nonlinear model circuits.

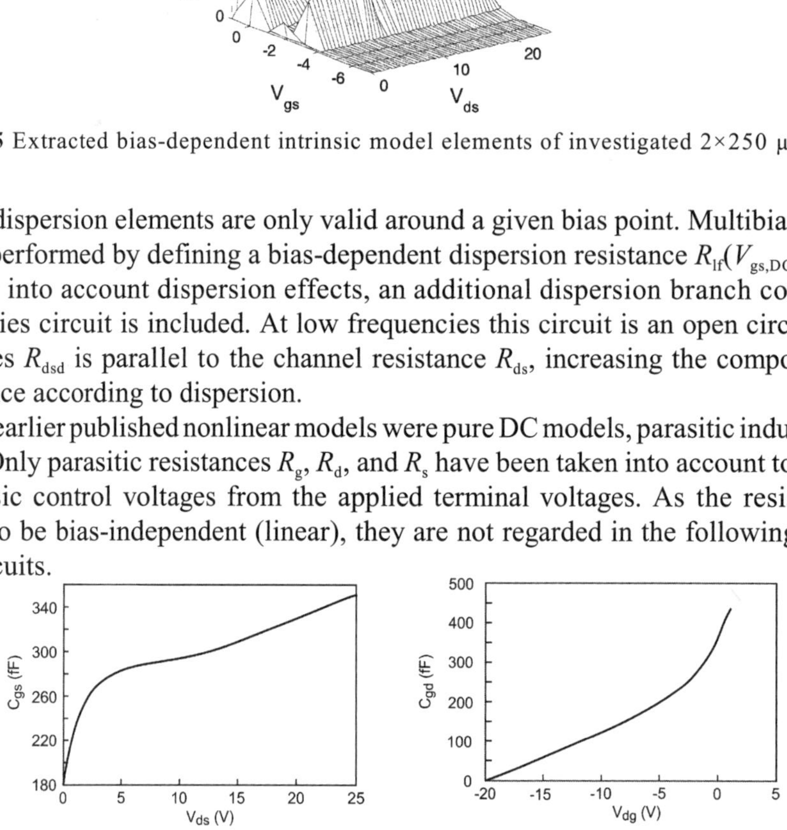

(a) (b)

Figure 25.6 (a) Extracted bias-dependent $C_{\mathrm{gs}}(V_{\mathrm{gs}} = 0\mathrm{V}, V_{\mathrm{ds}})$ and (b) extracted $C_{\mathrm{gd}}(V_{\mathrm{gd}})$ for investigated 8×125 μm gate GaN HEMT.

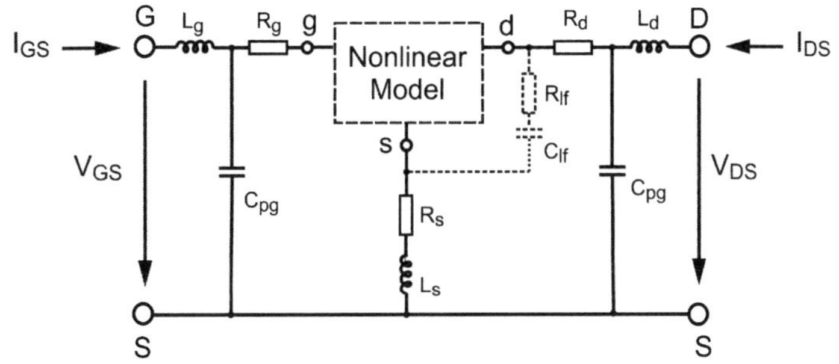

Figure 25.7 Nonlinear device model embedded in widely used linear parasitic network. Dotted R_{lf}- C_{lf} branch considers dispersion effects.

25.2 POPULAR NONLINEAR FET MODELS

25.2.1 Curtice Quadratic Nonlinear Model (1980)

The Curtice quadratic model, proposed by Curtice in 1980 [7], was one of the first high-frequency nonlinear GaAs FET models to be implemented in commercial simulators. The application of the MESFET model was mainly focused on the design and development of (low-power) GaAs integrated circuits (IC's), covering drain-source voltages of only several volts.

The nonlinear intrinsic model, shown in Figure 25.8, comprises a voltage-controlled drain current source I_{ds}, three interelectrode capacitances (gate-source, gate-drain and drain-source capacitances C_{gs}, C_{gd}, and C_{ds}, respectively) and clamping diode D between gate and source. The only nonlinear elements are $I_{ds}(V_{gs}, V_{ds})$ and $C_{gs}(V_{gs})$. The gate-drain capacitance is assumed to be constant.

The drain current source is defined as [7]

$$I_{ds}(V_{gs},V_{ds}) = \begin{cases} \beta(V_{gs} - V_p)^2(1 + \lambda V_{ds})\tanh(\alpha V_{ds}) & \text{for } V_{gs} + V_p \geq 0 \\ 0 & \text{for } V_{gs} + V_p < 0 \end{cases} \tag{25.19}$$

with four parameters (α, β, λ, V_p) to be determined from experimental data. β is the transconductance parameter and λ is the slope of the drain current characteristic in the saturation region. V_p is the (extrinsic) pinch-off voltage, which is determined by plotting I_{ds} versus V_{gs}. Due to time delay between gate-source voltage and drain current for reason of

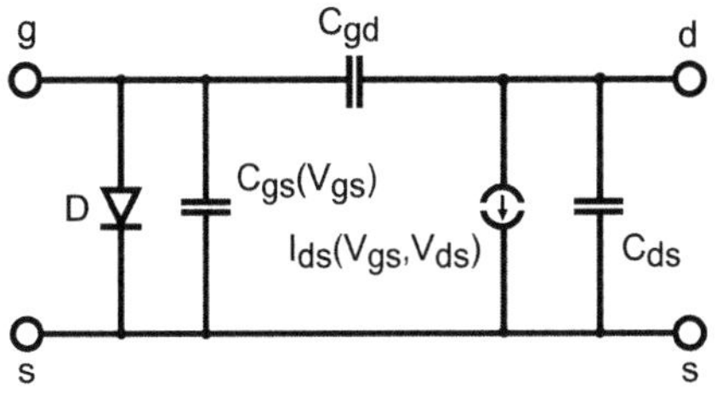

Figure 25.8 Curtice quadratic nonlinear intrinsic model [7].

transit time τ under the gate, the drain current source $I_{ds}(V_{gs}, V_{ds})$ should be altered to be $I_{ds}(V_{gs}(t-\tau), V_{ds})$ [6].

The gate-source capacitance C_{gs} is described as a voltage-dependent Schottky-barrier diode,

$$C_{gs}(V_{gs}) = \frac{C_{gs0}}{\sqrt{1 - V_{gs}/V_{bi}}} \tag{25.20}$$

where V_{bi} is the built-in voltage. C_{gs0} is the capacitance value for $V_{gs} = 0$.

25.2.2 Curtice-Ettenberg Cubic Nonlinear Model (1985)

In 1985, Curtice and Ettenberg proposed a nonlinear GaAs MESFET model for power applications. The nonlinear intrinsic model is shown in Figure 25.9. It comprises three nonlinear current sources. The current source $I_{ds}(V_{gs}, V_{ds})$ is the large-signal form of the usual small-signal transconductance. The new drain-gate voltage-controlled current source $I_{dg}(V_{gs}, V_{ds})$ represents the drain-gate avalanche current that can occur at large-signal operation. The new gate-source voltage-controlled current source $I_{gs}(V_{gs})$ represents gate current that occurs when the gate-source junction is forward-biased. Furthermore, a channel resistance R_i is considered.

The Curtice-2 model includes a square-law relationship between I_{ds} and V_{gs}. Curtice and Ettenberg had shown that real devices did not exhibit such a relationship and they assumed that it was more convenient to use a cubic approximation. The cubic Curtice-Ettenberg model (Curtice-3) provides an improved fitting of the DC characteristics, although sometimes a non-physical behavior can occur for voltages below pinch-off. As well, an interpretation of the model parameters is difficult.

The proposed cubic DC approximation of the drain-source current source reads

$$I_{ds}(V_{gs}, V_{ds}) = (A_0 + A_1 V_1 + A_2 V_1^2 + A_3 V_1^3)\tanh(\gamma V_{ds}) \tag{25.21}$$

with

$$V_1 = V_{gs}(t-\tau)[1 + \beta(V_{ds}^0 - V_{ds}(t))] \tag{25.22}$$

where (25.22) considers experimentally observed pinch-off voltage increase with drain-source voltage. β is a coefficient for pinch-off change. A_0, A_1, A_2, A_3, and γ are DC fitting parameters. In the saturation region A_0 is the saturation drain current for $V_{gs} = 0$. V^0_{ds} is the output voltage, at which A_0, A_1, A_2 and A_3 are evaluated. τ is the internal time delay of the

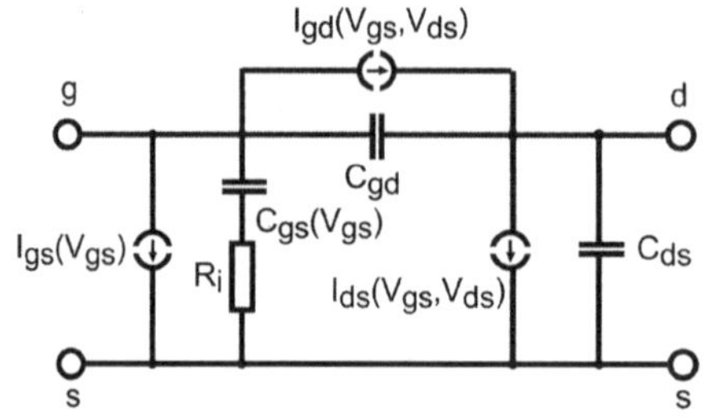

Figure 25.9 Curtice-Ettenberg cubic nonlinear intrinsic model [8].

device. A_1 can be preliminary evaluated as the transconductance parameter.

The drain current cannot be pinched off at large drain-source voltages due to the gate current produced by avalanche breakdown. In this model, the drain-gate avalanche current is taken as

$$I_{dg}(V_{dg}) = \begin{cases} \dfrac{V_{dg}(t) - V_B}{R_1} & \text{for } V_{dg} > V_B \\ 0 & \text{for } V_{dg} < V_B \end{cases} \tag{25.23}$$

with

$$V_B = V_{B0} + R_2 \cdot I_{ds} \tag{25.24}$$

where R_1 represents the approximate breakdown resistance and R_2 is the resistance that relates breakdown voltage to the channel currents. V_{B0} describes the gate drain junction reverse bias breakdown voltage.

The forward-biased gate current is taken to be

$$I_{gs}(V_{gs}(t)) = \begin{cases} \dfrac{V_{gs}(t) - V_{bi}}{R_1} & \text{for } V_{gs}(t) \geq V_{bi} \\ 0 & \text{for } V_{gs}(t) < V_{bi} \end{cases} \tag{25.25}$$

with the built-in voltage V_{bi}.

Gate-source capacitance C_{gs} and gate-drain capacitance C_{gd} are similar as in Curtice quadratic nonlinear.

25.2.3 Materka-Kacprzak Model (1983, 1985)

The intrinsic nonlinear equivalent circuit of the Materka-Kacprzak model is shown in Figure 25.10. It was developed for the analysis of GaAs MESFET power amplifiers with typical output power values of a few tens of milliwatts.

The voltage-controlled drain current source $I_{ds}(V_{gs}, V_{ds})$ is described by the relation

$$I_{ds}(V_{ds}, V_{gs}) = I_{dss}(1 - V_{gs}/V_p)^2 \tanh[\alpha V_{ds}/(V_{gs} - V_p)] \tag{25.26}$$

with

$$V_p = V_{p0} + \gamma V_{ds} \tag{25.27}$$

where I_{dss}, V_{p0}, α, and γ are model fitting parameters. I_{dss} represents the saturation current at $V_{gs} = 0$V. V_{p0} is the pinch-off voltage V_p for $V_{ds} = 0$V. To take into account the time delay

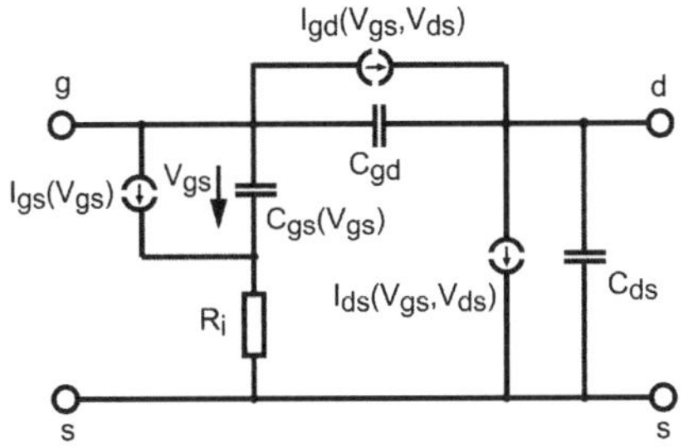

Figure 25.10 Materka-Kacprzak nonlinear intrinsic FET model [9,10].

between drain current and gate voltage, the instantaneous current $I_{ds}(t)$ is calculated from (25.26) with $V_{gs} = V_{gs}(t-\tau)$ and $V_{ds} = V_{ds}(t)$, where τ is a model fitting parameter.

The current of the gate-diode I_{gs} and gate-drain current I_{gd} is described by the relations

$$I_{gs}(V_{gs}) = I_s(\exp(\alpha_s V_{gs}) - 1) \tag{25.28}$$

$$I_{gd}(V_{gd}) = I_{sr}(\exp(\alpha_{sr} V_{gd}) - 1) \tag{25.29}$$

where I_s, I_{sr}, α_s, and α_{sr} are model fitting parameters. It should be noted that the current source I_{gd} does not represent any forward biased Schottky barrier junction connected between gate and drain electrodes; it only approximates the breakdown current.

The relation for the nonlinear gate-source capacitance corresponds to the Schottky barrier capacitance (25.20) for $V_{gs} < 0.8V_{bi}$

$$C_{gs}(V_{gs}) = \frac{C_{gs0}}{\sqrt{1 - V_{gs}/V_{bi}}} \quad \text{for} \quad V_{gs} < 0.8V_{bi} \tag{25.30}$$

For $V_{gs} \geq 0.8V_{bi}$, C_{gs} is approximated by a straight line with a slope equal to the derivative dC_{gs}/dV_{gs} obtained from (25.30) at $V_{gs} < 0.8V_{bi}$.

25.2.4 Statz Model (1987)

The Statz model [11] has been developed for SPICE simulation. The circuit configuration is similar to the Curtice-2 model. The differences are in the junction capacitances and in the current source equation. The model indicates that the square law approximation of drain-source current as function of gate-source voltage is only valid for small ($V_{gs} - V_p$) values, where V_p is the pinch-off voltage. The drain current I_{ds} becomes almost linear for larger values of ($V_{gs} - V_p$).

The drain-source current $I_{ds}(V_{gs}, V_{ds})$ is given by

$$I_{ds}(V_{gs},V_{ds}) = \begin{cases} \dfrac{\beta(V_{gs} - V_p)^2}{1 + b(V_{gs} - V_p)}(1 + \lambda V_{ds})\left[1 - \left(1 - \dfrac{\alpha V_{ds}}{3}\right)^3\right] & \text{for } 0 < V_{ds} < \dfrac{3}{\alpha} \\ \dfrac{\beta(V_{gs} - V_p)^2(1 + \lambda V_{ds})}{1 + b(V_{gs} - V_p)} & \text{for } V_{ds} \geq \dfrac{3}{\alpha} \end{cases} \tag{25.31}$$

where V_p, α, β, λ, b are DC model parameters, V_p is the pinch-off voltage, α signifies the slope of drain characteristics in the linear region, β is the tranconductance coefficient, λ is the channel length modulation parameter, and b is the doping profile extending parameter.

As a result, the approximation in (25.31) is quadratic for small values of $V_{gs} - V_p$; for their large values it becomes almost linear in $V_{gs} - V_p$, which is more realistic behavior of I_{ds} as a function of V_{gs} for real MESFET devices. A tanh function used in previous models below saturation was approximated by a simple polynomial with $n = 3$ that gives best fit.

In the Statz model both the gate-source capacitance C_{gs} and gate drain capacitance C_{gd} are derived from the total gate charge Q_g. In this way the charge-conservation condition is satisfied. The charge under the gate Q_g is composed of Q_{gs} and Q_{gd},

$$Q_g = Q_{gs} + Q_{gd} \tag{25.32}$$

The small-signal capacitances are related to the charge partial derivatives through the following expressions:

$$C_{gs} = \frac{\partial Q_{gs}}{\partial V_{gs}} + \frac{\partial Q_{gd}}{\partial V_{gs}} \tag{25.33}$$

$$C_{gd} = \frac{\partial Q_{gs}}{\partial V_{gd}} + \frac{\partial Q_{gd}}{\partial V_{gd}} \tag{25.34}$$

The expressions for gate charge and derived gate-source and gate-drain capacitances are as follows.

For $V_{\mathrm{new}} > V_{\mathrm{max}}$,

$$Q_g = C_{gs0}\left[2V_{bi}\left(1 - \sqrt{1 - \frac{V_{\max}}{V_{bi}}}\right) + \frac{V_{new} - V_{\max}}{\sqrt{1 - \frac{V_{\max}}{V_{bi}}}}\right] + C_{gd0}V_{eff_2} \tag{25.35}$$

$$C_{gs} = \frac{\partial Q_g}{\partial V_{gs}} = \frac{C_{gs0}F_1F_2}{\sqrt{1 - \frac{V_{\max}}{V_{bi}}}} + C_{gd0}F_3 \tag{25.36}$$

$$C_{gd} = \frac{\partial Q_g}{\partial V_{gd}} = \frac{C_{gs0}F_3F_1}{\sqrt{1 - \frac{V_{\max}}{V_{bi}}}} + C_{gd0}F_2 \tag{25.37}$$

$$Q_g = 2C_{gs0}V_{bi}\left(1 - \sqrt{1 - \frac{V_{new}}{V_{bi}}}\right) + C_{gs0}\frac{V_{new} - V_{\max}}{\sqrt{1 - V_{\max}/V_{bi}}} \tag{25.38}$$

For $V_{\mathrm{new}} \le V_{\mathrm{max}}$,

$$Q_g = 2C_{gs0}V_{bi}\left(1 - \sqrt{1 - \frac{V_{new}}{V_{bi}}}\right) + C_{gd0}V_{eff_2} \tag{25.39}$$

$$C_{gs} = \frac{\partial Q_g}{\partial V_{gs}} = \frac{C_{gs0}F_1F_2}{\sqrt{1 - \frac{V_{new}}{V_{bi}}}} + C_{gd0}F_3 \tag{25.40}$$

$$C_{gd} = \frac{\partial Q_g}{\partial V_{gd}} = \frac{C_{gs0} F_3 F_1}{\sqrt{1 - \dfrac{V_{new}}{V_{bi}}}} + C_{gd0} F_2 \tag{25.41}$$

where

$$V_{new} = \frac{1}{2}\left[V_{eff_1} + V_p + \sqrt{(V_{eff_1} - V_p)^2 + \delta^2}\right] \tag{25.42}$$

$$V_{eff_{1,2}} = \frac{1}{2}\left[V_{gs} + V_{gd} \pm \sqrt{(V_{gs} - V_{gd})^2 + \Delta^2}\right] \tag{25.43}$$

$V_{max} = 0.5$, $\delta = 0.2$, and

$$F_1 = \frac{\partial V_{new}}{\partial V_{eff_1}} = \frac{1}{2}\left[1 + \frac{V_{eff_1} - V_p}{\sqrt{(V_{eff1} - V_p)^2 + \delta^2}}\right] \tag{25.44}$$

$$F_2 = \frac{\partial V_{eff_1}}{\partial V_{gs}} = \frac{\partial V_{eff_2}}{\partial V_{gd}} = \frac{1}{2}\left[1 + \frac{V_{gs} - V_{gd}}{\sqrt{(V_{gs} - V_{gd})^2 + \Delta^2}}\right] \tag{25.45}$$

$$F_3 = \frac{\partial V_{eff_1}}{\partial V_{gd}} = \frac{\partial V_{eff_2}}{\partial V_{gs}} = \frac{1}{2}\left[1 - \frac{V_{gs} - V_{gd}}{\sqrt{(V_{gs} - V_{gd})^2 + \Delta^2}}\right] \tag{25.46}$$

Equations (25.43) are employed to achieve a gradual transition of capacitance values near $V_{ds} = 0$. V_{eff1} and V_{eff2} represent the larger and smaller value between V_{gs} and V_{dg}, respectively. Parameter Δ produces a smooth transition width in the value of V_{eff1} and V_{eff2} as a function of V_{gs} and V_{dg}. The transformation in (25.42) is employed to model capacitance beyond pinch-off. In the pinch-off region, the gate-source capacitance drops to small value that is normally determined by the fringing capacitance in the depletion region. The smooth transformation of (25.42) would set V_{new} to V_{eff1} before pinch-off, and to V_p when V_{eff1} is biased beyond pinch-off. Parameter δ stands for the voltage range over which the transition between V_{eff1} and V_p occurs, and is set to 0.2.

V_{max} is introduced to solve the singularity in junction capacitance when V_{eff1} becomes positive and is equal or greater than V_{bi}. The value of V_{eff1} is limited to a maximum value of V_{max} when $V_{eff1} \geq V_{max}$ (V_{max} sets to 0.5V). This limits the value of junction capacitance. The choice of V_{max} determines the maximum capacitance value. For voltage beyond V_{max}, the junction capacitances are assumed to remain constant.

C_{gs0} is the gate-source Schottky-barrier capacitance at $V_{gs} = 0$, C_{gd0} is the gate-source Schottky-barrier capacitance at $V_{gd} = 0$, V_{bi} is the built-in barrier voltage, and δ is the voltage width with smooth transition from V_p to V_{eff}, which can be chosen to be 0.2V.

The model proposed in [11] should be completed with gate-drain and gate-source diodes. These are very important for power amplifier operation to model forward current for the gate-source diode and the reverse gate current for the gate-drain diode. The forward gate current can be modeled by the standard diode current equation, which is generally given as

$$I_{gs} = I_s\left(e^{V_{gs}/(nkT/q)} - 1\right) \tag{25.47}$$

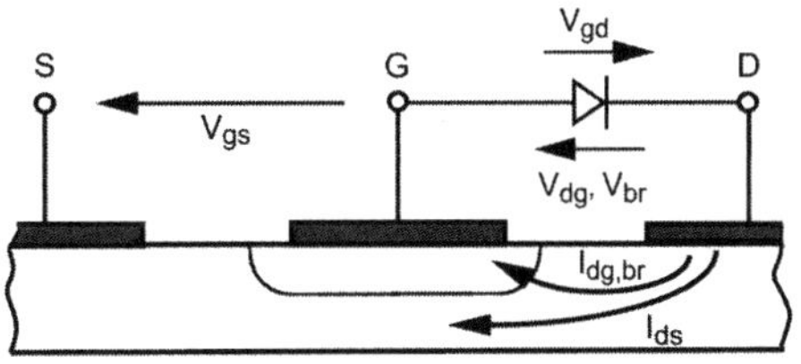

Figure 25.11 Illustration of gate-drain breakdown in a FET device.

where V_{gs} is the (forward) voltage across the diode, kT/q is the thermal voltage, n is a constant, and I_s is the saturation current and is a function of temperature. Since (25.47) is valid only for forward diode current, it is only applicable when $V_{gs} > 0$.

A variety of equations exist for modeling the reverse gate current in the junction diodes. A relatively simple equation that is used to model the reverse gate current, and which is very similar to the standard equation for forward current, is as follows. The condition for breakdown between drain and gate is with the notation in Figure 6.1 (see Section 4.2)

$$V_{dg} = V_{ds} - V_{gs} = V_{br} \tag{25.48}$$

Taking the standard diode equation,

$$I_{dg,br} = I_s\left(e^{(V_{dg} - V_{br})/nV_{th}} - 1\right) \tag{25.49}$$

and taking into account that $V_{gd} = -V_{dg}$ and $I_{gd,br} = -I_{dg,br}$, (25.49) reads

$$I_{gd,br} = -I_s\left(e^{-(V_{gd} + V_{br})/nV_{th}} - 1\right) \tag{25.50}$$

Frequently dispersion is considered by adding a series-RC network in parallel to the drain current source, which provides a correction to the channel conductance. At low frequencies the RC-network is an open so that the output conductance equals the DC channel conductance $G_{ds.DC}$. At sufficiently high frequencies, $1/R_{dis}$ is in parallel, increasing the output conductance. Thus, Figure 25.12 shows the completed Statz model.

25.2.5 Angelov Model (1992/96)

Angelov et al. [12,13] published in 1992 a nonlinear model for HEMTs and MESFETs,

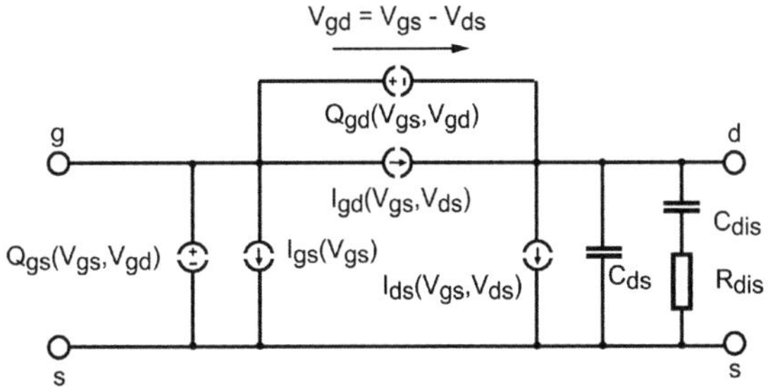

Figure 25.12 Statz model completed by gate-source and gate-drain current sources, and series RC-network for dispersion effects. Frequently the charging resistance R_i is considered in series with the charge source Q_{gs}.

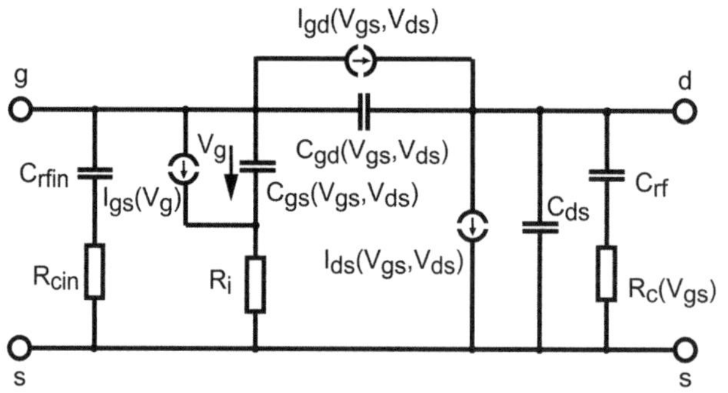

Figure 25.13 Nonlinear intrinsic Angelov FET model [14].

which has focused on the description of the nonlinear drain current source I_{ds} and the gate- and drain-source capacitances. Charging resistance R_i was already introduced. In 1996 the model was extended taking into account temperature, dispersion, and soft-breakdown effects [13]. Figure 25.13 illustrates the model topology as given in ADS library from 2008. In their original publication the model is more complex, containing an additional series RC-network in parallel to the source resistance (not shown) to describe a frequency-dependent decrease of R_s as reported in [15], thus increasing the G_m value at RF [13]. The model is capable of modeling the current-voltage characteristic and its derivatives, including the characteristic transconductance peak, gate-source, and gate-drain capacitances.

The decrease of the transconductance for MESFETs can be described by adding a parallel branch, nonlinear $R_c(V_{gs})$, and C_{rf} in series between the internal gate and source terminals. At frequencies higher than the corner frequencies for the dispersion effects the resistor R_c shunts the input and effectively decreases the transconductance.

The drain current function for this model is expressed

$$I_{ds}(V_{gs},V_{ds}) = I_{dA}(V_{gs})\cdot I_{dB}(V_{ds}) \tag{25.51}$$

where the first factor is dependent only on the gate voltage and the second only on the drain voltage. The $I_{dB}(V_{ds})$ term is the same as the one used in other models (Curtice model, Curtice-Ettenberg model, Materka model, etc.).

For $I_{dA}(V_{gs})$, Angelov et al. proposed a function whose first derivative has the same bell shaped structure as the measured transconductance function $G_m(V_{gs})$. The hyperbolic tangent (tanh) function describes the gate voltage dependencies and its derivatives well and is normally available in commercial HB simulators.

Expressions for the nonlinear model elements of the Angelov model; that is, $I_{ds}(V_{gs}, V_{ds})$, gate-source capacitance $C_{gs}(V_{gs}, V_{ds})$ and gate-drain capacitance $C_{gd}(V_{gs}, V_{ds})$ are listed below:

$$I_{ds}(V_{gs},V_{ds}) = I_{pk}(1 + \tanh(\Psi))(1 + \lambda V_{ds})\tanh(\alpha V_{ds}) \tag{25.52}$$

with

$$\Psi = P_1(V_{gs} - V_{pk}) + P_2(V_{gs} - V_{pk})^2 + P_3(V_{gs} - V_{pk})^3 \tag{25.53}$$

and

$$V_{pk} = V_{pk0} + \gamma V_{ds} \tag{25.54}$$

$$C_{gs}(V_{gs},V_{ds}) = C_{gs0}[1 + \tanh(\Psi 1)][1 + \tanh(\Psi 2)] \tag{25.55}$$

$$C_{gd}(V_{gs},V_{ds}) = C_{gd0}[1 + \tanh(\Psi 3)][1 - \tanh(\Psi 4)] \tag{25.56}$$

with

$$\Psi 1 = P_{0gsg} + P_{1gsg}V_{gs} + P_{2gsg}V_{gs}^2 + P_{3gsg}V_{gs}^3 \tag{25.57}$$

$$\Psi 2 = P_{0gsd} + P_{1gsd}V_{ds} + P_{2gsd}V_{ds}^2 + P_{3gsd}V_{ds}^3 \tag{25.58}$$

$$\Psi 3 = P_{0gdg} + P_{1gdg}V_{gs} + P_{2gdg}V_{gs}^2 + P_{3gdg}V_{gs}^3 \tag{25.59}$$

$$\Psi 4 = P_{0gdd} + P_{1gdd}V_{ds} + P_{2gdd}V_{ds}^2 + P_{3gdd}V_{ds}^3 \tag{25.60}$$

When an accuracy on the order of 5-10% of C_{gs} and C_{gd} is sufficient, then (5.25) - (5.28) can be simplified to:

$$C_{gs}(V_{gs},V_{ds}) = C_{gs0}[1 + \tanh(P_{1gsg}V_{gs})][1 + \tanh(P_{1gsd}V_{ds})] \tag{25.61}$$

$$C_{gd}(V_{gs},V_{ds}) = C_{gd0}[1 + \tanh(P_{1gdg}V_{gs})][1 - \tanh(P_{1gdd}V_{ds})] \tag{25.62}$$

where P_1, P_2, P_3, α, γ, λ and P_{1gsg}, P_{1gsd}, P_{1gdd} are DC model fitting parameters of I_{ds} and the capacitances C_{gs} and C_{gd}, respectively. V_{pk0} and I_{pk} are the gate voltage and drain current for maximum transconductance. C_{gs0} and C_{gd0} are the depletion capacitances at zero bias.

For small-signal applications the drain current reduction caused by thermal heating effects can be neglected in most cases, but for medium- and high-power applications the dissipated power P_d is large enough to heat the transistor to temperature T and thus significantly change the model parameters. The temperature effect on the device performance is simply expressed as

$$I_{ds_{term}} = \frac{I_{ds}}{1 + K_t P_d} \tag{25.63}$$

where K_t is the thermal constant for the transistor that summarizes the temperature changes of the different large-signal model parameters influenced by the temperature rise under the total dissipated power $P_d \approx I_{ds}(t) \cdot V_{ds}(t)$.

25.2.6 TOM Model (TriQuint Nonlinear Model) (1990)

The TOM model [16] is a modification of the Statz model. The first modification concerns an improvement of the poor fit at near pinch-off values of V_{gs}. This is accomplished by making V_T a function of drain voltage:

$$V_T = V_p - \gamma V_{ds} \tag{25.64}$$

The second modification relates to the way in which I_{ds} decreases at higher values of current and voltage, showing a smaller slope than would be predicted by the Statz model. As a result, the following drain current model for $V_{gs} > V_p$ is proposed:

$$I_{ds}(V_{gs},V_{ds}) = \frac{I_{ds0}}{1 + \delta I_{ds0} V_{ds}} \tag{25.65}$$

with

$$I_{ds0}(V_{ds},V_{gs}) = \begin{cases} \beta(V_{gs} - V_T)^Q[1 - [1 - (\alpha V_{ds})/3]^3] & \text{for } 0 < V_{ds} < 3/\alpha \\ \beta(V_{gs} - V_T)^Q & \text{for } V_{ds} \geq 3/\alpha \end{cases} \tag{25.66}$$

where α, β, δ, γ, Q are DC model parameters. V_p is the pinch-off voltage (fitting parameter). The approximations for C_{gs} and C_{gd} are practically the same as for the Statz model, except that V_p in the Statz relations must be replaced by V_T in accordance with (25.64).

25.2.7 Tajima Model (1981/84)

The equivalent circuit of the Tajima model is shown in Figure 25.14. Expressions for the nonlinear model elements of the Tajima model; that is, drain-source current $I_{ds}(V_{gs}, V_{ds})$, gate-source capacitance $C_{gs}(V_{gs}, V_{ds})$, gate-drain capacitance $C_{gd}(V_{gs}, V_{ds})$ and gate-forward current $C_{gs}(V_{gs})$ are listed below:

$$I_{ds}(V_{gs},V_{ds}) = I_{d1}(V_{gs})\cdot I_{d2}(V_{ds}) \tag{25.67}$$

with

$$I_{d1}(V_{gs}) = \frac{1}{k}\left\{1 + \frac{V'_{gs}}{V_p} - \frac{1}{m} + \frac{1}{m}\exp\left[-m\left(1 + \frac{V'_{gs}}{V_p}\right)\right]\right\} \tag{25.68}$$

$$I_{d2}(V_{ds}) = I_{dsp}\left[1 - \exp\left\{\frac{-V_{ds}}{V_{dss}} - a\left(\frac{V_{ds}}{V_{dss}}\right)^2 - b\left(\frac{V_{ds}}{V_{dss}}\right)^3\right\}\right] \tag{25.69}$$

$$k = 1 - \frac{1}{m}\{1 - \exp(-m)\} \tag{25.70}$$

$$V_p = V_{p0} + pV_{ds} + V_{bi} \tag{25.71}$$

$$V'_{gs} = V_{gs} - V_{bi} \tag{25.72}$$

where V_{p0} (> 0) is the pinch-off voltage at $V_{ds} \approx 0$, V_{dss} is the drain current saturation voltage, V_{bi} is the built-in potential of the Schottky barrier, I_{dsp} is the drain current for $V_{gs} = V_{bi}$, and a, b, m, p are DC model fitting parameters.

Breakdown currents are expressed by the following equations with parameters V_b (breakdown voltage) and R_b (series resistance) as dependent variables of I_{ds}:

$$I_{dg} = \begin{cases} 0, & V_{dg} < V_b \\ \frac{1}{R_b}(V_{dg} - V_b), & V_{dg} > V_b \end{cases} \tag{25.73}$$

with

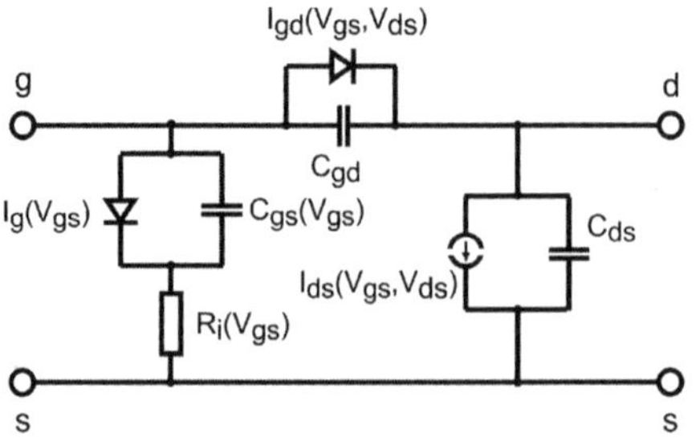

Figure 25.14 Equivalent circuit of the Tajima FET model [17,18].

$$V_b = V_{b0} + R_1 I_{ds} \tag{25.74}$$

$$R_b = R_{b0} + R_2 \frac{I_{ds}}{I_{dss}} \tag{25.75}$$

where V_{b0} is the breakdown voltage when $I_{ds} = 0$. Thus, four parameters, V_{b0}, R_{b0}, R_1, and R_2 can describe the breakdown current at any bias condition.

25.2.8 ROOT Model (1991)

In contrast to previously discussed nonlinear DC models, the Root model was one of the earliest models that took into account frequency dispersion of the drain current source. The intrinsic device was described by means of nonlinear state functions directly extracted from small-signal differential parameters. In frequency domain the large-signal drain current is formulated as

$$I_d(V_{gs},V_{ds}) = H(\omega)I_{ds}^{DC}(V_{gs},V_{ds}) + j\Omega Q_D(V_{gs},V_{ds}) + (1 - H(\omega))I_{ds}^{high}(V_{gs},V_{ds}) \tag{25.76}$$

where I_{ds}^{DC} is the measured nonlinear DC I- V characteristics and I_{ds}^{high} the nonlinear high-frequency I- V characteristics measured at a frequency above the inverse thermal and trap time constants (ω_t). The transition between DC and high-frequency currents is modeled by the low-pass-filter function $h(\omega)$, described in the frequency domain by the diagonal (dispersion) matrix $H(\omega)$, whose elements are $h(n\omega_0)$ ($n = 1, 2, ...$), being ω_0 the fundamental frequency. $h(\omega)$ characterizes the dispersion between DC and high-frequency currents.

The term $j\Omega Q_d$ corresponds, instead, to the time derivative of the nonquasi-static charge, where Ω is a diagonal matrix with elements $n\omega_0$, $n = 0, 1, 2, ...$ In this model, the nonquasi-static charges are identified by numerical integration of the intrinsic differential parameters, as well as the above-cutoff drain current; that is:

$$\nabla Q_g(V_{ds},V_{ds}) = \frac{ImY_{11}(V_{gs},V_{ds},\omega)}{\omega}\hat{V}_{gs} + \frac{ImY_{12}(V_{gs},V_{ds},\omega)}{\omega}\hat{V}_{ds} \tag{25.77}$$

$$\nabla Q_d(V_{ds},V_{ds}) = \frac{ImY_{21}(V_{gs},V_{ds},\omega)}{\omega}\hat{V}_{gs} + \frac{ImY_{22}(V_{gs},V_{ds},\omega)}{\omega}\hat{V}_{ds} \tag{25.78}$$

$$\nabla I_d^{high}(V_{ds},V_{ds}) = \frac{ReY_{21}(V_{gs},V_{ds},\omega)}{\omega}\hat{V}_{gs} + \frac{ReY_{22}(V_{gs},V_{ds},\omega)}{\omega}\hat{V}_{ds} \tag{25.79}$$

where $\nabla = (\partial/\partial V_{gs})\hat{V}_{gs} + (\partial/\partial V_{ds})\hat{V}_{ds}$ and $\hat{V}_{gs}$, $\hat{V}_{ds}$ are unit vectors along the orthogonal intrinsic voltage axes. All the model nonlinear functions are then stored into tables.

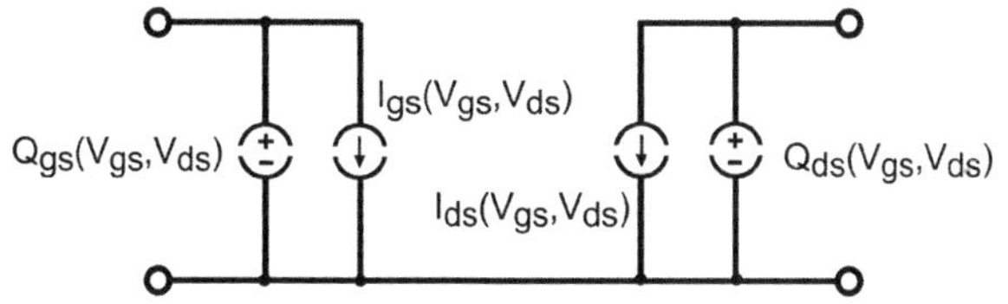

Figure 25.15 Nonlinear intrinsic Root FET model [19,20].

The model has the advantages of being fully compatible with harmonic balance simulations, and is fast and unambiguously identifiable and charge-conservative.

25.2.9 TOPAS Model (1996)

Transistor Parameter Scalable (TOPAS) model is a table-based model [21–25]. It is used for the simulation of any FET-type structure based upon different material systems such as GaAs, InP, or even Si. The TOPAS model has been developed by the Institute of Mobile and Satellite Communication Techniques (IMST). The small- and large-signal equivalent circuits of the TOPAS model is shown in Figure 25.16.

For parameters extraction of the model only small-signal S-parameters and forward IV-curves are needed. The determination of the intrinsic parameters is a bit more complicated, as can be seen in the equations below:

$$C_{gd}(V_{gs},V_{ds}) = \frac{\sum_{f=f_1}^{f_2} -\frac{1}{w}\mathrm{Im}(Y_{12}(V_{gs},V_{ds}))}{n(f_2 - f_1)} \tag{25.80}$$

$$C_{gs}(V_{gs},V_{ds}) = \frac{\sum_{f=f_1}^{f_2} \frac{1}{w}\left(\mathrm{Im}(Y_{11}(V_{gs},V_{ds})) + \mathrm{Im}(Y_{12}(V_{gs},V_{ds}))\right)}{n(f_2 - f_1)} \tag{25.81}$$

$$C_{ds}(V_{gs},V_{ds}) = \frac{\sum_{f=f_1}^{f_2} \frac{1}{w}\left(\mathrm{Im}(Y_{22}(V_{gs},V_{ds})) + \mathrm{Im}(Y_{12}(V_{gs},V_{ds}))\right)}{n(f_2 - f_1)} \tag{25.82}$$

$$R_i(V_{gs},V_{ds}) = \frac{\sum_{f=f_1}^{f_2} \left(\frac{\mathrm{Re}(Y_{11}(V_{gs},V_{ds}))}{\left(\mathrm{Im}(Y_{11}(V_{gs},V_{da})) + \mathrm{Im}(Y_{12}(V_{gs},V_{ds}))\right)^2}\right)}{n(f_2 - f_1)} \tag{25.83}$$

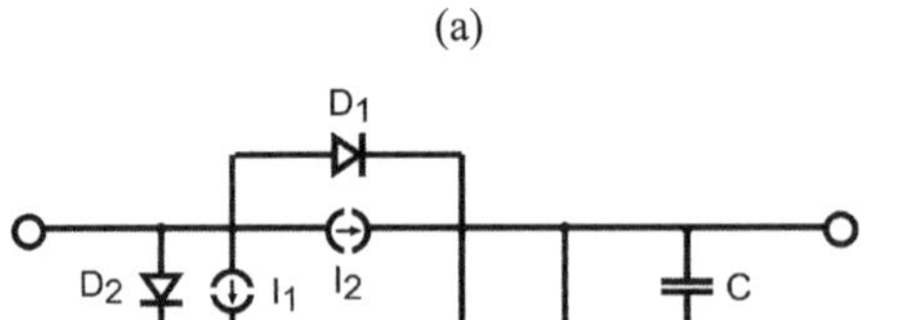

(a)

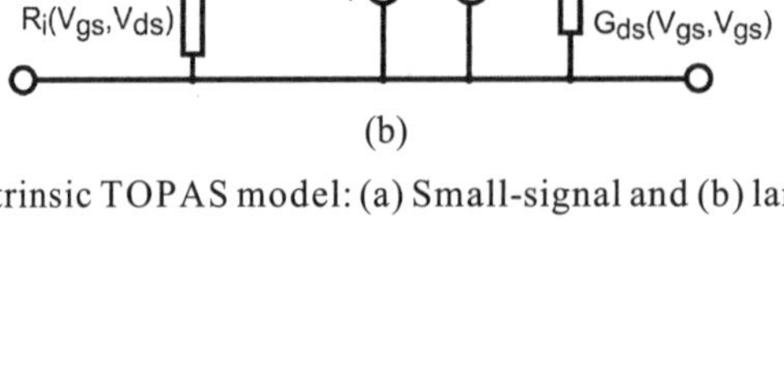

(b)

Figure 25.16 Nonlinear intrinsic TOPAS model: (a) Small-signal and (b) large-signal equivalent circuit.

where Y_{11}, Y_{12}, $Y_{21,}$ and Y_{22} in (25.80) - (25.83) are the de-embedded intrinsic small-signal admittance parameters.

Conventional techniques are employed to determine the extrinsic and intrinsic parameters [3, 5, 26]. Once these parameters have been determined, an optimization procedure is carried out to improve the fit between the deembedded S-parameters and the inner device model. This optimization procedure involves refining the intrinsic conductance parameter G_{ds}. It also introduces an additional empirically determined capacitance C [27].

The current source $I_{ds}(V_{gs}, V_{ds})$ is extracted from the measured DC IV-curves. Gate-source and gate-drain diodes currents can be extracted using small-signal scattering parameters:

$$-\mathrm{Re}(Y_{12}) = \frac{\partial I_{D1}}{\partial V_{gd}} \tag{25.84}$$

$$\mathrm{Re}(Y_{11}) = \frac{\partial I_{D2}}{\partial V_{gs}} \tag{25.85}$$

The large-signal equivalent circuit of the TOPAS model is shown in the Figure 25.16(b), where

$$I_1 = C_{gs}(V_{gs}, V_{ds}) \cdot v'_{gs} \tag{25.86}$$

$$I_2 = C_{gd}(V_{gs}, V_{ds}) \cdot v'_{gd} \tag{25.87}$$

$$I_3 = C_{ds}(V_{gs}, V_{ds}) \cdot v'_{gs} \tag{25.88}$$

$$I_4 = I_{ds}(V_{gs}, V_{ds}) \tag{25.89}$$

with $v' = \mathrm{d}v/\mathrm{d}t$ being the time derivative of voltage.

25.3 MODEL IMPLEMENTATION IN COMMERCIAL SIMULATION SOFTWARE

This chapter briefly describes the implementation of discussed analytical and table-based standard nonlinear FET models in commercial software.

25.3.1 Implementation of Analytical Models

The discussed analytical DC models (Curtice -2 model, Curtice-3 model, Materka model, Statz model, Angelov model, Tajima model, TOM model) are available in the Advanced Design System (ADS) [2]. The implementation of these models needs the complete knowledge of all model parameters. The extrinsic model parameters are found by conventional techniques. The nonlinear model parameters are obtained by optimization. Table 25.1 and Table 25.2 illustrate the initial and final values of the parameters for the analytical models.

Table 25.1
Initial Values of the Parameters for the Analytical Models

Modell ⇒ Parameters ⇓	Curtice 2	Curtice 3	Materka	Statz	Angelov	Tajima	TOM
V_{to}	-4.5	------	-4.5	-4.5	------	-4.5	-4.5
β	0.017	0.001	-0.04	2	------	------	0.25
λ	0.01	------	------	0.02	0.01	------	------
α	2	------	2.5	3	2	------	2
I_{DSS}	------	------	0.82	------	------	0.9	------
I_{pk}	------	------	------	------	0.9	------	------
V_{DSS}	------	------	------	------	------	2	------
V_{bi}	------	------	------	------	------	1.8	------
V_{out}	------	2.5	------	------	------	------	------
V_{pko}	------	------	------	------	2.3	------	------
γ	------	2.6	------	------	0.002	------	1.5
P	------	------	------	------	------	0.03	------
a	------	------	------	------	------	2	------
b	------	------	------	3.5	------	7	------
m	------	------	------	------	------	11	------
A_0	------	0.7	------	------	------	------	------
A_1	------	-0.025	------	------	------	------	------
A_2	------	-0.2	------	------	------	------	------
A_3	------	-0.01	------	------	------	------	------
P_1	------	------	------	------	0.25	------	------
P_2	------	------	------	------	0.1	------	------
P_3	------	------	------	------	0.01	------	------
Q	------	------	------	------	------	------	1
δ	------	------	------	------	------	------	0.01
Initial error	0.08	0.023	0.073	0.054	0.027	0.042	0.034

------: Parameter not applicable

Table 25.2
Final Values of the Parameters for the Analytical Models after Optimization

Modell ⇒ Parameters ⇓	Curtice 2	Curtice 3	Materka	Statz	Angelove	Tajima	TOM
V_{to}	-5.577	------	-5.329	-3.974	------	-3.914	-4.473
β	0.029	0.004	-0.019	1.472	------	------	0.413
λ	0.004	------	------	0.006	0.002	------	------
α	1.24	------	5.36	8.442	1.578	------	1.344
I_{DSS}	------	------	0.928	------	------	1.165	------
I_{pk}	------	------	------	------	1.244	------	------
V_{DSS}	------	------	------	------	------	1.624	------
V_{bi}	------	------	------	------	------	1.216	------
V_{out}	------	1.909	------	------	------	------	------
V_{pko}	------	------	------	------	2.856	------	------
γ	------	2.009	------	------	0.00036	------	1.225
P	------	------	------	------	------	0.015	------
a	------	------	------	------	------	1.719	------
b	------	------	------	6.89	------	8.227	------
m	------	------	------	------	------	19.67	------
A_0	------	0.825	------	------	------	------	------
A_1	------	-0.016	------	------	------	------	------
A_2	------	-0.155	------	------	------	------	------
A_3	------	-0.024	------	------	------	------	------
P_1	------	------	------	------	0.401	------	------
P_2	------	------	------	------	0.161	------	------
P_3	------	------	------	------	0.024	------	------
Q	------	------	------	------	------	------	0.607
δ	------	------	------	------	------	------	0.0002
Final error	0.0042	0.00025	0.004	0.0021	0.00035	0.002	0.0007
no. of iter.	273	421	326	286	392	275	314

------: Parameter not applicable

25.3.2 Implementation of Table-Based Models

Root Model

Regarding the Root model, the nonlinear model elements Q_{gs}, I_{gs}, Q_{ds}, and I_{ds} (Figure 25.15) have been determined. These elements are written in a CITI-file versus V_{gs} and V_{ds}. The extrinsic bias-independent elements L_g, L_d, L_s, R_g, R_d, and R_s are represented by lumped passive elements. The intrinsic nonlinear part is represented by a three-port symbolically defined device (SDD) component. The first port corresponds to the gate-source current and gate-source charge, the second port represents the gate-drain current and gate-drain charge, and the third port denotes the drain-source current and drain-source charge. During simulation the values of the nonlinear elements are read from the data file using a data access component (DAC).

TOPAS Model

Extracted nonlinear elements C_{gs}, I_{gs}, C_{gd}, I_{gd}, C_{ds}, G_{ds}, I_{ds}, and R_i are written in a CITI-file versus V_{gs} and V_{ds}. The extrinsic elements are represented by lumped passive elements. The intrinsic nonlinear part is represented by a nine-port SDD component. The first port represents the gate-source current and gate-source charge, the second port represents the gate-drain current and gate-drain charge, and the third port represents the intrinsic resistance R_i. The drain-source current and drain-source charge are represented in port 4 and the intrinsic conductance G_{ds} in port 5. The time derivatives of voltages ($\dot{v}_{gs}$, $\dot{v}_{ds}$ and $\dot{v}_{gd}$) are at the ports 6, 7, and 8, respectively.

25.4 SIMULATION RESULTS

The popular FET models have been verified with measurements of a GaN power HEMT (w_g = 8×125 μm) made by FBH, Berlin. Model verification includes static DC (Figure 25.17), pulsed DC (Figure 25.18) and large-signal measurements such as single- (Figure 25.19) and two-tone (Figure 25.20) signals.

25.4.1 Static DC Simulations

Static DC simulations have been done for gate-source voltages (V_{GS}) varying from −7V to +1Vin steps of 1V. Maximum drain-source voltage (V_{DS}) was 25V and maximum drain current (I_{DS}) about 0.9A.

GaN devices are typically affected by conspicuous kink phenomena in the DC characteristics (see 'D.3 Trap-induced Kink' in [28]). This phenomenon is less noticeable under pulsed operation conditions. As can be seen all older popular models cannot simulate this special effect (i.e., Curtice-2, Curtice-3, Statz, Materka, Tajima, and TOM). But the more recent version of the Angelov model [2] also does not include this pronounced property.

Addressing the overall simulation capability with respect to measurement, the hyperbolic tangent function works well in all cases. Differences are clearly seen regarding the V_{GS}-dependency. Curtice-2 and Materka, both with quadratic gate-source voltage function, show larger deviation from measurements. Slight improvement can be observed with the Statz

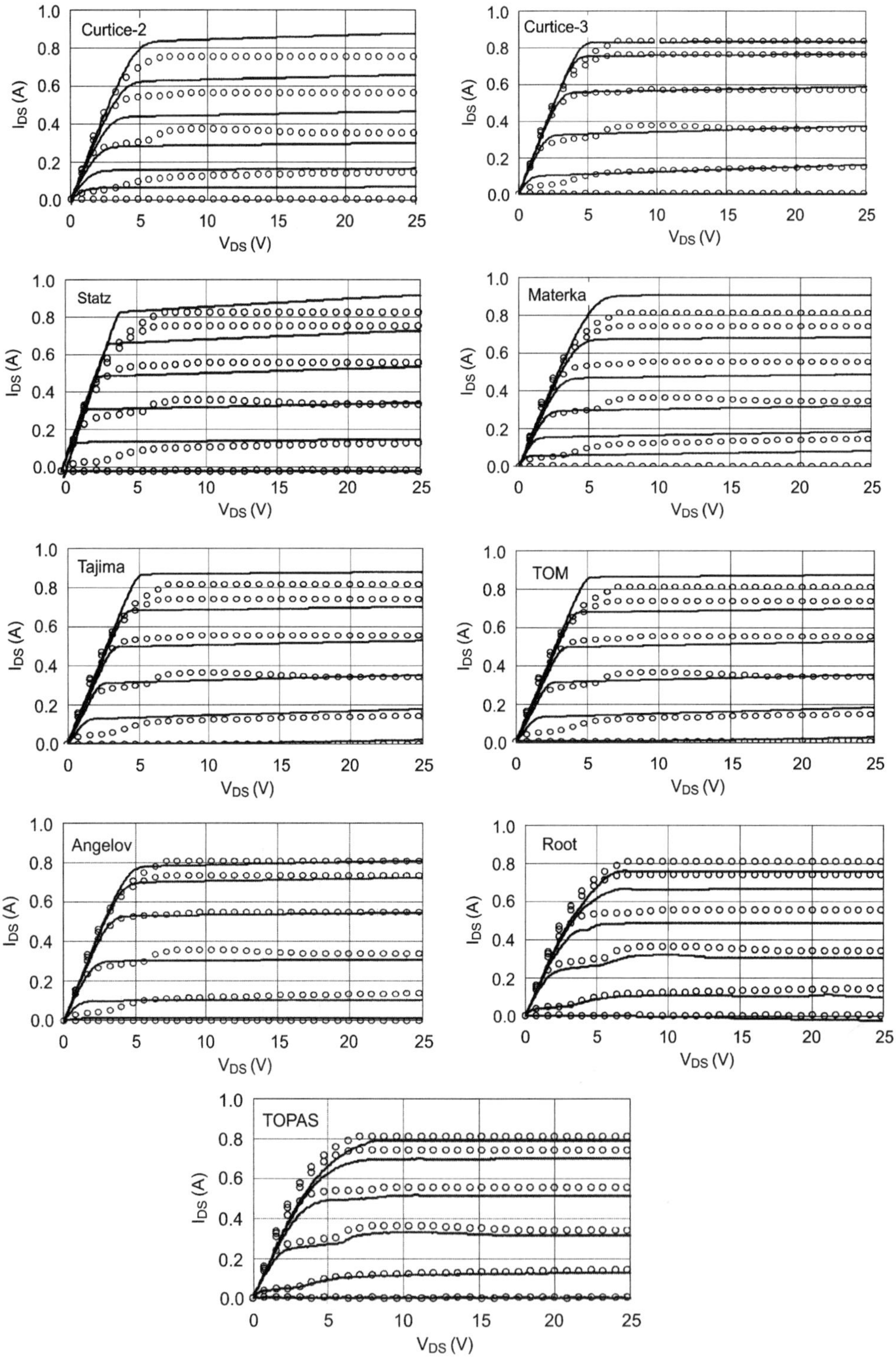

Figure 25.17 Simulated static DC characteristics of popular FET models (solid lines) in comparison with measurements (circles) of a GaN power HEMT ($w_g = 8 \times 125$ μm). V_{GS} varies from −7V to +1V, step 1V.

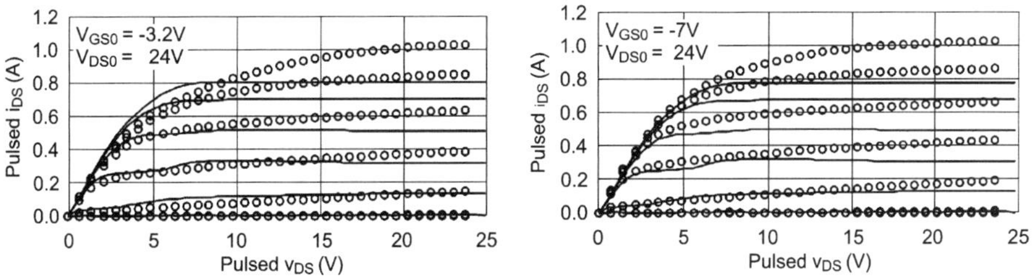

Figure 25.18 Simulated pulsed DC characteristics of TOPAS FET model (solid lines) in comparison with measurements (circles) of a GaN power HEMT ($w_g = 8\times125$ µm). V_{GS} varies from −7V to +1V, step 1V. Bias voltage for pulsing: $V_{GS0} = -3.2$V, $V_{DS0} = 24$V. (left), $V_{GS0} = -7$V, $V_{DS0} = 24$V (right).

model, providing a transition from quadratic to linear behavior. Models, which are measurement-based (i.e., Angelov, Root and TOPAS models) show very good overall agreement with measurements. This meets excellently with the knee voltages. Excellent fitting with measured data is also given for the Curtice-3 model.

25.4.2 Pulsed-DC Simulations

The tested models include the DC I−V characteristics. DC measurements are strongly affected by (static) self-heating and (static) dispersion effects. Thus, it is not surprising that this pulsed mode of operation cannot be simulated correctly. Therefore, the simulation results of pulsed DC stimulus is exemplarily given for the TOPAS model in comparison with measurements in Figure 25.18.

25.4.3 Single-Tone Input Power Sweep

Figure 25.19 presents simulation results of output power (P_{out}) and power-added efficiency (PAE) of single-tone input power sweep. These simulations have been done in the bias point ($V_{GS0} = -3.5$V, $V_{DS0} = 24$V). The operation frequency was 2 GHz. Source and load impedances were 50 ohms.

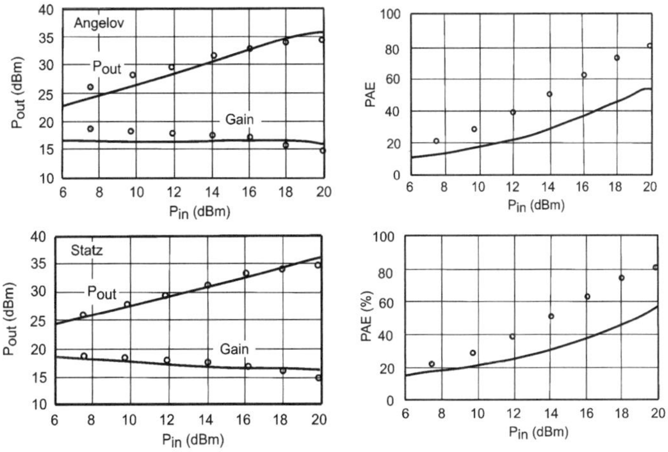

Figure 25.19 Single-tone power sweep: Measured and simulated P_{out} and PAE for tested nonlinear FET models.

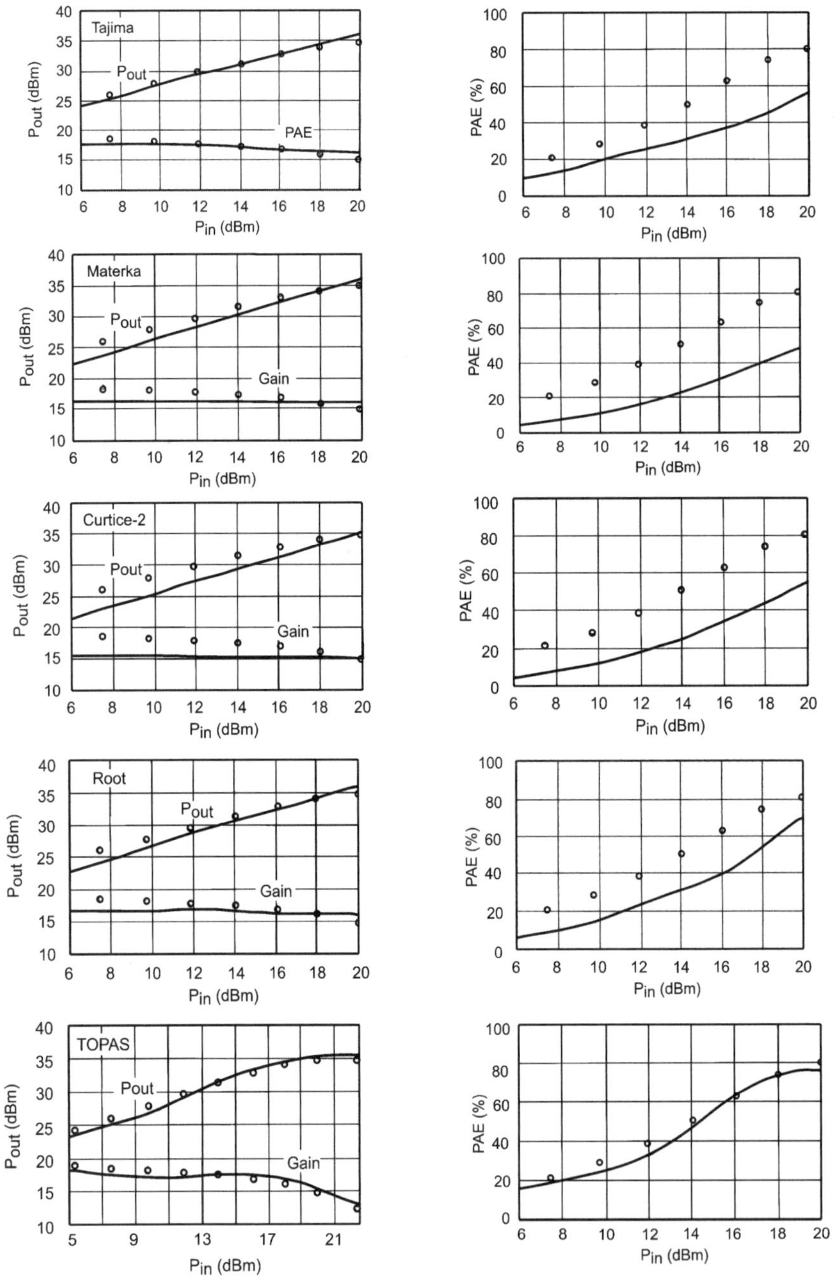

Figure 25.19, cont'd

Summarizing the simulation results in comparison with measurements, it can be observed that the Curtice-2 model exhibits the lowest accuracy. All other models deliver rather good agreement. It should be noted that the TOPAS model simulates the measured data most closely. This, in particular, also applies to the excellent agreement with the measured efficiency PAE.

25.4.4 Two-Tone Input Power Sweep

Two-tone simulation is typically used for device nonlinearity analysis because it describes the behavior of the device under realistic modulated large-signal conditions [29]. The simulations were performed under two-tone excitation centered at 2 GHz and separated by 100 kHz at a bias point (V_{GS0} = −2V, V_{DS0} = 24V).

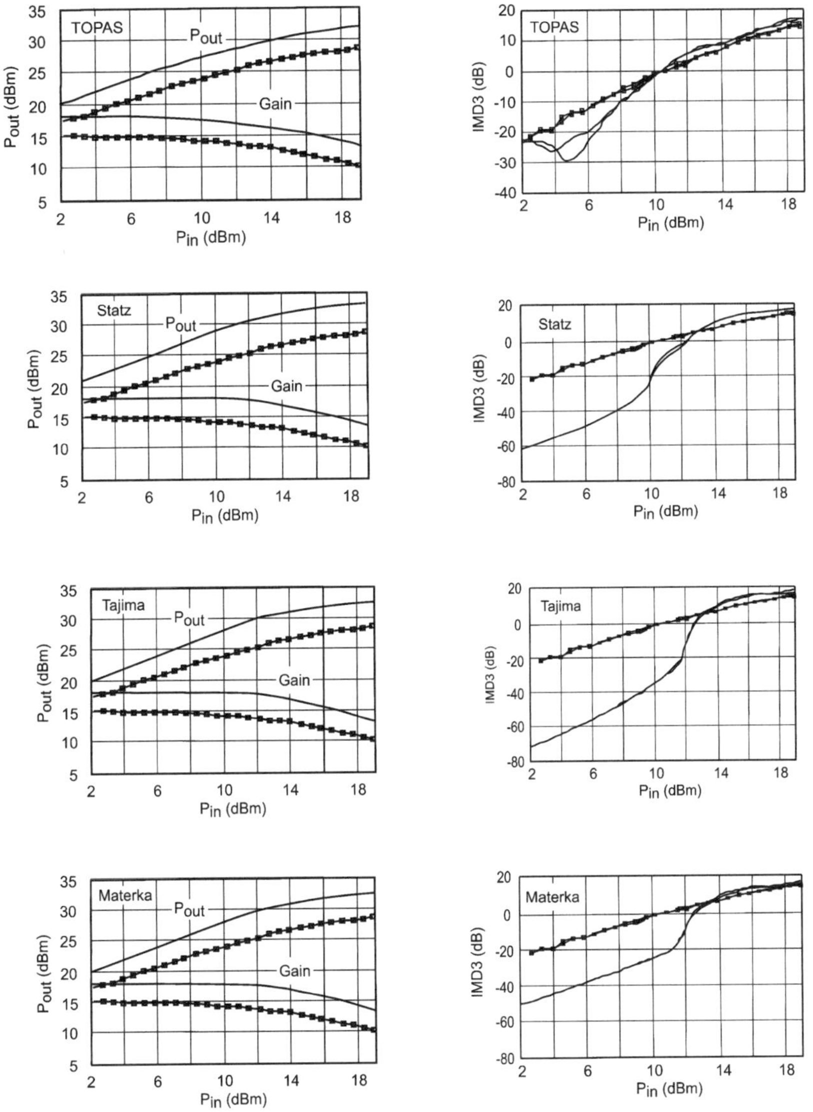

Figure 25.20 Two-tone input power sweep simulation of tested nonlinear FET models (solid lines) in comparison with measurements (symbols) of a GaN power HEMT (w_g = 8×125 μm) at bias point (V_{GS0} = −2V, V_{DS0} = 24V).

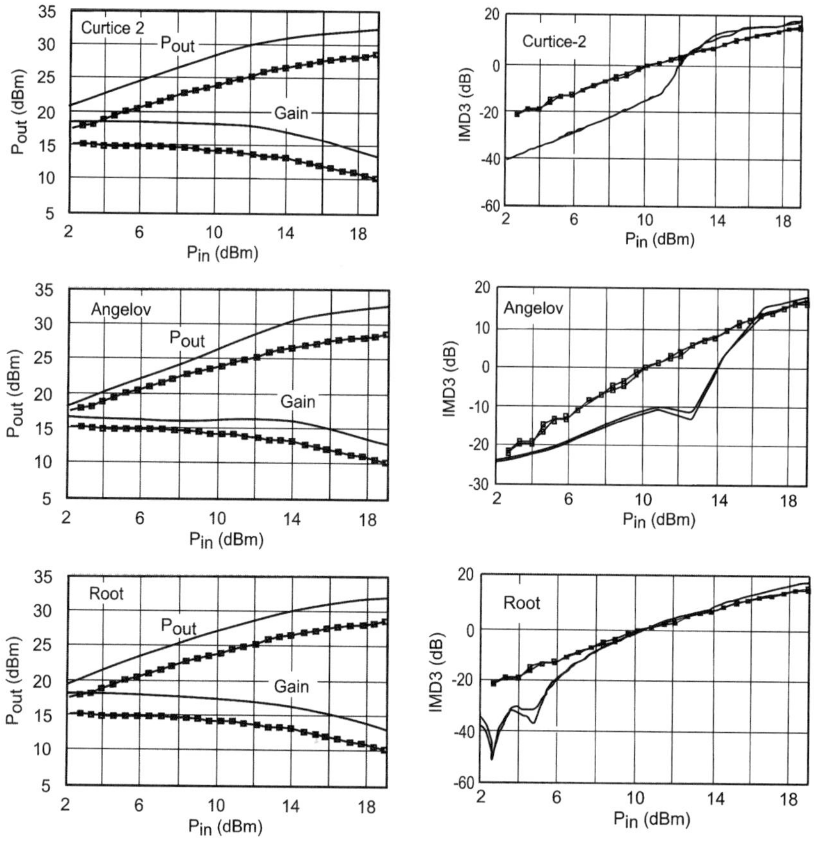

Figure 25.20, cont'd.

The simulated results of two-tone device excitation shown in Figure 25.20 are rather modest. It seems that the Angelov model is slightly superior to the others concerning output power and gain. All models simulate IMD3 very well at higher input power levels. Mostly the deviation between simulation and measurement increases with decreasing power. Notice that the root model is the best result over a wide input power range. However, at very low signal levels it degrades. This problem is well understood resulting from spline interpolation issues. However, with recently increasingly devoted artificial networks (ANN) this weakness can be principally eliminated. Surprisingly, the Angelov model fits the IMD3 measurement perfectly, but in the moderate power range it shows some weakness, which would need some interpretation. Note that the TOPAS model again performs very well with the overall IMD3 measurement.

25.5 SUMMARY

In this chapter popular nonlinear FET models were presented and described. All the models tested so far, were originally developed for gallium-arsenide field-effect-transistors, mostly MESFETs. Later on some models were extended to encompass GaAs HEMT applications.

Only recently has the Angelov model been specifically modified for GaN power devices.

The capacitance-based Curtice models do not take into account the charge conservation principle. The capacitance models are simple and described by single-valued bias-dependent junction capacitance of a Schottky-barrier diode. Note that Curtice has recently published an advanced Curtice model (CFET model) [30] that focuses on subthreshold-valid modeling, which may be important for designers who wish to operate in Class B, C,D, and E high-efficiency operating modes. It is noted that CFET model is well behaved in this region [31].

The Root model is a measurement-based nonlinear model that is capable of simulating the dispersion of microwave devices. The development of the large-signal model is based on DC and RF measurements. In addition the model is table-based, which means that the measured data is directly stored in a data file. Intermediate data points are provided by spline interpolation [32].

Simulated intermodulation distortion IMD3 shows excellent agreement with measurement at higher power levels. With decreasing input power, larger deviation is observed, which can be traced back to simple spline limitations. However, more recently, ANNs have become very attractive. They offer unique properties, being capable of fitting the two-dimensional state functions with utmost accuracy. This method also provides a smoother surface with continuous higher-order derivatives that is important for modeling nonlinear behavior, which is a prime concern in RF power amplifiers [33].

Regarding the measured DC characteristics in Figure 25.17, the kink effect, which is typical for GaN devices, is well simulated. This is also given for the simulation of output power and gain.

As is pointed out in [34], although the model allows very accurate simulation of both DC and RF characteristics, modeling of frequency transition effects is not quite up to the same standard. The root model uses a first-order function to interpolate between DC and drain current models [35]. The drain current expression of the Root model rewritten for conductance modeling reads

$$G(\omega) = \frac{1}{1 + \omega^2\tau^2} \cdot G_{low} + \frac{\omega^2\tau^2}{1 + \omega^2\tau^2} \cdot G_{high} \tag{25.90}$$

As τ is supposed to be independent of drain-source voltage, it implies that dispersion is attributed to one particular trap level. This is similar to other models using a series RC network with constant parameters in parallel to the drain current source. As was shown in [35], the conductance value at low and high frequencies were simulated correctly; however, the characteristic frequency $f = 1/(2\pi\tau)$ was found to depend quite strongly on the drain-source voltage, especially at high drain-source voltage V_{DS}.

The Angelov model was the first nonlinear model both for GaAs HEMT and MESFET devices in 1992. As noted in [32] the published Angelov model is capacitance-based. In general, the bias-dependent capacitor models are known to be less robust, which sometimes leads to nonconvergence problems. Charge-based models are normally more robust. ADS implementation of the Angelov model is enhanced by providing a charge-based model that is alternatively offered in ADS. The model is obtained from DC measurements and small-signal S-parameters.

Recently, the Angelov model has been improved in many respects for GaN applications, concerning, among other things, enhanced current and capacitance models, temperature- and biasdependent R_d and R_s resistances, enhanced breakdown model, and new breakdown models for gate-source and gate-drain junctions [36–38].

The EEHEMT model developed by Keysight Technologies for GaAs HEMT for RF applications [32,38] is modeled in a similar way as the Angelov model. However, V_{gs}- and V_{ds}-drain current terms are sophisticated piecewise empirical functions based on the piecewise regions of transconductance without physical explanations. Therefore it has not been treated in more detail, although it is taken as an industry standard model.

More recently, in 2014, Keysight Technologies proposed the DynaFET (Dynamic FET) model, which follows a modeling strategy that is totally different to known conventional approaches [39,40]. It is based on ANNs and subcircuits to model the behavior of the device. The ANNs are used to formulate gate charge, gate current, drain charge, and drain current while subcircuits model specific behaviors of GaN HEMTs such as self-heating and trapping effects [39,40]. The ANN takes the nodal voltage of the subcircuits (V_{gs}, V_{ds}, T_j, ϕ_1, ϕ_2, $\bar{V}_{gs}$, $\bar{V}_{ds}$) as input to the neuron networks and use measured large-signal waveform to train the weight of each neuron in the network. Weights of all neurons in the ANNs are iteratively adjusted during the training process until the output of the ANN network (such as terminal current) agrees with measured data within an acceptable tolerance.

The DynaFET model has achieved good results in DC, S-parameter, harmonic, intermodulation distortion and load-pull simulation over a very wide range of bias conditions, complex loads, powers, and frequencies [40]. Although the DynaFET model is computationally efficient in simulation, large-signal waveform measurement and the model training take a large amount of time and effort. As well, although trapping and self-heating equivalent circuits are included for modeling the trapping and self-heating effects, the relation between terminal voltages, currents, and charges are still modeled by empirical ANNs without physical explanations, which inherits all the limitations empirical models have [41].

REFERENCES

[1] *IEEE Workshop on Experimentally Based FET Device Modelling & Related Nonlinear Circuit Design*, Kassel, Germany, July 17–18, 1997, Workshop Proceedings available under https://archive.org/stream/DTIC_ADA331468/

[2] ADS2005A Manual, *Nonlinear Devices*, Chapter 3: Devices and Models, GaAs.

[3] Dambrine, G., A. Cappy, F. Heliodore, and E. Playez, "A New Method for Determining the FET Small-Signal Equivalent Circuit," *IEEE Trans.*, Vol. MTT-36, 1988, pp. 1151–1159.

[4] Hower, P. L., and N. G. Bechtel, "Current Saturation and Small-Signal Characteristics of GaAs Field-Effect Transistors," *IEEE Trans.*, Vol. ED-20, 1973, pp. 213–220.

[5] Fukui, H., "Determination of the Basic Device Parameters of a GaAs MESFET," *The Bell System Technical Journal*, Vol. 58, No. 3, 1979, pp. 771–797.

[6] Yang, L., and S. Long, "New Method to Measure the Source and Drain Resistance of the GaAs MESFET," *IEEE Electron Device Lett.*, Vol. EDL-7, 1986, pp. 75–77.

[7] Curtice, W. R., "A MESFET Model for Use in the Design of GaAs Integrated Circuits," *IEEE Trans.*, Vol. MTT-28(5), 1980, pp. 448–456.

[8] Curtice, W. R., and M. Ettenberg, "A Nonlinear GaAs FET Model for the Use in the Design of Output Circuits for Power Amplifiers," *IEEE Trans.*, Vol. MTT-33, 1985, pp. 1383–1393.

[9] Kacprzak, T., and A. Materka, "Compact DC Model of GaAs FET's for Large-Signal Computer Calculation," *IEEE J. Solid-State Circuits*, Vol. SC-18(2), 1983, pp. 211–213.

[10] Materka, A., and T. Kacprzak, "Computer Calculation of Large-Signal GaAs FET Amplifier Characteristics," *IEEE Trans.*, Vol. MTT-33, 1985, pp. 129–135.

[11] Statz, H., P. Newman, I. W. Smith, R. A. Pucel, and H. A. Haus, "GaAs FET Device and Circuit Simulation in SPICE," *IEEE Trans.*, Vol. ED-34(2), 1987, pp. 160–169.

[12] Angelov, I., H. Zirath, and N. Rorsman, "A New Empirical Nonlinear Model for HEMT and MESFET Devices," *IEEE Trans.*, Vol. MTT-40, 1992, pp. 2258–2266.

[13] Angelov, I., L. Bengtsson, and M. Garcia, "Extensions of the Chalmers Nonlinear HEMT and MESFET Model," *IEEE Trans.*, Vol. MTT-40(10), 1996, pp. 1664–1674.

[14] ADS2008A Manual, *Nonlinear Devices*, Chapter 3: Devices and Models, GaAs.

[15] Roblin, P., L. Rice, S. B. Bibyk, and H. Morkoç, "Nonlinear Parasitics in MODFET's and MODFET IV Characteristics," *IEEE Trans.*, Vol. ED-35(8), 1988, pp. 1207–1214.

[16] McCamant, A. J., G. D. McCormack, and D. H. Smith, "An Improved MESFET Model for Spice," *IEEE Trans.*, Vol. MTT-38(6), 1990, pp. 822–824.

[17] Tajima, Y., Wrona, B., and Mishima, K., "GaAs FET Large Signal Model and Its Application to Circuit," *IEEE Trans.*, Vol. ED-28(2), Feb. 1981, pp. 171–175.

[18] Tajima, Y., and Miller, P. D., "Design of Broad-Band Power GaAs FET," *IEEE Trans.*, Vol. MTT-32(3), March 1984, 261–267.

[19] Root, D. E., S. Fan, and J. Meyer, "Technology-Independent Large-Signal FET Models: A Measurement-Based Approach to Active Device Modeling," *Proc. 15th ARMMS Conference*, Bath, England, 1991.

[20] Root, D. E., and S. Fan, "Experimental Evaluation of Large-Signal Modeling Assumptions Based on Vector Analysis of Bias-Dependent S-Parameter Data from MESFETs and HEMTs," *IEEE MTT-S Int. Microwave Symp. Digest*, 1992, pp. 255–258.

[21] Institute of Mobile and Satellite Communication (IMST) GmbH, *TOPAS-Nonlinear Transistor Model and Extraction Software*, Germany, 1996.

[22] Follmann, R., W. Lütke, R. Tempel, and I. Wolff, "A New Method for Large-Signal FET Simulation for Use in MMIC Frequency Multiplier," *IEEE Workshop on 'Experimentally Based FET Device Modeling and Related Nonlinear Circuit Design'*, Kassel, 1997, https://archive.org/ stream/DTIC_ADA331468/.

[23] Follmann, R., R. Tempel, T. Sporkmann, and I. Wolff, "A New Spline Based FET Model for MESFETs and HEMTs," *27th European Microwave Conf. Symp. Dig.*, Jerusalem, September 1997

[24] Follmann, R., R. Tempel, and I. Wolff, "State-of-the Art FET Modeling," Institute of Mobile and Satellite Communication (IMST) GmbH, Germany, 1999.

[25] Follmann, R., J. Borkes, P. Waldow, and I. Wolff, "Extraction and Modeling Methods for FET Device," *IEEE Microwave Magazine*, Vol. 1, 2000.

[26] Lee, K. W., K. Lee, M. S. Shur, T. T. Vu, P. C. T. Roberts, and M. J. Helix, "Source, Drain, and Gate Series Resistances and Electron Saturation Velocity in Ion-Implanted GaAs FET's," *IEEE Trans.*, Vol. ED-32, 1985, pp. 987–992.

[27] McPherson, D. S., "Ultra-Broadband Nonlinear PHEMT Modelling Using TOPAS," *8th IEEE Int. High-Performance Electron Devices for Microwave and Optoelectronic Applications Symp.*, 2000, pp. 85–88.

[28] Kompa, G., *Basic Properties of III-V Devices–Understanding Mysterious Trapping Phenomena*, Kassel, Kassel University Press GmbH, 2014.

[29] Jarndal, A. H., *Large-Signal Modeling of GaN Device for High Power Amplifier Design*, Doctoral Thesis (2006), University of Kassel, published by Kassel University Press GmbH, 2007.

[30] Curtice, W. R., *User's Guide for the C_FET Model for Agilent's Advanced Design Simulator*, Washington Crossing, PA: W. R. Curtice Consulting, June 2004.

[31] Dunleavy, L., C. Baylis, W. Curtice, and R. Connick, "Modeling GaN: Powerful but Challenging," *IEEE Microwave Magazine*, October 2010, pp. 82–96.

[32] Agilent Technologies, *MESFET Models*, IC-CAP 2012.01, January 2012.

[33] Aaen, P. H., J. A. Plá, J. Wood, *Modeling and Characterization of RF and Microwave Power FETs*, Cambridge University Press, 2007

[34] Remley, K. A., D. F. Williams, D. M. M.-P. Schreuers, and J. Wood, "Simplifying and Interpreting Two-Tone Measurements," *IEEE Trans.*, Vol. MTT-52(11), 2004, pp. 2576–2584.

[35] Van den Bosch, S., and L. Martens, "Investigation of the Frequency Dispersion Effect in the Root-Model Applied to Conventional and Floating-Gate MESFET's," *IEEE Trans.*, Vol. ED-44(12), 1997, pp. 2312–2313.

[36] Angelov, I., V. Desmaris, K. Dynefors, P. A. Nilsson, N. Rorsman, and H. Zirath, "On the Large-Signal Modelling of AlGaN/GaN HEMTs and SiC MESFETs," *European Gallium Arsenide and Other Semiconductor Applications Symposium, GAAS 2005*, Paris, 2005, , pp. 309–312.

[37] Angelov FET (Angelov Nonlinear GaAsFET), *http://edadocs.software.keysight.com/.../ads2008U1/Nonlinear+Devices.*

[38] Tinti, R., J. Xu, and D. E. Root: *The Agilent Device Modeling Flow for Nonlinear RF Applications*, Amplifier Design Flow Seminar, Basking Ridge, NJ, 4/2/2014.

[39] Chen. Q., "Latest Advances in Gallium Nitride HEMT Modeling," *12th IEEE International Conference on Solid-State and Integrated Circuit Technology (ICSICT)*, Guilin, 2014, pp. 1–4.

[40] Xu, J., R. Jones, S. A. Harris, T. Nielsen, and D. E. Root, "Dynamic FET Model - DynaFET - for GaN Transistors from NVNA Active Source Injection Measurements," *2014 IEEE MTT-S International Microwave Symposium (IMS2014)*, Tampa, FL, 2014, pp. 1–3.

[41] Zhang, H., *Physics Based Virtual Source Compact Model of Gallium-Nitride High Electron Mobility Transistors*, Thesis, University of Waterloo, Ontario, Canada, 2017.

Chapter 26

Nonlinear Transistor Model Verification

Modeling of multifinger power devices with more complex gate peripheries have not yet been discussed. Source air bridges that are in close proximity to the drain and gate fingers may contribute to significant additional extrinsic capacitances [1]. As has been pointed out throughout this book, such complex equivalent circuits are also needed for field-effect transistors with smaller gate periphery operating at high frequencies taking into account interelectrode coupling by stray capacitances.

In this chapter the developed nonlinear power HEMT model (see Chapter 24) is verified using static-/pulsed-DC, small-signal S-parameter and large-signal single-/two-tone measurements.

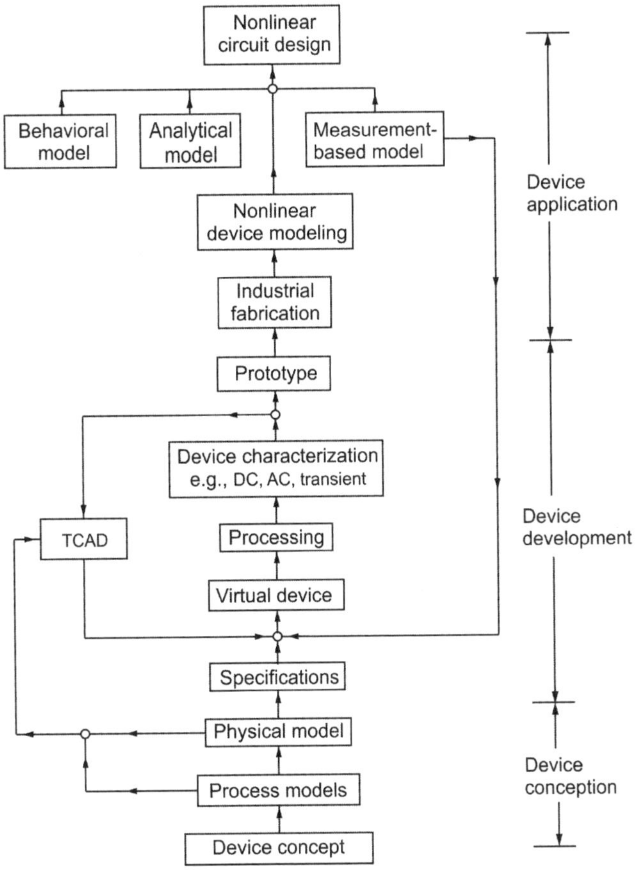

Figure 26.1 Device design and simulation flow from basic device conception to virtual device creation, device fabrication, and final device application in circuit design.

Figure 26.1 indicates schematically the device design flow from the basic device conception (e.g., new semiconductor material, novel device concept, new device structure creation, new device application) followed by device development based on technology computer-aided design (TCAD) device simulation and optimization, and prototype manufacturing. TCAD is capable of modeling semiconductor process and device physics through numerical computation. It can be used to develop and optimize semiconductor process and simulate device electrical behaviors with design of experiments (DoE) studies. By using simulation methods technology, specialists can observe what is going on inside the devices (useful in the development phase), which is impossible with measurement techniques applied directly to real devices [2]. The introduction into market requires matured, stable, and reproducible device performance. Finally, successful nonlinear circuit design is reliant on accurate nonlinear device models.

In the previous chapter popular nonlinear FET models were reviewed and tested with measurements on a GaN power HEMT. Analytical models such as the Angelov model rely on transcendential functions of nonlinear current and charge sources with formal fitting parameters, which are not always simple to be directly linked to physical device parameters. Such analytical models are primarily targeted at circuit designers for computer-aided design of nonlinear microwave and RF circuits. As the number of model parameters has permanently increased to enhance device performance prediction, model parameter extraction has become very sophisticated and needs intelligent extraction programs [3,4]. The extraction process of a large number of model parameters relies on optimization and entails the risk getting stuck in local minima and nonconvergence of the model for certain sets of parameter values at certain bias conditions [5].

Behavioral models, also termed black-box model, consider the device performance exclusively at the device terminals. Intrinsic device structure, the electrical network with distributed charge and current sources, resistances, and capacitances, is of minor importance. This kind of model is mainly focused on highly efficient circuit design. As already pointed out in the previous chapter, much progress has recently occurred extracting the nonlinear device model parameters directly from measurements with a nonlinear vector network analyzer (NVNA). The model parameter values are adjusted until the measured and simulated large-signal data agree [5]. However, as the models are artificial neural-network based, physical interpretation of the model parameters is difficult.

In contrast, measurement-based modeling exhibits advantages in various directions. Indeed, the main purpose is to create a reliable nonlinear model for microwave and RF circuit design. However, because the equivalent circuit is directly linked with device physics, it can give a feedback to the device modeler to improve the device performance. A good example has been given in connection with linearity improvement of GaN power devices. Based on physics-based equivalent circuit analysis and measurement-based characterization, intermodulation distortion sources inside of the device could be identified and localized so that convenient compensation measures could be proven by device structural modifications [6-8]. Thus, improved device linearity property simplifies external linearization techniques applied to the designed amplifier to fulfill communication standards.

26.1 COMPLETE LARGE-SIGNAL DEVICE MODEL

The developed large-signal transistor model has been implemented in the ADS® software of Agilent® to carry out simulations and obtain model predictions to compare them with the device measured behavior. The complete circuit representation of the large-signal model is shown in Figure 26.2. The extrinsic linear part of the device model comprises 12 elements, C_{pga}, C_{pda}, C_{gda}, C_{pgi}, C_{pdi}, C_{gdi}, R_g, R_d, R_s, L_g, L_d, and L_s, which are implemented in the ADS® schematic as lumped elements. The remaining elements form the intrinsic part of the circuit, which has been implemented with a SDD (symbolically defined device) block, is described in the following section.

26.2 MODEL IMPLEMENTATION

Figure 26.3 illustrates the large-signal model implementation in the ADS electrical circuit simulator. The SDD (symbolically defined device) block represents the intrinsic part of the device model. I_{gs} and Q_{gs} are represented in port 1, and I_{gd} and Q_{gd} in port 2. The current sources are directly entered as the current of the ports. The current component due to the charge sources is indicated entering the charge source in the corresponding port and applying the first weighting factor: The "1" in the second space inside the brackets in the definition F[n,1] by default is equivalent to the time derivative in the simulator. Simultaneously, v_{gs} is

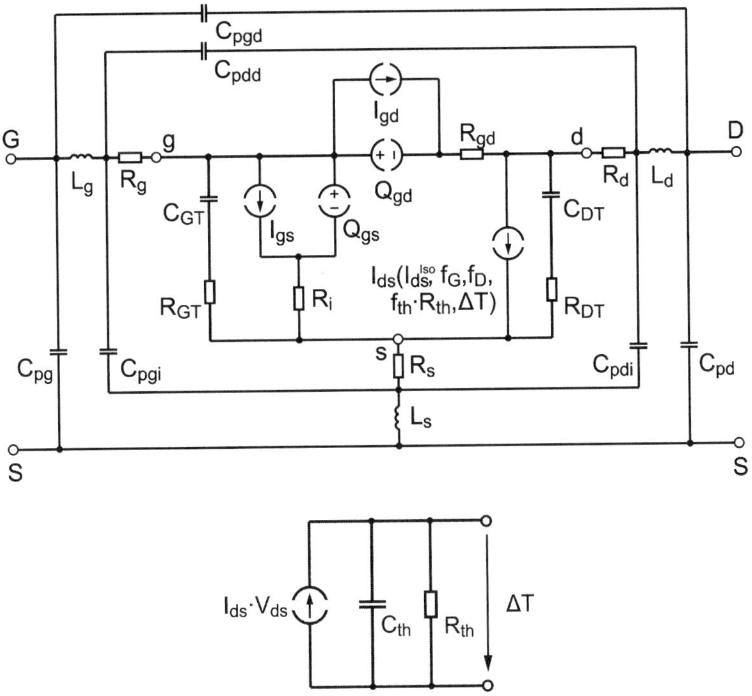

Figure 26.2 Complete electrothermal large-signal model for the analyzed 3.2-mm gate width AlGaN/GaN power HEMT (above) with subcircuit thermal model (below).

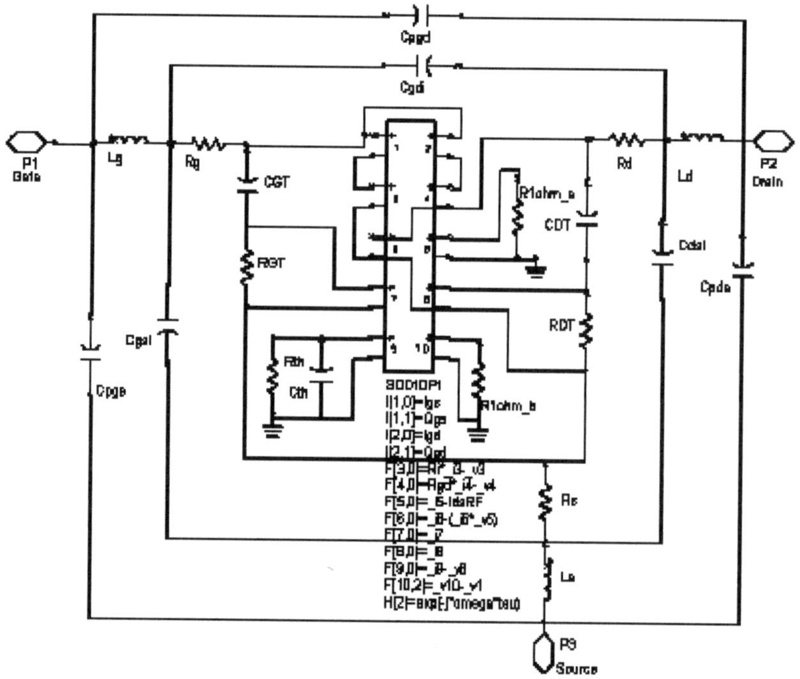

Figure 26.3 Implementation of the intrinsic part of the large-signal model with a 12-port SDD in the ADS® schematic [9].

given by the voltage sensed by the simulator in port 1,_v1. The ports 3 and 4 are used for the parameters R_i and R_{gd}. The I_{ds} current is implemented in a separate expression according to the I_{ds}-model equation derived in Section 23.5. Then, the instantaneous I_{ds} value calculated by the simulator is entered as the current of this port 5 and v_{ds} is given by the voltage sensed in port 5,_v5. The remaining five ports perform auxiliary functions:

- The product (_v5_i5) is calculated in port 6 and is entered as current, wherein an 1Ω resistor gives the dissipated power p_{ds} as _v6.
- The function of ports 7 and 8 is sensing the time-varying components (v_{gs} - V_{GS}) and (v_{ds} - V_{DS}) that are factors in the I_{ds} model, and which are available as _v7 and _v8.
- The power p_{ds} is entered as current in port 9, wherein the electrothermal subcircuit is implemented, and then, ΔT is obtained as _v9.
- The second weighting function is defined as H[2] = $e^{-j\omega\tau}$ (i.e., as the time delay of the transconductance). Then, the gate voltage denoted by _v1 is entered as voltage of port 10, applying the user-defined second weighting function with the nomenclature F[10,2]. As a result, the delayed gate voltage is obtained as _v10.

The extracted values of the bias-dependent intrinsic parameters R_{gd}, R_i, Q_{gs}, I_{gs}, Q_{gd}, I_{gd}, τ, and the bias-dependent I_{ds}-model parameters I_{ds}^{iso}, f_D, f_G, f_{th} are stored in text files. Then, these parameters are implemented in the ADS® schematic as look-up tables that point to the text files. This is done with data-access component (DAC) blocks. For every instantaneous value of controlling voltages, the simulator reads the value of the required parameter from the look-up table files.

The v_{ds} and v_{gs} values, given by _v5 and _v1, are used as controllers for the reading of the file that contains the parameters R_{gd}, R_i, Q_{gs}, I_{gs}, Q_{gd}, I_{gd}, τ. The values of v_{ds} and of the delayed v_{gs}, given by _v5 and _v10, are the controllers for the reading of the file that contains the parameters I_{ds}^{iso}, f_D, f_G, f_{th} [9].

26.3 SIMULATION AND COMPARISON WITH MEASURED DEVICE DATA

In the following, large-signal simulations have been performed and compared with measurements carried out on a large-size 3.2-mm GaN power device. The large-signal model has been tested for small-signal stimulus in Section 26.3.1, for pulsed-DC operation in Section 26.3.2, and for large-signal single- and two-tone operation in Section 26.3.3.

26.3.1 Simulation of Bias-Dependent S-Parameters

S-parameter measurements were compared with simulations using the large-signal model implemented in ADS®. The test device was a 3.2-mm AlGaN/GaN power HEMT, biased in the saturation region ($V_{GS} = -3.2$V, $V_{DS} = 30$V) and near the pinch-off region ($V_{GS} = -3.6$V, $V_{DS} = 55$V).

With regard to evaluation it should be noted that the parameters R_{gd}, R_i, Q_{gs}, I_{gs}, Q_{gd}, and I_{gd} have strong influence on the simulation of the device input impedance of the large-signal model. Those parameters were extracted on the basis of S-parameter measurements. On the other hand, the I_{ds}-model parameters have strong influence on the simulation of the device transconductance and output impedance of the large-signal model. Those parameters were extracted on the basis of pulsed-DC IV measurements, and thus, are less connected to S-parameter measurements.

In consequence, simulations of the large-signal models were expected to produce closer

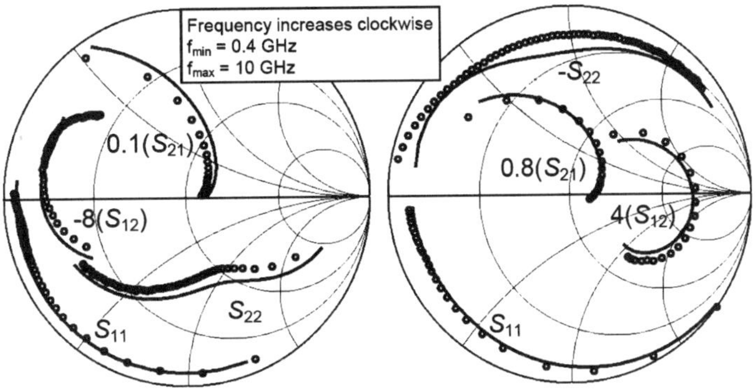

Figure 26.4 S-Parameter measurements of a 3.2-mm AlGaN/GaN HEMT (circles) and simulations with the large-signal model (lines) for bias points (a) in the saturation region ($V_{GS} = -3.2$V, $V_{DS} = 30$V), and (b) near the pinch-off region ($V_{GS} = -3.6$V, $V_{DS} = 55$V).

model agreements with measurements for S_{11}, related to the simulated input impedance than for S_{22} and S_{21} related to the simulated output impedance and transconductance. These expectations correspond to the results shown in Figure 26.4, with comparatively better agreement of simulations with measurements for S_{11} than for S_{22} and S_{21}.

The general good agreement of simulations and measured S-parameters verify the consistency between large-signal and small-signal models discussed in previous chapters.

26.3.2 Simulation of Pulsed-DC IV Characteristics

Comparisons of pulsed-DC IV measurements and simulations are shown in Figure 26.5 for the same GaN power HEMT as before. Results are presented only for quiescent bias points that were not employed to develop the I_{ds}-model. The excellent agreements of simulations with measurements are exemplified by the results shown in Figure 26.5(a) and Figure 26.5(c). These agreements indicate the ability of this large-signal model to accurately characterize low-frequency dispersive phenomena for quiescent bias points in the pinch-off and saturation regions, especially for operating points that are used in power amplifier design of classes A, AB, B or C (I_{DS} < 75% of I_{DSS}).

For bias points with larger quiescent currents as shown in Figure 26.5(b) and Figure 26.5(d), the agreements of simulations with measurements decrease as the pulsed-to v_{GS} approaches 0V (i.e., the highest i_{DS} values). This is explained by the residual effect of the stabilization resistance (see Section 9.2.3) that increases proportionally to the drain current flowing in the device.

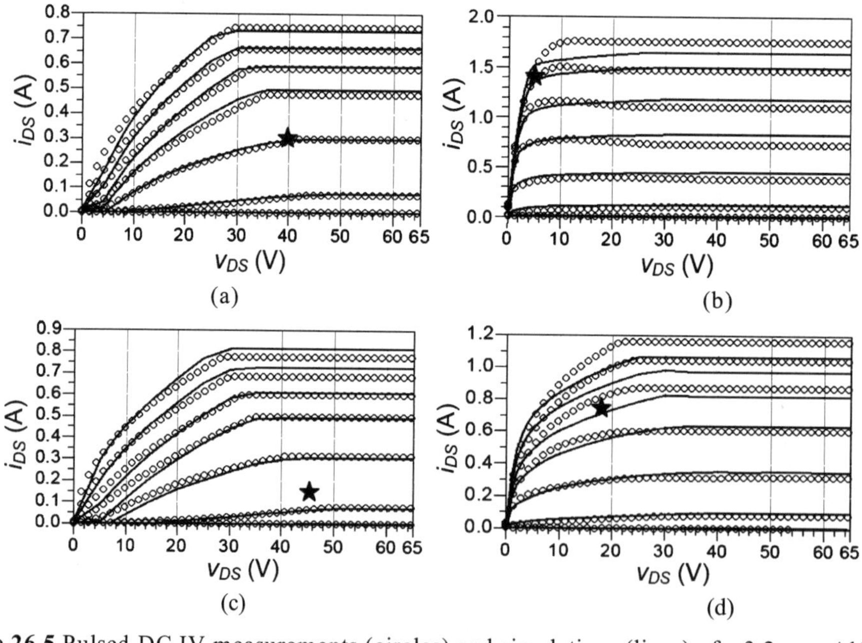

Figure 26.5 Pulsed-DC IV measurements (circles) and simulations (lines) of a 3.2-mm AlGaN/GaN HEMT on the quiescent bias points (stars): (a) $V_{GS} = -2.7$V, $V_{DS} = 40$V, $I_{DS} = 0.3$A, (b) $V_{GS} = -1$V, $V_{DS} = 5$V, $I_{DS} = 1.4$A, (c) $V_{GS} = -3.1$V, $V_{DS} = 45$V, $I_{DS} = 0.15$A, and (d) $V_{GS} = -1.7$V, $V_{DS} = 18$V, $I_{DS} = 0.76$A. v_{GS} curves from -3.6V to 0V in steps of 0.6V.

26.3.3 Simulation of Single- and Two-Tone Device Response

The large-signal models have been developed for the design of power amplifiers used for UMTS base station transmitters. The operating bias points were chosen in the regions as indicated in Figures 26.5(a) and 26.5(b) (i.e., V_{DS} between 45V and 50V and I_{DS} up to 300 mA for the analyzed 3.2-mm GaN device). The frequency of the fundamental tone for the single-tone test was selected as 2.15 GHz. This value is in the center of the frequency band assigned by ITU for the FDD variation of the UMTS communication channel in the base station. For the tests with two-tone stimuli, the middle point between the tones is also 2.15 GHz. The tone spacing was selected as 200 kHz, the channel raster of FDD UMTS.

Figure 26.6 shows single-tone measurements and simulations of the 3.2-mm GaN HEMT in a 50Ω environment. In the simulations, the power level of a sinusoidal source was swept and for each swept value model predictions were obtained for the output power at the frequencies f_0, $2f_0$, and $3f_0$ (fundamental, second- and third-harmonics), for the power gain and for the power-added efficiency (calculated with model predictions of I_{ds} current at each sweep step).

In order to assure that the simulated condition agreed with the real situation presented by the measurement setup, the term P_{out} in Figure 26.6 was calculated as equal to the power absorbed in the 50Ω load, and the term P_{in} was calculated as equal to the difference between incident and reflected powers at the transistor input.

For the model of the 3.2-mm GaN HEMT, accurate simulations of fundamental output power are reported in Figure 26.6(a) and Figure 26.6(b), for bias points of class-AB and C power amplifier design (I_{DS} values nearly 20% and less than 5% of I_{DSS}, respectively). In such

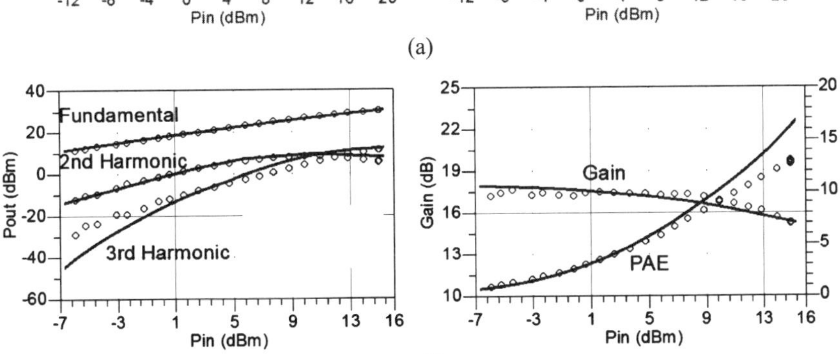

(b)

Figure 26.6 Single-tone measurements (circles) and simulations of a 3.2-mm AlGaN/GaN HEMT for bias points (a) $V_{GS} = -2.7V$, $V_{DS} = 45V$, $I_{DS} = 300$ mA and (b) $V_{GS} = -3.475V$, $V_{DS} = 40V$, $I_{DS} = 60$ mA.

a bias point, P_{out} was correctly predicted up to 36.5 dBm, approximately to 4W and 1.25 W/mm.

As shown in Figure 26.6(a) accuracy of large-signal predictions decreases for the third harmonic at low output powers (i.e. P_{out} lower than −30 dBm). These inaccuracies are a drawback of the look-up table implementation of the model parameters, as has been documented in [5,10]. When the power levels are small enough, the voltages controlling the reading of look-up table files are lower than the point-spacing of the grid of the stored parameter values, and then, the simulator must interpolate. Combination of interpolating highly nonlinear data and the higher-order derivatives required to calculate output power of the higher-order harmonics produces model predictions that show a high ripple response. The problems are diminished if the parameters are interpolated in a denser grid before being stored, but this is direct trade-off with the simulation time and computing effort, since the managed files are larger (a twice-denser grid means four-times larger files) and the look-up table process turns slower. However, the general agreement of the simulation results with the measured data is very good at the medium- and high-power level. As pointed out in [5], the main benefits of measurement-based nonlinear device models can be preserved while removing most of the limitations by replacing the tables and interpolation schemes by ANNs.

Two-tone verification tests shown in Figure 26.7 demonstrate the ability of the 3.2-mm device model to accurately simulate third-order intermodulation products, IMD3, for bias points of different power amplifier classes: in (a) I_{DS} was less than 5% of I_{DSS}, corresponding to class-C operation, in (b) V_{GS} was approximately V_{th}, corresponding to class-B operation, and in (c) I_{DS} was nearly 20% of I_{DSS}, corresponding to class-AB operation.

In the measurements depicted in Figure 26.7(a) the large-signal model predicts accurately the input powers and bias ranges where sweet spots were measured. Although small, but 10% uncertainty of the bias points (see Section 9.2.3: calibration of the stabilizing element by the built-in feature of the dynamic IV analyzer (DiVA) introduced a residual measurement error; correction of this residual error was demonstrated with tuning of the simulated bias voltages within a 10% range) is considered to decrease the agreement between measured and simulated power of the IMD3 products, especially for bias points related to sweet spots, because they are in the low-I_{DS} region, as is the case in Figure 26.7(a).

On other bias points (without appearance of sweet spots), simulated power of IMD3 products shows a close agreement with measurements, as exemplified in Figure 26.7(b) and Figure 26.7(c). A maximum deviation of 5 dB between IMD3 simulations and measurements appears only at extreme P_{in} and IMD3 values. That is shown in Figure 26.7(b) where P_{in} = −5 dBm and IMD3 about −35 dB. Similar deviation is observed in Figure 26.7(c) with P_{in} = 8 dBm and IMD3 about 0 dB. Decreased accuracy of simulated power of IMD3 products appearing in Figure 26.7(b) at low input power levels is attributed to the related low v_{GS} and v_{DS} values that control the reading of look-up table files. If those voltage values are lower than the point-spacing of the voltage grid of the stored parameter values, then interpolation of the simulator decreases the model prediction accuracy. This decrease of accuracy related to low input powers was explained in [5], and a similar effect was already discussed previously concerning the simulation of third harmonic output powers.

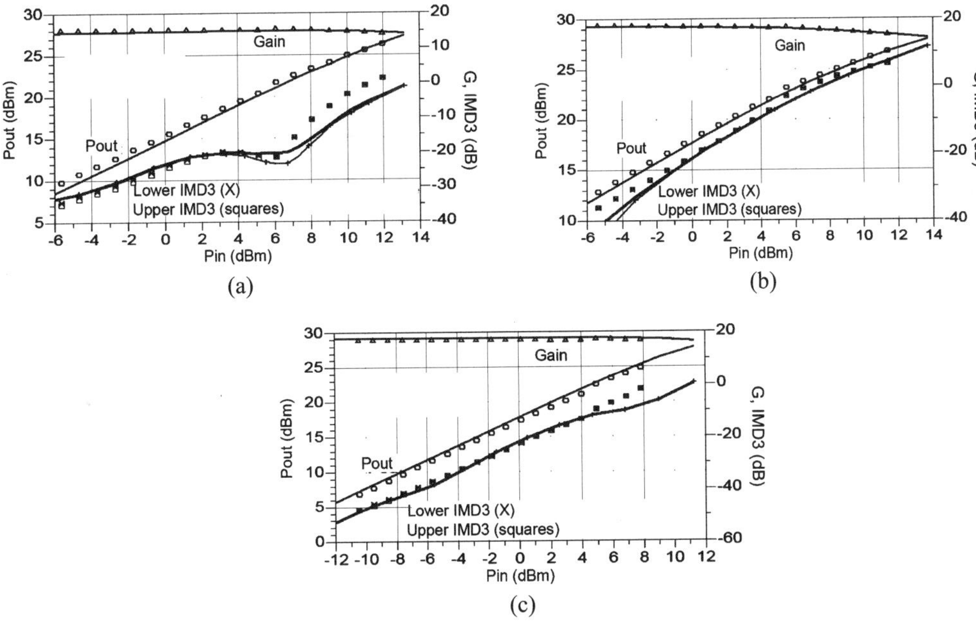

Figure 26.7 two-tone measurements (markers) and simulations of a 3.2-mm AlGaN/GaN HEMT for bias points (a) $V_{GS} = -3.475$V, $V_{DS} = 40$V, $I_{DS} = 25$ mA, (b) $V_{GS} = -3.1$V, $V_{DS} = 45$V, $I_{DS} = 190$ mA, and (c) $V_{GS} = -2.7$V, $V_{DS} = 40$V, $I_{DS} = 300$ mA. The tone-spacing is 200 kHz.

26.4 SUMMARY

Multifinger GaN power transistors comprise large-size complex gate peripheries with manyfold capacitive coupling effects. Similar parasitic capacitance effects can also be observed with smaller gate-peripheries comprising only one or two gate fingers at high operating frequencies. As has been demonstrated, accurate modeling of devices which take into account such distributed capacitive effects, requires a more elaborated model parameter extraction strategy than traditional methods can be provided.

The discussed complex transistor model comprises a distributed extrinsic parasitic network. The intrinsic nonlinear model includes thermal and frequency dispersion, and considers also low-frequency trapping transients at the gate and drain side. Bias dependent charging resistance R_i is included. Furtherrmore gate-forward and drain-gate breakdown is taken into account.

The model has been implemented in ADS as symbolically defined device (SDD). Various verification tests have been performed using a 3.2-mm AlGaN/GaN power HEMT on SiC substrate (8 gate finger device, fabricated by the Fraunhofer Institute for Applied Solid-State Physics (IAF) in Freiburg i. Br.). The proposed nonlinear model was verified on the basis of small-signal S-parameter, pulsed-DC and large-signal single- and two-tone measurements showing very good results for the 3.2-mm device. Modeling results for a 1-mm AlGaN/GaN power HEMT on SiC substrate were presented in [11]. A 2-mm AlGaN/GaN power HEMT on Si (with 10 gate fingers), manufactured by Nitronix® Corporation, has been treated in [9].

The complex transistor model described in this book and its model parameter extraction procedure has been developed over years supported partly by national funding [12], and has proven very successful in cooperative projects with partners from industry and universities.

REFERENCES

[1] Rudolph, M., C. Fager, and D. E. Root, *Nonlinear Transistor Model Parameter Extraction Techniques*, Cambridge: Cambridge University Press, 2011.

[2] Feng, P., *Design, Modeling and Analysis of Non-Classical Field-Effect Transistors*, Dissertation, Syracuse University, December 2012.

[3] Dunleavy, L., C. Baylis, W. Curtice, and R. Connick, "Modeling GaN: Powerful But Challenging," *IEEE Microwave Magazine*, October 2010, pp. 82–96.

[4] Tajima, Y., "Progress in RF Device Modeling: From MESFETs to GaN PHEMTs," *The Annual RF & Microwave Solutions Update*, 2006, pp. 40–44; *www.rfglobalnet.com*.

[5] Root, D. E., "Future Device Modeling Trends," *IEEE Microwave Magazine*, November/December 2012, pp. 45–59.

[6] Vuolevi, J., and T. Rahkonen, *Distortion in Power Amplifiers*, Norwood, MA: Artech House, 2003.

[7] Srinidhi, E. R., and G. Kompa, "Investigation of IMD3 in GaN HEMT Based on Extended Volterra Series Analysis," *Proc. of the 2nd European Microwave Integrated Circuits Conference (EuMIC)*, Munich, Germany, October 2007.

[8] Srinidhi, E. R., *Intermodulation Distortion Modelling and Measurement Techniques for GaN HEMT Characterization*, Doctoral Thesis, University of Kassel, 2008, Kassel: Kassel University Press GmbH, 2008.

[9] Zamudio Flores, J. A., *Device Characterization and Modelling of Large-Size GaN HEMTs*, Doctoral Thesis, Kassel: Kassel University Press GmbH, 2012.

[10] Jarndal, A., and G. Kompa, "Large-Signal Model for AlGaN/GaN HEMT Accurately Predicts Trapping- and Self-Heating Induced Dispersion and Intermodulation Distortion," *IEEE Trans.*, Vol. ED-54, 2007, pp. 2830–2836.

[11] Jarndal, A., and G. Kompa, "Large-Signal Model for AlGaN/GaN HEMT Accurately Predicts Trapping- and Self-Heating Induced Dispersion and Intermodulation Distortion," *IEEE Trans.*, Vol. ED-54, 2007, pp. 2830–2836.

[12] Federal Ministry of Education and Research (BMBF), Funding Priority "GaN-Electronics", 2003–2006.

Appendix A

Generic Two-Port Matrix Transistor Model

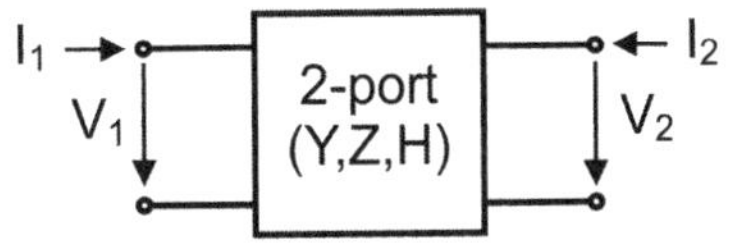

Figure A.1 General two-port network.

A general two-port network is shown in Figure A.1. V_1 and I_1 are input voltage and current, respectively. V_2 and I_2 are output voltage and current, respectively. A two-port can be described by Y-, Z-, and H-parameters. In addition ABCD parameters are used for the cascading of two-port networks.

We use the Y-parameter representation because these are appropriate for transistor equivalent circuit description:

$$I_1 = Y_{11}V_1 + Y_{12}V_2 \quad \text{(A.1)}$$

$$I_2 = Y_{12}V_1 + Y_{22}V_2 \quad \text{(A.2)}$$

where V_1 and V_2 are independent variables and I_1 and I_2 are dependent variables. In matrix form (A.1) and (A2) can be rewritten as

$$\begin{bmatrix} I_1 \\ I_2 \end{bmatrix} = \begin{bmatrix} Y_{11} & Y_{12} \\ Y_{21} & Y_{22} \end{bmatrix} \begin{bmatrix} V_1 \\ V_2 \end{bmatrix} \quad \text{(A.3)}$$

or in short form

$$I = Y \cdot V \quad \text{(A.4)}$$

Figure A.2 shows the common small-signal pi-type equivalent network of a transistor. The determination of the transistor parameters from the two-port Y-parameters is done as follows.

Y_{11} is calculated as

$$Y_{11} = \left. \frac{I_1}{V_1} \right|_{V_2 = 0} \quad \text{(A.5)}$$

and describes the short-circuit input admittance. Similarly, Y_{22} denotes the short-circuit output admittance,

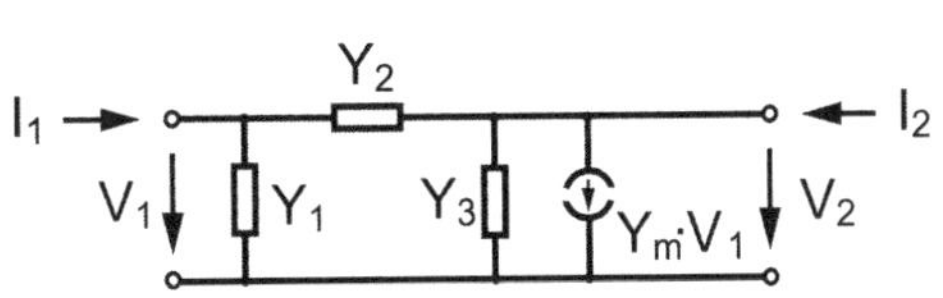

Figure A.2 Generic pi-type small-signal transistor model.

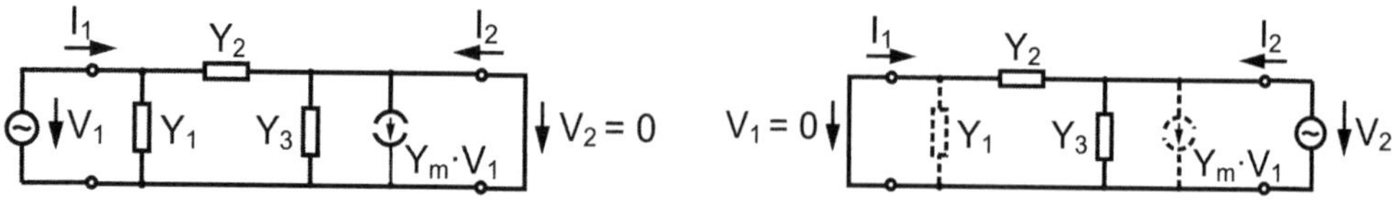

Figure A.3 Circuit arrangement to calculate (a) input and (b) output admittance parameters.

$$Y_{22} = \left.\frac{I_2}{V_2}\right|_{V_1 = 0} \tag{A.6}$$

Y_{21} is the forward transconductance and is defined as

$$Y_{21} = \left.\frac{I_2}{V_1}\right|_{V_2 = 0} \tag{A.7}$$

The reverse transconductance reads

$$Y_{12} = \left.\frac{I_1}{V_2}\right|_{V_1 = 0} \tag{A.8}$$

Figure A.3 illustrates the terminations of the two-port to determine (a) input and (b) output admittance parameters. Then, applying (A.5) - (A.8), the following relations can be found between the transistor elements Y_1, Y_2, Y_3, Y_m, and the Y-parameters:

$$\begin{aligned} Y_{11} &= Y_1 + Y_2 \\ Y_{12} &= -Y_2 \\ Y_{21} &= -Y_2 + Y_m \\ Y_{22} &= Y_2 + Y_3 \end{aligned} \tag{A.9}$$

Solving (A.9) for the transistor admittances Y_1, Y_2, Y_3, Y_m, we get

$$\begin{aligned} Y_1 &= Y_{11} + Y_{12} \\ Y_2 &= -Y_{12} \\ Y_3 &= Y_{22} + Y_{12} \\ Y_m &= Y_{21} - Y_{12} \end{aligned} \tag{A.10}$$

Figure A.4 shows the transistor model with the model elements represented by the Y-parameters.

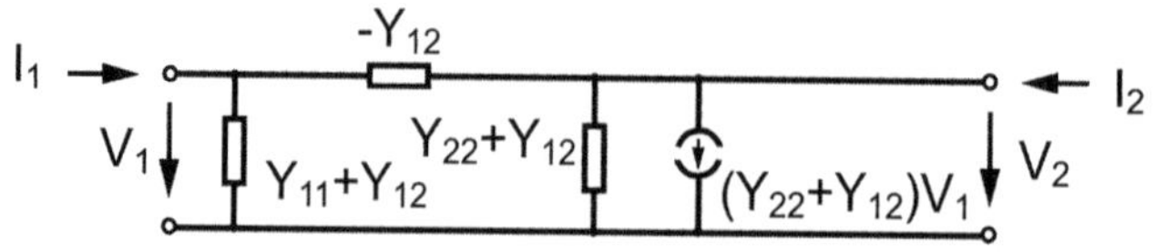

Figure A.4 Transistor equivalent network notation with generic two-port Y-parameters.

Appendix B

Direct Measurement of Series Resistances

B.1 INTRODUCTION

This chapter briefly reviews classical measurement methods to characterize the series resistances of R_s, R_d, and R_g (e.g., [1–5]). They are based on DC or low-frequency RF measurements.

B.2 DC METHOD AFTER WILLIAMS

In [5], Williams proposed a simple measurement method to determine the series resistances of the equivalent network of a field-effect transistor. Figure B.1 shows the typical DC characteristic of a Schottky diode of a FET device in forward conduction. At low gate-source voltage V_{GS} the gate current I_{SG} increases exponentially due to Schottky barrier property. With further increase of the gate voltage the resistances become noticeable so that the characteristic becomes linear. At higher voltages, saturation is observed. From the slope of the curve, the sum of the gate and source resistance, R_{GS}, can be measured.

The measurement of R_s is illustrated in Figure B.2. V_{GS} is applied to the gate electrode and V_{DS} to the drain electrode, which results in a small to moderate forward current I_{GS} and very small (compared to the gate current) drain current I_{DS}. The forward current is increased in precise current steps. The current reaching the drain is so small (compared to the source-gate current) that the drain voltage is essentially equal to the voltage between the source pad and the gate stripe. Hence, the source resistance is found by

$$R_s = dI_{GS}/dV_{DS}, \quad I_{DS} \ll I_{GS} \tag{B.1}$$

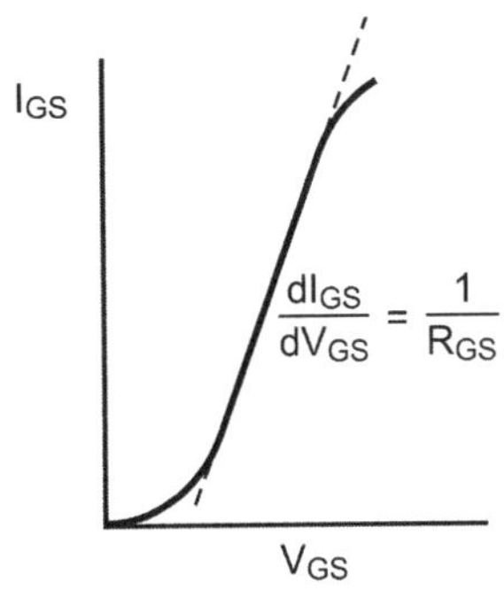

Figure B.1 DC forward gate characteristics of a field-effect transistor. The linear region is used to determine the sum of the gate and source resistance, R_{GS} [5].

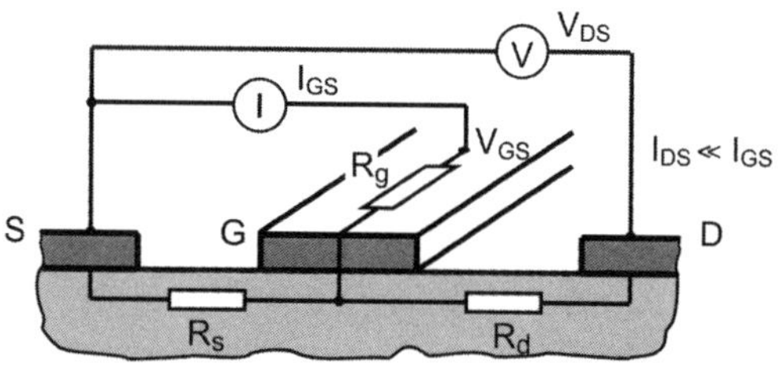

Figure B.2 Measurement of source resistance R_s. After [5].

The requirement that $I_{DS} \ll I_{GS}$ may be relaxed if the measurement is made under the condition of $\Delta I_{DS} = 0$. The drain resistance R_d is measured in a similar way. If R_s and $R_{GS} = R_g + R_s$ are known, then R_g is calculated as $R_g = R_{GS} - R_s$.

B.3 DC METHOD AFTER FUKUI

Fukui [2] proposed to measure the gate current of the FET device under analysis in forward conduction for different grounding combinations of source and drain electrodes (Figure B.3): (1) grounded source with open drain, (2) grounded drain with open source, and (3) both drain and source grounded. From the measured data the following relations can be derived:

$$R_a = R_g + R_s \tag{B.2}$$

$$R_b = R_g + R_d \tag{B.3}$$

$$R_c = R_g + \frac{R_s R_d}{R_s + R_d} \tag{B.4}$$

The series resistances are then calculated as

$$R_g = R_c - \sqrt{R_c^2 - R_c(R_a + R_b) + R_a R_b} \tag{B.5}$$

$$R_d = R_a - R_g \tag{B.6}$$

$$R_s = R_b - R_g \tag{B.7}$$

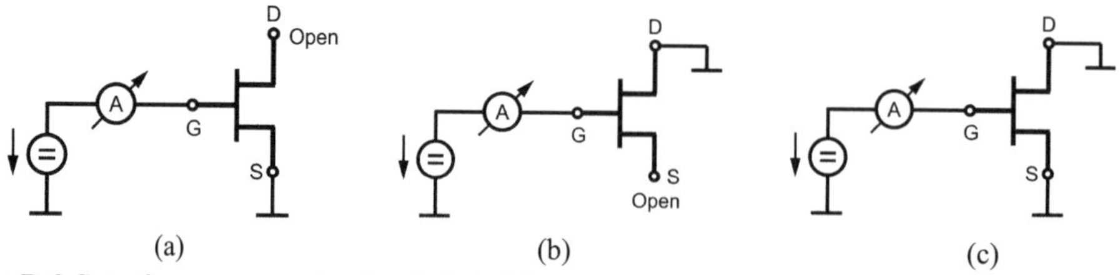

Figure B.3 Set of measurements after Fukui: (a) source grounded, drain open, (b)source open, drain grounded, and (c) source and drain grounded.

1. Determination of the Difference of the Series Resistances $R_s - R_d$

From (B.5) - (B.7) the three resistances can be calculated. However, these results would be inaccurate due to the additional resistive channel region under forward conduction. Thus, we should avoid relations that include the addition of measured resistance values. However, assuming same increase in value for R_s and R_d, the determined difference value will describe the situation under real operation (reverse operation of gate-source diode) sufficiently accurate. Thus, it follows from measurements (*a*) and (*b*) in Figure B.3

$$R_s - R_d = R_a - R_b \tag{B.8}$$

For the determination of R_s and R_d we need a further relation.

2. Determination of the Sum of the Series Resistances $R_s + R_d$

We consider the $I(V)$ characteristics in the ohmic region. The channel conductance G_{ds} in (4.42) reads (Shockley model with homogeneous doping, see Section 4.2)

$$G_{ds} = G_{d0}\left(1 - \sqrt{\frac{V_{ds} - V_{gs} + V_d}{V_{pi}}}\right) \tag{B.9}$$

where

$$G_{d0} = \frac{w\mu_0 q N_D a}{L} \tag{B.10}$$

In the linear region with $V_{ds} = 0.05$V; that is, $V_{ds} \ll (-V_{gs} + V_d)$ and $V_{ds} \ll V_{dss}$, (B.9) is approximated as

$$G_{ds} \approx G_{d0}\left(1 - \sqrt{\frac{-V_{gs} + V_d}{V_{pi}}}\right) \tag{B.11}$$

or with $V_{pa} = -V_{pi} + V_d$ in (4.18),

$$G_{ds}^{-1} \approx G_{d0}^{-1}\frac{1}{1 - \sqrt{\dfrac{-V_{gs} + V_d}{V_{pa} + V_d}}} \tag{B.12}$$

Defining

$$X = \frac{1}{1 - \sqrt{\dfrac{-V_{gs} + V_d}{V_{pa} + V_d}}} \tag{B.13}$$

it follows

$$G_{ds}^{-1} = G_{d0}^{-1} X(V_{gs}) \tag{B.14}$$

That means, G_{ds}^{-1} is a linear function of $X(V_{gs})$. For the external channel resistance G_{DS}^{-1} the series resistances R_s and R_d have to be taken into account,

$$G_{DS}^{-1}(V_{GS}) = R_s + R_d + G_{d0}^{-1} X(V_{gs}) \tag{B.15}$$

Note that indices with upper and lowercase characters indicate reference to the extrinsic and

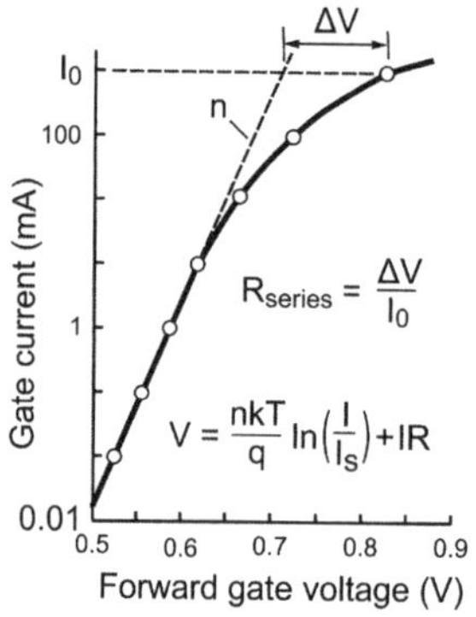

Figure B.4 Measured $I_G(V_G)$ characteristics.

intrinsic reference plane, respectively. However, as the drain current is relatively small for small drain-source voltages ($V_{GS} \leq 50$ mV) the voltage drop over R_S in negligible ($I_{DS} \cdot R_S \ll V_{GS}$) so that $V_{GS} \approx V_{gs}$. For the application of (B.15) the diffusion voltage V_d and the pinch-off voltage V_{pa} must be known.

Figure B.4 illustrates the forward $I_G(V_G)$ characteristic of a gate junction in a semi-logarithmic scale. It is seen that the log I_G versus V_G characteristic is linear at lower gate current values. However, with higher values of forward gate bias the gate current tends to saturate due to series resistance effect. At sufficiently high forward conduction of the gate junction, the resistance value is determined by $\Delta V/I_0$ as indicated in the figure.

The forward current density J of a Schottky barrier junction for $V > 3kT/q$ is approximately written as [6]

$$J = A^*T^2 \exp\left(-\frac{qV_d}{kT}\right) \exp\left(\frac{qV}{nkT}\right) = J_s \exp\left(\frac{qV}{nkT}\right) \tag{B.16}$$

with

$$J_s = A^*T^2 \exp\left(-\frac{qV_d}{kT}\right) \tag{B.17}$$

where A^* is the effective Richardson constant. T, k and n are the junction temperature in °K, the Boltzmann constant, and the ideality parameter, respectively. V_d is the diffusion or built-in voltage, and V is the forward bias voltage. The extrapolated value of current density to zero bias ($V = 0$) gives the saturation current density J_s. With the total gate width z and gate length L we get

$$J_s = \frac{I_0}{zL} \tag{B.18}$$

From (B.17) the diffusion voltage V_d is obtained as

$$V_d = \frac{kT}{q} \ln\left(\frac{A^*T^2}{J_s}\right) \tag{B.19}$$

After Fukui V_d in (B.19) can be approximated by

$$V_d \approx 0.768 - 0.06 \log(J_s) \tag{B.20}$$

with V_d in volts and J_s in 10^{-7} A/cm^2. For example, with $z = 260$ µm and $L = 0.5$ µm, the saturation current density reads $J_s = 189.23 \times 10^{-7}$ A/cm^2 and the diffusion voltage $V_d =$

0.631V.

The ideality parameter n describes the slope of the forward gate current characteristic. Taking the logarithm of (B.16), we get with (B.18)

$$\log(I) = \log(I_s) + \log(e)\frac{V}{nkT/q} = \log(I_s) + m \cdot V \tag{B.21}$$

with the slope m,

$$m = \frac{\log(I_2) - \log(I_1)}{V_2 - V_1} = \frac{\log(e)}{nkT/q} \tag{B.22}$$

The linear relationship does not hold for very small and very large diode voltages. The ideality factor n is given as

$$n = \frac{1}{m}\frac{\log(e)}{kT/q} \tag{B.23}$$

Figure B.5(a) illustrates a measured transfer function $I_{DS}(V_{GS})$. Linear extrapolation of the linear part of the transfer function intersects the x-axis in V_{pa}. Because V_{pa} can only be determined approximately, it must be optimized in such a way that a linear relationship between G_{DS}^{-1} and X is obtained as shown in Figure B.5(b). The extrapolation of the function for $X \to 0$ (intercept point with the y-axis) then delivers $G_{DS}^{-1}(X \to 0) = R_s + R_d$. It turns out that slight errors in the determination of V_{pa} have strong influence on the G_{DS}^{-1} curve progression. For example, an uncertainty of $\Delta V_{pa} = 10\ \%$ exhibits a substantial error of $\Delta(R_s + R_d) = 65\%$. In contrast, due to the weighting factor of 0.06 and log-function, uncertainty in J_s exhibits only minor error, e.g., $\Delta J_s = 100\ \% \to \Delta(R_s + R_d) = 2\%$. Thus, with (B.5), (B.8) and (B.15) all resistances R_g, R_s and R_d are known.

B.4 RF METHOD AFTER DAMBRINE ET AL.

The extraction method proposed by Dambrine et al. refers to the direct determination of both the extrinsic and the intrinsic small-signal elements of a 15-element FET model from measured S-parameters at rather low frequencies ($f < 5$ GHz) [4].

With the proposed method the authors wanted to eliminate the general problems known

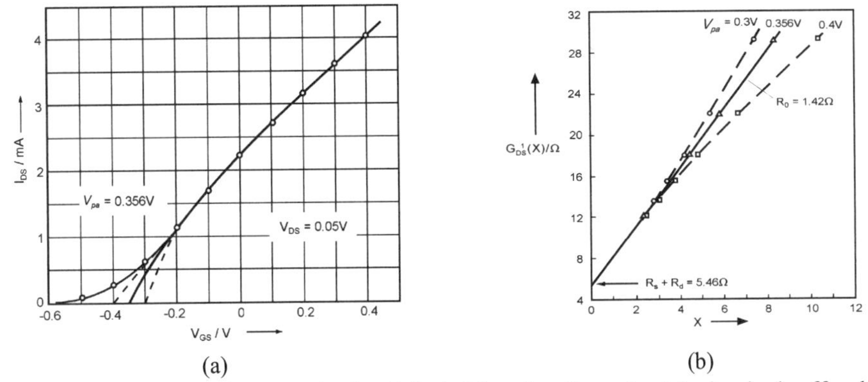

Figure B.5 Determination of $(R_s + R_d)$ after Fukui: (a) estimation of extrinsic pinch-off voltage V_{pa}, and (b) correction of V_{pa} by linearization of $G_{DS}^{-1} = f(R_s, R_d, X)$. (After [2].)

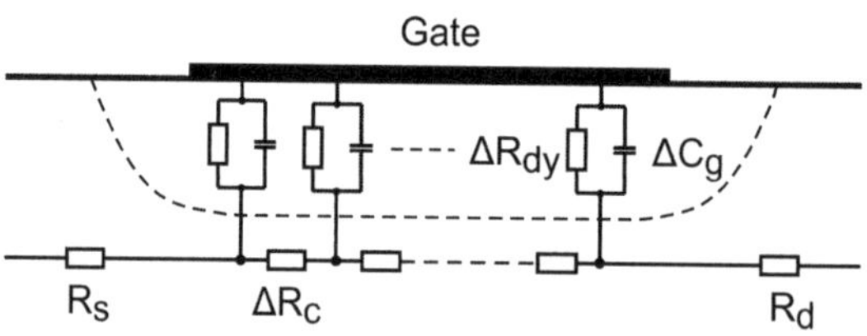

Figure B.6 Distributed RC-network as equivalent circuit of a field-effect transistor in gate-forward conduction ($V_{DS} = 0$).

from conventional extraction procedures, which involve an optimization process. In general, the local minima problem occurs, when the model element values of a 15-element model should be fitted to wideband-measured S-parameters using an optimization process. Commonly, as pointed out, the optimized element values are not unique, but depend on the used extraction algorithm, the starting values, and the truncation condition. Furthermore, it is stated that a physically relevant parameter extraction requires a preceding determination of certain parameters (e.g., gate-resistance or series inductances). In the following we will focus on the determination of the resistances as proposed by Dambrine et al.

All parasitic elements are determined from cold *S*-parameter measurements ($V_{DS} = 0$). In this case the equivalent circuit in Figure 8.1 simplifies and thus the determination of the model element values [2]. Figure B.6 shows the FET equivalent circuit for the gate-forward operation with $V_{DS} = 0$. For any gate voltage the following equations are valid [4]

$$Z_{11} = R_c/3 + Z_{dy} \tag{B.24}$$

$$Z_{12} = Z_{21} = R_c/2 \tag{B.25}$$

$$Z_{22} = R_c \tag{B.26}$$

where R_c denotes the channel resistance beneath the gate and Z_{dy} the impedance of the Schottky contact. For Z_{dy} it reads

$$Z_{dy} = \frac{R_{dy}}{1 + j\omega C_g R_{dy}} \tag{B.27}$$

with

$$R_{dy} = \frac{nkT}{qI_G} \tag{B.28}$$

where n and k are the ideality factor and Boltzmann constant, respectively. C_g is the distributed gate-channel capacitance and I_G the DC gate current.

With growing gate current, R_{dy} decreases exponentially with V_{GS}, but C_g increases only linearly. Thus, the term $\omega R_{dy} C_g$ becomes negligible for a sufficiently large current density of approximately $(5 \times 10^7 - 10^8)$ A/m^2. In this case it is

$$Z_{dy} \approx R_{dy} = \frac{nkT}{qI_G} \tag{B.29}$$

and Z_{11} in (B.24) becomes

$$Z_{11} = R_c/3 + \frac{nkT}{qI_G} \tag{B.30}$$

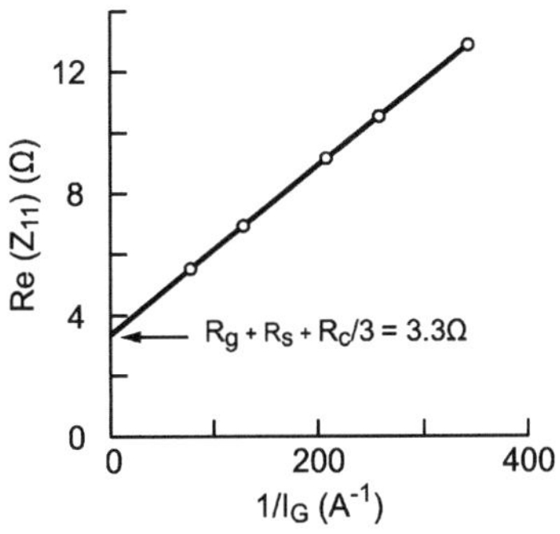

Figure B.7 Real part of the extrinsic Z_{11}-parameters as function of $1/I_G$.

The extrinsic parasitic capacitances C_{pg} and C_{pd} are neglected so that the extrinsic Z-parameters can be determined simply by addition of the parasitic resistances R_s, R_g, R_d and inductances L_s, L_g, L_d to the intrinsic Z-parameters,

$$Z_{11} = R_s + R_g + \frac{R_c}{3} + \frac{nkT}{qI_g} + j\omega(L_s + L_g) \tag{B.31}$$

$$Z_{12} = Z_{21} = R_s + \frac{R_c}{2} + j\omega L_s \tag{B.32}$$

$$Z_{22} = R_s + R_d + R_c + j\omega(L_s + L_d) \tag{B.33}$$

It is seen that the imaginary part of all extrinsic Z-parameters are dependent linearly on the frequency. Conversely, the real part is frequency-independent. However, the real part of Z_{11} increases with $1/I_G$.

Extrapolating the function $Z_{11} = f(1/I_g)$ linearly as indicated in Figure B.7, then the crossing point with the y-axis delivers the value $R_s + R_g + R_c/3$. Thus, the real parts of the Z-parameters deliver three equations for the four unknown resistances R_s, R_g, R_d, and R_c. A further fourth relation is needed. As pointed out by the authors this can be found in different ways: (1) the sum $(R_s + R_d)$ can be determined for example by additional measurements as proposed in [1] or [2], (2) R_g can be determined by a resistance measurement between two pads, (3) the values for R_s and R_d can be determined from DC measurement as described in [7], and (4) the value for R_c can be determined from technological data of the channel.

B.5 COMPARATIVE EXPERIMENTAL RESULTS

The following field-effect transistors were analyzed: MGFC1423/Mitsubishi, CFY19/Siemens, NE70083/NEC [8]. The extracted results are shown in Table B.1 and Table B.2. It can be clearly recognized that the resistance values based on the method after Fukui are essentially smaller in comparison with those obtained with Williams' approach. That is because in the latter case, part of the voltage controlled channel resistance is included in the measurements. The ratio of the resistances R_d/R_s is approximately the same for both methods.

Table B.1

Series Resistances R_s and R_d after the Method of Williams

Transistor	R_s (Ω)	R_d (Ω)	$(R_s + R_d)$ (Ω)	R_d/R_s
MGFC1423	2.6	3.1	5.7	1.19
CFY19	3.9	4.1	8.0	1.05
NE70083	2.6	2.6	5.2	1.00

Table B.2

Series Resistances R_s and R_d after the Method of Fukui

Transistor	R_s (Ω)	R_d (Ω)	$(R_s + R_d)$ (Ω)	$(R_s - R_d)$ (Ω)	R_d/R_s
MGFC1423	1.08	1.36	2.44	- 0.28	1.26
CFY19	2.51	2.54	5.04	- 0.03	1.01
NE70083	1.30	1.54	2.84	- 0.24	1.18

REFERENCES

[1] Hower, P. L., and N. G. Bechtel, "Current Saturation and Small-Signal Characteristics of GaAs Field-Effect Transistors," *IEEE Trans.*, Vol. ED-20(3), 1973, pp. 213–220.

[2] Fukui, H., "Determination of the Basic Device Parameters of a GaAs MESFET," *The Bell System Technical Journal*, Vol. 58, No. 3, 1979, pp. 771–797.

[3] Williams, R. E., *Gallium Arsenide Processing Techniques*, Dedham: Artech House, Inc., 1984.

[4] G. Dambrine, A. Cappy, F. Heliodore, and E. Playez, "A New Method for Determining the FET Small-Signal Equivalent Circuit," *IEEE Trans.*, Vol. MTT-36, pp. 1151-1159, July 1988.

[5] Williams, R. E., *Modern GaAs Processing Methods*, Norwood, MA: Artech House, Inc., 1990.

[6] Kompa, G., *Basic Properties of III-V Devices–Understanding Mysterious Trapping Phenomena*, Kassel: Kassel University Press GmbH, 2014.

[7] Yang, L., and S. Long, "New Method to Measure the Source and Drain Resistance of the GaAs MESFET," *IEEE Electron Device Lett.*, Vol. EDL-7, 1986, pp. 75–77.

[8] Bangert, A., *Realization of a Computer-Controlled Measurement Setup for the DC Characterization of Microwave Field-Effect Transistors and Development of Software for Setup Control and for the Extraction of Relevant DC Parameters* (in German), Project Work, University of Kassel, Dept. High-Frequency Engineering, 1988.

Appendix C

Parameter Extraction Relations for Inner FET Branch Topologies

In the following, all necessary relations for the quasi-analytical extraction of the intrinsic model elements are derived. This concerns the $R-L-C$ series connections of the FET model under cold pinch-off condition used for the generation of starting values of extrinsic elements and the saturation model comprising $R-C$ series and $R-C$ parallel branches as well as the voltage controlled current source. Furthermore, the extraction equations for the gate-source and gate-drain branches ($G \| C-R$) including Schottky diode conduction are presented.

C.1 $R-L-C$ SERIES CIRCUIT

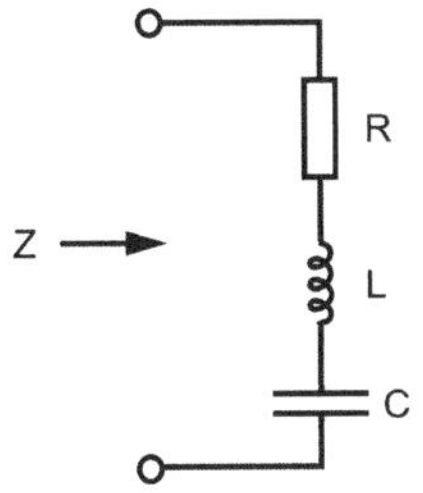

Figure C.1 $R-L-C$ series circuit.

Figure C.1 shows the circuit under analysis. The calculated response Z_k at a frequency point k reads

$$Z_k = R - j(\omega_k L - \frac{1}{\omega_k C}) \tag{C.1}$$

Denoting the measured response at the frequency point k by $Z_{k,m}$, the error ε_k between measured and calculated response can be written as

$$\varepsilon_k = w_k(Z_{k,m} - Z_k) = w_k[Z_{k,m} - R - j(\omega_k L - \frac{1}{\omega_k C})] \tag{C.2}$$

with the weighting factor w_k. Choosing $w_k = \omega_k C$ and decomposing the measured response $Z_{k,m}$ into real and imaginary part, $Z_{k,m} = Z'_{k,m} + jZ''_{k,m}$, (C.2) reads

$$\varepsilon_k = w_k C(Z'_{k,m} + jZ''_{k,m}) - w_k CR - j(\omega_k^2 LC - 1) \tag{C.3}$$

We introduce dimensionless normalized variables $z_k = Z_{k,m}/Z_0 = z'_k + jz''_k$ (e.g. $Z_0 = 50\Omega$) and $\tilde{f}_k = f_k/f_0$ (e.g., $f_0 = \max\{f_k\}$). Substituting the normalized impedance z_k and normalized frequency $\tilde{f}_k$ in (C.3) gives

$$\varepsilon_k = \frac{Z_{k,m}}{Z_0}\frac{f_k}{f_0}2\pi f_0 Z_0 C - \frac{f_k}{f_0}2\pi f_0 RC - j\left[(f_k/f_0)^2(2\pi f_0)^2 LC - 1\right] \tag{C.4}$$

With $x_1 = 2\pi f_0 Z_0 C$, $x_2 = 2\pi f_0 RC$, and $x_3 = (2\pi f_0)^2 LC$ we obtain

$$\varepsilon_k = \tilde{f}_k z_k' x_1 - \tilde{f}_k x_2 + j(\tilde{f}_k z_k'' x_1 - \tilde{f}_k^2 x_3 + 1) \tag{C.5}$$

The error function of l_2-norm reads

$$\|\varepsilon\| = \sum_k(\tilde{f}_k z_k' x_1 - \tilde{f}_k x_2)^2 + \sum_k(\tilde{f}_k z_k'' x_1 - \tilde{f}_k^2 x_3 + 1)^2 \tag{C.6}$$

Minimum value of the error function ($\|\varepsilon\| = \min$) is obtained by differentiation the error function with respect to the variables $\{x_i\}$ and put the expressions to zero,

$$\frac{\partial\|\varepsilon\|}{\partial x_i} = 0 \qquad (i = 1, 2, 3) \tag{C.7}$$

$\partial\|\varepsilon\|/\partial x_1$ yields

$$\sum_k \tilde{f}_k z_k'(\tilde{f}_k z_k' x_1 - \tilde{f}_k x_2) + \tilde{f}_k z_k''(\tilde{f}_k z_k'' x_1 - \tilde{f}_k^2 x_3 + 1) = 0 \tag{C.8}$$

and after sorting,

$$\sum_k \tilde{f}_k^2 |z_k|^2 x_1 - \sum_k \tilde{f}_k^2 z_k' x_2 - \sum_k \tilde{f}_k^3 z_k'' x_3 = -\sum_k \tilde{f}_k z_k' \tag{C.9}$$

$\partial\|\varepsilon\|/\partial x_2$ gives

$$-\sum_k \tilde{f}_k(\tilde{f}_k z_k' x_1 - \tilde{f}_k x_2) = 0 \tag{C.10}$$

and $\partial\|\varepsilon\|/\partial x_3$ results in

$$-\sum_k \tilde{f}_k^2(\tilde{f}_k z_k'' x_1 - \tilde{f}_k^2 x_3 + 1) = 0 \tag{C.11}$$

$$-\sum_k \tilde{f}_k^3 z_k'' x_1 + \sum_k \tilde{f}_k^4 x_3 = \sum_k \tilde{f}_k^2 \tag{C.12}$$

Equations (C.9), (C.10), and (C.12) can be summarized in form of the following linear (3 × 3)-matrix equation:

$$\begin{bmatrix} \sum_k \tilde{f}_k^2 |z_k|^2 & -\sum_k \tilde{f}_k^2 z_k' & -\sum_k \tilde{f}_k^3 z_k'' \\ -\sum_k \tilde{f}_k^2 z_k' & \sum_k \tilde{f}_k^2 & 0 \\ -\sum_k \tilde{f}_k^3 z_k'' & 0 & \sum_k \tilde{f}_k^4 \end{bmatrix} \begin{bmatrix} x_1 \\ x_2 \\ x_3 \end{bmatrix} = \begin{bmatrix} \sum_k \tilde{f}_k z_k'' \\ 0 \\ \sum_k \tilde{f}_k^2 \end{bmatrix} \tag{C.13}$$

From the solution for x_1, x_2 and x_3 the branch model elements read

$$R = Z_0\frac{x_2}{x_1}, \qquad L = \frac{Z_0}{2\pi f_m}\frac{x_3}{x_1}, \qquad C = \frac{1}{2\pi f_m Z_0}x_1 \tag{C14}$$

In the same, way derivative computations of further simple branch circuits can be performed.

C.2 R-C SERIES CIRCUIT

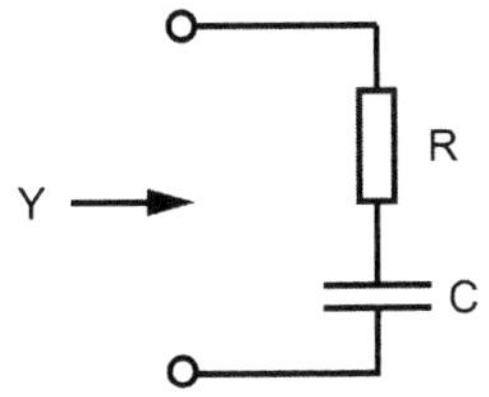

Figure C.2 R- C series circuit.

The calculated response Z_k at a frequency point k reads

$$Y = \frac{1}{R + 1/j\omega_k C} = \frac{j\omega_k C}{1 + j\omega_k RC} \tag{C.15}$$

$$\varepsilon_k = w_k(Y_{k,m} - Y_k) = w_k\left(Y_{k,m} - \frac{j\omega_k C}{1 + j\omega_k RC}\right) \tag{C.16}$$

Choosing $w_k = R + 1/(\omega_k C)$ the error function ε_k reads

$$\varepsilon_k = (R + 1/j\omega_k C)(Y'_{k,m} + jY''_{k,m}) - 1 \tag{C.17}$$

$$\varepsilon_k = [Y'_{k,m}R + Y''_{k,m}/(2\pi f_k C) - 1] + j[Y''_{k,m}R - Y'_{k,m}/(2\pi f_k C)] \tag{C.18}$$

With $x_1 = R$ and $x_2 = 1/(2\pi C)$, (C.18) reads

$$\varepsilon_k = (Y'_{k,m}x_1 + Y''_{k,m}/f_k x_2 - 1) + j(Y''_{k,m}x_1 - Y'_{k,m}/f_k x_2) \tag{C.19}$$

With (C.19) the l_2-norm is given as

$$\|\varepsilon\| = \sum_k (Y'_{k,m}x_1 + Y''_{k,m}/f_k x_2 - 1)^2 + (Y''_{k,m}x_1 - Y'_{k,m}/f_k x_2)^2 \tag{C.20}$$

$\|\varepsilon\| = \min$ is obtained by setting $\partial\|\varepsilon\|/\partial x = 0$ with $x = \{x_1, x_2\}$. $\partial\|\varepsilon\|/\partial x_1$ yields

$$\sum_k Y'_{k,m}(Y'_{k,m}x_1 + Y''_{k,m}/f_k x_2 - 1) + \sum_k Y''_{k,m}(Y''_{k,m}x_1 - Y'_{k,m}/f_k x_2) = 0 \tag{C.21}$$

which reduces to

$$x_1\sum_k |Y_{k,m}|^2 - \sum_k Y'_{k,m} = 0 \tag{C.22}$$

Similarly, $\partial\|\varepsilon\|/\partial x_2$ applied to (C.20) delivers

$$x_2\sum_k |Y_{k,m}|^2/f_k^2 - \sum_k Y''_{k,m}/f_k = 0 \tag{C.23}$$

From (C.22) and (C.23) the circuit elements $R = x_1$ and $C = 1/(2\pi x_2)$ are calculated as

$$R = \frac{\sum_k Y'_{k,m}}{\sum_k |Y_{k,m}|^2}, \qquad C = \frac{1}{2\pi}\frac{\sum_k (|Y_{k,m}|/f_k)^2}{\sum_k Y''_{k,m}/f_k} \tag{C24}$$

C.3 *G*-*C* PARALLEL CIRCUIT

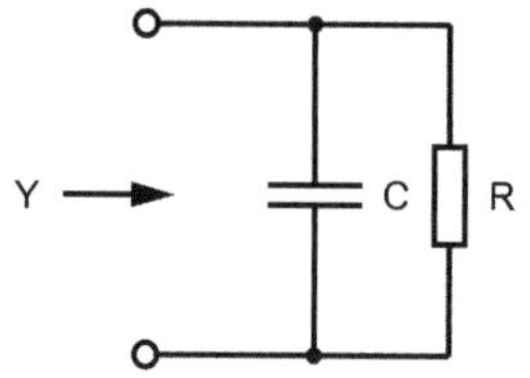

Figure C.3 *G*- *C* parallel circuit.

Referring to the preceded sections the error component ε_k can be written as

$$\varepsilon_k = w_k(Y_{k,m} - Y_k) = w_k(Y_{k,m} - G - j\omega_k C) \quad \text{(C.25)}$$

Accentuating lower frequencies, the weighting factor is chosen as $w_k = 1/\sqrt{f_k}$. With $x_1 = G$, $x_2 = 2\pi C$ and $Y_{k,m} = Y'_{k,m} + jY''_{k,m}$ the l_2-norm of (C.25) reads

$$\|\varepsilon\| = \sum_k (Y'_{k,m} - x_1)^2/f_k + \sum_k (Y''_{k,m} - f_k x_2)^2/f_k \quad \text{(C.26)}$$

Differentiation with respect to x_1 and, respectively, x_2 and putting to zero provide the branch elements G and C,

$$G = \frac{\sum_k Y'_k/f_k}{\sum_k 1/f_k}, \qquad C = \frac{1}{2\pi}\frac{\sum_k Y''_k}{\sum_k f_k} \quad \text{(C.27)}$$

C.4 R_2-*C* PARALLEL CIRCUIT CONNECTED TO SERIES R_1

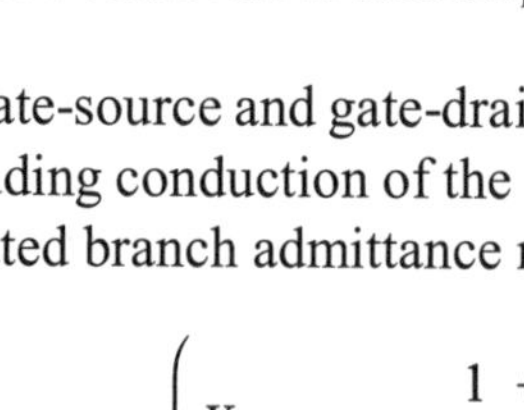

Figure C.4 R_2-*C* parallel circuit connected to series R_1.

Figure C.4 illustrates the gate-source and gate-drain branch topology of the intrinsic FET in the saturation region, including conduction of the Schottky diodes. The error function term ε_k of measured and calculated branch admittance reads

$$\varepsilon_k = w_k\left(Y_{k,m} - \frac{1 + j\omega_k CR_2}{R_1 + R_2 + j\omega_k CR_1R_2}\right) \quad \text{(C.28)}$$

With the introduction of weighting factor

$$w_k = R_1 + \frac{R_1 + R_2}{j\omega_k CR_2} \quad \text{(C.29)}$$

normalized branch admittance and frequency,

$$
\begin{aligned}
y_k &= Y_k Z_0 = y_k' + j y_k'', \qquad Z_0 = 50\Omega \\
\tilde{f}_k &= f_k / f_0, \qquad f_0 = \max_k \{f_k\}
\end{aligned}
\tag{C.30}
$$

and new variables $x_1 = R_1/Z_0$, $x_2 = (R_1 + R_2)/2\pi f_0 C R_2 Z_0$, $x_3 = 1/2\pi f_0 C R_2$, (C.28) gives

$$
\underline{\varepsilon_k} = y_k' x_1 + \frac{y_k''}{\tilde{f}_k} x_2 - 1 + j\left(y_k'' x_1 - \frac{y_k'}{\tilde{f}_k} x_2 + \frac{1}{\tilde{f}_k} x_3 \right)
\tag{C.31}
$$

The l_2-norm of (C.31) reads

$$
\|\varepsilon\| = \sum_k \left(y_k' x_1 + \frac{y_k''}{\tilde{f}_k} x_2 - 1 \right)^2 + \sum_k \left(y_k'' x_1 - \frac{y_k'}{\tilde{f}_k} x_2 + \frac{1}{\tilde{f}_k} x_3 \right)^2
\tag{C.32}
$$

Calculation of $\partial\|\varepsilon\|/\partial x_i = 0$ with $i = 1, 2, 3$ results in the following matrix equation for the variables x_i, $i = 1, 2, 3$:

$$
\begin{bmatrix}
\sum_k |y_k|^2 & 0 & \sum_k y_k''/\tilde{f}_k \\
0 & \sum_k |y_k|^2/\tilde{f}_k^2 & -\sum_k y_k'/\tilde{f}_k^2 \\
\sum_k y_k''/\tilde{f}_k & -\sum_k y_k'/\tilde{f}_k^2 & \sum_k 1/\tilde{f}_k^2
\end{bmatrix}
\begin{bmatrix} x_1 \\ x_2 \\ x_3 \end{bmatrix}
=
\begin{bmatrix}
\sum_k y_k' \\
\sum_k y_k''/\tilde{f}_k \\
0
\end{bmatrix}
\tag{C.33}
$$

Then the branch elements can be derived as $R_1 = x_1 Z_0$, $R_2 = x_2/x_3 Z_0 - R_1$, $C = 1/[2\pi f_0(x_2 Z_0 - x_3 R_1)]$.

C.5 VOLTAGE-CONTROLLED CURRENT SOURCE

The admittance Y is calculated as $Y = I_m/V_{gs} = G_m e^{-j\omega\tau} V_g/V_{gs}$. With $V_g/V_{gs} = (j\omega C_{gs})^{-1}[R_i + 1/(j\omega C_{gs})]^{-1}$ the relationship of the model elements G_m and τ to the intrinsic branch admittance Y_{gm} reads

$$
G_m e^{-j\omega_k \tau} = Y_{gm,k}(1 + j\omega_k \tau_m)
\tag{C.34}
$$

with $\tau_m = C_{gs} R_i$. The error term ε_k of measured and calculated branch admittance is

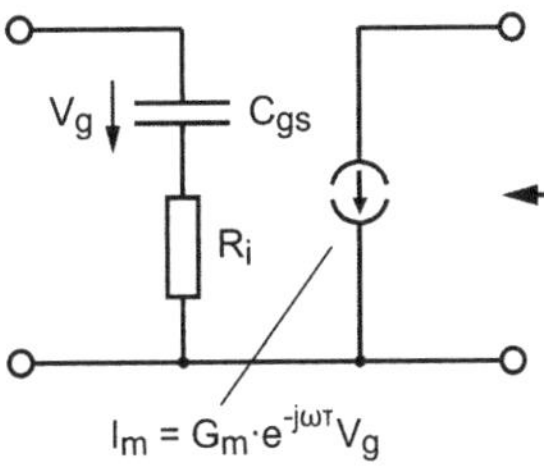

Figure C.5 Voltage controlled current source.

$$\varepsilon_k = w_k(Y_{k,m} - Y_k) = w_k\left(Y_{k,m} - \frac{G_m e^{-j\omega_k\tau}}{1 + j\omega_k\tau_m}\right) \tag{C.35}$$

Selecting the weighting factor as $w_k = (1 + j\omega\tau_m)/\sqrt{f_k}$, (C.35) yields

$$\varepsilon_k = \frac{1}{\sqrt{f_k}}(1 + j\omega_k\tau_m)Y_{k,m} - \frac{1}{\sqrt{f_k}}G_m e^{-j\omega_k\tau} \tag{C.36}$$

Expressing the complex terms in (C.36) by magnitude and phase gives

$$\varepsilon_k = \frac{|Y_{k,m}(1 + j\omega_k\tau_m)|}{\sqrt{f_k}}\measuredangle[Y_{k,m}(1 + j\omega_k\tau_m)] - \frac{1}{\sqrt{f_k}}G_m e^{-j\omega_k\tau} \tag{C.37}$$

With the magnitude error ε_{k1},

$$\varepsilon_{k1} = \frac{1}{\sqrt{f_k}}(G_m - |Y_{k,m}(1 + j\omega_k\tau_m)|) \tag{C.38}$$

and phase error ε_{k2},

$$\varepsilon_{k2} = \frac{1}{\sqrt{f_k}}\{\omega_k\tau_m + \measuredangle[Y_{k,m}(1 + j\omega_k\tau_m)]\} \tag{C.39}$$

the l_2-norm of (C.38) and, respectively, (C.39) read

$$\begin{aligned} \|\varepsilon_{k1}\| &= \sum_k \frac{1}{\sqrt{f_k}}(G_m - |Y_{k,m}(1 + j\omega_k\tau_m)|)^2 \\ \|\varepsilon_{k2}\| &= \frac{1}{\sqrt{f_k}}\{\omega_k\tau_m + \measuredangle[Y_{k,m}(1 + j\omega_k\tau_m)]\}^2 \end{aligned} \tag{C.40}$$

From $\partial\|\varepsilon_1\|/\partial G_m = 0$ and $\partial\|\varepsilon_2\|/\partial\tau = 0$ the elements of the current source are calculated as

$$G_m = \frac{\sum_k |Y_{gm,k}(1 + j\omega_k\tau_m)|/f_k}{\sum_k 1/f_k}, \qquad \tau = \frac{1}{2\pi}\frac{-\sum_k \measuredangle[Y_{gm,k}(1 + j\omega_k\tau_m)]}{\sum_k f_k} \tag{C.41}$$

Appendix D

Embedding the Intrinsic Model into an Extrinsic Network

D.1 EMBEDDING INTO AN IMPEDANCE NETWORK

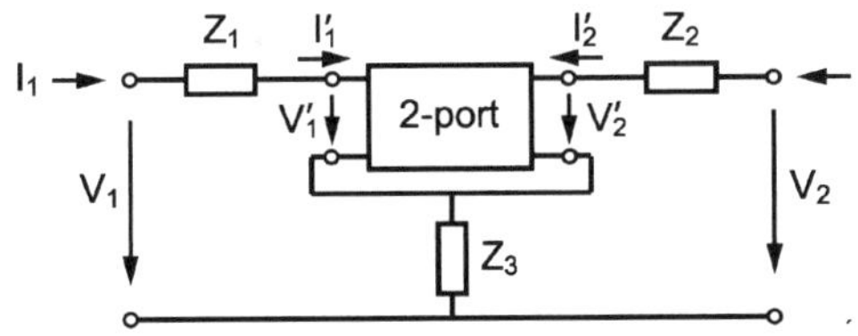

Figure D.1 Two-port embedded in an impedance network.

The extended constitutive matrix equation (17.8) of the two-port reads

$$[A']\boldsymbol{I}' = [B']\boldsymbol{V}' \tag{D.1}$$

Extended equation of embedded two-port in Figure D.1,

$$[A]\boldsymbol{I} = [B]\boldsymbol{V} \tag{D.2}$$

Voltage and current interconnection relations,

$$\boldsymbol{I} = \boldsymbol{I}' \tag{D.3}$$

$$\boldsymbol{V} = \boldsymbol{V}' + [Z]\boldsymbol{I} \tag{D.4}$$

Then, substituting $\boldsymbol{V}$ and $\boldsymbol{I}$ in (D.2),

$$[A]\boldsymbol{I}' = [B](\boldsymbol{V}' + [Z]\boldsymbol{I}') \tag{D.5}$$

With $[B] = [B']$ (D.5) yields

$$[A] = [A'] + [B'][Z] \tag{D.6}$$

and

$$[Z] = \begin{bmatrix} Z_1 + Z_3 & Z_3 \\ Z_3 & Z_2 + Z_3 \end{bmatrix} \tag{D.7}$$

D.2 EMBEDDING INTO AN ADMITTANCE NETWORK

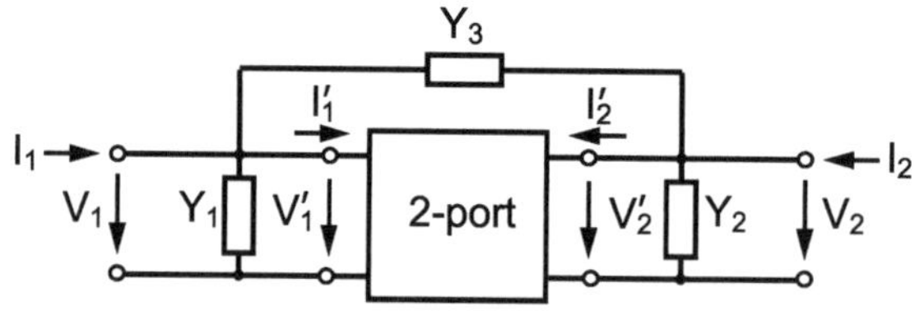

Figure D.2 Two-port embedded in an admittance network.

Voltage and current interconnection relations read now

$$\boldsymbol{V} = \boldsymbol{V}' \tag{D.8}$$

$$\boldsymbol{I} = \boldsymbol{I}' + [Y]\boldsymbol{V} \tag{D.9}$$

Then, substituting $\boldsymbol{V}$ and $\boldsymbol{I}$ in (D.2) gives

$$[A](\boldsymbol{I}' + [Y]\boldsymbol{V}') = [B]\boldsymbol{V}' \tag{D.10}$$

With

$$[A] = [A'] \tag{D.11}$$

it follows from (D.10)

$$[B] = [B'] + [A'][Y] \tag{D.12}$$

with

$$[Y] = \begin{bmatrix} Y_1 + Y_3 & Y_3 \\ Y_3 & Y_2 + Y_3 \end{bmatrix} \tag{D.13}$$

Appendix E

Derivation of Riccati Equation

In this appendix it is shown that Equation (21.99), Chapter 21, is equivalent to the relation (21.100). We introduce the following two definitions:

$$Z_1 \equiv A_1 + jB_1 = (Y_0 - jP)^{-1} \tag{E.1}$$

$$Y_2 \equiv A_2 + jB_2 = (Z_1 - j\Lambda)^{-1} = [A_1 + j(B_1 - \Lambda)]^{-1} \tag{E.2}$$

As explained in [1], any complex-valued matrix Z, which is defined in terms of two real-valued matrices A and B according to $Z \equiv A + jB$, has the real and imaginary parts of its inverse, Z^{-1}, expressed as

$$Re\{Z^{-1}\} = (Re\{Z\} + Im\{Z\}Re\{Z^{-1}\}Im\{Z\})^{-1} \tag{E.3}$$

$$Im\{Z^{-1}\} = -Re\{Z\}^{-1}Im\{Z\}Re\{Z^{-1}\} \tag{E.4}$$

Alternatively, (E.3) and (E.4) can be written as

$$M_Z = -Re\{Z^{-1}\}Im\{Z\} \tag{E.5}$$

$$Re\{Z^{-1}\} = [Re\{Z\}(I + M_Z^2)]^{-1} \tag{E.6}$$

$$Im\{Z^{-1}\} = M_Z Re\{Z^{-1}\} \tag{E.7}$$

Employing (E.5) to (E.7) in (E.2) results in

$$M_2 = -A_1^{-1}(B_1 - \Lambda) \tag{E.8}$$

$$A_2^{-1} = A_1(I + M_2^2) \tag{E.9}$$

$$B_2 = M_2 A_2 \tag{E.10}$$

Now combining (21.99) with (E.2) gives

$$Im\{[A_2 + j(B_2 - F)]^{-1}\} = E \tag{E.11}$$

Applying (E.3) and (E.4) in (E.11) leads to

$$-A_2^{-1}(B_2 - F)][A_2 + (B_2 - F)A_2^{-1}(B_2 - F)]^{-1} = E \tag{E.12}$$

Rearranging (E.12) into a homogeneous equation and then performing a right-hand side multiplication with the inverse of $-A_2$ results in

$$E^{-1}A_2^{-1}(B_2 - F)A_2^{-1} + I + (B_2A_2^{-1} - FA_2^{-1})^2 = 0 \tag{E.13}$$

The combination of (E.13) with (E.9) and (E.10) produces

$$E^{-1}A_1(I + M_2^2)[M_2 - FA_1(I + M_2^2)] + I + [M_2 - FA_1(I + M_2^2)]^2 = 0 \tag{E.14}$$

Performing the operations implied by the square brackets, applying a left-hand side multiplication by $(I + M_2^2)$, and factorizing, results in

$$[M_2 + A_1^{-1}(E^{-1} - F)^{-1}][M_2 + A_1^{-1}F^{-1}] + I = 0 \tag{E.15}$$

Substituting M_2 from (E.8) in (E.15) produces

$$A_1^{-1}[\Lambda - B_1 + (E^{-1} - F)^{-1}]A_1^{-1}(\Lambda - B_1 - F^{-1}) + I = 0 \tag{E.16}$$

With the following two definitions

$$Q \equiv -(E^{-1} - F)^{-1} - F^{-1} \tag{E.17}$$

$$X \equiv \Lambda - B_1 - F^{-1} \tag{E.18}$$

(E.16) transforms to

$$A_1^{-1}(X - Q)A_1^{-1}X + I = 0 \tag{E.19}$$

Finally, rearrangement of (E.19) delivers

$$X + \Lambda X^{-1}A_1^T = Q \tag{E.20}$$

Note that the transposition of the second A_1 factor in (E.20) is only a question of form, without actual effect, since A_1 is symmetric by definition.

Thus, it has been proven that (21.99) and (21.100) are actually equivalent.

REFERENCE

[1] Smith, W. W., and S. Erdman, "A Note on the Inversion of Complex Matrices," *IEEE Trans. on Automatic Control*, Vol. 19(1), 1974, p. 64.

Appendix F

General N-Port Network and Two-Port Admittance Matrix

With $\boldsymbol{I} = (I_1, I_2, ..., I_N)^T$, $\boldsymbol{V} = (V_1, V_2, ..., V_N)^T$ and $\boldsymbol{h} = (h^{(0)}, h^{(1)}, ..., h^{(K)})^T$ (23.2) can be written in matrix form as follows:

$$\boldsymbol{I}(\boldsymbol{V}(t)) = \boldsymbol{\Phi}(\boldsymbol{V}(t))\,\boldsymbol{h} \tag{F.1}$$

with

$$f\boldsymbol{\Phi} = \begin{pmatrix} \boldsymbol{\Phi}_1 \\ \boldsymbol{\Phi}_2 \\ \vdots \\ \boldsymbol{\Phi}_N \end{pmatrix} = \begin{pmatrix} \Phi_1^{(0)} & \Phi_1^{(1)} & . \; . \; . & \Phi_1^{(K)} \\ \Phi_2^{(0)} & \Phi_2^{(1)} & . \; . \; . & . \\ \vdots & \vdots & \vdots \; \vdots \; \vdots & \vdots \\ \Phi_N^{(0)} & . & . \; . \; . & \Phi_N^{(K)} \end{pmatrix} = \begin{pmatrix} \boldsymbol{\Phi}^{(0)} & \boldsymbol{\Phi}^{(1)} & ... & \boldsymbol{\Phi}^{(K)} \end{pmatrix} \tag{F.2}$$

Equation (12.26) reads in compressed matrix description:

$$d\boldsymbol{I}(\boldsymbol{V}(t)) = d(\boldsymbol{\Phi}(\boldsymbol{V}(t))\boldsymbol{h}) = \left.\frac{\partial \boldsymbol{\Phi}}{\partial \boldsymbol{V}}\right|_{\boldsymbol{V}_0} d(\boldsymbol{V}_h(t)) = \boldsymbol{\varphi}(\boldsymbol{V}_0)d(\boldsymbol{V}_h(t)) \tag{F.3}$$

where

$$\boldsymbol{\varphi} = \begin{pmatrix} \boldsymbol{\varphi}_1 \\ \boldsymbol{\varphi}_2 \\ \vdots \\ \boldsymbol{\varphi}_N \end{pmatrix} = \begin{pmatrix} \boldsymbol{\varphi}_1^{(0)} & \boldsymbol{\varphi}_1^{(1)} & . \; . \; . & \boldsymbol{\varphi}_1^{(K)} \\ \boldsymbol{\varphi}_2^{(0)} & \boldsymbol{\varphi}_2^{(1)} & . \; . \; . & \boldsymbol{\varphi}_2^{(K)} \\ \vdots & \vdots & \vdots \; \vdots \; \vdots & \vdots \\ \boldsymbol{\varphi}_N^{(0)} & . & . \; . \; . & \boldsymbol{\varphi}_N^{(K)} \end{pmatrix} = \begin{pmatrix} \boldsymbol{\varphi}^{(0)} & \boldsymbol{\varphi}^{(1)} & ... & \boldsymbol{\varphi}^{(K)} \end{pmatrix}$$

$$= \begin{pmatrix} \varphi_{1,1}^{(0)} & \varphi_{1,2}^{(0)} & \cdots & \varphi_{1,N}^{(0)} & \varphi_{1,1}^{(1)} & \varphi_{1,2}^{(1)} & \cdots & \varphi_{1,N}^{(1)} & \cdots & \cdots & \varphi_{1,1}^{(K)} & \varphi_{1,2}^{(K)} & \cdots & \varphi_{1,N}^{(K)} \\ \varphi_{2,1}^{(0)} & \varphi_{2,2}^{(0)} & \cdots & \varphi_{2,N}^{(0)} & . & . & \cdots & . & \cdots & \cdots & . & . & \cdots & . \\ \vdots & \vdots & \vdots\,\vdots\,\vdots & \vdots & \vdots & \vdots & \vdots\,\vdots\,\vdots & \vdots & \vdots\,\vdots\,\vdots & \vdots\,\vdots\,\vdots & \vdots & \vdots & \vdots\,\vdots\,\vdots & \vdots \\ \varphi_{N,1}^{(0)} & \varphi_{N,2}^{(0)} & \cdots & \varphi_{N,N}^{(0)} & . & . & \cdots & . & \cdots & \cdots & \varphi_{N,1}^{(K)} & \varphi_{N,2}^{(K)} & \cdots & \varphi_{N,N}^{(K)} \end{pmatrix} \tag{F.4}$$

or expressed by the partial derivatives of $\boldsymbol{\Phi}$:

$$\boldsymbol{\varphi} = \frac{\partial \boldsymbol{\Phi}}{\partial \boldsymbol{V}}$$

$$= \begin{pmatrix} \frac{\partial \Phi_1^{(0)}}{\partial V_1} & \frac{\partial \Phi_1^{(0)}}{\partial V_2} & \cdots & \frac{\partial \Phi_1^{(0)}}{\partial V_N} & \frac{\partial \Phi_1^{(1)}}{\partial V_1} & \frac{\partial \Phi_1^{(1)}}{\partial V_2} & \cdots & \frac{\partial \Phi_1^{(1)}}{\partial V_N} & \cdots & \cdots & \frac{\partial \Phi_1^{(K)}}{\partial V_1} & \frac{\partial \Phi_1^{(K)}}{\partial V_2} & \cdots & \frac{\partial \Phi_1^{(K)}}{\partial V_K} \\ \frac{\partial \Phi_2^{(0)}}{\partial V_1} & \frac{\partial \Phi_2^{(0)}}{\partial V_2} & \cdots & \frac{\partial \Phi_2^{(0)}}{\partial V_K} & \cdots\cdot & & & & & & & & & \\ \vdots & \vdots & & \vdots & \vdots & & & & & & & & & \\ \frac{\partial \Phi_N^{(0)}}{\partial V_1} & \frac{\partial \Phi_N^{(0)}}{\partial V_2} & \cdots & \frac{\partial \Phi_N^{(0)}}{\partial V_K} & \cdots & & & & & \cdots\cdot & \frac{\partial \Phi_N^{(K)}}{\partial V_1} & \frac{\partial \Phi_N^{(K)}}{\partial V_2} & \cdots & \frac{\partial \Phi_N^{(K)}}{\partial V_K} \end{pmatrix} \tag{F.5}$$

and the time-differential voltage vector V_h,

$$V_h = \begin{pmatrix} V_h^{(0)} \\ V_h^{(1)} \\ \vdots \\ V_h^{(K)} \end{pmatrix} = \begin{pmatrix} h^{(0)}V \\ h^{(1)}V \\ \vdots \\ h^{(K)}V \end{pmatrix} = \begin{pmatrix} h^{(0)}V_1 \\ \vdots \\ h^{(0)}V_N \\ h^{(1)}V_1 \\ \vdots \\ h^{(1)}V_N \\ \vdots \\ \vdots \\ h^{(K)}V_1 \\ \vdots \\ h^{(K)}V_N \end{pmatrix} \tag{F.6}$$

With the given schematic matrix representation of a generalized N-port the expressions for a two-port can simply be found as follows.The port currents $I_i(t)$ $(i = 1, 2)$ are derived from (F.1):

$$I_i(V_1,V_2) = \sum_{k=0}^{K} h^{(k)}\Phi_i^{(k)}(V_1,V_2) = h^{(0)}\Phi_i^{(0)}(V_1,V_2) + h^{(1)}\Phi_i^{(1)}(V_1,V_2) + h^{(2)}\Phi_i^{(2)}(V_1,V_2) + \ldots \qquad (i=1,2) \tag{F.7}$$

The incremental currents $dI_i(t)$ $(i = 1, 2)$ follow from (F.3) with the index notation of (F.4–F.6):

$$\begin{aligned} dI_i(V_{10},V_{20}) &= \sum_{k=0}^{K} \varphi_i^{(k)}(V_{10},V_{20})\,d(V_h^{(k)}) \\ &= \sum_{k=0}^{K} \varphi_{i,1}^{(k)}(V_{10},V_{20})\,d(h^{(k)}V_1) + \varphi_{i,2}^{(k)}(V_{10},V_{20})\,d(h^{(k)}V_2) \\ &= \sum_{k=0}^{K}\sum_{j=1}^{2} \varphi_{i,j}^{(k)}(V_{10},V_{20})\,d(h^{(k)}V_j) \\ &= \varphi_{i,1}^{(0)}(V_{10},V_{20})\,d(h^{(0)}V_1) + \varphi_{i,2}^{(0)}(V_{10},V_{20})\,d(h^{(0)}V_2) \\ &\quad + \varphi_{i,1}^{(1)}(V_{10},V_{20})\,d(h^{(1)}V_1) + \varphi_{i,2}^{(1)}(V_{10},V_{20})\,d(h^{(1)}V_2) \\ &\quad + \varphi_{i,1}^{(2)}(V_{10},V_{20})\,d(h^{(2)}V_1) + \varphi_{i,2}^{(2)}(V_{10},V_{20})\,d(h^{(2)}V_2) + \ldots \qquad (i=1,2) \end{aligned} \tag{F.8}$$

and analogous to (23.4) in the frequency domain

$$\begin{aligned} dI_i(\omega,V_{10},V_{20}) &= \sum_{k=0}^{K}\sum_{j=1}^{2} (j\omega)^k\,\varphi_{i,j}^{(k)}(V_{10},V_{20})\,dV_j(\omega) \\ &= (j\omega)^{(0)}\varphi_{i,1}^{(0)}(V_{10},V_{20})\,dV_1(\omega) + (j\omega)^{(0)}\varphi_{i,2}^{(0)}(V_{10},V_{20})\,dV_2(\omega) \\ &\quad + (j\omega)^{(1)}\varphi_{i,1}^{(1)}(V_{10},V_{20})\,dV_1(\omega) + (j\omega)^{(1)}\varphi_{i,2}^{(1)}(V_{10},V_{20})\,dV_2(\omega) \\ &\quad + (j\omega)^{(2)}\varphi_{i,1}^{(2)}(V_{10},V_{20})\,dV_1(\omega) + (j\omega)^{(2)}\varphi_{i,2}^{(2)}(V_{10},V_{20})\,dV_2(\omega) + \ldots \qquad (i=1,2) \end{aligned} \tag{F.9}$$

and after rearranging,

$$\begin{aligned} dI_i(\omega,V_{10},V_{20}) &= \left((j\omega)^{(0)}\varphi_{i,1}^{(0)} + (j\omega)^{(1)}\varphi_{i,1}^{(1)} + (j\omega)^{(2)}\varphi_{i,1}^{(2)} + \dots\right)dV_1(\omega) \\ &+ \left((j\omega)^{(0)}\varphi_{i,2}^{(0)} + (j\omega)^{(1)}\varphi_{i,2}^{(1)} + (j\omega)^{(2)}\varphi_{i,2}^{(2)} + \dots\right)dV_2(\omega) \end{aligned} \tag{F.10}$$

By comparison with the two-port admittance matrix the admittance coefficients Y_{ij} are as follows:

$$Y_{ij}(\omega,V_{10},V_{20}) = \sum_{k=0}^{K}(j\omega)^{(k)}\varphi_{i,j}^{(k)}(\omega,V_{10},V_{20}) \qquad (i,j=1,2) \tag{F.11}$$

About the Author

Günter Kompa received Dipl.-Ing. and Dr.-Ing. degrees in electrical communication engineering from the Technical University of Aachen (R.W.T.H. Aachen), Aachen, Germany, in 1970 and 1975, respectively.

In 1976, he joined Endress & Hauser, Germany, a renowned company in measurement and control engineering, focusing on microwave and laser radar research and development. In 1980, he became the head of the department of research and development of sensor systems for level control.

For his publication "Design of Stepped Microstrip Components" (*The Radio and Electronic Engineer*, Vol. 48, No. 1/2, January/February 1978, pp. 53–63), he was awarded the Heinrich-Hertz Award by the Institute of Radio and Electronic Engineers, London.

Since 1984, he has been a professor heading the Department of Microwave Engineering at the University of Kassel, Germany.

From 1990 to 2002, Dr. Kompa was a referee for the German Federal Ministry of Research and Technology (BMFT) (since 1994 BMBF, Federal Ministry of Education and Research). After that, he participated with his research group in the MobileGaN joint project funded by the BMBF. He also participated from 2004–2007 in the European FP6 Network of Excellence called TARGET (Top Amplifier Research Groups in a European Team).

From 1997 to 2001, he was a referee of the German Academic Exchange Service (DAAD).

In 1997, after a highly competitive selection process among German universities, he received a state subsidy to establish the international masters course "Electrical Communication Engineering (ECE)" at the University of Kassel, which is among the first eight established international postgraduate degree courses in Germany funded by the Federal Government.

In 1990, Dr. Kompa organized the International IEEE Workshop on Measurement Techniques for Microwave Device Characterization and Modeling in Stuttgart, Germany, and in 1997 the International Workshop on Experimentally Based FET Device Modeling and Related Nonlinear Circuit Design at the University of Kassel, Germany, supported by the DFG (German Research Foundation), the European Research Office (ERO, London), the European Office of Aerospace Research and Development (EOARD, London), and the Office of Naval Research-Europe (ONR-EUR, London).

Dr. Kompa was invited to the inaugural address of the International Conference on Modern Design, Manufacturing and Measurement at the Tsinghua University, Beijing, China, in 1993.

He was coorganizer of the DFG doctorate program Devices and Materials for Microsystem Engineering at the University of Kassel for six years, starting in 1995.

Dr. Kompa is a member of the professional societies of IEEE and VDE. He was on the honorary international advisory board for the Asia-Pacific Microwave Conference at Singapore in 1999.

His research activities have covered circuit and design of ultrawideband (UWB) microwave and laser radars, and multitone characterization and modeling of high-power devices and amplifiers.

He is the author and coauthor of more than 150 technical papers in the fields of microstrip technique, laser radar technology, microwave and millimeter-wave measurements, and circuit design at microwave and millimeter-wave frequencies for communications and sensor applications. He has filed numerous patents in Europe, the United States, and Japan.

In 2005, Dr. Kompa published with Artech House a book about microstrip technique titled *Practical Microstrip Design and Applications*. In 2014, he published with Kassel University Press GmbH, Kassel, a book titled *Basic Properties of III-V Devices—Understanding Mysterious Trapping Phenomena*.

He has offered international lectures on high-frequency technology, with emphasis on microwaves and millimeter waves, design of linear and nonlinear microstrip circuits, wave propagation on waveguiding structures, and the high-frequency sensor technique.

Although Dr. Kompa retired in 2008, he is still involved in research activities alongside his book writing.

Index

Artech House Microwave Library

Behavioral Modeling and Linearization of RF Power Amplifiers, John Wood

Chipless RFID Reader Architecture, Nemai Chandra Karmakar, Prasanna Kalansuriya, Randika Koswatta, and Rubayet E-Azim

Control Components Using Si, GaAs, and GaN Technologies, Inder J. Bahl

Design of Linear RF Outphasing Power Amplifiers, Xuejun Zhang, Lawrence E. Larson, and Peter M. Asbeck

Design Methodology for RF CMOS Phase Locked Loops, Carlos Quemada, Guillermo Bistué, and Iñigo Adin

Design of CMOS Operational Amplifiers, Rasoul Dehghani

Design of RF and Microwave Amplifiers and Oscillators, Second Edition, Pieter L. D. Abrie

Digital Filter Design Solutions, Jolyon M. De Freitas

Discrete Oscillator Design Linear, Nonlinear, Transient, and Noise Domains, Randall W. Rhea

Distortion in RF Power Amplifiers, Joel Vuolevi and Timo Rahkonen

Distributed Power Amplifiers for RF and Microwave Communications, Narendra Kumar and Andrei Grebennikov

Electric Circuits: A Primer, J. C. Olivier

Electronics for Microwave Backhaul, Vittorio Camarchia, Roberto Quaglia, and Marco Pirola, editors

EMPLAN: Electromagnetic Analysis of Printed Structures in Planarly Layered Media, Software and User's Manual, Noyan Kinayman and M. I. Aksun

An Engineer's Guide to Automated Testing of High-Speed Interfaces, Second Edition, José Moreira and Hubert Werkmann

Envelope Tracking Power Amplifiers for Wireless Communications, Zhancang Wang

Essentials of RF and Microwave Grounding, Eric Holzman

Frequency Measurement Technology, Ignacio Llamas-Garro, Marcos Tavares de Melo, and Jung-Mu Kim

FAST: Fast Amplifier Synthesis Tool—Software and User's Guide, Dale D. Henkes

Feedforward Linear Power Amplifiers, Nick Pothecary

Filter Synthesis Using Genesys S/Filter, Randall W. Rhea

Foundations of Oscillator Circuit Design, Guillermo Gonzalez

Frequency Synthesizers: Concept to Product, Alexander Chenakin

Fundamentals of Nonlinear Behavioral Modeling for RF and Microwave Design, John Wood and David E. Root, editors

Generalized Filter Design by Computer Optimization, Djuradj Budimir

Handbook of Dielectric and Thermal Properties of Materials at Microwave Frequencies, Vyacheslav V. Komarov

Handbook of RF, Microwave, and Millimeter-Wave Components, Leonid A. Belov, Sergey M. Smolskiy, and Victor N. Kochemasov

High-Efficiency Load Modulation Power Amplifiers for Wireless Communications, Zhancang Wang

High-Linearity RF Amplifier Design, Peter B. Kenington

High-Speed Circuit Board Signal Integrity, Second Edition, Stephen C. Thierauf

Integrated Microwave Front-Ends with Avionics Applications, Leo G. Maloratsky

Intermodulation Distortion in Microwave and Wireless Circuits, José Carlos Pedro and Nuno Borges Carvalho

Introduction to Modeling HBTs, Matthias Rudolph

An Introduction to Packet Microwave Systems and Technologies, Paolo Volpato

Introduction to RF Design Using EM Simulators, Hiroaki Kogure, Yoshie Kogure, and James C. Rautio

Introduction to RF and Microwave Passive Components, Richard Wallace and Krister Andreasson

Klystrons, Traveling Wave Tubes, Magnetrons, Crossed-Field Amplifiers, and Gyrotrons, A. S. Gilmour, Jr.

Lumped Elements for RF and Microwave Circuits, Inder Bahl

Lumped Element Quadrature Hybrids, David Andrews

Microstrip Lines and Slotlines, Third Edition, Ramesh Garg, Inder Bahl, and Maurizio Bozzi

Microwave Circuit Modeling Using Electromagnetic Field Simulation, Daniel G. Swanson, Jr. and Wolfgang J. R. Hoefer

Microwave Component Mechanics, Harri Eskelinen and Pekka Eskelinen

Microwave Differential Circuit Design Using Mixed-Mode S-Parameters, William R. Eisenstadt, Robert Stengel, and Bruce M. Thompson

Microwave Engineers' Handbook, Two Volumes, Theodore Saad, editor

Microwave Filters, Impedance-Matching Networks, and Coupling Structures, George L. Matthaei, Leo Young, and E. M. T. Jones

Microwave Imaging Methods and Applications, Matteo Pastorino and Andrea Randazzo

Microwave Material Applications: Device Miniaturization and Integration, David B. Cruickshank

Microwave Materials and Fabrication Techniques, Second Edition, Thomas S. Laverghetta

Microwave Materials for Wireless Applications, David B. Cruickshank

Microwave Mixer Technology and Applications, Bert Henderson and Edmar Camargo

Microwave Mixers, Second Edition, Stephen A. Maas

Microwave Network Design Using the Scattering Matrix, Janusz A. Dobrowolski

Microwave Power Amplifier Design with MMIC Modules, Howard Hausman

Microwave Radio Transmission Design Guide, Second Edition, Trevor Manning

Microwave and RF Semiconductor Control Device Modeling, Robert H. Caverly

Microwave Transmission Line Circuits, William T. Joines, W. Devereux Palmer, and Jennifer T. Bernhard

Microwaves and Wireless Simplified, Third Edition, Thomas S. Laverghetta

Modern Microwave Circuits, Noyan Kinayman and M. I. Aksun

Modern Microwave Measurements and Techniques, Second Edition, Thomas S. Laverghetta

Modern RF and Microwave Filter Design, Protap Pramanick and Prakash Bhartia

Neural Networks for RF and Microwave Design, Q. J. Zhang and K. C. Gupta

Noise in Linear and Nonlinear Circuits, Stephen A. Maas

Nonlinear Microwave and RF Circuits, Second Edition, Stephen A. Maas

On-Wafer Microwave Measurements and De-Embedding, Errikos Lourandakis

Parameter Extraction and Complex Nonlinear Transistor Models, Günter Kompa

Passive RF Component Technology: Materials, Techniques, and Applications, Guoan Wang and Bo Pan, editors

Practical Analog and Digital Filter Design, Les Thede

Practical Microstrip Design and Applications, Günter Kompa

Practical Microwave Circuits, Stephen Maas

Practical RF Circuit Design for Modern Wireless Systems, Volume I: Passive Circuits and Systems, Les Besser and Rowan Gilmore

Practical RF Circuit Design for Modern Wireless Systems, Volume II: Active Circuits and Systems, Rowan Gilmore and Les Besser

Principles of RF and Microwave Design, Matthew A. Morgan

Production Testing of RF and System-on-a-Chip Devices for Wireless Communications, Keith B. Schaub and Joe Kelly

Q Factor Measurements Using MATLAB , Darko Kajfez

QMATCH: Lumped-Element Impedance Matching, Software and User's Guide, Pieter L. D. Abrie